19th International Symposium on Aviation Psychology (ISAP 2017)

Dayton, Ohio, USA
8-11 May 2017

ISBN: 978-1-5108-4214-4

Printed from e-media with permission by:

Curran Associates, Inc.
57 Morehouse Lane
Red Hook, NY 12571

Some format issues inherent in the e-media version may also appear in this print version.

Printed by Curran Associates, Inc. (2017)

For permission requests, please contact Wright State University – Conferences & Events at the address below.

Wright State University – Conferences & Events
3640 Colonel Glenn Highway
Dayton, Ohio 45435
USA

Phone: (937) 775-1000

http://www.wright.edu

Additional copies of this publication are available from:

Curran Associates, Inc.
57 Morehouse Lane
Red Hook, NY 12571 USA
Phone: 845-758-0400
Fax: 845-758-2633
Email: curran@proceedings.com
Web: www.proceedings.com

19th INTERNATIONAL SYMPOSIUM ON AVIATION PSYCHOLOGY

MAY 8 - MAY 11, 2017 | DAYTON, OHIO

PAPERS

What's Next for NextGen?

Unmanned Aircraft Operations

Wednesday Posters

A TREND ANALYSIS OF HUMAN FACTORS ISSUES IN UK MILITARY AVIATION

Sarah Weedon, Victoria Cutler, and Saskia Revell
Royal Air Force Centre of Aviation Medicine
Bedfordshire, UK

Human factors issues in UK military aviation are identified and mitigated through a combination of proactive support and reactive investigations, both of which employ qualitative data collection and analysis methods. Each piece of work to identify human factors issues is performed on a standalone basis, but a regular review is undertaken to identify common trends. In the most recent review of trends, thematic analysis was used within the framework of the Accident Route Matrix to determine the most prevalent and qualitatively important human factors issues. The thematic analysis identified a wide range of human factors issues, including resource availability, training, documentation, and fatigue. By applying a qualitative approach throughout the data collection and analysis, it was possible to develop a rich understanding of each trend.

A combination of proactive support (examining normal flying operations) and reactive investigations (analysing air accidents) are used to identify and mitigate Human Factors (HF) issues in UK military aviation.

Proactive support is provided through the conduct of Operational Events Analysis (OEA, Revell, Harris, and Cutler, 2014). The OEA is a proactive and preventative approach, which examines typical military aviation operations and uses that information to identify HF issues which are influencing the work of the unit. OEA involves an HF specialist attending a unit for a period of time, typically between five and ten days. During that visit the specialist will conduct semi-structured interviews with a cross section of personnel on their experience of working on the unit. The specialist will also observe work on the unit such as flight planning, debriefing, engineering tasks, and team meetings. The information gathered during the visit is then analysed qualitatively to identify HF issues which could influence flight safety and specify the role those issues may play in an accident.

UK military air accidents are investigated by a Service Inquiry (SI) panel. Each SI panel is supported by a number of advisors, including an HF specialist. The HF specialist supports the panel in the collection of HF evidence and throughout the analysis phase. The HF specialist also determines where HF issues could have contributed to the accident. The HF specialist then prepares a report for the SI panel which characterises each relevant HF issue and their role leading up to, during and immediately post-accident (Harris, 2011).

The proactive support and reactive investigations use a common framework to analyse HF issues, which is known as the Accident Route Matrix (ARM). The ARM was developed by Harris (2016), by adapting the Human Factors Analysis Classification System (Wiegmann and Shappell, 2003) into an investigation matrix. As shown in Figure 1, the ARM allows HF issues to be presented by both the type of issue (on the y-axis) and time of effect (on the x-axis). The ARM also identifies the links between the HF issues and demonstrates how each HF issue is connected to its role in an (actual or potential) accident sequence (shown by the boxes hazard entry, recovery, escape, and survival).

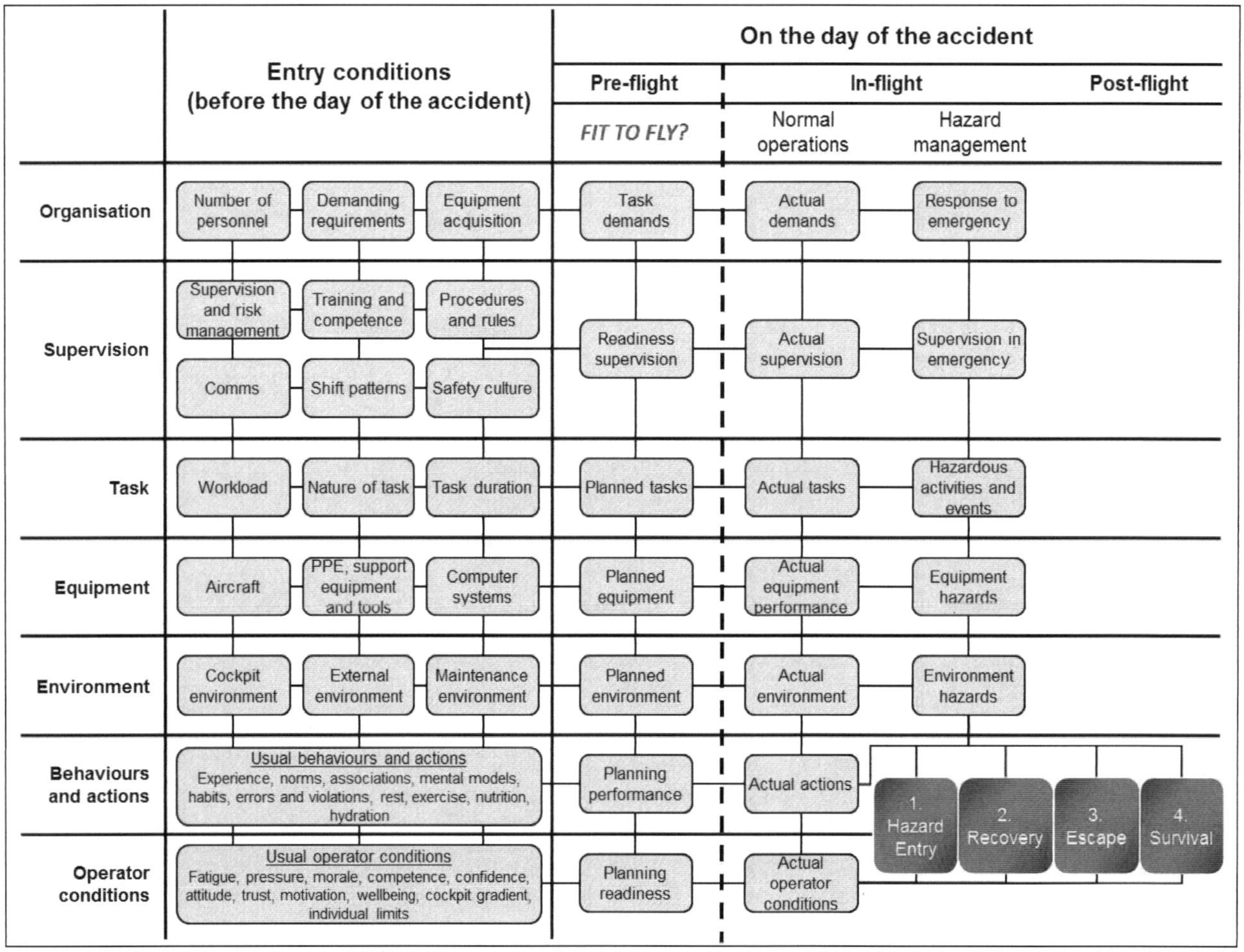

Figure 1. Accident Route Matrix.

The ARM is a fundamentally qualitative approach, as it is driven by the content and meaning of the information gathered. The benefit of such a qualitative approach in accident investigation is that the investigator can remain flexible during evidence collection and adapt to the nature of evidence available regarding the accident. The use of qualitative information reflects the richest available information about the accident, and so maximises the scope to understand why the accident happened. In applying the same process, used to investigate an accident, in the OEA immediately communicates the value of the OEA approach and means the OEA has good face validity. Applying a qualitative approach in a proactive safety investigation also offers benefits in terms of the depth of understanding of the HF issues and the links between those issues. Such an understanding assists in communicating the findings, demonstrating the credibility of the work, and in developing recommendations to address the issues identified.

Aviation safety incidents often share root causes and so analysing accidents and normal operations using the same framework (the ARM) enables common issues to be identified. However, each accident investigation and OEA is carried out on a standalone basis to ensure the HF input is appropriately tailored to the context. This enables targeted recommendations to be provided to the unit to improve safety but raises the risk that common issues and opportunities to address issues at the organisational level could be missed. Accordingly, a periodic trend analysis is undertaken with the aim of identifying the most critical trends.

Method

Data collection

Scope. Twenty reports were identified to act as the data set for the trend analysis. The data set comprised all the reports produced by the Aviation Psychology Team at the Royal Air Force Centre of Aviation Medicine (RAF CAM) between March 2013 and May 2015 inclusive. The type of reports included in this data set are shown in Table 1, the "other" reports refer to parachuting and Air Traffic Control (ATC).

Table 1.
Reports included in the trend analysis.

	Accident or incident investigation	Operational Events Analysis (OEA)	**Total**
Fixed-wing	2	4	**6**
Rotary-wing	4	7	**11**
Remotely Piloted Air System	0	1	**1**
Other	1	1	**2**
Total	**7**	**13**	**20**

The reports all shared the common qualitative investigative and data analysis procedure, as summarised in the introduction. The reports each presented the results of that analysis in the form of a series of descriptions of HF issues. Each description included the nature of the issue and, where possible, the causes of that issue and its impact on safety.

Analysis

Once the data set had been collated, a thematic analysis was carried out to "identify, analyse and report the patterns within the data" (Braun and Clarke, 2006). Thematic analysis was chosen as patterns within the data could be identified and reviewed in an iterative manner until the most prevalent themes emerged. As such, the process was driven by the qualitative information contained in the reports, but it also allowed a framework (the ARM) to be applied to the findings.

Data familiarisation and generating initial codes. Initially the reports were reviewed fully. Once fully immersed in the report contents, the HF issues were identified from the reports and collated so that very similar issues are grouped into a theme. A theme was defined as the highest level description of the issue and allowed for grouping later. Where similar but different issues were identified they were given a high level theme, but that theme was divided into sub-themes. The sub-theme provided more detail on the nature of the HF issue. For instance, a theme may be at the level of "number of personnel", which could be associated with sub-themes of "not enough supervisors" and "not enough instructors". During the analysis the titles of the themes and sub-themes were refined to reflect the whole body of information in the reports.

Categorisation of themes. Once all of the issues had been considered and the themes and sub-themes were drafted, they were compared against the ARM and categorised into one of the seven HF categories used in the ARM: organisation, supervision, task, equipment, environment, behaviours and actions, and operator conditions.

Reviewing themes and categorisation. After the ARM categorisation was completed, a full review was performed of the themes, sub-themes, and ARM categories. This comparison was undertaken by a different HF specialist, providing both an independent check of the initial identification of themes and a check of the suitability of the themes and sub-themes.

Defining trends. The ARM was then scrutinised in terms of the prevalence of each theme and sub-theme across the reports and its importance to flight safety. From this process, a number of themes and sub-themes were drawn out from the analysis to form the trends. A description of each trend was then prepared which was derived from the relevant descriptions in the twenty reports which comprised the data set.

Results

A total of thirty-one HF trends were identified from the thematic analysis and presented using the framework of the ARM, as shown in Figure 1.

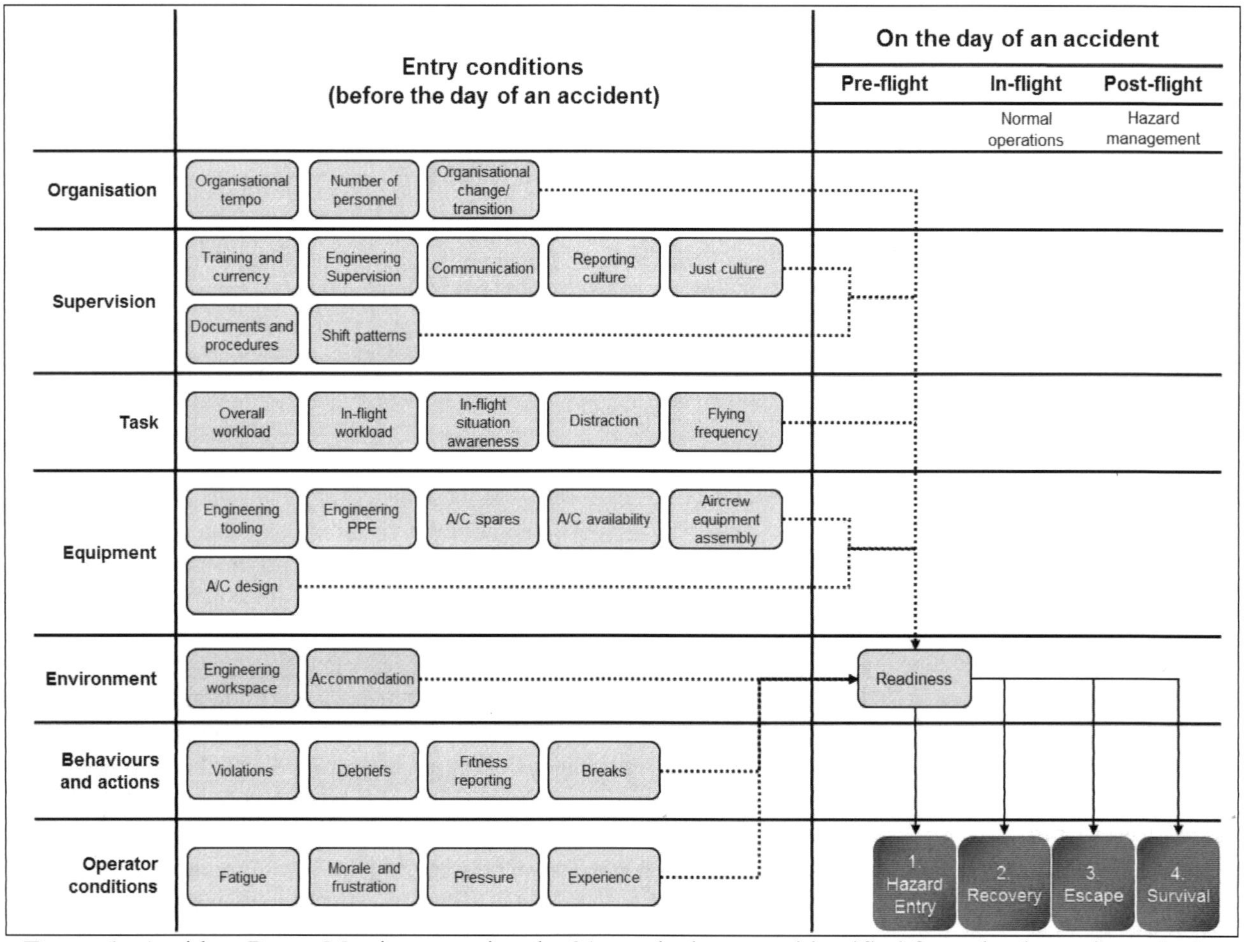

Figure 1. Accident Route Matrix presenting the 31 trends that were identified from the thematic analysis.

Descriptions were prepared for all thirty-one trends. The descriptions comprised a narrative of the issue, an actual example from the evidence, its causes, and the anticipated consequences of that issue for flight safety risk.

Amongst the thirty-one trends it was noted that there were a small number of critical trends which both prevalent and qualitatively important, and these were highlighted for particular focus and mitigation:

1. **Number of personnel**. Areas were highlighted where there were limited personnel in specific roles or with specific qualifications.

2. **Training and currency**. In all cases, training was provided to personnel to enable them to perform effectively in their role. However, some limitations were found in the content of training and in the opportunities to practice skills once trained.

3. **Documents and procedures**. A common issue in aviation is the high volume of rules, procedures and regulations. This issue was identified as a trend in the analysis, as it increased the risk of information being forgotten and so contributing to a procedural violation. There were also issues identified with the content of documentation – such as errors within the documents, unclear information, or information spread across multiple documents.

4. **Overall workload**. Rather than an issue with on-task workload, the critical trend was that personnel had a large number of tasks to perform during their working day which was challenging to achieve in the time available.

5. **In-flight Situation Awareness (SA)**. Difficulties in developing and maintaining SA in-flight were identified across a number of accident investigations. In OEA, limitations were identified with the cockpit equipment which could reduce SA in-flight.

6. **Distraction**. Two types of distraction were noted: In-flight distraction, most commonly linked to equipment discomfort, and general distraction, linked to uncertainty and frequent task changes.

7. **Fatigue and pressure**. There were very few reports of overt pressure being imposed in personnel, but personnel were highly motivated to achieve their tasks which, when combined with issues such as lack of personnel and high workload, was acting to impose a perceived pressure which could also contribute to a risk of fatigue.

8. **Experience**. Declining experience levels was identified as a critical trend, sometimes linked to new platforms where experience was naturally low, but also linked to limited opportunities to practice skills after training.

Discussion

Using qualitative analysis allows a large amount of contextual data, collected in various forms, to be examined in such a way that the feelings, values and perceptions underlying and influencing behaviours can be recognised. The language and imagery used by personnel can be captured to further understand the issues and factors being described in a way that statistical analysis cannot. Using thematic analysis allows for the identification of patterns and meanings across the data. Themes are developed from within the data and supported with assertions from grounded theory.

In the current study, combining the use of the ARM framework alongside thematic analysis has identified the HF issues which are critical trends for UK military aviation units. The analysis generated a wide range of HF issues which were then examined and explored before identifying the most critical eight. Each trend was identified based on qualitative data collection and analysis, which enabled an in-depth understanding of each issue to be developed, beyond what could have been achieved with a purely quantitative approach.

The nature of the qualitative approach used ensured that the results were evidence based, which was particularly important when presenting the findings to senior stakeholders to provide confidence in the conclusions. The nature of the analysis then allowed a detailed and descriptive set of results to be produced which could be easily and clearly explained to non-aviation psychologists. This clarity is vital in enabling action to be taken to address the issues identified and to guide decision making regarding the operational risks in military aviation.

The results of the trend analysis have been presented to senior personnel within the UK military to further aid their understanding of HF risks. Recommendations have been developed to address each of the eight critical trends at the organisational level, and to develop the use of OEA to support continual improvement in aviation safety.

Acknowledgements

The authors acknowledge the contribution of the relevant military units for which the original reports used in this paper were written.

References

Braun, V. and Clarke, V. (2006) Using thematic analysis in psychology. *Qualitative Research in Psychology,* 3 (2), 77-101.

Harris, S. (2011). Human factors investigation methodology. *Paper presented at the 16th International Symposium on Aviation Psychology.* Dayton, OH.

Harris, S. (2016) Errors and accidents. In *Ernsting's Aviation and Space Medicine* (707-722). Boca Raton, CA: Taylor and Francis Group.

Revell, S.D., Harris, S.L. and Cutler, V.J. (2014). A preventative approach to identifying and addressing flight safety human factors issues. *Paper presented at the 31st Conference of the European Association of Aviation Psychology.* Valetta, Malta.

Wiegmann, D.A. and Shappell, S.A. (2003). *A Human Error Approach to Aviation Accident Analysis.* Ashgate: Aldershot, UK.

Exploring Facilitated Debriefing Techniques Using a Diary Study

Shawn Pruchnicki, MS RPh ATP MEI
The Ohio State University, Department of Integrated Systems Engineering
290 Baker Systems, 1971 Neil Ave., Columbus, OH 43210 USA
Arjun H. Rao, Ph.D
The Ohio State University, Center for Aviation Studies
228 Bolz Hall, 2036 Neil Ave., Columbus, OH 43210 USA

Diary studies, when used as a qualitative research instrument, provide numerous advantages not possible with other methods. These differences become even more apparent when comparing diary study results to the vastly different quantitative type paradigms. Although less commonly used, their unique benefit to the researcher is both the volume and nature of the open-ended data captured. This underutilized method offers the researcher an opportunity to discover a rich first person account of the experiences, feelings, observations, and solutions to challenges. Here we present the beginning of a research study where we employed a diary study method to discover specific behaviours and observations from the perspective of aircraft simulator instructors. Specifically, during the post-simulator debriefing we examined first-person insights on how simulator instructors utilize facilitated debriefing techniques in addition to both the recognition and mitigations to learning barriers.

Understanding human behavior in natural settings offers both challenges and benefits simultaneously. For example, external factors can intrude unpredictably into your observations and can be both problematic and insightful as to how work is performed in complex environments. Traditionally, early in the study design process, researchers have a general idea if their methods could be categorized as either qualitative or quantitative in nature. Additionally, a researcher's field of study may also help with this categorization. Unfortunately, the social sciences have previously struggled with claims that, typically, qualitative methods in general lack the ability to find the provable "truth" or statistically supported findings. Historically, this claim has positioned the social sciences on the defensive, resulting in a consuming focus on trying to legitimize their research by following the lead of the more quantitative physical science research paradigms. Here we choose to not focus on language that invites argumentative discourse as this oration has gone on long enough and the arguments' relative merits are thoroughly contrasted elsewhere (Kunh, 1962; Flyvbjerg, 2001; Blaxter, Hughes, & Tight, 2006).

The question remains, how do we decide what methods of data collection are best for natural settings where work is complex, messy, and frequently does not follow a predictable script? Woods (1992) lends some insight by stating that in complex and dynamic systems we "must use a different subject population than the typical subject of psychology experiments." These environments include, but are not limited to, aircraft cockpits, nuclear power control rooms, and various health care settings. The nature of the study we describe here lends itself to one where we take a more holistic approach to sampling data from the context of real work as done. Described in the literature by Hutchins (1995a) as "cognition in the wild", this approach reminds us to be cautious as to not disturb the work process since this has the very real potential to truncate or alter our ability to understand work in context (Bartlett, 1932; Hollnagel & Woods, 2005; Hollnagel, Woods, & Leveson, 2007; Hollnagel et al., 2008; Hollnagel et al., 2009; Dekker, 2016; Hollnagel, 2017).

Diary Study

Here, we describe how we are employing a diary study method of data collection for research examining how professional simulator instructors discover and mitigate challenges to post-simulator facilitated debriefings. Diaries, in either the written or audio format, are self-reported instruments used to examine specific experiences. Diary studies in particular offer researchers the opportunity to investigate social, psychological, and physiological processes, with events that can be unpredictable (Bolger et al., 2003). Effectively, this underutilized research method offers us an opportunity to study cognition in natural settings from a more observational perspective. That is, to capture a very rich first person account of the experiences, feelings, observations and solutions to problems.

A researcher's specific research goals and questions will dictate if a diary study will be a good choice as a research tool. For example, when considering your research goals, a more open-ended examination of contextually rich environments lend themselves well to employing a diary method. In general, three broad types of research goals are possible using diary designs: (a) obtaining reliable person-level information; (b) obtaining estimates of within-person change over time, as well as individual differences in such change; and, (c) conducting a causal analysis of within-person changes and individual differences in these changes (Bolger et al., 2003). These are not trivial considerations as the methods and questions chosen for data collection will effect both the nature of your results and how they are interpreted (Bolger, et al., 2003). For the study we describe here, we are gathering reliable person-level information since we are amassing descriptions of specific events identified ahead of time (post-simulator debriefing) for each of the simulator instructors. These descriptions are not compared against each other but rather collected and analyzed as aggregate data.

Diary studies, when used as an ethnographic research instrument, provide numerous advantages not possible with other methods. Additionally, they can also support a more grounded theory approach—that is, one which is more exploratory in nature and later may reveal a potential hypothesis. The freedom provided by a diary study includes the ability of a researcher to explore the data and understand the unique complexities of work from different perspectives. These differences become even more apparent when comparing diary study results to the vastly different quantitative-type paradigms and their focus on a specific hypothesis and statistical justification. Other known benefits to the research community, diary studies offer unique research benefits. Some of these include both the volume and potential depth of the open-ended data captured which is simply not possible with other more rigid study design constructs. This rich contextual pool of data is possible by the unique flexibility and characterization of a diary study design.

Although diary studies provide the researcher a plethora of contextual rich data to examine, like any other research instrument, there are limitations and challenges unique to each.

For example, diary studies can suffer from being too tedious for the subject and they can invoke a "Heisenberg-style" challenge: that is, the process of influencing the observations by intruding upon and interfering with the very flow of the events being examined (Czerwinski et al., 2004). In the study presented here, we addressed each of these by providing recording pens so that they can verbally report their discussion as opposed to the more laborious task of writing out the details of their experience. As for the "Heisenberg-style" consideration, we addressed this by having the instructors make their recordings right after the post-simulator debriefing. This has the additional benefit of helping to prevent or at least reduce any memory recall problems with those that are captured later.

Facilitated Debriefings

Many safety-sensitive domains utilize advanced forms of simulation to capture learning objectives for both initial and recurrent training programs. Research has shown that these simulator sessions are more meaningful when followed by a structured debriefing session (Helmreich & Foushee, 1993). Precision flying skills are considered by many as easier to evaluate since they are based on specific quantitative flight parameters (i.e., airspeed ± 10 kts.). Instructors can easily debrief these training aspects as the performance is evaluated as being either within the allowable range or not.

Teamwork and collaborative constructs are much harder to evaluate for both the students and instructors as these events unfold due to either the more subjective nature of how these terms are defined or the lack of a measurable quantity. The evaluation of these collaborative teamwork constructs requires a more effortful discourse where students are the central focus. Post-simulator debriefings are more meaningful when conducted in a facilitated manner—that is, where the students through self-discovery discuss their non-technical performance (e.g., flight deck communication and collaboration), and as a team review the training event to discover areas of both strengths and weaknesses. If the debriefing is conducted correctly, the students will be able to better take their perspective of their performance back to the real aircraft and with reflection make changes to their day-to-day flying and collaborative abilities. Adult learning literature also suggests improvements in day-to-day performance is where a student-centered approach will lead to deeper understanding, better memory retention and later skill application (Duval & Wicklund, 1972; Gow & Kember, 1993; Jones, 1982; Dismukes, Jobe, & McDonald, 1997).

Although the adult learning literature discusses why facilitation is beneficial to promote a deeper understanding of the material and increased retention, there is limited guidance as how to conduct a facilitated debriefing. In other words, what are the essential components of these sessions, and how should they be conducted? Furthermore, there appears to not be, or at least not published, a serious research attempt to capture as many barriers to learning discovered in a simulator-training environment using ethnographic techniques. Even less available is guidance addressing any of these barriers and more importantly the successful strategies used to overcome obstacles to learning. Our research study presented in part here addresses these absences and the diary study method gives us the freedom to capture rich contextual data.

Data Collection

This study will utilize a group of professional simulator instructors who will conduct facilitated debriefings once we complete a literature review, subject matter expert (SME) consultation and standardized pre-study training. When the study begins, they will first answer four predetermined questions that are specific to facilitation methods and encountered barriers to learning. After these are addressed, they are encouraged to share all thoughts on the experience regardless of how pedestrian they may seem. The goal of this study, which is why the diary method is particularly effective, is that it offers subjects many degrees of freedom in both how and what they chose to report.

However, prior to data collection, literature searches for facilitation barriers to learning and previous aerospace research on debriefing facilitation was reviewed to see how this line of research could be further explored. Once completed, we met several times with subject matter experts (SME) that are simulator instructors and training captains who were able to provide significant insight into post-simulator debriefing challenges, in addition to how, in their experience they have seen facilitation both work successfully and fail. Thus, they were able to help us craft definitions of what facilitation means in this application and how that connects to the last of the aerospace research from the late 1990s (Dismukes et al., 1997). We were also fortunate to speak with the foremost NASA researcher who led this effort during that time.

All of these perspectives allowed us to establish several foundational components to our study: a) we developed a solid understanding of what facilitation is and what it is not, b) we established challenges and benefits commonly experienced (including known barriers to learning in the debriefing environment) by SMEs who use facilitation methods regularly (weekly basis) and, c) established the specific questions that we required instructors to include in their diary entries (see below). Once answers to these questions were established and prior to data collection, we provided a "standardization" class for the instructors. This class was used to ensure that they understood the meaning and goals of the study, their individual responsibilities, and satisfying Institutional Review Board (IRB) protocols. Materials covered included answers to what the SMEs felt was effective facilitation, operation of the recording pen, downloading and submitting their diary audio files, and a discussion on what a diary study is including history, advantages/disadvantages, and how to specifically make an audio diary entry. The specific required diary entry was initially structured around four questions that as a group with the help of the SMEs and the previous literature search we felt should be addressed in each diary entry. The questions are:

1. Over all, how well did the facilitation attempt work? Offer a high-level perspective of the experience as a whole.
2. What were the indications noticed that the crew arrived ready for self-discovery, or not?
3. What were the barriers to facilitation that you noticed? How were you able to discover them?
4. Were there any mitigation strategies attempted to any of the barriers experienced? If so, what were they and how well did they work? What would you do differently in other training events?

Otherwise, as part of diary study methods, instructors had free rein to discuss their observations, concerns, successes and failures while trying to conduct facilitated debriefings.

This type of study design would normally imply a retrospective analysis (the subject completes their diary entry once after the debriefing) complete with all of its biases and limitations. However, from a timing consideration this approach was our only opportunity for data collection since the instructors were not allowed to make their diary entries during the actual debriefing as requested by management. We agreed with their concern to the potential disruptive nature of trying to capture this data from the debriefing in real time. Some researchers would argue that this delayed capture may seem to shift the timing of the data collection from a prospective to a retrospective format. However, despite this apparent challenge, we felt that our data collection is actually far more prospective then many would appreciate. The instructors were guided to make their diary entries immediately after the post-simulator debriefing. This immediate entry would reduce biases and memory challenges, and we would be capitalizing on the learning principles of primacy and recency to significantly reduce the extent of retrospection bias and memory challenges (Bolger, et al., 2003). We felt that this approach was a reasonable balance between usual diary study methods and real world constraints and trade-offs that make this operational space challenging. We realized that there would be times when making the diary entry immediately after was not possible (for example during the middle of night while fatigued or when personal schedules are prohibitive). In those cases, the instructors were advised to make the diary entry as soon as practicable.

Conclusions

By using an ethnographic type research design (diary study), we were able to discover specific behaviors and observations from the first person prospective view from simulator instructors. For this specific work context, they are the best source of information which supports our understanding of both their challenges and opportunities when conducting facilitated debriefings. By our design, the simulator instructors offered truly a first person perspective that is captured in a prospective manner. This first perspective or first story has high ecological value because these experiences are carried out *in situ* or in the users' real environment (Czerwinski et al., 2004). In our research discussed here, using a diary study method allowed us insight to a contextual process that has not been previously explored and captured. This approach provided a much deeper and richer understanding of the challenges professional simulator instructors face. In our case, no other research instrument would have provided the balance between a comparatively less rigid method while yet still offering rich contextual data that will drive the next phase of our research on improving and standardizing the facilitating debriefing process.

References

Bartlett, F. C. (1932). Remembering: An experimental and social study. *Cambridge: Cambridge University.*

Blaxter, L., Hughes, C., & Tight, M. (2006). How to Research 3rd. Open University Press. NY.

Bolger, N., Davis, A., & Rafaeli, E. (2003). Diary methods: Capturing life as it is lived. *Annual review of psychology*, *54*(1), 579-616.

Czerwinski, M., Horvitz, E., & Wilhite, S. (2004, April). A diary study of task switching and interruptions. In *Proceedings of the SIGCHI conference on Human factors in computing systems* (pp. 175-182). ACM.

Dekker, S. (2016). *Drift into failure: From hunting broken components to understanding complex systems*. CRC.

Dismukes, R., Jobe, K. K., & McDonnell, L. K. (1997). *LOFT Debriefings: An analysis of instructor techniques and crew participation* (NASA Tech. Memo. No. 110442). Moffett Field, CA: NASA Ames Research Center.

Duval, S., & Wicklund, R. A. (1972). *A theory of objective self-awareness*. New York: Academic.

Flyvbjerg, B. (2001). *Making social science matter: Why social inquiry fails and how it can succeed again*. Cambridge university press.

Gow, L., & Kember, D. (1993). Conceptions of teaching and their relationship to student learning. *British journal of educational psychology, 63*(1), 20-23.

Helmreich, R. L., & Foushee, H. C. (1993). Why crew resource management? Empirical and theoretical bases of human factors training in aviation. In E.L. Wiener, R.L. Helmreich, & B.G. Kanki (Eds.), *Cockpit Resource Management* (pp. 3-41). San Diego, CA: Academic Press.

Hollnagel, E. (2017). Developing the Resilience Potentials: Safety II in Practice. Elsevier. Amsterdam.

Hollnagel, E., Nemeth, C. and Dekker, S.W. (2008). Resilience Engineering: Remaining Sensitive to the Possibility Of Failure. Aldershot: Ashgate Publishing Co.

Hollnagel, E., Nemeth, C. and Dekker, S.W. (2009). Resilience Engineering: Preparation and Restoration. Aldershot: Ashgate Publishing Co.

Hollnagel, E. & Woods, D.D. (2005). *Joint cognitive systems: Foundations of cognitive systems engineering*. Boca Raton, FL: CRC Press.

Hollnagel, E., Woods, D., & Leveson, N. (2007). *Resilience engineering: concepts and precepts*. Ashgate.

Hutchins, E. (1995a). *Cognition in the Wild.* MIT press.

Hutchins, E. (1995b). How a cockpit remembers its speeds. *Cognitive science, 19*(3), 265-288.

Jones, P. G. (1982). *Adult learning in your classroom: The best of training magazine's strategies and techniques for managers and trainers.* Minneapolis, MN: Lakewood Books.

Kunh, T. (1962). The structure of scientific revolutions. The University Press. London.

Woods, D. (1992). Cognitive Systems in Context: Joint Cognitive Systems and Research on Human-Machine Systems, Cognitive Systems Engineering Laboratory, Department of Industrial and Systems Engineering, The Ohio State University.

IS RHO THE KEY TO HAZARDOUS WEATHER AVOIDANCE?

William R. Knecht
Federal Aviation Administration
MMAC/CAMI/AAM510, Oklahoma City, OK USA

Current in-cockpit looping Next-Generation Radar (NEXRAD) is inadequate to maintain safe (20 nm) aircraft separation from heavy weather (> 40 dBZ reflectivity). This assertion is supported by mathematical information analysis and an empirical study (Knecht, 2016), as well as numerous previous empirical studies. The current work revisits the ecological analysis by examining the putative affordance ρ (rho) specifying when weather-avoidance maneuver should begin, as suggested by General Tau Theory (Lee, 2009). With "gap" defined as the distance between the on-screen aircraft icon and the weather hazard, ρ is specified by the ratio $((dg/dt)/g)(t)$, the instantaneous gap contraction rate divided by the instantaneous gap size. In current looping NEXRAD, ρ clearly does not reach perceptible threshold until too late to facilitate 20 nm separation from hazard. The addition of a range ring plus future-predicted weather and aircraft position could remedy this deficiency, enabling safe, efficient navigation around heavy weather.

Introduction

Background

Adverse weather remains a perennial challenge for all aviation, particularly for the smaller aircraft of general aviation (GA) and, therefore, a high priority for the U.S. Federal Aviation Administration (FAA). One important focus area involves pilot interpretation and use of color-coded weather-risk displays. In the U.S., the best known of these is the National Weather Service (NWS) NEXRAD. GA pilots are now being offered NEXRAD capability in the cockpit, for instance via XM satellite radio, and on handheld devices like tablet computers and smartphones. From a human-factors perspective, NEXRAD is effectively a *risk-proxy gradient*—a graphical representation of relative weather-related risk. Such gradients contain important perceptual information pilots can use to make hazard-avoidance decisions (Knecht & Frazier, 2015a; Wiggins, Azar, & Loveday, 2012)—particularly, how close their flight plan may take them to hazardous weather.

Normally, NEXRAD images are updated only about once every five minutes. But, rapid playback of about an hour's worth of individual frames is enough to create a time-lapse movie of precipitation. Repeating ("looping") such a movie conveys a strong sense of apparent motion (Wertheimer, 1912), enhancing the perception of where a storm is heading.

Nevertheless, looping NEXRAD ultimately shows a movie of where precipitation *used to be*. At issue is whether that information can be used to predict where both the aircraft and hazardous weather *will be* in the near future.

We know that pilots can estimate closest point of approach to storms on NEXRAD to a degree. Psychophysical studies by Bootsma & Oudejans (1993) have mathematically verified both the presence of detectable information in "an object moving toward a designated position," as well as the ability of observers to detect that information. Nonetheless, in virtually every aviation-related NEXRAD study to date (all *in simulo*), a substantial proportion of pilots seemed to overestimate closest point of approach (CPA), meaning they overestimated eventual minimum separation from heavy weather, and ended up approaching too closely (ATSC, 2013; Beringer & Ball, 2004; Burgess & Thomas, 2004; Hua, 2014; Knecht, 2016; Knecht & Frazier, 2015a,b; Lemos & Chamberlain, 2004; Novacek, Burgess, Heck, & Stokes, 2001; Wu, Duong, Koteskey, & Johnson, 2011; Wu, Gooding, Shelley, Duong, & Johnson, 2012; Wu, Luna, & Johnson, 2013; Yuchnovicz, Novacek, Burgess, Heck, & Stokes, 2001). In no study did all pilots consistently maintain the 20 nm separation from heavy weather advised in FAA AC 00-24-C (Table 1, FAA, 2013, p. 10, Sec 9c)

In previous investigation (Knecht, 2016) we took a theory-based look at the visual information present in looping NEXRAD. The current work revisits that investigation and suggests possible avenues of further research. The approach is that of *ecological psychology* (Gibson, 1979), *neurocomputation* (Marr, 1982), and *ecological interface design* (Dinadis & Vicente, 1999, Borst, Flach, & Ellerbroek, 2015), namely examination of the visual elements of a scene's "ecology" to determine *affordances*—information capable of "affording" completion of a given task in the sense of providing, supplying, facilitating, or enabling it in a way mathematically describable and computationally plausible by structures of neurons. Of particular concern to us in this discussion are the visual affordances in a NEXRAD display that would allow keeping an aircraft icon 20 scale miles away from "heavy" weather.

Summary of Key Findings to Date

The search for task-relevant information. Figure 1a represents an idealized map display of an aircraft moving NW in straight-line motion for 35 minutes with constant velocity $V_{aircraft}$ = 120 kt. Imagine a single point on the nose of the aircraft icon approaching a single designated point on the edge of a storm that does not change shape, but moves ENE in straight-line motion with constant velocity V_{storm} = 30 kt.

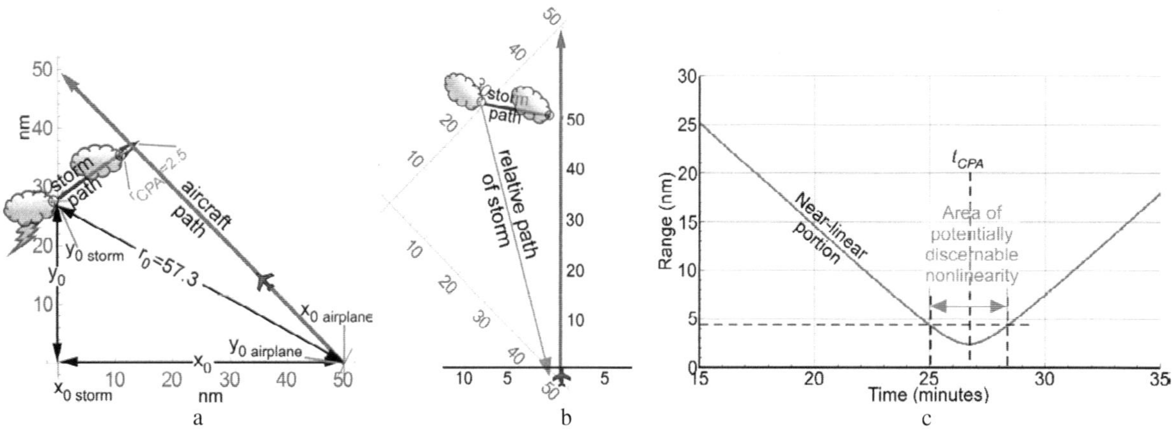

Figure 1, a) Cartesian geometry of a "pass-by" situation with 57.3 nm initial separation and CPA = 2.5 nm, b) the same situation rotated (45° clockwise), now depicting an aircraft-centered, moving-map display showing the storm's resulting relative motion (the logic of Eqs. 1 and 2 (below) is based on 1b), c) the gap function plotted over time produces a "rounded-V" shape having zero slope at time-of-CPA (t_{CPA}).

Avoiding a single point on such a storm's edge is arguably the simplest possible case of "weather avoidance." In reality, there would be many such points to consider along that edge, but we can consider just one because their mathematical logic will be similar.

Figure 1b shows the same weather situation, but transformed into the perspective of *relative* motion (Lenart, 1983) such as you would see in a *moving-map* format, centered on the aircraft, with the world rotated (here, 45° clockwise) to show the aircraft path headed straight up. The aircraft appears to stand still while objects around it move.

For a looping NEXRAD display without future-projection of weather, Figure 1c shows Figure 1a's *gap function*—the parametric (time-based) equation describing the instantaneous range r_t, or gap, between the tip of the aircraft icon and that single, moving point on the storm at time *t:*

$$r_t = \sqrt{(x_0 + v_x t)^2 + (y_0 + v_y t)^2} \qquad (1)$$

where x_0 and y_0 are initial relative separation distances (e.g., $x_0 = x_{0\,aircraft} - x_{0\,storm}$), and v_x and v_y are relative-velocity components (e.g., $v_x = v_{x\,aircraft} - v_{x\,storm}$), all of which can be estimated by comparing at least two views of the situation, separated by a known amount of time.

Solving Equation 1 for slope $d_r/d_t=0$ gives us CPA—the task-relevant information we need (see Knecht, Smith, & Murphy (2000), Appendix 1 for derivation). This shows that—at least in the absolute simplest case—looping NEXRAD theoretically contains sufficient information for pilots to estimate how close they will approach a storm boundary.

$$CPA = \sqrt{\frac{(x_0 v_y - y_0 v_x)^2}{(v_x^2 + v_y^2)}} \qquad (2)$$

Implausible vs. plausible solutions. We have retinal structures sensitive to position, various sizes of gap, angular orientation (Hubel, 1988), and motion (van Santen & Sperling, 1985). So, it may be plausible to detect the individual components of Equation 2. However, it is not plausible to imagine noisy neurons accurately executing all the delicate mathematical operations in the exact fashion specified by Equation 2.

We therefore look for a "hack"—some clever feature of the situation that might sidestep complicated computation, allowing what Gibson called *direct perception*. For instance, pilots have a hack to directly perceive if a distant airplane will collide with theirs. They just look out the window. If the relative position of the approaching aircraft on the windscreen never changes, but it keeps getting bigger and bigger—that represents an eventual collision (Bootsma & Oudejans, 1993).

The challenge is finding such a hack. Examining Figure 1c, we might, for instance, monitor the V-shaped gap function in non-future-projected looping NEXRAD to look for a sudden *change* in its slope (i.e., the second derivative). However, that approach seems implausible. As Figure 1c clearly shows, a "V" gives nearly no change-in-slope information until the time $t \approx 25$ minutes, where the aircraft is practically at CPA, and already dangerously close to the storm.

Ecological Enhancements for a Better Display

Rho as a potential cue to triggering avoidance maneuvering. Lee (2014) has considered ecological situations analogous to ours, namely ones where a viewer sees a gap changing size over time. The *way* the gap changes can serve as a trigger stimulus for actions such as an avoidance maneuver. The information that forms this potential trigger stimulus is called ρ (rho), and is defined (Eq. 3) as the *relative rate of change of the size of the gap.*

$$\rho_t = \frac{dg/dt}{g_t} = \frac{\text{instantaneous change in gap size}}{\text{instantaneous gap size}} = \frac{\text{slope of the gap function at time } t}{\text{size of the gap at time } t} \tag{3}$$

Readers may recognize ρ as essentially the inverse of τ (tau, that is *time-to-contact*), which is the basis of General Tau Theory (Lee, 2009). Regardless, the concept itself is simple enough. Given, say, a shrinking gap between an onscreen aircraft icon and a storm cell, the faster the gap is shrinking (bigger numerator)—or the smaller the gap itself is (smaller denominator)—the bigger ρ will be. The ratio forming ρ changes over time, and Bootsma & Oudejans (1993) suggest mathematical approximations that could be plausibly implemented by neurons without the need for implausibly extensive or delicate computation.

Figure 2a below is merely 1c repeated for convenience. Figure 2b shows how, in an onscreen conflict situation such as looping NEXRAD, the value of ρ would grow large enough to exceed a fixed threshold and trigger a neural circuit sufficiently far ahead of time to cover reaction and maneuver times. And, because any gain made in early alert translates directly into *available maneuver time*, ρ might constitute a key element in hazard avoidance.

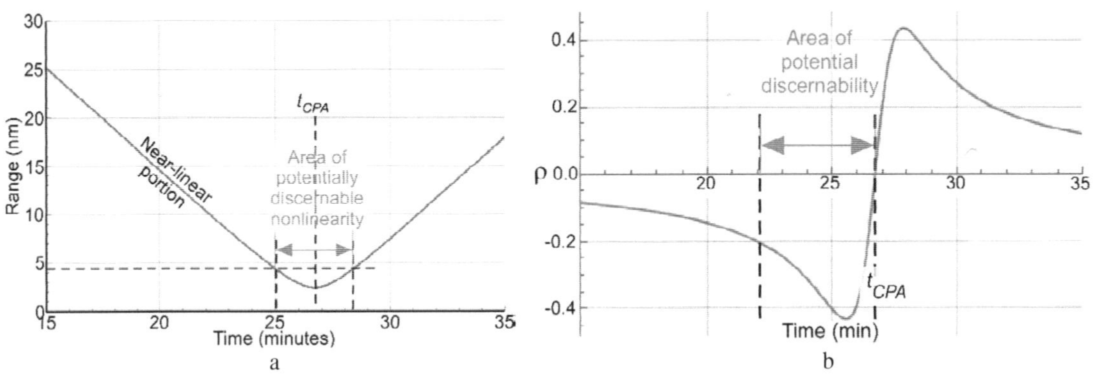

Figure 2a. The gap function of Fig. 1c, b) the time-evolution of ρ. Note that the threshold for earliest-time-of-discernability could be lower than that of mere slope change detection (Fig. 2b, t ≈ 22 minutes, about 3 minutes sooner than in Fig. 1c).

Addition of a range ring to the display. The addition of a range ring around the aircraft icon (Fig. 3a) should theoretically add even more benefit to a looping display.

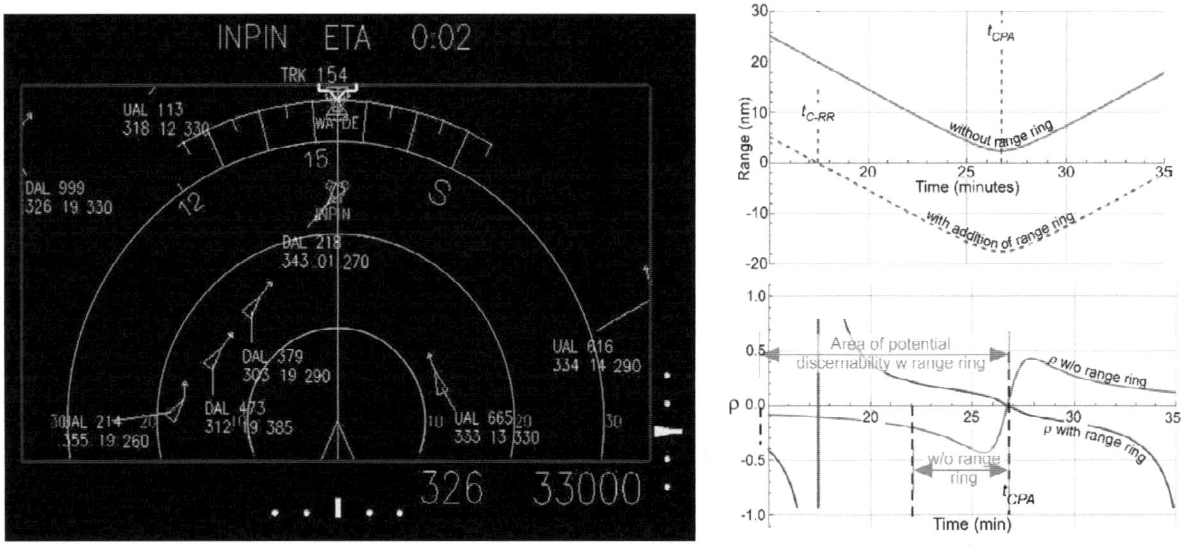

a b

Figure 3a. A cockpit multifunction display showing range rings centered around the aircraft icon, b, upper) The gap function with and without a 20-nm range ring, b, lower) time-evolution of ρ, and areas of potential discernability, with and without 20-nm range ring.

Figure 3b (upper) shows how a 20-nm range ring changes the gap function by effectively decreasing the instantaneous distance-to-hazard by 20 nm. If "Plan A" for hazard avoidance is based on perception of ρ in a looping display, then Figure 3b (lower) shows a marked decreased in earliest time-of-potential-discernability, from about 22 minutes without the range ring down to less than 15 minutes with it. In other words, having a range ring gains could provide 7 minutes additional maneuver time in this particular case.

Moreover, Figure 3a (upper) shows that the range ring itself will ultimately directly contact the edge of the hazard at time $t_{C\text{-}RR} \approx 17.4$ minutes, while the aircraft is still 20 nm distant. This constitutes a "Plan B" backup alert for even the least-attentive pilot.

Addition of a range ring *and* future-projected weather. Obviously, accurate estimation of the positions of both the aircraft icon and weather—even with as short as 30 minutes lookahead—would be a major step forward in tactical weather avoidance. This would eliminate having to depend on perception of an early-warning stimulus such as ρ. The display could either be looped, or simply "time-scrolled" ahead to see if the range ring itself contacted any hazard.

At issue, of course, is the accuracy of the convective weather forecasts themselves. Conversations with Keith Brewster (personal communication, July 30, 2015), Associate Director of the Center for Analysis and Prediction of Storms (CAPS) lead us to believe that 45 minutes lookahead appears feasible with current supercomputers running 3-km-resolution storm modeling. About 15 minutes of that lookahead would be needed to compensate for processing and data-broadcasting time, leaving the net 30-minute gain envisioned as necessary.

Conclusions

The Importance of Ecological Information Design

As human factors researchers, we need to be able to determine how task-critical in-formation from technological systems is detected by the user (Vicente, 1999). If we begin with the information present in the stimulus, we can then imagine how that information could be detected or derived by simple neural circuits. If these exist, then there may be the possibility for accurate, efficient, effortless Gibsonian direct perception, and the technology may function efficiently with little modification.

On the other hand, if we can logically show that either no easily detectible task-critical information exists in the stimulus, or no such simple neural detector of that information is plausible, then we can deduce that perception and/or cognition must be constructed. Constructed cognition is almost by definition less efficient, more error-prone, and is therefore an opportunity for augmented perception and augmented cognition, such as the theory and method of ecological interface design, which seeks to "make visible the invisible" (Vicente & Rasmussen, 1990.

Naturally, no cockpit display, no matter how advanced, can guarantee 100% freedom from weather hazard. Human factors issues always remain (e.g. "get-home-it is," fatigue, training issues, and so forth). Nonetheless, we feel compelled to support all efforts regarding the art and science of ecological interface design. To analyze the information available in the visual stimulus, to discern which tasks rely on hard-to-derive information, and to find creative ways of making visible the invisible are things clearly worth our effort. Ecologically enhanced displays have already shown considerable success in tactical aircraft collision avoidance (Ellerbroek, Visser, van Dam, & van Paassen, 2011; van Dam, Mulder, & van Paassen, 2008). Since weather is more or less a "large flying object," similar ecological approaches could, and should, be developed and tested.

Future research

Future research should center, first, on testing "the rho hypothesis" in a simplified psychophysical setting, for instance testing human ability to detect impending onscreen collisions between small moving dots. If psychophysical research confirms ρ as a likely stimulus capable of triggering avoidance maneuvering, then it would make sense to pursue the investigation, examining looping-NEXRAD displays with range rings and, ultimately, with range rings and future-projected storm displays.

Acknowledgements

This study was sponsored by FAA ANG-C61, Weather Technology in the Cockpit Program, and supported through the FAA NextGen Human Factors Division, ANG-C1.

References

Atmospheric Technology Services Company. (2013). *Demonstration comparing the effects of probabilistic and deterministic forecast guidance on pilot decision-making and performance.* (Report no. FAA-20130902.1). Washington, DC: Federal Aviation Administration.

Beringer, D.B., & Ball, J.D. (2004). *The effects of NEXRAD graphical data resolution and direct weather viewing on pilots' judgments of weather severity and their willingness to continue a flight.* (Report no. DOT/FAA/AM-04/5). Washington, DC: Federal Aviation Administration. Retrieved 1 Dec., 2014 from http://www.faa.gov/data_research/research/med_humanfacs/oamtechreports/2000s/media/0405.pdf .

Bootsma, R.J., & Oudejans, R.D. (1993). Visual information about time-to-collision between two objects. *Journal of Experimental Psychology, Human Perception and Performance, 19*(5), 1041-1052.

Borst, C., Flach, J.M., & Ellerbroek, J. (2015). Beyond ecological interface design: Lessons from concerns and misconceptions. IEEE Transactions on Human-Machine Systems, 45(2), 164-175.

Burgess, M.A., & Thomas, R.P. (2004). *The effect of NEXRAD image looping and National Convective Weather Forecast Product on pilot decision making in the use of a cockpit weather information display.* (Report no. NASA/CR-2004-212655). Hamp-ton, VA: NASA Langley Research Center.

Dinadis, N., & Vicente, K. J. (1999). Designing functional visualizations for aircraft systems status displays. International Journal of Aviation Psychology, 9(3), 241-269.

Ellerbroek, J., Visser, M., Van Dam, S.B.J., Mulder, M., & van Paassen, M. M. (2011). Design of an airborne three-dimensional separation assistance display. IEEE Trans-actions on Systems, Man, and Cybernetics-Part A: Systems and Humans, 41(5), 863-875.

Federal Aviation Administration (Feb. 19, 2013). Advisory Circular 00-24C. Download-ed 14 Aug., 2013 from http://www.faa.gov/documentlibrary/media/advisory_circular/ac%2000-24c.pdf .

Gibson, J.J. (1979). The ecological approach to visual perception. New York: Houghton-Mifflin.

Hua, L. (2014). *Effect of age of radar information on the pilot's weather-related judgment.* Unpublished doctoral dissertation, University of Oklahoma, Norman.

Hubel, D.H. (1988). Eye, Brain, and Vision. New York: Freeman.

Knecht, W.R. (2016). Current NEXRAD cannot reliably enable safe flight around heavy weather. *International Journal of Aviation Psychology, 26*(1-2), 46-61.

Knecht, W.R., & Frazier, E. (2015a). *Pilots' risk perception and risk tolerance using graphical risk-proxy gradients.* (Report no. DOT/FAA/AM-15/9). Washington, DC: Federal Aviation Administration. Retrieved 2 July,

2015 from http://www.faa.gov/data_research /research/med_humanfacs/oamtechreports/2010s/media/201509.pdf .

Knecht, W.R.,& Frazier, E. (2015b). *The influence of looping Next-Generation Radar on general aviation pilots' flight Into adverse weather.* (Report no. DOT/FAA/AM-15/16). Washington, DC: Federal Aviation Administration. Retrieved 2 July, 2015 from http://www.faa.gov/data_research/research/med_humanfacs/oamtechreports/2010s/media/201516.pdf .

Knecht, W.R., Smith, K., & Murphy, L. (2000). Sector Congestion Analytical Modeling Program. (Report no. KSU HFEEL 99-G-020-4).

Lee, D. (2009). General Tau theory: Evolution to date. *Perception, 38*, 837-858.

Lee, D.N. (2014). Moving to make contact. *Ecological Psychology, 26*(1-2), 47-59.

Lemos, K., & Chamberlain, J. (2004). *In-flight weather trending information: Optimal looping characteristics for animated NEXRAD images.* (Article no. 5.D.4). Proceed-ings of 23rd Digital Avionics Systems Conference.

Lenart, A.S. (1983). Collision threat parameters for a new radar display and plot tech-nique. Journal of Navigation, 36(3), 404-410.

Marr, D. (1982). Vision: A computational investigation into the human representation and processing of visual information. San Francisco: W.H. Freeman.

Novacek, P.F., Burgess, M.A., Heck, M.L., & Stokes, A.F. (2001). *The effect of ownship information and NexRad resolution on pilot decision making in the use of a cockpit weather information display.* (Report No. NASA/CR-2001-210845). Hampton, VA: NASA Langley Research Center.

Van Dam, S.B.J, Mulder, M., van Paassen, M.M. (2008). Ecological interface design of a tactical airborne separation assistance tool. IEEE Transactions on Systems, Man & Cybernetics, Part A, 38(6), 1221-1233.

van Santen, J.P.H., & Sperling, G. (1985). Elaborated Reichardt detectors. J. Optical Soc. Of America, 2(2), 300-321.

Vicente, K.J. (1999). Cognitive work analysis. Mahwah, NJ: Erlbaum.

Vicente, K.J., & Rasmussen, J. (1990). The ecology of human-machine systems II: Mediating "direct perception" in complex work domains. Ecological Psychology, 2(3), 207-249.

Wertheimer, M. (1912). *Experimentelle Studien über das Sehen von Bewegung. Zeitschrift für Psychologie, 61*, 161–265.

Wiggins, M.W., Azar, D., & Loveday, T. (2012). The relationship between pre-flight decision-making and cue utilization. *Proceedings of the Human Factors & Ergonomics Society, 56*(1), 2417-21.

Wu, S-C., Duong, C.G., Koteskey, R.W., & Johnson, W.W. (2011). Designing a flight deck predictive weather forecast interface supporting trajectory-based operations. *Ninth USA/Europe Air Traffic Management Research and Development Seminar.* Retrieved 27 Oct, 2015 from http://www.atmseminar.org/seminarContent/semi nar9/papers/151-Wu-Final-Paper-4-15-11.pdf .

Wu, S-C., Gooding, C.L., Shelley, A.E., Duong, C.G., & Johnson, W.W. (2012). Pilot convective weather decision making in en route airspace. *28th International Congress of the Aeronautical Sciences.* Retrieved 19 Oct, 2015 from http://ntrs.nasa.gov/archive/nasa/casi. ntrs.nasa.gov/20120016868.pdf .

Wu, S-C., Luna, R., & Johnson, W.W. (2013, October). *Flight deck weather avoidance decision support: Implementation and evaluation.* Paper presented at the 32nd Digital Avionics Systems Conference, E. Syracuse, NY. Retrieved 3 Nov, 2015 from http://human-factors.arc.nasa.gov/publications/Wu_etal_2013 Flight deck_weather_avoidance_decision_support.pdf .

Yuchnovicz, D.E., Novacek, P.F., Burgess, M.A., Heck, M.L., & Stokes, A.F. (2001). *Use of a data-linked weather information display and effects on pilot decision making in a piloted simulation study.* (Report no. NASA CR 2001-211047). Langley, VA: NASA Langley Research Center

OBJECTIVE AND SUBJECTIVE EVALUATION OF A NEW LIDAR-BASED SPEED PREDICTION AND ADVISORY DISPLAY

Jorg Onno Entzinger
The University of Tokyo (UTokyo)
Tokyo, Japan
Tomoko Iijima
Japan Aerospace eXploration Agency (JAXA)
Tokyo, Japan
Tsuneharu Uemura (UTokyo), Jun Matsumoto (JAXA),
Naoki Matayoshi (JAXA), Shinji Suzuki (UTokyo)

We evaluate a newly developed symbology that provides the pilot predicted and advisory airspeed information. This information is not only based on the current state of the aircraft, but also takes into account the wind field ahead of the aircraft measured by an onboard LIDAR system. Airline pilots flew landing approaches in wind shear scenarios that demanded careful consideration of whether to land on go-around in JAXA's full flight simulator. We obtained both subjective evaluations and objective data including flight performance, eye recoder data, electrocardiogram (ECG), electroencephalogram (EEG), and performance on a simple visual secondary task. The pilots considered all newly proposed systems useful during the landing approach, and reported better performance and lower workload compared to the conventional display, particularly in challenging situations. The objective data supported the subjective evaluation results.

The Japan Aerospace eXploration Agency (JAXA) is developing an onboard Doppler Light Detection and Ranging (LIDAR) system able to measure the wind velocity field up to several miles ahead of an aircraft when flying in clear air (Inokuchi et al. 2009, Inokuchi 2012). In a collaborative research between JAXA and The University of Tokyo we investigate how this information can be used to support the pilot's situational awareness and to reduce accidents, incidents, or inconvenience caused by strong turbulence and wind shear.

Proposed Systems

Figure 1 shows how the wind data measured by the onboard LIDAR can be used. This paper focuses on the LIDAR-based predictive wind shear (L-PWS) warning system, the predicted airspeed indicator (L-PSPD) and the target airspeed indicator (L-TSPD). Figure 2 shows an impression of the current implementation. We assume manual operations. Readers interested in future connections to the autopilot and auto throttle systems are referred to the paper by Kamo et al. (2016).

LIDAR-based predictive wind shear (L-PWS)

Closest to the raw LIDAR data is the use of LIDAR as a clear air extension to the weather radar system. The LIDAR data is displayed on the navigation display (ND) and a LIDAR-based Predictive Wind Shear (L-PWS) advisory, caution, or warning is generated analogous and in addition to the current radar-based system. An addition to the radar version is that the higher accuracy of the LIDAR allows us to provide a countdown timer until the expected wind shear occurrence ("ETA 5sec" on the ND in Figure 2).

The LIDAR data display is intended to support situational awareness on the perception level, while the warning system should facilitate decision making (i.e., preparing for or initiating a Go-Around).

LIDAR-based predicted airspeed indicator (L-PSPD)

The higher resolution and accuracy of the LIDAR system makes it possible to provide the pilot with predictions of airspeed changes up to several tens of seconds or even a minute ahead. The L-PSPD consists of 3 ovals ("bubbles") between the speed tape and the artificial horizon (Figure 2). The oval centers represent the predicted average speeds, their heights are a measure for the speed variation (i.e., short in calm air and tall in strong turbulence).

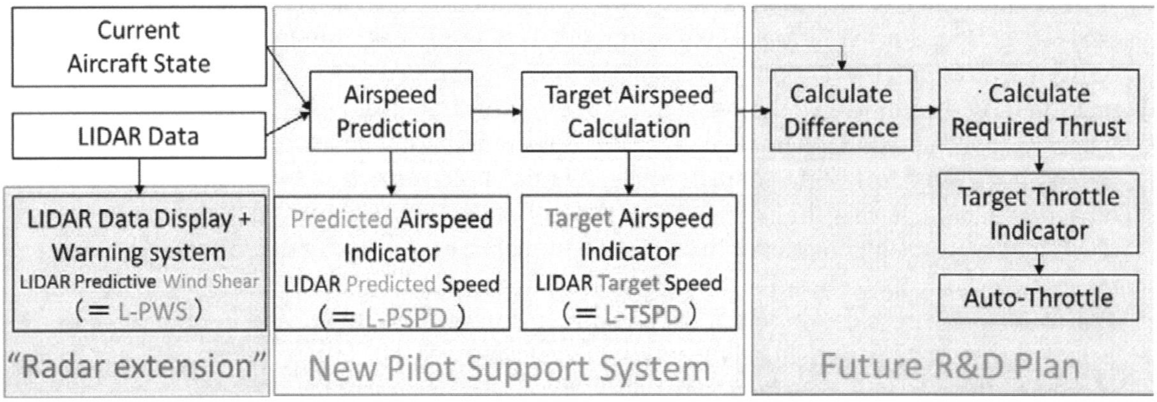

Figure 1. Overview of the proposed "SafeAvio" systems using data measured by the onboard LIDAR.

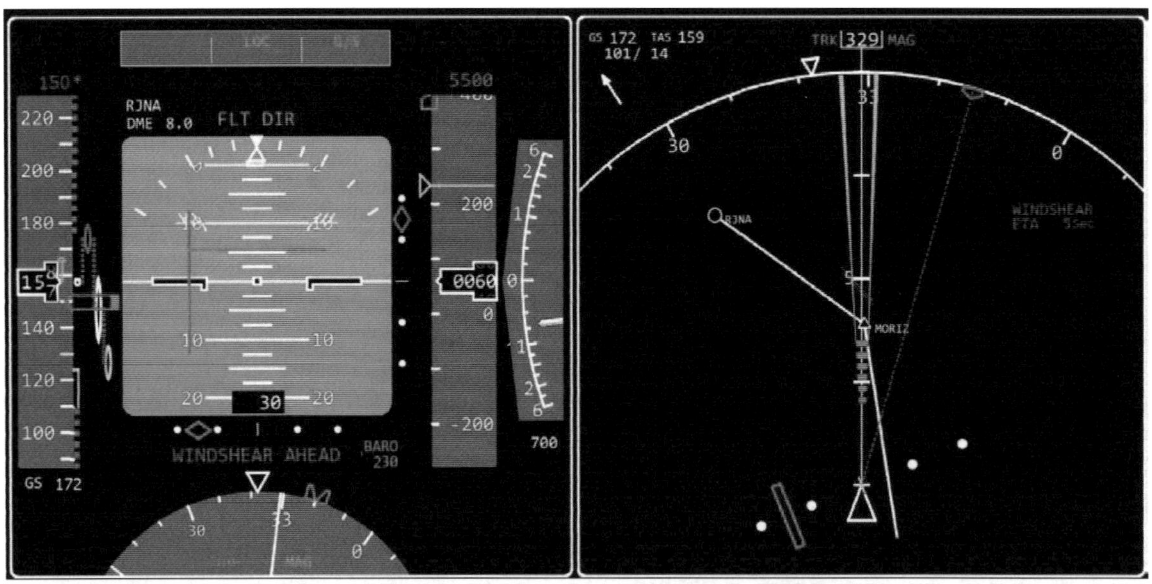

Figure 2. Impression of the L-PWS, L-PSPD ("Bubble") and L-TSPD ("Shell") additions to the Primary Flight Display (PFD, left) and Navigation Display (ND, right).

This indicator is similar to the speed trend vector on conventional displays, however:

1. it takes into account the future wind speed changes measured LIDAR (not only current wind),
2. it provides multiple predictions (e.g., 5, 10, and 20s ahead), (10s for the speed trend vector),
3. it shows the (in)stability of the airspeed through a variable height of each of the 3 "bubbles".

The predicted airspeed indicator is intended to support situational awareness on the (comprehension and) projection level. In particular, it is expected to be helpful for speed control, since the prediction could compensate the delays of several seconds the jet engines need to spin up and translate that additional thrust (acceleration) into speed.

LIDAR-based target airspeed indicator (L-TSPD)

In addition to the future airspeed predictions we propose a variable target airspeed (L-TSPD) that may temporarily deviate from the selected speed in order to prepare for large upcoming changes in wind speed. The symbology is similar to and replaces the "speed bug" or "selected speed" on conventional displays.

The predicted airspeed indicator is intended to directly support decision making, i.e., increasing or decreasing trust to keep following the advised target speed. It is expected to offer better protection of the minimum and maximum speed limits at all times.

Materials & Methods

To evaluate the proposed systems we carried out a series of simulator experiments with professional airline pilots. The experiment protocol was approved in advance by the ethics committees of JAXA and The University of Tokyo's School of Engineering. Each subject provided written informed consent before participating.

Primary Task

Subjects were asked to fly manual precision landing approaches to Tokyo Haneda airport runway 34L in JAXA's full flight simulator. The aircraft model was a generic model similar to the Boeing 737. Motion simulation was turned off. All experiments starting trimmed and on glide slope from 1400 or 2000ft altitude to 100ft altitude. Wind conditions were based on the longitudinal wind components from the FAA windshear database (Switzer et al., 1993) with additional light random turbulence. For the evaluation of the L-PSPD and L-TSPD the wind speeds were weakened to a level where a continued landing was possible, but a decision to go-around would not be unrealistic either.

Trials

In the morning subjects were given time to familiarize with the simulator, the various displays, and the various wind scenarios until they felt sufficiently familiar with them. They also practiced several trials with the secondary task.

Experiment 1A in the afternoon tested 4 different display combinations: the conventional display, and either or both the L-PSPD and L-TSPD display additions. All data except the secondary task were measured.

Experiment 1B was the same as 1A, but with a different wind profile. In addition, subjects performed the secondary task.

Experiment 2 compared the L-PWS system with a reactive windshear warning system under 2 different wind scenarios. For these trials only simulator data and subjective evaluations were recorded.

Some trials were duplicated to test for repeatability or by the subject's request (in order to develop a better basis for the subjective evaluation). After each experiment (1A, 1B, 2) the subjects filled out the corresponding subjective evaluation questions, followed by a short break.

Participants

8 male subjects participated in this experiment. 7 of them were captain pilots from 2 major Japanese airlines and 1 was a retired airline captain. The participants recruited through a contract with the airlines on the basis that they had significant flight experience and were able to provide a critical and detailed evaluation (e.g., have experience as instructor or examiner). The participants were compensated for their travel expenses.

Measuring Equipment

The simulator states were logged at 16.67Hz. Eye-mark data (pupil diameter, gaze direction, blink detection) was taken at 30Hz using the Takei TalkEyeLite. Electrocardiogram (ECG) data was recorded at 256Hz using the ParamaTech EP-301. Brainwave data (electroencephalogram, EEG) were recorded using the eMotiv EPOC+ at 128Hz.

In addition, a secondary task was administered in some of the trials. The task was a simple choice response time task, where the subject had to press one of two buttons attached to the control column depending on the change of either of two pictures displayed immediately left of the PFD within the pilot's peripheral field of view. The response time, error rate, and time-out rate were recorded.

The subjective evaluations consisted of a checklist based on the FAR 25.1301 and FAR 25.1322 Human Factors Considerations and a questionnaire focusing on perceived workload, situational awareness, and general usefulness or issues of each of the proposed systems.

Results

Experiment 1 (Conventional vs. L-PSPD vs. L-TSPS vs. LPSPD & L-TSPD)

Objective evaluation. Analysis of the simulator data showed that averaged deviations from the target airspeed (root of the mean square error, RMSE) were smaller when the L-PSPD indicator was present. This difference was significant at 5%-level compared to the conventional display and at 1% level compared to the L-TSPD only display. In particular deviations below the target airspeed were smaller. The addition of the L-TSPD indicator, on the other hand, significantly raised the all-time minimum airspeed throughout the trials. Also other flight performance parameters such as minimum and maximum pitch angles, pitch rate, and glideslope deviations showed small improvements, although differences did not reach the 5% significance level. In conclusion, the combination of both indicators helps the pilot to effectively stabilize the aircraft and guarantee a safe minimum airspeed.

From the ECG data we calculated the heart rate, which can be interpreted as a measure of arousal or stress, and an index of mental effort based on the heart rate variability (Vicente et al., 1987). The average and maximum heart rates were slightly lower for the L-TSPD only display. The mental effort, on the other hand, was slightly higher for the L-TSPD only display.

Mid-frontal brain wave activity in the theta band is said to correlate with the need for cognitive control (Cavanagh & Frank, 2014). The order from high to low activity was the conventional display, the L-TSPD, then the L-PSPD and finally the display with both L-PSPD and L-TSPD. However, the differences were not statistically significant. A more detailed analysis per phase (before wind, during wind, after simulation stop) may reveal clearer results, although the large measurement noise will still remain a problem.

The response times in the secondary task similar for the conventional, L-PSPD only and L-TPSD only displays, and only slightly slower for the display with both LPSPD and L-TSPD. For the L-PSPD display there were no timeouts, but the number of mistakes increased. One explanation could be that the subject sees the secondary task in his peripheral view when looking at the L-PSPD, noticing all changes, but not taking more effort to carefully check it.

Subjective evaluation. Pilots ranked their subjective performance best and workload lowest for the L-PSPD in Experiment 1A. In Experiment 1B with the secondary task and different wind pattern, they found the L-TSPD and combined L-PSPD&L-TSPD displays equally good (with slightly lower perceived workload for the combined display). They concluded that any of the newly proposed displays would be a valuable addition to the conventional display during the landing approach, although the preference among the new displays differed per person. Pilots commented that the lack of future wind information in the conventional display case made them initiate a go-around, and that knowing the wind changes ahead made it easier to control the airspeed.

Some subjects liked L-TSPD because it is simple. Others did not like it for the same reason: they wanted more raw information (L-PSPD) and draw their own conclusions. In case of higher workload (more difficult wind, additional tasks, etc.) these pilots would fall back on the L-TSPD, therefore the combined system proved effective.

General comment. We found significant differences between Experiment 1A and 1B for a number of physiological parameters (in particular the ECG related parameters), some flight performance parameters, and the subjective evaluations. Unfortunately, the current experiment design makes it impossible to know whether this was due to the different wind profile, the presence of the secondary task, the fact that it was the second round of trials, or a combination of these.

Experiment 2 (L-PWS versus reactive wind shear warning system)

Objective evaluation. In this experiment no psychophysiological and secondary task data were recorded. The objective (simulator) data showed highly significant flight performance improvements when using the proposed L-PWS system (smaller airspeed and glideslope deviations, lower pitch rates and sink rates, less extreme pitch angles, etc.). This is not surprising, since the Go-Around is initiated much earlier (46s on average), before the large wind speed changes occur.

Subjective evaluation. Subjects indicated that the proposed L-PWS enables them to plan ahead, and therefore reduced their workload. They noted it was compatible with the current radar based PWS system (which would detect noting in clear air) with equivalent workload and equal or even better performance. The main point of criticism was that the size of the detected wind shear area is narrow and therefore difficult to confirm on the ND. We believe this is partly mitigated by the added "Estimated time of arrival (ETA)" indication which counts down the time in seconds until wind shear. Since there is also no outside visual cues to confirm the existence of the wind shear (such as a cloud front), a clear instruction manual and training will be needed (similar to for example the Ground Proximity Warning System).

Conclusion

The proposed systems proved effective in supporting the pilot to maintain safe airspeeds and generally resulting in more stabilized landing approaches or earlier go-around decisions. Additional workload measurements did not indicate any problems, and may even be interpreted as showing reduced workload in some cases.

Subjective comments were also positive. Pilots found the newly proposed systems useful and reported lower workload because they were better able to plan ahead. There seem to be personal differences in the preference for the L-PSPD and the L-TSPD. In some cases pilots reported confusion when both were available, but overall the combined display proved best in both the objective and subjective evaluations.

Acknowledgements

We like to thank all subjects for their participation in the experiments. This research is carried out in as a collaborative research project between the University of Tokyo and the Japanese Aerospace eXploration Agency (JAXA) under JAXA's SafeAvio research program. The University of Tokyo received financial support from JAXA under their publicly funded research program.

References

Cavanagh, J.F., Frank, M.J. (2014). Frontal theta as a mechanism for cognitive control. *Trends in Cognitive Sciences*, August 2014, Vol. 18, No. 8. (pp 414-421). doi: 10.1016/j.tics.2014.04.012

Inokuchi, H., Tanaka, H., and Ando, T. (2009). Development of an Onboard Doppler LIDAR for Flight Safety. *Journal of Aircraft*, Vol. 46, No. 4. (pp. 1411-1415). doi: 10.2514/1.41738

Inokuchi, H. (2012). Detection of the Clear Air Turbulence by an Airborne Doppler LIDAR. In *Proceedings of the Asia-Pacific International Symposium on Aerospace Technology*, Jeju, Korea. (Paper #6.4.2).

Kamo, S., Entzinger, J.O., Uemura, T., Suzuki, S. (2016). Development of Control Advisory System Using Future Wind Information. *Asia-Pacific International Symposium on Aerospace Technology (APISAT)*, Toyama, Japan, Oct. 25-27.

Switzer, G.F., Proctor, F.H., Hinton, D.A., Aanstoos, J.V. (1993). *Windshear Database for Forward-Looking Systems Certification*. NASA Technical Memorandum 109012. Retrieved from https://ntrs.nasa.gov/archive/nasa/casi.ntrs.nasa.gov/19940012810.pdf

Vicente, K.J, Thornton, D.C., Moray, N. (1987). Spectral Analysis of Sinus Arrhythmia: A Measure of Mental Effort. *Human Factors*, Vol. 29, No. 2. (pp. 171-182).

DEVELOPMENT OF INCREASINGLY AUTONOMOUS TRAFFIC DATA MANAGER USING PILOT RELEVANCY AND RANKING DATA

Lisa R. Le Vie
Vincent E. Houston
NASA Langley Research Center
Hampton, VA

NASA's Safe Autonomous Systems Operations (SASO) project goal is to define and safely enable all future airspace operations by justifiable and optimal autonomy for advanced air, ground, and connected capabilities. This work showcases how Increasingly Autonomous Systems (IAS) could create operational transformations beneficial to the enhancement of civil aviation safety and efficiency. One such IAS under development is the Traffic Data Manager (TDM). This concept is a prototype 'intelligent party-line' system that would declutter and parse out non-relevant air traffic, displaying only relevant air traffic to the aircrew in a digital data communications (Data Comm) environment. As an initial step, over 22,000 data points were gathered from 31 Airline Transport Pilots to train the machine learning algorithms designed to mimic human experts and expertise. The test collection used an analog of the Navigation Display. Pilots were asked to rate the relevancy of the displayed traffic using an interactive tablet application. Pilots were also asked to rank the order of importance of the information given, to better weight the variables within the algorithm. They were also asked if the information given was enough data, and more importantly the "right" data to best inform the algorithm. The paper will describe the findings and their impact to the further development of the algorithm for TDM and, in general, address the issue of how can we train supervised machine learning algorithms, critical to increasingly autonomous systems, with the knowledge and expertise of expert human pilots.

Air traffic within the National Airspace System (NAS) is ever-increasing and "although humans today remain more capable than machines for many tasks, natural human capacities are becoming increasingly mismatched to the enormous data volumes, processing capabilities, and decision speeds that technologies offer or demand" (United States Air Force, 2010) Recognizing these challenges, NASA's Safe Autonomous Systems Operations (SASO) project's objectives are focused on developing technologies that enhance the safety and efficiency of civil aviation. Increasingly Autonomous Systems (IAS) are one avenue that could prove vital in decreasing a crew's workload, while enhancing safety and efficiency during the NextGen and other possible future airspace environments.

Utilizing IAS within the cockpit begins with understanding what an increasingly autonomous system is and what technologies are needed to profoundly improve a flight crew's overall awareness while maintaining or even decreasing workload. Autonomy allows an agent, human or machine, to act independently within a circumscribed set of goals; delegating responsibility to the agent(s) to achieve the overall system objective(s). (National Research Council, 2014). IAS lie within the sophisticated progression of current automated systems toward full autonomy. These systems, working together with humans, are expected to improve the safety, reliability, costs and operational efficiency of civil aviation. Implementation of IAS is imminent, which makes the development and proper performance of such technologies vital. The challenge is to develop these human-autonomy teams/systems where the combination of machine learning and human expertise exceeds the performance of either system alone.

To that end, an effort to develop cutting edge technology addressing an emerging airspace need as well as to serve as an IAS testbed for development and evaluation was created. The Traffic Data Manager (TDM) is an application that parses and displays traffic of interest while eliminating the clutter of insignificant surrounding traffic data. The application arises from an optimized Data Comm environment end-state where operations will become "net-centric" - as the transmittance of command, control, state, and intent information is passed autonomously between computers (agents) for efficient operational coordination and execution. Voice communication, between humans, will become non-existent as they can become a bottleneck to capacities. Nonetheless, to provide requisite human oversight, awareness, and intervention, an IAS is needed to effectively and concisely inform humans of "ownship-relevant" information (traffic, intent, messaging) being passed within this net-centric environment. The TDM application becomes an "intelligent party-line" process, only presenting (visual, aural, etc.) the information that the human must know to maintain the requisite awareness for possible subsequent action or intervention.

This technology relies on Machine Learning (ML) algorithms to parse all nearby traffic data, displaying only relevant data to the pilot. The primary component of the system is the TDM algorithm. TDM currently uses a supervised learning algorithm that relies on an Ensemble Learning framework (DeCoursey, 2003; Hastie, Tibshirani, & Friedman, 2009) where there are several methods for blending results into a very high-quality ensemble predictor. The fundamental challenge of IAS design and of this TDM application, in particular, is how to capture human expertise and knowledge and then effectively implement this knowledge within a machine learning architecture.

Data Collection Effort

Essential to the development of the machine learning algorithms was the collection of data needed to train the algorithm. A Dynamic Air Traffic Application (D.A.T.A) was developed and integrated into an EFB-like framework. Real-time flight data was randomly assigned a latitude and longitude and placed within a range of 20 or 40 nautical miles from the ownship. These data points were then displayed to the pilot in groups of 20, as shown in Figure 1a, and they were asked to rate the relevancy of the selected aircraft in relation to their own. When an aircraft was selected, a box appeared in the lower left hand corner (enlarged to enable viewing in Figure 1b) giving the selected aircraft's identification, type, altitude, speed, heading and vertical trend.

a. b.

Figure 1. D.A.T.A. Screenshots

Thirty-one pilots, all current or recently retired with an Airline Transport Pilot rating, were asked to choose a relevancy (relevant, maybe relevant, or not relevant) for each selected aircraft. This was repeated for each aircraft and each scenario. Each pilot saw 36 scenarios with 20 aircraft per scenario. Over 22,000 data points, with their selected relevancies, were collected from the pilots to be used in training the TDM algorithm.

Training TDM

The TDM supervised learning algorithm was initially trained using 75% of the 22,000 data points. These data points consist of aircraft state data that the pilots considered important to determining relevancy (i.e., course, heading, airspeed, altitude, range, bearing, etc.) as well as the pilot reported relevancy of the aircraft to the ownship's position. Testing was done with the remaining 25%, with the pilot-reported relevancy removed. The relevancy determination is only necessary to TDM in its training phase. Algorithm training took place using a MatLab Machine Learning, Tree-Bagger ensemble. The algorithm utilizes an embedded supervised learning algorithm to eliminate insignificant surrounding traffic, highlight traffic of interest or note, and identify operational significance autonomously.

Further Considerations

At the end of data collection, pilots were asked which of the two, heading or range, being the two salient parameters of the Navigation Display, was their primary consideration when choosing an aircraft in the scenario. Twenty pilots (64.5%) stated that heading was the first thing considered, while 11 pilots (35.5%) chose range. Pilot comments included:

"Heading aspect is most important between these two options. Also, altitude and trend are important considerations. Ultimately, if our flight paths will cross, then I am more likely to be concerned with a conflict, regardless of range."

"Based on range, you could rule many out quickly, regardless of climb/descent rate or speed. From there, the heading of the relevant aircraft ultimately highlighted the aircraft that posed true threats."

"Range was most important to those aircraft within several thousand feet. However the factors of descending or climbing in regards to heading are very important as well, making both very equal considerations in regards to converging traffic."

The pilots were also asked to rank order the importance of the secondary flight information given in the inset box for each aircraft (choices: aircraft's identification (ID), aircraft type, altitude, speed, course, and vertical trend; ranking of 1 being most important; 6 being least important). The results are shown in the box plot in Figure 2 showing the central tendency (mean rating (circle) and median), 25th and 75th percentile by the box height, 1.5 times the interquartile range by the whiskers, and asterisks denote outliers. Altitude was ranked of highest importance by 22 pilots (71%). The rankings of course and vertical trend were not statistically significant (T-Value = -1.19, P-Value = 0.245) enough to distinguish between a ranking of two or three. However, pilots generally agreed that speed, aircraft type, and aircraft ID were of lesser importance with rankings of four, five and six, respectively. Eleven pilots were in agreement as to a ranking order of: altitude, vertical trend, course, speed, aircraft type, and aircraft ID. This left 20 pilots choosing another nine separate ranking orders.

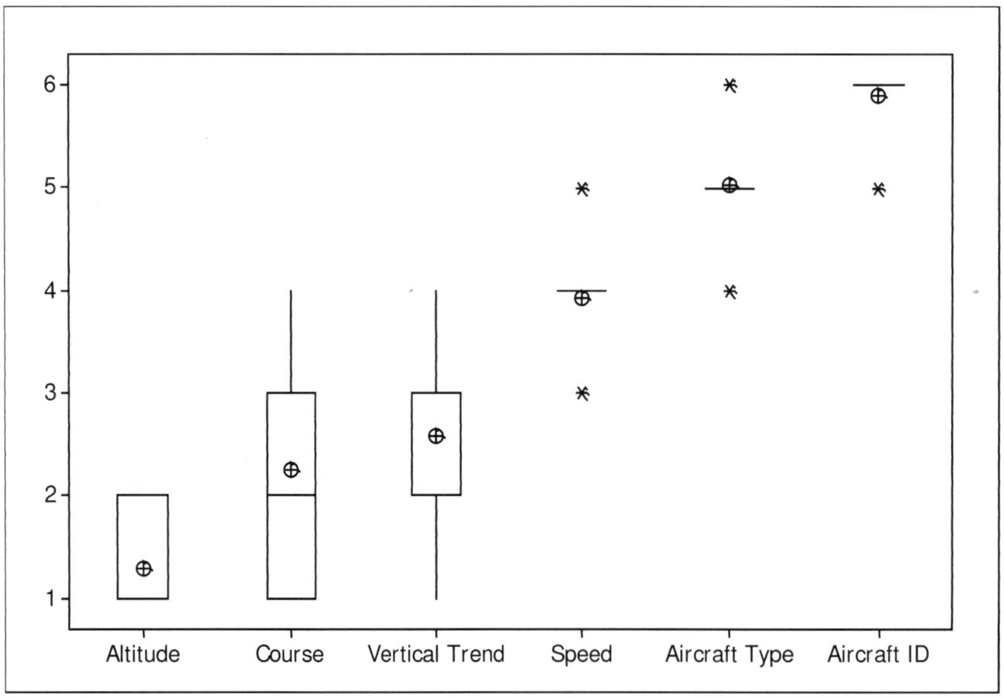

Figure 2. Rank Ordering of Secondary Aircraft Information (1: Most Important)

The pilots were asked to fill out a System Usability Scale (SUS) and ranked D.A.T.A.'s usability at 85.75 (Best). The SUS uses a five-point Likert scale (from Strongly Agree to Strongly Disagree) for a 10-item questionnaire. The scoring ranges from 0-100 and is seen as a reliable measure of usability. (Brooke, 1996) Pilots were also encouraged to give additional feedback of the system, if they wanted. Several pilots suggested the addition of ownship vertical speed to the data collection application, as they find it an important component when deciding relevancy. Others commented that knowing the selected aircraft's final climb or descent altitude would be helpful in determining the difference between "Maybe Relevant" and "Relevant/Not Relevant" These suggestions will be addressed in future work.

Discussion

We have now collected over 22,000 data points, from 31 ATPs, of traffic state data relative to ownship (i.e., altitude, course, range, bearing, speed) with pilot-derived relevancy labels (Relevant, Maybe relevant, and Not Relevant) to ownship. In addition, we have pilot-reported data on whether heading or range was the first consideration when choosing an aircraft to select. Pilots also ranked the order to which they used the aircraft information given to them when considering relevancy to ownship. After data collection, any discussion and additional pilot comments were also captured. These data were used to train machine learning algorithms that are designed to mimic human experts and their expertise. The detailed results of this work are reported elsewhere (Houston, Le Vie, in press) but overall, the algorithms are showing between 70% and 80% classification accuracy to the training. The results look promising, but not without challenges. One challenge experienced was making sure the machine learning algorithm was given not only enough data to train, test and learn from, but also enough of the "right" data. Through talking with pilots, a better understanding was gained in how they used the aircraft state data shown to make assumptions and predictions, which added extra information to their decision making. This was critical information that had not been considered and was not being provided to the machine learning algorithm. Having an expert walk through their decision-making process and selection method was fundamental in making sure that all of the processes that go into making a decision on whether an aircraft was relevant or not was captured and included. This effort continues to be a work-in-progress and is being used as a learning platform for the researchers to better gather this expertise in the future.

Future Work

This paper describes a data collection effort for training machine learning algorithms that will determine traffic relevancy, as the first-step in developing an intelligent party-line application and as a testbed for IAS development and evaluation. The TDM application is now running, using these training data, and shows promise in creating an intelligent decluttering and parsing agent.

The next immediate step is to assess the accuracy of the training data to the "expert" pilot population in general. An algorithm may never perfectly match the relevancy rating of every user. In fact, there is frequent disagreement among expert users about the "threat" of any individual aircraft. (St. John, Smallman, Manes, Feher & Morrison, 2005) In a study of six teams evaluating a threat management display, the interest level of an aircraft was agreed upon for only 41% of the aircraft. (Marshall, Christensen, & McAllister, 1996); Additional data collection up-coming will assess this training data against a new pilot population and assess the robustness of capturing expert pilot data for IAS development.

Future efforts will include real-time evaluation of the TDM algorithm performance and its ability to accurately predict air traffic relevancy in reference to ownship, the latency of its predictions, and its integration with other technologies being developed. As a learning/adaptive system, the stability of the algorithm as it adapts to the environment and changes will be assessed as this behavior may be a critical element in trust and human-autonomy teaming. Further, the system will be used for metrics development and evaluation as a Data Comm environment tool, assessing if relevancy changes as the operational context changes. The labels or markers for relevancy will also be expanded to include contextual or communication markers such as airport or runway identifiers, company names, routes of flight, etc.

Acknowledgments

This research was funded by NASA's Aeronautics Airspace Operations and Safety Program's Safe Autonomous Systems Operations (SASO) project. The contributions of Jason Neuhaus and Randall Bailey (NASA LaRC) were especially critical to the success of this work.

References

Brooke, J. (1996). SUS: A "quick and dirty" usability scale. In P. W. Jordan, B. Thomas, B. A. Weerdmeester, & A. L. McClelland (Eds.), *Usability Evaluation in Industry.* London: Taylor and Francis.

DeCoursey, W. J. (2003) *Statistics and probability for engineering applications: With Microsoft Excel.* Boston, MA: Newnes.

Hastie, T., Tibshirani, R., Friedman, J. (2009) *The elements of statistical learning: Data mining, inference, and prediction, second edition.* New York: Springer-Verlag. doi: 10.1007/978-0-387-84858-7.

Houston, V. E. & Le Vie, L. R. (in press). *Autonomous System Technologies for Resilient Airspace Operations*

Marshall, S.P., Christensen, S.E., & McAllister, J.A. (1996). Cognitive differences in tactical decision making. In *Proceedings of the 1996 Command and Control Research and Technology Symposium* (pp. 122-132). Monterey, CA: Naval Postgraduate School.

National Research Council (2014). *Autonomy Research for Civil Aviation: Toward a New Era of Flight.* Washington D.C.: National Academies Press. doi: https://doi.org/10.17226/18815.

St. John, M., Smallman, H. S., Manes, D. I., Feher, B. A., & Morrison, J. G. (2005). Heuristic Automation for Decluttering Tactical Displays. *Human Factors, 47*, 509-525. doi: https://doi.org/10.1518/001872005774860014

United States Air Force. (2005). Heuristic Automation for Decluttering Tactical Displays. *Human Factors, 47*, 509-525. doi: https://doi.org/10.1518/001872005774860014

United States Air Force, "Technology Horizons - A Vision for Air Force Science & Technology During 2010-2030" AF/ST-TR-10-01, 15 May 2010.

Four-Year Follow-Up of Intensive, Simulator-Based Pilot Training

Paper to be presented at the 19th International Symposium on Aviation Psychology
May 8 - 11, 2017, Dayton, Ohio - Proposal ID: 5780

Maxine Lubner, Ph.D.
Vaughn College of Aeronautics and Technology, New York, NY
Deb Henneberry, M.A.
Vaughn College of Aeronautics and Technology, New York, NY
Sharon DeVivo, Ed.D.
Vaughn College of Aeronautics and Technology, New York, NY
Andrew R. Dattel, Ph.D.
Embry-Riddle Aeronautical University, Florida
Emerson Allen, B.S.
Vaughn College of Aeronautics and Technology, New York

Sixteen students, who began with 0-20 flight hours, enrolled in an intensive, simulator based, collegiate training program. They completed their training with fewer flight hours than the US average: FAA Private Pilot within 4-6 weeks; instrument ratings in 3-4 weeks; Commercial within an average of 20 weeks and CFI ratings in an average of 40 weeks and all graduated with Bachelor's degrees (ISAP, 2013). The students had met selection criteria. At the time, indicators of success included variables associated with simulator based training, camaraderie, shared learning and opportunities to reflect on training. Four years later, 81% are now flying professionally: eight as flight instructors, four as first officer airline pilots, one commercial pilot, and three employed elsewhere in the aviation industry. This qualitative follow-up study suggests that initial, rapid learning was neither shallow nor short-lived. Most are still friends. They became active alumni and mentors for incoming flight students.

Simulator training for pilots is widely used and well regarded, particularly for advanced pilots, usually flying for airlines (McLean, Lambeth, Mavin, 2016). In the U.S., the role of simulator based training (SBT) is used for training ab initio pilots, but with varying times and methods (McLean et al., 2016). The reasons for inconsistent use of SBT for beginner pilots are numerous, including regulatory, such as the limited time that the Federal Aviation Administration (FAA) allows to be entered in pilots' log books (14 CFR Part 61 or Part 141), or because the protocols for this training and evidence of its effectiveness are still being examined (Goetz, Harrison, Voges, 2015).

Advantages of simulator training are well understood and include cost savings in terms of equipment requirements, particularly as moderate fidelity of simulators can be almost as effective as high fidelity trainers; tuition can take place regardless of weather or flying conditions; and that dangerous or unusual maneuvers can be taught without risk to the people and equipment involved (Harris, 2011; Salas, Bowers, & Rhodenizer, 1998; Taylor et al., 1999).

Techniques for flight training and general comprehension of learning processes have been fruitfully explored with the use of simulators. It is clear that certain scenarios, such as freezing a situation for detailed examination, can only be created in a simulator. Cognitive process have been examined, including questions about how part-task training can increase conceptual learning appropriate for complex situations (Dattel, Durso & Bedard, 2009); how procedural memory is acquired (Koglbauer, Riesel & Braunsting, 2016); when positive transfer of skills occurs (Koglbauer et al., 2016); and in situations where the level of expertise and the amount of detailed instruction do not correctly match, then negative transfer of training occurs (Hsu, Gao, Liu & Sweller, 2015). Simulator training can easily address a variety of styles of learning, such as conceptual, procedural, scenario, collaborative and individual styles of training (Dattel, et al., 2009, Dattel, Kossuth, Sheehan, & Green, 2013).

Complex cognition and communication and management within the social and technical context of flying with a crew or other actors in the aviation context have also been studied. In these more complex situations, scenario based instruction to teach aeronautical judgement and decision making and crew resource management (CRM) has been used in the simulators (FAA, 2008; Johnston, McDonald, and Fuller in Harris, 2011). It is important that pilots learn to operate in a multi-engine, multi-crew environment. Pilots must learn how to operate multifaceted technology and automation related to the aircraft, airport and airspace systems, and manage to fly in a complex, demanding and dynamic environment. Scenario centric training and SBT are complementary because the simultaneous demands of complex operations can be taught. The social and psychological components of instruction and flying in the real world, such as cognitive advancement of skills to manage flight operations, decision making, and ways to increase motivation, create useful attitudes, or uncover gaps in comprehension, can – and should - be taught.

The effectiveness of SBT in comparison to conventional training in the aircraft is supported (McLean et al., 2016), especially when specific cognitive processes such as those listed above are properly implemented. However, the duration of simulator centric learning is hard to assess. Multiple and hard-to-control variables as well as expenses of tracking pilots over time make duration questions difficult to test. Similarly, although cost savings because of reduced time required for SBT is accepted (Goetz et al., 2015), the training's longitudinal cost-effective benefits are not well documented. In this qualitative cohort study, the effectiveness of SBT over time was explored.

A descriptive examination of the effectiveness of a simulator-based training program for pilots was conducted. Of 55 students of varying backgrounds, but mostly with limited flight experience, 16 enrolled in an intensive, simulator-based flight training program. Within two years the remainder had enrolled in conventional collegiate flight training, supplemented with some simulator training. The students in the intensive program completed their FAA Private Pilot certificates in an average of 5 weeks (not including simulator time). Moreover, the intensive program group earned their private pilot's certificate in statistically significantly fewer hours (M=46.03) than the conventional collegiate flight training group (M=76.06). The intensive group returned to conventional training and completed their Commercial certificates in an average of 20 weeks and CFI qualifications in an average of 40 weeks. The potentially useful aspects of the intensive program are discussed, including type of training such as intensive classroom, simulator and traditional in-aircraft instruction in addition to the psychosocial impacts of camaraderie and shared learning experiences (Lubner, Dattel, Henneberry, & DeVivo 2105).

Aviation simulators have been a part of flight training since 1909, shortly after the Wright Brothers' first flights. The precursor to the modern aviation simulator, the Link Trainer, was developed as a cost effective and efficient form of flight training that could improve instrument flying skills from the early days of flying and during World War II (Wicks, 2003). When designed correctly, a training program that includes the appropriate use of simulators will provide facets of instruction that may not be otherwise possible (Harris, 2011).

Simulator centric training (SCT) offers several advantages. Firstly, depending on the equipment used and scenario being taught, costs can be significantly reduced when simulators instead of in-aircraft training are utilized. Capital investment in aviation simulators is becoming increasing affordable because high fidelity simulation is not required for positive transfer of training (Salas, Bowers, & Rhodenizer, 1998; Taylor et al., 1999). Secondly, overall training time can be used more efficiently because simulator training can take place when inclement weather prohibits in-aircraft training. Thirdly, many effective training scenarios can be created in a simulator. Learning objectives can be implemented in a deliberate manner to ensure that all performance criteria are satisfied. Fourth, by freezing the simulator during performance evaluation, deficiencies can be discussed as they occur. Full attention can be given to the analysis without devoting the resources needed to fly the airplane.

Fifth, the simulator offers many opportunities for part-task training, where the instructor can break a complex task into smaller parts so that the student can concentrate on mastering those and then re-incorporate the components into the larger task (Dattel, Durso, & Bedard, 2009; Harris, 2011). By evaluating performance at the time of action, flight instructors can better assess students' conceptual understanding of situations when part-task training is implemented. A greater conceptual understanding is particularly important for complex aviation maneuvers, non-routine conditions, and situation awareness (Dattel, Durso, & Bedard, 2009). One example of part-task training is allowing students to control the aircraft's yoke while the instructor handles the task of using the throttle. Another less commonly employed example is to have the student use only the throttle while the instructor operates the other airplane controls. Performing these exercises in a simulator allows the additional and important opportunity to return to the smaller building blocks making up those tasks, while engaging the student's conceptual understanding of the procedure. In this example, the simulator records the student's actions, thereby allowing analysis and reflection of each task component by the student and the instructor.

Sixth, by incorporating scenario-based training, students are able to develop mental models that permit them to hone judgment and decision-making skills for a variety of situations (FAA, 2008). Other factors have been examined in relation to SBT. Complex skill sets, such as crew resource management, have been positively transferred in even the most commonplace desktop simulators (Johnston, McDonald, and Fuller in Harris, 2011).

Comprehensive instruction in a simulator must use conceptual and procedural methodologies, both of which are independent of simulator fidelity (Hawkins, 1997). Conceptual training is accomplished by using scenario-based instruction as a part of the pilot's decision making process. Scenario-based instruction also assists teaching other skills, including traffic pattern operations. Simulator training can easily incorporate conceptual, procedural, scenario, collaborative and individual styles of training (Dattel, et al., 2009, Dattel, Kossuth, Sheehan, & Green, 2013). While flight simulators are generally considered an enhancement to the training process, a multi-factorial, instructional model should be followed by instructors and program designers. Simulator training should avoid excessive reliance on simulation-centric training. Certainly, individual instructor effectiveness is reported as necessary to ensure positive and satisfying pilot training (AOPA, 2010). Cognitive, and possibly psychosocial variables related to the students should also be included in a comprehensive flight training program. Several individual level variables have been found to influence training outcomes before and during training, including motivation, self-efficacy and attitudes (Alvarez, Salas, & Garofano, 2004). Scenario centric training should enhance SBT because scenarios require use of social and psychological skills, such as collaboration, communication, decision making, develop useful attitudes, ways to increase motivation, and address gaps in comprehension.

This qualitative paper describes the progress of three cohorts of Vaughn College sixteen students who participated in a simulator based, intensive flight training program four years ago. These students began their flight by traveling to the southwest US, stayed near a small airport and undertook a short duration, intensive simulator-based, ab initio flight training program. Later, the students returned to New York and completed the remaining flight qualifications required for their Bachelors' degrees in Aircraft Operations. Back in New York, they followed conventional training that offered some simulator practice. Lessons in New York were spaced over time and students had conventional opportunities for group interactions. A larger group of students who were not selected for the intensive program had remained in New York, where they had conventional flight training with some simulator practice too. In this follow-up, qualitative study, the progress of the sixteen students who participated in the intensive SBT program is reported. Questions are explored regarding the duration and efficiency of obtaining initial flight qualifications; predictors of training effectiveness; motivators; and duration of knowledge and skills acquired during initial learning.

Method and Program Description

Four years ago, starting in January 2012, three cohorts, totaling sixteen students, participated in the intensive, simulator based flight training program in the southwest United States. Each cohort of five to eight students traveled and studied together, following an intensive, simulator based program. The students had to meet several criteria, including having a G.P.A. of 3.0 or better, possessing an FAA Class III Medical Certificate, taken a demonstration flight, successfully passed the FAA private pilot knowledge exam, obtained financial counseling and agreed to remain substance free during the training period.

The students were expected to travel between the Texas and New York. In the Texas, they were to undergo intensive SBT, then return to New York to complete their academic studies and finish their FAA flight qualifications (private, instrument and commercial) as needed. The students stayed in the Texas for 4-6 weeks at a time, undergoing training in simulators and aircraft six days per week. Students lived in a hotel and dined together. As the program unfolded, the second cohort group could only travel to the Texas flight school twice – for private pilot and instrument training. The third group only participated in the Texas, SBT for their private pilot training. The conventional training in New York was conducted at a Part 141 flight school, located about an hour's drive from Vaughn College. By fall 2013, all students attended the conventional flight training at this Part 141 flight school. Students had limited access to simulators at the flight school and at Vaughn College.

In February and March 2017, semi-structured interviews were conducted with the three cohorts of students. The interviews were coded and examined for themes related to a priori questions of predictors of learning and impact on careers. The authors met to discuss results and conclusions to ensure agreement of interpretations. This follows accepted qualitative methods of analysis (Creswell, 2013). Outreach to each student included one to several contacts by one or more of the program instructors and administrators. Most students expressed delighted willingness to participate in the interviews, but two of the cohort members were not interviewed. One of the non-responders agreed to the interview, but did not participate. The second did not respond to any of the contacts. Some information on the progress of these two non-responders was obtained by looking up publicly available records, including the FAA airmen database, Linked-In and Facebook. The career paths of the two non-responders appear similar to those of their cohort members' paths (see below). The non-responder who did not participate in the interview obtained some flight qualifications and is working at a local, large airport and has recently returned to flight training. The second non-responder obtained flight qualifications up to ATP Instrument and two type ratings, and is flying as a first officer for a regional airline.

Analyses

Fourteen interviews were completed (11 m, 3 f). All but two of the interviewees had obtained a bachelor's degree. Chosen undergraduate major was equal between aeronautical science and aircraft operations (See Figure 1).

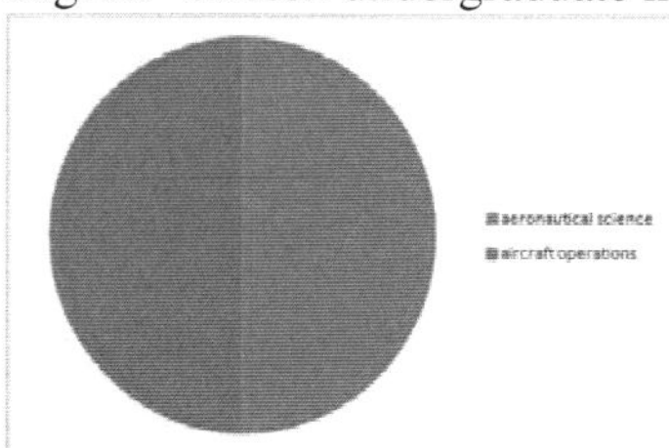

Figure 1. Cohort's undergraduate major:

The cohorts were interviewed about their experiences in the intense aircraft and simulator training. Private pilot training, instrument training, and commercial pilot training were all conducted in both the airplane and the flight simulator. Cohorts commented on their experience with the flight instruction, the mentor while training, their experience with the full motion flight simulator, and how effective the training was in skill development and knowledge retention. Cohorts were specifically asked about their experiences with their peers and the camaraderie that developed. Finally, the cohorts were asked about if they felt like their career goals were met, and if they were now mentors.

All cohorts seemed to be happy in their current position. Of the 12 interviewees who have obtained their BS, 11 are currently employed in paid pilot positions (See Figure 2 for a breakdown of employment positions). All interviewees said that their career goals have been met, or they were approaching their goals. One interviewee is on a hiatus from obtaining additional flight licenses and ratings due to medical reasons.

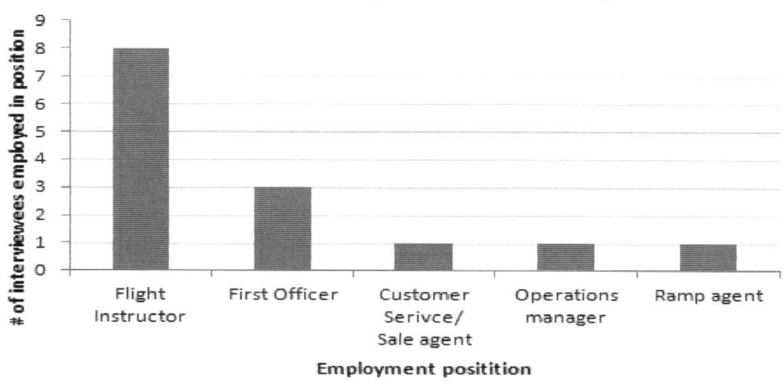

Figure 2. Employment positions:

Perception of instruction

Overall, the interviewees perceived the flight instruction as beneficial — better than they could have received at a traditional flight training program. However, which aspects were most beneficial varied. Everyone seemed to like the accessibility to practice on the flight simulator. As one interviewee commented, "24 hour access to the flight simulator helped me really learn." Another cohort admitted to not taking advantage of practicing in the flight simulator until the end of the training period. Practicing scenarios in a flight simulator without supervision has the potential for using improper flight skills that could lead to negative transfer. Some students commented that the flight instructors were better in the simulator, while others commented that the flight instructors were better in the airplane. Because the flight simulator was new to the instructors too, there may have been a learning curve that was developing with the transition from teaching in the airplane to teaching in the simulator. The consensus from the cohorts is that the flight simulator was most beneficial during the instrument training portion. One interviewee commented that it was easier to ask the flight instructor a question in the flight simulator than it was in the airplane.

Mentor

Having a mentor on site was considered a great advantage. The students felt that they could ask questions of the mentor that they were not comfortable asking of the flight instructors. Selected comments about the mentor were: "I felt that the mentor was my advocate"

- "Having the mentor there made me feel less nervous"
- "You need someone to give you insight"
- "Very knowledgeable. I felt like I could go back to my mentor for information"

Cohort as mentors

All but one of the interviewees said that they are now mentors. Being a mentor is not only rewarding, but the interviewees recognize that they learned from being mentors. Selected comments about being a mentor included:

- "When teaching (as a CFI), I love to see a student's progress. It's magical!"
- "Teaching others teaches you"
- "I learned from teaching others"
- "Mentoring helped me to use my knowledge"
- "Teaching helps me to prepare and move forward"
- "Mentoring provides a sense of satisfaction"
- "Mentees follow the lead of the mentor, so you have to always perform at your best"

Camaraderie

The most important aspect of the flight training program may be the camaraderie that was developed and how it contributed to the learning. The students spent several months (in various time periods) thousands of miles from their home university. For one student, the initial trip to the flight training program was the first experience flying on a commercial flight. Although each member of a cohort may just have been an acquaintance at the onset of the flight training program, they returned as lifelong friends.

Every interviewee stated that they still keep in contact with almost every person in their cohort. The cohorts professed a range of benefits from the camaraderie, including able to share concerns, learn from another, or just to socialize with a familiar friend who shares the same passion. One interviewee indicated meeting up with a cohort to "practice flying together." All interviewees claimed that the cohort helped facilitate their flight training. As one student stated, "we can discuss and learn from each other," while another student stated, "we share the same passion and support each other." Although the cohorts were supportive of each other, one did admit that "healthy competition builds motivation." Nonetheless, the one absolute consistency in the interviews was the interviewees perceived the importance of the relationships that were developed in the cohorts.

Conclusion

In January 2012, Vaughn College, New York City, launched a flight training program in partnership with a new training entity in Texas. Three cohorts of students participated over the next 18 months. While in Texas, these students flew twice a day five to six days a week, had constant access to simulators and were encouraged to use them to practice beyond their two flight lessons per day. The simulators provided ample opportunity to practice their emerging skills, but could have provided greater assistance if the instructors had been trained in a teaching pedagogy that provided reinforcement to flight lessons in a scenario-centric, structured and goal-oriented format. Once students acquired a baseline of knowledge and skill, the simulators were more helpful to the training process.

As stated by several members of the group, this was an intense form of training that required commitment, focus and a strong desire to achieve their goals. As demonstrated in the interviews, the aid of an on-site mentor, someone who had been both a flight instructor and was a current commercial airline pilot, supported student learning by providing additional information, advocacy and encouragement through the process. Another key finding is the role that camaraderie played in further supporting learning in terms of flight knowledge, skills and in sustaining their passion for flying. That sense of connectedness formed deep bonds between the students that have continued almost five years later and continue to be a source of information, career advice and friendship.

As compared with conventional flight training, where students are not flying 10 to 12 times per week but possibly one to three times a week, the advantage for these students was the ability to build flight skills in a focused setting. However, several students stated that a real deepening and understanding of those skills did not occur until they pursued their Certified Flight Instructor rating, which occurred in a conventional setting. This would seem to indicate that while intensive training has a role, in the case of these students it may not have been appropriate for every level of training. With roughly 67 % of the students currently flying as a profession and 100 % involved in the aviation industry, the results indicate that the program assisted students in achieving their goals. Finally, what cannot be understated is the passion that these students brought to their training and continue to bring to their pursuit of their goals in aviation. There was a sense across the interviews that the drive, commitment and focus required to fly is transformative and produces a student who becomes a teacher, while always keeping a "student mindset" to stay current and maintain their knowledge and skills.

In terms of further study, as part of the interviews students were also asked to rate their experiences on a Likert scale which will be analyzed later by comparing students to each other and to students pursuing conventional flight training. In terms of the simulator instruction, this study suggests that further work and training can be conducted to develop objectives for each simulator lesson tied to the stated outcomes of a particular certificate or rating to deepen learning and, potentially, reduce the learning in the aircraft. Additional study could be focused on the efficacy of that work and its subsequent impact on student training.

References

Aircraft Owners and Pilots Association (AOPA) (2010) The Flight Training Experience: A Survey of Students, Pilots and Instructors. http://www.aopa.org/

Alvarez, K., Salas, E., & Garofano, C. M. (2004). An integrated model of training evaluation and effectiveness. *Human Resource Development Review,* 3, 385-416.

Creswell, J.W. (2013). *Qualitative Inquiry and Research Design (3rd ed.)* Los Angeles: Sage

Dalgarnoa, B., Kennedy, G. and Bennett, S. (2014). The impact of students' exploration strategies on discovery learning using computer-based simulations. *Educational Media International*, Vol. 51, No. 4, 310–329, http://dx.doi.org/10.1080/09523987.2014.977009

Dattel, A. R., Kossuth, L., Sheehan, C. C., & Green, H. J. (2013). Poster presented at the *84th Annual Meeting of the Eastern Psychological Association*. New York, NY.

Dattel, A. R., Durso, F. T., & Bédard, R. (2009). Procedural or conceptual training: Which is better for teaching novice pilots landings and traffic patterns? *Proceedings of the 53rd Annual Meeting of the Human Factors and Ergonomics Society*. San Antonio, TX.

FAA(2008). Aviation Instructor's Handbook. Retrieved from http://www.faa.gov/about/office_org/headquarters_/avs/offices/afs/afs600

FAA (2008) Advisory Circular (AC) 00.2-15. Retrieved from www.faa.gov.

Goetz, S., Harrison, B. and Voges, J. (2015). The use of FAA flight training and aviation training devices at UAA institutions. *Collegiate Aviation Review*, 44 – 59.

Gopher, D., Weil, M., Bareket, T. (1994). Transfer of skill from a computer game trainer to flight. *Human Factors*, 36, 387-405.

Gopher, D., Sivan, R., & Iani, C. (2001). Comparing learning curves of experts and novices. *Proceedings of the Human Factors and Ergonomics Society*. 2, 1805.

Harris, D. (2011) *Human Performance on the Flight Deck.* Brookfield, VT: Ashgate.

Hawkins, F.H. (1997). *Human Factors in Flight. (2nd ed).* Brookfield, VT: Ashgate.

Koglbauer, I. (2016). Simulator training improves pilots' procedural memory and generalization of behavior in critical flight situations. *Cognition, Brain and Behavior*. XX:5 (December), 357-366.

Koglbauer, I., Riesel, M. and Braunstingl, R. (2016). Positive effects of combined aircraft and simulator training on the acquisition of visual flight skills. *Cognition, Brain and Behavior.* XX:5 (December), 309-318.

Hsu, Y., Gao, Y., Liu, T. C., & Sweller, J. (2015). Interactions Between Levels of Instructional Detail and Expertise When Learning with Computer Simulations. *Educational Technology & Society*, 18 (4), 113–127. 113 ISSN 1436-4522 (online) and 1176-3647 (print).

Lubner, M.E., Dattel, A.R., Henneberry, D., DeVivo, S. (2105) Follow-Up Examination Of Simulator-Based Training Effectiveness *18th International Symposium on Aviation Psychology*, Dayton, Ohio.

McLean, G. M. T., Lambeth, S. and Mavin, T. (2016). The Use of Simulation in Ab Initio Pilot Training. *The International Journal of Aviation Psychology*, 26:1-2, 36-45, DOI:
10.1080/10508414.2016.1235364 http://dx.doi.org/10.1080/10508414.2016.1235364

Salas, E., Bowers, C. A., & Rhodenizer, L. (1998). It is not how much you have but how you use it. *The International Journal of Aviation Psychology*, 8, 197-208.

Taylor, H. L., Lintern, G., Hulin, C. L., Talleur, D. A., Emanuel, T. W., & Phillips, S. I. (1999). Transfer of training effectiveness of a personal computer aviation training device. *The International Journal of Aviation Psychology*, 9, 319-335.

Wicks, F. (2003). Trial by flyer. *Mechanical Engineering*, 4-10.

THE EFFECT OF SURPRISE ON UPSET RECOVERY PERFORMANCE

Annemarie Landman
Delft University of Technology / TNO
Delft, The Netherlands / Soesterberg, The Netherlands

Eric L. Groen
TNO
Soesterberg, The Netherlands

M.M. (René) van Paassen
Delft University of Technology
Delft, The Netherlands

Adelbert W. Bronkhorst
TNO
Soesterberg, The Netherlands

Max Mulder
Delft University of Technology
Delft, The Netherlands

Introducing the element of surprise is one of the main challenges in simulator training of in-flight emergencies. In this simulator study, we investigated the differences in performance between predictable and surprising circumstances, in order to obtain insight into the transfer of training between predictable training settings and surprising circumstances in operational practice. This was done by testing twenty airline pilots who recovered from an aerodynamic stall in two conditions: one anticipation condition and one surprise condition. All pilots practiced beforehand using predictable, or non-surprising scenarios. The results show that pilots had significantly more difficulties in adhering to components of the FAA-commissioned recovery template in the surprise condition compared to the anticipation condition. These results suggest that predictable training may not be enough to prevent serious performance decrements under surprise.

As surprise and startle are considered to play an important role in a significant proportion of airplane safety events, aviation authorities have mandated the introduction of surprise and/or startle in upset prevention and recovery training (EASA, 2015; FAA, 2015). Both surprise and startle may occur in response to unexpected events, although the former relates specifically to a cognitive mismatch between new information and expectations (Meyer, Reisenzein & Schützwohl, 1997), while the latter refers to a highly physiological, sudden increase in stress (Martin et al., 2015). In the case of surprise, solving the cognitive mismatch (i.e., sensemaking) may be a mentally taxing and difficult task if one is unfamiliar with similar situations. It may require an adaptation, or switch, of one's cognitive "frame" (i.e., reframing; Rankin, Woltjer & Field, 2016). Frames are mental structures within which knowledge and procedures are grouped, and through which information is processed and understood (Klein, Phillips, Rall, & Peluso, 2007). Surprise is an indicator signaling that one's presently active frame is unable to fit with the emerging situation. This mismatch may cause the sensation of a loss of "grip" on the situation, and the desire to explain and understand it. If the surprising situation is also startling or threatening, the concomitant stress can be expected to impede the top-down, or goal-directed process of reframing (Eysenck, Derakshan, Santos & Calvo, 2007), which may further impair one's ability to respond quickly and appropriately (Landman, Groen, Van Paassen, Bronkhorst & Mulder, submitted). In contrast, when an upcoming event is anticipated, sensemaking can occur beforehand. The event is then immediately understood and less stress-evoking, which facilitates a quick and appropriate response.

It follows that if a procedure is only trained in highly anticipated conditions, the sensemaking activities that would be needed to identify and understand the situation before the procedure can be applied in operational practice, are never really practiced. The current simulator study aimed to test whether performance on a learned recovery procedure indeed suffers in surprising compared to anticipated conditions. In addition, the experiment aimed to test whether surprise can be used in simulator training to provide more challenging and realistic scenarios.

Several simulator studies have been published in which the effect of surprise was tested on pilot performance of learned procedures. One study (Schroeder, Bürki-Cohen, Shikany, Gingras & Desrochers, 2014) showed that adherence to a recovery template suffered when pilots were unexpectedly exposed to a previously practiced upset (aerodynamic stall). However, the experiment did not include a control condition, in which the pilots' performance was re-tested in a non-surprising scenario. Another relevant study showed that response times were longer when a stall was pilot-initiated versus when it was unannounced (Casner, Geven & Williams, 2013). However, this study did not include a detailed analysis of performance. The current study adds to these previous studies by comparing the effect of surprise to that of anticipation (manipulation check), while measuring several aspects of adherence to the recovery template.

Method

Participants

Twenty male airline pilots participated in the study (mean age = 36.3 years, SD = 7.88; mean flying experience: 12.4 years, SD = 5.05; 6986 flight hours, SD = 3804). Experience in operating medium-size twinjet aircraft types was required. Eight pilots had mainly experience with the A330, five with the B737, six with the E190 and one with the A320. All pilots were employed at the time of the experiment, and had been on duty at least once in the week prior to the experiment. Five were currently employed as captains, eleven as first officers and three as second officers. The pilots provided written informed consent prior to participation and the ethics committee of the TNO Soesterberg research institute approved the experiment.

Apparatus

The experiment was performed in the DESDEMONA flight simulator (manufactured by AMST Systemtechnik), located at TNO Soesterberg. The cockpit mockup was styled after the Boeing 737NG, and included the primary flight display, navigation display, engine-indicating and crew-alerting system, and a partial mode control with flight director and autopilot mode controls. There was no overhead panel or flight management system. Controls consisted of a yoke with control loading on pitch, rudder pedals with rudder limiter, throttles and a stabilizer with electric trim. Flaps and speed brake were not used. The aerodynamic model was derived from the SUPRA project (Groen et al., 2012), which extended the aerodynamic envelope of transport category aircraft (e.g., Boeing 737NG, Airbus A321, into high angles of attack.

Tasks and Conditions

Pilots were instructed beforehand that the simulator session would comprise two subsequent sections of circa 20 minutes: one upset recovery section and one spatial disorientation section. They were told that both sections were aimed at testing the simulator fidelity. In reality, the first section was used for practice of stall recoveries, while the second section did not take place as described: it was made up to manipulate the pilots' expectation before the test conditions. Figure 1 displays an overview of the experimental design. The order of the conditions was counterbalanced between pilots, and the two resulting groups were analyzed as one.

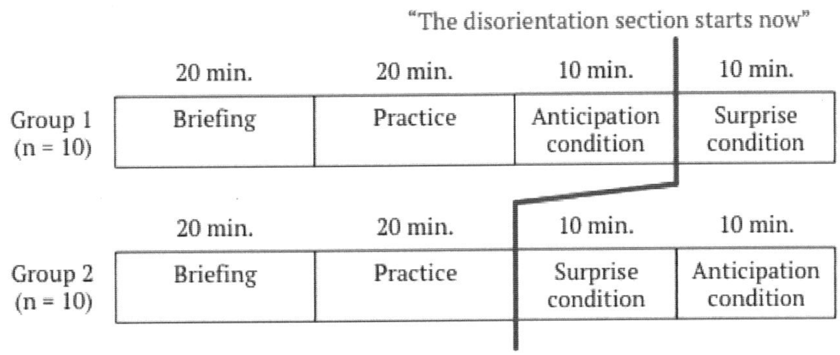

Figure 1. The experimental design, counterbalancing the two test conditions between two pilot groups. The groups were added together for analysis.

Pilots received verbal instructions about the simulated aircraft model and the stall recovery template as advised by the FAA (2015, p. 2. This involves the following steps: "1. Disconnect the autopilot and autothrottle/autothrust systems. 2a. Apply nose down pitch control until impending stall indications are eliminated. 2b. Use nose down pitch trim as needed. 3. Bank wings level. 4. Apply thrust as needed. 5. Retract speed brakes or spoilers. 6. Return the aircraft to the desired flight path." All pilots indicated that they were familiar with these steps. The practice session, aimed at decreasing inter-individual differences in skill level, consisted of the recovery of four different aircraft upsets and four different aerodynamic stall events. All scenarios in this training session were presented in a highly predictable and non-surprising manner, i.e., announced and explained beforehand. The final scenario of these was repeated until the pilot was able to push down quickly and forcefully enough to avoid stick shaker events, while also avoiding overspeed or excessive g-loads.

Unbeknownst to the pilot, the practice session transitioned into the testing section in which the same aerodynamic stall was presented in a surprise condition and in an anticipation condition. Each test condition was preceded and followed by three minutes of manual straight and level flight, with autothrottle on, at 5,000 ft. and with 220 knots. In the anticipation condition, pilots were told beforehand that the simulator operator would bring them into an aerodynamic stall when a certain landmark was crossed. The stall was induced by creating a strong "tailwind" (decreasing the calibrated airspeed (CAS) with 15 knots per second for five seconds), and by simultaneously adjusting the pitch trim up with 24 percent in 3 seconds time. None of the pilots reported afterwards that they were aware of any changes in pitch trim. In the surprise condition, exactly the same stall event was induced about five seconds before another landmark was crossed. In this case, however, pilots were instructed that the spatial disorientation section of the experiment had started, and that they would have to do a climb-out above the landmark while paying attention to any over-pitch sensation. In addition, their attention was taken away from the display at the moment the stall was initiated, as they were asked to give a rating on a sickness scale displayed to their lower right.

Outcome Variables

Flight parameters were logged from the simulator at a sample rate of 100 Hz. These flight parameters were twice (forth and back) low-pass filtered using a 2nd order Butterworth filter with a cut-off frequency of 2 Hz. A script was used to determine the moment of tailwind onset and the start of several control actions in terms of autothrottle, trim, pitch, aileron and rudder control inputs. Since pitch was continuously adjusted before stall onset, a deviation above 5 SD from the mean (taken from the preceding period of straight and level flight) was determined to be a significant pitch control input. Using these data, we checked whether pilots did or did not meet each of the four performance criteria (see, Table 1).

Table 1.

Description of the four measured performance criteria, with the corresponding FAA (2015, p. 2) recovery template principles.

Variable	Corresponding FAA principle	Description
C1. Disengage autothrottle first	1	Disengage the autothrottle at least 2.0 s before significant yoke or pedal inputs.
C2. Start with pitch down control	2a, 3	Give priority to pitch down control by starting the recovery with pitch down control inputs. Strong aileron inputs (> 50% of max) may not occur at around the same moment (within 2.0 s) of pitch down control to meet this criterion.
C3. Unload sufficiently	2a, 6	Respond (within 2.0 s) to stick shaker events with significant pitch down control and maintain significant pitch down control during stick shaker activation. Or, apply sufficient pitch down control to avoid any stick shaker events. Keep the aircraft sufficiently unloaded until CAS increases in order to avoid secondary stick shaker events. Stick shaker events were defined as secondary if they occurred subsequent to an earlier stick shaker event, or if they occurred after the first unloading action, i.e., following the first peak of pitch down control.
C4. Apply pitch down trim	2b	Using the pitch trim to aid in pitch down control during the recovery.

Besides measuring adherence to the recovery template, we performed a manipulation check by asking the pilots to rate their level of surprise caused by the tailwind on a 0-10 point Likert type scale. This was done after both conditions had ended, so as to not cause suspicion about the goal of the experiment in the second condition.

Statistical Analysis

The effect of Condition (anticipation or surprise) on the binary performance variables, i.e. meeting the criteria, was tested using generalized estimating equations (GEE) models of logistic regression. To protect against an overestimation of significant differences, the outcomes were corrected using Holm's sequential Bonferroni correction for multiple comparisons.

The effect of Condition on the pilots' subjective level of surprise was tested with a paired-samples T-test.

Results

Table 2 provides an overview of the statistical differences between conditions for each of the four performance criteria that were measured. All differences are statistically significant, with effect sizes (*d*) varying from medium to large in strength, i.e., in or above the range of 0.5 to 0.8. Despite the verbal instructions beforehand, it seems that meeting the criteria in the anticipated condition was already quite difficult, as the proportion of pilots who adhered to the criteria was around 50-80%. Nevertheless, in the surprise condition the proportion of pilots who met the criteria decreased with 20-30%, and with 50% in the case of 'start with pitch down control'. In sum, the surprise manipulation caused a significant decrease in adherence to several aspects of the recovery template.

The manipulation check confirmed that surprise was significantly higher in the surprise condition, 8.44 points, $SD = 1.50$, than in the anticipation condition, 1.39 points, $SD = 2.00$, $\Delta = 7.06$, $t = 12.35$, $p < .001$. This difference constituted a large effect size, i.e., Cohen's $d = 3.99$.

Table 2.
Criteria met in the two conditions. Effect sizes (d) are calculated by transforming the odds ratio, i.e. exp(B) from the GEE analysis, conform: Chinn (2000).

	Anticipation (met/unmet)	Surprise (met/unmet)	N[a]	Δ[a]	X^2	*p*	Cohen's *d*
C1: Disengage autothrottle early	11/9	6/14	20	-5*	5.10	.024	.69[b]
C2: Start with pitch down control	16/4	6/14	20	-10*	13.41	< .001	1.23[b]
C3: Unload sufficiently	10/10	5/15	20	-5*	3.94	.047	.61[b]
C4: Use trim	9/11	3/17	20	-6*	7.07	.008	.85[b]

* Significant after Holm-Bonferonni correction.

Discussion

The results of this simulator experiment show that the unexpectedness of an simulated aerodynamic stall effectively surprised the pilots, and negatively affected their recovery performance as measured using four criteria derived from the FAA stall recovery template. Our results suggest that significant decreases in adherence to learned procedures can be expected in operational practice compared to predictable training conditions, which is in line with previous similar studies. Our control condition shows that there is indeed a decrease in performance when pilots are surprised, thereby adding to the study of Schroeder et al., 2014. Also, adding to the study of Casner, Geven and Williams, 2013, our detailed measuring of performance indicates that several aspects of the recovery procedure were not followed in the surprise condition, while our manipulation check suggests that this was caused by our manipulation of surprise.

In the surprise condition, pilots were particularly more likely to incorrectly start their recovery with aileron control inputs. Perhaps this was influenced by an increase in lateral instability (wing drop) due to later responses in the surprise condition. Nevertheless, it suggests that ignoring a change in roll angle and giving priority to pitch control (step 1 in the FAA template) may be highly counter-intuitive and difficult to suppress, meaning that proper adherence to the FAA template in surprising situations may be very difficult.

Overall, the results suggest that skills trained under predictive conditions may not transfer to conditions containing an element of surprise. Hence, it may be useful to induce an element of surprise or unpredictability when designing training methods, which allows pilots to practice with sensemaking and switching frames and to increase their "cognitive flexibility" (Kochan, 2005). This conclusion is in line with ICAO's recommendation for scenario-based recurrent training: "Wherever possible, consideration should be given towards variations in the types of scenario, times of occurrences and types of occurrence, so that pilots do not become overly familiar with repetitions of the same scenarios." (2013, II-1-5). By practicing procedures under less predictable conditions, trainees learn to use the information presented by the situation itself to identify problems, and to apply solutions. Indeed, in the training domain it has been shown that experiencing examples of a concept in different contexts may strengthen the understanding (frame) relating to this concept and is thought to increase the trainee's ability to apply the concept in similar or novel situations (Van Merriënboer, 1997).

References

Casner, S. M., Geven, R. W., & Williams, K. T. (2013). The effectiveness of airline pilot training for abnormal events. *Human Factors: The Journal of the Human Factors and Ergonomics Society, 55*(3), 477-485.

European Aviation Safety Authority. (2015). Loss of control prevention and recovery training (Notice of proposed amendment 2015-13). Retrieved from https://easa.europa.eu/system/files/dfu/NPA%202015-13.pdf.

Eysenck, M. W., Derakshan, N., Santos, R., & Calvo, M. G. (2007). Anxiety and cognitive performance: attentional control theory. *Emotion*, *7*(2), 336-353.

Federal Aviation Administration. (2015). Stall Prevention and Recovery Training. (advisory circular no. 120-109A). Retrieved from https://www.faa.gov/documentLibrary/media/Advisory_Circular/AC_120-109A.pdf.

Groen, E., Ledegang, W., Field, J., Smaili, H., Roza, M., Fucke, L., ... & Grigoryev, M. (2012). SUPRA-Enhanced upset recovery simulation. In *Proceedings of the AIAA Modeling and Simulation Technologies Conference,* (pp. 2-5).

International Civil Aviation Organization. (2013). Manual of Evidence-based Training. (Doc 9995 AN/497). Retrieved from http://www2010.icao.int/SAM/Documents/2014-AQP/EBT%20ICAO%20Manual%20Doc%209995.en.pdf.

Klein, G., Phillips, J. K., Rall, E. L., & Peluso, D. A. (2007). A data-frame theory of sensemaking. *Expertise out of context: Proceedings of the sixth international conference on naturalistic decision making*, (pp. 113-155). New York, NY, USA: Lawrence Erlbaum.

Kochan, J. A. (2005). *The role of domain expertise and judgment in dealing with unexpected events* (Doctoral dissertation, University of Central Florida Orlando, Florida).

Landman, A., Groen, E. L, Van Paassen, M. M., Bronkhorst, A. W., Mulder, M. (submitted). *Dealing With Unexpected Events on the Flight Deck: a Conceptual Model of Startle and Surprise.*

Martin, W. L., Murray, P. S., Bates, P. R., & Lee, P. S. (2016). A Flight Simulator Study of the Impairment Effects of Startle on Pilots During Unexpected Critical Events. *Aviation Psychology and Applied Human Factors, 6,* 24-32.

Merriënboer, van, J. J. G. (1997). *Training complex cognitive skills: A four-component instructional design model for technical training.* Educational Technology Publications, Englewood Cliffs, New Jersey.

Meyer, W.-U., Reisenzein, R., & Schützwohl, A. (1997). Toward a process analysis of emotions: The case of surprise. *Motivation and Emotion, 21*(3), 251-274.

Rankin, A., Woltjer, R., & Field, J. (2016). Sensemaking following surprise in the cockpit—a re-framing problem. *Cognition, Technology & Work, 18*(4), 623-642.

Realer Than Real: The Quest for Immersive Realism in RPA Virtual Training

Matt J. Martin, Lt Col, USAF (ret)

L3 Link Simulation and Training

Arlington, TX

Aviation simulation, including that used for military aircrew training, has typically focused on stick-and-rudder tasks. But while great strides have been made over the years in the fidelity of virtual aviation environments, enterprise training efforts have lagged in areas of environmental immersion (physical, social, cognitive, etc.). Specifically, more work is needed on aircrew interaction with command and control, supporting ground units in near-pear scenarios, and building the skills needed to maintain a high level of situational awareness in complex scenarios. This has often left practitioners in the position of having to rely on live-fly training and accumulated experience to fill in the gaps—even when that experience can only be gained during the actual employment of the weapons system with all of its associated risks.

Recent developments in Live-Virtual-Constructive (LVC) training, Distributed Mission Operations (DMO), and high-fidelity mission simulation could change that. Focusing on the USAF MQ-1/9 training enterprise as a case study, this paper outlines the promise, potential, and reality of integrated RPA training.

Since the RQ-1 Predator first took flight to support military operations in Kosovo, Iraq, and other places, the public perception of remotely-piloted aircraft (RPA) has been repeatedly quick to invoke the video game analogy. The common presumption is that the experience of piloting an aircraft from thousands of miles away must obviously create a sense of detachment such that the pilot experiences no physical, mental, or emotional connection to the aircraft or to the players on the ground. This attitude can quickly lead the layperson to the conclusion that the RPA pilot might have a cavalier approach to civilian casualties or other tragic outcomes. What's more, the video game notion may lead one to assume that training and cognitive skill is not a particular problem for RPA crews, since the aircraft must employ a high level of automation with little human input.

However the experience of practitioners—now with over two millions hours logged of combat time—has been the exact opposite. The massive expansion of military RPA capability and capacity over the past 15 years has borne out the following conclusions: that effective employment of RPA systems at the tactical level requires a high level of physical, mental, and emotional engagement on behalf of the crew; that a certain set of mental and physical skills must be well-developed for success, and that the ability to produce an immersive and complex training environment in order to reproduce the experience of operational employment has proven difficult using live-fly training and elusive in the simulator.

This paper will focus on the third conclusion of the RPA employment experience—the need for a truly immersive and effective training experience. It will begin with a review of the training approaches and practices used to date, discuss the various experiments to move the employment experience to the range and the sim, and suggest improvements to future of immersive virtual training for the military RPA enterprise.

From its inception, the RQ-1 Predator (which later took on the multi-role designation of 'MQ-1' after the addition of weapons and a combat laser), was unlike the procurement of other Department of Defense major weapons systems. Initially acquired off-the-shelf as an Advanced Concept Technology Demonstrator (ACTD), its development and refinement over time took place in the midst of operational employment. In fact the tactics needed to employ every major innovation—from the use of satellites for beyond-line-of-sight (BLOS) operations, to the use of Hellfire missiles, to the global distribution of video and data for intelligence exploitation, were developed and inculcated during combat sorties (Whittle 2014). This meant that even crews employed these tactics in the field, there was no codified set of tactics, techniques, and procedures (TTPs) that could be incorporated into a syllabus or training courseware for the expanding Initial Qualification Training (IQT) effort. In fact even though Predator combat operations began in 1995 (Whittle 2014), there was no formal system of advanced tactics development, incorporation, and standardization until 2003 (McCurley 2015).

While a dedicated IQT forum was soon established in the form of the 11th Reconnaissance Squadron schoolhouse at Indian Springs Auxiliary field, there was no figh-fidelity simulator available until the acquisition of the L3 Predator Mission Aircrew Training System in 2006 (L3 2006). Prior to the time there were part-task trainers that allowed for familiarization and "switchology," but all mission training had to be done via live-fly on the Nellis range subject to the normal challenges and attrition factors of live-fly training such as weather, aircraft maintenance, and airspace availability (Colucci 2004).

During those live-fly events in the early days of the 11 RS, integrated mission training was a rare event. The procedures for conducting multi-ship and dissimilar live training had not yet been established. There was no formal system of identifying and synchronizing training requirements between Predator crews and crews of manned aircraft. And due to ongoing operations in Iraq and Afghanistan, there were no forward air controllers or other complementary training audiences available conduct integrated training with Predator crews. It was therefore up to instructors to “role-play” these roles while simultaneously performing instructor duties (McCurley 2015).

There was one integrated training event that was a regular feature of early MQ-1 IQT—Killer scout weekend (Martin 2010). Taking advantage of the training cycle of various F-16 Air National Guard units, the instructor cadre at the 11RS were able to arrange quarterly—sometimes monthly—visits by F-16 aircraft to conduct the “Killer Scout” Strike Coordination and Reconnaissance training events in the IQT syllabus. In this aerial interdiction scenario, MQ-1 students could locate targets on the range, conduct “talk-one” to verbally guide the F-16 pilots to the targets, practice the integration and deconfliction techniques to conduct a coordinated attack, and then debrief those simulated engagements after the mission alongside the F-16 pilot who had participated. These events provided an invaluable opportunity for MQ-1 crews to work with other strike assets prior to meeting them on real-world combat missions. Figure 1 below shows an operational view of a typical MQ-1/F-16 killer scout training event.

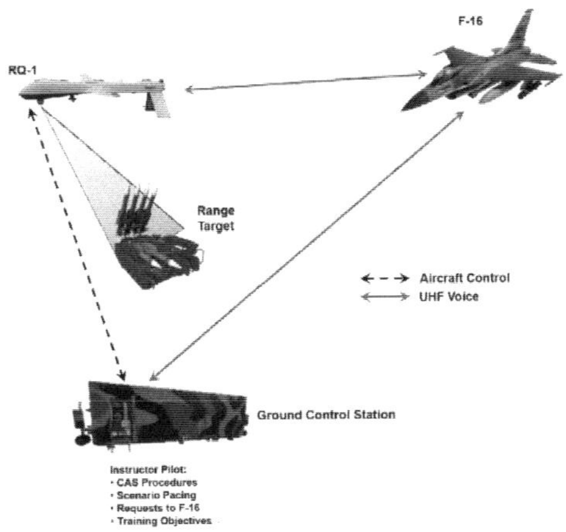

Figure 1 – Early Live-Fly "Killer Scout" Integrated Training Mission

But these types of integrated events were quite rare—as were all other types of live tactical events in MQ-1 IQT training. In fact it was not uncommon for MQ-1 crews to become qualified to fly in combat, transition to an operational squadron, began flying missions, and then be called upon to employ live ordnance from another aircraft against live targets, having never had the opportunity to perform a live weapons employment in training. In fact, due to safety rules on the range and the lack of a high-fidelity training system, many crews never even conduct the switch actuation in training (Martin 2010).

To help alleviate this lack of live integrated training opportunities, Air Combat Command hired several companies since 2008 to act as live forward air controllers, range targets, and stand-ins for live aircraft during IQT and Continuation Training Events. As contractors, these role-players bring the advantage of having no training objectives of their own. Likewise they can be dedicated to the RPA schoolhouse so that they are always available to insure all student crews have the opportunity to train with them. Since the inception of this contracted, live, role-playing capability—while there have been some limitations in capacity—every IQT crew has had the benefit of multiple tactical training events with live controllers and love-role players prior to graduating from the course. (Moore 2009)

With the advent of a high-fidelity simulator as part of the MQ-1/9 program of record—which is designed and managed to provide a certified, software concurrent, and technically realistic compliment to live-fly training—it became possible starting in 2010 to mirror this live role-playing capability in a virtual environment. The same contract forward air controllers who supported live-fly events on the range were brought into the simulator to perform those same roles for virtual training events. This brought the advantage of not only expanding student exposure to live-quality integrated training, but allowed for the transition of live-fly events into a virtual environment enabling savings in cost and efficiency by reducing attrition due to weather, aircraft maintenance, and airspace availability. It further allowed the inclusion of other operational elements in the virtual space including command and control, intelligence, and blue forces, to immerse the students in a highly-complex operational environment. Figure 2 below depicts an operational view of integrated virtual training in the PMATS simulator via role-playing.

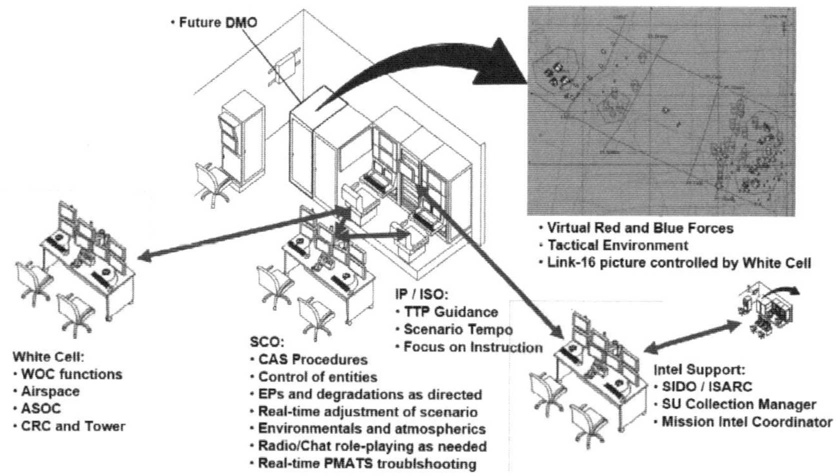

Figure 2 – The PMATS Role-Playing Set-up for RPA IQT

Beyond the intuitive, there is good data to back up the notion that virtual mission elements, and virtual scenario players, which place realistic limitations on mission execution when transposed to the real world, would lead to the development of the same cognitive skills that are importation to mission success in the real world. In fact, both the data and the experience of practitioners show that training to develop skills related to judgement, decision-making, and tactical leadership, can be trained to, and practiced in, a virtual environment with a high degree of effectiveness. Training these types of skills in simulators may be even more important than developing the physical stick-and-rudder skills that are the focus of traditional flight simulators.

For example, numerous studies of medical students have shown that scenario-based simulation training is superior to interactive problem-based training to develop critical assessment, coordinated team action, and management skills (Steadman et. al. 2006). Likewise, in civilian aviation, the Federal Aviation Administration has long recognized that scenario-based training is a far more effective way to develop judgement and decision-making skills among pilots than the old approach of focusing and physical aircraft control skills. The FAA has also accepted the concept—and helped implement it within commercial aviation—that pilots and crews can train entirely within simulators so long as they are of sufficient fidelity and complexity (FAA 2014).

The USAF has also moved more mission training into virtual environments, owing to the rising costs and restraints of live-fly training using 4th- and 5th-generation aircraft the shrinking size of the fleet, and the difficulty of reproducing sophisticated threats in a live environment. Even premier large-force training events such as Red Flag and the Weapons School integration phases, have relied more and more on virtual and constructive elements to replicate the true complexity and sophistication of high-level conflicts. This has allowed the integration of intelligence, cyber, and electronic warfare elements to make live-fly training even more complex and challenging (Bultman 2017).

What would such an approach look like for the MQ-1/9 enterprise? For starters the basic elements of software concurrency, accurate aeromodel, replication of malfunctions, weather, imagery, and weapons effects have to be of the highest possible quality. Any inaccuracy or "simism" will not just distract students from the realism of the scenario, but has the danger of teaching contrary or negative skills.

With a realistic weapons system baseline, automated and manual mission tools can be added such as command and control links, weaponeering tools, communications tools, and a datalink picture that would replicate picture used by crews during operations. Integration of these tools are essential to train not just their operation, but to train the cognitive skills of situational awareness, 3-diminension comprehension, attention and focus management, and crew coordination.

Finally, there has to be a wealth of red, blue, green, and white forces available so that students will be immersed in a truly complex mission environment. These can be provided by live players in networked sims, automated entities, or white cell role-players. The use of live players is particularly useful since it allows student to cooperatively, plan, brief, and debrief training events in the same manner and with the same emotional potency of real-world operations.

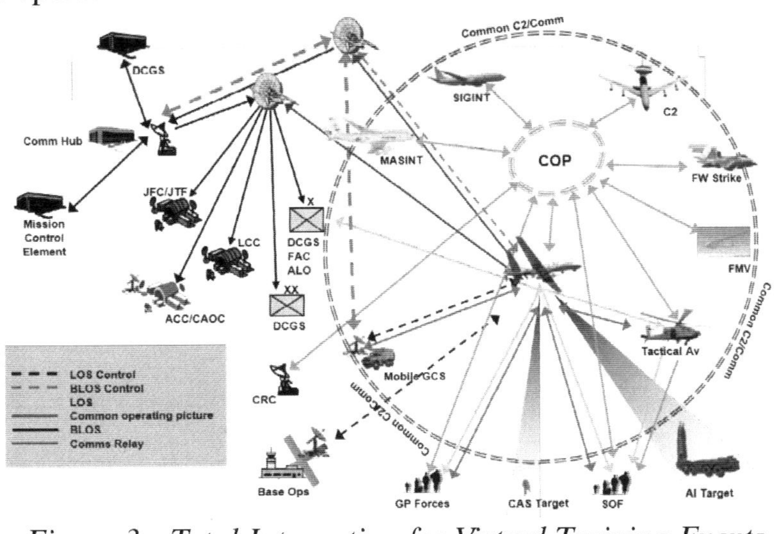

Figure 3—Total Integration for Virtual Training Events

In conclusion, highly complex mission training for RPA crews has been long in the making and is still not quite a reality. But based on the lessons of the past, both from USAF experience employing the still-growing RPA force, and from other virtual training applications, the goal can be achieved. And when virtual training is "realer than real," the RPA crew force can declare victory and focus on what really matters—the mission.

References

Whittle, Richard. 2014. *Predator: The Secret history of the Drone Revolution.* New York: Henry Holt and Company, LLC

L3 Technologies. 2006. Training and Simulation. *Annual Report,* Page 15.

Colucci, Frank. 2004. Air Force Refines Training Programs for UAV Operators. *National Defense Magazine,* retrieved from http://www.nationaldefensemagazine.org/archive/2004/May/Pages/Air_Force_Refines3555.aspx

Martin, Matt. 2010. *Predator: The Remote Control Air War over Iraq and Afghanistan: A Pilot's Story.* New York: Zenith Press

Moore, Alice. 2009. Train as we Fight; Close Air Support SMEs integrated into Predator Training. *Air Force Print News Today,* retrieved from www.rallypointmanagement.com/wp-content/uploads/2014/07/trainaswefight.pdf

TOWARD IDENTIFYING RISK IN OPERATIONAL PERFORMANCE OF ATC PROCEDURES

Paul Krois
Julia Pounds
Federal Aviation Administration
Washington, D.C.

Managing operational performance to reduce risk in the execution of air traffic control (ATC) procedures depends on understanding human performance in relation to patterns established in procedures. Procedures specify performance requirements meant to minimize unintended variation that is outside expected performance criteria and tolerances. We define human performance risk as unintended variation outside the envelope of applicable procedures.

Managing human performance to reduce risk in the execution of air traffic control (ATC) operations is predicated, in part, on understanding the human performance prescribed by operational procedures. A procedure is defined as "an established or official way of doing something", and as "a series of actions conducted in a certain order or manner" (Oxford Dictionary). Procedures specify requirements for how work is to be done. Procedures represent the standard technique or techniques to be followed and applied each time a particular operation or situation occurs.

Procedures are the cornerstone for a high reliability organization to provide safe services. Procedures establish air navigation requirements such as for pilot-controller communications (e.g., phraseology), airspace design, and airport arrival and departures. Similarly, procedures are critical to the Technical Operations workforce, flight crew operations and aircraft maintenance.

Our goal for this paper is to propose an approach for identifying where risk is introduced in the human performance of ATC procedures. This approach identifies unintended variation in execution and uses unintended variation as an observable and measureable dependent variable.

In this approach, unintended variation in human performance can be considered a potential hazard to future operations and can be characterized using available information and data. For example, information that has been entered into voluntary safety reporting systems or recorded radar and voice data can be used to identify human performance that varies from expectations given the procedural requirements.

Procedures represent management controls that specify how work is to be done and prescribe work using the following dimensions:

- Either *required* (must be performed, i.e., steps in sequence) or *discretionary* (may be performed, i.e., judgment over which steps to perform)

  Required management controls prescribe the steps of a process to be performed, e.g., the sequence of steps for a procedure, how a decision support tool is to be used, etc. Discretionary management controls pertain to situations where judgment is allowed and execution is optional. Judgment is used to choose which steps are needed to best fit the operational situation.

- Either a *standard* (performance must meet a single point criterion) or a *tolerance* (performance must be within a range).

  Required management controls can stipulate that human performance must meet a particular standard or that it must be completed within certain tolerances, e.g., based on altitude or distance.

- Either a *process* (how to perform an operation), an *output* (the result of an operation) or an *outcome* (ensure safe separation).

  Required management controls can establish a standard for an outcome from a process, such as mitigating the risk of fatigue (outcome) as a result of shift scheduling (the minimum staffing requirements per shift, minimum number of hours between shifts, etc.).

Defining Unintended Variation

We discussed risk as unintended variation in more detail elsewhere (Davis and others, 2015, 2016). In brief, unintended variation in human performance means that performance is outside approved management controls for the procedure relative to any standard or criterion that human performance must meet. Unintended variation can occur despite management controls established to constrain or prevent it. Procedures are developed to mitigate unintended variation.

Human performance refers to "the performance of jobs, tasks, and activities by operational personnel – individually and together" (EUROCONTROL/FAA, 2010). The relationship between these terms is that human factors is the scientific discipline whose sole purpose is to enhance human performance, that is, human performance can be ensured by applying human factors science.

Unintended variation can occur in operational processes, outputs, or outcomes. A procedure can be specified as a process: a series of steps to be executed either in sequence or in parallel. Unintended variation is introduced when a step is skipped (an error of omission), a unrelated step as added (an error of commission), or steps are completed out of sequence.

Unintended variation can occur in the output or the outcome of the procedure or both. It may result, for example, because of a unique operational context the procedure did not adequately address. This can produce a result that is out of tolerance. In terms of human performance, this likely would be classified as a human error.

Common statements used by people to explain their actions can reveal unintended variation in a procedure and responses to it:

- "I wasn't trained for this situation so my first thought was this situation looked like A so I did B and expected it to work,"
- "I expected X but the situation turned out to be Y. I hadn't seen Y before so I did what I always do for X and expected it would work," and
- "When a situation like this occurs I know from experience that I can take a shortcut and get the expected result."

Defining risk as unintended variation is different from the traditional approach to defining risk that uses consequence and likelihood, or probabilities. Use of probabilistic human reliability analysis to assess human performance failure in aviation has proven to be difficult.

ATC Procedures

There are two key FAA Orders that contain operational ATC procedures, Order 7110.65 and Order 7210.3. From time to time these procedures are updated to accommodate changes and introduction of new procedures. Updates are currently denoted with a suffix letter and previously with a change number.

FAA Order 7110.65W (change 2 effective December 10, 2015), Air Traffic Control – This Order documents the ATC procedures and phraseology required to be used by controllers who have the necessary expertise as it pertains to their operational responsibilities, e.g., en route or terminal. Controllers are required to exercise their best judgment if they encounter situations not covered by it. The Order covers all aspects of ATC operations including flight plans and flight progress strips, communications, terminal procedures, Instrument Flight Rules (IFR), radar and nonradar procedures, visual operations, offshore/oceanic procedures, special flights, emergencies, and decision support tools.

FAA Order 7110.65W is 784 pages in length. The Order was first published on January 1, 1976. When Order 71110.65, Change 7, became effective (July 1, 1977) it was 343 pages in length. Comparison between the 1977 and current versions provides some general observations, as follows:

- The 1977 version was organized so as to contain 1,773 numbered procedures.
- There are paragraphs in the 1977 version that no longer appear in the current version, e.g., removing procedures associated with ATC software capabilities that have been replaced by new systems.

- The 2015 version added numerous procedures for new operations and capabilities such as related to wake turbulence applications (for aircraft category, intersection departures, and intersecting runway/intersecting flight path operations).

The current FAA Order 7210.3Z (effective December 10, 2015, change 2 effective November 10, 2016), Facility Operation and Administration – The order contains direction and guidance for everyday operations of facilities and offices. The order currently spans 628 pages. Topics include familiarization/currency requirements for en route, terminal, and system operations facilities, watch coverage and supervision, national automation programs, flight service stations, and the traffic management system.

Other procedures pertain to various aspects of the National Airspace System (NAS). FAA Order 3120.4P (effective December 10, 2015), Air Traffic Technical Training – The order has instructions, standards, and guidance for training. The order addresses Academy qualification training, on the job training (OJT) for position certification, training of OJT instructors, and controller-in-charge training.

Additional FAA Orders address topics important to the NAS including contractions (7340.2), flight services (7110.10), location identifiers (7350.9), airspace (7400.2), Notices to Airmen or NOTAM (7930.2), special military operations (7610.4), traffic counting for determining facility classification levels (7210.57), the voluntary safety report program (7200.20), and occurrence reporting (7210.632).

Required and Discretionary Procedures

Procedures specify required and discretionary actions for what is intended for a particular operational situation. Required procedures prescribe that when a particular situation occurs there are certain actions that must be taken. Discretionary procedures recognize there is more than one course of action that can be taken. With discretionary procedures, there can be more than one pattern that is expected to occur. Discretionary management controls typically use such phrases as "the operator may discontinue the alerts if …" and "the documentation should include …" compared to required management controls indicated by such phrases as "the operator must discontinue the alerts if …" and "the documentation must include …" Discretionary procedures allow use of judgment to select the action to be taken.

Both required and discretionary procedures place limits on the actions so the pattern is predictable and consistent. By specifying the actions to be taken and any tolerances that are permitted, procedures intend to eliminate the potential for errors of omission (e.g., leaving a step out of the procedure) or commission (e.g., adding a unexpected step in the procedure), or that no action will be taken. The difference between required and discretionary procedures is shown in Table 1.

Table 1.
Comparison of required and discretionary procedures

	Required Procedure	Discretionary Procedure
Performance Within Tolerance	Performance fits within one permitted pattern	Performance fits within one of multiple permitted patterns
Performance Outside Tolerance	Performance does not fit the one permitted pattern	Performance does not fit within any permitted pattern

Changes to procedures occur such as when new procedures are developed for changes to existing or implementation of new NAS capabilities. New procedures are developed in response to the emergence of operational conditions or situations not addressed in current procedures. New procedures can also be developed in relation to the occurrence of safety-related operational conditions.

Procedures are sometimes executed through use of control techniques. For purposes of this paper, control techniques are defined as local facility methods for executing procedures contained in FAA Order 7110.65. With control techniques, procedure steps are aligned with local agreements, airspace design, local software adaptation, and other potential considerations. The patterns expected with required and discretionary management controls in 7110.65 extend to the patterns provided by control techniques.

There are numerous ways human performance can involve unintended variation. Bias in decision making can interfere with correct identification of patterns such as through expectation bias, confirmation bias, association bias, frequency bias, and coincidence bias. These types of bias can change performance that goes outside of tolerance. During multi-tasking, attention shifts back and forth among tasks allowing unintended variation to occur. As attention shifts between tasks the potential for errors of omission and commission can increase. Unintended variation can result from tunnel vision in which attention is focused on a particular situation and other situations are not addressed according to procedures. Distractions detract from fully recognizing an operational situation and determining the appropriate procedure. Training intends to build knowledge and skills for consistent problem solving in applying the right procedures to operational situations.

Patterns of Variation

Conceptually, human factors studies examine patterns in operational performance and assess changes to these patterns from unintended variation. Patterns are established through procedures, airspace design, traffic flows, training, equipment design, staffing, and other human factors considerations. Patterns are measured through laboratory and field studies that show how advanced concepts and new capabilities intersect with human capabilities and limitations. Required and discretionary procedures define the steps to be followed so that patterns are maintained. Recognizing and establishing patterns within a system can provide predictability and insight into the relationship between performance and tolerance.

Examples demonstrate how unintended variation from procedures can occur. These examples include that facilities may show differences in use of new capabilities. Facilities may also show differences in operational practices. Controllers may use new capabilities when discretionary procedures permit judgment on how those capabilities are used.

Unintended variation with use of a new capability was demonstrated with initial implementation of the User Request Evaluation Tool (URET) at three facilities (Bolic & Hansen, 2005). Discretionary procedures permitted facilities to adopt its use during implementation. For example, one facility had a discretionary procedure that when both the Radar and Data controllers were trained on URET, they could use the tool and disregard paper strips. These facilities had been using URET prototypes for different numbers of year. Training on URET across facilities ranged from 36 to about 70 hours. Qualitative data were collected at each facility using exploratory open-ended interviews with Subject Matter Experts. Results showed that different sector teams used URET in different ways and in many instances URET usage differed from what was intended. The three facilities used URET display functions for electronic flight strips to replace paper strips and the associated manual workload from handling the paper strips. Two facilities found amending routes was useful when severe weather occurred. A key finding was that facilities developed their own control practices in relation to unique operational conditions. Discretionary procedures permitted this variation.

Variation in operational performance was shown through past ATO research trials of the Normal Operations Safety Survey (NOSS). NOSS is an observational technique for collecting safety data in everyday ATC operations. Controllers volunteered and were trained on conducting sector position observations and classifying data. Controllers also participated in aggregating data at the facility level. NOSS uses the threat and error management taxonomy to classify observations of external threats to the controller, errors the controller may make, and mismanaged threats and errors that may challenge safe operations. An example of unintended variation was demonstrated involving the Transfer of Position Responsibility (TPR). TPR involved a step-by-step process with controllers following a checklist in which the Relieving Specialist previewed the position, the Specialist Being Relieved provided a verbal briefing, and the two Specialists completed the assumption of position responsibility (reviewing the position including signing in and checking information and equipment). TPR data were collected at two facilities over a standardized one-hour observation period. As shown in Table 2, the two facilities varied in completing the TPR checklist. Unintended variation occurred when required management controls were not followed. At the time research was conducted with NOSS, data showed unintended variation occurred both at the individual controller level and at different frequencies across facilities. At Facility A, many instances of the TPR checklist not used and not completed were associated with airspace having seasonal effects with low traffic counts. By the time Facility B trialed NOSS, the ATO was using a challenge and response technique to reduce TPR checklist not used and incomplete checklist use. Also, controllers memorize the TPR checklist through repeated use and not manually refer to a printed checklist. It is important to note NOSS data showed that none of the unintended variation with TPR involved unsafe conditions. Also, NOSS research did not evaluate individual controller performance but rather intended to examine patterns of operational performance.

Table 2.
NOSS trends for Transfer of Position Responsibility

Facility	Total One-Hour Observations	Number of TPRs	Checklist Not Used	Checklist Not Completed
A	90	96	47%	10%
B	147	220	10%	4%

Unintended variation has been examined in the laboratory with use of new capabilities. Kraut and others (AIAA, 2013) conducted a simulation studying how controllers applied a route planning tool to manage arrival traffic. Controllers used discretion to begin aircraft descent from cruise altitude before handing off the aircraft to a low altitude sector. A primary goal was to deliver aircraft to the meter fix within a parameter time of the scheduled time while maintaining standard separation. Results showed controllers used the tools and automation in both strategically and tactically different ways and this diverged further during high traffic demand situations. For example, when demand increased, some controllers deferred to manual control because it resulted in quicker action than the strategy of using the route planning tool. A conclusion from this paper was that tool designers should be concerned with how and when tools will be used and the training needed for their use.

Studies of advanced concepts and new capabilities often focus on assessing system benefits compared to baseline operational conditions. Unintended variation can be a useful perspective to examine how performance may vary from intended use with consequent limitations on intended benefits.

Future Directions

In the future, specific methods should be developed to more closely examine when and where unintended variation occurs and the circumstances associated with it. Once identified, the effects of unintended variation on operations can be examined and better understood. Methods could include use of radar and voice tapes, and systematic observation of ATC operations.

Patterns in unintended variation can also be examined in use of advanced capabilities. For example, simulation studies could assess patterns and variance in operational performance using within subjects experimental designs. Also, training and human-centric design of ATC automation should be examined for their effect on mitigating unintended variation.

Key challenges that can be addressed include whether and how unintended variation in operational performance can be identified. This includes whether unintended variation in operational performance would be minimized by limiting use of discretionary procedures. Another challenge is whether differences in control techniques for the same procedure can lead to unintended variation.

In a laboratory setting, safety is sometimes considered and measured using such measures as losses of separation, controller ratings, and anecdotal evidence. In contrast, examining variation in human performance provides increased understanding and insight into unintended variation and how it influences safety in the design and use of prototype capabilities. Consistency and predictability of ATC operations can be complicated by individual control techniques for executing a procedure. That is, a controller may adapt a control technique they have used successfully many times before to a new capability even if that technique is not well suited to the capability and operational condition. This can introduce unintended variation and may result in performance outside tolerances for that procedure.

Conclusion

Unintended variation must be identified before it can be managed. The evidence for recognizing unintended variation can be derived through field and laboratory measurements. Unintended variation acts as a marker of performance risk. Establishing management controls for required performance in executing an operational procedure creates shared expectations of consistent and predictable performance. Management controls keep human performance within expected performance tolerances.

This paper proposes to use the concept of unintended variation to better understand risks to NAS safety. Unintended variation in human performance can create risk to the safety of NAS operations. Examples of three areas were provided where unintended variation in human performance creates potential for risk. Further work is needed to develop and validate measures as indications of human performance risk.

In future studies, researchers and engineers should consider methods to identify unintended variation in human performance and its relationship to operational processes and outcomes. As a result of this approach, implications for training and human-centric design of ATC automation can be identified along with potential risks in system design.

Acknowledgments

The opinions expressed are those of the authors alone, and do not necessarily reflect those of the Federal Aviation Administration, the U.S. Department of Transportation, or the government of the United States of America.

References

Bolic, T. and Hansen, M., 2005, "User Request Evaluation Tool (URET) Adoption and Adaptation, Three Center Case Study," FAA/EUROCONTROL 2005 Air Traffic Management Seminar.

Davis, B.C., Pounds, J., Krois, P., and Wishy, M., November, 2015, "Can Risk Be Managed If it Can't Be Measured," AEROSAFETYWORLD, pp. 10-12.

Davis, B.C., Pounds, J., Krois, P., and Wishy, M., March, 2016, "Next Generation Management of Risk: System Patterns," System Engineering Conference SEDC.

EUROCONTROL/FAA Action Plan 15 Safety (2010). Human Performance in Air Traffic Management Safety: A White Paper. Paris, France: EUROCONTROL.

Kraut, J.M., Mercer, J., Morey, S., Homola, J., Gomez, A., and Prevot, T., 2013, "How do Air Traffic Controllers Use Automation and Tools Differently During High Demand Situations," AIAA Aviation Technology, Integration, and Operations Conference.

Oxford Dictionary, 2017. Retrieved from https://en.oxforddictionaries.com/definition/procedure.

OPERATIONALIZING THE DEFINITION OF RISK AS VARIABILITY[1]

Julia Pounds[1], Paul Krois[2], Barry C. Davis[1] and Melissa Wishy[1]
[1]Air Traffic Safety Oversight Service
[2]NextGen Human Factors Division
Federal Aviation Administration[2]
Washington, DC

Maintaining safety requires acknowledging risk. However, one's definition of risk can depend on whether the word is being used in everyday conversation or by safety practitioners or by domain experts. Not having a commonly agreed-upon definition poses problems for those charged with identifying, reducing, and communicating about risk. In an effort to standardize the definition, the International Organization for Standardization defined risk simply as the effect of uncertainty on objectives. Still, this general definition lacks enough specificity to describe uncertainty's positive or negative effects. Relevant information can reduce uncertainty's potential effects if it's not ambiguous, unreliable, incomplete, or unavailable.

The global aviation community strives to continually increase the level of safety and relies on a highly procedural system of systems to maintain flight safety.Rules and regulations for all airspace users specify procedures for executing safe operations. For example, specific words distinguish required performance: must, must not, shall, shall not.[i] For example, "At tower-controlled airports where radar coverage does not exist to within 1/2 mile of the end of the runway, arriving aircraft must be informed when radar service is terminated" (FAA, 2012). This mandatory procedure is executed to reduce risk. If 10 out of 100 arriving aircraft are not informed, then there is variation from the required standard.

Maintaining safety requires acknowledging risk. However, "risk" is used as a very plastic concept. One's definition can depend on whether the word is being used in everyday conversation or by safety practitioners or by statistical experts. Not having a generally agreed-upon definition can pose problems for those charged with identifying, reducing, and communicating about risk. As either noun or verb, it can be used to characterize a variety of situations. Aviation safety professionals seem to have developed an unspoken requirement to use "risk" concepts in every document.

Variability and Risk

In aviation, risk characterizes a future when consequences are unknown. "In general, risk considers uncertain and undesired future occurrences and it is typically assessed by combining probability and severity levels of future occurrences. Note that there is no risk involved in runway incursion events as such, since they did occur and their consequences are known" (Stroeve, S., van Doorn, B., Bakker, B., & Som, P. 2015).

The opinions expressed in this paper are those of the authors and do not reflect FAA policies or positions.

Efforts to identify future safety risks typically rely on stochastic methods but the complexity of situations may disguise risk, making its prediction difficult. Jaeger (2000) proposed that prediction methods may reveal financial risk but not describe the nature of it and too much reliance on quantitative forecasting tools can lead to trouble. For example, financial risk industries predict price volatility but risk is uncertainty, not volatility. Jaeger believed that difficulty measuring risk could lead to improvements in one's ability to manage risk. To quantitatively oriented financial experts, variance is a commonly used substitute for risk. (Chang, Lin, & Zhu, 2008).

"Given the ubiquity of risk in almost every human activity, it is surprising how little consensus there is about how to define risk" (Damodaran, A. 2008). In an effort to standardize the definition, the International Organization for Standardization (ISO) defined risk simply as the effect of uncertainty on objectives (ISO 31000, 2009; ISO Guide 73). Uncertainty is generally viewed as undesirable vagueness, or ambiguity, that is unintended and to be avoided. An overview how uncertainty relates to variability was discussed in Davis et al, (2015).

We discuss variability as a simple solution for the complex problem of defining, identifying, measuring and mitigating safety risks by defining risk as unintended variation. This can be operationally useful when defined using recognized indicators of deviation from a standard. We hypothesize that safety can be supported by recoginzing and reducing sources of unintended variation.

Rather than trying to forecast potential adverse events, a method we examine to address these considerations is to re-conceptualize uncertainty, risk and outcome in terms of variability. We are examining methods from other areas as means to operationalize, measure, and mitigate current safety hazards. Early detection and mitigation of hazards can block risk in future operations.

Variability and Safety Controls

The concept of variability is not new. This definition can be tested using methods to measure variance that are familiar to most safety professionals. Reducing variation is a standard manufacturing tool for improving consistency in production. However, to our knowledge it has not been used to re-conceptualize uncertainty. This definition can be tested using methods to measure variance that are familiar to most safety professionals using frequencies distributions. We don't presume that this is a predictive, stochastic method for forecasting outcome, only that it is a way of identifying deviation from an expected standard.

Safety compliance with the appropriate standards is required to control known risks.

- Variability that is known to the controlling organization and permitted is classified as "intended" and thus is an acceptable uncertainty that the controlling organization has judged does not need to be controlled.
- Variability that is known to the controlling organization and not permitted is classified as "unintended" and thus is unacceptable uncertainty that the controlling organization has judged needs to be controlled.

Risk management with appropriate mitigations in place as controls is needed to control unknown risks.

- Variability that is unknown to the controlling organization but could be present is classified as "unintended but discoverable." Because it is unknown to the controlling organization, it is classified as "unintended."
- Variability that is unknown to the controlling organization and cannot be imagined is classified as "unintended and unimagined." This type is difficult to discover and control, if needed, because it *is* unimagined.

By using present variability instead of estimated future states, this approach can describe how and when known and unknown variability occurs, whether it is intended or unintended, what influences it, whether a mitigation reduces it, etc. In short, using this approach can provide an objective method for managers of safety organizations in government and industry to address the present potential for future risk by recognizing the hazard of unintended variability.

To explore the implications of this, we used the metaphor of an iceberg metaphor, with intended safe outcomes at its top (Figure 1). Whatever the desired performance (e.g., an accurately executed procedure), the organization's safety controls can reduce variability and increase the potential of a process achieving its intended outcome (i.e., to maintain separation) or objective (e.g., safety). Moreover, any practitioner will recognize that there are situations when variability occurs as expected; but in some cases, variability occurs and is unexpected.

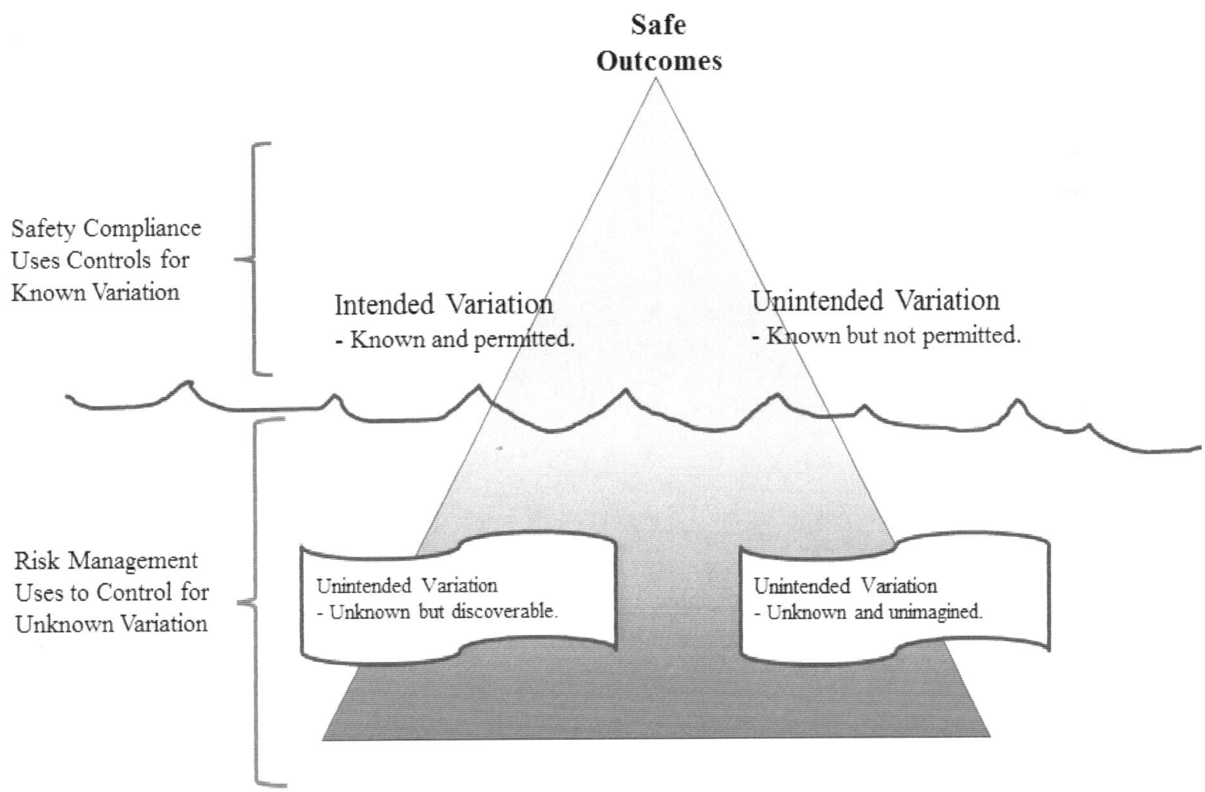

Figure 1. Types of intended and unintended variation in safety compliance and risk management.

This approach can be used to recognize how and when variability occurs, what influences it, if a mitigation is effective in reducing it, etc. In short, using this approach provides an objective, measurable method for safety practitioners to address potential risk.

References

Chang, K. J., Lin, C., Zhu, T. (2008). *Risk Measurement and Management.* New York: Cambridge University Press.

Damodaran, Aswath, (2008). *Strategic Risk Taking: A framework for risk management.* Indianapolis, Indiana. Pearson Prentice Hall.

Davis, B. C., Pounds, J., Krois, P., Wishy, M. (2015, Nov.). *Can Risk Be Managed If It Can't Be Measured?* AeroSafetyWorld, 10-12. Flight Safety Foundation.

Federal Aviation Administration. (2012). *Air Traffic Control*, JO 7110.65.

International Organization for Standardization (ISO). (2009). *ISO 31000/ISO Guide 73. Risk management – Vocabulary.*

Stroeve, S., van Doorn, B., Bakker, B., & Som, P. (2015). *A risk-based framework for assessment of runway incursion events.* Presented at 11th USA Europe ATM R&D Seminar, Lisbon, Portugal.

[i] In FAA documents the practice of using directives such as "shall" and "shall not" is changing to forms such as "must" and "must not."

COGNITIVE CONSTRAINTS FOR AUTOMATION ON FLIGHT TESTING

Diogo Silva Castilho
Massachusetts Institute of Technology
Boston – MA

The first flight of a new aircraft is still a dangerous event. Despite all simulations and software predictions, test pilots face many unknowns when a prototype leaves the ground for the first time. The cultural celebration of first flights masks the concerns of many stakeholders about the technical challenges of the new equipment. The pilot extensively prepares to react properly to unexpected situations and often bring a new story to tell, but in a time when remotely piloted and autonomous aircraft fly every day, the question about how to use their technologies to save a test pilot life arises. This study investigates the technical advantages of using specific autopilot modes and remote or autonomous controls. It also discusses the disadvantages of relying on airborne sensors instead of using pilots cognitive capabilities and judgment. The analysis on the data collected by students in a Flight Testing Course supports that there are clear advantages of the suggested new approach. The control stick input technique to explore the longitudinal stability of an aircraft is used as an example of human limitations on measuring quantitative variables. The results are extended to the critical phases of the campaign and the analysis points to new safety constraints that cannot be ignored.

Humans and machines progressively share the control of numerous vehicles and this symbiosis has reached a level of maturity that makes it a main topic of research. Aviation pioneered the use of automation for decades, but human pilots are still manually controlling aircraft in dangerous and precise situations.

Despite the fact that machines execute many tasks more precisely and faster than humans, according to Fitts List, humans are more versatile, innovative and better for error correction and judgment (de Winter and Dodou, 2014). Automation is not able to autonomously support strategies to manage complexity, anticipate the dynamics of cross-adaptive processes or to deal with tradeoffs and dilemmas (Woods and Sarter, 2000). However, as algorithms become more reliable and versatile, increases on the levels of automation in flight controls happen through improvements on autopilots modes.

The activity of Flight Testing (FT) must safely verify the accomplishment of requirements and validate the product for certification. If automation can reduce costs, time of development, or make it safer, then manufacturers must explore related technologies using new devices and techniques.

FT campaigns of fixed-wing aircraft designed to be piloted by humans can improve the use of automation in two different ways. The first is a human-machine partnership providing a more efficient investigation of aircraft characteristics with inputs precise in timing, frequency, and amplitude. The second is the remote operation of the aircraft to execute dangerous test events.

The Benefit of Precision

Handling qualities (HQ) events test for longitudinal and latero-directional stabilities. The FT crew measures the natural frequency of an aircraft oscillation and the damping after an input on that axis (IPEV, 2015). Test pilots are trained to investigate many different frequencies and amplitudes on each axis trying to induce a pilot-aircraft coupling. If it ever happens, it causes changes in the design or reduced operational limits.

An analysis of data collected during the 2016 Brazilian FT Course, focusing on the technique applied to the excitation of the Short Period mode, proved that automation would provide a more efficient investigation of the handling qualities than traditional methods. There are several different techniques to investigate the Short Period. All of them have an input followed by the observation of the aircraft's reaction.

Figure 1 shows the stick input followed by the natural response of the aircraft in angle of attack (AoA) and pitch angle. The first input that the pilot provided is a frequency sweep followed by multiple short duration inputs to investigate the Short Period. In this sweep, the pilot starts cycling the control stick longitudinally in a low frequency while maintaining a relatively constant altitude (+/-20ft) and air speed (+/-2kt). For doublet applications, the pilot memorizes the frequency in which reactions have more amplitude. The two inputs before 100s in the x-axis of figure 1 are longitudinal doublets.

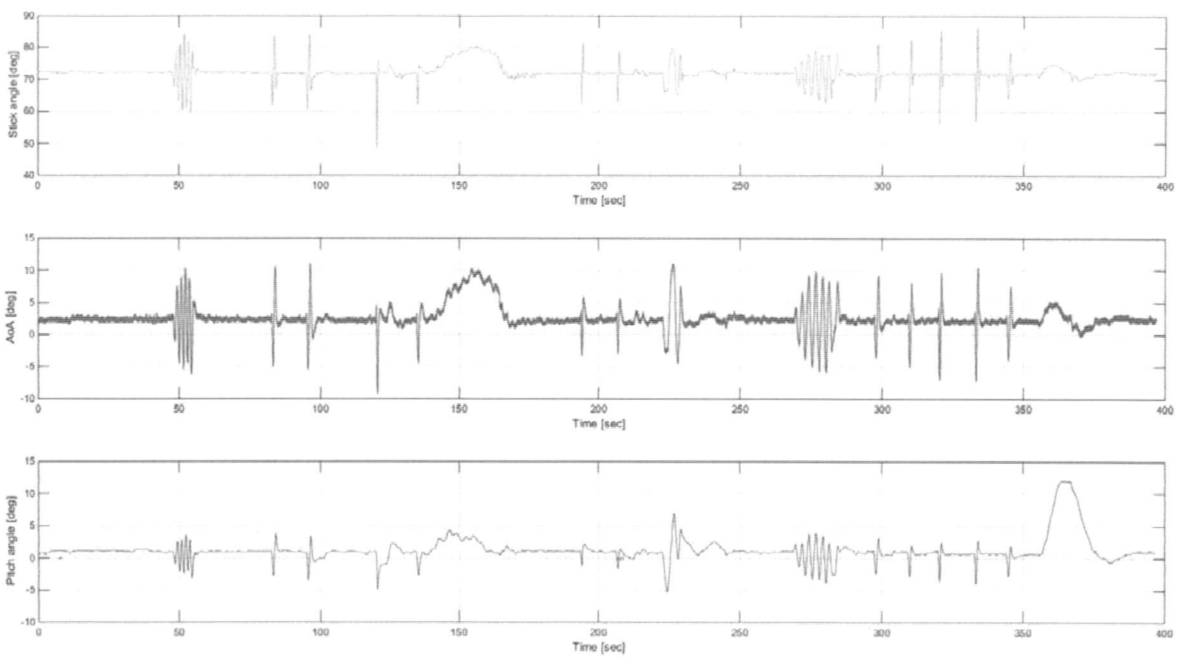

Figure 1. Recorded flight data of short period test events

The pilot performs the first doublet at a small amplitude to avoid exceeding load factor limits. If the response in pitch is safe, i.e., not resulting in pilot-aircraft coupling, the next must have more amplitude, but with the same frequency.

Flight test engineers analyze the relation between the input on the flight stick and the reaction of the aircraft to build aerodynamic models of the aircraft considering the technique applied by the pilot as standard. The same issues observed in this analysis happen directionally for the investigation of the Dutch Roll.

The data was collected from the flights of three pilots. As a result of training, in theory, there should be only a small variation in input frequency among them. However, pilots applied doublets with frequencies varying from 0.45 to 1.25Hz and this variability on the input, just after the frequency sweep, leads to uncertain conclusions made by engineers. The analysis of pilot's inputs showed a cognitive pattern as they all have a similar but wide spread in input frequencies. This proved that it is incorrect to assume that the test pilot input frequency is always precise. When the precision of input is critical, a specific autopilot mode could provide inputs on a flying stick or yoke that are precise in amplitude and frequency. It would eliminate the uncertainties and characterize the response of the aircraft more efficiently. Other phases of FT campaigns could also take advantage of the precision provided by automation, including fixed decelerations for stall (e.g.: 1 kt/s) and windup turns (e.g.: dg/dt = 1g every 5s).

Merging the natural advantages of humans and machines in a cockpit, in 2016, Aurora Flight Sciences developed for DARPA a concept program called ALIAS (Aircrew Labor In-Cockpit Automation System). This system has portable hardware and software that can be configured in less than one month to operate any different type of aircraft. The goal of this system is to reduce crew requirements by the robot replacing the copilot in the right seat. The machine reads the gauges using machine vision and its arms operate the yoke and the throttle levers. The combination of the strengths of humans and robots in the cockpit supposedly provide less workload for the pilot and enforce the execution of all procedures. The system has no Artificial Intelligence to be predictable and reliable for the pilot.

Similar systems might be developed to test aircraft that have mechanical flight controls, including the hydraulically boosted ones. The idea of applying the inputs on the yoke and pedals is to include all the looseness and inflections of the control system on the analysis. For aircraft with fly-by-wire controls and autopilot, a simple autopilot mode might be implemented exclusively for testing.

All of these experimental solutions would be applied before certification. Thus, the risk analysis of the FT campaign must address malfunctions and systemic issues. Woods and Sarter (2000) named the following new problems caused by automation: over automation, human error and bad man-machine coordination. The solution to deal with all of these is training because the use of intermediate levels of automation with inaccurate mental models are a source of new unsafe control actions (Leveson, 2012). All mode confusion in regular operation characterize scenarios that were not sufficiently explored during FT and other development phases.

Risk in FT

Robots are easier to replace than humans. Cost and time to train a professional limit the replacement of highly qualified human operators. After manufacturing new hardware and loading the latest version of the software, the robot is ready to face the unknown again. Moreover, technology is getting cheaper and more accessible. Thus, the replacement of human operators by automation for dangerous tasks becomes more attractive.

Some activities could already use robots as the preliminary tester. However for social purposes, the presence of the human is still essential, despite the implicit danger. For example, sending an astronaut to explore Mars is more socially relevant than doing the same with robots receiving orders from earth. Similarly, facing the unknown behavior of a new aircraft, test pilots face adversities and human presence on the first flight of a prototype is still the target for historical pictures and the cover of magazines.

To open the flight envelope of an experimental piloted aircraft means flying at extreme speeds, altitudes and load factors to explore and enhance the performance of the aircraft. The pilot must work together with the FT engineer to determine the operational limits that will be followed during the entire life of the product. The exploration of these limits often provides amazing stories about things that went wrong, including losing the control of the aircraft or losing its parts. Most test pilots learn about mishaps that marked the development of important aircraft. In each of these stories, there was one common fact: the life of the pilot was in danger.

There is a big expectation for the first flights of new aircraft. When test pilots and FT engineers execute a long campaign plan, they are so concerned about not adding complexity to the first flight that many times they don't even retract the landing gear or change the flaps position. That happens not only because it is a technical milestone. First flights are also a media event loaded with cultural celebration.

On the other hand, the subsequent flights explore critical features, such as the aerodynamic flow during the stall, handling characteristics or the aircraft controllability when aborting a takeoff run. On many of these flights, the pilots know from the risk analysis that the probability of finding undesired vibrations or controllability issues is higher during specific events. The pilot prepares his mindset to react to surprises using emergency procedures that might include ejecting from the prototype. He flies because he was taught to do it. He flies because the adrenaline of facing the unknown makes him feel good. He flies because he seeks personal glory (O'Mara, 2011) by being the main character of stories to be told.

Each test has a piloting technique and its execution has a series of cognitive demands; all of which have safety impacts. The first type of investigation deals with finding operational performance limits, including stalling and maximum speed, the maximum load factors, and the maximum altitude. The second type relates to the handling qualities, e.g. sources of pilot-aircraft coupling and spins. Finally, system testing also has critical events, like the weapons separation from pylons, launch rails, and bomb bays.

The Performance phase extends the flight envelope using the build-up approach[1] to investigate unknown behaviors. The exploration of high **speeds** might find a buffet on the structure with potential loss of parts. At the other end, lower speeds explored during stall investigations might cause the pilot to lose control of the aircraft. Both situations require complex sensing, diagnosis and judgement of the test pilot to determine operational limits.

While chance of structural issues due to excess **load** is very small, if it happens there is no time to react. The signals that the structure is about to collapse are cognitively perceived by the pilot as noises and vibrations different than usual. If the positive or negative limit is high, as in fighter aircraft, the senses of the pilot are affected by the g-force and his or her judgment is compromised. Automation would not suffer such restrictions and the combination of acceleration and vibration sensors, and microphones would provide the recording of the phenomena and a basic reaction.

For high **altitudes**, risk is related to pressurization issues, such as noises and the increase in cabin altitude. Pilots have a better judgment than automation about diagnosing off-nominal situations with the structure or sub-systems when the aircraft struggles with huge differential pressures. But for first climbs, the effects of hypoxia and decompression explosions are

[1] Build-up approach means that the event starts at a safe initial condition and the parameter is increased gradually and at a constant rate. This rate depends on how unpredictable the behavior of the system is to that extreme condition.

extremely dangerous to humans. In this case, a first flight to the maximum altitude without a human on board would reduce drastically the severity of this test event on the risk analysis.

Adding external payloads in pylons require an investigation of the effects of **flutter**, a resonant vibration of wing tips and stabilizers. If the oscillation is divergent, the test might be catastrophic. Even after using software and wind tunnels, this test is still important. Devices designed to produce these oscillations are installed on the trailing edge of the surfaces and the aircraft take off with a chase aircraft to record videos of the test. Accidents caused by flutter are not common, but their severity is often high. Thus, automation would be welcome for the same reasons as in load factor.

For handling qualities, the **spin** is one of the most critical maneuvers on a FT campaign of training and combat aircraft. The pilot must explain the behavior of the aircraft while reading speed, attitude and altitude. After the recovery, the pilot classifies the spin according to a metric chosen for the test. The maneuver itself take less than one minute, but the workload and dizziness are close to the human limitations. The remote control of a spin would be challenging because its implicit delay in communication interfere with the successful exit from spinning.

Finally, among all sub-systems tests, first-time **weapon separations** on military aircraft is critical because it might cause damage that interferes with the aircraft's controllability[2] and demands a fast decision about ejection. Remote operation would provide better chances of recovering the prototype, but as with spins, the delay would limit a proper reaction on controls.

The technology necessary to remotely control an aircraft with a seat and controls for a human pilot already exists. The QF-16 is an adaptation on the flight controls of a regular F-16 that enables it to be controlled remotely. The system has been flying since 2013 and reached operational capability as aerial target in 2016.

The challenge of building a machine to autonomously react properly and timely to all of these dangerous situations is the core of a cognitive paradigm, because the sum of methodological and theoretical approaches to all aspects of human psychology such as instincts, motor skills, memory, speech, values, personality, and problem-solving is too complex to be reproduced by software.

The safety improvement on aviation statistics with automation leads us to believe that, little by little, Intelligence Augmentation (IA) will reduce workload and increase autonomous properties up to the moment that Artificial Intelligence (AI) will take over and reduce the remote control to emergency modes of operation. In the light of the hexagon of cognitive sciences (Miller, 2003), the use of IA and the safe substitution of the human by AI is a multidisciplinary endeavor. This evolution must respect social phenomena and consider user's behavior when reacting to scenarios of automation failures and mode confusion.

In October 27th, 2016, Uber Technologies Inc. released a white paper picturing an aerial vision for urban transportation, envisioning that "pilot aids will evolve over time into full autonomy, which will likely have a marked positive impact on flight safety" (Uber, 2016). The initial certification process and operation of these new machines will be as piloted aircraft. Thus,

[2] When a bomb is launched from a pylon, one or two explosive charges are used to initiate the movement of the bomb away from the aircraft. Depending on many aspects, such as the charges sizing, sideslip, angle of attack, speed, attitude, and altitude, the bomb might present unstable separation and collide with the aircraft. Self-propelled weapons, such as missiles and rockets, use launch rails or launchers, but they are equally susceptible to separation issues like limited time for reaction.

the first generation of this urban VTOL (Vertical Takeoff and Landing) will pass on a regular FT campaign with the same performance and HQ issues discussed on this paper. It it will be an opportunity to prove the value of using higher levels of automation building statistical proof for users and regulators.

Conclusion

This paper discusses the main advantages and challenges of using different levels of automation on piloted and remote/autonomous control on fixed-wing aircraft designed to be piloted by humans. The use of machines acting on flight controls or devoted autopilot modes for precision on FT techniques along with the remote operation on dangerous events are new applications of automation with potential to make FT campaigns safer and more efficient. These applications are restricted to the FT events that do not require complex perception or judgment.

Intermediary levels of automation as a new autopilot mode require the adjustment of the test pilot's mental models to the limitations of the system. This means that more preparation is necessary to avoid surprises. When the aircraft is fully autonomous or remotely operated for dangerous events, those sources of confusion diminish and the risk comes from the incompleteness of software or delayed communications link. The development of such systems will bring unforeseen accidents that must be addressed in the FT campaign risk analysis.

References

de Winter, J.C.F., Dodou, D., 2014. Why the Fitts list has persisted throughout the history of function allocation. Cogn. Technol. Work 16, 1–11. doi:10.1007/s10111-011-0188-1

Elevate, U., 2016. Fast-Forwarding to a Future of On-Demand Urban Air Transportation.

Fitts, P. M. (1954). "The information capacity of the human motor system in controlling the amplitude of movement". Journal of Experimental Psychology 47(6), 381–391.

Instituto de Pesquisas e Ensaios em Voo (IPEV), Divisão de Formação em Ensaios em Voo. Manual de Técnicas de Ensaios em Voo, Estabilidade Estática Longitunal EQ-02-TEV. São José dos Campos, 2015.

N. G. Leveson, Engineering a Safer World: Systems Thinking Applied to Safety. The MIT Press, 2012.

Miller, G.A., 2003. The cognitive revolution: A historical perspective. Trends Cogn. Sci. 7, 141–144. doi:10.1016/S1364-6613(03)00029-9

O'Mara, Raymond P. 2011. The Sociotechnical Construction of Precision Bombing : a Study of Shared Control and Cognition by Humans, Machines, and Doctrine During World War II / by Raymond P. O'Mara. c2011.

N. B. Sarter, D. D. Woods, and C. E. Billings, "Automation surprises," in Handbook of Human Factors and Ergonomics, vol. 2, John Wiley and Sons, 1997, pp. 1926– 1943

Acknowledgements

This research was possible due to the scholarship provided by CNPq/Brazil.

EVALUATING SPATIAL-AUDITORY SYMBOLOGY FOR IMPROVED PERFORMANCE IN LOW-FIDELITY SPATIAL AUDIO DISPLAYS

Griffin D. Romigh
U.S. Air Force Research Labs
Wright-Patterson AFB, OH

Jose Villa
U.S. Army Natick Soldier RD&E Center
Natick, MA

Jason Ayers
Ball Aerospace
Dayton, OH

For decades, spatial auditory displays have been considered to be a promising technology to help fight pilot disorientation and loss of SA. Inherently heads-up, these displays can provide time-critical spatial information to pilots about navigational targets, air and runway traffic, wingman location, and even the attitude of one's aircraft without placing additional demands on the already over-tasked visual system. Unfortunately, currently-fielded auditory displays often suffer from poor spatial fidelity, particularly in elevation, due to their use of a one-size-fits-all (i.e., non-personalized) head-related transfer function (HRTF), the set of filters responsible for creating the spatial impression. The current study investigated the utility of combining a spatial cue (non-personalized HRTF) with one of two auditory symbologies, one providing both object and location information, and the other only location information. In one case, ecologically-valid sounds were paired with a particular class of visual object, and spatial cues indicated a plausible target elevation (e.g., a squeak indicated the target was a rat on the floor). In the other condition, the cue was a broadband sound, the repetition rate of which indicated target elevation (i.e., the cue provided only location information, not object information). Results indicate that target acquisition times were lower when meaningful (i.e., ecologically-valid) cues were added to non-personalized spatial cues when compared to the case in which the source-based cues provided no information about the target source. These results indicate that careful construction of auditory symbology could improve performance of cockpit-based spatial auditory displays when personalized, high-fidelity spatial processing is not practical.

Background

Because of its natural function as the body's "early-warning system," the auditory system provides an intuitive channel for portraying time-critical information. Many auditory displays leverage a listener's natural ability to rapidly identify different sound sources, and use source identification (ID) as way to alert a user to not only when, but also what type of event has occurred (e.g., different alerts for low altitude vs. traffic warnings).

Several experiments have also shown the benefits of spatial audio cues provide in visual search tasks, specifically, a reduction in visual search times compared to visual-only search conditions (Bolia et al., 1999, Perrot et al., 1996). In general, these studies have also shown that auditory-aided visual search is largely unaffected by the number of visual distractors, leading to large performance benefits for more complex visual scenes.

Displays that aim to take advantage of this spatial cueing are referred to as Virtual Audio Displays (VADs) or sometimes referred to as Spatial or 3D-Audio Displays. These displays rely on the creation of a perceptual illusion that headphone-based sounds actually originate from real-world locations in 3D space. If properly designed, VADs can have application to aircraft threat avoidance, station keeping, and navigation (Simpson et al., 2005), as well as, the more traditional use as a tool for radio speech intelligibility improvement. Despite their promise, VADs have not yet made a large impact in the aviation market, due mostly to the difficulty of achieving robust, high-fidelity spatial audio imagery on a commercial scale.

The signal processing that underlies a VAD is done by filtering a single-channel, non-spatial sound source with a pair of head-related transfer functions (HRTFs). That filtering operation results in left- and right-ear signals, which when presented over headphones can result in the perceptual illusion that the sound source was presented from a physical location out in space. Unfortunately, HRTF filters are both position- and listener- specific, meaning high-fidelity virtual auditory space can only be achieved by making electro-acoustic measurements on each listener from a large number of spatial directions. This means that commercial VAD systems, which often need to have one-size-fits-all convenience, typically have poorer fidelity than a personalized system. While lack of personalization is the major drawback of most commercial VAD technology, other compromises have also been made to save on processing power and/or battery life in some resource-constrained, real-world systems.

When non-personalized or low-fidelity HRTFs are used in a VAD, typical problems include: the perception that sources originate from inside your head (a.k.a., lack of externalization), a compression of perceived sound source elevation, and an increase in the rate of front-back reversals (the perception that sources presented in the front came from the back and vice versa) (Wenzel et al., 1993). In most applications, these perceptual shortcomings result in a decrease in the effectiveness of the VAD to accurately support its intended purpose.

The current study was designed to investigate whether the robustness of a listener's sound source identification ability could be leveraged to improve performance in an auditory-aided visual search task, when the fidelity of the spatial rendering was low.

Methods

In order to investigate whether source-ID cueing could provide a benefit for VADs with low-fidelity spatial rendering, an auditory-aided visual search task was conducted in a virtual environment with varying levels of spatial rendering quality and two auditory display symbologies that provided ID-based spatial cues.

Experimental Conditions

Ten paid listeners with normal hearing and vision participated in 24 experimental blocks over the course of three weeks. Each block consisted of 120 auditory-aided visual search trials with a fixed audio cueing condition. The audio cueing condition for each block was selected randomly from a 2 by 3 by 4 condition matrix composed of cue duration (250ms single burst, continuous), audio source type (Noise, Ecological, Click Train) and spatial rendering type (Enhanced, KEMAR, Panning, Diotic).

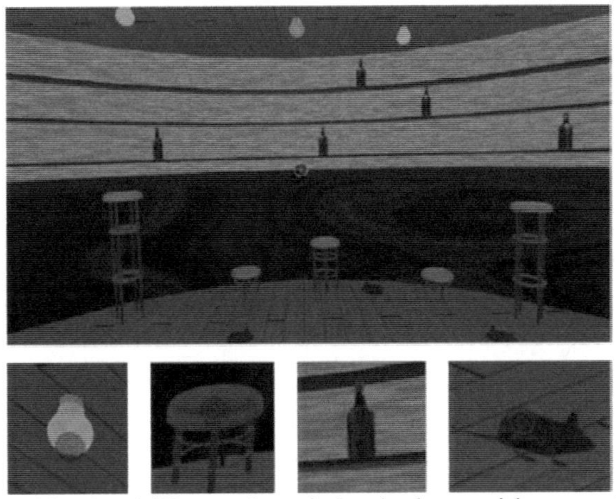

Figure 1. Virtual saloon scene used in the A/V search task (top) along with example targets from each object class (bottom).

The broadband noise stimuli were bandpass filtered between 200 Hz and 16 kHz and were independent, but statistically identical, for all target object types. This type of cue therefore provides no source-ID-based spatial information, in contrast to the ecological and click train cue types that follow. The ecological stimuli were constructed to resemble the type of auditory event a listener might expect from each of the four visual objects; electrical sparking of a light, rattling of a bottle, clanking of a barstool, squeaking of a rat. In general, all of the ecological stimuli contained spectro-temporal features sufficient to provide localization accuracy on par with the broadband noise stimuli. Conversely, random-phase click train stimuli were constructed to allow audio identification of target object classes without having any ecological validity, meaning subjects would have to learn the association between each click-train type and visual object class. The click trains were constructed by modifying the random phase click rate (100, 141, 200, and 283 Hz) and a $\sin^2$ temporal modulation window (2, 4, 6, and 8 Hz) for each class of object (rats, stools, bottles, lights, respectively). In general, these random phase click trains contain sufficient information to allow good localization; however, due to an implementation error, the click-train stimuli were lowpass filtered at 8 kHz, meaning some of the important cues for sound source localization above 8 kHz were not available.

To generate spatial audio cues, an HRTF specific to the current spatial rendering condition was loaded into slab3D and a pre-generated .wav file of the appropriate source type and duration was played through the engine. The KEMAR condition utilized a conventional non-individualized HRTF, recorded on the KEMAR mannequin as described in Romigh et al. (2015). The Enhanced condition utilized the same KEMAR HRTF after being pre-processed to exaggerate spectral cues as described in Brungart & Romigh (2009). The Panning HRTF was constructed to provide stereo panning between the left and right headphone signals based on the sound source's head-relative lateral angle. This setup means only the inter-aural level difference (ILD) cues that were relevant to sound source lateralization were present, without any high frequency monaural spectral cues, which are critical for elevation. The Diotic HRTF was constructed so that the original source signal was passed directly to both ears without any processing. This manipulation provided no spatial information, since in this condition all sound sources should appear as though they originate from the center of the listener's head.

Task Environment

Listeners were seated on a rotating stool in the Spatial Hearing Anechoic Research Chamber (SHARC) at Wright Patterson AFB, OH. An audio-visual virtual environment was presented via an HTC Vive VR headset and a pair of Sennheiser HD280 headphones. The Vive allows 6-DOF motion and also includes a tracked wand to enable cursor-based pointing within the scene. Spatial audio rendering was accomplished using slab3D (Miller & Wenzel, 2002)an open-source audio rendering engine that allows incorporation of custom HRTFs and has been shown to produce virtual sound sources that permit localization accuracy on par with free-field sources (Romigh et al., 2015).

The virtual environment was created in Unity3D, a game engine for developing interactive 3D virtual environments. The environment resembled a 360° saloon scene (top panel of Figure 1) and consisted of a cylindrical room partitioned into four distinct regions in elevation (i.e., floor, bar, wall, ceiling). In each elevation region, 30 instances of a single class of object were scattered randomly throughout the region at all azimuths; light objects occupied the ceiling region from +36 to +18 degrees in elevation, bottle objects occupied the wall region from +18 to 0 degrees in elevation, stool objects occupied the bar region from 0 to -18 degrees in elevation, and rat objects occupied the floor region from -18 to -36 degrees in elevation.

Experimental Task

Each trial started when the listener pulled the trigger to "shoot" the large bullseye in the front of the visual scene by aiming a wand-slaved crosshair cursor. Then, a visual target was presented in the form of a semi-transparent bullseye placed randomly in front of one of the 120 scene objects. The transparency of the target bullseye was manipulated to subjectively equalize the salience of all target objects and make it less likely that a visual target could be identified in the visual periphery. Simultaneously, a virtual audio cue was presented from the location of the visual target, and the task of the subject was to locate and shoot (aiming a wand-slaved crosshair) the visual target as quickly and as accurately as possible. The first shot aimed within 10 degrees of the visual target scored as a hit and the timing of the shot was recorded as the response time.

Results and Discussion

Average response times for all conditions are shown in Figure 2. Results for short duration, "Burst" stimuli and continuous stimuli and shown in the left and right panels, respectively. In general, response times for the burst and continuous stimuli were similar. The biggest differences appear to be for Noise stimuli and/or the Panning rendering condition. This suggests that when some cues for sound source elevation are available (i.e. in the Enhanced and KEMAR rendering conditions and/or with Ecological or Click Train stimuli) the additional information provided by dynamic head-motion cues and a longer observation window do not reduce search times.

With the Noise stimuli, response times increased with decreasing spatial rendering quality, as expected, rising from 2 seconds in the Enhanced condition to over 6.5 seconds in the Diotic condition. In contrast, the Ecological stimuli were less affected by rendering condition, increasing from just under 2 seconds in the Enhanced condition to roughly 4 seconds in Diotic condition. The Click Train stimuli fell in between the Noise and Ecological conditions, which could have resulted from both decrease in localizability caused by its reduced bandwidth, or because the mapping of the click train parameters to elevation (or source type) was less intuitive that the Ecological stimuli. The fact that a difference is seen between the Ecological and Click Train stimuli in the Diotic condition suggests the latter.

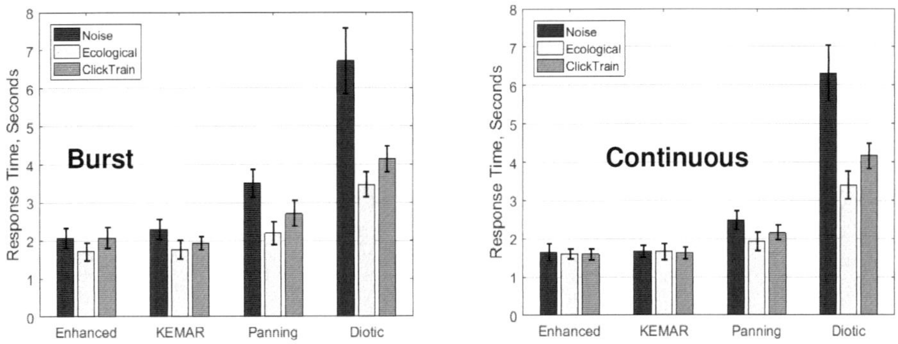

Figure 2. Average response times for each experimental condition. Error bars represent their 95% confidence intervals.

Comparing the results across the conditions, it appears that providing source-ID based elevation cues can provide increasing benefit in terms of reduced search times as the fidelity of the spatial rendering cues goes down. The largest benefit is therefore found when no rendering-based elevation cues are available (e.g. in the Panning and Diotic conditions); however, since the performance benefit between Noise and Ecological stimuli goes up from the Panning to the Diotic condition, it suggests another non-spatial cue is being used (e.g. a benefit from a reduced valid set-size).

Figure 3 shows head-tracking elevation data for all Burst trials. Each panel represents a different experimental condition, as indicated, and colors are used to identify the target object type and target elevation range (Purple-Lights, Cyan-Bottles, Yellow-Stools, Red-Rats). Dramatic differences are apparent for the Ecological and Noise stimuli. Even in the Enhanced rendering condition where average response times are fairly similar, the ecological stimuli clearly resulted in more definitive head movements, as can be seen by the clear separation of tracks with different target object types (i.e., tracks with different colors). This suggests very different search strategies are employed when different sources of spatial information are available.

Conclusion

The current study investigated the benefit of providing source-ID based spatial cues in addition to traditional spatial rendering cues in an auditory-aided visual search task. Response time results indicate that the benefit of adding source-ID cues goes up with decreasing fidelity of the spatial rendering, and may not be influenced by stimulus duration and/or presence dynamic head-motion cues. Head tracking results indicate that different search strategies are employed when source-ID based cues are available.

Acknowledgements

This work was partially supported through funding from the Air Force Office of Scientific Research (Grant #16RHCOR366).

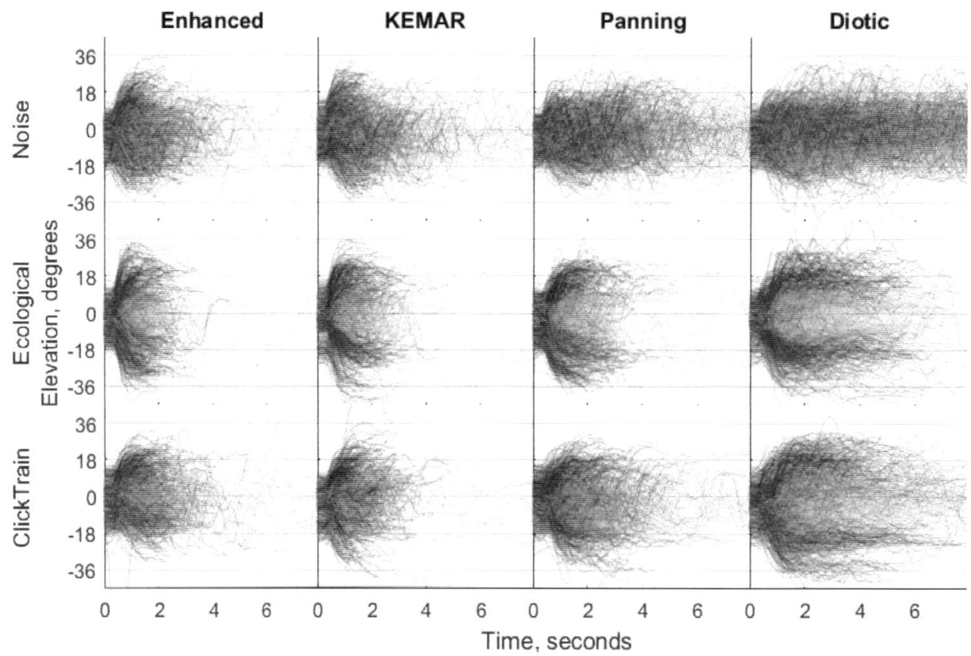

Figure 3. Headtracking data for all trials from the short duration "Burst" trials. Tracks show the elevation component of the head-orientation as a function of time. Each panel shows a single experimental condition. Colors indicate the target object type (Purple – lights, Cyan – Bottles, Yellow – Stools, Red – Rats).

References

Bolia, W. S., D'Angelo, W. R., & McKinley, R. L. (1999). Aurally aided visual search in three-dimensional space. Human Factors, 41, 664-669.

Brungart, D. S. and Romigh, G. D.(2009) "Spectral HRTF enhancement for improved vertical-polar auditory localization," *2009 IEEE Workshop on Applications of Signal Processing to Audio and Acoustics*, New Paltz, NY, pp. 305-308. doi: 10.1109/ASPAA.2009.5346479

Miller, J. D., Wenzel, E. M. (2002) Recent Developments in SLAB: A Software-Based System for Interactive Spatial Sound Synthesis, Proceedings of the International Conference on Auditory Display, ICAD 2002, Kyoto, Japan, pp. 403-408

Perrott, D. R., Cisneros, J., McKinley, R. L., & D'Angelo, W. R. (1996). Aurally aided visual search under virtual and free-field listening conditions. Human Factors, 38, 702–715.

Romigh, G. D., Brungart, D. S., Simpson, B. D. (2015). Free-field localization performance with a head-tracked virtual auditory display. IEEE Journal of Selected Topics in Signal Processing, 9, 943-954

Simpson, B.D.; Brungart, D.S.; Gilkey, R.H.; McKinley, R.L. (2005) Spatial Audio Displays for Improving Safety and Enhancing Situation Awareness in General Aviation Environments. In New Directions for Improving Audio Effectiveness, pp. 26-1 – 26-16.

Wenzel, E.M., Arruda, M., Kistler, D.J., Wightman, F.L. (1993) Localization using non-individualized head-related transfer functions. J. Acoust. Soc. Am., 94, pp. 111–123

VIRTUAL REALITY AND 2D INTERFACES: A COMPARISON OF VISUAL SEARCH TASK PERFORMANCE

Matthew Brown, Kathleen Van Benthem, James Howell,
Jonathan Poisson[1], Scott Arbuthnot, and Chris Herdman
Carleton University
[1]General Dynamics Mission Systems Canada
Ottawa, Canada

Airborne surveillance operations present challenging environments for tactical operators and for the technologies that support these activities. Information from multiple sources is currently presented on 2D displays, but the influx of data has made it difficult to represent this information using traditional technologies. Recent innovations in VR have laid the groundwork for a promising solution to this problem by allowing users to immerse themselves in 3D representations of the real world with embodied tracking capabilities. The present research examined the feasibility of transitioning two common tactical operator tasks from a 2D to a 3D/VR user interface. Naive participants searched for targets amongst a set of non-targets on a traditional 2D interface and on a custom-built VR interface rendered on an Oculus Rift. Participants reported a target's geographical coordinates or the distance between two targets. Search difficulty and search specificity were manipulated. Results and future directions are discussed.

Airborne surveillance operations require the visual integration of multiple streams of data from ground, air, and maritime sources. The ever-increasing availability of real-time sensor data, fused track data, and environmental data has surpassed the capability of traditional 2D displays to provide the operator with a coherent visual representation of the operational environment. Consequently, the operator must devote considerable mental effort to navigate multiple layers of cluttered displays in order to maintain effective situation awareness. The limitations of 2D displays have accelerated the need to develop human machine interfaces that can leverage recent innovations in virtual reality (VR) and augmented reality (AR) technologies.

A potential benefit of VR/AR interfaces is that the user can view and interact with one-to-one mappings of an environment in virtual 3D space (VR) or with synthetically rendered/enhanced objects in the environment itself (AR). This offloads the operator's task of mentally re-mapping a 2D plan-view to encompass a vertical dimension (e.g., Carswell & Wickens, 1987; Wickens, Merwin, & Lin, 1994). The benefits of 3D over 2D visualization have also been shown in the context of "tunnel in the sky" displays (Haskell & Wickens, 1993) and for conflict avoidance on air traffic displays (Ellis, McGreevy & Hitchcock, 1987). There is, however, evidence showing that 2D displays are *better* than 3D displays in certain contexts (e.g., Boyer, Campbell, May, Merwin, & Wickens, 1995; O'Brien & Wickens, 1997; Tham & Wickens, 1993; Wickens & May, 1994). One limitation of many 3D displays is that they only provide the user with one viewpoint, which can result in closer objects obscuring distant objects (e.g., Ellis et al., 1987). The immersiveness of VR interfaces circumvents this problem by providing the user with a potentially infinite number of viewpoints.

The purpose of the present work was to examine the impact of a 3D/VR user interface on tasks that are representative of what a tactical operator would commonly perform using a 2D interface. To this end, participants performed a visual search task in which they were to locate target objects amongst distractors. Participants reported a target's location (latitude, longitude or altitude) or the distance between two targets. Participants performed these tasks using an in-house prototype 3D/VR interface and a commercial-off-the-shelf 2D interface. The difficulty of the search task was manipulated by having either 18 or 36 objects in the search environment. Further, search specificity was manipulated by providing the target object's domain (airborne, surface, sub-surface), its classification (friendly, neutral, enemy), or by not providing any domain/classification information. An object's domain was visually represented in 2D/3D as a triangle/pyramid, square/cube, or circle/sphere for airborne, surface, or sub-surface objects, respectively. An object's classification was visually represented by the object's color – green, yellow, or red for friendly, neutral, or enemy objects, respectively.

It was hypothesized that the one-to-one mapping of the search environment provided by the 3D display, coupled with the ability to change viewpoints in VR (i.e., participants could move along the x, y and z-axes) would yield better performance than the 2D display. It was further hypothesized that the performance benefit when using the 3D/VR interface would be magnified for difficult searches. It was also hypothesized that the anticipated benefits of the 3D/VR would be more evident when the target object's domain (air, surface, sub-surface) was specified because the vertical separation of the objects can be visually represented in 3D/VR, but not in 2D.

Method

Participants

A total of 17 Carleton University undergraduate students (12 females) participated in exchange for $20. All participants had normal or corrected-to-normal visual acuity and normal color vision. Three participants were unable to complete the experiment due to VR-induced motion sickness and were therefore excluded from the sample.

Design

A 2 (Interface: 2D vs. 3D/VR) x 2 (Search Difficulty: 18 objects vs. 36 objects) x 3 (Search Specification: No Specification vs. Domain Specified vs. Classification Specified) repeated measures design was used. Interface was blocked and counterbalanced across participants. Search difficulty and search specification were mixed factors, with the six conditions created by crossing these two factors randomly presented with the constraint that there were an equal number of trials per condition. A total of 72 trials were presented – 36 in the 2D condition and 36 in the 3D/VR condition.

Apparatus and Stimuli

2D interface. The operational enviroment – a surface area of approximately 150 km^2 off of the coast of Halifax – and search instructions were displayed on two LCD monitors with a 1920 x 1200 resolution. An overhead plan-view of the search environment (see Figure 1, left panel) was shown on one monitor while the search instructions (e.g., "What is the altitude of object ID #1?") and a countdown timer were displayed on the other monitor. Input devices were a standard Microsoft keyboard and mouse. The visuals and user interface were driven by VR Forces (Version 4.4) software produced by VT MÄK. The environment was populated with 18 or 36 objects, depending on the search difficulty for that trial, that were represented as icons created by crossing three shapes (triangle, square, circle) with three colors (green, yellow, red). Each object was labeled with a unique numerical identifier (i.e., the digits 1 to 18/36), which was located adjacent to the icon. The countdown timer appeared with the search instructions and started at a predetermined time based on the task and search difficulty. If time elapsed, the message "TIMEOUT" was displayed and a buzzer sounded.

3D/VR interface. The same computer used in the 2D interface condition was used to render the search environment on an Oculus Rift CV1 head-mounted VR display (see Figure 1, right panel), which tracked participants' head movements such that the environment was always in view. The field of view was approximately 110° vertically and horizontally. A Leap Motion hand tracker was affixed to the front of the Oculus Rift and used IR tracking technology to fit a kinematic model to the user's hands in order to track and visually represent hand/finger movement in real time. Input devices consisted of a SpaceNavigator 3Dconnexion 3D mouse, which allowed users to move along the x, y, and z-axes in 3D space and a virtual number pad. The visuals and user interface were controlled by custom in-house software built on the Unreal gaming engine platform (Version 4.13). The objects in the search environment were volumetric equivalents of the icons in the 2D condition. The size of the objects was scaled according to the distance between the participant's current location and the object. The search instructions and countdown timer were identical to those in the 2D condition, but were displayed on a virtual screen that was located on the right side of the search environment and maintained a set size and position relative to the participant's current location.

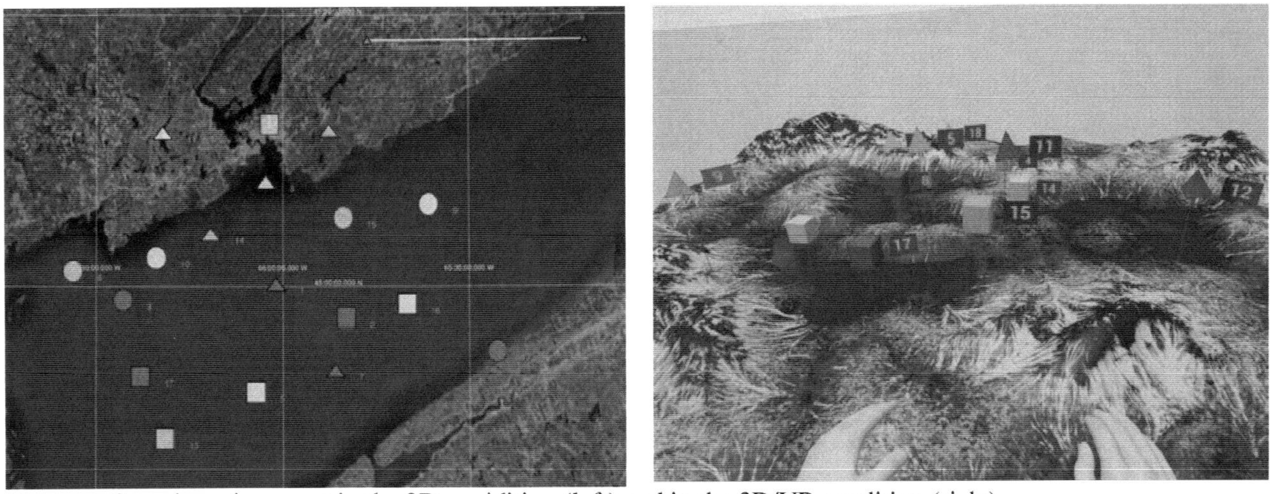

Figure 1. Search environment in the 2D conidition (left) and in the 3D/VR condition (right)

Procedure. The 3D/VR condition consisted of a 10-minute training session to familiarize participants with the VR-specific apparatus, 12 practice trials, and 36 experimental trials. The 2D condition consisted of 12 practice trials and 36 experimental trials. Half of the participants received the 3D/VR condition followed by the 2D condition and the other half received the reverse order. Trials began with the presentation of the search task instructions, which specified the target object's unique identifying number, the target object's domain or classification (except on no-specification trials), and whether the participant was to report the target object's latitude, longitude, altitude or distance from another target object. Participants entered their responses on the keyboard's number pad in the 2D interface condition or on the virtual number pad in the 3D/VR interface condition. Correct responses always consisted of four digits.

In the 2D interface condition, participants accessed an object's location by clicking on the target object, which activated a drop-down menu. Participants selected an option on this menu that activated a secondary menu that displayed the target object's location. To find the distance between two objects, participants clicked on the two target objects and then used the mouse to drag and drop the end points of a distance measurement tool onto the activated targets. Participants then right clicked the measurement line, which activated pop-up menu that displayed the line's current length (i.e., distance between the two targets).

In the 3D/VR interface condition, participants accessed an object's location by fixating on an object, which activated a blue halo that surrounded the object, and then clicked the 3D mouse to activate a pop-up menu that displayed the object's location. To find the distance between two objects, participants fixated on the first target and clicked the 3D mouse to activate it and then fixated on the second target object and activated it. Participants then made a "pinch" gesture with their left hand on a target, which activated a distance finder tool, signalled by the appearance of a blue sphere that was displayed in the participant's virtual left hand. While maintaining the pinch gesture, participants used the mouse to move to the second target. Participants then "dropped" the distance finder (i.e., the blue sphere) on the target by releasing the pinch gesture. A measurement line connecting the two target objects then appeared, with the distance displayed above the line.

Results

Three participants were unable to complete the experiment due to VR-induced motion sickness. Their data were eliminated from all further analyses, which reduced the sample to n=14. Additionally, 3.5% of the trials were flagged as mistrials due to data collection failure and were therefore eliminated from the analyses.

The remaining data were analyzed using a 2 (Interface) x 2 (Search Difficulty) x 3 (Search Specification) repeated measures ANOVA.

Activation Response Times

Activation response times were measured as the time between the onset of the search instructions and the activation of the target (location task) or targets (distance task). Only correct response times were included in the analysis.

Location task. The main effect of interface was not significant, $F(1, 13) = 2.17$, $p > .15$, nor was the main effect of search specification, $F(2, 26) = 1.03$, $p > .35$. The main effect of search difficulty was significant, $F(1, 13) = 27.56$, $p < .001$, with faster responses on 18-object trials (M=5.67 s) than on 36-object trials (M=8.82 s). The interface by search difficulty interaction was not significant ($F<1$). As shown in Figure 2 (left panel), the interface by search specification interaction was significant, $F(2, 26) = 7.62$, $p < .005$. This interaction was driven by domain-specified targets being activated significantly slower than non-specified and classification-specified targets in the 2D interface condition, but being activated faster in the 3D/VR condition.

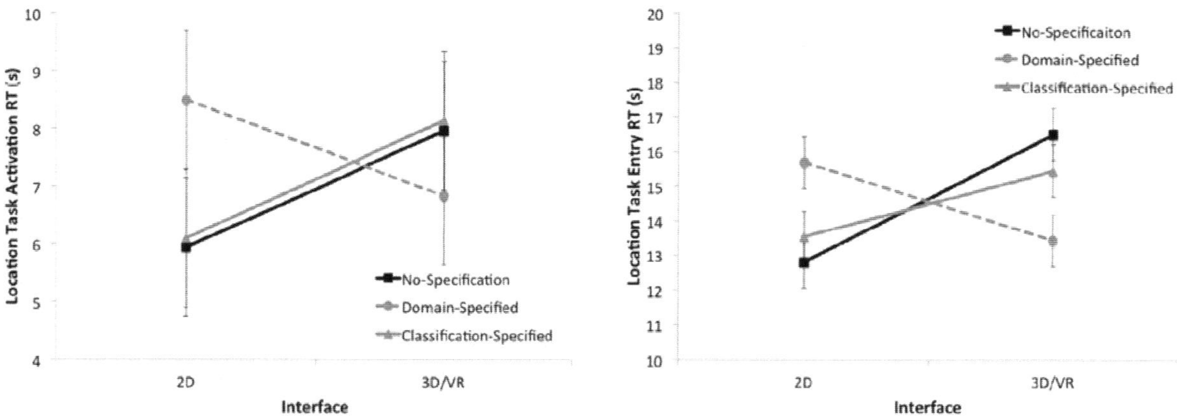

Figure 2. Location task activation response times (left panel) and entry response times (right panel) as a function of interface and search specification with 95% confidence intervals.

Distance task. There was a significant main effect of interface, $F(1,13) = 19.38$, $p < .005$, with faster responses in the 2D interface condition (M=9.72 s) than in the 3D/VR interface condition (M=13.82 s). The main effect of search difficulty was also significant, $F(1, 13) = 42.01$, $p < .001$, with faster responses on 18-object trials (M=8.71 s) than on 36-object trials (M=14.82 s). The main effect of search specification was not significant ($F<1$), nor were the interface by search specification or interface by search difficulty interactions ($Fs<1$).

Entry Response Times

Entry response times were measured as the time between the onset of the search instructions and the entry of the 4-digit target object location or distance. Only correct response times were included in the analysis.

Location task. Neither the main effect of interface ($F<1$) nor the main effect of search specification, $F(2, 26) = 2.26$, $p > .10$, were significant. The main effect of difficulty was significant, $F(1, 13) = 35.67$, $p < .001$, with faster responses on 18-object trials (M=13.26 s) than on 36-object trials (M=15.88 s). The interface by search difficulty interaction was not significant ($F<1$). As shown in Figure 2 (right panel), the interface by search specification interaction was significant, $F(2, 26) = 6.72$, $p < .005$. As in the location task activation response time data, this interaction is caused by significantly slower responses on domain-specified trials than on no-specification and classification-specified trials in the 2D condition, but significantly faster responses in the 3D/VR condition.

Distance task. The main effect of interface was not significant, F(1, 13) = 1.54, p > .20. The main effect of search specification was marginally significant, F(2, 26) = 3.28, p < .06, with slower responses on domain-specified trials (M=29.14 s) than on no-specification (M=27.04 s) or classification-specified (27.34 s) trials. The main effect of difficulty was also significant, F(1, 13) = 127.74, p < .001, with faster responses on 18-object trials (M=23.83 s) than on 36-object trials (31.84 s). There was a marginally significant interface by search specification interaction, F(2, 26) = 3.24, p < .06 (Figure 3, left panel). Entry response times were significantly slower on domain-specified trials than on no-specification and classification-specified trials in the 2D condition, but did not differ in the 3D/VR condition. The interface by difficulty interaction was also marginally significant, F(1, 13) = 3.29, p < .10 (Figure 3, right panel), with faster entry response times in 2D than in 3D/VR for 18-object searches, but not for 36-object searches.

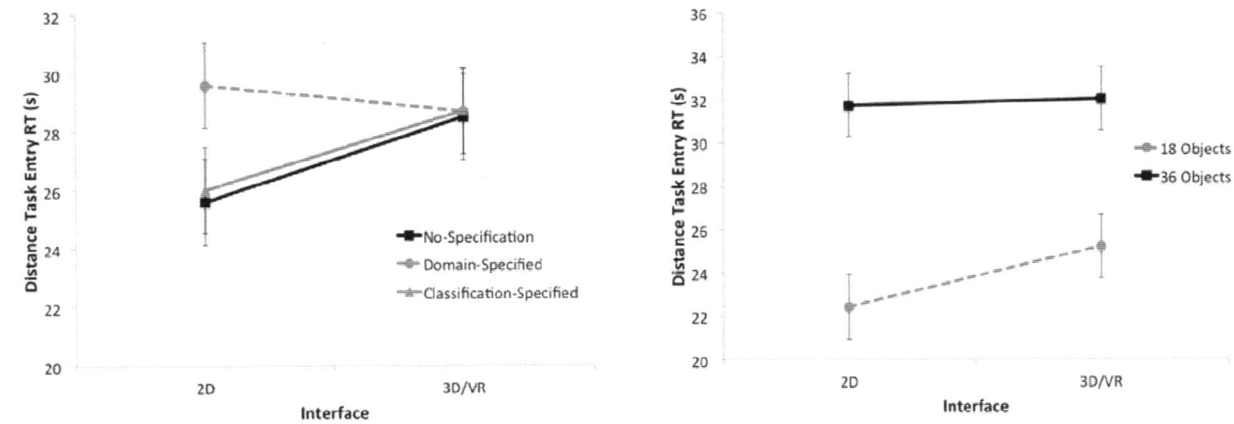

Figure 3. Distance task entry response times as a function of (left panel) interface and search specification and (right panel) interface and search difficulty with 95% confidence intervals.

Accuracy

Accuracy was recorded as binary data (correct vs. incorrect). In order for a trial to be deemed correct, the participant had to enter their 4-digit response before the trial timed-out and the response had to match the target's true location or distance value.

Location task. The main effects of interface and search specification were not significant (Fs<1). There was a marginally significant main effect of difficulty, F(1, 12) = 4.11, p < .07, with higher accuracy on 18-object trials (97.4%) than on 36-object trials (94.7%). Neither the interface by search specification interaction, F (2, 24) = 1.12, p > .30 nor the interface by difficulty interaction (F<1) were significant.

Distance task. The main effect of interface was not significant, F(1, 11) = 1.80, p > .20. The main effects of search specification and difficulty were not significant (Fs<1). Neither the interface by search specification interaction, F(2, 22) = 1.28, p > .25, nor the interface by difficulty interaction (F<1) were significant.

Discussion

The key finding is that the current implementation of a 3D/VR user interface did not yield many performance advantages over a traditional 2D interface on a visual search task that required the user to report a target's location or the distance between two targets. However, one observed advantage of 3D/VR over 2D is that the 3D/VR interface allowed users to find and query targets faster when the target's domain (airborne, surface, sub-surface) was known (see Figure 2). This finding supports the hypothesis that the visual separation of vertically disparate objects in 3D/VR helps the user effectively constrain their search to include only relevant objects.

The search difficulty manipulation had a robust and consistent effect on performance. In contrast, search specification typically did not influence performance, which indicates that overall, participants were not using the additional information provided in domain-specified or classification-specified trials to help guide their searches. One explanation for this finding is that participants simply ignored this supplementary information when searching for the target because its unique numeric identifier was sufficient. In order to encourage participants to use this additional information, the search instructions in subsequent experiments will be modified such that the target's number will be enclosed in a circle, triangle, or square to indicate its domain or will be colored green, yellow, or red to indicate its classification.

The fact that there are many experimental differences between the 2D and 3D/VR interfaces besides the dimensionality of the search environment and how the user interacts with it makes it impossible to pinpoint why the 3D/VR interface did not yield the anticipated benefits. As a starting point, it was clear that many participants struggled with the 3D/VR interface apparatus, including the 3D mouse and the virtual number pad. Future experiments will therefore attempt to level the playing field by having participants in the 2D condition view the display on a Oculus Rift, navigate using the 3D mouse, and respond using a virtual number pad. It is anticipated that the continual refinement of the 3D/VR interface's usability, coupled with the elimination key experimental confounds between interface conditions will provide a clearer picture of the benefits of immersive visualization technologies.

Acknowledgements

This research is supported by General Dynamics Mission Systems Canada and by an Ontario Centres of Excellence grant awarded to C.H. We would like to acknowledge Danielle Krukowski for her assistance with data collection.

References

Boyer, B. S. and Wickens, C. D. (1994). 3D weather displays for aircraft cockpits. Technical Report ARL-94-11/NASA-94-4, Aviation Research Laboratory, Savoy, IL.

Carswell, C. M. and Wickens, C. D. (1987). Information integration and the object display: An interaction of task demands and display superiority. *Ergonomics*, 30, 511-527.

Ellis, S. R., McGreevy, M. W., and Hitchcock, R. J. (1987). Perspective traffic display format and airline pilot traffic avoidance. *Human Factors*, 29, 371-382.

Haskell, I. D. and Wickens, C. D., (1993). Two and three-dimensional displays for aviation: A theoretical and empirical comparison. *International Journal of Aviation Psychology*, 3, 87-109.

O'Brien, J. V. and Wickens, C. D. (1997). Free-flight cockpit displays of traffic and weather: Effects of dimensionality and database integration. In *Proceedings of the Human Factors and Ergonomics Society 41st Annual Meeting* (pp. 18-22). Albuquerque, NM: Human Factors and Ergonomics Society.

Tham, M. and Wickens, C. D. (1993). Evaluation of perspective and stereoscopic displays as alternatives to plan view displays in air traffic control. Technical Report ARL-93-4/FAA-93-1. Aviation Research Laboratory, Savoy, IL.

Wickens, C. D. and May, P. (1994). Terrain representation of air traffic control: A comparison of perspective with plan view displays. Technical Report ARL-94-10/FAA-94-2, Aviation Research Laboratory, Savoy, IL.

Wickens, C. D., Merwin, D. H., and Lin, E. L. (1994). Implications of graphics enhancements for the visualization of scientific data: Dimensional integrality, stereopsis, motion, and mesh. *Human Factors*, 36, 44-61.

A COLOR-CODE DESIGN TOOL

David L. Post
ϕ
Fairborn OH

William E. Goode
Goode Systems
Shawnee OK

The FAA is developing a standard set of colors for use in air traffic control (ATC) displays. The set will be defined in terms of CIE Yu′v′ values, corresponding sRGB values, and color names. A significant complication is that the ATC controller population includes people who have color-vision deficiencies (CVDs). We have written a software tool to assist the FAA in selecting and testing a suitable set of colors. It accepts a set of Yu′v′ values as input and: (1) Draws graphics and calculates color-related figures of merit to predict whether the set will be acceptable for color-normal and CVD users; (2) Flags colors and pairings that violate criteria; and (3) Allows the designer to adjust the colors and see the results immediately. The tool should be useful for designing other color sets, also.

The Federal Aviation Administration (FAA) is conducting experiments to develop a standard set of colors for use in air traffic control (ATC) displays in terminal approach, en route, and oceanic ATC systems. The colors must be discriminable, recognizable, and legible for ATC controllers, including those with color-vision deficiencies (CVDs). The FAA intends to incorporate these colors in the ATC displays within new and modified ATC systems.

We have written a Microsoft Excel-based[1] tool (named *Palette Designer*) to assist with selecting a suitable set of display colors. Palette Designer (PD) allows users to input a candidate set of colors, expressed as CIE luminances and chromaticity coordinates. It then draws graphics and calculates figures of merit, derived from human factors research on color perception and embodied in human factors standards regarding the use of color on electronic displays.

Palette Designer's Main Table

Figure 1 shows PD's main table. The first four columns allow the user to input a color name, CIE 1976 u′v′ chromaticity coordinates, and luminance for as many as 25 colors. (Excel's *Hide Rows* function has been used in the figures for the 11-color set shown herein for illustration.) Luminance is expressed as a percentage relative to the display's peak white luminance, i.e., the luminance produced when the red (R), green (G), and blue (B) tuple {255, 255, 255} is loaded to drive the computer's graphics card.

[1] Palette Designer uses features unique to Excel 2010 for Windows presently. We are eliminating them so it can be used with Excel 2011 for OSX also. It was developed using federal funds, so it is available to the public with unlimited distribution. We are developing a website to distribute it freely. Meanwhile, please contact davepost@woh.rr.com to obtain copies.

Color name	u'	v'	%Y	sR	sG	sB	Pred R	Pred G	Pred B
White	0.1978	0.4683	100.00	255	255	255	251	237	243
Mustard	0.2092	0.5350	57.60	213	203	90	234	198	100
Rose	0.2500	0.4100	40.64	232	138	224	242	147	223
Dark Green	0.1460	0.5560	42.00	96	196	34	136	193	38
Gray	0.1978	0.4683	22.50	130	130	130	162	139	145
Blue	0.1570	0.3675	35.10	67	164	255	102	170	242
Orange	0.3276	0.5308	31.66	244	115	42	246	120	57
Red	0.4100	0.4800	20.90	245	22	95	246	26	111
Green	0.1350	0.5300	73.00	71	253	134	114	235	149
Yellow	0.1770	0.5470	85.00	203	255	79	228	236	89
Magenta	0.2750	0.3100	24.00	218	45	254	234	65	242
Background	0.1978	0.4683	0.01	0	0	0	0	0	0
Ambient from screen	0.1978	0.4683	0.38						
Stimulus size (arcmin)	20.0								

Figure 1. Palette Designer's main table. User input goes in the green cells.

Toward the bottom of the main table, the user inputs the background color's luminance and chromaticity coordinates. For the case shown, that color is black, i.e., the color produced for RGB = {0, 0, 0}, which produces measurably non-zero output typically because most contemporary displays (i.e., LCDs) emit light even for {0, 0, 0}. If the viewing environment includes illumination reflecting off the display screen (as in the example shown here), the user inputs the resulting luminance and chromaticity coordinates produced on the screen. Finally, the user inputs the size of the alphanumerics, symbology, or other stimuli that will be color coded.

The next three columns show the colors' corresponding standard RGB (sRGB) values, which PD calculates according to IEC (1999). The last three columns show the RGB values that should reproduce the colors accurately on a secondary display for which a characterization file has been specified in another area of the spreadsheet (not shown). The file contains measurements of the luminances and chromaticity coordinates produced by the secondary display's R, G, and B channels for RGB values ranging from 0 to 255. The calculated RGB values are obtained using the PLVC method described in Post and Calhoun (1989, 2000). If a secondary display is connected to the computer that is running PD and a characterization file is provided, a color-swatch chart will be displayed there using the calculated RGB values so the user can see a colorimetrically accurate rendition of the current color set.

Color-Swatch Chart (Recognizability)

PD always displays the current color set in a swatch chart on the main display screen using the calculated sRGB values, as shown in Figure 2. The colorimetric accuracy of the colors shown there depends on how well the main display conforms to the IEC (1999) sRGB standard. Ordinarily, the rendition will be at least approximately accurate. The swatches include character strings so the user can judge legibility, also. Those strings, including their font and size, are user-specified in another area of the spreadsheet (not shown).

CIELUV Color-Difference Table (Search Time)

As shown in Figure 3, PD computes color differences between all pairings of the current color set, taking into account the user-specified ambient illumination and symbol size, using Equation 1, as presented by Carter (1989):

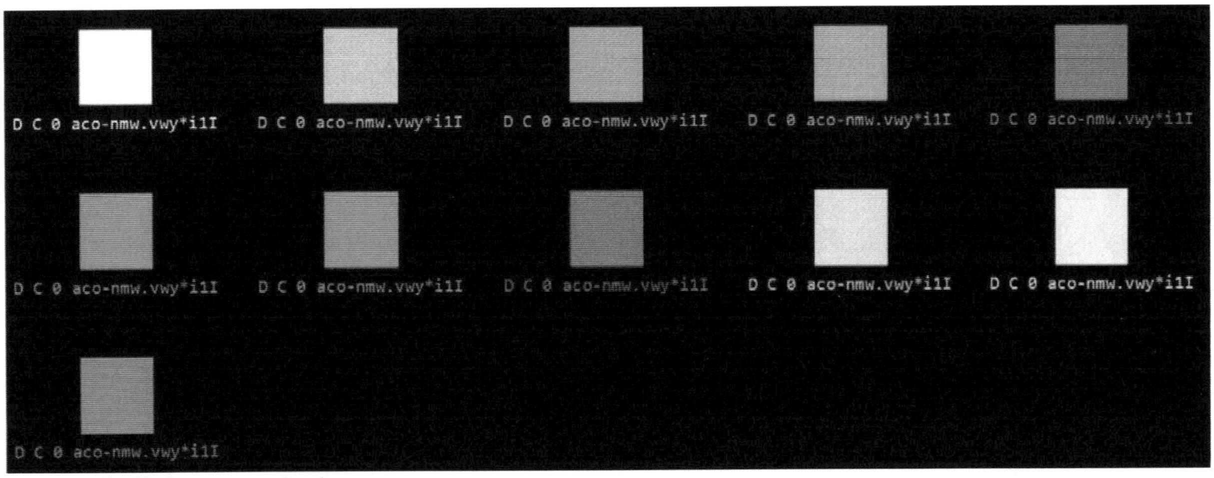

Figure 2. Color-swatch chart.

$\Delta E^*_{uv\text{-}sc}$ w/ambient	White	Mustard	Rose	Dark Green	Gray	Blue	Orange	Red	Green	Yellow
Mustard	13.94									
Rose	21.67	10.99								
Dark Green	21.69	12.05	17.54							
Gray	30.20	17.94	13.04	14.64						
Blue	24.19	16.35	15.12	11.81	11.14					
Orange	30.85	20.03	12.02	27.12	19.49	26.08				
Red	40.12	29.71	20.39	35.67	25.09	32.85	10.01			
Green	15.33	15.38	24.67	12.44	26.12	19.40	35.01	44.40		
Yellow	8.73	11.09	21.60	15.82	27.24	22.32	30.64	40.43	9.05	
Magenta	29.09	19.37	9.15	22.26	11.70	15.26	14.50	18.62	30.85	29.52

Figure 3. CIELUV color-difference table with values < criterion (28) highlighted.

$$\Delta E^*_{uv\text{-}sc} = \left((K_{L^*} * \Delta L^*)^2 + (K_{u^*} * \Delta u^*)^2 + (K_{v^*} * \Delta v^*)^2 \right)^{0.5}, \quad (1)$$

where $\Delta E^*_{uv\text{-}sc}$ is the size-corrected color difference, the coefficients K_{L^*}, K_{u^*}, and K_{v^*} are computed as shown below, and ΔL^*, Δu^*, and Δv^* are computed in accordance with the conventions of the CIE 1976 ($L^*u^*v^*$) color space (CIELUV) described in CIE (2004).

$$K_{L^*} = 1.0366 - e^{0.15263 - 0.05766A} \quad \text{for } 0 < A < 60, \quad (2)$$

$$K_{u^*} = 0.008991A - 0.0065 \quad \text{for } 0 < A \leq 32, \quad (3)$$
$$= 0.0257A - 0.5403 \quad \text{for } 32 < A < 60, \quad (4)$$

$$K_{v^*} = 0.005446A - 0.042 \quad \text{for } 0 < A \leq 32, \text{ and} \quad (5)$$
$$= 0.031A - 0.8594 \quad \text{for } 32 < A < 60, \quad (6)$$

where A is the visual angle subtended by the stimulus in arcmin. For $A \geq 60$ arcmin, $K_{L^*} = K_{u^*} = K_{v^*} = 1$.

Carter (1989) estimated that a difference ≥ 28 is needed to yield asymptotic search times for color-coded stimuli; therefore, values < 28 are highlighted in the table to alert the user. It can be seen that many pairs in Figure 3 fail the criterion, but this outcome predicts only that search times will be suboptimal – not that they will be unacceptable, necessarily. Nonetheless, the user should try adjusting the nearest pairs to increase their Equation 1 color differences. Ideally, experimental testing should follow, to ensure that the search times are acceptable.

It is worth noting that the use of Equations 1-6 and a criterion of 28 is different and more complex than one sees in human factors color standards. A simpler equation and criterion of 20, also from Carter (1989), is seen typically. (Sometimes, the simpler equation and a criterion of 40, based on Carter and Carter, 1981, is seen instead.) We suspect the choice of the simpler equation that Carter (1989) showed to yield a substantially inferior R^2 has been motivated by a bias toward ease of use, which underscores one of PD's advantages: It eliminates the need for users to perform or even understand more complex and accurate colorimetric calculations.

CIELAB Color-Difference Table (Discriminability)

As shown in Figure 4, PD also computes color differences between all pairings of the current color set plus the background color, taking into account the user-specified ambient illumination, using the equation:

$$\Delta E^*_{ab} = ((\Delta L^*)^2 + (\Delta a^*)^2 + (\Delta b^*)^2)^{0.5} , \qquad (7)$$

where ΔE^*_{ab} is the color difference and ΔL^*, Δa^*, and Δb^* are computed in accordance with the conventions of the CIE 1976 ($L^*a^*b^*$) color space (CIELAB) described in CIE (2004).

PD's criterion value in this case is 9.9, which is the maximum that Brainard (2003, p. 203) obtained after computing 95% confidence intervals for the lengths of the major and minor axes of MacAdam's (1942) ellipses in CIELAB. The 9.9 criterion is taken here to be a conservative estimate of the minimum acceptable color difference among spatially adjacent colors that must be discriminable. All values in Figure 4 meet the criterion; hence, none are highlighted.

ΔE^*_{ab} w/ambient	White	Mustard	Rose	Dark Green	Gray	Blue	Orange	Red	Green	Yellow	Magenta
Mustard	60.12										
Rose	46.71	78.40									
Dark Green	88.72	45.60	114.13								
Gray	45.14	62.45	38.96	85.39							
Blue	62.84	110.39	57.23	129.39	54.04						
Orange	83.15	58.76	67.12	100.35	75.06	120.37					
Red	93.06	98.95	53.39	140.17	80.43	106.73	51.20				
Green	82.90	60.70	117.72	30.56	88.72	121.87	118.73	153.14			
Yellow	84.53	37.21	113.16	28.91	92.91	136.49	92.19	135.70	42.62		
Magenta	105.93	147.66	69.27	181.49	98.16	75.45	124.83	82.79	182.06	182.09	
Background	96.49	95.81	75.61	107.64	51.35	81.88	95.57	94.50	118.16	123.60	113.08

Figure 4. CIELAB color-difference table.

Contrast-Ratio Table (Legibility)

As shown in Figure 5, PD computes the luminance-contrast ratio for each color against the background color, taking into account the user-specified ambient illumination. The criterion for this case is 3:1, which is the minimum needed to ensure symbol legibility against the background according to many sources, such as ANSI-HFES-100 (2007) and MIL-HDBK-87213A (2005). All values in Figure 5 meet the criterion; hence, none are highlighted.

Contrast ratio w/ambient	White	Mustard	Rose	Dark Green	Gray	Blue	Orange	Red	Green	Yellow	Magenta
Background	257.4:1	148.7:1	105.2:1	108.7:1	58.7:1	91.0:1	82.2:1	54.6:1	188.2:1	218.9:1	72.0:1

Figure 5. Luminance contrast-ratio table.

Protan, Deutan, and Tritan Confusion-Line Charts for CVD Viewers

As shown in Figures 6 and 7, PD draws a confusion line for each color for protanopic (red-weak), deuteranopic (green-weak), and tritanopic (blue-weak) viewers, using the copunctal points from Wyszecki and Stiles (1982, p. 464). It also shows the sRGB chromaticity gamut so users can see the colors' spacing within that gamut. The figures show that the Yellow and Orange lines are nearly colinear for protans, and the Dark Green and Orange lines are nearly colinear for deutans. These observations indicate that luminance differences must be provided between those color pairs so CVD viewers will be able to discriminate and recognize them.

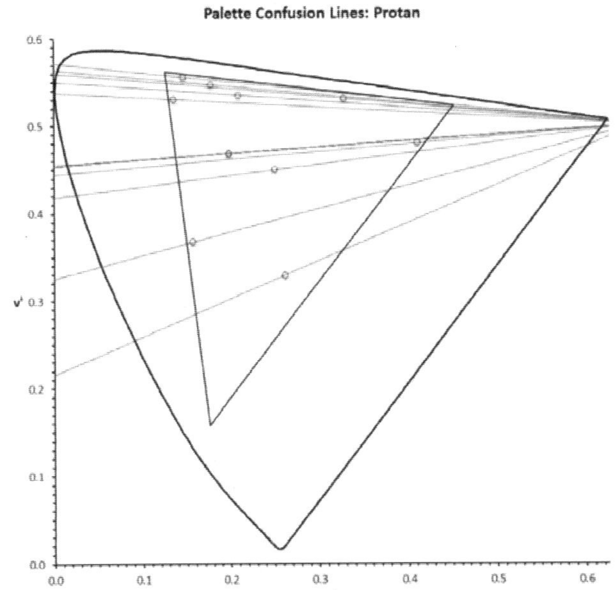

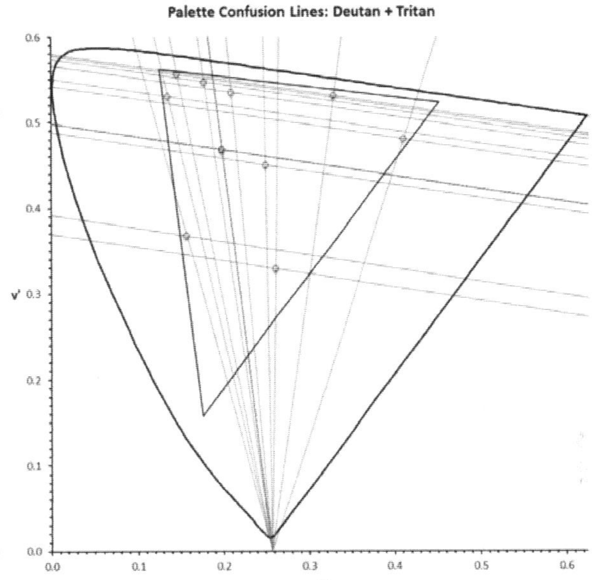

Figures 6 and 7. Protan, deutan, and tritan confusion lines for the color set, plotted on the CIE 1976 u′v′-chromaticity diagram with the sRGB chromaticity gamut (inset triangle) included.

Color-Adjustment Tools

PD allows users to adjust each color's luminance and chromaticity coordinates by making changes directly in the appropriate cells of its main table or by clicking a color name and then using the computer keyboard's arrow keys to change the color's luminance or move it on the CIE 1976 u′v′-chromaticity diagram. Either way, the results are reflected immediately in all the figures and tables. This interactive mechanism simplifies exploring ways to improve the discriminability, recognizability, and legibility of the colors under consideration.

General Utility

Palette Designer aids the design of color codes by automating the calculation of important figures of merit found in human factors design standards for color use on electronic displays and producing helpful graphical representations. Although we created the tool to facilitate development of a color palette for air traffic control displays, we believe that it could be useful for any project that involves designing color codes for electronic displays.

Acknowledgements

The Federal Aviation Administration provided funding for this work under contract DTFAWA-12-D-00011, FAA Cooperative Agreement 16-G-009 with Wright State University, and contract DTFAAC-15-P-05247. The views expressed herein are solely the authors' and do not necessarily reflect those of the FAA.

References

ANSI-HFES-100 (2007). *Human factors engineering of computer workstations.* Santa Monica, CA: Human Factors Society.

Brainard, D. H. (2003). Color appearance and color difference specification. In S. K. Shevell (Ed.), *The science of color* (2nd ed.). Oxford: Elsevier.

Carter, R. C. (1989). Calculate (don't guess) the effects of symbol size on usefulness of color. In *Proceedings of the Human Factors Society 33rd Annual Meeting* (pp. 1368-1372). Santa Monica, CA: Human Factors Society.

Carter, R. C., & Carter, E. C. (1988). Color coding for rapid location of small symbols. *Color Research and Application, 13,* 226-234.

CIE (2004) *Colorimetry* (3rd edition). Publication 15. Vienna, Austria: Commission Internationale de l'Eclairage.

IEC (1999). *Multimedia systems and equipment - Colour measurement and management - Part 2-1: Colour management – Default RGB colour space – sRGB* (1999). Publication 61966-2-1:1999. Geneva, Switzerland: International Electrotechnical Commission.

MacAdam, D. L. (1942). Visual sensitivities to color differences in daylight. *Journal of the Optical Society of America, 32,* 247-274.

MIL-HDBK-87213A (2005). *Electronically/optically generated airborne displays.* Washington, DC: US Department of Defense.

Post, D. L., & Calhoun, C. S. (1989). An evaluation of methods for producing desired colors on CRT monitors. *Color Research and Application, 14,* 172-186.

Post, D. L., & Calhoun, C. S. (2000). Further evaluation of methods for producing desired colors on CRT monitors. *Color Research and Application, 25,* 90-104.

Wyszecki, G., & Stiles, W. S. (1982). *Color science* (2nd ed.). New York: Wiley.

Proposed Scenarios for the Standardization of the Evaluation of New ATC Technologies and Procedures

Jerry M. Crutchfield
Federal Aviation Administration
Oklahoma City, Oklahoma
Angel Millan
Cherokee CRC, LLC
Oklahoma City, Oklahoma

This project provides Federal Aviation Administration acquisition program managers and system development integration contractors with a standard set of human-in-the-loop simulation scenarios against which new Air Traffic Control (ATC) technologies and procedures can be evaluated. We scripted 24 scenarios, eight scenarios for each of three different types of airspace, including a TRACON arrival sector and both low and high altitude en route sectors. The scenarios were scripted to re-create real world airspaces that analyses showed are associated with complex traffic situations. They included representations of severe weather and high traffic load for the purpose of demonstrating the performance of new ATC technologies and procedures when challenged by such real world events. The scenarios were vetted by retired controllers who had experience working the selected sectors and were provided in a format that allows for input into any ATC simulation platform.

The Federal Aviation Administration (FAA) evaluates proposed new Air Traffic Control (ATC) technologies and procedures (henceforth referred to as ATC tools) that have the potential to enhance safety and efficiency in the United States' air transportation system. Evaluations that examine human performance and other aspects of human factors are important in that they produce evidence for benefits that a new ATC tool may provide along with potential problem areas that will need to be addressed prior to deployment. The FAA has a long history of using human-in-the-loop (HITL) simulations to evaluate aspects of ATC (Anderson & Vickers, 1953). Although the use of simulations has limitations and is not without its own challenges (Buckley, DeBaryshem, Hitchner, & Kohn, 1983), it remains the primary means of evaluating ATC tools prior to using them with live air traffic and can provide cost savings when conducted prior to the completion of costly prototypes.

Additionally, new ATC tools should be evaluated against situations as close as possible to those that are likely to occur in real air traffic operations. Events that perturb the status quo but would not be considered rare in the context of ATC operations, such as convective weather, medical emergencies, equipment malfunction and air traffic compression, are sometimes referred to as off-nominal events (Burian, 2008). Researchers conducting HITL evaluations of new ATC tools should use scenarios containing off-nominal events, in addition to other scenarios that do not include off-nominal events, whenever possible.

Proposals to change the ATC system may come from a variety of developers and researchers both within and outside the FAA. These organizations coordinate some aspects of their development and evaluation work formally with the FAA, and infrequently with each other, and coordinate other aspects only occasionally. There are a variety of ATC simulation platforms, including those commercially available and those developed in house by the organizations. Furthermore, there are multiple accepted ways to measure many human factors variables such as workload (Stein, 1985, and Hart & Staveland, 1988). The outcome of the situation in the ATC industry has been that the different industry groups that develop and propose new ATC tools usually do not have access to the same ATC Subject Matter Expert

(SME) participants, use the same scenarios or even the same airspace, may not use the same ATC simulation platforms, and may not collect and report the same performance measures as other industry groups when conducting evaluations. Attempting to compare results from HITL simulations that use different airspace and air traffic situations and report different types of performance measures presents an additional challenge to ATC tool evaluators and decision makers at the FAA, in addition to those inherent to the HITL simulation evaluation method.

This project provides organizations, both within and outside the FAA, with a standard set of scenarios against which new ATC tools can be evaluated. If organizations that run HITL simulations to evaluate their proposed ATC tools use these scenarios, it will facilitate comparisons between various proposed tools going forward.

Methods

This section describes the methods we used to identify airspace suitable for evaluating new ATC tools, the traffic volume and pattern for the scenarios, the number and type of scenarios created, and the off-nominal events included in the scenarios.

Airspace Selection

The first step in the creation of scenarios was the selection of airspace in which the scenarios would take place. The use of generic (designed by researchers) airspace was considered as a possibility as it has certain advantages. Generic airspace allows researchers to build made-to-order challenges into sectors and, given that no controller would have encountered the airspace outside of a lab, controller participants would all have the same level of unfamiliarity with the sectors. However, we ultimately elected to use real world sectors. There is a finite number of sectors in the National Airspace System (NAS). It was decided that the benefits of allowing proposed ATC tools to be tested for their ability to solve real world air traffic issues and the face validity that accompanies the use of a real world sector would, for this project, outweigh the advantages of using a generic sector. But the benefits of simulating real world airspace could only be achieved if the airspace selected would provide enough real world challenges or opportunities to solve real air traffic issues. Therefore, we attempted to identify the busiest and most complex sectors in the NAS.

Finding a suitable Terminal Radar Approach Control (TRACON) airspace was fairly straightforward. We used the Air Traffic Activity Data System (ATADS) to identify the airport with the greatest number of operations annually for the year 2014. It follows that this facility's TRACON would also be the busiest. The airport identified was Hartsfield–Jackson Atlanta International Airport (ATL). We selected the TRACON arrival airspace due to the preponderance of tools proposed for this type of operation.

We attempted to identify complex en route sectors by contracting an analysis of sectors in the three busiest Air Route Traffic Control Centers (ARTCCs): Atlanta (ZTL), Chicago (ZAU), and New York (ZNY). The sector analysis examined air traffic characteristics across a two-year time span (2013 to 2015). Traffic characteristics considered included Average Number of Aircraft in the sector per hour, Number of Climbing or Descending Aircraft per hour, Number of Potential Aircraft Conflicts per hour, and Number of Adjacent Sectors with which that sector controller would have to coordinate. Sectors with an average of fewer than 25 aircraft per hour were eliminated as possible candidates for simulation because of insufficient activity. The remaining sectors were compared with regard to how many climbing and descending aircraft and how many potential conflicts occurred per hour. We decided that, to increase opportunities to evaluate a wide variety of en route ATC tools, it would be necessary to provide scenarios for both a low altitude and a high altitude en route sector. Low and high altitude sectors have different characteristics that may differentially affect the way new tools are used or may differentially affect their utility for resolving the problem the tools were created to resolve. Certain sectors in both ZAU and ZNY were comparable in complexity, depending upon how one weighted the selected traffic characteristics. In ZNY, however, a low altitude sector and a high altitude sector on the candidate list were adjacent to each other. Since the two sectors were adjacent to each other, selecting them would create the possibility of simulating traffic through the two sectors simultaneously and, thus, provide an opportunity to collect data regarding coordination between sectors. The candidate ZNY sectors adjacent to each other were presented to project sponsors at the FAA who concurred with their selection.

Defining Scenarios and Scenario Events

Events are occurrences of interest scripted to take place during a scenario. We began the identification of suitable off-nominal events by using an event list collected during a previous project (Crutchfield & Pfleiderer, 2009). This list was created from the input of controller, pilot, and weather SMEs across five knowledge elicitation sessions that occurred during 2008. We updated that list using a hazard analysis of new ATC tools associated with NextGen (Sawyer, Berry, & Blanding, 2010). As our scenarios are meant to be used during HITL simulations, any event that specified a scripted error on the part of controller participants was removed from the list although pilot errors or errors on the part of controllers for adjacent scenarios were retained. Other events that we dropped from the list were events which we deemed to occur too rarely to be considered off-nominal (e.g., special handling of Air Force 1) or that would result in such a significant change to operations that the situation might be considered a better measure of emergency procedures than of a new ATC tool for normal operations (e.g. aircraft hijacking). Some of the events from the 2008 list required highly similar responses from controllers encountering them. In these cases a single representative event was selected from the group of similar events.

It is not likely that ATC tool evaluators will have the resources necessary to see how well a new tool performs during all of the off-nominal events identified. Furthermore, it was beyond the scope of our project to provide the number of scenarios necessary to cover all of these events. Therefore, we decided to select three high profile off-nominal events that should be included, along with a time period with no off-nominal events, in the standardized scenarios. The scenario that includes a time with no off-nominal events allows the ATC tools to demonstrate the benefits they can provide under ideal traffic conditions. The off-nominal events selected were Pop-up Storm, High Traffic Load, and Equipment Failure. Severe weather occurs somewhere in the NAS on a frequent basis and has the potential to impact traffic flows across the NAS for many hours. Additionally, evaluators and decision makers are interested in knowing how new ATC tools will perform in the face of high air traffic loads predicted to occur years into the future. Concerns about how the ATC system recovers during an equipment failure make it important to include a failure-related event as well.

We determined that all scenarios should be designed to be 40 minutes in length to minimize the amount of time controller participants are needed, while providing sufficient time to collect a useful amount of performance data. We determined that two versions of each evaluation scenario should be developed for each sector so that one version could be used as a baseline condition while the other could be used with the new ATC tool(s).

Air Traffic

The scenarios developed for this study were created by a retired controller employed with the ATAC Corporation using an I-Sim simulator provided by Kongsberg Geospatial. This retired controller had no experience controlling traffic in any of the three selected airspaces. The SME was directed to develop 40-minute long scenarios from real world air traffic data recorded in the summer of 2014 for the specified sectors using Performance Data Analysis and Reporting System (PDARS). Once the draft scenarios were created, the SME used WebEx to run them for other retired controllers to review. During this review, a retired controller from ZNY, familiar with the two selected sectors, reviewed the corresponding en route scenarios and a retired controller from our selected Atlanta TRACON approach control sector reviewed the TRACON scenarios. While they watched the scenarios, the retired controllers noted which flights they believed needed to be changed in some way to achieve the desired degree of realism in the scenario. The SME who developed the scenarios made changes to the scenarios in response.

Next, we used a second set of retired controllers familiar with the respective airspace and sectors to run the scenarios again. These scenario runs were conducted using a high fidelity simulation of an En Route Automation Modernization (ERAM) workstation and of a TRACON (Standard Terminal Automation Replacement System (STARS) workstation, again provided by Kongsberg Geospatial. The scenario runs used live pseudo-pilots to perform the associated flight deck/controller communications. Comments on how to improve these scenarios were collected from this second set of retired controllers and the scenarios were changed accordingly.

Lastly, we used a third set of retired controllers, one experienced with the ZNY sectors and another experienced with the selected Atlanta TRACON airspace, to run the scenarios with pseudo-pilots and make comments. We used these comments to make any final adjustments.

The second and third set of retired controllers were also asked to help us create presentations to be used in familiarizing controllers naïve to the selected airspaces with Letters of Agreement (LOAs), traffic flows, sector boundaries, and other types of information necessary to be able to control the simulated traffic. The familiarization material was then presented to retired controllers naïve to those airspaces who subsequently controlled two scenarios from each airspace. These naïve controllers were interviewed afterwards to identify information in the presentations that needed further clarification or recommend additional information that controller participants would need to be able to successfully control traffic in these sectors.

Results

Airspace Materials

We identified airspace at Atlanta TRACON A80 and New York ARTCC ZNY10 (a high altitude sector) and ZNY27 (a low altitude sector) to represent in the standardized scenarios. We developed Microsoft Excel files that include the sector boundaries, altitude definition, waypoints and fixes, routes, airports, and winds for each airspace. We also developed materials to familiarize participants with the airspaces being represented.

Air Traffic Scenarios

We developed and validated six scenarios suitable for use in evaluating TRACON tools, six scenarios suitable for use in evaluating en route tools in low altitude airspace, and six scenarios suitable for use in evaluating en route tools in high altitude airspace.

Two moderate traffic load scenarios for each sector were scripted without any off-nominal events. Two scenarios for each sector represented a moderate traffic load with the addition of a severe weather system that impacts operations in the sectors. Two scenarios for each sector have a traffic load 15% higher than the average that occurred in 2014. This traffic level represents what is predicted to occur in the year 2025 (Federal Aviation Administration, 2016).

The first 20 minutes of the moderate scenarios without off-nominal events enable new ATC tools to demonstrate the benefits they can provide under ideal traffic conditions. Our intent is that evaluators add their own equipment failure event to the second 20 minutes of these two scenarios. It was not possible to predict all the types of tools that may be evaluated with these scenarios, and selection of an unrelated type of equipment failure event would result in a less meaningful evaluation. Therefore, we suggest that evaluators include their own customized equipment failure event directly related to the ATC tool being evaluated.

Additionally, two moderate traffic load scenarios for each sector were developed as examples of scenarios that can be used to familiarize controller participants who are naïve to a given sector with the sector operations and traffic flow and also to familiarize them with the new ATC tool(s) being evaluated. Evaluators are encouraged to create more familiarization scenarios given available time and resources.

Discussion

We intend the scenarios to be used by a variety of organizations, both within and outside of the FAA, when evaluating new ATC tools. In so doing, this will foster a greater opportunity to compare controller performance associated with a wide variety of new ATC tools. The scenarios and other materials provided by this project were designed so that evaluators could either use controller participants who are familiar with controlling traffic in the sectors represented without any additional training, or use other participants who are naïve to the sectors but who can learn about them through familiarization materials and training scenarios. When running the HITL scenarios, it is expected that the evaluators will use a repeated measures design where every participant runs every scenario in turn. It is expected that evaluators will run one of each type of scenario provided (moderate traffic, weather, busy traffic) as a baseline using current technologies and procedures and run a second scenario (moderate traffic, weather, busy traffic) in an experimental condition that includes the use of the new ATC tools. It is further expected that the order of the scenarios used (baseline vs. new ATC tool) will be counterbalanced across participants to further control for differences in difficulty level that may inadvertently exist in the scenarios.

Although our primary goal was to provide scenarios that would allow the evaluation of the effectiveness of new ATC tools under conditions that might stress them, another goal was to fashion the experimental scenarios to be independent of any new ATC tool being evaluated. In some cases, the change to be evaluated may have an impact on air traffic flows into the airspace or on the structure of airspace objects (such as routes) within the airspace being represented itself. In these cases, it is suggested that when running the baseline condition, evaluators use the airspace as provided. When running the condition that uses the new ATC tool, evaluators are justified in changing the sequencing or spacing of aircraft entering into and operating within the airspace or the routes and/or airspace objects in the airspace, if the changes are similar to changes that would be made to any airspace using the new ATC tool. The number, type, and destination of the aircraft should not be changed.

Comparisons of tools evaluated using these standardized scenarios would be facilitated if evaluators collect and report standardized performance measures as well. Preliminary work to identify appropriate performance measures was done as part of this project. Further work, however, is needed to provide evaluators with details required to assure the measures are fully comparable.

Acknowledgments

This research was supported and funded by the FAA NextGen Organization's Human Factors Division, ANG-C1.

References

Anderson, C. & Vickers, L. (1953). *Application of simulation techniques in the study of terminal area air traffic control problems.* (Report No. 192). Washington, DC: Civil Aeronautics Administration Technical Division.

Burian, B.K. (2008). Perturbing the system: Emergency and off-nominal situations under NextGen. *International Journal of Applied Aviation Studies*, *8*(1), 114-127.

Buckley, E. P., DeBaryshe, B. D., Hitchner, N., & Kohn, P. (1983). *Methods and measurements in real-time air traffic control system simulation* (DOT/FAA/CT-83/26). Atlantic City International Airport: Federal Aviation Administration Technical Center.

Crutchfield, J. M., & Pfleiderer, E. (2009). *CAMI Far-Term TBO Scenarios.* Oklahoma City, OK: Authors.

Federal Aviation Administration. (2016). *FAA Aerospace Forecast Fiscal Years 2016-2036.* Washington DC: Forecasts and Performance Analysis Division, APO-100. Rctricvcd from http://www.faa.gov/data_research/aviation/.

Hart, S. G., & Staveland, L. E. (1988). Development of NASA-TLX (Task Load Index): Results of Empirical and Theoretical Research. In Hancock, Peter A.; Meshkati, Najmedin. *Human Mental Workload* (PDF). Advances in Psychology. 52. Amsterdam: North Holland. pp. 139–183.

Sawyer, M., Berry, K., & Blanding, R. (2010). *NextGen Human Hazard Assessment Report.* Washington DC: TASC Inc.

Stein, E.S. (1985). *Air traffic controller workload: An examination of workload probe* (DOT/FAA/CT-TN84/24). Atlantic City International Airport: Federal Aviation Administration Technical Center.

AN INTERDISCIPLINARY APPROACH TO EVALUATING U.S. ARMY AVIATION TRAINING

Martin S. Goodwin
University of Central Florida, Institute for Simulation & Training
Orlando, Florida
Lauren Reinerman-Jones
University of Central Florida, Institute for Simulation & Training
Orlando, Florida
Brian F. Goldiez
University of Central Florida, Institute for Simulation & Training
Orlando, Florida
Robert A. Crapanzano
U.S. Army Program Executive Office for Simulation, Training, and Instrumentation (PEO STRI)
Orlando, Florida

The U.S. Army is seeking to update and expand its use of simulation-based aviation training to address operational and fiscal concerns that are driving the need for more efficient training solutions. This has created a need to evaluate whether lower-cost, game-based simulations may potentially augment higher-cost, traditional simulation-based training for specific aviation training tasks. However, current approaches to Training Effectiveness Evaluation (TEE) do not address the complete range of factors to adequately evaluate today's increasingly sophisticated simulation training environments. Leveraging recent research and drawing from the tools and techniques of human performance assessment, instructional science, and phenomenology, an interdisciplinary approach to performing TEEs is introduced and described in the context of evaluating UH-60A/L aviation collective mission training. This novel TEE approach optimizes a research-based evaluation methodology to more fully capture the range of factors that contribute to training effectiveness in interactive simulation training environments.

The United States continues to face uncertain and unprecedented threats around the world. Increasing acts of terror by both state and non-state actors, rising global instability, and the need to maintain readiness for both conventional and unconventional warfare are key strategic concerns. At the same time, technology innovations such as the expanding role of unmanned aircraft systems (UAS) and the emergence of the cyber-battlefield are changing the characteristics of modern warfare. Today's warfighters must be prepared to meet the challenges of highly dynamic, increasingly technological military operations. To help prepare warfighters to meet those challenges, the U.S. Army is seeking to update and expand its use of simulation-based aviation training. While the Army continues to rely on traditional simulation as a proven aviation training method, game-based simulation has become more sophisticated and may provide viable training options in some applications. The use of game-based simulation to augment traditional simulation-based training can potentially reduce costs, enhance return on investment, advance training objectives, and inform future training environment designs.

Operational imperatives are mandating training strategies that produce optimum levels of readiness for a wide range of mission scenarios. Simultaneously, fiscal concerns are driving the need for more efficient training methods. This need for optimized training can be addressed for the U.S. Army by investigating whether lower-cost, game-based simulations may potentially augment higher-cost, traditional simulation-based training for specific aviation training tasks. Such investigations are typically performed by conducting Training Effectiveness Evaluations (TEEs). The most popular and widely used methods for performing training evaluations are based on Kirkpatrick's Four-Level Training Evaluation Model (1959, 1976, 1994). However, the Kirkpatrick model does not adequately address the complete range of factors that exist in dynamic training simulations. Additionally, the model inherently limits the types of questions that need to be answered to effectively evaluate today's increasingly sophisticated simulation training environments. It also provides little guidance on how different simulated environments may be combined to meet evolving training requirements. This paper describes the structure of the Kirkpatrick model, the reasons for its popularity in the training community, and the contrast between its intended purpose and its use to address modern simulation training evaluation objectives. A novel, interdisciplinary approach to evaluating training effectiveness, called Assessing Simulated Systems Empirically for Training, or ASSET, is then introduced. ASSET addresses the limitations of TEE methods based on the Kirkpatrick Model by building on a methodology better aligned with the purpose of modern TEEs. The ASSET approach is then described in the context of a use case to evaluate whether game-based systems can potentially augment traditional simulation-based U.S. Army UH-60A/L Blackhawk helicopter collective training.

Training Effectiveness Evaluation Considerations

Kirkpatrick's Four-Level Training Evaluation Model (1959, 1976, 1994) seeks to evaluate training effectiveness through an assessment of four hierarchical levels (Figure 1).

- Level 1: Reaction – Evaluates trainees' reactions to the training event.
- Level 2: Learning – Evaluates changes in trainees' knowledge, skills, attitudes, and abilities as a result of the training event.
- Level 3: Behavior – Evaluates the change in behavior in trainees from the training context to the performance context to determine training transfer and application.
- Level 4: Results – Evaluates the degree to which specific targeted outcomes have been achieved.

Figure 1. Kirkpatrick's Four-Level Training Evaluation Model

The popularity of the Kirkpatrick Model can be traced to a number of factors: 1) it provides a multi-level approach to training evaluation; 2) it organizes the complexities of training evaluation into four distinct areas; and 3) it simplifies outcome measures by reducing the number of variables involved in the evaluation analysis (Bates, 2004). The Kirkpatrick Model is used to conduct TEEs in many different training contexts, but its use to evaluate modern simulation training is problematic. The original purpose of the Kirkpatrick Model was to gain information on the *value* of training programs to help determine instructional improvements and decide if a program should be continued (Kirkpatrick, 1959). As such, it follows a traditional evaluation methodology and has utility in evaluation contexts where the intent is to determine whether the training is meeting desired objectives. In other words, the scope of the evaluation is limited to assessing a single training program in terms of the need it was designed to meet. Evaluating the effectiveness of training in today's simulation domains typically extends beyond this concern. While the imperative to determine if training is meeting its desired objective still exists, this is now generally part of a much larger evaluation goal that encompasses the need to inform decisions concerning how, what, when, and where simulation training will be used to meet specific training requirements. These decisions are typically based on factors unique to simulated environments, such as levels and types of fidelity, the affordances of instructional interfaces, and the dynamics of the environments themselves.

For simulation training then, TEEs are less concerned about *improving* a single training program and more concerned about *proving* the efficacy of specific individual factors that influence training effectiveness. This focus on proving instead of improving necessitates the use of a TEE approach based on a research methodology instead of a standard evaluation methodology. It is from this perspective that the interdisciplinary TEE approach called Assessing Simulated Systems Empirically for Training (ASSET) was deveoped.

Assessing Simulated Systems Empirically for Training (ASSET)

The ASSET approach draws on the tools and techniques of human performance assessment, instructional science, and phenomenology to establish a multidimensional, interdisciplinary perspective to performing TEEs. This approach increases the breadth of the evaluation to more fully capture the range of factors that contribute to training effectiveness in dynamic, interactive simulation training environments. ASSET follows the procedures and rigor of a research methodology, with some slight modification to optimize its use to conduct TEEs in simulation training environments. A condensed version of the ASSET approach is illustrated in Figure 2.

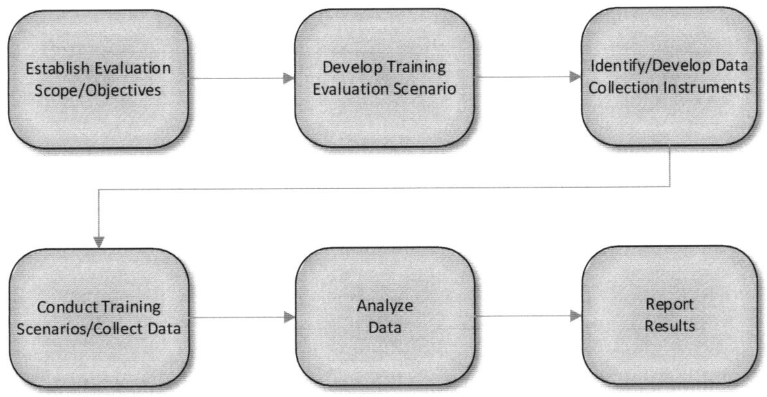

Figure 2. ASSET Evaluation Approach

The ASSET approach is described in the following sections in the context of a use case to evaluate Army Aviation training. The U.S. Army Aviation Combined Arms Training Strategy (2016) emphasizes the use of Training Aids, Devices, Simulations, and Simulators (TADSS) to prepare Army aviation forces for future combat. This strategy highlights multiple types of environments that encompass a wide range of fidelity and cost. Some broad examples include game-based systems, moderate-fidelity trainers, and high-fidelity flight simulators. Of these environments, there is a high level of interest in the training potential of game-based systems. However, the effectiveness of game-based simulations requires further investigation (Sotomayor & Proctor, 2009; Whitney, Tempby, & Stephens, 2014). In particular, the use of game-based training as an adjunct to traditional simulation-based training has not been adequately evaluated.

A TEE was performed using the ASSET approach to conduct evaluations of three simulated training environments to determine the potential of lower-cost, game-based simulations to augment higher-cost, traditional simulation-based training. The training environments evaluated in the study were the Aviation Combined Arms Tactical Trainer (AVCATT; the current U.S. Army Program of Record for aviation collective training), a moderate-fidelity training simulator that integrates augmented reality helmet mounted displays (HMDs) to blend the physical cockpit with the virtual environment; the Virtual Battlespace 3 (VBS3) low-fidelity, first-person, games-for-training system operated on a desktop computer with commercial-off-the-shelf (COTS) flight controllers; and Microsoft Flight Simulator (MSFS), a commercially available flight simulator game that provides a similar level of fidelity and operation as VBS3. An operational flight trainer (OFT), a full-motion FAA Level D flight simulator, served as a real-world analog and was used for evaluation of the training environments.

The ASSET approach began with an identification of the scope and objectives of the evaluation. This was an essential part of the process, as it established the parameters for performing the rest of the evaluation. For the present use case, it was determined that the primary objective was to determine how and where lower-cost game-based training could be used as an equally effective adjunct to higher-cost simulation-based training for a particular set of aviation collection mission training tasks. Based on this evaluation objective, the following three evaluation questions were identified to establish the scope of the TEE: 1) Are there differences among the three simulated training environments?; 2) Are there differences in a real-world analog environment (OFT) based on the preceding simulated training environment?; and 3) Are there differences in the degree to which each simulated training environment corresponds to the real-world analog environment (OFT)?

Once the scope and objectives were established, the training scenarios that formed the basis of the evaluation were developed. The training evaluation scenarios involved a flight of UH-60A/L Blackhawk helicopters engaged in a collective air assault mission and consisted of a set of operationally demanding tasks and cognitive decision-making points. Operational tasks focused on mission events that are part of standard operating procedures or explicit items covered in mission and crew briefings. Cognitive decision-making focused on the pilot's specific choices and reactions to changing conditions during the mission scenario. These tasks and decision points directly related to the ability of the investigated training environments to support their execution and were part of the mission performance rubrics for the study.

The next step was to identify, develop, and collect data using a set of specific measures and data collection instruments that supported the objectives of the evaluation. An interdisciplinary set of empirically validated measures that contribute to training effectiveness were used. These measures aligned within the disciplinary areas of psychology, physiology, and phenomenology.

Psychology

Psychological measures included a mission performance rubric and questionnaires. The mission performance rubric consisted of 12 individual tasks and 5 decision points. Questionnaires from the psychology discipline were used to record a variety of subjective measures related to immersion, presence, workload, stress, and simulator sickness.

An Immersive Tendencies Questionnaire (ITQ; Witmer & Singer, 1998), version 3.01, as revised by the Université du Québec en Outaouais Cyberpsychology Lab, was administered at the beginning of the experimental session. Immersive tendencies were scored across four subscales: Focus (paying attention to current tasks), Involvement (interacting with current tasks), Games (becoming engaged within a scenario), and Emotions (experiencing fear, excitement, or other feelings). A Presence Questionnaire (PQ; Witmer & Singer, 1998), version 3.0, as revised by the Université du Québec en Outaouais Cyberpsychology Lab, was administered at the completion of each experimental session. The PQ assessed the degree to which participants experienced presence in each of the simulated environments, as well as the intensity of this experience as influenced by seven individual factors (realism, possibility to act, possibility to examine, quality of interface, self-evaluation of performance, sounds, and haptic). A Simulator Sickness Questionnaire (SSQ: Kennedy, et. al., 1993) was used to assess the level of discomfort experienced by participants in each of the simulated environments. The SSQ consists of items related to symptoms of simulator and motion sickness (eyestrain, headache, dizziness, etc.), clustered into three factors: Oculomotor, Disorientation, and Nausea. The Dundee Stress State Questionnaire (DSSQ; Matthews, et. al., 2002) differentiates 11 primary state factors relating to affect, motivation, and cognition. These primary state factors support three broader second-order factors: engagement (qualities of interest, motivation, and energy), distress (feelings of confidence, tension, and control), and worry (levels of self-esteem, self-focus, and cognitive interference). A short version of the DSSQ was used in the described study (Matthews, Emo, & Funke, 2005). A pre-task questionnaire was administered at the beginning of the experimental session and a post-task questionnaire was administered at the completion of each experimental session. The NASA-Task Load Index (TLX; Hart & Staveland, 1988) was used to assess each participant's perceived workload during the performance of the mission scenarios. The TLX is composed of six subscales that measure workload across the dimensions of mental demand, physical demand, temporal demand, effort, frustration, and performance. A separate global workload score is computed as the unweighted averages of the six subscale scores. The TLX was administered at the completion of each experimental session.

Psychological measures provided important data related to training effectiveness that is often overlooked in traditional TEEs. Factors relating to immersion, presence, workload, stress, and simulator sickness all correspond to the ability of a simulated training environment to support the positive performance of training tasks. Performance measures may also provide indications of differences between training environments.

Physiology

Physiological measures consisted of electrocardiography (ECG) and galvanic skin response (GSR). Both of these measures were captured using a Procomp Infiniti system. ECG is a direct measure of cardiac activity and one of the most common physiological measures of workload and stress in response to task demands. ECG measures included Inter-Beat Interval (IBI), Heart Rate Variability (HRV) and Beats per Minute (BPM). Increases in BPM have been associated with increases in workload and this particular measure is more sensitive to physiological workload (Wilson & O'Donnell, 1988; Jorna, 1993). HRV is generally associated with cognitive workload rather than physiological workload. As such, it reflects engagement in effortful information processing (Jorna, 1993). Increases in cognitive workload of task demands are associated with decreases in HRV (an inverse relationship; Mulder, Waard, & Brookhuis, 2004).

GSR is a measure of emotional stress and nervous tension based on the electrical conductance of the skin (Mundell, Vielma, & Zaman, 2016). Increases in GSR are associated with increases in stress and tension (Shi, Choi,

Ruiz, Chen, & Taib, 2007). GSR drift, the difference between the upper and lower levels of galvanic skin response, is a measure of emotional arousal related to stress. Absolute drift, in particular, is the absolute change in raw GSR from the beginning to the end of a session (Mundell, Vielma, & Zaman, 2016). Absolute drift reveals slow variations in the GSR signal. GSR Maximum Increase Drift is the absolute difference in raw GSR from the minimum point to the end of the session (Mundell, Vielma, & Zaman, 2016). Maximum increase drift gives a measure of trends in the GSR signal existing at the end of the session.

These measures captured the direct, real-time physiological responses of study participant's as they were engaged in mission scenarios within the simulated training environments investigated in this study. This provided an additional dimension of training effectiveness that helped broaden the evaluation effort.

Phenomenology

Study participants were interviewed at the end of each experimental session to collect first-person experiential data for each simulated training environment. The interview method was based on Petitmengin (2006) and implemented following the guidance provided by Bockelman, Reinerman-Jones, and Gallagher (2013). Participant interviews consisted of questions designed to focus the participant's attention on the real-time subjective experience of performing the mission scenario in a particular simulation environment. Questions such as "Describe what it is like performing the mission in the [*type of simulator*] environment." and "Tell me your thoughts as you progress through the mission." provided opportunities for participants to relate their direct experiences with the simulated environments. Copilots were also interviewed after each experimental session. Although they were confederates in the study, data collected from copilot interviews provided an additional source of evaluation information. These interviews helped capture the ability of the simulated environments to support mission training tasks in terms of graphics, controls, responsiveness to inputs, and representation of flight and mission characteristics.

Summary

Evaluating the effectiveness of training in simulation domains cannot be adequately accomplished by standard TEE approaches and methods. The ASSET approach represents a novel method for conducting TEEs in simulation training environments that transcends the limitations of standard approaches. ASSET is based on the procedures and rigor of a research methodology, but is specifically optimized to conduct TEEs for simulation training. Its interdisciplinary focus on human performance assessment, instructional science, and phenomenology increases the scope of the evaluation effort to more fully capture the range of factors that contribute to training effectiveness in dynamic, interactive simulation training environments. Beyond its application in the described use case, the ASSET approach provides a powerful methodology for evaluating simulation training in any context. Its use becomes essential when the objective of the evaluation extends beyond a determination of the value of a training program and into the need to inform decisions concerning how, what, when, and where simulation training will be implemented to meet specific training requirements.

Acknowledgements

This paper was prepared in conjunction with work sponsored by the U.S. Army Program Executive Office for Simulation, Training, and Instrumentation (CA# W911NF-14-2-001). The U.S. Army Program Executive Office for Simulation, Training, and Instrumentation nor any agency thereof, nor any of their employees, makes any warranty, expressed or implied, or assumes any legal liability or responsibility for any third party's use, or the results of such use, of any information, apparatus, product, or process disclosed in this report, or represents that its use by such third party would not infringe privately owned rights. The views expressed in this paper are not necessarily those of the U.S. Army Program Executive Office for Simulation, Training, and Instrumentation.

References

Army Aviation Training Strategy. (January 2016). U.S. Army Aviation Center of Excellence. Fort Rucker, AL.

Bates, R. (2004). A critical analysis of evaluation proactice: the Kirkpatric model and the principle of beneficience. *Evaluation and Program Planning 27*(2004), 341-347.

Bockelman P, Reinerman-Jones L and Gallagher S (2013) Methodological lessons in neurophenomenology: Review of a baseline study and recommendations for research approaches. *Frontiers in Human Neuroscience,* 7:608.

Hart, S. G. & Staveland, L. E. (1988). Development of NASA-TLX (Task Load Index): Results of empirical and theoretical research. In P. A. Hancock and N. Meshkati (Eds.) *Human Mental Workload.* Amsterdam: North Holland Press.

Jorna, P.G.A.M. (1993). Heart-rate and workload variations in actual and simulated flight. *Ergonomics, 36*(9), 1043–1054.

Kennedy, R. S., Lane, N. E., Berbaum, K. S., & Lilienthal, M. G. (1993). Simulator sickness questionnaire: An enhanced method for quantifying simulator sickness. *The International Journal of Aviation Psychology, 3*(3), 203-220.

Kirkpatrick, D. L. (1959). Techniques for evaluating training programs. *Journal of American Society for Training and Development*, 11, 1-13.

Kirkpatrick, D. L. (1976). Evaluation of training. In R.L. Craig (Ed.), *Training and development handbook: A guide to human resource development.* New York: McGraw Hill.

Kirkpatrick, D. L. (1994), *Evaluating Training Programs: the Four Levels.* San Francisco, CA: Berrett-Koehler.

Matthews, G., Campbell, S. E., Falconer, S., Joyner, L. A., Huggins, J., Gilliland, K., et al. (2002). Fundamental dimensions of subjective state in performance settings: Task engagement, distress, and worry. *Emotion*, 2, 315–340.

Matthews, G., Emo, A. K., & Funke, G. J. (2005). *A short version of the Dundee Stress State Questionnaire.* Paper presented at the Twelfth Meeting of the International Society for the Study of Individual Differences, Adelaide, Australia.

Mulder, L. J. M., de Waard, D., & Brookhuis, K. A. (2004). Estimating mental effort using heart rate and heart rate variability. In N. Stanton, A. Hedge, K. Brookhuis, E. Salas, & H. Hendrick (Eds.), *Handbook of Human Factors and Ergonomics Methods* (pp. 201-208). Boca Raton, FL: CRC Press.

Mundell, C., Vielma, J. P., & Zaman, T. (2016). *Predicting performance under stressful conditions using galvanic skin response.* Retrieved from https://arxiv.org/ftp/arxiv/papers/1606/1606.01836.pdf.

Petitmengin, C. (2006). Describing one's subjective experience in the second person: An interview method for the science of consciousness. *Phenomenology and the Cognitive Sciences,5*(3-4), 229-269.

Shi, Y., Choi, E. H. C., Ruiz, N., Chen, F., & Taib, R. (2007). Galvanic skin response (GSR) as an index of cognitive workload. In *ACM CHI Conference Work-in-progress*.

Sotomayor, T., & Proctor, M. (2009). Assessing Combat Medic Knowledge and Transfer Effects Resulting from Alternative Training Treatments. *The Journal of Defense Modeling and Simulation: Applications, Methodology, Technology*, 121-134.

Whitney, S., Tempby, P., & Stephens, A. (2014). A Review of the Effcetiveness of Game-based Training for Dismounted Soldiers. *Journal of defenese Modeling and Simulation*, 319-328.

Wilson, G. F., & O'Donnell, R. D. (1988). Measurement of operator workload with the neuropsychological workload battery test. *Advances in Psychology, 52*, 63-100.

Witmer, B.G. & Singer. M.J. (1998). Measuring presence in virtual environments: A presence questionnaire. *Presence: Teleoperators and Virtual Environments*, 7(3), 225-240.

QUANTIFYING PILOT CONTRIBUTION TO FLIGHT SAFETY DURING AN IN-FLIGHT AIRSPEED FAILURE

Timothy J. Etherington
Rockwell Collins
Hampton, Virginia
Lynda J. Kramer
Randall E. Bailey
Kellie D. Kennedey
NASA Langley Research Center

Accident statistics cite the flight crew as a causal factor in over 60% of large transport fatal accidents. Yet a well-trained and well-qualified crew is acknowledged as the critical center point of aircraft systems safety and an integral component of the entire commercial aviation system. A human-in-the-loop test was conducted using a Level D certified Boeing 737-800 simulator to evaluate the pilot's contribution to safety-of-flight during routine air carrier flight operations and in response to system failures. To quantify the human's contribution, crew complement was used as an independent variable in a between-subjects design. This paper details the crew's actions and responses while dealing with an in-flight airspeed failure.

Accident statistics like Baker (2001) often cite flight crew error as the primary contributor in accidents and incidents in transport category aircraft. However, the Air Line Pilots Association (2011) suggests "a well-trained and well-qualified pilot is acknowledged as the critical center point of the aircraft systems safety and an integral safety component of the entire commercial aviation system." This is generally acknowledged but cannot be verified because little or no quantitative data exists on how or how many accidents/incidents are averted by crew actions. Anecdotal evidence suggest crews handle failures on a daily basis and Aviation Safety Action Program (2003) data generally supports this assertion, even if the data is not released to the public. However without hard evidence, the contribution and means by which pilots achieve safety of flight is difficult to define. Thus, ways to improve the human ability to contribute or overcome deficiencies are ill-defined.

Method

The pilot contribution to flight safety was investigated by experimentally manipulating crew complement (single pilot and crewed conditions) during normal and increasingly challenging non-normal airline operations.

Experiment Design

To assess human performance and safety, the experiment contrasted normal two-crew operations to conditions when one of the crew was absent from the flight deck. If the condition included a temporary absence, it was designated as reduced crew operations (RCO). If the condition included a permanent absence, it was designated as single pilot operations (SPO). The experimental independent variables were crew complement and scenario. The three crew complement conditions were: Two-crew, RCO, and SPO. Two normal scenarios and six non-normal scenarios were flown. The six non-normal scenarios were grouped into three categories; Category A featured failures initially unannunciated with the autopilot available, B featured annunciated failures with autopilot available, and C featured annunciated failures with autopilot not available. Failures were triggered near top of climb (TOC) or top of descent. This paper details one Category A failure, unreliable airspeed. Etherington et al (2016) details the entire experimental matrix and details one Category C failure.

The data shown here is taken from the 18 nominal Two-crew and SPO runs, 6 nominal RCO runs (with the Captain resting), and 6 unreliable airspeed non-normal runs in each of the SPO, Two-Crew, and RCO crew conditions. For the RCO condition, the non-normal started out with one pilot flying and the other resting in the seat, isolated in sight and sound from the cockpit. Two minutes after the failure, the resting pilot returned to flying duties in the cockpit.

Participants

Thirty-six pilots (18 crews), representing five airlines, participated. Each pilot held a current Airline Transport Pilot certificate and was current in the Boeing 737-800 aircraft. Crews were paired by function (Captain and First Officer) and by employer to minimize conflicts in training, standard operating procedures, and crew resource management techniques. Crews were instructed to bring their company's paper and/or electronic charts and checklists with them to further reduce conflicts in standard operating procedures or training.

Apparatus

The research was conducted using the Boeing 737-800 simulator operated by the FAA AFS-440 at Oklahoma City, OK (See Figure 1). The simulator is Level D certified and yet fitted with experimental controls, modifications, and recording capability to support research operations. The fidelity of the simulator and the recording capability were both critical to this research effort. The scenario was an air carrier flight from Denver International to Albuquerque. Dispatch paperwork was provided and constituted the flight release. Simulated weather en-route consisted of convective activity along the mountain range to the west of Denver, and weather and visibility were designed to affect any diversion decisions. Live Air Traffic Control (ATC) and pseudo-pilots provided interactive clearance procedures, realistic pilot workload, and a level of realism to the scenario. Dispatch could be contacted on the radio.

Figure 1. FAA OKC Boeing 737-800 Simulator.

Results

The results detailed here describe the major findings of only one of the Category A failure conditions, unreliable airspeed. This failure emulated an iced-over pitot tube at the cruise altitude of Flight Level (FL) 350 which caused erroneous airspeed readings on the corresponding side. When the pitot tube became blocked, the airspeed indicator then performed like an altimeter such that increasing or decreasing altitude from FL 350 would also appear as an increase or decrease in airspeed. The failure is latent and cannot be detected until the aircraft deviates from the altitude at which the blockage occurred.

At approximately 15 minutes after TOC, the failure scenario was triggered by ATC instructing the crew/pilot(s) to climb to FL370 from FL350. As the aircraft climbed, the airspeed indicated an increase on the failed side. At a difference of 5 knots airspeed, the "IAS DISAGREE" amber warning would appear on both pilots' primary flight displays under the airspeed tape. Eventually the failed side airspeed reached an overspeed condition and the overspeed warning clacker triggered.

Failure Identification

As a Category A failure scenario, inconspicuous symptoms of failures, impending failures, or non-normal conditions were evident in the cockpit before a warning triggered; in this case, airspeed would diverge side to side and the IAS Disagree amber message illuminated before the overspeed warning clacker.

In the total of 21 non-normal runs, only 11 pilots noticed the IAS Disagree light before the overspeed warning clacker started. Only 33% of the SPO pilots recognized the failure before the clacker, while more than half (56%) of the RCO crews and two-thirds (67%) of the two pilot crews did so. One two-crew noticed the failure before the IAS DISAGEE annunciation.

Flight Path Control and Failure Handling

The average time between IAS DISAGREE and overspeed was 8-10 seconds so even for those that detected the IAS DISAGREE light, all crews/pilot(s) experienced an overspeed warning clacker. The clacker is extremely loud and distracting and continues until the overspeed condition is cleared. The clacker sounded for an average of 1.5 minutes with a range of 30 seconds to 15 minutes.

Because of some high profile accidents, this failure has been extensively trained for the past few years. Prior to referring to the appropriate checklists, nearly all crews immediately disconnected the autopilot and autothrottle from memory due to this training and to ensure that the automation was not causing the problem.

A few pilots found the clacker so distracting that they attempted to locate the circuit breaker before attempting to troubleshoot. As this clacker sounded immediately following an ATC command to climb, the majority of pilots sought to reverse the most recent action and requested a descent to the previous altitude. When the aircraft returned to that altitude the majority of the non-normal indications cleared and the aircraft behaved normally until the crew initiated the descent.

In the midst of the failures, all pilots alerted ATC to an airspeed problem but only 14 crews declared an emergency. All but two crews requested a descent or block altitude clearance from ATC with an average time of approximately 45 seconds. If this occurred during the two minute delay in an RCO configuration, the failure effects were no longer apparent when the resting pilot re-engaged.

Some pilots were erroneous in conceptualization of the flight control warning system and indicated a concern that the failed pitot would trigger the stick shaker as they descended. The stick shaker system that warns of aircraft stall conditions is based on an angle of attack sensor and not just airspeed.

Typically, the autopilot was re-engaged to the non-failed side within a minute or less.

Even for this short period of time, there were many control difficulties. At high altitude, there is a small airspeed range between stall warning and over speed which requires only a small pitch excursion to go from over speed to stall warning. Eight of the 18 crews experienced one or more stick shaker events that precede a stall. Five of the 18 crews experienced an actual overspeed because of inappropriate pitch control during the event. At least one crew received a bank angle warning. Three crews experienced both stick shaker and overspeed during recovery. Although all crew configurations had at least one event, 67% of single pilot crews experienced a stick shaker or actual overspeed. Approximately 50% of the total stick shaker events occurring during SPO and the majority of the stick shaker events during the RCO conditions occurred before the resting pilot was re-engaged. Therefore, approximately 90% of the total stick shaker events occurred when the pilot flying was essentially performing SPO. The crew resource management when the other pilot indicates “watch your airspeed” occurs long before the aircraft warnings.

Checklist Usage

Time to first correct checklist is an indicator of crew understanding of the problem. This data is shown in Figure 2. Time to complete the checklist is another indicator, as well as how closely the crew follows the checklist and if they complete additional checklists that apply to the failure.

Checklist use for this failure was complicated by the fact that the first annunciation, IAS DISAGREE, points to a checklist with the only action “Refer to the unreliable airspeed checklist”, which is the required checklist for this failure. The checklist has recently been re-designed to handle multiple failures as well as other failures. The checklist requires qualitative decisions and some of the indications disappear before completing the checklist.

The time to start the checklist was significantly faster in the two-crew condition. The time for the resting pilot to re-engage in the flight was fixed at two minutes and that is the approximate difference in times between two-crew and RCO condition. On average, SPO pilots took 50% longer to start the checklist than Two-Crew pilots did.

This time is essentially a reflection of not being able to delegate any tasks like talking to ATC, gathering weather information, talking to dispatch, and maintaining aircraft control.

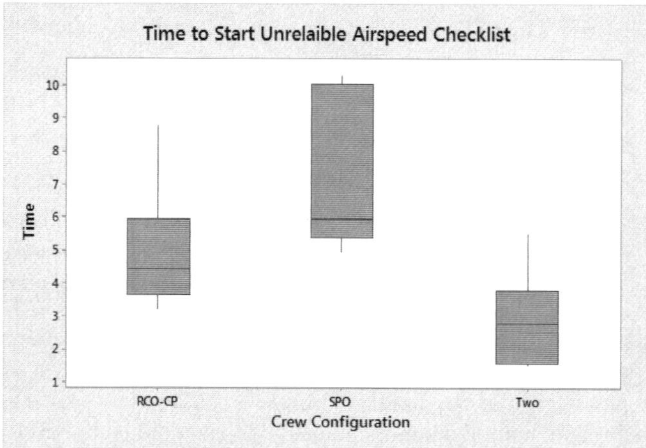

Figure 2. Time to first starting the Unreliable Airspeed Checklist.

Diversion Decision

The diversion decision is an indicator of how well the non-normal was handled and includes many factors in the decision making process. Airspeed failures, especially while in the clouds, can become critical. The diversion decision includes factors like icing potential and location of the nearest suitable airport with good weather.

The weather at Denver (DEN) and Colorado Springs (COS) was configured to be similar and relatively bad, with DEN being especially problematic. Weather at Albuquerque (ABQ) was okay and weather at Grand Junction (GJT) was good. Crews, in general, knew what the weather was like at Denver and Albuquerque but had to ask specific questions of ATC or dispatch to get other weather. The flight management system was already configured for a landing at ABQ.

Figure 3 shows diversions by airport and crew configuration. For RCO where the first officer is initially flying (RCO-CP) and doesn't make any diversion decisions until the captain is back active on the flight deck, only one of six crews diverted to other than the destination. When the Captain was flying the SPO condition (SPO-CP), they always diverted and the majority found the good weather at Grand Junction, but for First Officers flying an SPO flight (SPO-FO), only one in three diverted. For two-crew condition, half diverted to Denver. These data suggest a correlation between the perceived criticality and crew experience in the diversion decision. Not all crews considered the airspeed failure a critical problem and when flying two crew decided the risk of weather less important than the expediency of the closer airport, Denver. For single pilot Captains, the failure was critical enough that they all diverted and they felt that getting to better weather was a priority.

Workload

Overall workload was measured using the NASA Task Load Index (TLX) presented to the pilots immediately after completion of each run (Figure 4). There was an increase in workload for airspeed failure compared to the nominal runs but this difference was not found to be significant. Overall workload increased more for the first officer. Analysis of the TLX components found a significant difference in the temporal subscale ($F=3.24$, $p= 0.035$) likely due to time pressure of the first officer while completing the checklist items.

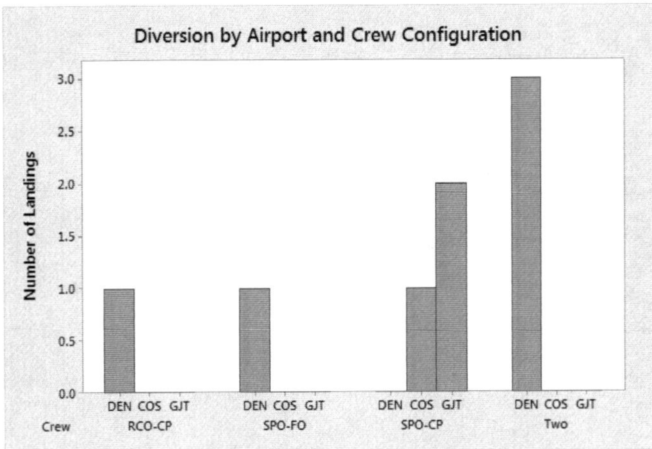

Figure 3. Diversions by Airport and Crew Configuration.

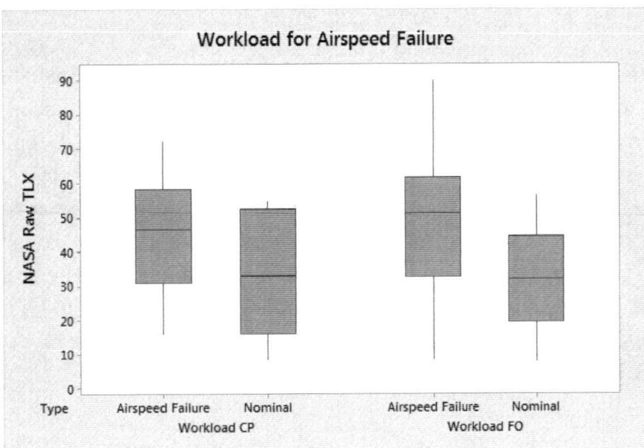

Figure 4. NASA Task Load Index Workload

Perceived Level of Safety

Post-run the crews used a Likert scale to self-assess their perceived level of safety for the airspeed failure by crew configuration and crew member, captain or first officer. The first three data columns are crew configuration as perceived by the Captain (CP) and the last three columns are crew configuration as perceived by the first officer (FO). A safety level of 1 is completely acceptable, 4 is neutral and 7 is completely unacceptable. Although a blocked pitot tube is a simple failure, RCO and SPO crew configurations rate this as unacceptable.

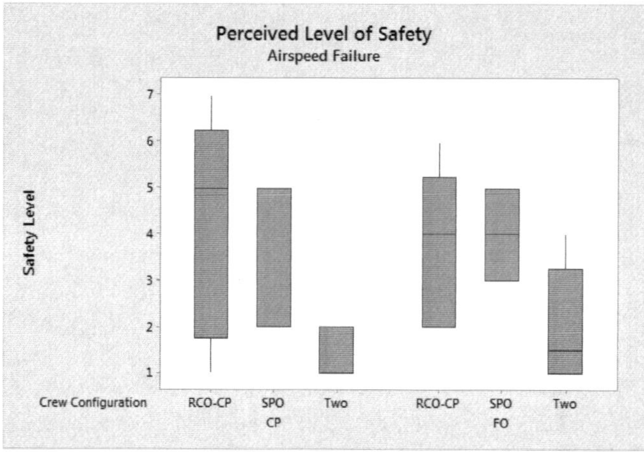

Figure 5. Perceived Level of Safety for the Airspeed Failure by Crew Configuration and Crew Member.

Conclusion

Although relatively benign, the Pitot tube failure presented some challenges that were especially problematic as the crew complement was reduced from the nominal two-crew condition. Unreliable airspeed is a well-trained event. Many crews had the initial procedures memorized; however, crews still had difficulty with aircraft control as stick shaker, overspeed, and overbank warning were common. Time to identify and begin to work the correct checklist was double for SPO compared to two-crew. Data analysis for this failure (and the other five, including nominal runs) is being used to establish quantitative baseline levels of performance and safety during nominal crew configuration. These data are being used to assess the performance and safety decrement in reduced crew and single pilot crew configurations using current-day flight deck design and certification. From this baseline, technology requirements will be identified that may inform future normal two crew operations and may eventually help enable reduced crew or possibly even commercial single pilot operations.

References

Air Line Pilots Association, "Air Line Pilots Association White Paper on Unmanned Aircraft Systems," April 2011.

Li, G., Baker, S.P., Grabowski, J.G., and Rebok, G.W., "Factors Associated with Pilot Error in Aviation Crashes," Aviation, Space, and Environmental Medicine, Vol. 72, No. 1, January 2001.

Emergency and Abnormal Situations in Aviation Symposium, LOSA Data and Emergency and Abnormal Situations http://human-factors.arc.nasa.gov/flightcognition/download/EAS_Symposium_Presentations/LOSA.pdf , June 10-11, 2003

Etherington, T.J., Kramer, L.J., Kennedy, K.D., Bailey, R.E., and Stephens, C.L., *Quantifying Pilot Contribution To Flight Safety For Normal And Non-Normal Airline Operations*. Paper presented at AIAA Digital Avionics Systems Conference.

Acknowledgements

This work was sponsored by NASA's Airspace Operations and Safety Program, Safe Autonomous Systems Operations project, led by Parimal Kopardekar. It was conducted under an interagency agreement between NASA and the FAA. This work could not have been possible without the hard work, diligence, assistance and support of Mr. Harry Hodges (AFS-440), Mr. Doug Rodzon (AFS-440), the entire AFS-440 Flight Simulation Branch, and Mr. Bruce DeCleene (AFS-400). The technical assistance of Dr. Robert "Buck" Joslin and Mr. Bryan Burks, Alaska Airlines, was greatly appreciated.

DEVELOPMENT OF A MODEL OF 'SEE AND AVOID' IN PARACHUTING

Victoria Cutler and Saskia Revell
Royal Air Force Centre of Aviation Medicine
Henlow, UK

The UK military undertakes in-depth investigations of serious parachuting accidents, which have recently included two mid-air collisions. The analysis of these accidents identified that collision avoidance in parachuting uses similar processes to the see-and-avoid task performed by aircraft pilots. However, no research was identified that had explored see-and-avoid when parachuting. Accordingly, a model of parachuting see-and-avoid was developed which consisted of six stages which must be performed in sequence for a collision to be avoided successfully. Each stage of see-and-avoid was associated with key errors, the likelihood of which was influenced by a range of factors within the individual, their operating environment, and equipment. The model of see-and-avoid can be applied to identify human factors influences in a parachute accident and in the development of initiatives to improve parachuting safety.

Parachute jumping represents an area of injury and fatality risk in aviation. Accordingly, the Royal Air Force Centre of Aviation Medicine (RAF CAM) was tasked to provide Human Factors (HF) support to the investigation of parachuting accidents involving UK military personnel. Two recent accidents involved a collision between two parachutists while under canopy control in the late stages of the descent. These collisions were unintended and led directly to the injuries sustained in the accident. Therefore, the investigation undertaken by RAF CAM aimed to identify why the parachutists collided, what HF issues may have increased the likelihood of the collision, and what could be done to reduce the likelihood of such collisions in future.

The British Parachuting Association (BPA) Operations Manual (1998) states that "*throughout the descent parachutists should be aware of other parachutists and, if necessary, take avoiding action*". As such, collision avoidance relies on the parachutists maintaining adequate look out, which reflects the pilot's task to see-and-avoid other air traffic.

While a number of detailed studies have characterised the pilot's see-and-avoid task, no research has been identified that has explored see-and-avoid during parachuting. Therefore, the aim of this work was to review the applicability of see-and-avoid aviation research to collision avoidance during parachuting, and use this to develop a model of Parachuting See-and-Avoid (PSA).

Method

Literature regarding see-and-avoid in aviation was reviewed in relation to the parachuting environment. The literature review included journal articles, aviation accident investigations, technical reports, and advice from regulators. The stages of see-and-avoid that have been specified for aviation were identified from the literature, and compared against the parachuting task to provide an initial model of PSA. For each stage in the initial model, key errors and HF issues which would prevent that stage from being effective were described. The completed model was reviewed informally within the team, and further refined through application in two parachuting accident investigations.

Results

The PSA model is presented in Figure 1. The left hand column in Figure 1 outlines the stages of see-and-avoid, and the right hand column outlines key errors which would prevent that stage from being effective. Each stage in the left hand column must occur successfully for a collision to be avoided.

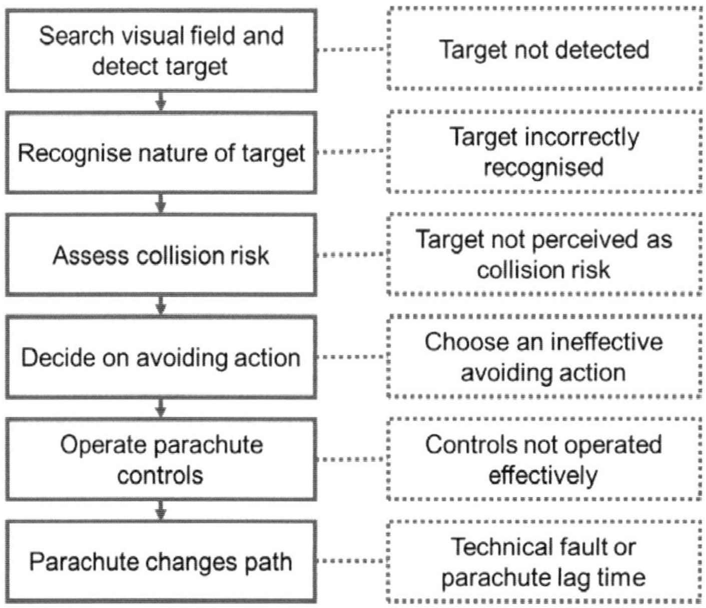

Figure 1. Model of see-and-avoid in parachuting.

Search, detect, and recognise target

As shown in Figure 1, the first stage of PSA is to search the visual field to detect the target. The target may be another parachutist, but could refer to any other collision hazard. In the second stage the nature of the target is recognised. These initial stages are fundamental for see-and-avoid to be successful, but can be influenced by a wide range of factors.

Position within visual field. For a target to be detected, it must be present within the visual field. The ability to detect a target that is within the visual field will then depend on its position at the centre or periphery of the visual field, the apparent size of the object, and any relative movement (Scott and Wright, 2016).

Visual contrast. The contrast between a target and the background against which it is viewed is a key determinant of the ease with which it is detected (Scott and Wright, 2016). Particular considerations in parachuting are the canopy colours, differentiation between the canopy and the sky/ground, and the background complexity. Environmental conditions such as light levels, visibility, and glare can all influence the visual contrast of the target and so the ease with which a target can be detected (Kroemer and Grandjean, 1997).

Alerting equipment. Where there are no tools available to alert the individual to a target, visual search will be non-directed and associated with a lower level of success (Australian Transport Safety Bureau (ATSB), 1991). While alerting systems are common place in commercial aviation, such systems are not used in parachuting. In parachuting, an alert may be provided by another parachutist giving a warning or an instructor or safety officer giving talk down instructions.

Equipment obstructions. The parachute canopy and risers, and the parachutist's goggles and helmet could limit field of view. Scratches or marks on the goggles could also reduce visibility.

Time to search visual field. Given the range of movement possible in parachuting, potential targets could be in a large proportion of the airspace and so even with the application of a highly efficient visual search strategy it could take considerable time to perform a complete search of the visual field.

Attention and distraction. Attentional resources are limited and so if a parachutist's attention is targeted at one particular area this is likely to be at the expense of other areas; so if attention is directed away or distracted from the target, then it may not be detected.

Workload and stress. Parachuting is perceived as a high stress task and so has been used as an experimental context for research on physiological responses to stress. This research has shown that physiological stress responses are found before, during, and after parachute jumps (Chatterton, Vogelsong and Hudgens, 1997) and that stress from parachuting could reduce cognitive performance (Taverniers, Smeets, Lo Bue, Syroit, Van Ruysseveldt, Pattyn and von Grumbkow, 2011). However, there have been few studies into the effect on performance and there is no evidence to indicate if workload and stress varies through the jump, between different types of parachuting, or between parachutists.

Environmental stressors. Stressors such as hypoxia, noise/wind rush, vibration, and temperature could influence target detection during parachuting. However, there has been relatively little research to characterise the impact of these factors in the parachuting environment. For instance, while it is known that hypoxia can lead to impairments to decision making, reaction times, and vision, as well as changes in attitude to risk (Hodkinson, 2011; Petrassi, Hodkinson, Walters, and Gaydos, 2012) there have been no studies exploring the effect of hypoxia on performance during the dynamic environment of a parachute jump. Overall noise levels have been measured in civilian parachuting, indicating a noise level of approximately 105dB across all phases of the jump (aircraft flight, free fall and under canopy; Penman and Epstein, 2011); however, no study has been identified which measures the noise levels at each stage of the jump. No research has been identified to determine the levels of vibration found when parachuting.

Diffusion of responsibility. Diffusion of responsibility is *"The process by which individual's may fail to act in a situation requiring intervention as a result of the presence of other people"* (Stratton and Hayes, 1999). The scope for diffusion of responsibility to influence see-and-avoid behaviour has been considered in relation to piloted aircraft (ATSB, 1991) and may be applicable to parachuting.

Assess collision risk

Once detected, the parachutist must determine if the target poses a collision risk. This task involves assessment of the other's trajectory and speed in relation to the parachutist's own flight path. Many of the same factors which reduce the likelihood of detecting and recognising the target could also influence this task – particularly the visibility of the target, workload, and environmental stressors. However, there are also HF issues specific to the decision process.

Nature of the flight paths. The parachutists may only be on a conflicting path for a short period of time which could limit the opportunity for the collision risk to be assessed. Unpredictable movements by either of the parachutists would also make it difficult to assess the risk of collision.

Judgement of speed, trajectory, and size. Assessment of own and other's speed, trajectory and size during a parachute descent is purely visual and so subject to a range of visual illusions and misjudgements which could lead the parachutist to believe that a collision was less likely than it was.

Gathering additional information to reduce the likelihood of misjudgement takes time and so reduces the time available to take avoiding action.

Decision time. The time required to recognise a collision will vary depending on a wide range of factors including the clarity of information regarding the collision risk, training, and experience. FAA Advisory Circular 90-48D suggests five seconds is required to recognise a mid-air collision risk, although the extent to which this applies to decision making while parachuting is not known.

Training and experience. It is not known if training in assessing collision risk could improve parachutist's judgement. However, a greater level of training and experience at the tasks being performed could improve decision making at critical times.

Decide on avoiding action

Having identified that there is a risk of a collision, the parachutist must choose an appropriate avoiding action. Workload, environmental stressors, training, experience, and decision time could influence this task, alongside three factors specific to this stage of PSA.

Procedures. Where procedures are available for the parachutist to adopt to avoid the collision, and the parachutist is aware of those procedures, it may be possible to achieve a reliable level of performance.

Diffusion of responsibility. Responsibility for collision avoidance is placed with the upper parachutist during a descent (BPA, 1998). This clear specification of roles is beneficial in preventing confusion when deciding on an avoiding action, but could lead the lower parachutist to delay making a necessary avoidance decision due to a perception (conscious or otherwise) that the upper parachutist would take the action.

Freezing response. Although response to emergency situations has often been characterized as 'fight or flight', the response to freeze - or take no action - has also been observed. Leach (2004) characterizes the freezing response as reflecting a situation in which *"no behavioural schema"* exists and the person perceives that the time to choose the appropriate behaviour is longer than the time available. Such a response could be anticipated in a PSA if the parachutist does not feel able to select and implement a suitable response in the perceived time available before the collision.

Implement avoiding action

The final two stages of the PSA model reflect the human and system tasks to implement the avoiding action. As with the previous stages, the operation of the controls could be influenced by workload, environmental stressors, training, and experience. However, the nature of the tasks associated with implementing avoiding action introduces novel influences.

Equipment usability. The design of the controls for adjusting the canopy could impact on the ease with which actions are implemented.

Reaction time. Simple reaction times (for a simple motor action to a stimulus) can be as short as 0.15 secs, and for a simple choice between three options it has been estimated that reaction time could be as fast as 0.4 secs (Kroemer and Grandjean, 1997). This figure is in line with the FAA Advisory Circular 90-48-D which uses a reaction time of 0.4 secs once a course of action has been selected. However, no data was identified which recorded the time taken to operate parachute controls.

Physical characteristics. The size, shape, and other physical characteristics of the parachutist can impact on their ability to operate the parachute controls. In particular, the anthropometry of the upper body and the parachutists' strength are relevant to the task of taking action to avoid a collision. Injury, either pre-existing or sustained during the jump may also impact on the parachutist's ability to implement the required actions.

Lag time. There will be a lag time between the parachutist operating the control and the parachute changing direction which could influence the ability to avoid the collision.

Discussion

RAF CAM has developed an initial model of PSA which has been adapted from models of similar tasks undertaken by aircraft pilots. The PSA model has been applied during the investigation of two parachuting accidents. In these investigations, the use of the model enabled the investigator to characterise which stages of PSA had been successful, and where shortfalls may have occurred. As such, the model contributed to understanding what happened during the accident.

The inclusion of the factors that could influence each stage of PSA within the model was particularly beneficial during the investigation process as it enabled a wide range of factors to be considered and provided a framework against which the available evidence could be compared. By reviewing the evidence against these factors, the PSA model contributed to understanding why the collision avoidance process was unsuccessful. In doing so, the analysis identified changes which could be made to improve safety and reduce the likelihood of recurrence. The two HF investigations produced a total of 23 recommendations which covered issues including training, parachuting equipment, and talk down practices.

In developing and applying the PSA model it has become apparent that there were a number of areas where adequate research was not available to determine if a factor could influence the risk of a collision. In particular, further research is required to determine the impact of workload and stress, hypoxia, noise, and vibration on performance during parachuting tasks. Further work is also required to conduct full scrutiny of the PSA model. To date the model has only been applied to two parachuting accidents, both involving military personnel; therefore further assessment would be required to determine the suitability of the model for both military and civilian parachuting. However, the initial work presented in this paper suggests that the application of the PSA model could be beneficial to improving safety and reducing the risk of mid-air collision in parachuting.

Acknowledgements

The authors acknowledge the contribution of the relevant Service Inquiry and Defence Accident Investigation Branch personnel to the analysis presented in this paper.

References

Australian Transport Safety Bureau (1991). *Limitations of the See and Avoid Principle*. Retrieved from https://www.atsb.gov.au/publications/1991/limit_see_avoid.aspx

British Parachute Association (1998). *Operations Manual (third re-write)*. Retrieved from http://www.bpa.org.uk/assets/Operations-Manual.pdf/Operations-Manual-July-2016.pdf

Chatterton, R.T., Vogelsong, K.M., Lu, Y., and Hudgens, G.A. (1997). Hormonal response to psychological stress in men preparing for skydiving. *Journal of Clinical Endocrinology and Metabolism*, 82(8), 2503-2509.

FAA Advisory Circular 90-48D (2016). *Pilots' Role in Collision Avoidance*. Retrieved from https://www.faa.gov/documentLibrary/media/Advisory_Circular/AC_90-48D.pdf

Hodkinson, P.D. (2011). Acute exposure to altitude. *Journal of the Royal Army Medical Corps,* 157(1), 85-91.

Kroemer, K.H.E. and Grandjean, E. (1997). *Fitting the Task to the Human. Fifth Edition*. London: Taylor and Francis.

Leach, J. (2004). Why people freeze in an emergency: Temporal and cognitive constraints on survival responses. *Aviation, Space and Environmental Medicine*, 75(6), 539-542.

Penman, T. and Epstein, M. (2011). Skydiving Noise Levels. *Poster presentation at Joint Defense Veterans Affairs Audiology Conference (JDVAC)*, San Diego, CA.

Petrassi, F.A., Hodkinson, P.D., Walters, P.L. and Gaydos, S.J. (2012). Hypoxic hypoxia at moderate altitudes: review of the state of the science. *Aviation Space, and Environtal Medicine*, 83(1) 1 – 10.

Scott, R.A.H. and Wright, P. (2016). Ophthalmology. In Rainford, D.J. and Gradwell, D.P. (Eds). *Ernsting's Aviation Medicine*. London: Hodder Education.

Stratton, P. and Hayes, N. (1999). A *Student's Dictionary of Psychology Third Edition*. London: Arnold Publishers.

Taverniers, J., Smeets, T., Lo Bue, S., Syroit, J., Van Ruysseveldt, J., Pattyn, N. and von Grumbkow, J. (2011). Visuo-spatial path learning, stress, and cortisol secretion following military cadets' first parachute jump: The effect of increasing task complexity. *Cognitive, Affective, and Behavioural Neuroscience*, 11, 332-343.

A STUDY OF ACCIDENTS AND INCIDENTS OF LANDING ON WRONG RUNWAYS AND WRONG AIRPORTS

Linfeng Jin

Purdue University

West Lafayette, Indiana

Edmund Lo

Purdue University

West Lafayette, Indiana

This paper addresses the landings on wrong runways/ at wrong airports happened between 03/26/1992 and 05/08/2012. The visibility, intended landing runway heading, accident landing runway heading, pilots' flight hours, and the ages of those pilots are studied to test whether they have correlations with the number of personnel injury, the number of personnel death, and the degrees of aircraft damage. Some significant findings are: the most likely angular difference between the supposed landing runway headings and wrong runway headings among wrong runway/airport landings is 180 degrees, and there is a weak negative correlation between aircraft damage and pilot flight hours. All the data used in the paper was collected from the National Transportation Safety Board (NTSB) database.

Introduction

Ever since first manned flight, the people have been enjoying the freedom of flight, and they must land back on earth. However, landing is not always so easy. The legendary pilot Chuck Yeager once said, "If you can walk away from a landing, it's a good landing. If you use the airplane the next day, it's an outstanding landing (Yeager, 2016, p.1)." The researchers of this research analyzed total of 84 accidents and incidents due to landings on wrong runways or wrong airports between 1992 and 2012 in the States. There are many studies that have been focused on landings at wrong airports regarding spatial disorientation and aviation physiology, but few existing research studies available based on empirical data based on the perspectives of pilots. In the study, the researchers reviewed the accident and incident reports from the NTSB database to conclude what pilot-related factors may cause landings on wrong runways or wrong airports. After a series of search and discussions, the researchers narrowed down these variables in this study: visibility, supposed landing runway heading, accident/incident landing runway heading, pilot in command flight hours, the age of the pilots, the degree of aircraft damage, andthe numbers of injuries or deathes and their injury conditions.

Literature Review

Mismatches between external world reality and the "internal world" aircrew mental picture relative to the real world would cause landings at wrong airports. In other words, the pilots misjudged the time, speed and distance, finally misidentified the wrong airport as the correct ones through the distortion of the facts of the reality (Antuano & Mohler, 1989). Landings on wrong runways or at wrong airports can be considered as the instances of disorientation on the part of the pilot. The pilots who become disoriented, are inadequately informed by the the external visual environment, deceived by the force environment, or both effects (Stott, 2012). The main cause of such accidents or incidents can be listed as perceptual error under errors under unsafe acts of pilots by Human Factors Analysis and Classification System (HFACS) (Shappell & Wiegmann, 2000). It is noteworthy that the NTSB is calling on the FAA to issue new rules requiring controllers to withhold a landing clearance until after an aircraft has passed all other airports that may be confused with the destination airport (Croft, 2015). This terse news shows that the aviation safety investigation body come to realize that other than pilot errors, there are more things can be done outside the cockpit. From renowned researcher Dr. Douglas Wiegmann, it has been learn that aviation accidents; especially general aviation accidents, usually happen when pilots fly VFR (Visual Flight Rules) into instrument meteorological conditions (IMC), and the reason behind that can be from the pilots don't realize the dangerous transition between VFR and IMC during the flights, or they are overconfident in their piloting skills and don't fully appreciate the risks of flying into the adverse weather (Wiegmann & Goh, 2002). According to the

Aircraft Owners and Pilots Association (AOPA), "since 2002, more than 86% of all fixed-wing VFR-into-IMC accidents have been fatal" (AOPA, 2016). Even before this alarming number found in 2016, there was a research paper suggested ground all VFR (Visual Flight Rules) flights when there are Marginal VFR weather conditions because it had shown that restricted visibility was the leading cause or a contributing factor in the fatal accidents when those accidents materialized in Marginal Visual Flight Rule (MVFR) or Instrument Flight Rule (IFR) (Pearson, 2002).

On average, there are more than ten incidents of commercial operations involved with landings on the wrong runways every decade domestically and internationally combined since 1960s (Silversmith, 2016). Even though commercial pilots are better trained compared with general aviation counterparts; however, they would land on the wrong runways, and even wrong airports in perfect weather conditions. A research team from Purdue University found that runway incursions occur at a more frequent rate for airports with intersecting runways compared to airports with no intersecting runways after they analyzed the data from the 30 busiest airports with intersecting runways and the 30 busiest airports without intersecting runways were compared in USA between 2009 and 2013 (Johnson, Zhao, Faulkner, & Young, 2016). In the research, two independent variables: the flight hours of the pilots in command, and the age of pilot, which could be counted as liveware in the SHELL (Software, Hardware, Liveware, Liveware) model proposed by the International Civil Aviation Organization (ICAO) in ICAO Circular 216-AN31. The model put emphasis on the connection between liveware and either one of rest four components, and it shows that any breakdown of two or more components can lead to human performance problem (Australian Government, 2014).

Mr. Voogt and Mr. Doorn recommended comparison of airports near the destination airport and the use of GPS to the identification procedure to prevent landing at wrong airports after they did an analysis of 54 incidents and 11 incidents happened between 1981 and 2004 (De Voogt & Van Doorn, 2007).

Research Questions

The researchers of this study addressed the following questions:

What is the most likely angular difference between the supposed landing runway headings and wrong runway headings among wrong runway/airport landings? (descriptive statistics)

What is the flight hour distribution of the pilots in command from the wrong runway/airport landing?

What is the most likely visibility when the wrong runway/airport landing? (descriptive statistics)

What is the correlation between the degree of aircraft damage and the flight hours of the pilot, age of the pilot? (MLR)

What is the correlation between the degree of personnel injury and the flight hours of the pilot, age of the pilot? (MLR)

What is the correlation between the number of personnel loss and the flight hours of the pilot, age of the pilot? (MLR)

Methodology

The data is from NTSB (National Transportation Safety Board) aviation accident database, and the database is accessible to the public. The query is constrained to wrong runway or wrong airport, broad phase of flight: landing, and Injury Severity: all. There are entirely 82 NTSB accident or incident reports generated. In other words, there is a total valid sample size of 82 (n=82). The accidents/incidents happened between March 26th, 1992 and May 8th, 2012.

Most of the independent variables in the research are directly from the NTSB report, the visibility, the mislanded runway heading, and the flight hours of the pilot, and the age of pilot.

For the dependent variables in the research, there are aircraft damage, the number of people injured, and the fatality number directly from the NTSB report. And the researchers can either determine the supposed runway the pilot should land at from reading the description of accident or incident in the NTSB reports or get it directly from the description from the NTSB report. In order to do a multilinear regression, the degrees of aircraft damage have been converted to numerical variables like 0, 1, 2, 3 in respect to none, minor, substantial destroyed. And for the same reason, the researchers combined the degree and the number of personnel injury together, and the new indications are 0, 1, 2 in respect to nobody injured, one person with minor injury, one person with serious injury.

Results

The angular difference between supposed landing runway heading and mislanded landing runway heading distribution is shown below:

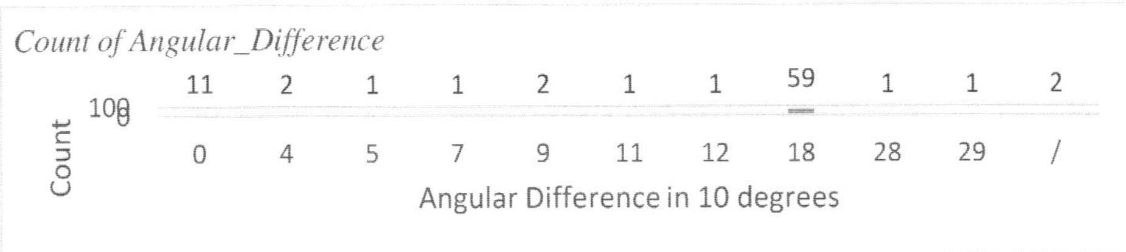

The graph shows that the most frequent (59/82) angular difference between the supposed landing runway headings and wrong runway headings among wrong runway/airport landings is 180 degrees.

The flight hour distribution of the pilots in command having the wrong runway/airport landing is presented in histogram shown below:

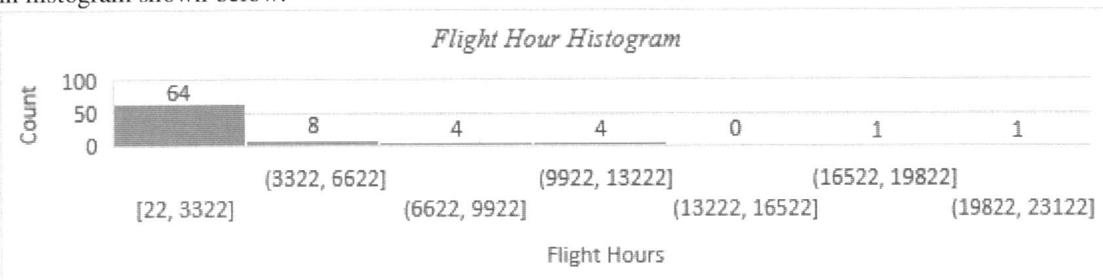

The histogram indicates that the absolute majority of the pilots in command from wrong airport/runway landings are in low flight hour range.

The visibilities of accidents and incidents are recorded below:

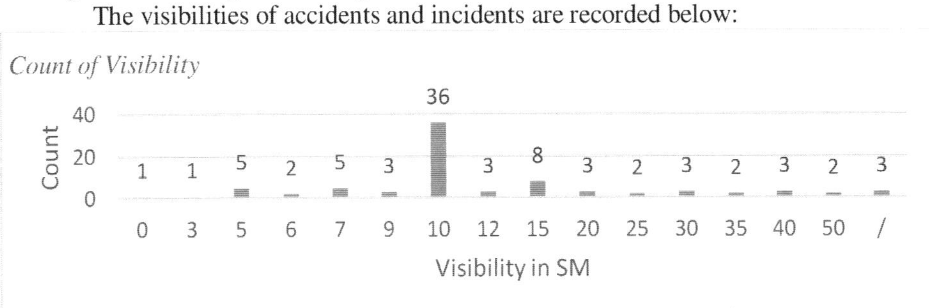

From the graph, the researchers found out the most likely visibility in the wrong runway/airport landings are 10 statute miles (SM). And the average visibility in the wrong runway/airport landing is 14 SM.

The Multilinear Linear Regression (MLR) is used to test the correlation between the degree of aircraft damage and the flight hours of the pilot, age of the pilot.

By using MLR, the researcher found out that there is a correlation between the degree of aircraft damage and the flight hours of the pilot, age of the pilot because the whole model Pr value is 0.002, which is smaller than the alpha value 0.05. However, after a further examination, it has been found out that Pr value of age of pilot is greater than 0.05, hence the researchers used Bonferroni method to remove the independent variable age of pilot, and made a new single linear regression (SLR) to show the correlation between the degree of aircraft damage and hours of flight.

By using the SLR analysis, it shows that there is a correlation between the degree of aircraft damage and the flight hours of the pilot. And it can be expressed as the following equation:

$$Y = -0.00003330\ X + 2.13139$$

Y: Degree of aircraft damage, and X: Hours of Flight.

The MLR is again used to test the correlation between the degree of personnel injury and the flight hours of the pilot, age of the pilot. By using the MLR, the researchers found out that there is no correlation established between the degree of personnel injury and the flight hours of the pilot, age of the pilot because Pr value of the model is 0.3963 and it is bigger than the alpha value.

Finally, the MLR is used to test the correlation between the number of personnel loss and the flight hours of the pilot, age of the pilot. The researchers found out that there is no correlation established between the degree of personnel injury and the flight hours of the pilot, age of the pilot because Pr value of the model is 0.2639, and it is greater than the alpha value 0.05.

Discussion

There were total 82 observations of the angular difference between mislanded runway heading and supposed runway heading, but there are only 80 valid observations because two observations are lack of supposed landing runway headings and mislanded runway headings. The absolute majority of the angular difference is 180 degrees. In other words, most of the pilots in the wrong runway/airport landing accidents landed in the tail wind. They were either unware of tail wind situations or making the risky landings with the knowledge of the tailwind, all of those corresponded with the fact that most of the pilots were low-flight-time pilots in the research. And the next most frequent angular difference is 0 degree. It entails the following situations: landings on the parallel runways in the same airports, landings on the parallel runways in different airports, or landings on the same runways in the same airports.

Against the findings of many previous spatial disorientation studies, the accidents/ incidents in this research happened on excellent weather condition in terms of visibility (average visibility 14 SM). One possible reason is that the accidents/incidents pilots are low hour pilots and they were apt to flying in Visual for Reference (VFR) weather.

There is only one weak negative correlation existed between the degree of aircraft damage and the flight hours of the pilots in command, which makes sense because more experienced pilots can make better judgments so that minimize the damage to aircraft in accidents/incidents.

There is no correlation between the degree of personnel injury and the flight hours of the pilot, age of the pilot. And there is no correlation between the number of personnel loss and the flight hours of the pilot, age of the pilot. One reason is that there were not enough samples with the personnel injury or loss so that it is impossible to establish statistical significance.

According to the NTSB, most of the accidents are attributed as "WRONG RUNWAY - SELECTED - PILOT IN COMMAND". However, this study tends to find out what has caused these pilots, even the more experienced ones to commit such error? It is deemed that to choose a right runway regarding wind condition is the basics of pilotage skills. As stated by Wiegmann and Shappell (Wiegmann & Shappell, 2001), various human factors could play a role in these accidents and NTSB's report may not be comprehensive enough for covering these factors behind.

Human Factors Analysis and Classification System (HFACS) is a framework developed by Dr. Weigmann and Dr. Shappell based on Reason's Swiss cheese model in 1990. HFACS framework allows investigators to identify Human causes in aviation accidents. Developed from the Swiss cheese model, HFACS has gone further from just identifying latent and active failures. The framework was initially developed for the U.S. military use, but it has also now been applied outside the military. HFACS categorized human error into four levels of failure, namely I) Unsafe Acts of operators, II) Preconditions for unsafe acts, III) unsafe supervision and IV) organizational influences.

From studying the final report of each case in the NTSB's database, it revealed that rarely did NTSB indicate how the pilots achieve weather information. Most of the information is factual and does not disclose why and how the pilots make such mistake in selecting the wrong runway. According to HFACS classification, selecting the wrong runway is probably a decision error or a skill-based error, if not a violation. Regrettably, the factual information in NTSB's report is too scarce to deduce what are the human factors that caused pilots errors, as all of the 78 cases showed no mechanical failures in the aircraft.

The results of the regression analysis revealed that there is a correlation between flight hours of pilot and aircraft damage: the higher the flight hours obtained by the pilot, the lesser damage is to a plane. Such result is in line with previous similar research on crash rate (Li, et al., 2003). Although the rate of wrong runway landings does not change significantly across pilots' age, flight experience as measured by total flying hours does have a correlation with wrong runway landings. Li, et al. (2003) in a similar study found that flight experience has a protective effect against the risk of crash decreases as flight experience increases until a certain threshold, which is 5000 hours of flight. It can be regarded that as pilots built their flight experience, they are exposed to a variety of risks and probably have learned how to handle different risk through experiencing them. Such risk handling skills enable pilots with more flight experience to choose the runway correctly when wind information is presented.

On the other hand, the age of pilots may not necessary be associated with the degree of aircraft damage. Although pilots are more vulnerable to health problems through aging, FAA's rigorous health standards on the pilost are not interrelated with applicants' age. If a pilot can obtain FAA's airman medical certificate, it denotes that his/her health is up to an airman standard and are deemed fit to fly and not jeopardizing flight safety. If a pilot's health function is disqualifying, he/she would be declined in the issuance of airman medical certificate, leaving only those who are medically fit in the skies. Our research, in fact, reflects that a wrong runway landing is more related to the pilots' ability to make a right decision rather than aircraft maneuvering skills. Such ability is part pf the pilots's situaton awareness which is not correlated with age, but experience.

From the study statistics, most accidents being studied happened at day time rather than night time. In fact, only 5 cases (6.7%) and 1 case happened at night and dusk respectively. It is suggested that it is because pilots are more vigilant at night and more likely to verify information in a meticulous manner. At night, windsocks can hardly be identified and may cause pilots to stay cautious to weather information. This results are in line with FAA's (Lee, 2012) findings in 2011. Pilot's attitude toward risk and their risk perception should not be overlooked. This research shows that pilots committing wrong runway landing varied drastically in terms of flight experience, ranging from student pilots to airline transport pilot. The least experienced student pilot clocked 22 hours, while the most experienced pilot had 19306 hours of flight. This suggested that with ample of flight experience, a pilot can still make basic mistakes if they are not vigilant enough.

Based on NTSB data for U.S. rotorcraft accidents from 2001-2010, over 88% of the accidents occurred in daylight conditions and over 95% occurred in visual meteorological conditions. Interestingly, 60 of all 78 (77%) accidents happened in between 1992-2002, and only 18 (23%) of them happened in between 2002-2012. Such huge difference in accident rate has aroused our attention. One probable reason for such discrepancy may be attributed to the technology advancement and a deeper understanding on aviation human factors. Glass cockpit on light aircraft wasn't something new. In 2003, Cirrus Design Corporation began to use glass cockpits in Federal Aviation Administration (FAA)-certified light aircraft as electronic primary flight displays (PFD). The company quickly standardized electronic PFDs on their SR20 and SR22 models subsequently. These electronic PFDs also as known as glass cockpit displays, are integrated with a lot of functions which are necessary to pilots, such as terrain and traffic avoidance, synthetic vision, and autopilots and global positioning systems(GPS) (National Transportation Safety Board, 2010). Glass cockpit has proven very useful in commercial aviation by relieving pilot's workload. It is thus deduced that the glass cockpit when coupled with integrated weather update service, have greatly reduced the chances of wrong runway landing, particularly the 180 degree ones (i.e. landing on the opposite direction) either by directly navigating or giving cues to pilots to land on the right runway.

The other reason for less accidents reported during 2002-2012 perhaps may be the deeper understanding and emphasizes of human factors and safety in aviation. The airlines started to look into human factors and employed crew resources (CRM) training after the 1977 Tenerife airport disaster after two Boeing 747 collided. Human factors have been studied greatly. In 1980, several authorities including the European Joint Aviation Authority (now EASA) have incorporated quality assurance program in their management system, and FAA quickly followed them as well.

10-15 years later, in the 1995 aviation safety summit which comprises of about a thousand representatives from the aviation sector, FAA's administrator David Hinson advocated to work towards a goal of zero accidents in a large, international scale. In the summit meeting, FAA formed a new office of system safety and issued an aviation safety action plan with 173 initiatives. These are regarded as stepping stone of the current safety management system(SMS) which is required by ICAO to be implemented by aviation service providers in 2006 (Britton, 2016).

The introduction of the notion of safety in human factors and implementation of SMS in general aviation have been raised the pilots' concern for safety training. Now, it is an order from FAA (FAA, 2016) to develop and implement SMS in certain providers, such as flight schools in the United States. Under the implementation of SMS on flight schools, training organizations and airline operators, it is thought that pilots' attitude and understanding towards aviation safety have been improved. Pilots may have been benefited by going through a more rigorous training and thus lowered the chance by committing errors.

In Weigmann and Shappell's study (2001), it revealed that skill based and decision errors have accounted for over forty percent of all the accidents which associated with human errors. In this study, 62 out of 75 (83%) cases have landed on the opposite runway. Presume these errors are made by pilots out of decision error by wrongfully selected the opposite runway, it is believed that decision errors and skill-based error are easily being made in the landing phrase. In fact, according to FAA, there are more than 250 accidents happened during the landing phase in 2014 and were accounted for the largest portion of accidents by flight phases. Decision errors denote when an operator implements a plan which is unsatisfactory in achieving the desired outcome and results in an unsafe situation and skill-based error entails an error which occurs in the execution of a routine procedure by an operator (Wiegmann & Shappell, 2001).

In our study, most pilots select the wrong runway under daylight with good visibility and weather conditions. Although NTSB did not provide sufficient factual information, we suspected that the pilots committed these errors either by not understanding wind components and intentionally chose the wrong runway (decision error) or simply mixed up the runway and landed wrongly (skill-based error). Landing phase is the busiest phase of a flight and pilots would be busy setting up an aircraft for landing configuration. Such high workload could distract pilots' concentration on their decision-making process and makes them feel tired and disorientated, especially when the pilots are landing at somewhere with which they are not familiar. We, therefore, advocate that pilots are easier to commit skill based and decision error during the landing phase.

It must be stressed that HFCAS is much more than just skilled-based and decision errors. There are also other latent factors, such as environmental factors and physical conditions of operators and supervision factors. However, because of not having enough the factual information from the NTSB, we propose the most probable reason for wrong runway landing by cross referencing Weigmann and Shappell's data.

Conclusion

Wrong runway landings can be lethal if not handled properly. When pilots are flying into an uncontrolled airport, it is pilots' duty to ensure flight safety by choosing the most suitable runway based on factors inside and outside cockpits. This paper studies the correlation between flight experience, the age of pilots, personnel injury, aircraft damage and visibility and the results are tabulated and presented. Findings show that there is a correlation between flight experience and degree of personal injury and aircraft damage, and the possible cause of such correlation is discussed. It is hoped that through such study, more attention would be given to pilots' training in Aeronautical Decision-Making(ADM) and wrong runway landing awareness in the future.

References

Antuano, M. J., & Mohler, S. R. (1989). Geographical disorientation: Landing at Wrong Airport. *Aviation, space, and environmental medicine*, 996-1004.

AOPA. (2016). *VFR INTO IMC ACCIDENTS 08/14/2015 - 08/14/2016*. Retrieved from AOPA Organization Web site: https://www.aopa.org/asf/ntsb/vfrintoimc.cfm?window=1

Australian Government. (2014). *SMS 6 SMS For Aviation- A Practical Guide 2nd Edition*. Retrieved from Australian Government Civil Aviation Safety Authority: https://www.casa.gov.au/sites/g/files/net351/f/_assets/main/sms/download/2014-sms-book6-human-factors.pdf

Croft, J. (2015, November 15). NTSB: Controllers, Software Complicit In Wrong - Runway Landings. *Aviation Daily*. Washington, United States of America: Penton Media, Inc., Penton Business Media, Inc.

De Voogt, A. J., & Van Doorn, R. R. (2007). Approach and Landings at Wrong Airports: Analysis of 54 Incidents and 11 Accidents, 1981-2004. *Aviation, Space, and Environmental Medicine*, 117-120.

Johnson, M. E., Zhao, X., Faulkner, B., & Young, J. P. (2016). Statistical Models of Runway Incursions Based on Runway Intersections and Taxiways. *The Journal of Aviation Technology & Engineering*, 15-26.

Lee, R. (2012, Jan). *U.S. Rotorcraft Accident Data and Statistics*. Retrieved from Aircraft Electronics Association: https://www.aea.net/events/rotorcraft/files/US_Rotorcraft_Accident_Data_And_Statistics.pdf

Pearson, D. C. (2002). *"VFR Flight Not Recommended" A Study of Weather-Related Fatal Aviation Accidents.*

Shappell, S. A., & Wiegmann, D. A. (2000). *The Human Factors Analysis and Classification System–HFACS.* Springfield, MO: National Technical Information Service.

Silversmith, J. A. (2016, July 10). *The Legacy of Douglas Corrigan: "Wrong Way" Landings by Commercial Airliners*. Retrieved from Third Amendment: http://www.thirdamendment.com/WrongWay.pdf

Stott, J. R. (2012). Orientation and Disorientation in Aviation. *Extreme Physiology and Medicine*.

Wiegmann, D., & Goh, J. (2002). Human Factors Analysis of Accidents Involving Visual Flight Rules Flight into Adverse Weather. *Aviation, Space, and Environmental Medicine* , 817-822.

Yeager, C. (2016, August 27). *Chuck Yeager Quotes*. Retrieved from Brainy Quote: http://www.brainyquote.com/quotes/quotes/c/chuckyeage598656.html

A CASE STUDY USING THE HUMAN FACTORS ANALYSIS AND CLASSIFICATION SYSTEM FRAMEWORK

Flavio A. C. Mendonca, Ph.D. Candidate
Chenyu Huang, Ph.D. Candidate
Richard O. Fanjoy, Ph.D.
Julius Keller, Ph.D.
Purdue University
West Lafayette, Indiana

Aircraft accidents are generally the end result of a number of latent conditions arising in the organizational and managerial sectors. These conditions frequently permit or even motivate the unsafe acts by the flight crew. The Human Factors Analysis and Classification System (HFACS) is a system safety tool for the investigation and analysis of underlying human causal factors in aircraft accidents. Using the HFACS framework, four researchers classified the human factors identified by the Brazilian Aeronautical Accidents Investigation and Prevention Center (CENIPA) during the investigation of a mishap (PR-AFA) that happened in Brazil in 2014. CENIPA argued that errors and violations by both pilots contributed to the accident. Results of this study indicate that inappropriate decision making by upper-level management had an adverse effect on the performance of the PR-AFA pilots. Most importantly, safety strategies to mitigate unsafe acts by crewmembers should receive significant attention from the highest managerial levels of the organization.

Approximatelly 80% of aircraft mishaps are associated with human errors (Wiegmann & Shappell, 2003). The terms human error and procedural violations may have limited value in preventing future accidents (Reason, 1997, 1998). These factors could indicate where the breakdown occurred, but provide no guidance as to why an accident occurred or how to prevent one from occurring in the future (ICAO, 2013; Reason, 1998; Wiegmann & Shappell, 2003). Several accident causation models have been developed to assist in mitigating human errors and violations. The Human Factors Analysis and Classification System (HFACS) describes four levels of failure (Li, Harris, & Yu, 2008; Wiegmann & Shappell, 2003) proposed in the Reason model (Reason, 1997, 1998). HFACS is a system safety tool that can be used within aviation sectors to systematically and effectively examine underlying human causal factors during the investigation of aircraft accidents. This tool facilitates the development of data-driven investment safety strategies to enhance aviation safety addressing areas where the benefits will be the highest.

A Cessna Citation CE-560XLS+, registered as PR-AFA, crashed in Brazil in August 2014, claiming the lives of seven people, including a Brazilian presidential candidate during the political campaign. The Brazilian Aeronautical Accidents and Prevention Center (CENIPA) thoroughly investigated this accident (CENIPA, 2014) in accordance with the ICAO Standards and Recommended Practices (SARPs) (ICAO, 2016). Weather conditions were below flight minimums at the destination airport. The crewmembers performed an instrument flight rules (IFR) procedure and missed approach with a profile different from the one prescribed in the aeronautical chart. In addition, CENIPA (2014) presented other human factors issues that could have contributed to the accident, such as fatigue, spatial disorientation, and poor team dynamics. Using the HFACS framework, the purpose of this case study was to analyze the human factors elements, including errors and violations, which may have contributed to the accident. Findings were expected to suggest new insights to mitigate the risk of aircraft accidents due to human factors.

The Human Factors Analysis and Classification System

Safety professionals have used organizational and systemic models during the investigation of aircraft accidents as well as the development of the ensuing mitigation strategies since the 1990s (Reason, 1997, 1998). Human factors models such as the "Swiss Cheese", also known as Reason's model (Reason, 1997; 1998), and the HFACS model (Wiegmann & Shappell, 2003) provide a better capture of the complexity of organizational and social-technical systems. Therefore, they enable safety professionals to have a greater understanding of the factors that may contribute to aircraft mishaps (Shappell et al., 2007). Reason's model, the most popular accident causation framework, describes the interactions between active failures by frontline personnel and latent conditions. According to Reason (1997, 1998), it is inadequate to attribute accidents to individual operator performance. Human errors and violations are the end result rather than the cause of mishaps, and just the starting point of the safety investigation process. Accident investigators must focus on events beyond the Unsafe Acts by pilots to latent preexisting conditions, which are usually induced by fallible decisions made on managerial levels.

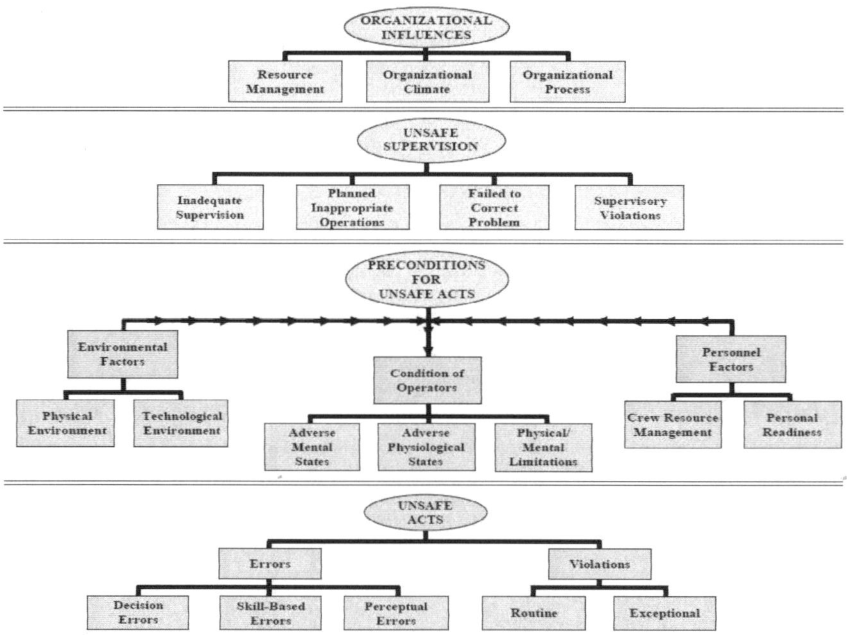

Figure 1. The HFACS Framework. Adapted from the "Human error approach to aviation accident analysis: The human factors analysis and classification system" by Wiegmannn, D. A., & Shappell., S. A. (2003). Burlington, VT: Ashgate Publishing Limited.

The HFACS framework was drawn upon the concept of latent conditions and active failures by Reason (1997). It bridges the gap between theory and pratice by providing safety professionals with a scientifically tested framework designed to investigate the active failures by operators. Additionally, it also encourages safety experts to investigate the latent conditions upstream in the organization (Shappell et al., 2007; Wiegmann & Shappell, 2003). The HFACS model succesfuly describes human errors at four levels: Unsafe Acts of Operators, Preconditions for Unsafe Acts, Unsafe Supervision, and Organizational Influences. The HFACS framework is presented diagrammatically in Figure 1.

Each lower level is impacted by the higher levels in the HFACS framework (Li et al., 2008). The HFACS model goes beyond the identification of unsafe acts by frontline employees, and provides a better understanding of the latent conditions that permited or even prompted Unsafe Acts by human operators. Human errors and violations are viewed as consequences of systemic failures, and are the starting point of an investigation process (Wiegmann & Shappell, 2003). The use of the HFACS framework during the investigation of mishaps facilitates the identification of the contributing factors to the accident, the elaboration of hypotheses, and the development of safety recommendations designed to mitigate latent conditions and Unsafe Acts, greatly improving aviation safety.

The PR-AFA Accident

The PR-AFA, a Cessna Citation CE-560XLS+, was on a non-scheduled flight from Santos Dumont Airport (SBRJ) bound for Santos Aerodrome (SBST), in Brazil, on August 13, 2014. At the time of the accident, the destination airport was operating under severe weather conditions with mist and rain significantly affecting both visibility and operational ceiling. The crewmembers informed the Aerodrome Flight Information Service (AFIS) their intention to perform a non-directional beacon (NDB) instrument flight rules (IFR) approach procedure to land on runway 35. However, they did not follow the profile of the Echo 1 IFR procedure. CENIPA raised the hypothesis that the captain used the aircraft flight management system (FMS) to intercept a direct approach to land at SBST, even though the aircraft manual warned the crew that the FMS visual approach mode must not be utilized in instrument meteorological conditions (IMC) as a substitute for IFR approaches. The pilots discontinued their approach, but did not follow the profile prescribed in the aeronautical chart. The PR-AFA crashed into the ground at a high negative pitch angle and at a high speed, killing two pilots and five passengers, including a well-known Brazilian politician who was campaigning for president. The mishap was thoroughly investigated by CENIPA (CENIPA, 2014).

In addition to the aforementioned factors, CENIPA (2014) posited in its final report that both pilots had not had the adequate and prescribed training while transitioning to the CE-560XLS+ (they were not qualified in that aircraft model). CENIPA (2014) also argued that other human factors issues could have contributed to this mishap. For example, at the time of the accident, there was a self and organizational pressure on the pilots relative to flight schedule due to the political campaign of a passenger. Analysis of the copilot's voice, speech, and tone indicated compatibility with fatigue and somnolence. Moreover, both pilots had difficulties in applying crew resource management concepts. CENIPA (2014) also postulated that the first officer operational capabilities (e.g., cockpit and operational routine management, provision of support as a pilot-not-flying [PNF], effectiveness in the execution of procedures) were inadequate. Those conditions degraded the crewmembers' aeronautical decision making process (ADM).

Following CENIPA (2014), the captain had previously utilized the FMS resources (visual mode) for making direct approaches and very likely used the FMS for reducing the time spent in the Echo 1 IFR procedure. Because the pilots did not follow the profile of the Echo 1 IFR procedure, and due to a tailwind, the crewmembers had difficulty in maintaining a stabilized approach. Thus, they had to perform a missed approach. Yet, after the missed approach the flight crew attempted to maintain visual meteorological conditions (VMC), despite the bad weather conditions. CENIPA (2014) also claimed that the inadequate training, the conflicting relationship and synergy between crewmembers, and the pilots' personal characteristics (e.g., captain authoritarian, first officer passive) hindered the dynamics of the crewmembers, and greatly increased their workload. Moreover, such conditions favored the onset of spatial disorientation of an incapacitating type during a high-risk flight-condition.

Methods

CENIPA is a Brazilian Air Force organization responsible for the investigation of aircraft accidents and incidents involving civil and Brazilian Air Force aircraft in Brazil, all in accordance with the ICAO SARPs. The final report of the PR-AFA, the unit of this case study, was available at the CENIPA website. Using both tabular and narrative data from the PR-AFA final report, each human causal factor was classified using the HFACS framework (Wiegmann & Shappel, 2003). One researcher, who had previous HFACS training and experience using the model during the investigation of aircraft mishaps, made the initial classification. After that, the remaining members of the research team, all with experience in aviation safety and human factors, reviewed potential classifications independently until all researchers reached an agreement. Considering the high inter-rater reliability found in previous studies using the HFACS model (Li et al., 2008; Shappell et al., 2007; Wiegmann & Shappell, 2003), consensus classification was deemed appropriate for the study.

Findings and Discussions

The current study presents an analysis of the accident involving the PR-AFA, a Cessna Citation CE-560XLS+, using the final report by CENIPA (2014) and the HFACS tool (Wiegmann & Shappell, 2003). The HFACS model provides safety investigators with an empirically tested framework that bridges the gap between theory and practice, and assists in identifying and classifying human errors and violations in aircraft mishaps. In addition, it helps safety professionals to focus on latent conditions, active failures, and their interrelationships (Wiegmann & Shappel, 2003). Most importantly, it permits the identification of the underlying causes of Unsafe Acts by crewmembers.

The analysis of this accident started with the level most closely tied to the mishap: Unsafe Acts of operators (Wiegmann & Shappell, 2003). In the first level, researchers agreed that the following actions by the crewmembers could be classified as:

1. Execution of the Echo 1 IFR procedure by the flight crew even though the weather was below the minimums for the procedure (Exceptional Violation);
2. Probable use of the aircraft FMS by the pilots to make a direct approach (Routine Violation);
3. Nonconformity with the profile established in the aeronautical chart during the procedure (Routine Violation) and ensuing missed approach (Exceptional Violation);
4. Attempt to maintain VMC during the missed approach (Decision Error); and
5. Inadequate response to spatial disorientation (Percpetual Error).

Latent conditions, arising in the managerial and/or organizational levels, such as failing to provide crews with proper training, are unavoidable components of the aviation system. They could combine with local triggering conditions and allow or even induce unsafe acts by frontline personel (Reason, 1997, 1998). Unsafe acts of crewmembers can reduce safety margins and lead to mishaps. However, it is paramount to investigate the second level of the HFACS framework, Preconditions for Unsafe Acts, in order to better prevent future accidents. For example, both pilots had not received the prescribed training to transition to the Citation CE-560XLS+ (Personal Readiness). Therefore, they did not have the adequate knowledge and skills to safely operate the aircraft, or the adequate experience for the complexity of the situation (Mental Limitations). In addition, such conditions reduced the pilots' situational awareness (SA) and demanded more cognitive efforts during the IFR procedure, especially the missed approach. The copilot's fatigue and somnolence were Adverse Physiological States that also reduced the crewmembers' SA, thus precluding their ADM process and the safe operation of the aircraft (CENIPA, 2014). In the final report, CENIPA argued that both pilots had difficulty in applying CRM concepts. Even more, they had an unfriendly relationship before the accident. Hence, this situation led to poor coordination, confusion, low SA, and inadequate ADM by both pilots (FAA, 2016). Moreover, these factors most likely contributed to the spatial disorientation of the flight crew. The researchers agreed that loss of SA, complacency, and overconfidence (Adverse Mental States) were factors that adversely influenced the pilots' performance and ADM. The operational environment, the deteriorating weather before and during the time of the accident, also had an adverse effect on the Unsafe Acts by the flight crew. First officer operational weaknesses as a crewmember (Mental Limitations) also was a precondition for the unsafe acts committed by the flight crew.

The Unsafe Supervision level of the HFACS framework connects Unsafe Acts by pilots to the level of the front-line supervisors. The role of front-line supervisors is to provide their personnel leadership, training, guidance, and the adequate tools to perform their jobs efficiently and safely (ICAO, 2013; Shappell et al., 2007). At the supervisory leadership level, researchers identified actions and inactions that had an adverse effect on the safety of the PR-AFA. For instance, both pilots were neither provided with nor required to undergo the adequate and prescribed training before operating the aircraft. Leadership also failed to provide proper CRM training for both crewmembers. Middle management failed to identify and correct risky behaviors by the captain (e.g., inappropriate use of the aircraft FMS; poor CRM skills), by the first officer (e.g., lack of aptitude and skills to act as a crewmember), and the unfriendly relationship of the crewmembers. Additionally, front-line supervisor(s) failed to provide adequate rest in order to mitigate fatigue (Inadequate Supervision). The fourth level of the framework describes the contributions of fallible decisions in upper-levels of management that have a negative effect on the lower levels of the model. Corporate-level decision-making for organization resources, including monetary and human resource management (e.g., inadequate CRM training), played a role in this accident (Resource Management). A poor safety culture (Reason, 1997, 1998), and ill-defined safety policies (ICAO, 2013) contributed to the mishap (Organizational Climate). Finally, organizational pressures due to the presidential campaign (e.g., time; schedule), and inadequate safety programs to mitigate safety hazards were latent conditions that allowed and prompted unsafe acts by the crewmembers (Organizational Process).

Conclusion

Human errors and violations in aviation are elusive and complex to investigate. The accident involving the PR-AFA was analyzed using the HFACS framework. This analysis, demonstrated that actions and inactions at the highest organizational levels can promulgate throughout lower levels. Moreover, those actions and inactions could allow or even motivate Unsafe Acts by crewmembers on the aircraft flight deck. Furthermore, it indicated that the HFACS framework could provide accurate information that should be used for the development, implementation, and the quantifiable assessment of effective safety intervention and mitigation strategies addressing the highest organizational levels. The most cost-effective strategies with the greatest improvement in safety should target these areas (Li et al., 2008; Reason, 1997, 1998; Shappel et al., 2007; Wiegmann & Shappell, 2003).

References

Brazilian Aeronautical Accident Prevention and Investigation Center (CENIPA). (2014). *Final report PR-AFA* (A - 134/CENIPA/2014). Retrieved from http://www.cenipa.aer.mil.br/cenipa/paginas/relatorios/relatorios-finais

Federal Aviation Administration (FAA). (2016). *Pilot's handbook of aeronautical knowledge.* Retrieved from https://www.faa.gov/regulations_policies/handbooks_manuals/aviation/phak/media/pilot_handbook.pdf

International Civil Aviation Organization (ICAO). (2013). *ICAO safety management manual* (Doc. 9859-AN/474) (3rd ed.). Montreal, Canada: Author.

International Civil Aviation Organization (ICAO). (2016). *Annex 13 to the Convention on International Civil Aviation, Aircraft Accident and Incident Investigation* (11th ed.). Montreal, Canada: Author.

Li, W. C., Harris., D., & Yu, C. S. (2008). Routes to failure: Analysis of 41 civil aviation accidents from the Republic of China using the human factors analysis and classification system. *Accident Analysis and Prevention, 40*(2), 426-434.

Reason, J. (1997). *Managing the risks of organizational accidents.* Aldershot, England: Ashgate Publishing Limited.

Reason, J. (1998). Achieving a safe culture: Theory and practice. *Journal of Work and Stress, 12*(3), 293-306, doi:10.1080/02678379808256868

Shappell, S., Detwiler, K. H., Holcomb, K., Hackworth, C., Boquet, A., & Wiegmann, D. A. (2007). Human error and commercial aviation accidents: An analysis using the human factors analysis and classification system. *Human Factors, 49*(2), 227-242

Wiegmann, D. A., & Shappell, S. A. (2003). *A human error approach to accident analysis: The human factors analysis and classification system.* Burlington, VT: Ashgate Publishing Company

IMPROVING AVIATION STUDENTS' TEAMWORK, PROBLEM SOLVING, COORDINATION, AND COMMUNICATIONS SKILLS DURING A HIGH-FIDELITY SIMULATION

Evan Lester, Andrea Georgiou, Michael Hein, Glenn Littlepage, Richard Moffett III, & Paul Craig
Middle Tennessee State University
Murfreesboro, Tennessee

The National Aeronautics and Space Administration (NASA) Flight Operations Center – Unified Simulation (FOCUS) lab at Middle Tennessee State University (MTSU) is a high-fidelity simulation of a regional airline's flight operations center. During a simulation, a team of senior undergraduate aerospace students must work together across disciplines to manage 24 simulated Canadair Regional Jet– 200 aircraft and resolve real-world scenarios. After the simulation, the lab's staff evaluates the team's performance, which is discussed during its After Action Review (AAR). The AAR allows the team to establish strategies and an action plan to improve its performance and skills during subsequent simulations. Overall, as the lab continuously increases in standardization and fidelity through various ways, such as the utilization of WSI Fusion and WSI Fusion Replay, the lab's simulations help MTSU's aerospace students improve their problem solving, teamwork, coordination, and communication skills while also helping the lab's staff conduct reliable research on teamwork.

Every position in the aviation industry, such as flight dispatchers and pilots, operates in a complex environment requiring effective teamwork, communication, coordination, and problem-solving (Helmreich, 2000). Without these elements, accidents and incidents can occur, which can lead to a tremendous loss of life (Helmreich, 2000). Unfortunately, a gap still exists between the information and skills that students learn in the classroom and the effective application of the skills and information in the real-world setting (Sleeper & Thompson, 2008). As a result of the massive increase in computing power and various types of technology over the last decade, there has been an increase in the development and use of high-fidelity simulations in aviation to resolve this issue (Beaubien & Baker, 2004; Miller, Crandall, Washington, & McLaughlin, 2012).

A simulation is technology that is designed in such a way as to virtually reproduce one aspect of the working environment (Maran & Glavin, 2003; Sleeper & Thompson, 2008). Simulations can be classified into three different types of fidelity (Beaubien & Baker, 2004; Maran & Glavin, 2003). Fidelity is defined as the extent to which a simulation's behavior and appearance match the appearance and behavior of the replicated aspect of the working environment. The first type of fidelity is called physical fidelity (Maran & Glavin, 2003). Physical fidelity is the extent to which a simulation replicates the physical aspects of the actual working environment. Although increasing the physical fidelity of the simulation helps participants slightly improve their performance and skills, an increase in physical fidelity can cost a significant amount of money. The second type of fidelity is called equipment fidelity, which means the extent to which a simulation replicates the sensory information, such as the motion and visual cues, of the actual work environment (Beaubien & Baker, 2004). The third type of fidelity is called psychological fidelity (Maran & Glavin, 2003). This type of fidelity means the extent to which a simulation replicates the actual work environment's tasks and responsibilities. The level of psychological fidelity of a simulation depends on the task being replicated and the skills that participants need to be able to transfer to the actual working environment. For example, simulations replicating complex work environments, like the aviation industry, need to have a high level of psychological fidelity to help participants improve their skills and prevent them from experiencing a negative transfer of training (Maran & Glavin, 2003).

There are several advantages of using simulations that are high in all three types of fidelity. First, teams and individuals in simulations can practice the knowledge and skills that they have learned (Beaubien & Baker, 2004). Second, after applying or not applying their knowledge and skills to a situation in a simulation, teams and individuals can observe the positive or negative consequences of their action or inaction while in a safe environment. Third, simulations allow teams and individuals to face and respond to emergency scenarios that are impossible to train for in actual work environments (Beaubien & Baker, 2004). Fourth, simulations provide an opportunity to train teams and individuals on human interaction skills, such as coordination, communication, problem-solving, and teamwork (Shapiro et al., 2004). Finally, simulations that are immediately followed by a debriefing process allow participants to understand how they performed, identify their strengths and weaknesses, and learn how they can improve their teamwork and human interaction skills in subsequent simulations and the real-world environment (Fanning & Gaba, 2007; Hunt, Shilkofski, Stavroudis, & Nelson, 2007; Shapiro et al., 2008)

However, there are also some negative aspects of simulations that are high in all three types of fidelity. For example, the teams and individuals participating in the simulation can be reluctant to participate, which can cause them to put little effort into the simulation's scenarios and tasks (Sleeper & Thompson, 2008). Also, if unrealistic scenarios are implemented into a simulation, the teams and individuals participating in the simulation can learn inappropriate skills and information. In addition, teams and individuals participating in a simulation may feel overwhelmed by the stress, time pressure, and scenarios if they have never had an opportunity prior to the simulation to practice their knowledge and skills (Beaubien & Baker, 2004).

In conclusion, high-fidelity simulations are critical tools for training teams and individuals in the aviation industry. Although there are possible weaknesses associated with high-fidelity simulations, they have many strengths. For example, high-fidelity simulations allow teams and individuals to apply their knowledge and skills toward real-world scenarios in safe environments and learn how to improve their skills and performances during debriefing sessions. However, one of the most important strengths of simulations is that they train teams and individuals on human interaction skills, such as coordination, communication, teamwork, and problem-solving skills, that are crucial for working in the aviation industry.

History and Concept of the NASA FOCUS Lab

Before 2010, Middle Tennessee State University's (MTSU) Aerospace Department was teaching students in their specific aerospace concentrations, or educational "silos." This means that students in one aerospace concentration at MTSU were only taught the skills and information that they needed to succeed in the aviation industry and never truly interacted with students from the other aerospace concentrations. For example, the students in the flight dispatch concentration at MTSU only took classes and interacted with students in the flight dispatch concentration. This was a major problem for the department because aviation professionals from every aerospace concentration must effectively communicate, coordinate, problem-solve, and work together across disciplines 24 hours a day, seven days a week in the industry in order to conduct legal, safe, and efficient operations. In addition, several experts in the aviation industry have found that it can take up to 10 years for newly-hired aviation professionals, such as recent aviation graduates, to truly understand the big picture of the aviation industry and how their decisions and performances impact the aviation company that they work for and, ultimately, the aviation industry. In response to these issues, Dr. Paul A. Craig, an MTSU aerospace professor, decided to apply for two National Aeronautics and Space Administration (NASA) grants to build a simulation lab that could bring all of MTSU's aerospace students from every aerospace concentration together in order to break down the aerospace department's educational silos, reduce the amount of time it takes for recent MTSU aerospace graduates to understand the big picture of the aviation industry and how their performances impact the industry, and help MTSU's aerospace students enhance their teamwork skills that are critical for working in the aviation industry. In 2010, Dr. Craig was awarded both NASA grants, and he used them to create the simulation lab called the NASA Flight Operations Center – Unified Simulation (FOCUS) lab.

The NASA FOCUS lab is a high-fidelity simulation of a Part 121 regional airlines' flight operations center. Every MTSU senior undergraduate aerospace student enrolled in the "Aerospace Senior Capstone Lab" are placed into teams of 10 to work a three-hour "shift" in the flight operations center for the virtual airline called "Universal E-Lines." Also, each team's students are placed into one of the following positions that are directly related to their aerospace concentrations: Flight Operations Coordinator (FOC), Maintenance Control, Maintenance Planning and Scheduling, Flight Operations Data 1 (FOD 1), Flight Operations Data 2 (FOD 2), Crew Scheduling, Weather and Forecasting, Nashville International Airport (BNA) Ramp Tower / Duty Pilot, and Canadair Regional Jet (CRJ) – 200 Flight Crew. For example, if one student on a team is in the flight dispatch concentration, then that student will be placed in the FOC position. During a three-hour simulation, the students on a team must coordinate, communicate, problem-solve, and work together across concentrations to manage Universal E-Lines' 24 simulated CRJ-200 aircraft. These aircraft operate approximately 80 flights in the southeastern United States along a hub-spoke system to 14 spoke airports, such as McGhee Tyson Airport and Tampa International Airport, and two hub airports, which are Nashville International Airport and Jacksonville International Airport.

However, the previously-mentioned positions are not all in a single location at MTSU. These positions are in one of three locations that are utilized during every simulation. The first location is the NASA FOCUS lab, which is home to Universal E-Lines' flight operations center. The positions located in the flight operations center are the FOC, FOD 1, FOD 2, Weather and Forecasting, Maintenance Control, Maintenance Planning and Scheduling, and Crew Scheduling. The second location is the BNA Ramp Tower, which is in a room adjacent to the lab. The BNA Ramp Tower is home to the BNA Ramp Tower / Duty Pilot position where he or she manages Nashville International Airport's arriving and departing aircraft and reroutes all cargo and passengers that missed their connecting flights at Nashville International Airport. The third location is the MTSU Simulator Building at Murfreesboro Municipal Airport (KMBT). The MTSU Simulator Building is home to MTSU's Federal Aviation Administration (FAA)-certified Level 5 CRJ-200 flight training device (FTD). During every simulation, two students from a team are sent to the MTSU Simulator Building to fly the CRJ-200 FTD as three of Universal E-Lines' simulated flights. One additional location that is not home to a student's position is the office across the hall from the NASA FOCUS lab. In this office, a lab staff member plays the role of the pilot-in-command for every Universal E-Lines' simulated flight, except the flights operated by the CRJ-200 Flight Crew.

In addition, during a three-hour simulation, the NASA FOCUS lab staff, which consists of professors, graduate students, and undergraduate students from MTSU's Aerospace Department and Industrial and Organizational Psychology program, implements real-world scenarios into the simulation, which gives students on a team the opportunity to enhance their skills and apply the knowledge that they have gained throughout their undergraduate aerospace education to resolve the scenarios. After a team creates and carries out a solution for each scenario, the team immediately learns how its solution impacted Universal E-Lines through simulated financial data and immediate feedback from the lab's staff. Also, while a team is managing Universal E-Lines' flights and resolving real-world scenarios, the lab's staff conducts various measures and takes detailed notes about the team's performance. These measures and notes are used to give the team constructive and concrete feedback on their performance at the After Action Review (AAR), which is a facilitated debriefing process that helps a team identify how it can improve its performance and skills in subsequent simulations.

Implementation and Standardization of Real-World Scenarios

During a simulation, the NASA FOCUS lab staff implements real-world scenarios, or triggers, that vary in difficulty into the simulation. Before 2016, the lab's staff would implement a different number of triggers that varied in difficulty into the simulations for each team. However, during 2016, the lab's staff decided to standardize the triggers. This means that every team now faces the same number and difficulty of triggers, which increases the reliability of the data collected during each simulation. Also, after the lab's staff implements the triggers into a simulation, the team must resolve the triggers in a legal, safe, and efficient manner. After the team creates and implements a solution into the simulation, the lab's staff evaluates and determines the effectiveness and appropriateness of the solution. If the solution was not legal, safe, and efficient, then the team faces negative downstream consequences, such as a simulated financial penalty and missed passenger connections. Overall, the implementation of negative downstream consequences allows MTSU's aerospace students to understand how their decisions and performances impacts the virtual airline and, ultimately, the aviation industry.

High-Fidelity Components of the NASA FOCUS Lab

The NASA FOCUS lab relies on both specially developed and commercially available technology and software to provide MTSU's aerospace students a realistic simulation that will enhance their teamwork knowledge, skills, and abilities (KSAs) and prepare them for working in the aviation industry.

At every position in the NASA FOCUS lab and the BNA Ramp Tower, there are new Dell Optiplex 7040 desktop computers with 22-inch dual monitors that provide each student the capability to access multiple sources of information without any delays and the necessary space to organize the multiple sources of information. In addition, interactive Microsoft Excel documents have been developed for each position in the lab and BNA Ramp Tower that students can manipulate on their computers to gather the information that they need to complete their positions' tasks and responsibilities. The students also manipulate the Excel documents to gather the information that they need to share with their team in order to resolve real-world scenarios. Every position in the lab and BNA Ramp Tower has Plantronics headsets connected to their computers for direct verbal communications with any team member or lab staff member.

In addition, each position and the lab's staff utilize two computer applications called Join.Me and Skype in order to manage and communicate information to one another. "Join.Me" is a screen-sharing application that allows the lab's staff to view each position's desktop screen and analyze and record each student's performance during a simulation on a mobile-device that has internet capabilities, such as a laptop computer or tablet. Also, Skype allows every student in the lab and BNA Ramp Tower to communicate information verbally and through text messages to one another in order to manage Universal E-Lines fleet. The staff also uses Skype to communicate with the students at the positions in the lab and BNA Ramp Tower to provide them with technical assistance and respond to the team's solutions to the real-world scenarios that are implemented into the simulation.

Every position in the lab and BNA Ramp Tower has access to three new 65-inch Sony Ultra High Definition televisions on both sidewalls of the lab that display three specific sources of information. The first source is the Universal E-Lines' flight schedule. The flight schedule displays every simulated flight's number, departure airport using its International Civil Aviation Organization (ICAO) identifier, departure time, destination airport using its ICAO identifier, and arrival time. The flight schedule also has five status lights for each simulated flight to inform the students in the lab and BNA Ramp Tower when the flights are about to reach their scheduled departure time, due, in progress, delayed by more than 15 minutes, and delayed by more than 30 minutes. These five status lights are automatically updated throughout the simulation by comparing each flight's departure time to the current Zulu time. The flight schedule also displays each team's average arrival and departure performance, total time of delays, delay loss, and daily revenue, which are automatically updated when a Universal E-Lines' flight is released from its departure airport or arrives at its destination airport. The second source of information that is displayed on the televisions in the lab is the radar screen. The radar screen displays all of Universal E-Lines' simulated flights and the flights being operated by the CRJ-200 Flight Crew that are in progress, so the students in the lab and BNA Ramp Tower can monitor the progress of each flight. The third source of information displayed on the televisions in the lab is the weather radar, which displays the weather in the southeastern United States.

In the room adjacent to the lab, the BNA Ramp Tower consists of three 55-inch LG televisions, four control stations, and 12 servers. The 12 servers operate the Computer Sciences Corporation and Frasca software that provides the graphical display of each Universal E-Lines' flight on the radar screen in the lab and the programs at each control station in the BNA Ramp Tower. The software also generates a 150-degree view of Concourse C at Nashville International Airport on the three televisions that Universal E-Lines' simulated aircraft utilize. The student in the BNA Ramp Tower / Duty Pilot position uses this view to safely and efficiently manage and monitor the movement of Universal E-Lines' simulated aircraft at Nashville International Airport.

Located in MTSU's Simulator Building at the Murfreesboro Municipal Airport (KMBT), the FAA-certified Level 5 CRJ-200 FTD, or simulator, is used during every NASA FOCUS lab simulation. Two students from the team in the lab are sent to KMBT to fly three flights for Universal E-Lines. While the students are flying these flights for Universal E-Lines, the team in the lab can track the flights on the radar screen due to several network connections. Students in the lab can also communicate with the students in the CRJ-200 simulator using Voice Over Internet Protocol (VOIP) connections. Specifically, the students in the Maintenance Control, BNA Ramp Tower / Duty Pilot, FOC, and Weather and Forecasting positions will use the VOIP connections to verbally communicate information to the students in the CRJ-200 simulator that is pertinent to their flights, such as the dispatch release and weather information. Also, the students in the CRJ-200 simulator will use the VOIP connections to verbally communicate information about their flights to the students in the lab.

In March 2016, The Weather Company, an IBM business, and Southwest Airlines donated five licenses for the aviation analysis and flight tracking software called WSI Fusion and the weather replay software called WSI Fusion Replay. With WSI Fusion, the lab's staff has and continues to capture the weather on the radar; weather data; Air Traffic Control demand, ground stops, and weather reports (i.e., METARs) from the airports that Universal E-Lines services; and various weather charts (i.e., Winds Aloft charts) from any day. After capturing the data and charts, the lab's staff saves them in WSI Fusion Replay and Microsoft Word documents. Then, one week prior to a team's simulation, the Chief Meteorologist for the NASA FOCUS lab provides the student in the Weather and Forecasting position the weather data and charts to analyze and create a weather briefing that he or she will give to the team before its simulation begins. The student also creates a briefing for each flight operated by the CRJ-200 Flight Crew, which he or she must give to the CRJ-200 Flight Crew before each flight using the VOIP connections.

Once the simulation begins, the student in the Weather and Forecasting position uses WSI Fusion Replay to view the saved weather and analyze how it will impact Universal E-Lines' operations in the southeastern United States. The student also has to utilize WSI Fusion to ensure that the winds at each airport do not exceed each simulated CRJ-200's maximum crosswind or tailwind components. Based on the weather shown on WSI Fusion Replay, the student in the Weather and Forecasting position must pick the most legal, safe, and efficient route for each Universal E-Lines' flight. The student has three options for each flight that are already programmed into WSI Fusion Replay. The first option is called CP1, or company preferred 1. This route is the most direct route from a flight's departure airport to its destination airport. The other two options are called CO1 and CO2, or company option 1 and company option 2. These routes are not direct routes between a flight's departure airport and destination airport. These routes are intended to be used when a flight must fly around hazardous weather, such as thunderstorms or icing conditions. After gathering the wind and route information for each Universal E-Lines flight, the student in the Weather and Forecasting position must give that information to the Flight Operations Coordinator (FOC), who relays the information to the pilot-in-command of each Universal E-Lines flight.

Overall, WSI Fusion and WSI Fusion Replay are providing many benefits to both MTSU aerospace students and the lab's staff. One of the benefits of this software is that MTSU's aerospace students are given the opportunity to use software that current regional and major Part 121 airlines are using daily to monitor the progress of their flights, ensure that their flights do not fly into hazardous weather, and make determinations on whether to release their flights. By having experience using this software, MTSU's aerospace students will have a significant competitive advantage over other aerospace graduates. Second, the software allows the lab's staff to use the same weather scenarios across teams. Before WSI Fusion and WSI Fusion Replay, students in the Weather and Forecasting position would have to analyze live weather data and weather charts; however, if there was not significant weather in the southeastern United States, the students would not have many tasks or responsibilities to complete, reducing the usefulness of the simulation training for that student. With WSI Fusion and WSI Fusion Replay, students in the Weather and Forecasting position on every team encounter the same weather scenarios, which increase in difficulty in the teams' subsequent simulations. As a result, the software keeps the students in the Weather and Forecasting position engaged in the simulation and provides them the opportunity to enhance their problem-solving, communication, coordination, and teamwork skills. Third, by using the same weather scenarios across teams, the lab's staff can collect more valid and reliable data than ever before. Finally, since this software is used by both regional and major Part 121 airlines, this software ultimately enhances the fidelity of the NASA FOCUS lab's simulations.

When the NASA FOCUS lab was created, there were no documents that could accurately track each student's performance in the lab's simulations. Therefore, over the last four years, the lab's staff has created more than 20 documents that determine whether or not the students on a team are completing their tasks legally, safely, and efficiently. The documents also help the lab's staff determine the simulated financial penalties a team should receive due to not following federal regulations and standard operating procedures, not dispatching flights in a safe or efficient manner, or not resolving downstream consequences. After a simulation has ended, the documents used by the lab's staff are gathered and used during the lab's After Action Review to give every team member constructive and concrete feedback about how they performed in the simulation and how the team can improve its performance in subsequent simulations. Overall, the purpose of the documents is to help students realize that their decisions and performances do affect the success of the virtual airline.

After Action Review

After a team's simulation ends, the students on the team must complete an After Action Review (AAR) form, which asks the students about the strengths and weaknesses of the team, along with ways in which the team can improve in subsequent simulations. One week after the team's simulation, the students bring their AAR forms to the lab's AAR. The AAR is facilitated by MTSU's Industrial and Organizational (I/O) Psychology professors and graduate students. During an AAR, the I/O professors and graduate students provide the team feedback on how they performed during their simulation. Also, the I/O professors and graduate students ask the team members to discuss the team's strengths, weaknesses, and areas of improvement. This allows the team to learn from their mistakes, reinforce their strengths, and build new strategies that can improve their weaknesses during subsequent simulations. In addition, the I/O professors and graduate students discuss the team's decisions that violated Federal Aviation Regulations and standard operating procedures to ensure that the students on the team do not make the same decisions again in subsequent simulations and in the actual aviation industry. Overall, the lab's AAR provides

MTSU's aerospace students the opportunity to create strategies that can combat their weaknesses; enhance their strengths; and improve their teamwork, coordination, communication, and problem-solving skills that they need to become successful aviation professionals.

Summary

In conclusion, the NASA FOCUS lab is an important training and research tool for MTSU's aerospace students and the lab's staff. By participating in the lab's high-fidelity simulations and AARs, MTSU's aerospace students can enhance their problem-solving, communication, coordination, and teamwork skills that they need for working in the aviation industry and reduce the amount of time needed for fully understanding the big picture of the aviation industry. Also, the lab's high-fidelity simulations provide the lab's staff the opportunity to conduct valid and reliable research on various aspects of teamwork, which is used to publish articles in highly-respected academic journals, such as *Human Factors*. As the NASA FOCUS lab continues to improve, it will continue to be an important tool for MTSU's aerospace students, the lab's staff, and the aviation industry.

Acknowledgements

This research was supported in part by a contract (NNX09AAU52G) from NASA awarded to the Middle Tennessee State University Center for Research on Aviation Training. We are grateful to Dr. Paul A. Craig at the Department of Aerospace and the Center for Research on Aviation Training at Middle Tennessee State University. Correspondence concerning this article should be addressed to Evan M. Lester via email (eml3d@mtmail.mtsu.edu).

References

Beaubien, J. M., & Baker, D. P. (2004). The use of simulation for training teamwork skills in health care: How low can you go?. *Quality and Safety in Health Care, 13*, 51-56. doi: 10.1136/qshc.2004.009845

Fanning, R. M., & Gaba, D. M. (2007). The role of debriefing in simulation-based learning. *Simulation in Healthcare, 2*(1), 1-11. Retrieved from https://www.ncbi.nlm.nih.gov/pubmed/19088616

Helmreich, R. L. (2000). On error management: Lessons from aviation. *BMJ, 320*, 781-785. Retrieved from http://www.jstor.org.ezproxy.mtsu.edu/stable/pdf/25187424.pdf

Hunt, E. C., Shilkofski, N. A., Stavroudis, T. A., & Nelson, K. L. (2007). Simulation: Translation to improved team performance. *Anesthesiology Clinics, 25*, 301-319. doi: 10.1016/j.anclin.2007.03.004

Maran, N. J., & Glavin, R. J. (2003). Low- to high-fidelity simulation – a continuum of medical education?. *Medical Education, 37*(1), 22-28. doi: 10.1046/j.1365-2923.37.s1.9.x

Miller, D., Crandall, C., Washington, C., & McLaughlin, S. (2012). Improving teamwork and communication in trauma care through in situ simulations. *Society for Academic Emergency Medicine, 19*, 608-612. doi: 10.1111/j.1553-2712.2012.01354.x

Shapiro, M. C., Gardner, R., Godwin, S. A., Jay, G. D., Lindquist, D. G., Salisbury, M. L., & Salas, E. (2008). Defining team performance for simulation-based training: Methodology, metrics, and opportunities for emergency medicine. *Academic Emergency Medicine, 15*, 1088-1097. doi: 10.1111/j.1553-2712.2008.00251.x

Shapiro, M. J., Morey, J. C., Small, S. D., Langford, V., Kaylor, C. J., Jagminas, L., Suner, S., Salisbury, M. L., Simon, R., & Jay, G. D. (2004). Simulation based teamwork training for emergency department staff: Does it improve clinical team performance when added to an existing didactic teamwork curriculum?. *Quality Safe Health Care, 13*, 417-421. doi: 10.1136/qshc.2003.005447

Sleeper, J. A., & Thompson, C. (2008). The use of hi fidelity simulation to enhance nursing students' therapeutic communication skills. *International Journal of Nursing Education Scholarship, 5*(1), 1-12. doi: 10.2202/1548-923X.1555

TEAMWORK AND EMERGENT COGNITIVE STATES AS PREDICTORS OF ROUTINE AND ADAPTIVE PERFORMANCE IN FLIGHT DISPATCH CENTERS

Glenn E. Littlepage and Megan Wertheimer
Middle Tennessee State University
Murfreesboro, TN

This study examines relations between the emergent cognitive state of transactive memory, the emergent affective state of collective efficacy, teamwork processes, and team performance. Mediation is examined as well as comparison of states and processes related to performance in routine and non-routine situations.

Meta-analytic findings indicate that teamwork processes are related to team performance (Lepine, Piccolo, Jackson, Mathieu, & Saul, 2008). Meta-analysis has also shown that the emergent states of collective efficacy and transactive memory are known to relate to team performance (DeChurch & Mesmer-Magnus, 2010; Stajkovic, Lee, & Nyberg, 2009). Salas, Sims, & Burke, 2005 theorized that cognitive states provide a basis for teamwork processes. This suggests that teamwork processes serve as a mediator between cognitive states and team effectiveness. Based on previous research and theories of Marks, Mathieu, & Zaccaro (2001), and Salas and colleagues (Salas, Rosen, Burke, & Goodwin, 2009; Salas et al., 2005) we propose the following hypotheses:

H1: Teamwork processes are positively related to team performance
H2: Emergent states are positively related to teamwork. Specifically, H2a: Collective efficacy is related to effective teamwork, H2b: Transactive memory is related to effective teamwork.
H3: Emergent states are positively related to team performance. Specifically, H3a: Collective efficacy is related to team performance, and H3b: Transactive memory is related to team performance.
H4: Teamwork mediates relations between emergent states and team performance.

Routine vs. Adaptive Performance

Teams in aviation conduct many routine activities, but also must adapt to unexpected situations. Airline operations offer a context in which to study team performance in both routine and non-routine contexts. Performance during routine tasks and non-routine tasks can be markedly different. Cognitive ability is more critical on non-routine tasks and dependability (closely following existing protocols) facilitates performance on routine tasks while it may be dysfunctional on non-routine tasks where adaptation is needed (LePine, 2003). Likewise, effective teams modify interaction patterns (Stachowski, Kaplan, & Waller, 2009) and strategies (Randall, Resick, & DeChurch, 2011) to cope with non-routine situations. In addition, non-routine task contexts may require changes to the role relations between members (LePine, 2003). Furthermore, LePine (2005) observed only a moderate relationship ($r = .38$) between team performance on routine tasks and performance on non-routine tasks requiring adaptation. These findings suggest the relations between both emergent cognitive states and teamwork behaviors with team effectiveness may differ across routine and adaptive performance. These theories and findings suggest important differences between routine and adaptive performance. Based on the literature on team adaptation, we explore relations between emergent states and team processes with team performance separately for routine and non-routine tasks.

Method

Participants

Forty teams of senior-level aerospace students comprised this study. These students came from different aerospace specialties including flight dispatch, professional pilot, maintenance management, aerospace administration, and aerospace technology. Participants were assigned to specific positions in the flight operations center simulation (described below).

Procedure

The study was conducted in a high-fidelity simulation of an airline's flight operations center (FOCUS Lab). Positions in the lab include Flight Operations Coordinator, Flight Planning, Flight Scheduling, Maintenance Planning and Control, Crew Scheduling, and Weather and Forecasting. The center also externally coordinates with a Ramp Tower Coordinator, and Pilots. The team works together to release flights and overcome problems (i.e., "triggers") during the simulation. Each team completed two or three simulations, or "work shifts." Each simulation lasted approximately two and a half hours. During each simulation, participants came into the lab, worked on their specific job duties, and coordinated with team members to solve problems as they arose. During each simulation, the team collectively managed approximately 60 flight elements (takeoffs and landings). See Littlepage, Hein, Moffett, Craig, & Georgiou (2016) for a complete description of the lab and positions and duties within the simulations.

Measures

Participants completed a 10-item *collective efficacy* scale based on Quinones (1995). Items were rated on a five-point scale from strongly disagree to strongly agree. *Transactive memory* was assessed using a 15 item self-report scale (Lewis, 2003). These items were rated on a five-point scale from strongly disagree to strongly agree.

Self-rated teamwork was assessed using a teamwork scale developed by Mathieu and Marks based on Marks et al. (2001). The scale comprised 30 items rated from 1 (not at all) to 10 (a very great extent) and measured the extent to which team members engaged in certain teamwork behaviors. The scale assesses a broad array of teamwork behaviors including planning activities such as analysis of goal specification, coordination, backup behavior, and conflict management. *Observer-rated teamwork* was assessed using a locally developed ten-item behaviorally anchored scale. This scale included three subscales: problem solving, coordination, and information utilization. These items were measured on a seven-point scale from 1 (trainee level) to 7 (professional level).

Both objective and subjective measures of team performance were collected for this study. *Delay loss* was measured in dollars and is a consequence of flight delays during the simulation. Delay loss is conceptualized as a measure of routine performance with lower delay loss representing more effective performance. *Trigger effectiveness* measured the team's performance solving problems that arose during the simulation. Examples include: pilot illness, mechanical issues, severe weather, and passenger issues. Triggers not only require adaptation to the current situation, they may also require actions to prevent or minimize disruptions to other flight segments and to accommodate passengers who are stranded or miss connections. Trigger Effectiveness was measured on a seven-point scale from 1 (highly ineffective) to 7 (highly effective) by observers upon the completion of a simulation.

Results

Data analyses were conducted at the team level. See Table 1 for means, standard deviations, Cronbach alpha, and correlations.

Table 1.
Descriptive statistics and correlations.

Measure	Mean	SD	α	1	2	3	4	5
1. Collective Efficacy	2.91	.19	.70					
2. Transactive Memory	3.76	.12	.73	.08				
3. Observer-rated Teamwork	4.54	.90	.97	-.12	.37*			
4. Self-rated Teamwork	4.10	.30	.97	-.20	.61**	.58**		
5. Delay Loss	$25K	$14K		.10	-.00	-.51**	-.14	
6. Trigger Effectiveness	4.93	.67		.07	.45**	.66**	.48**	-.01

Note. *$p < .05$; **$p < .01$

Teamwork and Team Effectiveness (H1).

Observer-rated teamwork was significantly related to both team effectiveness measures: delay loss, where low scores indicate better performance (r (37) = -.51, p <.01) and trigger effectiveness (r (31) = .66, p <.01). Self-rated teamwork was significantly related to trigger effectiveness (r (31) = .48, p <.01) but not to delay loss. These findings provide partial support for H1. Self-rated teamwork was related to observer-rated teamwork, r (38) = .58, p <.01, providing some evidence of the construct validity of these measures of teamwork behavior.

Emergent States and Teamwork (H2).

Collective efficacy was not significantly related to either teamwork measure, thus hypothesis 2a was not supported. Transactive memory was positively related to both self-rated teamwork, r (38) = .61, p <.01, and observer-rated teamwork, r (38) = .37, p <.05. These results provided support for hypothesis 2b.

Emergent States and Team Performance (H3).

Collective efficacy was not related to either delay loss or trigger effectiveness. Transactive memory was related to trigger effectiveness, r (31) = .45, p = .01, but not to delay loss. This pattern of results does not support hypothesis 3a (collective efficacy), but provides partial support for hypothesis 3b (transactive memory).

Teamwork as a Mediator (H4).

Based on the pattern of correlations, three relationships showed the potential for mediation. Direct and indirect effects were identified as described by Hayes (2013). See Figure 1-3 for a visual depiction of the mediation analyses and standardized regression coefficients. In each figure, the top relationship represents the overall effect of the predictor on the criterion and the lower relationships represent direct and indirect paths. As indicated in Figure 1, observer-rated teamwork mediated relations between transactive memory and trigger effectiveness (z = 2.09).

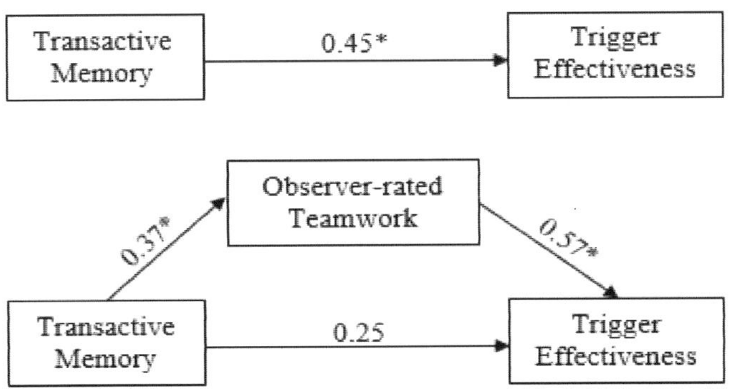

Figure 1. Testing observer-rated teamwork as a mediator between transactive memory and trigger effectiveness.

Although zero-order correlations indicated that self-rated teamwork was related to both transactive memory and trigger effectiveness, additional analyses did not confirm mediation (See Figure 2). Both transactive memory and self-rated teamwork were directly related to trigger effectiveness, but no mediation was observed.

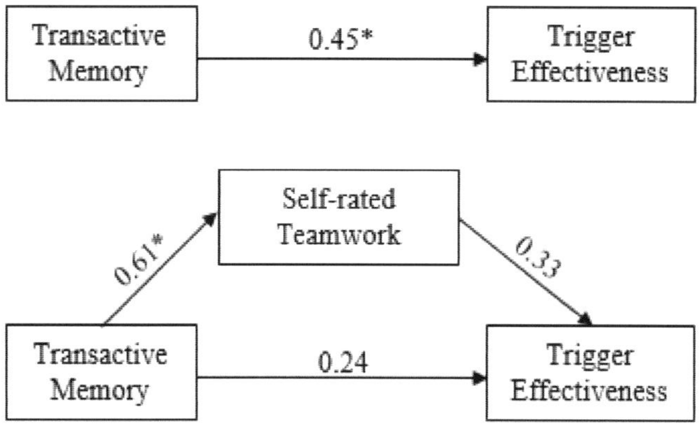

Figure 2. Testing self-rated teamwork as a mediator between transactive memory and trigger effectiveness.

The correlation between transactive memory and delay loss was extremely small (r = -.004). Nevertheless, mediation analysis indicated an indirect effect of transactive memory on delay loss. (z = - 2.06). The effect of transactive memory on delay loss was indirect and operated through observer-rated teamwork. This is explained by the existence of a suppressor variable. A variable with a positive relationship with the predictor and a negative relationship with the criterion can suppress the observed relationship between the independent and dependent variable (Schwab, 2005, p. 57). This is the pattern observed in the indirect relationship between transactive memory and delay loss. Transactive memory was positively related to observer ratings of teamwork and those ratings were negatively related to delay loss. Note that delay loss is an indicator of poor team performance. This relationship is shown graphically in Figure 3.

The overall pattern of mediation results provides mixed support for hypothesis 4. Two of the four analyses suggested that teamwork mediated the relationship between transactive memory and team performance. In both cases, teamwork was operationalized via observer ratings of teamwork. Mediation was not observed when self-ratings were used to operationalize teamwork.

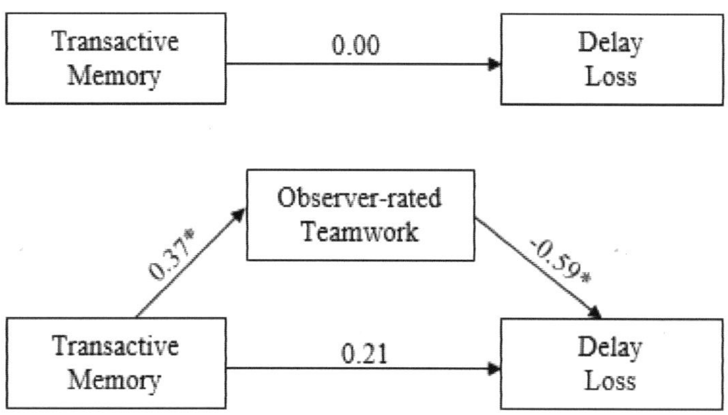

Figure 3. Testing observer-rated teamwork as a mediator between transactive memory and delay loss.

Routine and Adaptive Performance

Delay loss was conceptualized as a measure of routine performance while trigger effectiveness was conceptualized as a measure of adaption to non-routine conditions. Results indicated that the two performance measures were uncorrelated (r = -.01). Predictors such as transactive memory and teamwork ratings showed a more consistent pattern of relationships with trigger effectiveness than with delay loss. As previously indicated, trigger effectiveness was related to transactive memory (r = .45), observer-rated teamwork (r = .66), and self-rated teamwork (r = .48). Delay loss was related only to observer-rated teamwork (r = -.51).

Discussion

Results indicate partial support for hypothesis 1. Three of the four relations between measures of teamwork and measures of team performance were significant and moderate or strong in magnitude. Positive relationships between teamwork and team performance are consistent with theory (Marks et al., 2001; Salas et al., 2009; Salas et al., 2005) and findings from LePine et al. (2008).

Hypothesis 2 and 3 received partial support. Consistent with hypothesis 2b, transactive memory showed moderate to large correlations with both measures of teamwork. In partial support of hypothesis 3b, transactive memory was related to one of the two measures of team performance (trigger effectiveness). The positive relationships with teamwork and performance are consistent with meta-analytic findings (DeChurch & Mesmer-Magnus, 2010). Contrary to hypotheses 2a and 3a, collective efficacy was not related to any measure of teamwork or performance. This is surprising given that meta-analysis indicates the importance of collective efficacy (Stakovic et al, 2009). While we cannot offer a definitive explanation, the varied roles in the simulation may provide a clue. Some positions within the simulation (e.g. flight operations coordinator) are especially critical and represent core roles (Humphrey, Morgeson, & Mannor, 2009). Perhaps on this task where one person performs a coordinating role, the level of efficacy possessed by the person in this core role is more critical than the overall degree of collective efficacy. While this potential explanation is speculative, it may be worth exploring.

Partial support was found for hypothesis 4 that teamwork processes mediate the relationship between emergent states and team performance. No support was found for the mediating role of collective efficacy. One of the two teamwork measures (observer-rated teamwork) mediated the relationship between transactive memory and measures of both routine and adaptive performance.

The two measures of teamwork were highly related (r = .58). In some cases they showed consistent patterns of relations with other variables. This includes significant relationships with transactive memory, and trigger effectiveness. But observer-rated teamwork was related to delay loss while self-rated teamwork was not. In addition, observer-rated teamwork mediated relationships between transactive memory and team performance while self-rated teamwork did not. It is unclear whether these differences represent differing perspectives of the two rating sources or whether they represent differences in the facets of teamwork rated by the two types of raters. Self-rated teamwork utilized scales designed to measure the Marks et al. (2001) teamwork model that reflects a broad range of teamwork behaviors including transition, action, and interpersonal processes. The observer-rated teamwork examined a narrower set of teamwork processes including problem solving, coordination, and information utilization.

The two measures of team performance, delay loss and trigger effectiveness, were not related. These measures were representative of two different aspects of team effectiveness. While delay loss tends to capture routine performance, trigger effectiveness captures adaptive performance. Our finding that the two performance measures are unrelated is consistent with theoretical positions that distinguish between routine and adaptive performance (e.g., Burke, Stagl, Salas, Pierce, & Kendall, 2006; Rosen et al., 2011) and are relatively consistent with previous findings that routine and adaptive performance are not highly related (LePine, 2005). The lack of a close relation between measures of routine and adaptive performance and the differing pattern of relations with predictors suggests the need for research that distinguishes between factors related to performance under routine and non-routine conditions. This is especially true for aviation research because aviation requires performance in both routine and non-routine situations.

References

Burke, C. S., Stagl, K.C., Salas, E., Pierce, L., & Kendall, D. (2006). Understanding team adaptation: A conceptual analysis and model. *Journal of Applied Psychology, 91*, 1189-1207.

DeChurch, L. A. & Mesmer-Magnus, J. R. (2010). The cognitive underpinnings of effective teamwork: A meta-analysis. *Journal of Applied Psychology, 95(1)*, 32-53

Hayes, A. F. (2013). *Introduction to mediation, moderation, and conditional process analysis: A regression-based approach.* New Yourk: Guilford.

Humphrey, S. E., Morgeson, F. P., & Mannor, M. J. (2009). Developing a theory of the strategic core of teams: A role composition model of team performance. *Journal of Applied Psychology, 94*(1), 48-61. doi:10.1037/a0012997

Lepine, J. A. (2003). Team adaptation and postchange performance: Effects of team composition in terms of members' cognitive ability and personality. *Journal of Applied Psychology, 88*(1), 27-39.

Lepine, J. A. (2005). Adaptation in response to unforeseen change: Effects of goal difficulty and team composition in terms of cognitive ability and goal orientation. *Journal of Applied Psychology, 90*(6), 1153-1167.

Lepine, J. A., Piccolo, R. E., Jackson, C. L., Mathieu, J.E., & Saul, J. R. (2008). A meta-analysis of teamwork processes: Tests of a multidimensional model and relationships with team effectiveness criteria. *Personnel Psychology, 61*, 273-307.

Lewis, K. (2003). Measuring transactive memory systems in the field: Scale development and validation. *Journal of Applied Psychology, 88*, 587-604. doi: 10.1037/0021-9010.88.4.587.

Littlepage, G. E., Hein, M. B., Moffett, R. I., Craig, P. A., & Georgiou, A. M. (2016). Team training for dynamic cross-functional teams in aviation: Behavioral, cognitive, and performance outcomes. *Human Factors, 58*(8), 1275-1288. doi:10.1177/0018720816665200

Marks, M. A., Mathieu, J. E., & Zaccaro, S. J. (2001). A temporally based framework and taxonomy of team processes. *Academy of Management Review, 26*, 356-376. doi: 10.2307/259182607.

Quinones, M. A. (1995). Pretraining context effects: Training assignment as feedback. *Journal of Applied Psychology, 80*, 226-238. doi: 10.1037/0021-9010.80.2.226

Randall, K. R., Resick, C. J., & DeChurch, L.A. (2011). Building team adaptive capacity: The role of sensegiving and team composition. *Journal of Applied Psychology, 96*, 525-540. doi: 10.1037a/0022622

Rosen, M. A., Bedwell, W., Wildman, J. L., Fritzsche, B. A., Salas, E., & Burke, C. S. (2001). Manaaging adaptive performance in teams: Guiding principles and behavioral markers for measurement. *Human Resource Management Review, 21*, 107-122.

Salas, E., Rosen, M. A., Burke, C. S., & Goodwin, G. F. (2009). The wisdom of collectives in organizations. In E.Salas, et al. (Eds.) *Team effectiveness in complex organizations* (pp. 39-79). New York: Routledge.

Salas, E., Sims, D. E., & Burke, C. S. (2005). Is there a 'Big Five' in Teamwork? *Small Group Research, 36*(5), 555-599. doi:10.1177/1046496405277134

Schwab, D. P. (2005). *Research methods for organizational studies.* Mahwah, NJ: Erlbaum.

Stachowski, A A., Kaplan, S. A., & Waller, M. J. (2009). The benefits of flexible team interaction during crises. *Journal of Applied Psychology, 94*(6), 1536-1544.

Stajkovic, A. D., Lee, D., & Nyberg, A. J. (2009). Collective efficacy, group potency, and group performance: Meta-analyses of their relationships, and test of a mediation model. *Journal of Applied Psychology, 94*(3), 814-828. doi:10.1037/a0015659

SELF-LEADERSHIP STRATEGIES & PERFORMANCE PERSPECTIVES WITHIN STUDENT AVIATION TEAMS

Christopher Ryan Bearden
Middle Tennessee State University
Murfreesboro, TN

This study uses a correlational-design to explore relationships between peer ratings of team member effectiveness, supervisor ratings of performance, and self-reported performance strategies associated with self-leadership. Team members that were perceived as effective by their peers were also favorably rated on job performance by their supervisors. Peer-ratings on possession of job-relevant knowledge, skills, and abilities increased with frequency of communication behaviors, as rated by supervisors. This finding replicates previous research that suggests talking leads to perceived expertise in teams. Finally, self-goal setting was found to be related to peer-rated teammate effectiveness, but not supervisor-rated performance.

Organization-level outcomes may be contingent upon individual-level performance strategies (Krokos, Baker, Alonso, & Day, 2009). As a normative theory, self-leadership strategies may be able to prescribe behaviors to individuals that would improve personal effectiveness at work (Andressen, Konradt, & Neck, 2011). Self-leadership entails both proactive behaviors and thought processes geared towards engineering productive and positive affective experiences. Bligh, Pearce, and Kohles (2006) suggest member-to-member interactions may be improved through individual cognitive-behavioral strategies associated with self-leadership, which may lead to overall enhanced team efficacy, trust, and commitment to the team. Still, supervisors perceive team behaviors differently than peers (Murphy & Cleveland, 1991). For example, talking may lead to perceived competence by team members (Littlepage, Schmidt, Whisler, & Frost, 1995); however, talking may not lead to increased job performance on individual taskwork or job duties. The aim of the present study is to explore the relationship between performance strategies, observer-rated individual performance, and perceived team member effectiveness in an aviation team work setting. The study will provide a comparison of the perspectives of work-role performance on various dimensions using a correlational-design.

Methodology

All participants ($N = 216$) were students enrolled in a southern university's Aerospace Seminar. Data was collected from participants enrolled between the Fall semester 2013 to Fall semester 2016. Participation in the lab portion of the class is required for graduation; however, participation in the research portion was voluntary. Institutional Review Board approval and informed consent were obtained before commencing data collection. Participants were assigned to teams of approximately 10 by the instructor of the aerospace seminar according to their major concentration within the aerospace program. Data from a total of 33 teams are included.

Each team completes a minimum of three 3-hour iterations in the lab during the course of the academic semester. The lab portion of the seminar incorporates multiple software components and technologies to simulate a regional flight dispatch center, the Flight Operations Center – Unified Simulation (FOCUS; see Littlepage, Hein, Moffett, Craig, & Georgiou, 2016). Dispatching flights within the lab requires coordination and information sharing from every student position. The positions held by students include: flight operations coordinator (FOC), weather and forecasting (WX), crew scheduling (CS), flight operations data - scheduling (FOD1), flight operations data - planning (FOD2), and maintenance (MX). Data from other student positions were not included in this study, namely pilots and ramp tower coordinators.

Measures

Abbreviated Self-Leadership Questionnaire (ASLQ). The ASLQ is a nine item scale published by Houghton and his colleagues (2012). Self-leadership is assessed using three 3-item subscales, each subscale is associated with performance strategies subsumed under self-leadership: behavior awareness and volition, constructive cognition, and task motivation. Lab participants self-reported on the ASLQ using a 5-point Likert scale from 1 (*rarely*) to 5 (*usually*) during the final class meeting of the semester.

Behaviorally-Anchored Comprehensive Assessment of Team Member Effectiveness (CATME-B). Each lab participant rates his or her team members (i.e., peers) using the CATME-B (Ohland et al., 2012) on a scale from 1 (*below average*) to 5 (*excellent*). Team members did not rate themselves because self-ratings tend to be overly biased (Holzbach, 1978), especially for poor-performers (Murphy & Cleveland, 1991). Each team member is rated by his or her peers using three dimensions: *contributions to the team's work*, *teammate interaction*, and *possession of related knowledge, skills, and abilities* (KSAs).

Individual Performance Measures (IPMs). A series of scales developed within the FOCUS lab were used to assess individual performance. Scales differ by student position and were created through the process of task analysis, in which essential work-role behaviors were identified for each position. Each scale contains three items related to communication that remain the same across positions; however, all other items are unique to the taskwork required by each respective position. A different subject matter expert, acting as a supervisor, rated each position on how often a participant engaged in work-role behaviors during the third simulation on a Likert-scale from 1 (*never*) to 7 (*always*).

Results

See Table 1 for descriptive statistics. The IPMs demonstrated acceptable levels of internal consistency, FOC ($\alpha = .95$), WX ($\alpha = .93$), CS ($\alpha = .96$), FOD1 ($\alpha = .88$), FOD2 ($\alpha = .89$), and MX ($\alpha = .93$). Confirmatory factor analysis (CFA) on each IPM provided support for a correlated two-factor model: *taskwork* and *communication*. The ASLQ did not have acceptable levels of internal consistency, further the CFA failed to support a one-factor model, $\chi^2(27, n = 85) = 48.82$, $p = .006$, CFI = .88, TLI = .83, and RMSEA = .10. Therefore, the individual ASLQ items associated with specific strategies were used when calculating correlations. An index of within-team agreement (r_{wg}) was calculated on each CATME-B item. See Table 2 for average within-team agreement per item and per position. On average, teams agreed the most on their members' possession of KSAs ($r_{wg} = .75$), and across all items teams agreed the most on the effectiveness of the maintenance position ($r_{wg} = .79$). Average scores for each participant on the ASLQ, IPMs, and CATME-B and their respective subscales were used to calculate correlations.

Table 1.
Descriptive Statistics

Measure	*n*	Min	Max	*M*	*SD*	Range
Individual Performance	181	1.78	7.00	5.29	0.99	8 - 10[1]
Task-work	181	1.33	7.00	5.29	1.05	1 - 7[1]
Communication	181	1.00	7.00	5.30	1.04	1 - 3
Team Member Effectiveness	198	2.50	5.00	4.26	0.49	1 - 3
Contributions to the Team's Work	198	1.00	5.00	4.00	1.09	1
Teammate Interaction	198	2.50	5.00	4.37	0.47	1
Possession of KSAs	198	2.50	5.00	4.41	0.46	1

Note. [1] = Scales for task-work behaviors in the individual performance measures varied in size across position, ranging from 5 items to 7 items.

The self-goal setting item of the ASLQ was positively correlated with perceived team member effectiveness as rated by his or her peers on both contributing to teammate interaction ($n = 51$, $r = .30$, $p = .032$) and to the team's work ($n = 51$, $r = .28$, $p = .045$), but not with supervisor ratings of performance. No other self-leadership performance strategy measured in this study was correlated with any performance outcomes. See Table 3 for all other correlations.

Table 2.
Average Within-Team Agreement (r_{wg}) for Peer-Rated Teammate Effectiveness

	Position						
CATME-B Dimension	FOC	FOD1	FOD2	CS	WX	MX	Average/Item
Team's Work	0.78	0.78	0.74	0.71	0.70	0.76	0.74
Teammate Interaction	0.75	0.71	0.72	0.76	0.68	0.79	0.73
Possession of KSAs	0.78	0.72	0.73	0.73	0.75	0.83	0.75
Average/Position	0.77	0.73	0.73	0.73	0.71	0.79	

Note. $N = 33$ teams. CATME-B = Behaviorally-anchored comprehensive assessment of team member effectiveness; FOC = Flight operations coordinator; FOD1 = Flight operations - scheduling; FOD2 = Flight operations - weight & balance; CS = Crew scheduling; WX = Weather & Forecasting; MX = Maintenance control. Team-level r_{wg} Min = .00 Max = 1.00.

While individual performance was moderately correlated with team member effectiveness ($r = .31$, $p < .001$), contributions to teammate interaction was only correlated with the communication subscale of individual performance ($r = .19$, $p = .010$). Further, teammate interaction was strongly correlated with possession of KSAs ($r = .75, p < .001$), while possession of KSAs was moderately correlated with the IPM subscale of communication ($r = .29, p < .001$).

Table 3
Correlations Between Performance Dimensions

Measure	1	2	3	4	5	6	7
1. Individual Performance[1]	1						
2. Taskwork	.97*	1					
3. Communication	.90*	.77*	1				
4. Team Member Effectiveness[2]	.31*	.31*	.27*	1			
5. Contributions to the Team's Work	.26*	.28*	.16‡	.81*	1		
6. Teammate Interaction	.13	.09	.19*	.61*	.08‡	1	
7. Possession of KSAs	.25*	.22*	.29*	.68*	.17‡	.75*	1

Note. * = Correlation is significant at the .01 level (2-tailed); ‡ = Correlation is significant at the .05 level (2-tailed). [1] = Ratings provided by lab researchers acting as job supervisors. [2] = Ratings provided by peers on the same team as the participant.

Discussion

Self-goal setting is a performance strategy that was found to be related to teammate perceptions of effectiveness. Other performance strategies that comprised the *constructive cognition* and *task motivation* dimensions, including self-observation, visualizing successful performance, self-reward, self-talk, and evaluating beliefs and assumptions, were not related to peer-perceived effectiveness or supervisor-rated performance. The strategies measured may not generalize to the aviation industry or perhaps only the specific research setting. Another explanation is that the items did not adequately capture these strategies.

Members of the student teams generally shared an acceptable level of consensus on member effectiveness across the three dimensions: *contributing to the team's work, contributing to the team's interaction, and possession of KSAs*. Teams shared the strongest level of agreement on the effectiveness of the flight dispatcher (FOC) and the maintenance control position, and the weakest level of agreement on the effectiveness of the weather and forecasting position. In other words, the participants generally agreed on the level of KSAs, contributions to the team's work, and contributions to the team's interaction of their peers within the lab.

On average, as team members were rated more favorably by their supervisor on job-related tasks and communication behaviors, such as the sharing and solicitation of information and coordination, they were also perceived as more effective by their peers. Further, team members seen by their peers as contributing to the team's work, were seen by supervisors as engaging in job-related tasks and behaviors. Team members perceived by their peers as contributing to team interaction through feedback seeking and providing encouragement, were also favorably rated by a supervisor on frequency of communication behaviors, but not on performing job-specific duties. Interestingly, as team members engaged in communication behaviors more frequently, they were perceived by their teammates as possessing superior knowledge, skills, abilities.

Acknowledgements

Portions of this research are derived from my undergraduate thesis, supervised by Dr. Michael Hein. Financial support for this research was provided directly by the Middle Tennessee State University Undergraduate Research Center, University Honors College, and Aerospace Department. This research could not have been completed without the initial 2003 NASA space grant. Further, it would not be possible without the contribution of many researchers, including Dr. Paul Craig, Dr. Glenn Littlepage, Dr. Andrea Georgiou, Dr. Richard Moffett III, Mr. Gerald Hill, Megan Wertheimer, Jordan Gunn, and Evan Lester.

References

Andressen, P., Konradt, U., & Neck, C. (2011). The relation between self-leadership and transformational leadership: Competing models and the moderating role of virtuality. *Journal of Leadership & Organizational Studies*. doi:1548051811425047

Bligh, M., Pearce, C., & Kohles, J. (2006). The importance of self-and shared leadership in team based knowledge work: A meso-level model of leadership dynamics. *Journal of Managerial Psychology*, *21*(4), 296-318. doi: 10.1108/02683940610663105

Holzbach, R. L. (1978). Rater bias in performance ratings: Superior, self-, and peer ratings. *Journal of Applied Psychology*, *63*(5), 579. doi: 10.1037/0021-9010.63.5.579

Krokos, K. J., Baker, D. P., Alonso, A., & Day, R. (2009). Assessing team processes in complex environments: challenges in transitioning research to practice. *Team effectiveness in complex organizations: cross-disciplinary perspectives and approach*, 383-408.

Littlepage, G. E., Hein, M. B., Moffett, R. G., Craig, P. A., & Georgiou, A. M. (2016). Team training for dynamic cross-functional teams in aviation behavioral, cognitive, and performance outcomes. *Human Factors: The Journal of the Human Factors and Ergonomics Society*, *58*(8), 1275-1288. doi: 10.1177/0018720816665200

Littlepage, G. E., Schmidt, G. W., Whisler, E. W., & Frost, A. G. (1995). An input-process-output analysis of influence and performance in problem-solving groups. *Journal of personality and social psychology*, *69*(5), 877. doi: 10.1037/0022-3514.69.5.877

Murphy, K. R., & Cleveland, J. N. (1991). *Performance appraisal: An organizational perspective*. Boston, MA: Allyn & Bacon.

STUDENT PERCEPTIONS ON THE USEFULNESS OF SIMULATION-BASED TRAINING

Andrea M. Georgiou, Megan E. Wertheimer, & Megan E. Snodgrass
Middle Tennessee State University
Murfreesboro, Tennessee

The NASA Flight Operations Center Unified Simulation (FOCUS) lab is a high-fidelity simulation of an airline operations center. Its purpose is to train senior aerospace students to collaborate and communicate effectively with team members in a highly interdependent environment that mirrors the airline industry. Data was collected from the participants on their perceptions and the lessons learned from running the lab. These results were analyzed across eleven semesters over the last five years. Specifically, the quantitative data captured student perceptions about whether the lab was helpful in preparing them for their future job demands. The qualitative questions assessed their most important lessons learned, the problems they encountered, and their recommended changes.While there were some variances in student perceptions, teamwork and communication were repeatedly cited as being the most crucial variables to their success in running the virtual airline.

Airline operations are complex and demand multi-level coordination and communication among multiple teams to ensure safety and efficiency (Zaccaro, Marks, & DeChurch, 2012). Even as some issues are outside one's immediate control, such as hazardous weather and in-flight equipment failures, people can control how they react and take action to resolve issues. Integral to safe operations, teamwork training is woven into the airline industry for pilots, dispatchers, flight attendants, and many other entities. The NASA Flight Operations Center Unified Simulation (FOCUS) lab provides the platform for undergraduate aerospace students to improve and refine their non-technical teamwork, aeronautical decision-making, communication, and situational awarness skills. With 5 years of collected data, we felt it was appropriate to assess how the simulation lab has been helpful in improving their teamwork KSA's (knowledge, skills, and abilities) and review participant suggestions for revising or upating the simulation design.

Simulation-based training (SBT) is an excellent way to allow individuals to practice their technical and non-technical skills in a nonconsequential environment (Alinier, Hunt, Gordon, & Harwood, 2006; Beaubien & Baker, 2004; Lazzara et al., 2010; Shapiro et al., 2008). While the training efficacy of the NASA FOCUS lab has been confirmed (Littlepage, Hein, Moffett, Craig, & Georgiou, 2016), the perceptions of the participants were not formally analyzed prior to this study. As part of quality control going forward with the simulation training, it was important to analyze participant feedback after completion of the lab. As participant reactions to training can have implications for learning and transfer of training, evaluating how they felt about the simulation experiences and lessons learned is a vital educational component of the training (Morgan & Casper, 2000). According to the FAA (2005), collecting participant feedback after training has proven helpful to determine areas that can be strengthened. With the rapidly evolving technological and regulatory changes in the aviation industry, it is important to continuously monitor the realism of the simulation design, scenarios, and debriefing procedures.

Method

Participants

572 senior-level aerospace students participated in this research while enrolled in their capstone course. These students came from different aerospace majors including professional pilot, flight dispatch,

maintenance management, aerospace administration, aerospace technology, and unmanned aircraft systems. They worked together in teams comprised of approximately ten students. Each student was assigned to a position in the flight operations center simulation (described below). These positions are similar to those typically found in airline operations.

Simulation Lab

The FOCUS Lab is a high-fidelity simulation of a true flight operations center. Upon entering the lab, students are onboarded to a simulated airline, Universal E-lines, and trained in their respective positions before participating in a simulation. Positions include the Flight Operations Coordinator, Flight Operations Data, Flight Operations Scheduling, Maintenance Planning and Control, Crew Scheduling, and Weather and Forecasting. Ramp Tower Coordinator is in an adjoining room. Pseudo Pilot is in a separate, nearby location and the CRJ Pilot Crew is off-site flying a simulator connected to the lab's software. During the simulations, teams work together to release flights and solve problems as they arise during their shift. They participate in three simulations throughout the duration of the semester and review their performance in an After Action Review (AAR) following each simulation. See Littlepage, Hein, Moffett, Craig, & Georgiou, 2016, for an in-depth description of the lab.

Procedure

Data were collected over the last five years across eleven semesters with three to six teams participating in the lab each semester. After being onboarded to Universal E-lines, students participate in three simulations that act as their "work shifts" lasting approximately two and a half hours. During the simulations, participants completed their position's job duties while coordinating with other team members to solve various problematic scenarios that arise. The overall goal is to release flights safely and efficiently. A week after each simulation, participants engage in an AAR (After Action Review) to discuss their performance in the lab including what went well, what did not go well, and what behaviors led to various outcomes. Following the third simulation and associated AAR, all students completed an evaluation of the lab wherein they were asked quantitative and qualitative questions regarding what they learned, problems they encountered, and what they would change about their experience.

Two researchers separately content coded the qualitative comments. The first rater content coded the comments and developed the overarching categories for each qualitative question. Then, these overarching categories were given to the second rater and the second rater content coded the comments according to those categories. Inter-rater reliability was assessed using Cohen's Kappa to adjust for chance agreement. Then, a third researcher assessed all of the comments for which coders disagreed and made an expert judgment as to the final codes for frequency calculations.

Measures

Although participants take many measures throughout the duration of their participation in the FOCUS Lab, the measure of interest for this study is the FOCUS Lab Evaluation. This measure consisted of five quantitative items and four qualitative questions. The five quantitative questions were rated on a scale from 1 (Strongly Disagree) to 6 (Strongly Agree) and were as follows: "The FOCUS Lab experience helped me learn how my aerospace specialization relates to other specializations," "The FOCUS Lab experience helped me understand the work of other specializations," "The FOCUS Lab experience helped me understand the need for good communication among specializations," "The FOCUS Lab experience helped me understand the need for coordination among specializations," and "The FOCUS Lab experience will help me with the job demands as I start my professional career." The qualitative questions were, "What is the most important thing you learned in the FOCUS Lab this semester," "What were some of the problems you encountered in the FOCUS Lab that prevented smooth operations,"

"What would you change about the FOCUS Lab and your experiences in the lab to help future students," and "Is there anything that should have been included in the previous classes that would have made you better prepared to work in the FOCUS Lab." As described above, the qualitative questions were coded for content and then recoded by a second coder to assess inter-rater reliability.

Results

Inter-rater agreement was .78 and Cohen's Kappa was .75. The average rating of each of the five quantitative items assessing the understanding of specialization relationships, the work of specializations, the need for communication, the need for coordination, and the perception that the lab prepared them for job demands were all relatively high ($M = 5.16$, $M = 5.18$, $M = 5.47$, $M = 5.45$, $M = 4.96$, respectively). See Table 1 for the breakdown of these average ratings across semesters. Overall, ratings were stable across time.

Table 1.
Average Ratings of Quantitative Items across Semesters.

Semester	How Specializations Relate	Understand Specializations	Communication	Coordination	Job Demands
Fall 2011	5.28	5.38	5.69	5.67	5.11
Spring 2012	5.04	5.02	5.36	5.33	4.78
Fall 2012	5.16	5.26	5.53	5.42	5.32
Spring 2013	5.21	5.30	5.51	5.49	5.10
Fall 2013	5.14	5.16	5.55	5.43	4.84
Spring 2014	4.96	5.12	5.24	5.23	4.96
Fall 2014	5.11	5.11	5.44	5.39	4.94
Spring 2015	5.34	5.28	5.45	5.45	5.15
Fall 2015	5.13	4.82	5.50	5.53	4.69
Spring 2016	5.37	5.46	5.54	5.63	4.89
Fall 2016	4.58	4.58	5.13	5.08	4.58

In order from highest to lowest frequencies, the categories derived for each question and examples of qualitative comments are described in Table 2. The frequency of responses in each category for each question are in Table 3. Results indicate that for question one, the two most frequently listed responses for the lessons learned were in relation to communication/coordination ($N = 189$) and teamwork ($N = 129$). For question two, encountering problems, many students indicated that miscommunication was an issue ($N = 124$); this result clearly mirrors the results in question one with the emphasis on communication. Other problems encountered during the simulation included the scenarios ($N = 92$) and lack of knowledge or deficit in training ($N = 80$). The most frequent comment for question three, recommended changes for the lab, was a request for more training ($N = 112$). Finally, in question four, which asks about whether they would include previous classes before the lab, most individuals indicated that no additional classes were needed ($N = 160$).

Table 2.
Comment Coding Categories and Example Comments for Each Qualitative Question.

Question 1: Most Important Lesson Learned *Categories 1-7*	Example Comment
1. Communication/Coordination	Communication is essential to a positive outcome
2. Teamwork	How to better my teamwork skills....
3. Airline Functions	I learned valuable information about flight operations...
4. Other	The operation system
5. Knowledge of Team Member Roles	The understanding of the work in other job areas
6. Staying Calm/Positive Atittude	Stay calm, trust your FOC, talk to someone when you need help
7. Attitude to Detail/Thinking Ahead	You have to pay close attention to every detail

Question 2: Problems Encountered *Categories 1-8*	Example Comment
1. Miscommunication / Lack of Communication	Lack of good communication. Some information was never received...
2. Scenarios / Workload	Weather delays and closures and emergencies during flight
3. Knowledge/Deficit Training	Lots of inexperience
4. Technical Difficulties	Glitches in the system, technology difficulties
5. Other	Poor planning from FOC
6. Attitudes / Stress	People becoming stressed and losing situational awareness
7. Lack of Resources/ Absences	Missing team members, people not arriving early
8. Situational Awareness/ Anticipating Problems	Not everyone was ahead of the SIM

Question 3: Changes That Could Improve the Lab *Categories 1-9*	Example Comment
1. Training	Maybe allow extra time to learn how each position works
2. Nothing	I would change nothing
3. Time in the Lab	More labs if time permitted
4. Resources	Warning lights when approaching deadlines
5. Other labs	Get students to interact with each other between labs
6. Position Specific	I would have the FOC and FOD sit beside one another
7. Communication	Standard way of communication will help
8. Pilot/Ramp More Involved	Allow pilots to preview another team's sim session
9. Technical	More reliable communication devices

Table 2. Continued
Comment Coding Categories and Example Comments for Each Qualitative Question.

Question 4: Anything That Should Have Been Provided in Previous Classes to Prepare for the Simulations *Categories 1-5*	Example Comment
1. No	No. Classes prepared me pretty well
2. Other	Some time to get to know everyone in the group
3. More Training	More training time and a longer intro sim
4. More Classes	Maybe a communication class…
5. Learning About Other Positions	A overview of each position

The frequencies of these comments were also analyzed across time, indicating that there were not substantial changes across semesters. Students consistently valued communication, coordination, and teamwork as important lessons and consistently reported miscommunication as a major problem. They also consistently highlighted the contribution of training in the lab and reported that no additional classes are needed for preparation. Although these are the most frequently occuring comments, the particpants made a variety of other significant comments that underscore other learning experiences including the value of staying calm, the necessity for adequate resources, and situational awareness.

Table 3.
Frequency of Comments for Each Qualitative Question.

Content Category	Q1Freq	Q2Freq	Q3Freq	Q4Freq
Category1	189	124	112	160
Category2	129	92	51	80
Category3	35	80	50	41
Category4	30	62	40	28
Category5	21	36	36	27
Category6	18	28	28	
Category7	15	24	27	
Category8		14	24	
Category9			23	

Conclusion

Overall, this research highlights students' perceptions of the lab's value in teaching them how to communicate, coordinate, and work as a team. In their future careers, they will need to break out of their educational silos to effectively work as a team and develop creative solutions to abnormal problems. Participants clearly see the value in the lab and its ability to prepare them for the workplace. Based on their feedback, the most important lessons learned were the criticality of teamwork, communication, and coordination. Further, the most frequent change request was for more training. In direct response to this qualitative feedback, job aids, Captivate training, and PowerPoint training modules were developed for individual positions. A downstream consequences training was also developed for students to better understand the larger impact of decisions made in response to an immediate problem. Overall, based on their quantitative and qualitative feedback, participants seem to value the lab along with its immediate educational benefits and its contribution toward students' future careers.

Acknowledgements

This research was supported in part by a 2008 contract (NNX09AAU52G) from NASA awarded to the Middle Tennessee State University Center for Research on Aviation Training. The success of the NASA FOCUS lab would not be possible without the contribution of many researchers including Dr. Paul Craig, Dr. Glenn Littlepage, Dr. Michael Hein, Dr. Richard Moffett III, Mr. Gerald Hill, and Evan Lester. Correspondence concerning this article should be emailed to Dr. Andrea M. Georgiou at andrea.georgiou@mtsu.edu.

References

Alinier, G., Hunt, B., Gordon, R. & Hardwood, C. (2006). Effectiveness of intermediate-fidelity simulation training technology in undergraduate nursing education. *Journal of Advanced Nursing 54*(3), 359–369.

Beaubien, J. M., & Baker, D. P. (2004). The use of simulation for training teamwork skills in health care: How low can you go? *Quality and Safety in Health Care, 13*(1), i51-i56, doi: 10.1136/qshc.2004.009845.

Federal Aviation Administration [FAA]. (2005). Advisory circular 121-32A Dispatch resource management. Retrieved from https://www.faa.gov/regulations_policies/advisory_circulars/index.cfm/go/document.information/documentID/22513

Lazzara et al. (2010). Team Medss: A tool for designing medical simulation scenarios. *Ergonomics in Design: The Quarterly of Human Factors Applications, 18*(11), 11-17. doi:10.1518/106480410X12658371678435.

Littlepage, G. E., Hein, M.B., Moffett III, R.G., Craig, P.A., & Georgiou, A.M. (2016). Team training for dynamic cross-functional teams in aviation: Behavioral, cognitive, and performance outcomes. *Human Factors, 58*(8), 1275-1288. doi: 10.1177/0018720816665200.

Morgan, R. B., & Casper, W. (2000). Examining the factor structure of participant reactions to training: A multidimensional approach. *Human Resource Development Quarterly, 11*, 301-317.

Shapiro, M.J., Gardner, R., Godwin, S.A., Jay, G.D., Lindquist, D.G., Salisbury, M.L., & Salas, E. (2008). Defining team performance for simulation-based training: Methodology, metrics, and opportunities for emergency medicine. *Academic Emergency Medicine, 15*, 1088-1097, doi: 10.1111/j.1553-2712.2008.00251.x.

Zaccaro, S. J., Marks, M. A., & DeChurch, L. A. (2012). *Multiteam systems: An organizational form for dynamic and complex environments.* Routledge: New York.

DESIGN CONSIDERATIONS FOR ATTITUDE STATE AWARENESS AND PREVENTION OF ENTRY INTO UNUSUAL ATTITUDES

Kyle K.E. Ellis, Lawrence J. (Lance) Prinzel III, Jarvis J. (Trey) Arthur, Stephanie N. Nicholas
NASA Langley Research Center
Hampton, Virginia

Captain Daniel Kiggins[1], Harry Verstynen[2], Captain Clay Hubbs[1], Captain James Wilkerson[3]
[1] National Institute of Aerospace, Hampton, Virginia
[2] Unisys Corp, Hampton, Virginia
[3] Boeing, Seattle, Washington

Loss of control – inflight (LOC-I) has historically represented the largest category of commercial aviation fatal accidents. A review of the worldwide transport airplane accidents (2001-2010) evinced that loss of attitude or energy state awareness was responsible for a large majority of the LOC-I events. A Commercial Aviation Safety Team (CAST) study of 18 worldwide loss-of-control accidents and incidents determined that flight crew loss of attitude awareness or energy state awareness due to lack of external visual reference cues was a significant causal factor in 17 of the 18 reviewed flights. CAST recommended that "Virtual Day-Visual Meteorological Condition" (Virtual Day-VMC) displays be developed to provide the visual cues necessary to prevent loss-of-control resulting from flight crew spatial disorientation and loss of energy state awareness. Synthetic vision or equivalent systems (SVS) were identified for a design "safety enhancement" (SE-200). Part of this SE involves the conduct of research for developing minimum aviation system performance standards (MASPS) for these flight deck display technologies to aid flight crew attitude and energy state awareness similar to that of a virtual day-VMC-like environment. This paper will describe a novel experimental approach to evaluating a flight crew's ability to maintain attitude awareness and to prevent entry into unusual attitudes across several SVS optical flow design considerations. Flight crews were subjected to compound-event scenarios designed to elicit channelized attention and startle/surprise within the crew. These high-fidelity scenarios, designed from real-world events, enable evaluation of the efficacy of SVS at improving flight crew attitude awareness to reduce the occurrence of LOC-I incidents in commercial flight operations.

Recent data indicate that Loss-Of-Control In-Flight (LOC-I) accidents are the leading cause of commercial aviation accidents and incidents today (Boeing, 2016). Recent analysis by the Commercial Aviation Safety Team (CAST, 2014a) showed that LOC-I is primarily comprised of two causal factors: Spatial Disorientation (SD) and Loss-of-Energy State Awareness (LESA). SD is defined as an erroneous perception of aircraft attitude that can lead directly to a LOC-I. LESA is typically characterized by a failure to monitor or understand energy state indications (e.g., airspeed, altitude, vertical speed, commanded thrust) and a resultant failure to accurately forecast the ability to maintain safe flight. The leading consequence of LESA is aircraft stall.

To address the safety concerns surrounding LOC-I, CAST formulated a Joint Safety Analysis Team (JSAT) to study 18 recent LOC-I events. The JSAT study determined that a lack of external visual references (i.e., darkness, instrument meteorological conditions, or both) was associated with flight crew loss of attitude awareness or energy state awareness in 17 of these events (see Figure 1). A Joint Safety Implementation Team (JSIT) was formed to address the safety concerns identified in the JSAT study (CAST, 2014b). CAST recommended that, to provide visual cues necessary to prevent LOC-I, manufacturers should develop and implement virtual day- visual meteorological condition (VMC) display systems, such as synthetic vision systems. In support, CAST requested the National Aeronautics and Space Administration (NASA) to conduct research and lead efforts to support definition of minimum aviation system performance standards (MASPS) for virtual day- VMC displays to accomplish the intended function of improving flight crew awareness of airplane attitude. CAST established Safety Enhancement 200 (SE-200) entitled, "Airplane State Awareness – Virtual Day-VMC Displays" to formalize this effort.

Airplane State Awareness – Virtual Day-VMC Displays

The purpose of SE-200 is to reduce the risk of LOC-I by having manufacturers develop and implement virtual day-VMC display systems (such as SVS) that will support flight crew attitude awareness similar to a day-VMC-like environment in applicable new transport category airplane programs. SE-200 includes a detailed implementation plan that defined specific research needs to support the design and implementation of these displays that will enable the necessary visual cues to prevent LOC-I due to flight crew SD/LESA and aid in detecting unusual attitude entry and performing recovery. In large transport aircraft, an unusual attitude is operationally defined as a nose-up pitch attitude greater than 25 degrees, a nose-down pitch attitude greater than 10 degrees, a bank angle greater than 45 degrees or flight within these parameters but with airspeeds inappropriate for the conditions.

Virtual day-VMC display standards are not currently in effect for this intended function and the NASA research will inform the development of MASPS under RTCA Special Committee (SC)-213, Enhanced Flight Vision Systems and Synthetic Vision Systems (EFVS/SVS).

Virtual Day-VMC Displays

Virtual day-VMC displays are intended to provide similar visual cues to the flight crew that are available when outside visibility is not restricted (i.e., often observed under VMC). Their intended function would be to improve continuous attitude, altitude, and terrain awareness, reducing the likelihood of unstable approach, inadvertent entry into an unusual attitude, spatial disorientation, and/or collision with terrain through a synthetic vision (SV) display. SV is a computer-generated image of the external scene topography from the perspective of the flight deck, derived from aircraft attitude, high-precision navigation solution, and database of terrain, obstacles, and relevant cultural features

Technologies for Airplane State Awareness

The SE-200 detailed implementation plan defined areas of research needs for design and implementation of virtual day-VMC displays to prevent loss-of-control accidents due to loss of attitude awareness and lack of external visual references. The NASA "Technologies for Airplane State Awareness" (TASA) project was created to address SE-200 and other safety enhancements. NASA research has been completed that evaluated design characteristics such as image minification, optical flow cues, and field-of-view (Nicholas, 2016). The present paper describes high-fidelity, large commercial transport simulation research that evaluated various types of SVS displays for their efficacy to improve attitude awareness and prevent unusual attitude (UA) conditions from developing during realistic flight operations scenarios.

Experimental Method

Research Pilots

Twelve current major commercial airline pilot crews participated in the research. The average experience was 22,000 hours. Pilots were required to have 737/A320 or larger aircraft type ratings from major domestic airlines, with preference given to those with glass cockpit experience.

Research Simulator

The research was conducted in the Research Flight Deck (Figure 1) at NASA Langley Research Center, which is a high-fidelity, 6 degrees-of-freedom motion-based large commercial aircraft simulator with full-mission capability and advanced glass, Boeing 787-like flight deck displays.

Figure 1. Research Flight Deck Simulator

Special Purpose Operations Training Scenarios

The research employed four special purpose operations training (SPOT) scenarios based on FAA training guidance (FAA Advisory Circular 120-35D). NASA and subject matter experts designed the four SPOT scenarios using a sequence of off-nominal events that create challenging flight and workload conditions that may ultimately lead to an unusual attitude without timely pilot intervention. The compound failures required pilots to address several issues, often unrelated, that saturated the pilot's/crew's attention. The SPOT scenarios stressed the crews' aircraft state awareness to evaluate the efficacy of the display system to maintain pilot attitude awareness and identify recognition of impending unusual aircraft attitude conditions.

The four SPOT scenarios were: (1) False-Glideslope with Radar Altimeter Fail; (2) Fuel Leak with Clear Air Turbulence; (3) Reduced Engine Performance/High-Alpha; and, (4) Missed Approach with Degraded Autopilot in the roll axis.

In addition to the SPOT scenarios, nearly identical distractor scenarios were created for each of the four SPOT scenarios but with the removal of one or several off-nominal events. These additional scenarios were challenging, requiring significant pilot interaction, but did not lead to an unusual attitude conditions.

Special Purpose Operations Training Experimental Method

Eight scenarios, four SPOTS and four distractor scenarios, were evaluated in an ordered sequence - the crews flew the SPOT scenario prior to the distractor scenario of similar type. Because the SPOT scenarios involved "black swan" events, they could only be presented once to successfully achieve the high level of task saturation and surprise required for the experiment (Taleb, 2007). The crews were assigned to one of four experimental blocks with each block given a different display condition for each scenario. The scenario order was fixed across all crews, randomized by display condition block in a between-subjects design. Scenarios lasted on average eight minutes.

The purpose of the test was to evaluate a flight crew's ability to maintain attitude awareness and prevent entry to unusual attitudes. Pre-experimental briefings provided instructions and training including FAA- (FAA, 2016) and Boeing- (Boeing, 2004) recommended UA recovery (UAR) techniques. Pilots were briefed about evaluations of the displays, not the off-nominal nature of the scenarios. This training is in addition to the training that the pilots have received with their respective airlines.

Display Concepts

The experimental display concept conditions are shown in Figure 2. The Baseline display emulated a Boeing 787-like primary flight display (PFD); this display does not include SV. Three virtual-day VMC (SVS) display concepts were used – one was representative of the MASPS, as defined by RTCA under DO-315A, for a synthetic vision display intended for terrain awareness (i.e., the so-called "SVS1- MASPS"). The other SV concept was representative of the Industry Standard virtual-day VMC (SVS) in operational use today (i.e., the so-called "SVS-2 - Industry"). The SVS3-Advanced display concept was the industry standard (SVS2) with an added innovative optical flow cue designed to aid situation awareness when the aircraft enters an unusual attitude. If no unusual attitude condition is present, the display is effectively the same as the industry standard type. All display concepts included roll arrow recovery guidance (Ewbank et al, 2016) and angle-of-attack indication (Cashman et al,

2000) (note: angle-of-attack indicator is standard on B-787 PFDs). The roll arrow guidance symbology is displayed when the aircraft attitude meets roll angle exceedance criteria (see Figure 3).

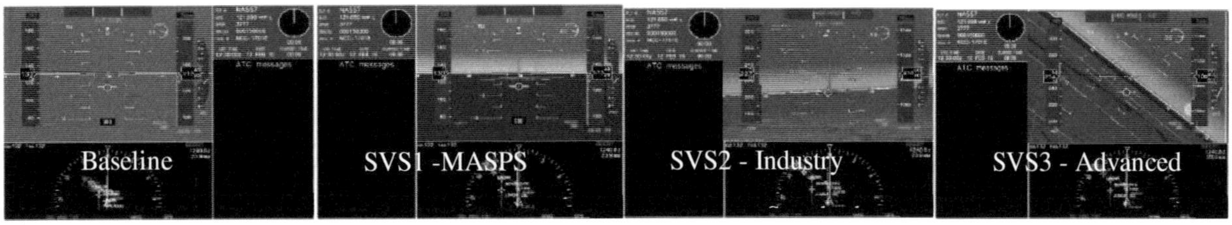

Figure 2. Experimental Display Conditions

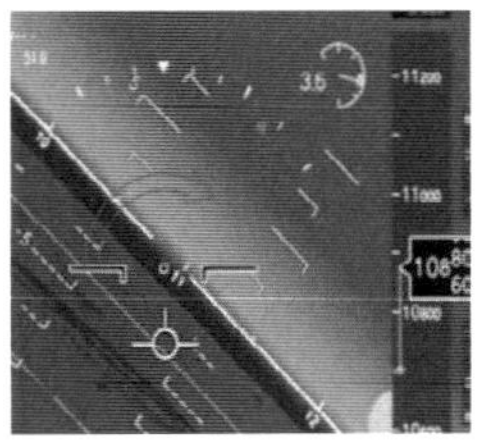

Figure 3. Roll Arrow Recovery Guidance and Angle-of-attack (Alpha) Symbologies

Experimental Results

The crew was informed of the initial flight condition and the display concept being flown. All flights were conducted with Memphis as the destination airport. The SPOT was orchestrated by pre-programmed non-normal events to induce the unusual attitude conditions. Once the recovery was completed, the trial ended and post-trial subjective scales were administered and pilot comments solicited. Post-scenario questionnaires were administered, including the NASA-Task Load Index (TLX) evaluation of workload, a three-question Situation Awareness Rating Technique evaluation of Situation Awareness (SA), and single score evaluation of crew-member workload.

Quantitative Results

Several dependent measures were assessed during specific time windows leading up to, during, and immediately after the unusual attitude events. These data revealed that SPOT-2 and SPOT-4 were the most effective in achieving pilot crew surprise and task saturation to properly evaluate the display conditions.

The SPOT-2 UA condition was induced by an autopilot disconnect (due to a fuel imbalance) followed by a near-simultaneous clear air turbulence event. Both events required pilot intervention to maintain attitude control. Time-to-first correct input distributions for SPOT-2 are shown below in Figure 4. Analysis show nearly significant results ($p < 0.05$) for time-to-first correct roll input across display condition, $F(3, 8) = 3.44$, $p = 0.072$. Results were not significant for time-to-first correct pitch input $F(3,8) = 2.15$, $p = 0.172$.

The SPOT-4 scenario involved a degradation in the roll-axis autopilot, occurring while the aircraft was turning following a missed approach vector from the tower. This resulted in pilots expecting the aircraft to turn based on the commanded heading setting on the autopilot, however, the aircraft would continue to roll beyond 45 degrees of bank without pilot intervention due to the un-annunciated degraded autopilot condition. Data was evaluated from the moment the autopilot was degraded in the roll-axis and the 15 seconds following that event. Time- to-first correct input distributions for SPOT-4 are shown below in Figure 5. No statistically significant results were observed across the four display conditions for time-to-first correct pitch input $F(3,8) = 2.36$, $p = 0.148$, or for the time-to-first correct roll input $F(3,8) = 1.48$, $p = 0.291$.

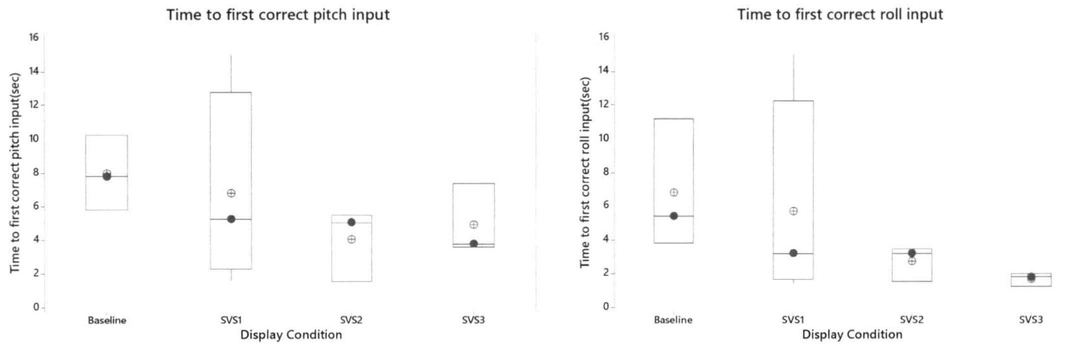

Figure 4. SPOT 2 Time-to-First Correct Pitch and Roll Input

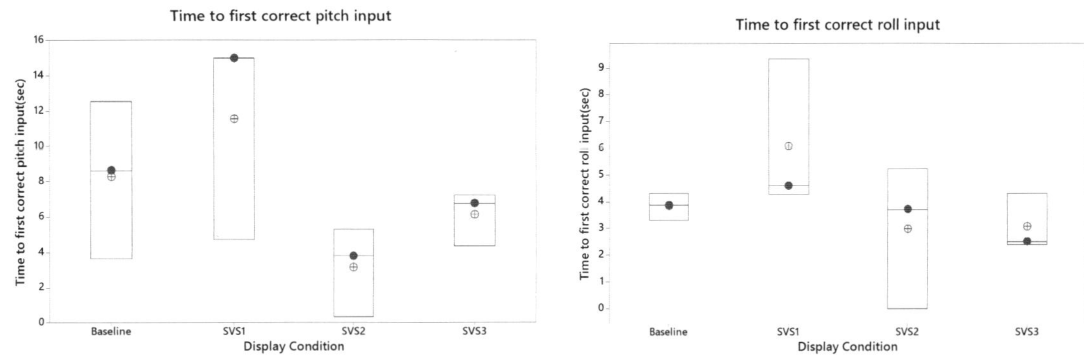

Figure 5. SPOT 4 Time-to-First Correct Pitch and Roll Input

Qualitative Results

<u>NASA-Task Load Index</u>. No significant differences were found across display concepts for NASA Task Load Index (NASA-TLX) for any of the presented SPOT scenarios.

<u>Situation Awareness Rating Technique</u>. No significant differences were found across display concepts for Situation Awareness Rating Technique (SART) for any of the presented SPOT scenarios.

<u>Paired Comparisons</u>. A mixed-factor ANOVA was conducted on the independent variables of display type (Baseline, SVS1, SVS2, SVS3) and pilot role (First Officer, Captain). The ANOVA revealed a significant main effect for display concept, $F(3,36) = 17.291$, $p < 0.001$ and display-role interaction, $F(3,36) = 3.15$, $p < 0.05$. The main effect of role was not significant, $F(1,12) = 0.143$, $p > 0.05$. Post-hoc simple effects analysis evinced that Baseline, SVS1, SVS2, and SVS3 were not significantly different. However, SVS2 and SVS3 were significantly different from Baseline and SVS1.

<u>Main Effect for Display Concept</u>. The results suggest that both the Captain and First Officer rated the advanced synthetic vision display concepts (SVS2 and SVS3) higher for attitude awareness than either the baseline or lower fidelity SVS display concept. However, the addition of "optical flow" (SVS3) did not enhance the SA ratings compared to the industry standard SV concept (SVS2).

<u>Interaction Effect for Display Concept x Role</u>. The significant interaction revealed that the First Officer provided significantly higher paired comparison ratings for the SVS2 and SVS3 concepts. Although the Captain rated the SVS2 and SVS3 significantly higher than the baseline or SVS1 concepts, the First Officer provided the most significant contrast in ratings as they tended to provide lower ratings than Captains for the baseline and SVS1 concepts but much higher ratings for SVS2 and SVS3. The results suggest that the advanced features of the SVS2 and SVS3 were more beneficial for SA for the monitoring pilot (First Officer). Although both pilots rated the SVS2

and SVS3 displays higher in terms of SA, the Captains did not statistically rate the SVS1 as higher than the SVS2 or SVS3, but did rate all three SVS concepts higher than baseline. The First Officers however, provided that the baseline and SVS1 concepts were statistically equivalent for SA but there was a substantial SA increase for SVS2 (highest) and SVS3 and the differential pattern of results accounts for the significant display-role interaction.

Conclusions

The pilots that participated in the research had substantial experience and training in recognizing and recovery from unusual attitudes. The pilot population was conservatively selected because it was hypothesized that, if significant differences were found across displays, it would be even more significant with less experienced commercial pilots (i.e., the identified risk group in the CAST report). The limited number of trials presented to each of the pilot crews does not allow for any statistical evidence to generalize to the commercial pilot population. However, these data do provide indications that are useful in evaluating pilot response in extremely rare circumstances such as presented in the SPOT scenarios.

The performance data suggest there may exist an operational improvement in UAR, as indicated by the pilot's time-to-first correct roll input when using the Industry Standard (SVS2) and Advanced (SVS3) SVS display concepts. These results show that pilots generally had faster correct control inputs while using SVS concepts that included higher definition details such as terrain texturing, shading, and terrain features. Additionally, pilot comments indicated that the inclusion of the roll arrow recovery guidance symbology and angle-of-attack displays helped the highly-experienced pilots to recover more easily from unusual attitudes and reduced reliance on external visual cues. The roll arrow was included because it is part of the SVS MASPS standard and there is significant likelihood of it being standard on all primary flight displays in the future.

Pilot preference was substantially biased toward the use of SVS, with top preference for the Industry Standard SVS2 condition. Feedback indicated that the awareness enhancement provided by the optical flow cues of the Industry Standard and Advanced virtual day-VMC displays was substantial (compared to Baseline and MASPS). Research evaluating SVS for UA recognition and recovery using comparative, repetitive testing techniques have also been performed, indicating no performance differences or preferences (Prinzel, 2017). These data, however, suggest that commercially trained pilots use SVS for attitude awareness with either comparable or improved performance to that of the existing baseline displays available today during operational flight profiles.

References

Boeing. (2004, August 1). Airplane Upset Recovery. Industry Solutions for Large Swept-Wing Turbofan Airplanes Typically Seating More than 100 Passengers. Boeing.

Boeing (2016), Statistical Summary of Commercial Jet Airplane Accidents Worldwide Operations, 1959–2015; retrieved at http://www.boeing.com/resources/boeingdotcom/company/about_bca/pdf/statsum.pdf

Cashman, J.E., Kelly, B.D., and Nield, B.N., Operational Use Of Angle Of Attack. Boeing Aero Magazine, Aero No. 12, Oct 2000.

Commercial Aviation Safety Team (2014a). Airplane State Awareness Joint Safety Analysis Team Final Report. Washington, D.C.: CAST.

Commercial Aviation Safety Team (2014b). Airplane State Awareness Joint Safety Implementation Team Final Report. Washington, D.C.: CAST.

Ewbank, C. E., Mumaw, R. J., & Snow, M. P. (2016). Development of the enhanced bank angle warning. In Digital Avionics Systems Conference (DASC), 2016 IEEE/AIAA 35th. pp. 1-9.

Federal Aviation Administration (2015). Upset Prevention and Recovery Training. Advisory Circular 120-35D. Washington, D.C.: FAA.

Nicholas, S.N. (2016). Evaluation of Synthetic Vision Display Concepts for Improved Awareness in Unusual Attitude Recovery Scenarios. Department of Defense Human Factors and Ergonomics Meeting 70 TAG. Presented May 10, 2016.

Prinzel, L.J., Ellis, K.E., Arthur, J.J., Nicholas, S.N., Kiggins, D., (in press). Synthetic Vision System Commercial Aircraft Flight Deck Display Technologies for Unusual Attitude Recovery (paper #4060). 2017 meeting of the International Symposium on Aviation Psychology. Dayton, OH: Wright State.

Taleb, N. T., The Black Swan: The Impact of the Highly Improbable. (2007) Random House Publishing Group

SYNTHETIC VISION SYSTEM COMMERCIAL AIRCRAFT FLIGHT DECK DISPLAY TECHNOLOGIES FOR UNUSUAL ATTITUDE RECOVERY

Lawrence (Lance) J. Prinzel III, Kyle E. Ellis, Jarvis (Trey) J. Arthur, Stephanie N. Nicholas
NASA Langley Research Center
Hampton, Virginia

Captain Daniel Kiggins
National Institute of Aerospace
Hampton, Virginia

A Commercial Aviation Safety Team (CAST) study of 18 worldwide loss-of-control accidents and incidents determined that the lack of external visual references was associated with a flight crew's loss of attitude awareness or energy state awareness in 17 of these events. Therefore, CAST recommended development and implementation of virtual day-Visual Meteorological Condition (VMC) display systems, such as synthetic vision systems, which can promote flight crew attitude awareness similar to a day-VMC environment. This paper describes the results of a high-fidelity, large transport aircraft simulation experiment that evaluated virtual day-VMC displays and a "background attitude indicator" concept as an aid to pilots in recovery from unusual attitudes. Twelve commercial airline pilots performed multiple unusual attitude recoveries and both quantitative and qualitative dependent measures were collected. Experimental results and future research directions under this CAST initiative and the NASA "Technologies for Airplane State Awareness" research project are described.

Recent accident and incident data suggests that Spatial Disorientation (SD) and Loss-of-Energy State Awareness (LESA) for transport category aircraft are becoming an increasingly prevalent safety concern in all domestic and international operations (Bateman, 2010). SD is defined as an erroneous perception of aircraft attitude that can lead directly to a Loss-Of-Control (LOC) event and result in an accident or incident. LESA is typically characterized by a failure to monitor or understand energy state indications (e.g., airspeed, altitude, vertical speed, commanded thrust) and a resultant failure to accurately forecast the ability to maintain safe flight. The leading consequence of LESA is aircraft stall.

A CAST study of 18 loss-of-control accidents determined that a lack of external visual references (i.e., darkness, instrument meteorological conditions, or both) was associated with a flight crew's loss of attitude awareness or energy state awareness in 17 of these events. The Airplane State Awareness Joint Safety Analysis (JSAT) and Implementation Team (JSIT) reports (CAST, 2014a; CAST, 2014b) recommended that, to provide visual cues necessary to prevent LOC resulting from a flight crew's SD/LESA, manufacturers should develop and implement virtual day-VMC display systems, such as synthetic vision systems. In support of this implementation, CAST requested the National Aeronautics and Space Administration (NASA) to conduct research to support definition of minimum requirements for virtual day- VMC displays to accomplish the intended function of improving flight crew awareness of airplane attitude; see CAST Safety Enhancement 200 (SE-200) entitled, "Airplane State Awareness – Virtual Day-VMC Displays".

Airplane State Awareness – Virtual Day-VMC Displays

A NASA project, entitled Technologies for Airplane State Awareness (TASA), has been developed which, in part, addresses the CAST request for research to support manufacturer design and implementation of virtual day-VMC displays that will enable the necessary visual cues to prevent SD/LESA and aid in detecting unusual attitude and performing recovery. In large transport aircraft, an unusual attitude is operationally defined as a nose-up pitch attitude greater than 25 degs, a nose-down pitch attitude greater than 10 degs, a bank angle greater than 45 degs or flight within these parameters but with airspeeds inappropriate for the conditions.

Virtual Day-VMC Displays

Virtual day-VMC displays are intended to provide similar visual cues to the flight crew that are available when outside visibility is unrestricted (i.e., observed under VMC). Their intended function is improve continuous attitude, altitude, and terrain awareness, reducing the likelihood of unstable approach, inadvertent entry into an unusual attitude, spatial disorientation, and/or collision with terrain through the use of a synthetic vision display; that is, a computer-generated image of the external scene topography from the perspective of the flight deck, derived from aircraft attitude, high-precision navigation solution, and database of terrain, obstacles and relevant cultural features. Virtual day-VMC display standards are not currently in effect for this intended function and the NASA research will inform the development of minimum aviation system performance standards under RTCA Special Committee (SC)-213, Enhanced Flight Vision Systems and Synthetic Vision Systems (EFVS/SVS).

Experimental Method

Technologies for Airplane State Awareness

SE-200 defined areas of research needed for design and implementation of virtual day-VMC displays to prevent loss-of-control accidents due to loss of attitude awareness and lack of external visual references. NASA research has been completed that evaluated virtual day-VMC display design characteristics, such as image minification, optical flow cues, and field-of-view for attitude awareness (Nicholas, 2016). The present paper describes high-fidelity, large commercial transport simulation research that evaluated various types of synthetic vision system displays and a symbology concept termed, "background attitude indicator", as they may promote aircraft attitude awareness as evident from pilot recognition of and in their ability to recovery from unusual attitudes.

Research Pilots

Twelve active major commercial airline pilots participated in the research. The average experience was 22,000 hours. All pilots had been trained on large transport aircraft unusual attitude recovery procedures.

Research Simulator

The research was conducted in the Research Flight Deck at NASA Langley Research Center, which is a high-fidelity, 6 degree-of-freedom motion-based large commercial aircraft simulator with full-mission capability and advanced glass, Boeing 787-like flight deck displays.

Unusual Attitude Recovery Scenarios

The research employed four unusual attitude (UA) initial conditions based on FAA training scenario guidance. The four UA scenarios were: (a) Nose-up 30 degrees, 90 degrees right roll; (b) Nose-up 30 degrees, 90 degrees left roll; (c) Nose-down 30 degrees, 60 degrees right roll; and (d) Nose-down 30 degrees, 60 degrees left roll. The initial starting altitude was 22,000 ft. mean sea level and each trial lasted an average of 30 seconds.

Unusual Attitude Recovery Trial Method

Twenty trials were conducted such that all display concepts (five) were evaluated in each of the four UA scenarios. Prior to data collection, pilots were provided detailed briefings on Boeing and FAA-recommended UA recovery techniques with subsequent discussion on each pilot's airline specific training; it was observed that there were not any substantive differences across pilots (US air carriers) in terms of UA recovery technique training. Training in the simulator followed with specific instruction and practice and with the display concepts, performing UA recoveries until the pilots demonstrated an asymptotic level of performance.

Each data collection trial began with the pilot being briefed on the display concept. When ready, the displays were blanked and real motion cueing was used while flying the simulator to the UA initial condition to keep the pilots unaware of the actual attitude. (Post-experimental briefings validated that the method was successful and all pilots confirmed they had no awareness of attitude prior to start of each trial.) Once the simulator reached the UA condition, a tone was sounded followed by the front panel displays unblanking and pilots were instructed to move from hands-in-lap, open their eyes, recognize the UA condition, and perform a successful UA recovery. Pre-experimental briefings provided instructions and training including FAA- (FAA, 2016) and Boeing- (Boeing, 2004) recommended UA recovery techniques (all pilots had been trained by their respective airlines), followed by in-simulator practice. Once the pilot judged the aircraft had been recovered (criteria being wings-level attitude; zero vertical speed), the trial ended and post-trial ratings and pilot comments were solicited.

Display Concepts

The 5 experimental display concept conditions are shown in Figure 1 below. The first concept was a baseline display emulating a Boeing 787 primary flight display; this display does not include SV. Two virtual-day VMC (SV) display concepts were used – one was representative of the minimum aviation system performance standards, as defined by RTCA under DO-315A, for a synthetic vision display intended for *terrain* awareness (i.e., the so-called "MASPS SV"). The other SV concept was representative of virtual-day VMC (SV) in operational use today (i.e., the so-called "Industry Standard SV"). The fourth display concept - Advanced virtual-day VMC (SV) display - added an innovative optical flow cue when the aircraft entered into an unusual attitude, that aided situation awareness in proper execution of recovery. The optical flow cue consisted of a series of yellow ball symbols that moved in the direction of the aircraft attitude (e.g., when nose-up and climbing, the cues would depict movement in direction of up in the primary flight display). Finally, the Industry Standard + BAI condition uses the Industry Standard SV but extended the presentation of the SV scene beyond the primary flight display (PFD) window across the entire display panel (see Bailey et al, 2013) using the Captain's PFD as the BAI reference point.

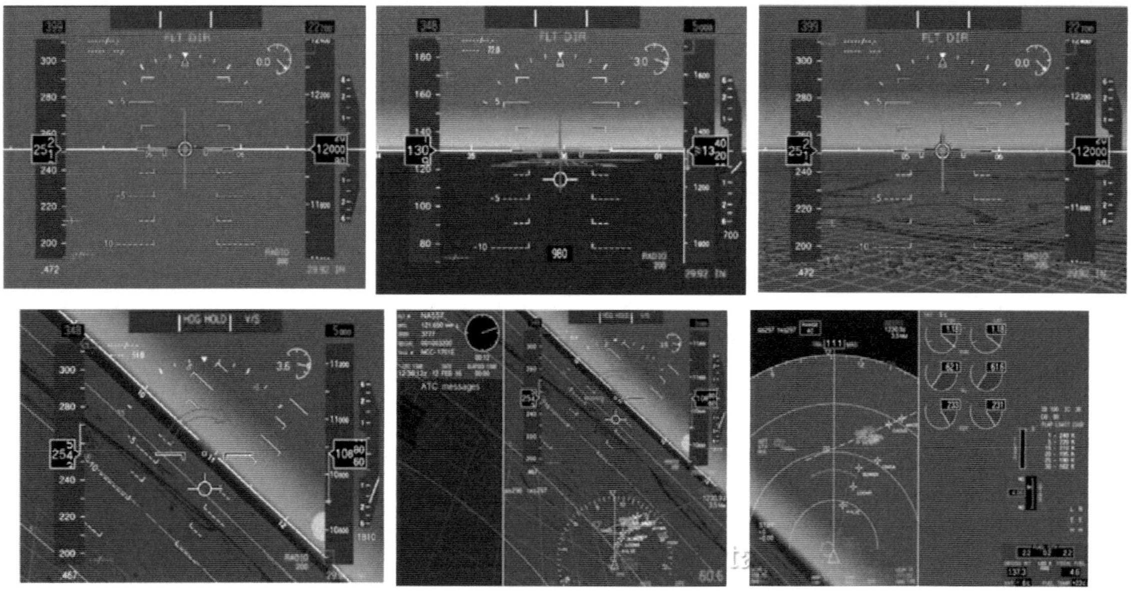

Figure 1. Experimental Display Conditions

Experimental Results

A number of dependent measures were collected and analyzed for attitude recognition and unusual attitude recovery. A UA recovery (UAR) score was also calculated based on whether the correct, incorrect, or neutral pitch, roll, and throttle input was made (using a score of +1, -1, or 0, respectively) for a total score that ranged from -3 (poor) to + 3 (excellent).

Quantitative Results

Scenarios. For scenario, the four UA scenarios were combined into either a pitch-up or pitch-down condition for analysis. The results showed significant main effect for time-to-first pitch input, $F(1, 23) = 37.599$, $p < 0.01$; time-to-first correct pitch input, $F(1, 23) = 9.130$, $p < 0.01$; time-to-first roll input, $F(1, 23) = 5.479$, $p < 0.05$; and time-to-first correct roll input, $F(1, 23) = 24.951$, $p < 0.01$. A significant main effect was found for UAR score for scenario, $F(1, 23) = 61.408$, $p < 0.01$. The 30 degree pitch-up UAR scenario condition was significantly poorer for dependent measures compared to the 30 degree pitch-down UAR scenario condition. No significant effect was found for number of control reversals, $F(1,23) = 0.04$, $p > 0.05$.

Displays. No significant differences were found for time-to-first pitch input, $F(4, 92) = 1.407$, $p > 0.05$; time-to-correct first pitch input, $F(4, 92) = 0.145$, $p > 0.05$; time-to-first roll input, $F(4, 92) = 2.131$, $p > 0.05$; time-to-first correct roll input, $F(4, 92) = 0.345$, $p > 0.05$; number of control reversals, $F(4, 92) = 1.100$, $p > 0.05$; and UAR score, $F(4, 92) = 0.063$, $p > 0.05$. The scenario *display interaction effects for all quantitative dependent measures were also not significant. Figure 2 presents boxplots of each of the quantitative dependent measures for the pitch-up (left side of figure) and pitch-down UAs (right side of figure) for each display concept. The boxes indicate the median value and $25^{th}/75^{th}$ percentiles with the whiskers extending to 1.5 times the height of the box or to the minimum or maximum values. The points beyond the whiskers are extreme values or outliers and are indicated by circles.

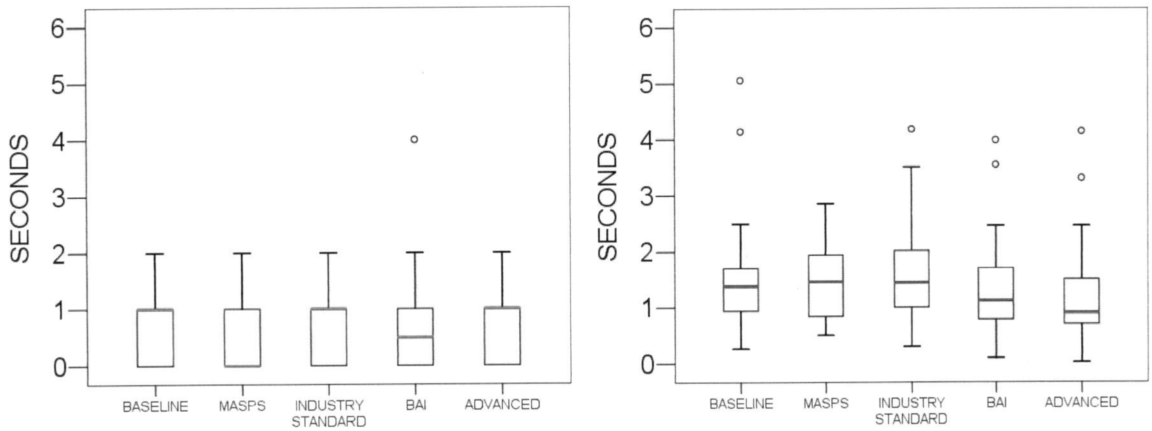

Time-to-First Pitch Input

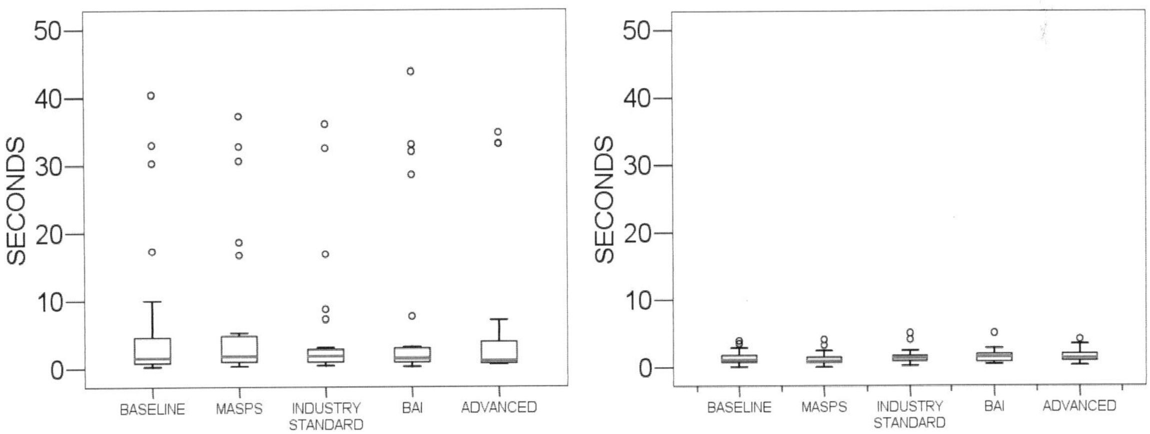

Time-to-First Correct Pitch Input

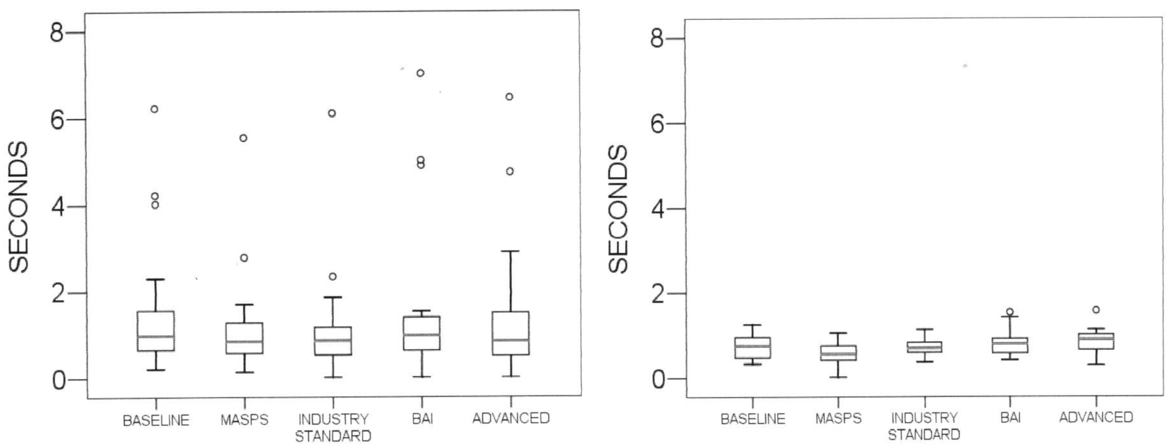

Time-to-First Roll Input

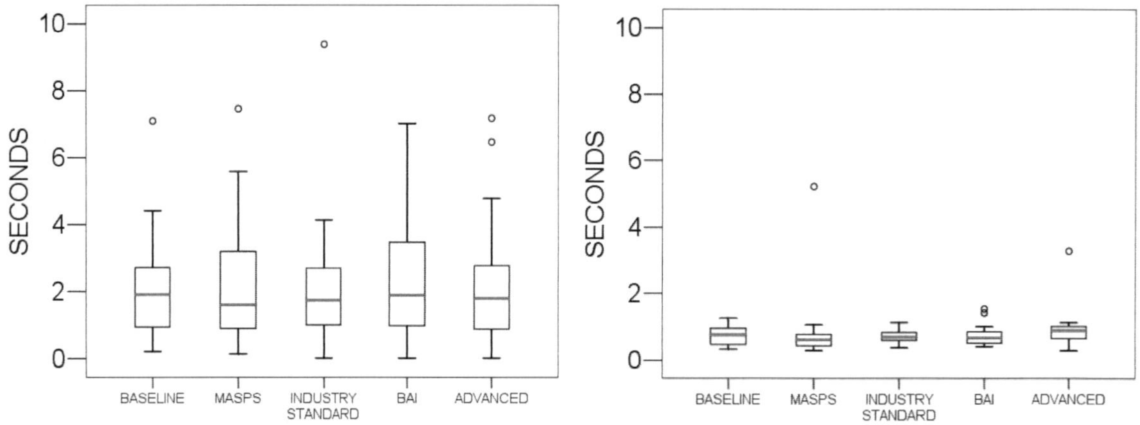

Time-to-First Correct Roll Input

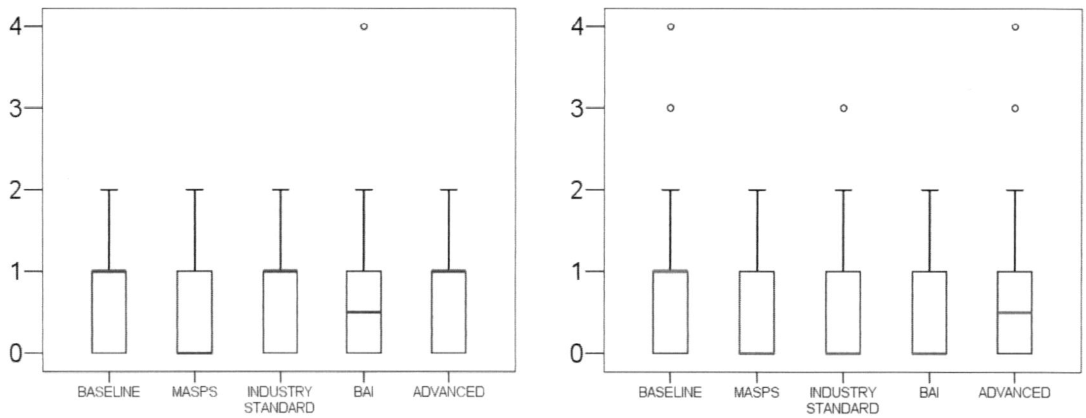

Number of Control Reversals

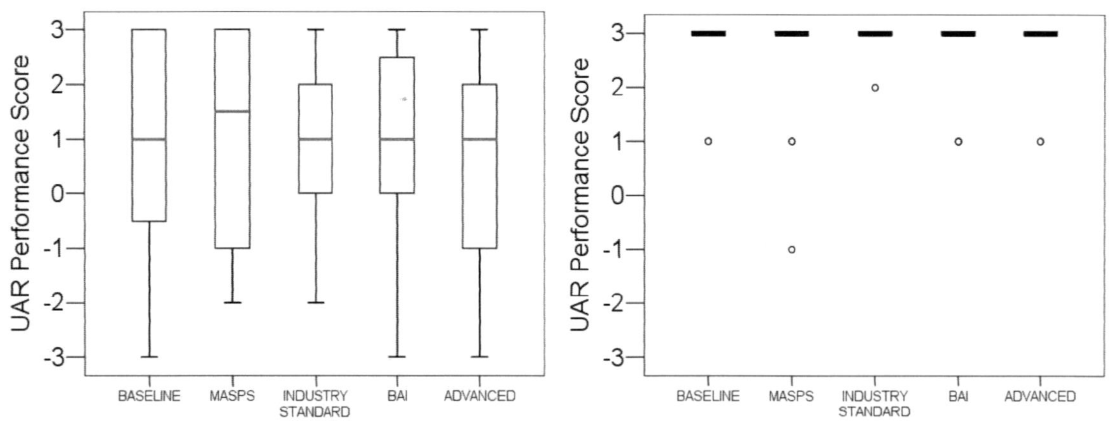

UAR Score

Figure 2. Boxplots of Quantitative Dependent Measures (left – 30 degrees up; right – 30 degree down)

Qualitative Results

Scenarios. Post-experimental analyses revealed that the nose-down, 60 degrees right roll UA had highest workload (using NASA-Task Load Index, TLX), $F(3, 177) = 26.15$, $p < 0.01$ and lowest situation awareness (using Situation Awareness Rating Technique, SART), $F(3, 177) = 26.15$, $p < 0.01$.

Displays. No significant differences were found across display concepts for NASA-TLX, $F(4, 204) = 0.565$, $p > 0.10$. No significant differences were found across display concepts for SART, $F(4, 204) = 0.847$, $p > 0.10$. The ANOVA revealed a significant main effect for paired comparison geomeans for display, $F(4,48) = 24.033$, $p < 0.0001$. Pilots rated the Industry Standard + BAI (0.34) as significantly higher for SA than all four other display concepts. The BAI was reported to significantly enhance attitude awareness. The results also showed that the Advanced (0.19) and Industry Standard (0.21) virtual day-VMC displays were not significantly different from each other. These concepts were significantly different from Baseline (0.13) and MASPS (0.12) concepts, and pilots reported that their enhanced synthetic vision presentations provided better situation awareness and more intuitive interpretation of aircraft attitude than the baseline (no SV) or MASPS (minimal SV).

Conclusions

The subject pilots had substantial experience and training in recognizing and recovery from UAs. This pilot population was conservatively selected because, if significant differences were found across displays, it would be even more significant with less experienced commercial pilots - the identified risk group in the CAST report. Although the pitch-up scenarios were found to be significantly different than nose-down, the differences are associated with the difficulty of quantifying nose-up transport aircraft UAR performance (see Gawron, 2009); no quantitative performance differences for displays were found and all pilots were well adept to recover from the UA conditions. The pilots subjectively rated the nose-down conditions as being the most difficult scenarios.

Although no performance differences were found, pilot comments revealed that the added situation awareness provided by the background attitude indicator and the terrain visual cues of the Industry Standard and Advanced virtual day-VMC (SV) displays was substantial (compared to Baseline and MASPS concepts). The BAI concept was rated significantly better than all other display concepts and will be further researched. The Advanced virtual day-VMC concept with optical flow cues was not found to be quantitatively or qualitatively different compared to Industry Standard, but pilot feedback suggests that modifications to the cues would substantially improve efficacy. Furthermore, pilots stated that the optical flow cues, as implemented, may not provide useful information for recovery but would be of value to help them recognize an impending unusual attitude.

In general, the results posit that virtual day-VMC displays have potential benefit to aid in recognition of, and recovery from, unusual attitudes. The next steps are to evaluate the SV display concepts with low-hour (< 1200 hours) international pilot populations and continue research and development of the BAI and optical flow cues.

References

Bailey, Randall E., Kyle KE Ellis, and Chad L. Stephens. (2013) Test and Evaluation Metrics of Crew Decision-Making And Aircraft Attitude and Energy State Awareness. AIAA Infotech@Aerospace 2013.

Bateman, D. (2010). Some Thoughts on Reducing the Risk of Aircraft Loss of Control. AIAA Guidance, Navigation, and Control Conference, Guidance, Navigation, and Control and Co-located Conferences, AIAA Paper No. 2010-8139. , http://dx.doi.org/10.2514/6.2010-8139.

Boeing. (2004, August 1). Airplane Upset Recovery. Industry Solutions for Large Swept-Wing Turbofan Airplanes Typically Seating More than 100 Passengers. Boeing.

Commercial Aviation Safety Team (2014a). Airplane State Awareness Joint Safety Analysis Team Final Report. Washington, D.C.: CAST.

Commercial Aviation Safety Team (2014b). Airplane State Awareness Joint Safety Implementation Team Final Report. Washington, D.C.: CAST.

Federal Aviation Administration (2016). Upset Prevention and Recovery Training. Advisory Circular 120-111 Change 1. Washington, D.C.: FAA.

Gawron, V., Bailey, R., and Randall, A., (2009) Differeneces in In-Flight Unusual Attitude Recovery Performance and Workload,

Nicholas, S.N. (2016). Evaluation of Synthetic Vision Display Concepts for Improved Awareness in Unusual Attitude Recovery Scenarios. Department of Defense Human Factors and Ergonomics Meeting 70 TAG. Presented May 10, 2016.

FATIGUING THE FORCE: USING OPERATIONAL DATA TO IMPROVE THE UNITED STATES AIR FORCE'S MISSION EFFECTIVENESS MODEL

Kelly M. Amaddio
Headquarters Air Mobility Command, Scott Air Force Base, IL
Jennifer P. Howland
Headquarters Air Mobility Command, Scott Air Force Base, IL

Air mobility pilots routinely fly multiple missions spanning several time zones, thereby disrupting their circadian rhythm. As a result, they consistently operate at a sub-optimal performance level. After several fatigue-related accidents, the Air Mobility Command (AMC) Safety Office incorporated the Sleep, Activity, Fatigue, and Task Effectiveness (SAFTE) model into its Aviation Operational Risk Management (AvORM) program to inform aircrew members of their fatigue levels during critical phases of flight. Further analysis indicated that aircrew members experience higher fatigue levels than predicted, which directly reduces flight safety. This study seeks to improve the underlying assumptions within the sleep model to more accurately predict aircrew member performance during critical phases of flight, thereby improving the predictive power of the mission effectiveness model within AvORM. This is the first study to collect operational data from the United States Air Force (USAF) C-17 pilot community using actigraph watches, self-report daily logs, and objective aircraft data to determine the relationship between fatigue and pilot mission effectiveness. Additionally, this study provides policy recommendations to enable aircrew, squadron leadership, and mission planners to mitigate some factors contributing to aircrew fatigue.

In response to numerous Class A mishaps where fatigue was deemed a contributing factor, Air Mobility Command (AMC) has been employing a version of the Sleep, Activity, Fatigue, and Task Effectiveness (SAFTE) model in the Aviation Operational Risk Management (AvORM) program since 2012. Class A mishaps are defined as mishaps resulting in: a total cost of $2 million dollars or more, a fatality or permanent total disability, and/or destruction of an aircraft (AFI 91-204). Schedulers are instructed to use the SAFTE model to plan missions with the goal of keeping the pilot's performance effectiveness level above 70%. Within AvORM, the performance effectiveness graphs indicate expected times for critical phases of flight (e.g. aerial refueling, landing, etc.). Pilots are given a print out of the performance effectiveness graph when starting a mission to aid in situational awareness and plan possible mitigation strategies. While the merits of the SAFTE model are well documented, it has limitations when employed with a unique population such as air mobility pilots.

Development of the SAFTE model was sponsored by the Department of Defense (DoD), and carried out by a collaboration among Walter Reed Army Institute of Research (WRAIR), Air Force Research Laboratory (AFRL) and Science Applications International Corporation (SAIC). The SAFTE model was accepted as the base model for continued DoD development in 2002. The role of the SAFTE model within AvORM is to make predictions of pilot performance effectiveness. Inputs to the SAFTE model within AvORM include scheduled sleep periods (duration and timing) as well as scheduled flight time and duration. The inputs are used to infer an individual's state in terms of his or her sleep reservoir and determine where he or she is within their circadian rhythm. The SAFTE model is a variant of the two-process model proposed originally by Borbely (Borbely, 1982; Daan, 1984). Although other models seek to predict fatigue, the SAFTE model distinguishes itself from others by considering reduced effectiveness due to sleep inertia immediately upon waking and interrupted sleep.

The SAFTE model is highly accurate when predicting the collective effectiveness level for a group of individuals in a laboratory setting over broad time ranges; however, external issues may limit the model's utility in an operational environment. Primarily, the SAFTE model has not been rigorously validated in an operational setting. Tuning the underlying assumptions within the model with field-relevant data will enhance the model's accuracy at planning and mission time.

As depicted in Figure 1, the SAFTE Model employs numerous physiological factors to predict performance effectiveness (Hursh et. al, 2004). The model is very sensitive to the input assumptions. The SAFTE model within AvORM holds the following factors constant: sleep intensity, sleep quality, rate in which the sleep reservoir is replenished and depleted, sleep inertia, and circadian rhythm. While the rate in which the sleep reservoir is replenished and depleted, sleep inertia, and circadian rhythm will not be addressed in this study due to the type of data collected, the sleep intensity and quality are addressed. Replacing those assumptions with measurement-based statistics should improve the accuracy of the predictions and guide planners in minimizing fatigue throughout all phases of the mission.

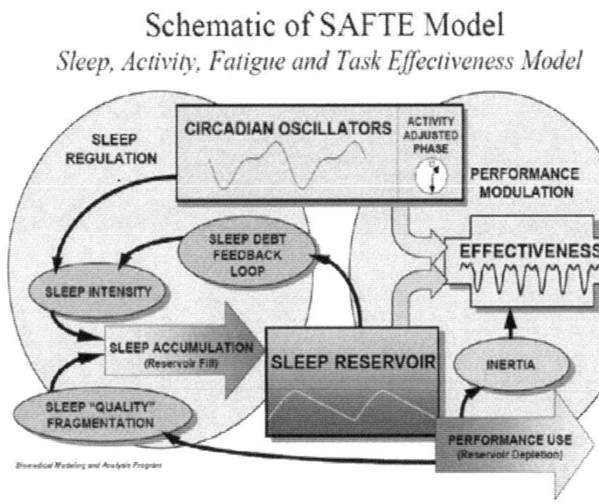

Figure 1. Schematic of the SAFTE Model

Strong assumptions are currently made about the rest-related status of pilots as they begin a mission and about the duration and quality of sleep that crew members obtain during real missions. Within the AvORM program, the SAFTE model assumes that the pilot will start a mission with a sleep reservoir at 90%, start sleeping two hours after landing, sleep for a total of eight hours, experience good quality sleep, experience the same quality sleep with in-flight napping as napping in a bed, and adjust to jet lag at a constant rate of 1.5 hours per day (Hursh, 2004). These factors may be optimized with operational data to reflect a more accurate representation of mobility pilots' operational and fatigue patterns. By using operational data to replace the general population baseline, the SAFTE model will more accurately predict aircrew fatigue and increase safety of flight.

Psychomotor Vigilance Task

This study utilizes the Psychomotor Vigilance Task (PVT) to measure an individual's behavioral alertness. The standard PVT is ten minutes in length; however, operational constraints require this study to use a three-minute PVT. Basner, Mollicone & Dinges found that the 22.7% decrease in effect size garnered by the three-minute verses the ten-minute PVT was an acceptable tradeoff in sensitivity, especially considering the test is 70% shorter in length (2011).

Performance measures on the PVT typically include: lapses of attention, false starts, and response time mean, median, and standard deviation. In this study, we measured performance by mean response time and range. The PVT is best taken in a quiet area free of distractions. However, when the pilots took their PVTs during a mission (whether in flight or on the ground), they were subject to radio calls and other various operational distractions. This led to numerous PVT data points that were outside a person's "normal" range (loosely defined as 100ms to 500ms is a laboratory setting, 100ms to 650ms in an operational setting). Therefore, it was difficult to determine whether a participant's response time increased due to distractions or fatigue. Our method for managing this was to only analyze the response times that were between 100 and 650ms, which comprised 96.8% of the data.

The hypothesis for the study was that mobility pilots are consistently operating at a performance level below the predicted level in the underlying SAFTE model within AvORM. Hence, mobility pilots were more fatigued at critical phases of flight than predicted during the flight planning phase, thereby decreasing the safety of flight. Improving the underlying assumptions within the sleep model would more accurately predict pilot performance during critical phases of flight. It was predicted that sleep duration and quality would be significantly shorter and degraded while on a mission compared to participants' sleep duration and quality at home. A corresponding relationship with the PVT data was expected. Specifically, the mean and range of reaction times should increase throughout a mission and PVTs taken on flying days should be slower than those taken on non-mission days.

While this study addresses the relationship between the predicted performance effectiveness level prior to mission execution and the actual performance effectiveness level, the focus of this paper is the actigraph and PVT analysis.

Method

Participants

Thirty Air Force C-17 pilots stationed at Joint Base Charleston volunteered to participate in this study. All were physically cleared to fly by a flight doctor; therefore nobody had a condition negatively affecting his or her ability to fly safely. Unfortunately, many participants failed to provide demographic information. Of the 30 recruited participants, eleven (nine male and two female) completed all parts of the study so the analysis only includes their data. The mean age of participants was 28.25 years (min 26, max 30). Participants were initially recruited through email and a unit level safety briefing. Interested participants met with the researchers to receive test materials. Pilots unable to attend were directed to the unit safety office to attain test materials.

Apparatus

Participants wore an Ambulatory Monitoring Inc. (AMI) Motionlogger watch which features a built-in accelerometer used to record active and sleeping activity during the duration of the testing period. This watch also has a three-minute PVT to assess reaction time, a proxy variable for performance effectiveness. Although participants were also given a daily log to provide information related to their sleep and flight parameters, the analysis of this data is beyond the scope of this paper.

Experimental Procedure

For 30 days, each pilot was instructed to continuously wear the actigraph and complete a series of activities. If not flying that day, they were instructed to complete the three-minute PVT at least three times daily (within 45 minutes after waking and prior to going sleep, and once anytime throughout the day). If flying that day, they were instructed to complete the PVT within 45 minutes after waking, prior to going to sleep, takeoff, and landing. If the participant flew multiple sorties in the day, they were instructed to only complete the PVT within 45 minutes of landing (not on takeoff).

Analysis

Sleep

The data on the actigraph watch was downloaded and cleansed using Ambulatory Monitoring Inc.'s ActionW software. Episodes when the watch was obviously not worn (e.g. taken off to take a shower, prior to the start of study period, etc.) were removed. After the sleep periods were highlighted within the dataset, the ActionW software scored the sleep episodes. The scored sleep was then imported in Fatigue Science's Fatigue Avoidance Scheduling Tool (FAST) to conduct the performance effectiveness analysis.

Eleven participants wore the actigraph watch during 248 total sleep episodes. Six participants flew on multi-day missions that required them to sleep at a hotel. The 155 sleep episodes from these six participants were used for this portion of the analysis. There were 101 bedtime sleep episodes at home ranging from 196 minutes to 612 minutes (μ=410.3, σ=75.5). There were 26 bedtime sleep episodes away from home ranging from 139 minutes to 704 minutes (μ=371.2, σ=130.1). There were 19 nap sleep episodes at home ranging 22 minutes to 126 minutes (μ=76.1, σ=36.9). There were 9 nap sleep episodes away from home ranging from 21 minutes to 114 minutes (μ=52.4, σ=29.0). Duration of sleep episode, quality of sleep (100*sleep minutes/0-0 period), and longest sleep period were analyzed for each sleep episode. The 0-0 period is defined to start when there are twenty minutes of continuous non-movement until the first continuous movement of twenty minutes; hence the period a participant would consider themselves asleep.

The first relationship analyzed was bed time sleep away from home versus at home. An independent samples t-test indicated that bedtime sleep durations away from home were statistically significantly shorter (μ= 371.2, σ = 130.1) than bedtime sleep durations at home (μ= 410.3, σ = 75.5), $t(127) = 1.47$, $p = .08$, $\alpha = .1$. An α of .1 was used due to the highly variable nature of human subjects research. Analysis of variance showed a main effect of sleep location on sleep duration, $F(1, 127) = 3.98$, $p = .048$, $\alpha = .1$. An independent-samples t-test indicated that the length of the longest sleep period away from home were statistically significantly shorter (μ= 151.1, σ = 79.1) than the longest sleep period when sleeping at home (μ= 171, σ = 92.0), $t(127) = 1.84$, $p = .04$, $\alpha = .1$. This suggests that participants were awoken more frequently while sleeping away from home compared to sleeping at home, and therefore not getting as much restorative deep sleep. An independent samples t-test indicated that the quality of sleep away from home was statistically significantly diminished (μ= 93.4, σ = 5.6) than the quality of sleep at home (μ= 95.2, σ= 4.6), $t(127) = 1.82$, $p = .07$, $\alpha = .1$.

The next relationship analyzed was nap time sleep duration away from home and at home. An independent samples t-test indicated that nap duration away from home was significantly shorter (μ= 52.4, σ =29.0) than nap duration at home (μ = 76.1, σ = 36.9) $t(26) = 1.84$, $p = .04$, $\alpha = .05$. The quality of nap sleep was not found to be significantly different away from home and at home.

A 2x2 ANOVA with sleep location (home, away) and sleep type (bed, nap) as between-subjects factors revealed a statistically significant main effects of sleep duration, $F(2,155) = 191.9$, $p < .0001$.

Finally, the last relationship analyzed was sleep after local flight days and non-flying days. This indicated whether pilots slept longer after flying compared to a non-flying day. Sleep duration, sleep quality, and longest sleep length after flying a local mission was not statistically different than normal bedtime sleep at home.

PVT

The PVT data was cleansed to remove obvious outliers and analyzed using Microsoft Excel. There were a total of 509 PVTs with 10841 individual button presses (and response times) analyzed. A one-tailed t-test indicated that mean response times for PVTs taken on flying days (μ = 257.74 ms) were significantly faster than those on non-flying days (μ = 265.20), $t(311) = 1.28$, $p = .048$. There was no statistically significant difference between mean PVT trial ranges on flying verses non-flying days.

Mission lengths ranged between one and six days. A positive correlation was found between mission day number and response time, $r(134) = .164$, $p = .028$ (small effect size), but there was no statistically significant correlation found between range of PVT trial scores and mission day. There also was no statistically significant correlation between response time and time of day.

Discussion

Sleep

The operational data collected in this study was incomplete in many aspects. There were three times more home station sleep episodes than mission sleep episodes. This was unavoidable due to the fact that many of the actively participating pilots spent more time at home than away on missions. In particular, the number of nap sleep periods was low which prevents the ability to make any real analytical conclusions.

There are numerous factors contributing to the shorter duration and lower quality of sleep experienced during a mission compared to at home. While policy requires crew rest to be at least twelve hours, there are multiple factors affecting a pilot's ability to get enough sleep during a mission. Transportation to the hotel or crew rest facility may take longer than the 30 minutes currently assumed in the AvORM model, especially when the mission ends after normal operating hours. If a mission ends while it's still daylight outside, pilots will have a harder time falling asleep. The quality of the lodging facilities (sound, light, new bed, etc.), location of the hotel, and mission constraints (aircraft commander being called to put the crew on alert) can also affect sleep duration and quality. Finally, crew rest may simply not be the ideal length to recover from the previous mission and prepare for the next mission. For example, if the crew rest length is too short, the pilots may not be able to fully recover from the previous mission. Conversely, a crew rest period can be too long to get in two sleep periods so pilots are reporting for their next mission with one long sleep period and possibly a short nap.

When a person is sleep deprived, their body will spend subsequent recovery sleep periods in deep sleep (Corsi-Cabrera, 1992; Borbély, 1981). It would have been expected that the pilots would have experienced increased longest sleep periods while on a mission due to the long duty day; however, the opposite was seen. This supports the hypothesis that the sleep environment was not conducive to restorative sleep. Further analysis is needed to determine if the pilots had increased longest sleep periods upon return to home station and for how many subsequent days to return to baseline performance effectiveness levels. Hursh et. al. (2004) found that for some individuals under extreme sleep deprivation, the assumed three days needed to recover was insufficient for performance to return to baseline levels.

Shorter length and lower quality naps when on a mission versus at home was also expected. While mission dependent, flight planners may expect pilots to nap in-flight to keep his or her performance level above 70%. The C-17 aircraft crew rest facilities are not conducive to high quality sleep since they are located underneath the stairs leading to the flight deck. In addition, if an aircraft commander is accompanied by other inexperienced pilots, they may sleep less or have lower quality sleep. It is necessary to note that there were less than half the naptime episodes when on a mission compared to at home. Unexpected though, were the low number of naps at the hotel prior to showing up for a flight while on a mission. When crew rest is approximately 24 hours, it is not conducive for two long sleep periods so many pilots reported that they would have one long sleep period with a nap. Further data collection is required to make any conclusions on this issue.

PVT

While cleaning the data, it was apparent that a possible confounding variable is the location where the individual is completing their PVT. On flying days, pilots are most likely doing their PVTs on the aircraft or around other crew members, possibly causing them to be distracted. When critical radio calls are required, they must respond in a timely manner. However, it is likely that on non-flying days, the pilots are taking the PVT in a quieter area since they have more control over where and when they take it. The data showed that faster response times on flying days verses nonflying days, which was unexpected due to the fatiguing effects of flying and circadian shift. Further analysis is needed to determine the root cause for this finding.

An expected result was the positive correlation between mission day number and response time. As the missions continued in length, the participants responded slower to the PVT stimulus. Long missions usually involve the crossing of multiple time zones and circadian rhythm shifts along with long duty days and short periods of sleep.

Conclusion

This study supports the hypothesis that air mobility pilots are not getting the same sleep duration and sleep quality on a mission than they do at home station. As evidenced by the PVT scores, it appears that there is a compounding effect of fatigue as a mission increases in duration. In the short-term, this may decrease flight safety. In the long-term, Pilcher (1996) found that repetitive sleep deprivation can negatively affect one's health and well-being. Additional data is required to make a more definitive algorithmic conclusion concerning the SAFTE model; however, it is clear that pilots experience shorter and lower quality sleep when on a mission than when at home. Additionally, since naps away from home are shorter and of lower quality, it is advisable that the SAFTE model not count naps as fully restorative sleep.

The authors recommend that the SAFTE model decrease the current eight hour crew rest sleep duration to the mean of six and a half hours of sleep. Next, the sleep quality should be decreased from good quality to poor. With further data collection, the decreased in sleep quality can be more accurately quantified. Finally, the authors recommend that naps not be used as a fatigue mitigation strategy by the flight planners to keep pilot performance effectiveness above 70% since the duration and quality of naps during a mission were of such low quality.

Acknowledgements

We would like to thank the pilots at Joint Base Charleston for participating in this study. The views in this paper do not represent the views of Air Mobility Command or the United States Air Force.

References

AFI 91-204 (2014). Department of the Air Force, Headquarters Air Force Safety Center. http://static.e-publishing.af.mil/production/1/af_se/publication/afi91-204/afi91-204.pdf.

Basner, M., Mollicone, D., & Dinges, D. F. (2011). Validity and sensitivity of a brief psychomotor vigilance test (PVT-B) to total and partial sleep deprivation. *Acta Astronautica,* 69(11-12), 949-959.

Mallis, M.M, Mejdal, S., Nguyen, T.T. & Dinges, D.F. (2004). Summary of the Key Features of Seven Biomathematical Models of Human Fatigue and Performance. *Aviation, Space, and Environmental Medicine* 75:A4-14.

Borbély A.A., 1982. A two process model of sleep regulation. *Human Neurobiology* 1:195-204.

Borbély A.A., Achermann, P. (1999). Sleep homeostasis and models of sleep regulation. *Journal of Biological Rhythms*14:557-568.

Borbély, A. A., Baumann, F., Brandeis, D., Strauch, I., & Lehmann, D. (1981). Sleep deprivation: effect on sleep stages and EEG power density in man. Electroencephalography and clinical neurophysiology, 51(5), 483-493.

Balkin, T., Thorne, D., Sing, H., Thomas, M., Redmond, D., Wesensten, N., Williams, J., Hall, S. & Belenky, G. (2000). Effects of sleep schedules on commercial driver performance. Washington, DC: U.S. Department of Transportation, Federal Motor Carrier Safety Administration; Report No.: DOT-MC-00–133.

Corsi-Cabrera, M., Ramos, J., Arce, C., Guevara, M. A., Ponce-de Leon, M., & Lorenzo, I. (1992). Changes in the waking EEG as a consequence of sleep and sleep deprivation. SLEEP-NEW YORK-, 15, 550-550.

Daan, S., Beersma, D.G.M., & Borbély, A.A. (1984). Timing of human sleep: Recovery process gated by a circadian pacemaker. *American Journal of Physiology* 246:R161-R178.

Hursh, S. R., Redmond, D. P., Johnson, M. L., Thorne, D. R., Belenky, G., Balkin, T. J., & Eddy, D. R. (2004). Fatigue models for applied research in warfighting. *Aviation, space, and environmental medicine*, *75*(3), A44-A53.

STATUS OF FAA AIR TRAFFIC CONTROL FATIGUE INTERVENTIONS 2013-2016

Darendia McCauley
And
Thomas Nesthus
FAA Civil Aerospace Medical Institute
Oklahoma City Oklahoma

The Federal Aviation Administration (FAA) has long been concerned with the impact of fatigue in Air Traffic Operations. Fatigue has been cited as a factor in operational incidents. The shift work and quick turn-around shifts contribute to this problem. In conjunction with the collective bargaining agreement, FAA management and the National Air Traffic Association (NATCA) agreed to jointly develop a series of interventions designed to mitigate some of the aspects of fatigue in the controller workforce. This resulted in a Fatigue Group comprised of FAA management, NATCA representatives, and fatigue scientists. Following 15 work-intensive meetings, the Fatigue Risk Management Group produced 12 fatigue mitigation recommendations. This Fatigue Risk Management Group also supported research conducted by NASA for ATC. The ATC research focused on two components, a fatigue survey of the ATC workforce and a field study with participant volunteers using wrist activity monitors. This research supported the 12 fatigue mitigation recommendations. To date, all recommendations have been fully or partially implemented.

The Federal Aviation Administration (FAA) presented a Fatigue Training Workshop for the 17th International Symposium on Aviation Psychology (Nesthus, Avers, & McCauley, 2013). Fatigue is an important human performance problem. The impact of fatigue, its risks, and mitigations have become key concepts managers and shift workers involved with aviation systems must acknowledge, understand, and manage. The understanding of and support for fatigue mitigation initiatives is critical. The FAA is working to maintain the safety of the National Airspace System (NAS) and ensure the health and well-being of its workforces as well as other workforces within the aviation industry through regulations with consideration of fatigue issues. Fatigue awareness and mitigation are important components of this effort. FAA has developed a full spectrum of fatigue awareness and mitigation programs designed to impact shift workers and managers within aviation systems including air traffic controllers and technical operations specialists, pilots, flight attendants, and maintenance workers.

The workshop presented in 2013 introduced the fatigue science background used in the development of various intervention materials and the modification of those materials to accommodate multiple vocational backgrounds for those involved with aviation systems. Along with the awareness of fatigue issues, maintainance and distribution of this knowledge set, the development and use of personal strategies to optimize sleep and maximize alertness, and the use of ergonomic scheduling principles (to the extent possible), a reduction in fatigue-related risks can be achieved and will contribute to safer operations throughout this industry.

During the 2013 workshop, an overview of the 12 fatigue mitigations mutually agreed upon by FAA and the National Air Traffic Association (NATCA) through the Article 55 Fatigue Risk Management (FRM) Work Group was presented. Article 55 of the NATCA Collective Bargaining Agreement of 2009 directed FAA management and NATCA to jointly develop recommended fatigue mitigation strategies. The NATCA and FAA management representatives appointed to make these recommendations became known as the Article 55 FRM Work Group.

The Work Group Charter established that FAA management and union (NATCA, PASS) representation had important and equal voting status for their recommendations. The Article 55 FRM

Work Group relied on fatigue science and research as an independent resource for the basis of establishing recommended mitigations. This resulted in a balanced approach and provided the inclusion of Civil Aerospace Medical Institute (CAMI) scientists and other research consultants with independent fatigue expertise for the transportation industry. The Office of Aerospace Medicine Medical Specialties Division was also directly involved with medical fatigue issues, obstructive sleep apnea, in particular. The resulting 12 fatigue mitigation recommendations were briefed to the FAA Administrator, the NATCA President, and AT Management shortly after the work group completed them. Also, the National Transportation Safety Board (NTSB) was officially briefed in order to meet the requirement of Safety Recommendation A-07-30 through -32 and A-07-34. Since that time all 12 recommendations have been addressed and implemented in several ways involving many areas of FAA.

Background

The Article 55 (FRM) Work Group promoted fatigue risk management, which relied on a basic Fatigue Risk Management System (FRMS) approach to promote an awareness of fatigue safety and minimize fatigue risks in ATC operations. The basis of the FRMS was first, that fatigue is a physiological state affecting everyone to varying degrees. Second, fatigue is inherent in all shift work environments. Third, fatigue can introduce a risk to the health and well-being of employees and the safe operation of the National Airspace System (NAS). Guidance for an FRMS should include the following elements:

- Must be data-driven and scientifically-based
- Must enable continuous monitoring and management of safety risks associated with fatigue-related hazards
- Must provide a means of measuring, mitigating, and reassessing fatigue risk
- Must include schedule assessment, data collection, and systematic analysis
- Provides scientifically guided fatigue mitigations—both proactive and reactive

The Article 55 FRM Work Group sponsored analyses of current scheduling practices (Orasanu, Parke, & Kraft, 2012) and fatigue modeling using the Sleep, Activity, Fatigue, and Task Activity Effectiveness (SAFTE)/Fatigue Avoidance Scheduling Tool (FAST; IBR, 2016) to identify fatigue-related issues in the ATC work environment. The FRM Work Group also reviewed International research on fatigue. This provided an informed and comparative basis for the developed recommendations.

Addressing fatigue is a shared responsibility. Fatigue countermeasures can help to mitigate fatigue safety risks and improve employee health and well-being. The Article 55 FRM Work Group focused on these themes as well as the components of an FRMS in the development of mitigation strategies reflected in the Work Group recommendations.

Objectives

The FRM Work Group objectives for fatigue mitigation efforts were developed to promote understanding of the basics of fatigue and its mental, physical, and emotional signs; recognition that fatigue can represent a hazard to the safety of FAA operational employees and the NAS; awareness of fatigue countermeasures that can be used to help reduce fatigue risks and increase both personal and NAS safety; and understanding that fatigue may represent a safety risk, depending on the likelihood and severity of the fatigue hazard.

Based on its tasking, the Article 55 FRM Work Group defined a set of guidelines to help focus its efforts. These guidelines provided a backdrop on which Work Group activity was based and included the following:

- Increase the safety of the NAS by reducing fatigue hazards and risks,

- Improve the health and wellbeing of the workforce through better fatigue management,
- Base findings and recommendations on science and data while leaving implementation issues for later discussions, and
- Collaborate with internal and external organizations.

Fatigue Risk Management Group Recommendationations and Implementation

Recommendations/Implementation

The Air Traffic Operations (ATO) Safety and Technical Training Fatigue Risk Management Team approved FRMS Work Group Findings and Recommendations by FAA management and NATCA. These recommendations are presented with implementation strategies in Table 1 below:

Table 1.
Article 55 Fatigue Risk Management Work Group Recommendations and Implementation

Component	Recommendation/Implementation
FRMS	1. Design and implement an FRMS within the ATO operational environment. FAA ATO established its Fatigue Risk Management system in January 2012, via a Charter, agreed to and signed by the FAA, NATCA and PASS. FAA ATO JO 1030.7A (2012) formally established the ATO Fatigue Risk Management program as the cornerstone to the ATO implementing a fatigue risk management system.
	2. Continue to support the post-recommendation work efforts by creating a transition team composed of Article 55 FRM Work Group members until the formal ATO FRMS is established. A Post-Article 55 FRM Workgroup met in early 2011, completed the agreements resulting in the July and August FAA/NATCA Fatigue MOU and Fatigue MOU Guidance (2011) and then collaborated to establish the ATO FRMS as referenced in recommendation 1 actions above.
Scheduling	3. Provide a minimum of nine hours between evening and day shifts. FAA/NATCA Fatigue MOU (2011): *The Parties recognize the need for watch schedules that meet operational needs and mitigate system risks due to fatigue. In response to the scientific data supplied by the Article 55 workgroup, the Parties agree that employees are required to have a minimum of nine (9) consecutive hours off-duty preceding the start of a day shift. For purposes of this document only, a day shift is generally defined as a schedule where the majority of hours fall between 7:00 a.m. and 4:00 p.m. This requirement applies to all shift changes, swaps, and overtime to include scheduled, call-in, and holdover assignments.*
	4. On a 2-2-1 counterclockwise (CCW) rotation, reduce the day shift preceding the first midnight shift from eight to seven hours, and begin that shift one hour later, to provide the opportunity for an extra hour of restorative sleep at the end of the nighttime sleep period. FAA/NATCA Fatigue MOU Guidance (2011): *Consistent with the Article 55 Workgroup recommendations, for those facilities that utilize 2-2-1 counterclockwise schedules, it is encouraged that schedules be constructed to reduce the day shift preceding the first midnight shift from eight to seven hours, and begin that shift one hour later, in order to provide the opportunity for an extra hour of restorative*

sleep at the end of the nighttime sleep period.

This reduced shift duration would be offset by adding the hour to a shift, or a combination of shifts, earlier in the workweek. It is recommended that the additional time be scheduled either at the beginning of a normal evening shift(s), or at the end of a normal day shift(s), so as to not infringe on nighttime sleep.

Such schedules would be constructed as an Alternative Work schedule (AWS) and would require employees to volunteer. In the event that there are insufficient volunteers, this AWS schedule cannot be implemented and existing 2-2-1 counterclockwise scheduling practices may be utilized.
(This recommendation was not implemented as a regulation, but is available as a component of AWS.)

Recuperative Breaks

5. Modify current policy, orders, etc., to permit naps during relief periods (breaks).

FAA JO 7210.3Y, Section 2-6-6, Relief Periods, Paragraph c., was modified with the following language:
Personnel performing watch supervision duties must not condone or permit individuals to sleep during any period duties are assigned. Any such instance must be handled in accordance with applicable Agency policy and the applicable collective bargaining agreement.
The above clarified that sleeping while on duty is prohibited. Notably, it did not explicitly prohibit controllers from sleeping while on a recuperative break.

6. In addition to normal breaks on midnight shifts, include a provision for a recuperative break for 2.5 hours, which incorporates time to overcome sleep inertia should an employee choose to nap.

FAA/NATCA Fatigue MOU Guidance (2011):
Employees are permitted to have break periods away from their assigned duties to sufficiently recuperate from the effects of fatigue, if needed, attend to personal needs, and rejuvenate their mental acuity.
Length of recuperative breaks on midnight shifts shall be longer than those normally provided during other shifts, to the maximum extent possible, considering staffing and workload, consistent with the recommendations of the Article 55 Workgroup.
The above recognizes the need for longer breaks on midnight shifts. Activities that rejuvenate mental acuity are not specified.

Sleep Disorders

7. Create policies and procedures that encourage self-initiated evaluation, diagnosis, and demonstration of initial treatment effectiveness of Sleep Apnea (SA) by removal or reduction of economic disincentives.

Obstructive Sleep Apnea was covered in an article titled *Obstructive Sleep Apnea: Know the Signs, Take Action,* in Focus FAA (2016). Sleep disorders (including sleep apnea) have been included in the ATO Fatigue Awareness and Countermeasures Training

Part 1 – Fatigue Basics, Section1, secondary contributors to fatigue – Sleep Disorders, and

Part II – Sleep basics, Section 6, Sleep Disorders.

8. Use AAM-prepared SA education to build sleep apnea awareness in the ATO workforce, include raising awareness of respiratory coaching to SA patients.

Sleep disorders (including sleep apnea) have been included in the ATO Fatigue Awareness and Countermeasures Training

9. Aerospace Medicine:
   - AAM to stay current with state of the art in sleep medicine
   - AAM to utilize AASM standards and practices for SA risk factor identification, diagnosis and treatment standards
   - AAM to document the process for medical qualification for individuals at risk for sleep apnea
   - AAM to develop educational materials for the workforce and AMEs
   - AAM to educate AMEs on SA

OSA materials for AMEs have been developed and published on the FAA Guide for Aviation Medical Examiners Website (2016).

Personal Fatigue Management

10. Develop policy and education for employees designed to minimize fatigue and report fit for duty, and action to be taken when they consider themselves too fatigued to safely perform their duties.

FAA/NATCA Fatigue MOU (2011) states the following:

Section 8. *All operational personnel are obligated by their significant safety duties and professional responsibilities to prepare for duty with consideration for being well-rested and mentally alert. It is the employees' responsibility to recognize and report to their supervisor when they are unable to perform operational duties due to fatigue. Upon request, employees that self-declare as unable to perform operational duties due to fatigue will be granted leave in accordance with the leave provisions contained within the 2009 CBA. Additionally, at his/her request, an employee that self-declares as fatigued, shall be assigned other facility duties, to the extent such duties are available. If no such duties are available, the employee will be granted leave as described above.*

The FAA's ATO Operational Supervisors Workshop, Fatigue Lesson, reviews scenarios when employees might self-declare fatigue, and the responsibilities of the manager in those situations.

11. In order to avoid on-the-job fatigue that threatens safety, develop policy and education for managers that incorporates emphasis on a non-punitive approach when an employee, in accordance with the developed policy, self-declares as too fatigued to safely perform operational duties.

The FAA's ATO Operational Supervisors Workshop, Fatigue Lesson, reviews scenarios when employees might self-declare fatigue, and the responsibilities of the manager in those situations.

Fatigue Education

12. Update existing fatigue awareness training to reflect current science and to provide applications specific to all people in certain occupations personalize the application of the training.

ATO Fatigue Awareness and Countermeasures Training programs have been developed for air traffic controllers and technicians. These electronic courses reflect current science and methods to personalize the training and make it relevant to the learner. Additional fatigue lessons are instructor-led and are delivered at the Mike Monroney Aeronautical Center in Oklahoma City, for Air Traffic Controllers. All of the content for these lessons reflects current science and is intended to allow learners to reflect on what they learn to make better choices regarding sleep and fatigue.

Summary

Fatigue presents an acknowledged hazard to the safety of the NAS and to the health and well-being of FAA employees. By raising awareness of fatigue and ways to reduce its impact, FAA will work to make the FAA a better and safer place to work, while improving the safety of the NAS. Keeping stakeholders informed of the FAA's efforts in fatigue safety is important to maintaining the public trust placed in the agency.

Acknowledgment

Research reported was conducted under the Air Traffic Program Directive / Level of Effort Agreement between the Human Factors Division (ANG-C1), FAA Headquarters, and the Aerospace Human Factors Division (AAM-500) of the Civil Aerospace Medical Institute.

References

FAA JO 1030.7A (2012)

FAA JO 7210.3Y, (2014).

FAA/NATCA Fatigue Memorandum of Understanding, (2011).

FAA ATO Fatigue Risk Management System (FRMS) Fatigue Safety Steering Committee (FSSC) Charter, (2012).

FAA Guide for Medical Examiners, Oklahoma City, OK: FAA Civil Aerospace Medical Institute. Retrieved from https://www.faa.gov/about/office_org/headquarters_offices/avs/offices/aam/ame/guide/dec_cons/disease_prot/osa/ref_materials/.

IBR, SAFTE/FAST. (2016). Fatigue Science, Baltimore, MD.

Nesthus, T., Avers, K., & McCauley, D. (2013). FAA Fatigue Training Workshop. For 17th International Symposium on Aviation Psychology, Dayton, Ohio, 2013.

Orasanu, J., Parke, B., & Kraft, N., (2012). Evaluating the Effectiveness of Schedule Change for Air Traffic Service (ATS) Providers: Controller Alertness and Fatigue Monitoring Study, NASA Ames Research Center, Moffett Field, CA.

SMS HAZARD ANALYSIS AT A UNIVERSITY FLIGHT SCHOOL

Don Crews
Middle Tennessee State University
Murfreesboro, Tennessee
Wendy Beckman
Middle Tennessee State University
Murfreesboro, Tennessee

For the last several years, the flight school of a mid-sized university has been working to implement a safety management system (SMS). As part of the effort, a robust self-reporting system has been developed, from which data has been used to effect changes in school policies and procedures. In this project, the safety reports that have accumulated over the life of the reporting system were classified based on the hazards experienced which caused the report generation. Non-use of standard procedures was found to be the leading hazard, with 90 of the 176 reports indicating improper procedure application. The traffic pattern at the non-towered airport where the flight school operates was the phase of flight found to be most prevalent in the safety reports, with non-standard pattern procedures, improper judgement/decision-making and communication issues cited as common hazards. Student knowledge/skill and instructor technique were also frequently reported hazards.

A 14 CFR Part 141 flight school within an aviation department at a mid-sized southeastern university initiated an anonymous safety reporting program in the spring of 2010. The first report was filed on 4/22/10, and the safety report data base at the time of analysis contained 176 total reports. The department Safety Committee "owns" the database, and as such, each report in the database has been reviewed and accepted by the committee. The primary role of the safety committee is to identify safety hazards, assess the risk associated with a given hazard, and recommend steps to mitigate the hazard. An additional role of the committee is to disseminate safety information to the flight school community to promote awareness of hazards and identification of risk factors, and to encourage the use of mitigation measures.

The safety report database is primarily a catalogue of reported safety related events. It has been used to identify several metrics including: events per year (see Table 1), weather conditions (93% VFR, 3.4% MVFR, 3.4% IFR), role of reporter (78.6% instructor, 9.8% student, 8.7% dispatcher, 2.9% other), as well as to record both initial actions and further actions taken as a result of reports.

Table 1.
Number of Safety Reports Per Year

Year	Total	Percentage
2010	7	3.98%
2011	12	6.82%
2012	20	11.36%
2013	24	13.64%
2014	20	11.36%
2015	39	22.16%
2016	54	30.68%

The safety report database has been used consistently to inform instructors and students of safety issues and promote safety awareness. A synopsis of each report and related recommendations is provided to the flight school community in a timely manner as reports are submitted. However, a lack of manpower has previously halted the systemic analysis of the safety report data available at this point.

While a full scale Safety Management System (SMS) is not currently required for Part 141 flight school operations, the goal of the department is to move towards that model in as many ways as feasible. The effort undertaken in this project was to assess the information in the database by identifying and categorizing hazards in a systematic fashion to aid the flight school and the overall airport community in which the flight school exists.

Literature Review

The Safety Management International Collaboration Group (SMICG) considers hazard identification the key element in safety risk management (2010). Likewise, the FAA defines the initial step in safety risk management (SRM) as conducting a thorough system description or analysis, to be able to "understand the aspects of the operation that might cause harm," and indicates that "in most cases, hazard identification flows from this system analysis," (Federal Aviation Administration [FAA], 2015, p.6). This includes the development of a hazard taxonomy and categorization process. Bahr (1997) suggests that an effective hazard analysis process should be "...a systematic, comprehensive method to identify, evaluate, and control hazards," (p.72).

The basic definition of a hazard from the SMICG is similar to those found in almost all general safety literature. "A hazard...is an object or condition with the potential to cause injuries to personnel, damage to equipment or structures, loss of material, or reduction of ability to perform a prescribed function," (SMICG, 2010, p.2). The FAA is more pointed in its definition of a hazard, indicating that it is, "a condition that could foreseeably cause or contribute to an aircraft accident," (FAA, 2015, p.7). The development of a comprehensive hazard taxonomy for each sector of the aviation industry is acknowledged as a challenge by the SMICG, as hazards may differ greatly between organizations, depending on their specific processes and procedures (SMICG, 2013, p.3). However, the need for organizations to attempt to identify the hazards within their activities, and to use this data to develop risk mitigation strategies, is also made clear (SMICG, 2013).

This project was an attempt at capturing the hazards that have been implied in the safety reports that have been filed at the subject flight school during the past 6 years. The development of a data driven understanding of the current condition of the system will lead to the ability to more appropriately apply accepted risk management techniques.

Methodology

Each reported event in the MTSU safety data base was reviewed by both of the researchers to determine the specific hazard(s) that was experienced, and to identify potential contributing factors. After a separate analysis, the safety reports were reviewed again by the researchers as a team, to further develop and clarify the hazards present in the submitted reports. As suggested by ICAO, this resulted in the development of a hazard categorization and identification process that was directly related to the available data. Cross referencing each category of safety concern with its contributing factors presented the data in a way that was more likely to identify the true nature of the hazard. The nature of a hazard was identified by the prevalence of certain kinds of events and/or behaviors found in the safety reports. These events and behaviors were related to the contributing factors in a reported safety event. Several additional passes though the data were made to clarify further the hazard categories to be utilized. To be consistent in identifying the nature of a hazard it was necessary to carefully define each type of contributing factor. The hazard categories developed following review of the data are described below:

Procedures – flight crew not following documented routines for a particular phase of flight

Judgement/Decision making – flight crew not exhibiting proper analysis of inputs, leading to failure to make a timely or correct decision

Situational Awareness – flight crew not aware of immediate circumstances or not able to project their circumstances into the future as appropriate

Checklist Use – check list not utilized; check list used but items not completed; non-optimal design of checklist

Communications – misunderstanding of communication; failure to communicate; communication not successfully transmitted

Air Proximity – when the PIC of either aircraft involved felt the need to take immediate evasive action to avoid a potential mid-air collision

Maintenance procedure discrepancy – an inoperative component was not properly reported by a previous crew, resulting in a flight taking place with this discrepancy; maintenance not being aware of a discrepancy report which has been completed; pilots not checking discrepancy reports prior to flight

Mechanical discrepancy – an inoperative aircraft component is identified by a pilot during flight operations

Student knowledge/skill – lack of student knowledge/skill that is expected, given the phase of training or experience level of the student

Instructor technique – lack of awareness of opportunity to allow students to learn from a situation; or, a lack of intervention when circumstances are beyond a student's skill level

In addition to coding the hazards, the phase of flight in which the hazard was reported was also recorded. These locations included traffic pattern (further coded as pre-flight, taxi, takeoff, departure, descent, approach, and landing).

Data Analysis

As described above, each safety report was ultimately coded with the hazards involved that led to the circumstances necessitating submission of a safety report. Multiple factors could be (and in most cases, were) found to be existent in each report. An overall analysis of contributing factors indicated revealed that non-compliance with standard procedures (90 instances) was by far the most prevalent factor found. Judgement/decision-making was the second highest factor found, with 72 instances. Student knowledge/skill (33 instances), instructor technique (29 instances) and communication issues (28) were the next three highest contributors (see Figure 1).

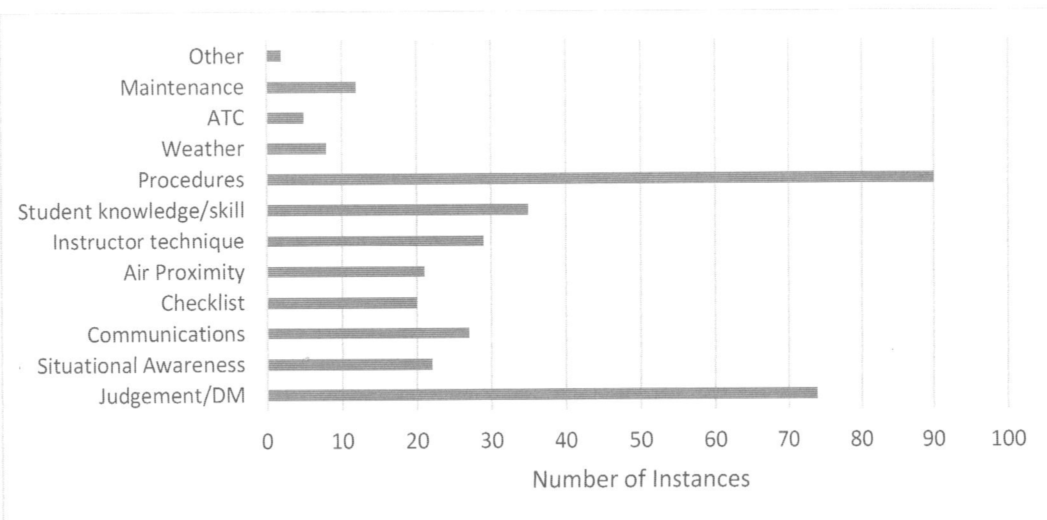

Figure 1. Contributing factors coded on safety reports

Hazard: Non-use of Standard Procedures

Given the high incidence of lack of use of standard procedures, the 90 safety reports coded with this hazard were scrutinized to determine the other hazards that existed in concert with non-use of standard procedures (see Figure 2). It was found that judgement and decision-making were also present in 57% of the safety reports that had procedures indicated. Student knowledge/skill and instructor technique were both also highly prevalent hazards in the safety reports that had lack of standard procedures cited as a hazard. Multiple reports with procedures indicated also specifically included improper use of checklists. These reports include items such as fuel mismanagement (landing with fuel imbalance side to side, or with less than flight school mandated one hour minimum reserve), forgetting to shut of magnetos (multi-engine aircraft) and forgetting to remove cowl plugs.

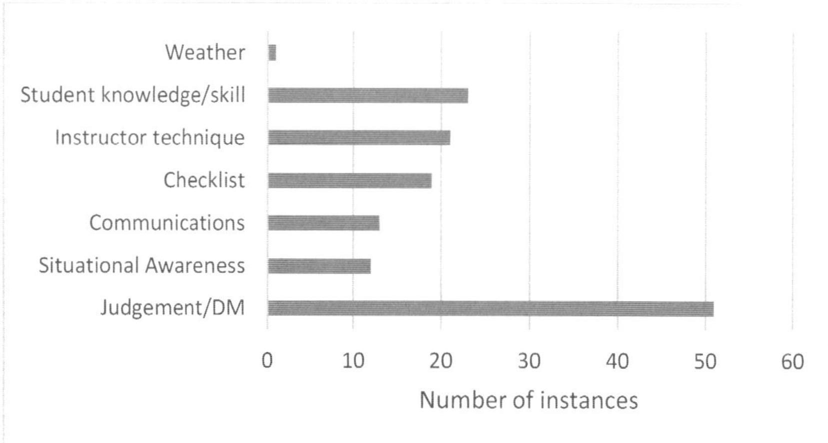

Figure 2. Frequency of occurrence of additional hazards in conjunction with lack of standard procedures usage

Hazard: Traffic Pattern

It is worth noting that 31 (34%) of the 90 reports that were found to have "procedures" as a hazard experienced were due to non-standard procedures conducted by aircraft in the traffic pattern. Similarly, 17 of the 54 (31%) of the reports with "judgement/decision-making" were from traffic pattern experiences. While a few cases involved flight school aircraft using non-standard procedures, the vast majority cited non-flight school aircraft which were not following standard traffic pattern procedures. The flight school is based at a non-towered public airport, where the traffic pattern is shared with another flight school and an active GA community, as well as significant itinerant traffic. The traffic pattern is often busy and just as often it is thought of as a hazard. Given the MTSU expectation of strict adherence to AIM recommended non-towered airport procedures, MTSU instructors and students have been quick to notice and file reports of aircraft that depart from those recommendations.

The traffic pattern issue was of concern to the researchers prior to beginning the formal analysis of hazards, as simply based on the anecdotal experience with safety reports over the years it was clear there was a high frequency of safety reports involving events in the local traffic pattern. An analysis of the safety reports revealed that 54 of the 176 total reports (31%) indicated the phase of flight in which the circumstances which caused the filing of the report was experienced was in the traffic pattern. While this is quite a large number, further analysis revealed that 14 of these "traffic pattern" reports indicated an air proximity (i.e. potential collision threat) danger. This means 26% of the traffic pattern reports were felt to be at the level of a potential mid-air collision threat, while the majority of the others cited lack of procedures (31 reports), lack of judgement/decision making (17 reports) and communication issues (16 reports). Figure 3 below gives a complete breakdown of the issues cited in the traffic pattern reports.

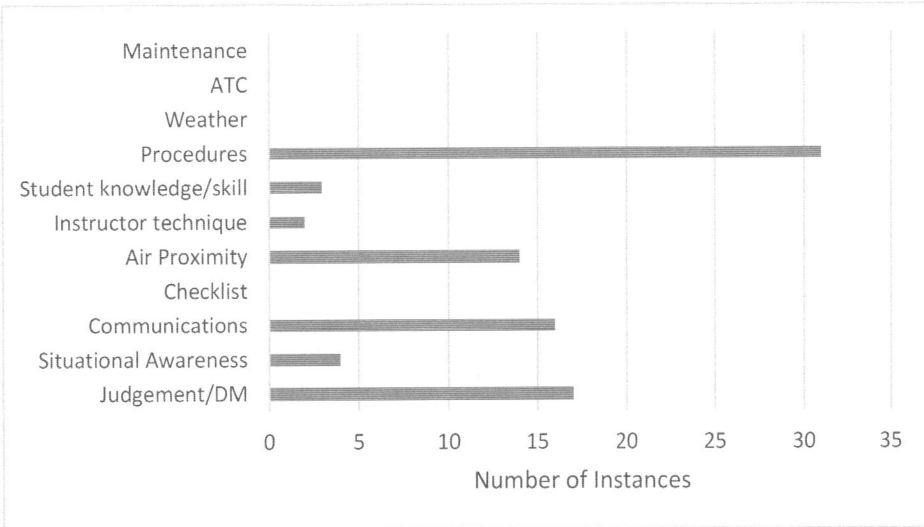

Figure 3. Hazards identified within traffic pattern operation safety reports

The phase of flight within the traffic pattern was also analyzed. The majority of the reports detailed circumstances within the landing phase (23 reports), with the approach phase (12 reports) next, followed by takeoff (10 reports) and departure (3 reports). Figure 4 below depicts the phase of traffic pattern reported.

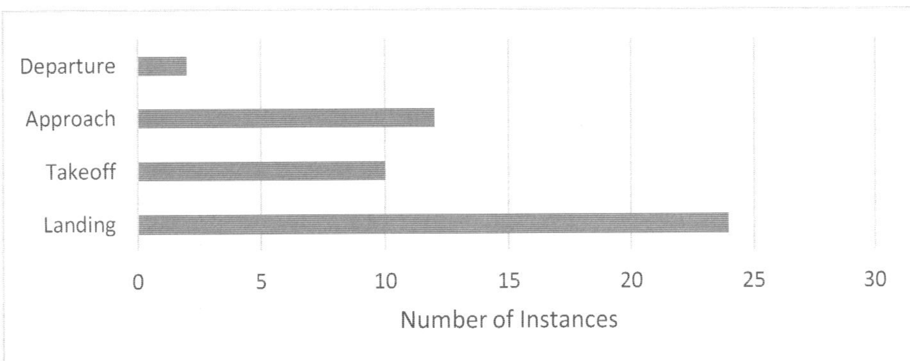

Figure 4. Phase of flight within traffic pattern when event was experienced

Hazard: Student Knowledge/Skill and Instructor Technique

Given the training environment inherent in a flight school, the hazards of "student knowledge/skill" and "instructor technique" were two specific items of interest. As indicated previously, 33 reports were found to have "student knowledge/skill" as a contributing factor, while 29 were found to have "instructor technique" as a contributing factor. When the overlap between these two hazards was evaluated, it was found that 14 of the reports indicating "instructor technique" were also found to have "student knowledge/skill" as a hazard. This was not surprising, as in these cases what caused the hazard was the instructor not realizing and responding to a lack of student knowledge until a situation warranting a safety report was encountered. In particular, 7 of the 14 reports indicating both instructor technique and student knowledge/skill occurred within the landing phase of flight, when instructor vigilance of and reaction to student actions is obviously much more time sensitive than in other phases of flight.

Conclusion

Unlike the hazards experienced by other aviation operations such as air carriers, flight schools operations by definition are associated with students in training. Even when the 31 traffic pattern procedures, mostly observed with non-flight school aircraft, are removed, 59 of the 176 reports (34%) indicate procedure issues, by far the largest category of hazards. While non-adherence to procedures is often cited by all aviation operators as a predominant hazard, this analysis of reports indicates the need to emphasize the importance of procedure use from the earliest days of flight training, even in relatively simple aircraft. This mitigation, in the form of specific communication to flight school students of the fact that non-adherence to standard procedures is the largest hazard, must continue to be a priority. Additional ways of making this point clear, such as during safety meetings and in academic classes, will be investigated. As a subset of procedures, checklist compliance must also continue to be emphasized.

If a student were asked what the most significant hazard experienced by students during flight training at this flight school would be, it is likely that operations in the traffic pattern would be cited. However, it is important to understand that the traffic pattern itself is simply a place and a phase of flight, not a hazard in itself. Analysis of the contributing factors suggest that the hazard(s) in this case are related to certain kinds of behavior in the traffic pattern. Aircraft non-compliance with recommended procedures in the traffic pattern is the hazard, coupled with lack of judgement/decision-making. Therefore an effort to improve procedural integrity, communication, and pilot judgment and decision making appears to be an avenue for effective mitigation. Mitigation in this example might involve providing all airport operators at this field with insight into the nature of the real hazard(s) in order to promote a common approach to traffic pattern procedures, communication, and pilot judgment and decision making.

An additional recommendation to come from this study is the need to further refine the safety reporting form that is currently in use. To assist in the continuing identification of trends in hazards, self-selection by reporters of the hazards experienced would be beneficial. While safety committee review, oversight, and coding of the reported hazards will be continued, this initial coding by users will greatly assist in the maintenance of an up to date hazard analysis database.

Acknowledgements

The authors would like to thank the MTSU Airport Safety Committee for their willingness to share this data for analysis and subsequent publication. And, a particular thank you is extended to Mr. Matt Torres for his diligence in maintaining the Safety Committee data base.

References

Bahr, N.J., (1997). *System safety engineering and risk assessment: A practical approach.* Taylor & Francis, NY.

Federal Aviation Administration (2015). AC No: 120-92B Safety management systems for aviation service providers. Retrieved from https://www.faa.gov/documentLibrary/media/Advisory_Circular/AC_120-92B.pdf

Safety Management International Collaboration Group (2013). Hazard taxonomy examples. Retrieved from http://www.skybrary.aero/bookshelf/books/2301.pdf

Safety Management International Collaboration Group (2010). Development of a common taxonomy for hazards. Retrieved from http://www.skybrary.aero/bookshelf/books/1779.pdf

LONGITUDINAL SAFETY CLIMATE ANALYSIS: MODELING FOR ENHANCED ORGANIZATIONAL RESPONSE

Erin E. Bowen
Behavioral & Safety Sciences Dept., Embry-Riddle Aeronautical University
Prescott, AZ

Brian J. Roggow
Aviation Safety Program Manager, Embry-Riddle Aeronautical University
Prescott, AZ

Much of the safety climate research captures only a transient state in the aviation environment, by extension limiting organizational responses to transactional approaches. The limits of the transient annual safety climate audit traps safety attitudinal/behavioral research in a static or reactive cycle. The present study takes advantage of a collegiate aviation environment with multiple training locations (each with its own culture), participating in regular safety climate audits across flight operations, to develop an enhanced safety culture model. Using longitudinal climate data collected from the organization, the authors present a mixed-methods trend analysis of safety climate changes to date, incorporating organizational structure and resource variables. The longitudinal model creates a more comprehensive evaluation of the long-term safety culture of the organization at all training locations and creates a new format for a more enhanced organizational response. The study utilizes the new longitudinal model as a framework for developing systems-based responses to climate concerns, and in turn documenting the impact of the organizational changes made in result. This paper presents initial findings based on the primary training location; final results are presented at the ISAP meeting and available after the presentation. Application across multiple aviation operation settings are discussed, including characteristics and strategies for improving organizational response to safety climate and culture evaluations.

Safety climate and safety culture have become nearly ubiquitous constructs in current discussions of both accident prevention and organizational performance (e.g., Block, et al, 2007; Gibbons et al, 2006; Karanikas, 2016). The ubiquity, rather than being a sign of a topic that has been over-researched, points to the criticality of these constructs and the acknowledgement that no research has yet completely tackled or resolved all of the challenges in the organizational safety climate field. The evolution of research into human error in aviation has continued to evolve from focus on the individual's error (e.g., Hunter, 2005) to crew/group level factors (e.g., Taylor & Thomas, 2003), and then to larger organizational influences (Mjos, 2004; Block, et al, 2007). This in turn has led to attempts to capture aspects of the the climate and/or culture within the organization that contribute to or impede 'safety' with regard to attitudes, policies, and behaviors (e.g., Bowen, et al 2011; Bowen, 2013).

Research conducted by Von Thaden, Wiegmann, and Shappell (2006) identified ten categories of organizational factors that appeared associated with commercial airline accidents investigated by the National Transportation Safety Board (NTSB). These factors included: training, surveillance, procedures/directives, standards, information, supervision, documentation, pressure, substantiation, and facilities. Their research indicated that inadequate procedures and directives were most commonly linked with aviation accidents. Both facets of their investigation provide strong evidence in favor of a systems theory approach to aviation safety. While the work of von Thaden, et al. and others (e.g., Soeters & Boer, 2000) in reviewing accident data for safety culture and organizational systems trends is extremely valuable for the creation of failure models of safety climate and culture, most aviation practitioners prefer to identify factors that will support safety in advance of incidents or accidents, rather than being forced to review and attempt to *post facto* address these failures.

One strategy to pre-emptively identify weaknesses or risk factors within an aviation organization is the implementation of an annual or semi-annual "safety climate audit". Employees at multiple levels of the organization may be asked to complete a written or oral questionnaire documenting their beliefs, behaviors, observations, or opinions regarding various categories of organizational factors and structure. Some of these questionnaires have been created by commercial designers and provided to the aviation organization, but many are self-created by a safety manager or other technical expert with safety responsibility. Many of these designing the questionnaires, however, lack any training on survey methodology, design, implementation, or analysis, leaving the organization with potentially incorrect or misleading data, or results that have been under-analyzed due to a lack of comprehension.

Unfortunately, the nature of organizational safety climate as residing heavily within the perceptions and beliefs of its members makes understanding of climate as anything more than a transient organizational state a challenge, particularly to the safety practitioner. Many practitioners as well as researchers focus on single-year findings or, at most, year-to-year changes in attitude or action as indicators of the health of the organization's safety climate, and by extension, its long-term culture (Schein, 2004). However, little work has been done to examine multi-year trends in safety climate audit data, nor to use such multi-year trends to begin an evaluation of the longer-term safety culture of the organization. The current research is an attempt to begin to fill this gap as well as provide insight into other scientist-practitioners faced with organizational questions and concerns about safety climate.

Methodology

In an attempt to begin to address the lack of multi-year data analysis within aviation organizations, the authors collected 4 years' worth of data (2012-2016) from the annual safety climate audit questionnaire at a U.S. university's collegiate aviation program, collected in the fall of each year. Flight instructors, dispatchers, office workers, and supervisors throughout the flight operation were requested to complete the audit survey each year; the organization has a nearly 100% response rate each year the survey was administered.

The safety climate audit questionnaire was created by the collegiate aviation program to evaluate potential safety concerns occurring at the individual, team, or organizational level. The 74-item questionnaire was designed by the organization and has been in use in various iterations since 2003. The most recent revision occurred in 2012; the present data set contains responses from 2013 to the present.

Respondent Demographics

Demographic data on respondents over the past 4 years can be seen in Table 1; as is apparent in the tables, respondents are primarily young (69.4% are age 30 or younger) instructors (88.1%) who are relatively new to the organization (77.7% have five or fewers years with the operation). A total of 175 respondents completed the audit questionnaire over the past four years.

Table 1: Respondent Demographics

Respondent Reported Age Ranges

20-30	118
31-40	36
41-50	5
51-60	0
60	11

Primary Job Responsibility

Flight Instructor	148
Supervisor/Manager	20

	Years in Organization	Years at Current Job
<1	38	68
1-5	94	76
6-10	19	16
11-15	14	5
16-20	3	3
20+	2	2

Certificates/Ratings Possessed

CFI	6
CFII	84
MEI	59
ATP	13
Other	3

The relative youth and short tenure of the majority of organization employees would suggest a safety climate that would be more likely to be transient from year to year based on turnover and developmental factors. To evaluate this, year over year comparisons for the safety climate audit were analyzed using univariate analysis of variance with Bonferroni correction. Results found that, of the 74 items on the safety climate questionnaire, only ten showed significant change in the past four years. These items can be seen in Table 2.

Table 2: Questionnaire Items with Significant Longitudinal Change

Safety Climate Item	F value	Pairwise Comparison Mean Scores (Significant)	
The Assistant Aviation Safety Program Manager has the power to make changes.	$F_{(3,169)}$ =3.613, p=.015	4.2766, s=1.28 5.1166, s=1.16	Year 1 Year 4
The Assistant Aviation Safety Program Manager has little or no authority compared to operations personnel.	$F_{(3,169)}$ =3.032, p=.031	3.978, s=1.39 3.113, s=1.29	Year 1 Year 4
Flight department management shows favoritism to certain pilots.	$F_{(3,170)}$ =4.635, p=.004	4.500, s=1.709 3.204, s=1.678	Year 3 Year 4
Pilots who call in fatigued fear being scrutinized by the chief pilot.	$F_{(3,171)}$ =4.164, p=.007	3.707, s=1.887 2.477, s=1.355	Year 3 Year 4
The chief pilot does not hesitate to contact instructor pilots to discuss safety issues.	$F_{(3,171)}$ =4.212, p=.007	4.553, s=1.47 5.302, s=1.26 4.553, s=1.47	Year 1 Year 2 Year 1

		5.463, s=1.24	Year 3
As long as there is no accident or incident, the chief pilot does not care how flight operations are performed.	$F_{(3,171)}$ =2.761, p=044	2.425, s=1.39 1.772, s=0.773	Year 1 Year 4
The chief pilot has a clear understanding of risks associated with flight operations.	$F_{(3,171)}$ =3.513, p=.017	5.872, s=1.11 6.418, s=0.663 5.872, s=1.11 6.454, s=0.588	Year 1 Year 2 Year 1 Year 4
Pilots often report safety concerns to their chief pilot rather than the safety department.	$F_{(3,169)}$ =2.952, p=.034	2.617, s=1.189 3.348, s=1.325	2013-2014
The flight supervisor consistently emphasizes information or details (e.g., weather requirements, NOTAMs) that affect flight safety.	$F_{(3,171)}$ =2.927, p=.035	4.634, s=1.71 5.545, s=1.021	Year 3 Year 4
The flight supervisor is responsive to pilots' concerns about safety.	$F_{(3,171)}$ =3.142, p=.027	5.439, s=1.449 6.09, s=0.603)	Year 3 Year 4

Factor Analysis

The survey was designed with items clustered around 14 theoretical constructs; however, no confirmatory analyses had been conducted to evaluate the extent to which questions actually mapped to the organizational factors. Given the high turnover rate of the primary respondent group (flight instructors), as well as the gap in time between each administration of the audit questionnaire, the decision was made for the purpose of preliminary analyses to treat the annual samples as independent for the purpose of evaluating the proposed factors. Even taking this liberty, principal components analysis (PCA) failed to provide a stable factor structure. PCA was attempted in order to reduce the number of survey items in use for subsequent analyses and provide recommendations to the flight training program for ways in which to reduce the length of the questionnaire. This failure is in part likely due to the questionnaire length (74 items) and relative overall sample size (N=175).

Table 3: Intended Factors of Safety Climate Audit

Reporting System
Aviation Safety Program Manager
Assistant Aviation Safety Program Manager
Accountability
Pilot Authority
Professionalism
Chief Flight Instructor
Training Managers
Flight Supervisor
Ramp Operations
Instructors
Safety Values
Going Beyond Compliance
Institution Safety Record

Discussion

The present study sought to increase understanding of longitudinal trends in organizational safety climate, in order to identify stronger leverage points for organizational change and enhanced safety performance. In addition, the study sought to evaluate the quality and utility of the annual safety climate audit questionnaire in use in a large-scale collegiate aviation training organization. Data presented here are based upon analysis of the initial training location under investigation; final results based upon multi-site comparison are presented at the International Symposium on Aviation Psychology and available after that meeting.

One of the key concerns to be discussed in final presentation of the data is the utilization of disparate safety climate audits at each flight training location within the institution. The authors strongly recommend that the institution identifies a single set of safety climate items for use at all training locations in order to facilitate future cross-analyses and the impact of larger institutional trends.

Data from Site 1 suggest that employee attitudes regarding the majority of safety climate components have remained consistent over the past four years. Only ten of the 74 items in the climate questionnaire showed significant differences in comparing data over time. This, when coupled with the high rate of turnover among front line flight instructor employees, suggests a remarkably consistent culture in existence within the training operation. This may be due in part to the highly-regulated structure of the FAA Part 141 training program, or due to other organizational factors. This finding may be one of the most significant of the study, as it indicates areas in which change may be initially occurring within the organization, with regard to employee attitudes. These ten items may be the indices of leverage points within the organization; future research to explore and clarify these results is planned.

The authors propose substantially reducing the number of items in the safety climate audit using a more theory-based factor structure. The current structure, with 74 items attempting to encompass 14 factors, contains a large degree of conceptual overlap and a lack of question clarity. This can be seen in the failure of the principal components analysis to provide a consistent factor structure.

References

Block, E. E., Sabin, E. J., & Patankar, M. S. (2007). The structure of safety climate for accident free flight crews. *International Journal of Applied Aviation Studies, 7,* 46-59.

Bowen, E. E. (2013). Predicting impact of maintenance resource management training utilizing individual difference variables. *Journal of Aviation/Aerospace Education and Research, 22(3),* 9-17.

Bowen, E. E., Sabin, E. J., & Patankar, M. S. (2011). Aviation maintenance human factors in a systems context: Implications for training. *International Journal of Applied Aviation Studies, 11*(1), 13-26.

Hunter, D.R. (2005). Measurement of hazardous attitudes among pilots. *The International Journal of Aviation Psychology, 15*(1), 23-42.

Karanikas, N. (2016). Critical review of safety performance metrics. *International Journal of Business Performance Management,* 17(3), 266-285.

Mjos, K. (2004). Basic cultural elements affecting the team function on the flight deck. *The International Journal of Aviation Psychology, 14*(2), 151-169.

Schein, E.H. (2004). Organizational Culture and Leadership (3rd edition). New York: Wiley & Sons, Jossey-Bass.

Soeters, J.L. & Boer, P.C. (2000). Culture and flight safety in military aviation. *Human Resource Development Quarterly, 8*(1), 23-40.

Taylor, J.C. & Thomas, R.L. (2003). Toward measuring safety culture in aviation maintenance: The structure of trust and professionalism. *The International Journal of Aviation Psychology, 13*(4), 321-343.

Von Thaden, T.L., Wiegmann, D.A., & Shappell, S.A. (2006). Organizational factors in commercial aviation accidents. *The International Journal of Aviation Psychology, 16*(3), 239-261.

RELATIONSHIP BETWEEN RELATIONAL ENERGY, EMOTIONAL LABOR, AND COGNITIVE FLEXIBILITY AMONG FLIGHT ATTENDANTS

Rithi Baruah

Christ University

Bangalore, India

The primary aim of the civil avaiation insdustry is to work for the safety and comfort of their clients and customers. This study concentrated on the frontline employees of the aviation industry, the flight attendants who are paid to smile. Energy at workplace is a fairly new concept and is an organisational resource which help employee attain their goals. The aim of the study was to establish the relationship between relational energy and the major issue of emotional labor (deep acting and surface acting) and cognitive flexibility among flight attendants. A correlational research design was used to study the relationship among 39 flight attendants in India. The study revealed that relational energy was positively related to deep acting. Also, relational energy proved to be a significant predictor of deep acting. However, no statistically significant relationships were found between relational energy and surface acting and also between relational energy and cognitive flexibility.

The work culture in general has seen a shift from working in industries to the ones that include working for people. The civil aviation industry is one such sectors whose primary aim is to work for the safety and comfort of their clients and customers. The crew members in the civil aviation industry include pilots, flight attendants, air traffic controllers, and baggage and maintenance personnel. In any airline industry the frontline workers are the flight attendants also called the cabin crew members. Although the job of the flight attendants seems to be glamorous and appealing, it is very strenuous and taxing. As the cabin attendants are the first source for the clients and customers to form an impression about the airline company, it becomes imperative for the cabin attendants to deliver the best possible services. In this bargain, the well-being of the cabin crew members are often neglected.

Issues Experienced by Flight Crew

Some of the leading issues in the aviation industry are interpersonal and communication errors (Avis, 2012). Emotional dynamics also lead to malfunction in communications (Brown & Moren, 2003). As reported by Avis in 2012, 37% of the aviation employees primarily the pilots, cabin crew, and ground employees do not communicate the relevant information to other crew members thereby resulting in major mishaps. Brown and Moren (2003) reported that the sterile cock-pit rule also leads to major frustrations between the pilot and the flight attendants. Some of the emotions that the crew members face are that of shame, excitement, awkwardness and inhibitions that seem to adversely affect their performance (Avis, 2012). Inconsistent work schedules, different time zones at work, food habits, variable altitudes, attitudes of the aviation employees, differences in culture and continuous interactions with clients and customers lead to mental and physical exhaustion.

Emotional Labor

Flight attendants belong to the niche group of population who are paid to smile (Hoschild, 2003). No matter what the flight attendant is going through on actuality their work situation demands them to smile and maintain a positive demeanour. Hoschild (1983) called this as Emotional Labor (EL) in 1983. She defined Emotional Labour as managing one's feelings to produce a publicly acknowledged facial and bodily demonstration of emotions. When there is an incongruence in the emotions felt and emotions exhibited, there is emotion dissonance (1983).

Forms of emotional labor. According to Hochschild (1983), there are two forms of emotional labor. They are surface acting and deep acting. Surface acting involves acting or expressing an emotion on the surface without actually feeling them (Hochschild, 1983). Deep acting involves modifying feeling to match the organizationally demanded emotion (Hochschild, 1983). Though both the types of emotional labour signify dissimilar intensions they are internally false. That is, surface acting involves managing the overt expressions to abide by the organizational display rules, while deep acting consists of managing the underlying emotions to genuinely feel the emotion demanded by the display rules (Grandey, 2000; Hochschild, 1983).

Relational Energy

Energy at workplace is a fairly new concept. As cited by Owen, Baker, Sumpter and Cameron in 2015, the capacity of the employee's motivation and action is influenced by the energy at work. They cited that energy is an organisational resource which help employee attain their goals. Absence of energy results in stress, burnout, and disengagement (Sonnentag, Kuttler, & Fritz, 2010; Schaufeli, Bakker, & Van Rhenen, 2009;

Demerouti, Bakker, Nachreiner, & Schaufeli, 2001). Research reveals that work performance levels are improved if individuals are surrounded by energised people (Baker, Cross, & Wooten, 2003; Cross & Parker, 2004). Owens et al. (2015) defined relational energy (RE) as a "heightened level of psychological resourcefulness generated from interpersonal interactions that enhances one's capacity to do work" (p 37).

Cognitive Flexibility

According to Martin and Rubin (1995, p 623), "Cognitive flexibility refers to a person's (a) awareness that in any given situation there are options and alternatives available, (b) willingness to be flexible and adapt to the situation, and (c) self-efficacy in being flexible". Individuals who acknowledge more possible adjustments are more cognitively flexible than the counterparts.

In 2005, Canas defined cognitive flexibility as the individual's ability to change and adapt the strategies of cognitive processing to face unpredictable and new situations in the environment. When faced with any new problem, individuals with higher cognitive flexibility will be able to consider various alternatives and will outperform the others with lower cognitive flexibility (Stewin & Anderson, 1974). It has been reported that more an individual is cognitively flexible, better will be his/her ability to optimise his/her potential (Bergland, 2015). According to Bergland (2015), previous studies showed that higher levels of cognitive flexibility are directly related to resilience in adulthood, better reading capabilities of children, and higher quality of life in older age. The neurological mechanics of cognitive flexibility are directly linked to multitasking executive functions.

Rationale of the Study

Relational energy being a relatively new concept of energy at work, the empirical studies are scanty. Therefore, the researcher tried to fill in the gap by carrying out further exploration of relational energy and contribute to the theory. According to conservation of energy theory, lack of resources at work lead to burnout (Owen et al., 2015). And enhanced psychological resources which result from relational energy at work would enhance coping with stressors at work, burnout and lead to well-being at work. Therefore, the researcher aimed to study and verify if better relational energy at work result in lower emotional labor which causes burnout, thereby, filling the research gap.

Also, previous research have shown the influence of emotions over cognitive flexibility but no published research has tried to find if emotional labor (surface and deep acting) has any relation with the flexibility of cognition. Cognitive flexibility has also shown to be effective in interpersonal and intrapersonal communication. This study tried to fill the gap by investigating if interaction with the human resources at work and their psychological exchange have any relation with cognitive flexibility. Positive affect has proved to be better predictor of cognitive flexibility. Therefore, the researcher aimed to study if emotional labor (surface and deep acting), which is related to positive and negative affect is also related to cognitive flexibility.

Objectives of the Study

To study the relationship between relational energy, emotional labor (surface acting and deep acting) and cognitive flexibility.

Hypotheses of the Study

H(1). There is no relationship between relational energy and surface acting.

H(2). There is no relationship between relational energy and deep acting.

H(3). There is no relationship between relational energy and cognitive flexibility.

H(4). There is no relationship between cognitive flexibility and surface acting.

H(5). There is no relationship between cognitive flexibility and deep acting.

Tools Used in the Study

Relational energy Scale. This scale was developed by Owens et al. 2015. There are five items which are measured on a 7-point Likert scale (1-strongly disagree to 7-strongly agree). The reliability of this scale is 0.96.

Emotional labor Scale. This scale was developed by Brotheridge and Lee in 2003. It is a 5 point Likert scale with 14 items. The reliability of this scale is 0.89.

Cognitive flexibility Scale. This scale was developed by Martin and Rubin in 1995. It is a 6-point rating scale with 12 items. The reliability of the ale is 0.83.

Procedure. An online survey was carried out. Individuals who gave their consent to participate in the study were included. The online questionnaire comprised of the consent form, demographic checklist, Relational Energy Scale, the Emotional Labor Scale, and the Cognitive Flexibility Scale. 39 participants responded to the online survey. The data gathered was subjected to appropriate statistical analysis using SPSS 21. Subject matter experts, academic experts and cabin crew members were asked to give their feedback on the same.

Results and Discussion

Table 1.

Descriptive Statistics of the Variables Under Study.

Variable	Total	Mean	Std. Deviation
Relational Energy	39	23.90	8.178
Cognitive Flexibility	39	51.56	6.648
Deep Acting	39	9.44	3.119
Surface Acting	39	44.46	10.918

Table 3.

Table 2.

Findings Based on Correlation Analysis.

Variable	1	2	3	4
1. Relational Energy				
2. Deep Acting	0.632**			
3. Surface Acting	0.221			
4. Cognitive Flexibility	0.027	0.038	0.230	

** Correlation is significant at 0.01 level (2-tailed)

Findings Based on Regression Analysis.

Variable	R	R2	F	Sig.
Relational Energy Deep Acting	0.585	0.343	19.294	0.000

The data gathered was subjected to Shapiro-Wilk normality test and it was found that the data for emotional labor (deep acting and surface acting) was normally distributed. However, the data for relational energy and cognitive flexibility was not normally distributed. The descriptive analysis of the data is displayed in Table 1.

To check hypotheses 1, 2, 3, 4, and 5, Spearman's correlation was used respectively. From Table 2 it is seen that a significant positive relationship was found between relational energy and deep acting. However no statistical significant relationship was found in testing hypotheses 1, 3, 4, and 5. Table 3 displays the Regression analysis between relational energy and deep acting. The table reveals that relational energy is a significant predictor of deep acting.

From the analysis we can conclude that the exchange of psychological resources (relational energy) that take place at work increases the flight attendants to deep act. Previous research has shown a positive relationship between deep acting and variables like job performance (Bursali, Bagci, & Kok, 2013), employee creativity and role prescribed customer service performance (C. Liu, X. Liu, & Geng, 2013). The result also supports the precious findings that work performance are improved when surrounded by energised people (Baker, Cross, & Wooten, 2003; Cross & Parker, 2004). Therefore, relational energy at work should be encouraged during Crew Resource Management in the Aviation industry and practised among colleagues so that a healthy environment at work is maintained and thereby help the flight attendants to ward off the negative consequences related to emotional labor at work.

Scope of the Study

The insignificant relationships with regard to the other hypotheses could be because of the fact that the data was gathered from only 39 flight attendants in India, which may not be adequate to represent the population. Also, all the tools used in the study were self-report measures and hence there is a probability of personal bias in their responses and also the accuracy of retrospective accounts are questionable in self reports. Hence, to gain more interesting insights, a qualitative analysis would be adopted by the researcher in the future.

From the feedback gathered by the subject matter experts it was observed that, the relational energy scale which had five items measured the psychological exchange that take place between two specific individuals at work. As the job of the flight attendants require them to go on rotation with no specific group of batch mates or colleagues on a regular basis, this may cause a discrepancy in the way they responded to the items. Therefore, there is a scope for adaptation of the relational energy scale to suit the specific sample of this study and come out with interesting findings that can contribute to the literature of Aviation Psychology and also help the Aviation industry at large.

Ethical Consideration

1. Participants were included in the study only after their consent was taken.
2. Participants had the liberty to withdraw from the study at any point they want.
3. Confidentiality and anonymity was maintained.
4. Participants were given the option to ask for their results if they were willing to know.
5. No injury or harm of any nature was meted out to the participants during the research.

Acknowledgements

I would like to extend my sincere gratitude to Dr Tony Sam George, HOD of Department of Psychology, Christ University, India, for his constant support and encouragement. I am indeed grateful to Dr K. Jayasankar Reddy, Associate Professor, Christ University, and my PhD research supervisor for his abled guidance and inputs. Also, I would like to thank Dr Anuradha Sathiyaseelan, Associate Professor, Christ University, and Ms Sowgandhi Chaturvedala, Aviation Psychologist, IAF for their expert opinion and invaluable feedback. I deeply appreciate and thank all the participants for taking time out to take this survey and for their valuable inputs. I would sincerely like to thank Dr Bradley P. Owens for graciously accepting my request and allowing me to adapt the Relational Energy Scale to suit the sample of my study. Last but not the least, I would like to thank the Almighty and my family for their never ending support and blessings.

References

Avis, J. (2012). The Social and Psychological Aspects Behind Flight. *Invoke: Undergraduate Sociology Journal, 1 (1).*

Baker, W., Cross, R., & Wooten, M. (2003). Positive organizational network analysis and energizing relationships. In K. S. Cameron, J. E. Dutton, & R. E. Quinn (Eds.), *Positive organizational scholarship: Foundations of a new discipline* (pp. 328–342). San Francisco, CA: Berrett-Koehler.

Bradley. J. R., & Cartwright. S. (2002). Social Support, Job Stress, Health, and Job Satisfaction Among Nurses in the United Kingdom. *International Journal of Stress Management. 9*(3), 163-182.

Brotheridge, C. M., & Lee, R. T. (2003). Development and validation of the Emotional Labour Scale. *Journal of Occupational and Organizational Psychology, 76,* 365.

Brown, N., & Moren, C. R. (2003). Background Emotional Dynamics of Crew Resource Management: Shame Emotions and Coping Responses. *International Journal of Aviation Psychology 13*(3):269-286. DOI: 10.1207/S15327108IJAP1303_05.

Bursalı, Y. M., Bağcı, Z., & Kök, S. B. (2014). The relationship between emotional labor and task/contextual/innovative job performance: A study with private banking employees in Denizli. European Journal of Research on Education, 2(2), 221-228.
Cole, M. S., Bruch, H., & Vogel, B. (2012). Energy at work: A measurement validation and linkage to unit effectiveness. *Journal of Organizational Behavior, 33*, 445–467. http://dx.doi.org/10.1002/job.759.
Cossette, M. & Hess, U. (2009). *How organizational factors impact emotional labour strategies.* Paper presented at the Conference of the Administrative Sciences Association of Canada, Niagara Falls.
Cross, R., & Parker, A. (2004). The hidden power of social networks: Understanding how work really gets done in organizations. Boston, MA: Harvard Business School Press.
Demerouti, E., Bakker, A. B., Nachreiner, F., & Schaufeli, W. B. (2001). The job demands resources model of burnout. *Journal of Applied Psychology, 86,* 499–512. http://dx.doi.org/10.1037/0021-9010.86.3.499.
Grant S. Shields, Brian C. Trainor, Jovian C. W. Lam & Andrew P. Yonelinas (2016): Acute stress impairs cognitive flexibility in men, not women. Stress, DOI: 10.1080/10253890.2016.1192603.
Goodwin, R. E., Groth, M., & Frenkel, S. J. (2011). Relationships between emotional labor, job performance, and turnover Journal of Vocational Behavior, 79 (2), 538-548 doi:10.1016/j.jvb.2011.03.001.
Hochschild, A. R. (1983). The managed heart: Commercialization of human feeling. Berkeley, CA: University of California Press.
Liu, C., Liu, X., & Geng, Z. (2013). Emotional labor strategies and service performance: the mediating role of employee creativity. Journal of Applied Business Research, 29(5), 1583.
Owens, B.P., Baker, W.E., Sumpter, D.M., & Cameron, K.S. (2016). Relational Energy at Work: Implications for Job Engagement and Job Performance. Journal of Applied Psychology, 101 (1), 35-49.
Martin, M., & Rubin, R. (1995). A New Measure of Cognitive Flexibility. Psychological Reports, 76, 623-626, DOI: 10.2466/pr0.1995.76.2.623.
Schneiderman, N., Ironson, G., & Siegel, S. D. (2005). Stress and Health: Psychological, Behavioral, and Biological Determinants. Annual Review of Clinical Psychology, 1, 607–628. http://doi.org/10.1146/annurev.clinpsy.1.102803.144141.
Sonnentag, S., Kuttler, I., & Fritz, C. (2010). Job stressors, emotional exhaustion, and need for recovery: A multi-source study on the benefits of psychological detachment. Journal of Vocational Behavior, 76, 355–365 http://dx.doi.org/10.1016/j.jvb.2009.06.005.
Schaufeli, W. B., Bakker, A. B., & Van Rhenen, W. (2009). How changes in job demands and resources predict burnout, work engagement, and sickness absenteeism. Journal of Organizational Behavior, 30, 893–917. http://dx.doi.org/10.1002/job.595.
Schneiderman, N., Ironson, G., & Siegel, S. D. Stress and Health: Psychological, Behavioral, and Biological Determinants. Annual Review of Clinical Psychology, (1), 607-628) DOI: 10.1146/annurev.clinpsy.1.102803.144141.
Suryavanshi, R. (2015). Exploring the Effects of Cognitive Flexibility and Contextual Interference on Performance and Retention in a Simulated Environment.

INNOVATIVE AIRPORT VISUAL AIDS TO ENHANCE SITUATIONAL AWARENESS AND FLIGHT TRAINING FOR GENERAL AVIATION

Wesley Major
Sarah Hubbard
Purdue University
West Lafayette, Indiana

Runway incursions are a threat to runway safety and have been increasing in recent years. Incursions are categorized into three categories, pilot deviations (PD), operational incidents (OI), and vehicle pedestrian deviations (VPD). At general aviation airports, PDs are the most prevalent runway incursion type. Inadequate situational awareness is one of the human factors associated with PDs. Student pilots, pilots flying to an unfamiliar airport, ground operations personnel, and emergency planning and emergency responders can benefit from the use of visual aids that extend beyond an airport diagram or static Google Earth imagery. More robust visual aids can potentially increase situational awareness and reduce the risk of a runway incursion, and increase airfield familiarity through 360-degree photographs of the airfield facilities, including markings, signage, and intersecting taxiways/runways. This educational and informational tool has the ability to increase familiarity of airfield characteristics and increase safety.

Safety is the top priority in aviation, and runway safety is a critical aspect of aviation safety. According to the International Civil Aviation Organization (ICAO), runway safety related events account for more than half of all accidents, and 14% of fatal accidents (International Civil Aviation Organization, 2015). For this reason, runway safety is a high priority for all aviation stakeholders, and reducing runway incursions is one way to improve runway safety. Since October 2001 there have been 19,184 runway incursions at United States airports (Federal Aviation Administration, 2017b). The Federal Aviation Administration (FAA) defines a runway incursion as, "any occurrence at an aerodrome involving the incorrect presence of an aircraft, vehicle or person on the protected area of a surface designated for the landing and take-off of aircraft" (Federal Aviation Administration, 2017c). FAA categorizes incursions based on the cause, resulting in the following three incident types (Federal Aviation Administration, 2012):

- *Operational incident (OI):* runway incursion caused by air traffic controller (ATC) error that violates the required minimum separation between two or more aircraft or between an aircraft and an obstacle,
- *Pilot deviation (PD):* runway incursion caused by pilot error that violates any Federal Aviation Regulation, such as entry onto runway without permission, and
- *Vehicle/pedestrian deviation (V/PD):* runway incursion caused by unauthorized entry of vehicles or pedestrians onto the airport movement areas, such as ground vehicle entry onto runway without ATC authorization.

Incursions have been increasing in recent years. As a result of this increase, the FAA announced the Runway Incursion Mitigation (RIM) program in 2015. The purpose of RIM is to identify airport risk factors that might contribute to a runway incursion. Examples of risk factors consist of unclear taxiway markings, unclear airport signage, and complex airfield geometries, including unusual runway or taxiway layouts, and runway intersections. Through RIM, the FAA is focusing on reducing runway incursions by addressing risks at specific locations at the various airports, especially those that have a history of runway incursions (Federal Aviation Administration, 2016). While the FAA continues to be proactive in taking steps to reduce runway incursions through mitigation measures and airfield development projects that reduce runway intersections, there are other potential opportunities to improve runway safety that would support the activities of the RIM program.

The proposed tool suggested in this research leverages technology to supplement and augment existing airfield diagrams and increase situational awareness for pilots and ground operations workers. Traditionally, airport diagrams (Figure 1) are used to familiarize pilots and other personnel with an airport's layout and geometry. While these diagrams meet basic needs and provide one frame of reference, they do not provide pilots and airport ground crews with a visual representation of the airport facilities, markings, signage, and intersecting taxiways/runways. Creating a more robust visual aid will fill this gap and potentially improve airport safety and pilot training.

Literature Review

Traditional airport diagrams reflect an aerial perspective and provide critical, but rudimentary information regarding how the runways, taxiways and terminal are oriented with respect to true north, and with respect to one another. Airport diagrams fill a critical need for pilots as they plan their trip and upon approach to an airport. Once on the ground, however, many pilots and ground vehicle operators would benefit from a more robust depiction of airport facilities. Specifically, one that is enhanced with actual photo images. The addition of visual references can be important to convey information, especially in complex environments such as airports where situational awareness is critical to safe operations.

Perhaps the need for enhanced tools is best evidenced by current runway incursion statistics. Since October 2001, there have been 6,288 runway incursions at general aviation (GA) airports, with a majority of these incursions classified as PD or V/PD (Figure 2) (Federal Aviation Administration, 2017b). Nearly two-thirds of GA incursions are a result of pilot error. Chang and Wong (2012), as well as Endsley and Garland (2000), identified a lack of situational awareness as one of the leading factors associated with pilot error. The proposed use of photo enhanced airfield diagram would be an appropriate intervention strategy to enhance situational awareness not only for pilots, but also for ground operators; together these two categories cause 96% of runway incursions.

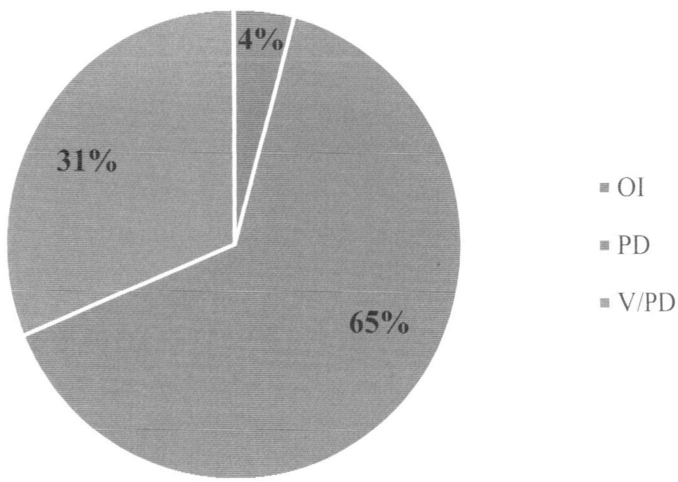

Figure 1. Classification of runway incursions at GA airports since October 2001 (Federal Aviation Administration, 2017b).

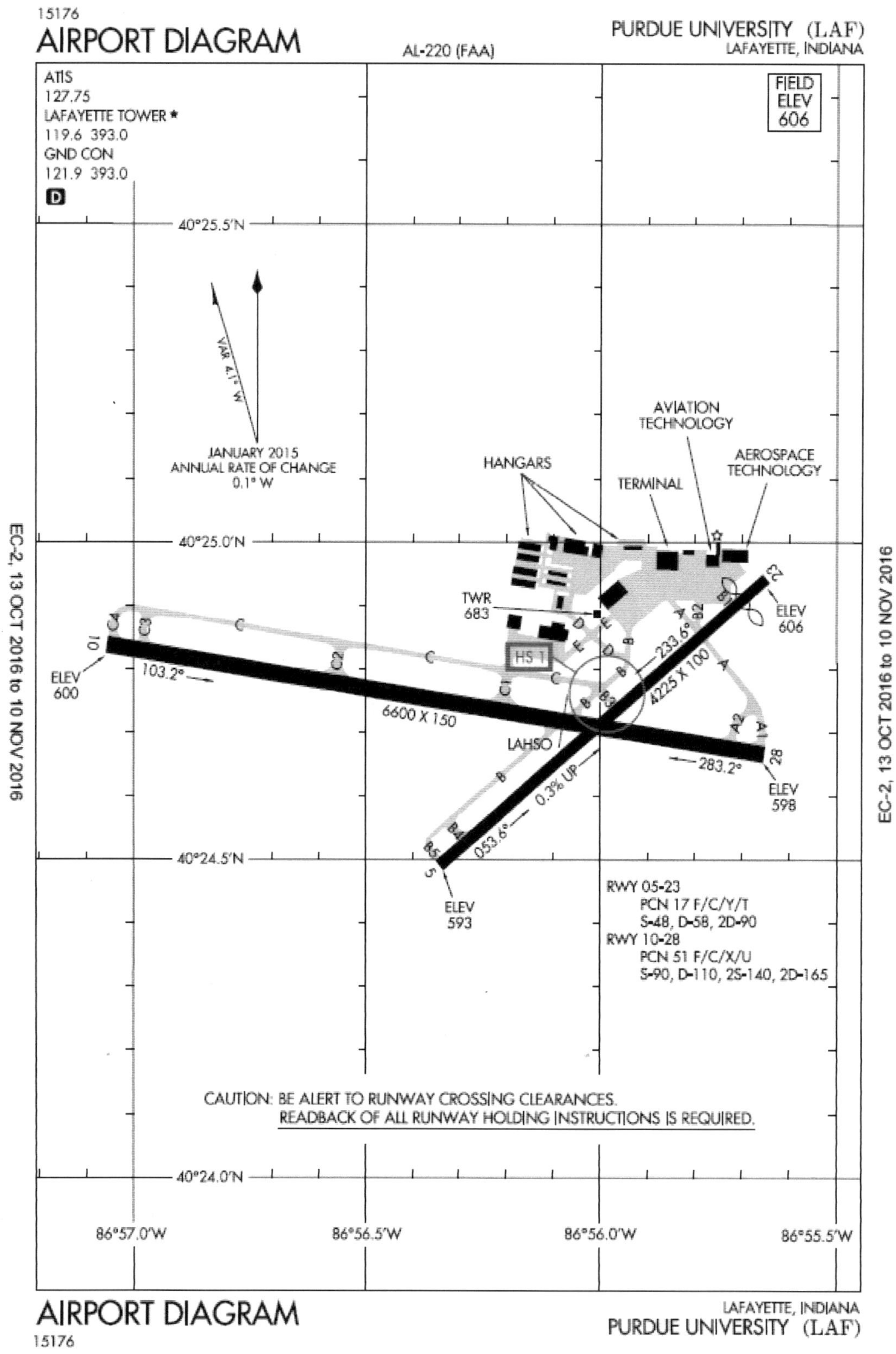

Figure 2. Airport diagram of Purdue University Airport (Federal Aviation Administration, 2017a).

Dublin International Airport was the first airport to address this need with the creation of a Google Street View perspective for their airfield. Vincent Harrison, Dublin Airport managing director, said, "these images will help the airport's Airside Safety Training department, as they will become an essential piece of the training suite in educating and familiarizing all airport employees" (Kennedy, 2016). The advent of virtual globe software, such as Google Street View, allows users to navigate and explore areas in three dimensions. This is very useful, as reported by Schultz, Kerski, and Patterson (2008), virtual globes can be used by educators to help students think spatially by investigating processes and places.

Other research has also demonstrated the value of this perspective. Oulasvirta, Estlander, and Nurminen (2009) compare 2D maps (similar to the traditional airport diagram) with 3D maps (similar to Google Street View), and state 3D imagery can provide realistic first-person perspective versus a use of flat symbolic illustrations to represent space. The information gathered through maps (2D) and actual navigation (analogous to Google Street View) is different. From a map, people acquire survey knowledge, from navigation people acquire procedural knowledge of the routes connecting diverse locations (Thorndyke & Hayes-Roth, 1982). The use of landmarks is often a key element in navigation (Raubal & Winter, 2002; Snowdon & Kray, 2009) and the integration of realistic visual cues to supplement existing airfield tools makes good sense.

Creating A More Robust Airport Visual Aid

Currently, through satellite images, a Google Earth view is available for airports, and is used by many GA pilots to provide additional information when landing at an unfamiliar airport. While the Google Earth view is helpful, it provides a top down perspective that is useful from the air (and similar to the perspective provided by the airport diagram), but less useful from the perspective of a taxiing pilot or a ground vehicle operator. Both Google Earth and the airport diagram lack the ability to convey important spatial cues, including airfield signs, markings, and views of intersecting taxiways and runways, as observed during taxiing and airfield operations. Through the use of emerging technologies, an improved and more robust visual aid can be created to allow an accurate representation of the sight picture pilots and operations personnel will encounter on the airfield.

This research explores the use of 360-degree photo enhanced airfield diagrams. Photo spheres, or 360-degree still photos, were taken of select locations on Purdue University's airport (KLAF). When paired with the airport diagram, the result is a more robust tool for training and for airport familiarization. Figure 3 shows how an airfield diagram can be enhanced with 360-degree photos. These photos can illustrate not only the upcoming pavement markings (both threshold markings and runway markings) and airfield signs, but also the upcoming intersection. Google Maps allows the photo spheres to be linked together to create a custom Street View, which is an application many users are familiar with from landside applications on city streets. Creation of an enhanced airport diagram allows for specific and unique areas of each airfield to be highlighted and emphasized to the wide variety of personnel that may need to operate on the ground.

Anticipated Uses and Benefits of the Enhanced Airport Diagram

This enhanced airport diagram could be utilized by many aviation stakeholders. In addition to pilots and ground operations personnel, airport managers, emergency planning and emergency response crews (including community partners who participate under a Memorandum of Agreement in an emergency) are some of the groups that could benefit from such a tool. Other users include student pilots, who can begin to orient themselves to the airport layout and airfield markings prior to beginning their flight training. Certified flight instructors (CFI) can walk a student through expected taxi procedures to provide virtual experience navigating an airfield for the first time; this virtual experience could allow student pilots to focus on aircraft operations rather than airfield orientation, especially in the beginning

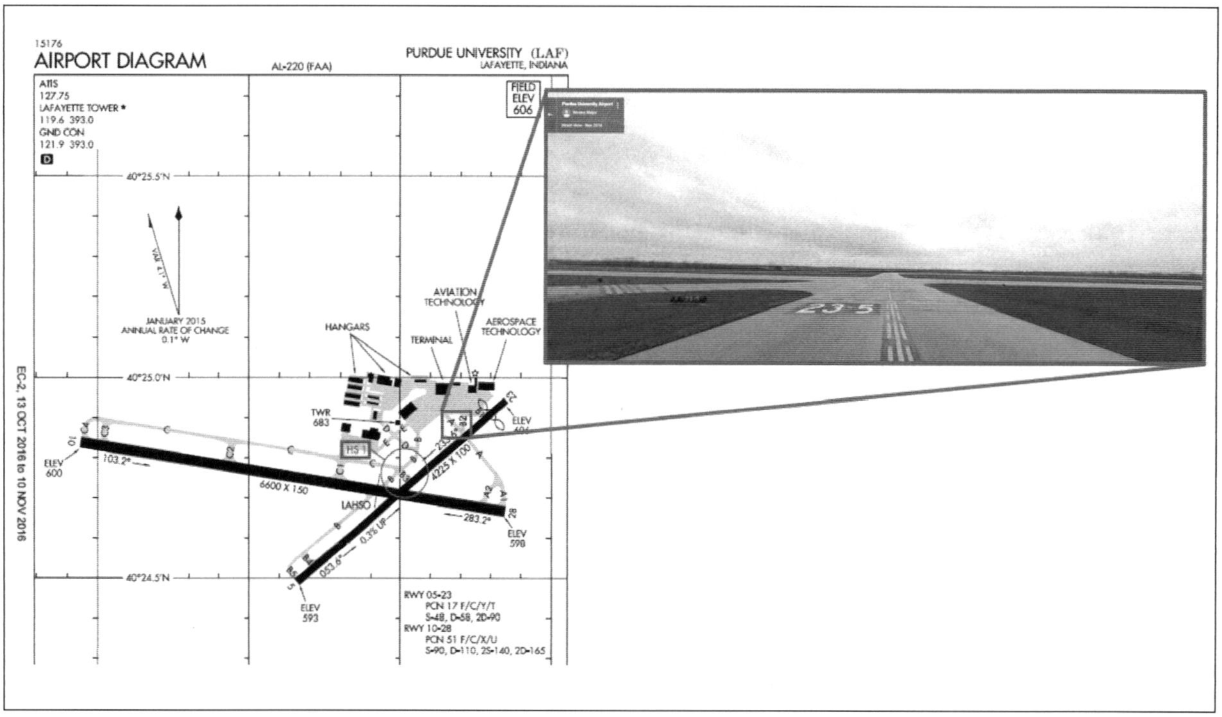

Figure 3. A 360-degree photo can provide an enhanced context for the airfield environment.

flight training. These are the kind of benefits that were substantiated by Schultz et al. (2008), in research that documented that virtual globes can be used to educate about the spatial environment. Similarly, pilots flying to an airport for the first time can familiarize themselves with the new environment and can incorporate the enhanced airport diagram into their pre-flight planning; familiarizing themselves with visual cues, and supporting the development of a movement plan, if desired. Ground personnel can also use this tool to train new team members on proper airfield navigation in a low risk environment. Emergency teams that do not normally operate on active airfields can utilize such a tool to maintain familiarity and support practice with airfield protocol. The enhanced airfield diagram would also be very useful as an aide during the table top exercises that are a required component for airport certification under Part 139. Perhaps most importantly, this tool can be used to illustrate hot spots and other potentially confusing areas on the airfield. Although airport diagrams label hot spots, they do not provide a strong visual context for hot spots. The use of the enhanced airport diagram, provides a means to examine airfield signage, runway markings, and other landmarks prior to experiencing them on the airfield.

Conclusion

Aviation safety is paramount, and the increase in runway incursions has prompted the FAA to created programs specifically to reduce incursions. This research sets forth a low cost method to familiarize airport users with the airfield, which will contribute to enhanced situational awareness and support a reduction in runway incursions. The traditional method of providing airfield information to aviation stakeholders via an airport diagram is useful, but a more modern version that incorporates photos is beneficial. A Google Street View style map of the airfield increases situational awareness, one of the risk factors identified by the RIM program, and results in a more robust visual aid, providing aviation stakeholders an accurate representation of the airfield procedures and conditions from a ground-based perspective, which has the potential to increasing safety and training efficiency.

References

Chang, Y.-H., & Wong, K.-M. (2012). Human risk factors associated with runway incursions. *Journal of Air Transport Management*, *24*, 25–30. https://doi.org/10.1016/j.jairtraman.2012.05.004

Endsley, M. R., & Garland, D. I. (2000). Pilot situation awareness tralnlng in general aviation. *Proceedings of the Human Factors and Ergonomics Society Annual Meeting*, *44*(11), 357–360. Retrieved from https://doi.org/10.1177/154193120004401107

Federal Aviation Administration. (2012). *Runway Safety Report 2011 – 2012*. Washington D.C. Retrieved from https://www.faa.gov/airports/runway_safety/publications/media/2012-AJS-475-FY2011-Runway-Safety-Annual-Report.pdf

Federal Aviation Administration. (2016). Runway Incursion Mitigation (RIM) program. Retrieved February 11, 2017, from https://www.faa.gov/airports/special_programs/rim/

Federal Aviation Administration. (2017a). KLAF airport diagram. Retrieved from http://aeronav.faa.gov/d-tpp/1701/00220ad.pdf#search=KLAF

Federal Aviation Administration. (2017b). RWS search form. Retrieved February 10, 2017, from http://www.asias.faa.gov/pls/apex/f?p=100:28:0::NO:28::

Federal Aviation Administration. (2017c). RWS system info page. Retrieved February 10, 2017, from http://www.asias.faa.gov/pls/apex/f?p=100:32:0::NO::P32_REGION_VAR:1

International Civil Aviation Organization. (2015). *ICAO safety report, 2015 edition*. Montreal, Canada. Retrieved from http://www.icao.int/safety/Documents/ICAO_Safety_Report_2015_Web.pdf

Kennedy, J. (2016). Dublin Airport's runways are now on Google Street View. Retrieved February 11, 2017, from https://www.siliconrepublic.com/play/dublin-airport-street-view

Oulasvirta, A., Estlander, S., & Nurminen, A. (2009). Embodied interaction with a 3D versus 2D mobile map. *Journal Personal and Ubiquitous Computing*, *13*(4), 303–320.

Raubal, M., & Winter, S. (2002). Enriching wayfinding instructions with local landmarks. *International Conference on Geographic Information Science*, 243–259. https://doi.org/10.1007/3-540-45799-2_17

Schultz, R. B., Kerski, J. J., & Patterson, T. C. (2008). The use of virtual globes as a spatial teaching tool with suggestions for metadata standards. *Journal of Geography*, *107*, 27–34. https://doi.org/10.1080/00221340802049844

Snowdon, C., & Kray, C. (2009). Exploring the use of landmarks for mobile navigation support in natural environments. *Proceedings of the 11th International Conference on Human-Computer Interaction with Mobile Devices and Services*. https://doi.org/10.1145/1613858.1613875

Thorndyke, P. W., & Hayes-Roth, B. (1982). Differences in spatial knowledge acquired from maps and navigation. *Cognitive Psyschology*, *14*, 560–589.

EFFECTS OF SIMULATOR TRAINING FOR UNMANNED AERIAL SYSTEMS IN UNDERGRADUATE EDUCATION

William Weldon
Purdue University
West Lafeyette, IN
Dr. Brian Kozak
Purdue University
West Lafeyette, IN

Unmanned aerial systems (UAS) are being added to the national airspace (NAS) in very large numbers. Many universities have seen this demand for UAS operators and begun to create programs in order to train undergraduate students in their use. The UAS industry currently lacks adequate training requirements for beginning operators. This fact makes university training important, because universities are one of the few places that offer structured training. While the UAS industry in the US is in its infancy it is possible to draw parallelisms the training history of manned flight in order to avoid pitfalls and offer training in the most efficient way possible. This study utilizes a mid-fidelity UAS simulation program in order to test the application of this technology. Preliminary results show that simulator technology is helpful in teaching UAS flight in an undergraduate setting.

Unmanned aerial systems (UASs) are being introduced into the United States national airspace system (NAS) by the hundreds of thousands. As this new technology is implemented, the question of how the operators of these vehicles will be trained is raised. With the implementation of code of federal regulations (CFR) 14 part 107, the Federal Aviation Administration (FAA) has set the framework for commercial operations of UASs inside the NAS. These regulations not only allow for commercial use of UASs in the NAS, it places limitations on the vehicles that can be used. The limitations are; the vehicle must weigh less than 55lbs, have a maximum speed of 100mph, stay at or below 400ft above ground level (AGL), and stay within line of sight (Federal Aviation Administration, 2016). CFR 14 part 107 was released in 2016 but in December 21, 2015 the FAA began requiring that all UASs between .55-55lbs be registered. February 5, 2016 saw the number of registered UASs surpass the number of registered manned aircraft, and by May 12, 2016 over 466,000 vehicles had been registered (Federal Aviation Administration, 2016). This rapid growth far surpassed any of the forecasting attempts done in the previous years.

In 2013, Darryl Jenkins and Dr. Bijan Vasigh, researchers for the Association for Unmanned Vehicle Systems International (AUVSI), published a report on the economic impact of these vehicles in the US. In this report, Jenkins and Vasigh used 100,000 vehicle sales as a benchmark for the number of vehicles sold per year, for commercial purposes. They went on to forecast that UASs would add over $82.1 billion to the economy by 2025 as well as add over 103,000 jobs paying on average $40,000 per year (Jenkins & Vasigh, 2013). However, in 2016, the FAA predicted that around 600,000 UASs would be sold for commercial use. This large difference in predictions will likely have an effect on the forecasted economic impact of UAS integration into the NAS, as well as the jobs they create. The incredible potential that UASs have to influence the US economy in a positive manner make it extremely likely that these vehicles will quickly become an integral part of the NAS.

Assuming that these vehicles will become a vital part of the NAS the question of how these operators will be trained is incredibly important. One training option is to have individuals follow CFR 14 part 107 requirements, as they exist at this time, which requires no experience flying these vehicles. While this follows the letter of the law, it may not be the safest way to train new UAS operators. Another approach is to implement UAS training at universities, and have this training mimic flight training already done at universities. This style of training is being used across the country by many different universities. One popular method is having students build and flight test a UAS from a kit. This student built vehicle is then utilized in that course and later courses. This method has a good amount of merit as it allows students to learn the components of a UAS, learn how the construction methods affect flight, and allows them inexpensive flight experience. However, this method alone allows for a wide range of hazardous errors.

This method generates two very prominent errors when used with students that have no prior experience with UASs; the first problem is the possibility of components being installed incorrectly leading to the vehicle to function incorrectly; the second problem is the increased risk of crashes during flight. The first error can be easily fixed by providing more comprehensive instructions to the students as they construct the vehicles, or by having a professor or teaching assistant oversee the construction. The second error however is much more difficult to mitigate, as any time an individual attempts to develop a new skill it is almost guaranteed that they will make mistakes as part of the learning process. While this outcome is expected during any new skill development, it becomes dangerous when teaching UAS operations. Kit built vehicles tend to weigh between 3-5lbs, but their performance capabilities increase the possible damage from a ground collision. Many are capable of maintaining speeds of 30 miles per hour or more, and climbing to altitudes over 400 feet. This factor accompanied by the risk of laceration by the propellers on the vehicle are the sort of risks that accompany an UAS flight. By solely training new UAS operators with "real world" flight experience the likelihood of personal or property damage is increased..

While there is no way to teach new operators without allowing them to fly their vehicles, these flights can be augmented with simulator training. Simulator training has been the standard of training in aviation for decades. This long usage offers a great deal of experience and refinement for the emerging UAS industry can use to create the most efficient training programs. One of these lessons is the idea that utilizing simulators that are extremely realistic, high fidelity, in order to give realistic flight experiences does not mean that the simulator training is effective. For many years aviation has placed a premium on how realistic a simulator is, because these simulators are being used as a replacement for using an actual aircraft for familiarization and recurrent training. In order to replace flying the aircraft with flying a simulator it appears that replicating the flight environment is of the upmost importance. In this setting, it would seem that a company would get the best training for their flight crews by spending money on the most recent and most high tech simulators available. However, these simulators are often used to train crews on day-to-day flying activities like standard operating procedures and crew roles (Dahlstrom, Dekker, Winsen, & Nyce, 2009). This style of training has come about because training evaluation has generally been done by having the trainees evaluate the training upon their completion, this idea has led to high fidelity simulators to be rated extremely high because they very flashy and include a great deal of "bells and whistles" (Salas, Bowers, & Rhodenizer, 1998).

This approach to simulator training evaluation has come about because of the lack of opportunity for research with high fidelity simulators. A high fidelity simulator is very costly, and in order to make sure this equipment is used in a cost effective manner they are generally in use for training continuously (Salas, Bowers, & Rhodenizer, 1998). In the place of high-fidelity simulation training for normal operations, it has been suggested that lower fidelity simulators that focus on adverse tasks and crew resource management (CRM) are more useful in practice. Dahlstrom, et al. (2009) tested the use of a mid-fidelity simulation of a ship's bridge, during this simulation the subjects were given time critical and event driven scenarios in order to see if the subjects could begin to develop skills useful in the target environment. This training experiment was conducted over two days and included; two runs of the simulation along with briefings, discussions, and lectures (Dahlstrom, Dekker, Winsen, & Nyce, 2009). These experiments showed the subjects adapting quickly to different situations, and breaking out of their predefined roles to better control the situation at hand (Dahlstrom, Dekker, Winsen, & Nyce, 2009). The participants of this study also requested more simulation training similar to the experiment, which shows that while they not only did better the subjects found the training enjoyable (Dahlstrom, Dekker, Winsen, & Nyce, 2009). Dahlstrom, et al. go on to state that high-fidelity simulations run the risk of reducing the imaginative and creative involvement of the participants. This in turn can lead to the "internalization of a series of highly contextualized instrumental stimulus-response relationships-putatively stress-resistant procedural response that may be insensitive to, or even make actors unprepared for, contingencies outside of rehearsed routines," (Dahlstrom, Dekker, Winsen, & Nyce, 2009, p. 311).

Currently the only large operator of UASs in the US is the US military, and all four branches utilize this technology. The United States Air Force (USAF) and the United States Army differ greatly in their UAS missions and in their training methods. Both of these organizations apply simulator technology in their training to differing degrees of success. The USAF requires that all UAS pilots be trained pilots that have flown a minimum of one tour of duty (Tvaryanas, Thompson, & Constable, 2005). A 2002 report by the United States Air Force Research Lab describes the kind high-fidelity simulation used to cross train these experienced pilots into the RQ1 Predator (Schreiber, Lyon, Martin, & Confer, 2002). By 2005, this simulation training had proven to have a great many flaws (Tvaryanas, Thompson, & Constable, 2005). A study conducted over the UAS mishaps in every branch of the military showed that despite the greater flight experience of USAF operators, and their simulation training, they had the highest rate of skill based errors (Tvaryanas, Thompson, & Constable, 2005). One of the main causes of these errors was that this "high-fidelity" simulation did not represent the handling characteristics of the vehicle (Tvaryanas, Thompson, & Constable, 2005). The US Army chooses UAS operators from enlisted personnel and gives these individuals UAS specific training (Tvaryanas, Thompson, & Constable, 2005). The UAS specific training that is given to US Army personnel consists of 88 simulator hours in a 20 day period as well as training with the actual system (Rosenberg, 2012). Unlike the USAF, the US Army uses many small, rugged, and relatively inexpensive vehicles, which allows them to utilize them for real world training at a lesser cost than if the USAF utilized it's vehicles for many training flights (Rosenberg, 2012). This training approach led the US Army having the lowest number of skill-based errors when compared with the other three branches of the US military (Tvaryanas, Thompson, & Constable, 2005).

Methodology

By utilizing commercially available simulation software, Real Flight 7.5, students enrolled in 300 level UAS courses were able to gain and practice UAS flight skills. These students had to complete 14 labs over the course of the semester, five simulator labs and nine outdoor flying labs. Each of these simulator labs was designed to present the students with a different aspect of UAS flight.

The first lab required the students to hover a very basic quadcopter, one without equipped stabilization assistance, in different orientations during a 10mph cross wind from a third person perspective. This task introduced students to the challenges of dealing with wind as well as the challenges of partial and complete control reversal. The second lab required students to operate the same vehicle from the previous lab in order to locate a missing item, this lab was done in first person. This lab introduced one of the functions of UAS, and forced the students to maintain constant situational awareness during the search. Lab three used the same quad-rotor vehicle to navigate a course of tubes placed at different altitudes throughout the flight area, this lab was also done in first person. Students performing this lab quickly learned that their vehicle's battery would deplete if they did not perform the course quickly enough. This challenge forced the students to learn how to quickly and accurately make flight corrections in a time critical environment. Labs four and five required the students to use a specific fixed wing vehicle instead of the quad-rotor vehicle used for labs 1-3. During lab four, students had to demonstrate their ability to land a fixed wing vehicle, from a third person perspective, consecutive times and in wind. Flying a UAS from a third person view presents a problem with depth perception when tracking the vehicle. The fifth lab required students to fly the same obstacle course as lab three with a fixed wing vehicle. For this lab, the battery life of the vehicle was still a factor, but the fixed wing vehicle was able to travel much more quickly, which mitigated this factor. Highlighting this difference between vehicles allowed students to learn that certain vehicles are better suited for certain applications.

The students who took this course also completed labs with quadcopters built from a kit in a previous class. After completing this course these students continued onto the following course that focuses on flying vehicles equipped with payloads. After this continued flight experience, the students were given a survey asking their perceptions of the UAS simulator and its usefulness. In this survey students were asked to state which simulator lab they felt were the most helpful, least helpful, easiest, most difficult and if the labs increased their confidence operating a UAS. Along with rating the labs the students were asked to describe the reasoning behind their ratings.

Real Flight 7.5

Real flight 7.5 is a mid-fidelity UAS simulator that utilizes a mock UAS controller for control inputs. This simulator contains over 140 aircraft of many configurations, and over 40 flight areas. Each of these aircraft is accurately modeled to the flight characteristics of their real counterparts, and each of the flight areas has controllable atmospheric conditions present. This simulator allows the user to operate their UAS from a third person view, as if they were looking at the vehicle and flying, or from a first person view, as if they are looking through a camera mounted on the vehicle.

Results

At the time of publication, this study is ongoing. However, preliminary results suggest that there is a correlation between gaining experience with the simulation equipment and skill with the UAS in flight operations.

Conclusions

By utilizing simulators alongside inexpensive vehicles, undergraduates can be professionally trained for safe operations in the NAS. This method mimics the training given to enlisted personnel in the US Army, and should be easily adapted to undergraduate education. In order to improve this study in the future, these surveys should be completed yearly and the results compiled.

References

Dahlstrom, N., Dekker, S., Winsen, R. v., & Nyce, J. (2009). Fidelity and validity of simulator training. *Theoretical Issues in Ergonomics Science*, 305-314.

Federal Aviation Administration. (2016). *FAA Releases 2016 to 2036 Aerospace Forecast.* Washington DC: U.S. Department of Transportation.

Federal Aviation Administration. (2016). *FAA Releases Drone Registration Location Data.* Washington DC: US Department of Transportation.

Federal Aviation Administration. (2016). *Fact Sheet – Small Unmanned Aircraft Regulations (Part 107).* Washington D.C.: Department of Transportation.

Jenkins, D., & Vasigh, D. B. (2013). *The economic impact of unmanned aircraft systems integration in the United States.* Arlington, VA: AUVSI.

Salas, E., Bowers, C. A., & Rhodenizer, L. (1998). It is not how much you have but how you use it: Toward a rational use of simulation to support aviation training. *The International Journal of Aviation Psychology*, 197-208.

CREW STATE MONITORING AND LINE-ORIENTED FLIGHT TRAINING FOR ATTENTION MANAGEMENT

Chad Stephens, Lawrence Prinzel, Angela Harrivel, Ray Comstock, Nijo Abraham, and Alan Pope
NASA Langley Research Center
Hampton, VA

James Wilkerson
Boeing Flight Services
Seattle, Washington

Daniel Kiggins
National Institute of Aerospace
Hampton, VA

Loss of control – inflight (LOC-I) has historically represented the largest category of commercial aviation fatal accidents. A review of worldwide transport airplane accidents (2001-2010) indicated that loss of airplane state awareness (ASA) was responsible for the majority of the LOC-I fatality rate. The Commercial Aviation Safety Team (CAST) ASA study identified 12 major themes that were indicated across the ASA accident and incident events. One of the themes was crew distraction or ineffective attention management, which was found to be involved in all 18 events including flight crew channelized attention, startle/surprise, diverted attention, and/or confirmation bias. Safety Enhancement (SE)-211, "Training for Attention Management" was formed to conduct research to develop and assess commercial airline training methods and realistic scenarios that can address these attention-related human performance limitations. This paper describes NASA SE-211 research for new design approaches and validation of line-oriented flight training (LOFT).

Recent accident and incident data suggests that Spatial Disorientation (SD) and Loss-of-Energy State Awareness (LESA) for transport category aircraft are becoming an increasingly prevalent safety concern in all domestic and international operations (Commercial Aviation Safety Team, 2014a). SD is defined as an erroneous perception of aircraft attitude that can lead directly to a Loss-of-Control Inflight (LOC-I) event and result in an accident or incident. LESA is typically characterized by a failure to monitor or understand energy state indications (e.g., airspeed, altitude, vertical speed, commanded thrust) and a resultant failure to maintain safe flight.

Commercial Aviation Safety Team (CAST) Analysis of LOC-I

A Commercial Aviation Safety Team (CAST) study of 18 LOC-I events determined that issues with flight crew attention were involved in all of the 18 events. CAST created a research "Safety Enhancement" (SE) specifically to address this problem state as identified in the CAST JSAT (Joint Safety Analysis Team) and JSIT (Joint Safety Implementation Team) analyses (CAST, 2014a). It was recommended that the aviation community (government, industry, and academia) should conduct research on methods for understanding the phenomena of flight crew channelized attention, startle/surprise, diverted attention, and confirmation bias. In response, NASA initiated a sub-project under the Airspace Operations and Safety Program (AOSP), "Technologies for Airplane State Awareness", to address this SE and others. The research described in this paper specifically addresses SE-211, "Training for Attention Management".

Training for Attention Management. CAST recommended research and training organizations develop methods to detect and measure attention-related human performance limiting states (AHPLS). Furthermore, research organizations should work with industry partners (air carriers, manufacturers, and commercial training providers) to develop methods and guidelines for creating training scenarios that induce AHPLS and develop and assess potential mitigations to these issues in the training environment. The "detailed implementation plan" for SE-211 (Commercial Aviation Safety Team, 2014b) described two keys tasks, assigned to NASA, which were: 1) the development of valid methods to detect and measure AHPLS in pilots; and, 2) the development of methods for creating realistic, high workload scenarios that can induce human performance limitations, including channelized attention, startle/surprise, diverted attention, and confirmation bias.

Scenarios for Human Attention Restoration Using Psychophysiology (SHARP)

The SHARP study was conducted at NASA Langley Research Center in the spring of 2016 and consisted of multiple facets to assess crew state monitoring measures (Harrivel, et al., 2017) and the induction of AHPLS via benchmark tasks and a line-oriented simulation (LOS) scenario. Data collection was performed in the Research Flight Deck in the Cockpit Motion Facility at NASA Langley. The simulator has full-mission, Level D type capabilities and the flight deck emulates a B-787, but with a B-757 aerodynamic model. The crew state monitoring (CSM) data will be used, post-test, in the development of classification methods for detecting AHPLS.

A LOS scenario was designed to provide a high-fidelity simulation of line operations with event sets designed to induce channelized attention and startle/surprise; the CSM and pilot qualitative and quantitative data collected during the event sets will be used for purpose of validating AHPLS classification algorithms during LOFT. This paper describes the LOS scenario results. The results of analyses on the crew state monitoring (CSM) measures captured during the benchmark tasks and during employment in the LOS scenario are reported elsewhere (Harrivel et al., 2017, and Harrivel et al., 2016).

NASA Langley Research Center subject matter experts (SMEs) and line-operational commercial airline pilots with combined experience of more than 30 years designed the LOS event set. The scenario was also developed by reference to FAA Advisory Circular (AC) 120-35D (Federal Aviation Administration, 2015a) which presents the guidelines for the design, implementation, and validation of LOFT. The LOS used a gate-to-gate (from pushback to taxi-in) scenario with multiple event sets designed to induce startle/surprise and channelized attention AHPLS.

Twelve flight crews were paired based on pilot role (Captain, First Officer) and same airline. Each flight crew averaged 22,000 hours of experience with both the B-757 and B-787 aircrafts. The LOFT scenario included debriefing, dispatch paperwork, and other materials and instruction that airlines typically provide for LOFTs (based on two major airlines and manufacturer that had partnered with NASA for this research).

LOFT Scenario Events

Wake Hazard Event. Following the taxi-out, the first major event consisted of a wake encounter which occurred at 700 ft. mean sea level (MSL) after take-off from Runway 36L at Memphis (KMEM). The event created a startle state due to an aircraft roll upset at low altitude. The simulated wake encounter aerodynamic behaviors were verified by SMEs and calibrated by line-operational commercial airline pilots who had each experienced similar low-altitude wakes.

Hydraulic System/Anti-Skid Failure Events. The second major event was a right hydraulic system pressure and antiskid failure approximately 20 nmi. from the LEOOO waypoint on the BBKNG 4 departure. The event set was designed to induce channelized attention on the part of the flight crews by requiring an extensive sequence of checklist items and decision-making considerations (e.g., alternate airports, systems integrity, landing/stopping distances, availability of controls and gear, etc.). The event set provided behavioral indicator checks to determine whether the flight crew was channelized on the basis of: (a) communication patterns and verbiage and attentional management toward other activities (e.g., Air Traffic Control, ATC, responses); and, (b) detection of "proximate" traffic that was also heading to the LEOOO waypoint. The potential incursion traffic was an aircraft that departed from Runway 18C (the scenario design allowed for both north and south traffic departure flows) and party-line communications were provided that indicated that the traffic was cleared to the LEOOO waypoint at altitude that conflicted with the own ship. The event set was designed to cause a proximate traffic encounter if pilots were channelized in attention, since the traffic was observable and appropriate mitigation responses could be performed (e.g., contact ATC) well before the encounter. The traffic was clearly visible on the navigation display for the entire duration of the event set and SMEs predicted that the traffic should be detected 100% of the time under normal operations and conditions (note: depending on how the flight crews navigated and managed the situation, the incursion traffic could become a Traffic alert and Collision Avoidance System, TCAS, "caution").

Trailing Edge Flap Asymmetry Event. The third major event was a trailing edge flap asymmetry (TE FLAP ASYM) which occurred after flight crews were directed back to KMEM for approach to Runway 36L following the hydraulic leak. Runway 36C is the longer runway at Memphis, but the scenario had the runway occupied and unavailable due to foreign object debris that was on runway. Because of the weather conditions and poor braking action reported, flight crews had significant cognitive overhead when deciding whether to accept the runway assignment or request to go an alternate airport due to the aforementioned hydraulic leak and antiskid failures. There were significant variations in how flight crews handled the decision and problem-solving and exhibited threat and error management. However, all flight crews eventually accepted an approach to Runway 36L.

During the approach, the trailing edge flap asymmetry event occurred; the flap asymmetric deployment was alerted to the crew on the Engine Indication and Crew Alerting System. The checklist allowed for a flight crew decision to continue the landing based on the flap configuration but most flight crews requested a go-around and executed the missed approach. For those that elected to continue, ATC issued a go-around (traffic was reported on the runway). The event, combined with the existing issues, was designed to induce channelized attention due to the temporal demands and decisional factors that needed to be considered once the event occurred (e.g., electronic checklist, decision to go-around or land, etc.) The amount of cognitive effort was high during the timing of the event (which went caution alert to the go-around and clean-up and climb to Hold), regardless of whether the pilots initiated the go-around and contacted ATC or ATC issued the go-around, to include the subsequent crew coordination, clean-up of aircraft, and discussion on option. The exception were the four flight crews that immediately executed the missed approach after the TE FLAP ASYM caution was presented on engine indication and crew alerting system (EICAS) display (see discussion below).

Missed Approach Event. After initiating the Runway 36L missed approach, the flight crews climbed and then leveled-off at 3000 ft. on the runway heading and then were turned to a heading of 330 and instructed to proceed to the KALIE waypoint to hold at 5000 ft. ATC then gave vectors to return to KMEM Runway 36C (the longer runway that all pilots preferred earlier was now available). Flight crews were provided speed and vectors to the ILS 36C approach. Due to the trailing edge flap asymmetry, the approach speed was significantly higher than normal (186 knots indicated airspeed).

Runway Incursion Event. The Runway 36C runway incursion event was designed to induce startle/surprise. The incursion was triggered by an aircraft that had erroneously crossed the active runway. Because the landing speed is higher than nominal approach, the reaction time to such an event was reduced creating the conditions for a startle/surprise response. The aircraft timing was intended to purposely not cause a collision on the runway but to simulate a Category B runway incursion event. Due to flight crew decisions or timing issues, in a few cases, the runway incursion aircraft was blocking the runway when the aircraft landed; in such cases, the pilots were briefed that the event was not as intended.

ATC Taxi Clearance Event. After the flight crew turned off the runway, ATC instructed the aircraft to hold on the taxiway and contact ground. Ground ATC issued a plausible and almost correct taxi clearance that would require the flight crews to carefully consider the path prior to execution to avoid an error in taxi. Depending upon their exit, they were either given a taxi clearance which crossed a runway (without a hold short of or clearance to cross the runway in the ATC taxi clearance) or were given a clearance that had a discontinuity (i.e., the cleared route omitted a taxiway). If the flight crews communicated that they had an issue with the clearance, ATC immediately corrected it. It is standard practice for pilots to immediately read-back the clearance to ATC verbatim (which in this case was an intentionally generated ATC error), or 'Roger' or call sign or other (which is not recommended SOP but this would not be marked as an error if done), but they then should review and verify the route on chart. Often, this is done while the aircraft is taxiing, but in this case the aircraft was stopped on taxi-way and there were no temporal pressures to begin taxi until the pilots were ready (due to the runway incursion event ATC had located the aircraft where they were a non-issue for other aircraft and ATC told the pilots they could begin when ready). However, if the flight crew did not identify the error and contact ATC before taxiing, this was not considered as an error; only, if the flight crews did not detect the routing deficiency prior to arriving at the route error was it marked error (recognizing that the original error was ATC).

Discussion

Quantitative and qualitative data were collected from the crews to assess the efficacy of the scenario to illicit realism, training effectiveness, and AHPLS. The qualitative data results for the LOFT scenario evince that the LOFT scenario was rated to be "excellent" / "very good" (82%) with 68% of pilots responding that NASA LOFT scenario was of higher quality than airline LOFT scenarios they had experienced. The NASA LOFT scenario was also judged "very good" to "excellent" for all pilots' responses in comparison of realism to actual commercial flight operations and these hazards encountered on the line.

Startle/Surprise

Wake Encounter Event. The LOFT scenario was found to be highly effective to producing startle/surprise responses for the wake encounter event set - 58% of Captains and 33% of First Officers exhibited behavioral indicators of startle/surprise (based on SME video analyses). Participant pilots rated the wake encounter as 4.5/5 on the Wake Vortex Encounter (WVE) questionnaire (Ahmad et al., 2014) for realism. The WVE data ranged from pilot ratings of 'Minor' (2 responses), 'Major' (18 responses), to 'Hazardous' (6 responses) in effect. Pilots reported that roll angle and roll rate (20 out of 26 responses) was the most significant parameter identifying the disturbances as a wake. Pitch angle and rate (6 out of 26 responses) was also indicated as significant parameter. Pilot comments validated that the simulated wake event was realistic and similar to those operationally encountered.

Runway Incursion Event. The LOFT scenario was also found to be highly effective to producing startle/surprise responses for the runway incursion event set; 42% of Captains and 33% of First Officers displayed behavioral indicators of startle/surprise. Jones and Prinzel (2011) reported on a set of standard dependent measures used in runway incursion research based on the Runway Incursion Severity Index (Federal Aviation Administration, 2015b). The LOFT scenario event was designed to be a "pilot deviation" event (cross hold line on active runway of other traffic) - a Category B type runway incursion - requiring the flight crews to make corrective/evasive action to avoid a collision but was not expected to result in a collision unless the flight crew exhibited poor attention management. Post hoc analysis, based on the FAA Runway Severity Index Rating, of video of the 10 crews who experienced an incursion showed that 4 were rated as Category A events, no Category B, 6 Category C, and 2 Category D events. These data support that these highly experienced flight crews were mostly effective at recognizing and preventing a more serious runway incursion situation.

ATC Taxi Clearance Event. The ATC taxi clearance error event set demonstrated that approximately half of the flight crews accepted the erroneous taxi-in clearance without cross-checking and verification. The error was not that the flight crews read-back of the erroneous ATC clearance and ATC confirmed the read-back, but that the pilots were told to stop on the taxiway after runway turn-off and to contact ground and, therefore, were given ample time to review route before starting taxi again. It is standard practice for pilots to read-back the clearance upon receiving it, but to then to after review the route on the charts to ensure it is correct (often while taxiing where the pilot-not-taxiing reviews the route on the chart) and that they know where they are going. There were no temporal demands on the pilots, as there often are at major airports, and the event was not meant to be a major safety event although one flight crew had taxied onto the active runway before stopping beyond the hold line before contacting Tower. The results evinced that those pilots that experienced the highest channelized attention and startle/surprise responses previously during LOFT did not review, or did so only cursorily, the taxi-in route before or during taxi; these flight crews only realized the ATC error when they came to the mistake in the route. The results suggest that the effects of startle/surprise and channelized attention can continue after the event even when pilots had substantial opportunity to stop and reset without significant temporal demands.

Channelized Attention

Hydraulic System/ Anti-Skid Event. The first channelized attention event set was highly effective to induce channelized attention. 92% (11/12 flight crews) did not detect the proximate traffic and in several cases, a TCAS 'caution' alert was generated due to the attentional focus required by the complex and lengthy electronic checklist.

Trailing Edge Flap Asymmetry. The second channelized attention event set was marginally effective owing largely to the highly variable nature of scenario segment which, to maintain realism, allowed degrees of freedom for pilot responses; as consequence, the trailing edge flap asymmetry and behavioral indicators did not always manifest themselves in the LOFT scenario. 42% (5/12 flight crews) showed evidence of channelized attention. Half of the flight crews did not complete the scenario event set segment as crafted so they did not encounter the event mechanisms designed to induce AHPLS.

Qualitative and Quantitative Pilot Performance

Overall, the flight crews exhibited acceptable threat and error management (e.g., Maurino, 2005) Human Factors Training Manual Doc 9683, and NOTECH or non-technical skills (e.g., Flin et al., 2003) and line/LOS behavioral markers (e.g., Kanki, Helmreich, and Anca, 2010) were found to be "acceptable" to "good" across all the commercial pilots (based on SME video analyses). Pilot technical standards were found to meet the FAA published standards (FAA-S-8081-5F, 2008) and were evaluated against the performance standards for each phase of operations during the LOFT scenario.

Pilot responses to an extensive and detailed final questionnaire provided a wealth of data in terms of current LOFT scenario implementation at airline training centers and substantial information for further work for SE-211. The questionnaire revealed significant and valuable data for how to enhance LOFT scenario and implementations and potential avenues to explore for further scenario development specific to construction of training for attention management scenario and related constructs (confirmation bias, diverted attention, startle/surprise, and channelized attention).

Future Directions

Although not discussed here, analyses are on-going to compare these AHPLS behavioral responses to the CSM classification data. The subjective data suggests that there are a number of potential other or additional opportunities to implement and assess the CSM data for AHPLS, including diverted attention, within the LOFT scenario. Communication analyses (Kanki, Lozito, and Foushee, 1989) are on-going for each event set and the overall LOFT to add additional behavior markers for this characterization/classification. A substantial amount of data cannot be fully described here within the space available, but the results show that LOFT scenarios can be effectively designed to induce AHPLS.

The data suggests that LOFT sessions may have more value if event sets were used with the goal of training pilots to combat AHPLS rather than focus on the event set itself (e.g., training on runway incursion mitigation). Results described in Harrivel et al. (2017) suggest that CSM methods and approaches may be useful in the validation of event sets and potentially for real-time analysis during LOFT sessions. Harrivel et al. (2017) describe the CSM benchmark classification results and similar analyses that are being conducted.

Future directions include additional LOFT scenario evaluation with events sets designed to induce other AHPLS, including diverted attention and confirmation bias. Airline and major aircraft manufacturer training centers have partnered with NASA to continue to improve the design of LOS design and training methods. The planned efforts include evaluation of scenario event sets and recommended approaches during actual airline training LOFTs for training AHPLS.

References

Ahmad, N., VanValkenburg, R., Bowles, R., Duparcmeur, F.L., Gloudemans, T., van Lochem, S., and Ras, E. (2014). Evaluation of fast-time wake vortex models using wake encounter flight test data. Atmospheric and Space Environments Conference. Atlanta, GA. June 16-20, 2014.

Commercial Aviation Safety Team. (2014a). Airplane State Awareness Joint Safety Analysis Team Interim Report. Retrieved from http://www.skybrary.aero/index.php/ Commercial_Aviation_Safety_Team_%28CAST%29_Reports.

Commercial Aviation Safety Team. (2014b). CAST Safety Enhancement SE 211 detailed implementation plan (DIP). Retrieved from http://www.skybrary.aero/bookshelf/books/2541.pdf.

Federal Aviation Administration. (2015a). AC 120-35D - Flightcrew Member Line Operational Simulations: Line-Oriented Flight Training, Special Purpose Operational Training, Line Operational Evaluation. Washington, DC.

Federal Aviation Administration (2015b), FAA Runway Safety Report 2013-2014. Department of Transportation, Federal Aviation Administration, Office of Runway Safety, Washington DC. May, 2015.

Flin, R., Martin, L., Goeters, K. Hoermann, J., Amalberti, R., Valot, C., Nijhuis, H. (2003). Development of the NOTECHS (Non-Technical Skills) system for assessing pilots' CRM skills. Human Factors & Aerospace Safety, 3, 95-117.

Harrivel, A. R., Liles, C., Stephens, C., Ellis, K., Prinzel, L., & Pope, A. (2016). Psychophysiological Sensing and State Classification for Attention Management in Commercial Aviation. Paper presented at the American Institute of Aeronautics and Astronautics, SciTech, San Diego, California.

Harrivel, A. R., Stephens, C. L., Milletich, R. L., Heinich, C. M., Last, M. C., Napoli, N. J., Pope, A. T. (2017). Prediction of Cognitive States during Flight Simulation using Multimodal Psychophysiological Sensing. Paper presented at the American Institute of Aeronautics and Astronautics, SciTech, Grapevine, Texas.

Jones, D., & Prinzel, L. (2011). Simulator evaluation of runway incursion prevention technology for general aviation operations. NASA-TP-2011-217045. Hampton, VA: NASA.

Kanki, B., Helmreich, R., & Anca, J. (2010). Crew Resource Management (2nd Ed). Amsterdam: Academic Press.

Kanki, B., Lozito, S, & Foushee, H.C. (1989). Communication indices of crew coordination. Aviation, Space, and Environmental Medicine, 60(1), 56-60.

Maurino, D. (2005). Threat and Error Management (TEM). Canadian Aviation Safety Seminar (CASS), Vancouver, Canada, 18-20 April 2005.

EFFECTS ON TASK PERFORMANCE AND PSYCHOPHYSIOLOGICAL MEASURES OF PERFORMANCE DURING NORMOBARIC HYPOXIA EXPOSURE

Chad Stephens, Kellie Kennedy
NASA Langley Research Center
Hampton, VA

Nicholas Napoli, Matthew Demas, Laura Barnes
University of Virginia
Charlottesville, VA

Brenda Crook, Ralph Williams, Mary Carolyn Last
Analytical Mechanics Associates
Hampton, VA

Paul Schutte
US Army - Aviation Development Directorate
Fort Eustis, VA

Human-autonomous systems have the potential to mitigate pilot cognitive impairment and improve aviation safety. A research team at NASA Langley conducted an experiment to study the impact of mild normobaric hypoxia induction on aircraft pilot performance and psychophysiological state. A within-subjects design involved non-hypoxic and hypoxic exposures while performing three 10-minute tasks. Results indicated the effect of 15,000 feet simulated altitude did not induce significant performance decrement but did produce increase in perceived workload. Analyses of psychophysiological responses evince the potential of biomarkers for hypoxia onset. This study represents on-going work at NASA intending to add to the current knowledge of psychophysiologically-based input to automation to increase aviation safety.

The Next Generation Air Transportation System (NextGen; FAA, 2011) is designed to improve airspace capacity and maintain, if not, improve flight safety. Within the NASA Aeronautics Research Mission Directorate (ARMD), the Safe Autonomous Systems Operations (SASO) project is supporting the NextGen development and directing research into increasingly autonomous systems-supported operations. Increasingly autonomous systems are envisioned as a future flight deck technology "building-block" where machine learning/artificial intelligence algorithms are aware of the vehicle, operator, and airspace system state and respond appropriately (Stephens et al., 2011). These future systems will sense internal and external hazards, evaluate them, and facilitate timely and suitable responses for mitigation.

A controlled method for inducing poor operator functional state (OFS) will further development of behavioral, psychophysiological, and performance indices to augment automation capabilities and potentially enable the creation of technologies necessary for reduced crew operations. Hypoxic hypoxia is a reduction of oxygen in the arterial blood with a resulting decrease in oxygen for diffusion into the tissues (Gradwell, 2006). Hypobaric hypoxia, a mechanism for hypoxic hypoxia, is caused by a reduction of oxygen partial pressure in inspired air at altitude, and has been an aeromedical and human performance concern since the dawn of aviation (Dille, 2002). Brief hypoxia in humans results in temporary cognitive impairment including lapses of attention or loss of situation awareness related to human error. The intentional use of hypoxia in human test subjects for the concomitant cognitive impairment has potential for understanding limitations of human operators and performance augmentation from autonomous systems.

A research team at NASA LaRC applied normobaric hypoxia induction in human subjects to study the impact on aircraft pilot performance. Voluntary subjects in the study experienced simulated altitudes of Sea Level (21% O_2) and 15,000 feet (11.2% O_2) induced by an Environics, Inc. Reduced Oxygen Breathing Device (ROBD-2). During non-hypoxic and hypoxic exposures each test subject performed a battery of written, computer-based, and flight simulation tasks. Task performance measures, NASA Task Load Index subjective self-report of workload, and physiological responses including: pulse oxygen saturation (SPO_2), electrocardiogram (ECG), respiratory effort, and electroencephalogram (EEG) were recorded. The performance, subjective, and physiological data were examined to understand cognitive impairment due to mild hypoxia exposure. The purpose of this study is to add to the current knowledge of psychophysiologically-based input to automation to increase aviation safety.

Technical Goals

- Stage and deploy hypoxia induction equipment, psychological testing batteries, and physiological recording equipment.
- Assess efficacy of cognitive impairment induced by mild hypoxia exposure.
- Determine safe and effective method for performing hypoxia induction in future studies to support OFS assessment.

Method

Experiment Subjects

The experiment sample (N = 57) included 8 women and 49 men. All subjects were screened for disorders and excluded from participating if they reported neurological, cardiovascular, prescription medication affecting cenral nervous system and autonomic nervous system activity, and smoking. Furthermore, a medical examination conducted at the NASA LaRC Clinic consisting of a 12-lead electrocardiogram, complete blood count with differential, and pulmonary function test with all results being within normal limits was required in order to participate.

Experiment Apparatus

An Environics, Inc. ROBD-2 (see Figure 1) portable computerized gas-blending instrument used to induce hypoxia, without changes in atmospheric pressure, through an aviator's oxygen mask worn by the human research subject. The ROBD-2 is capable of producing Sea Level (21% O_2) to 34,000 feet (4% O_2).

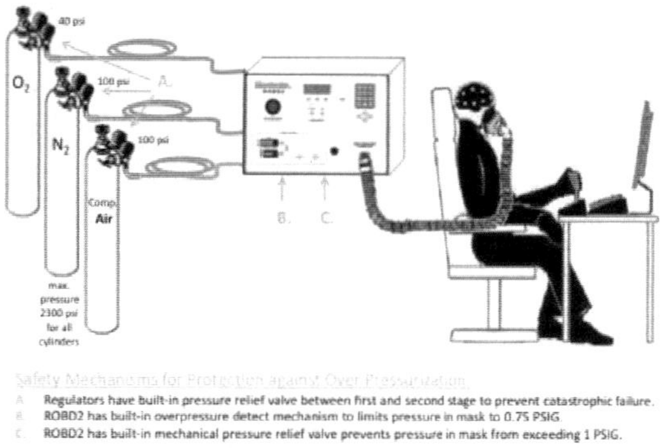

Figure 1. Schematic depiction of ROBD-2 configuration

Figure 2. Test subject in flight sim

Physiological Recording

EEG was recorded from 16 electrode sites (AF_1, AF_2, F_1, F_2, F_Z, C_1, C_2, C_Z, P_1, P_2, PO_Z, T_7, T_8, O_1, O_2, O_Z ; International 10-20 system; Jasper, 1958) through a gTEC gUSBAmp amplifier (see Figure 2). SPO_2, ECG, respiration effort, and galvanic skin response (GSR) were also recorded through a gTEC gUSBAmp amplifier. All signals were digitized at 256 Hz. SPO_2 and pulse rate were redundantly monitored and recorded from the ROBD-2.

Experiment Tasks

Each test subject performed a battery of written, computer-based, and flight simulation tasks. A Cognitive Function Test (Westermann, 2004) battery of written tasks was completed by half of the subjects including: simple computational problems – addition, subtraction, or multiplication; serial 7 subtraction; eye-hand coordination drawing; semantic memory and visual–motor coordination; working memory digit and address recall; and trail-making A and B. The other half of the subjects completed the CogScreen Hypoxia Edition (Kay, 1995). The Multi-Attribute Task Battery-II was performed involving tasks analogous to activities performed by aircraft crewmembers in flight (Santiago-Espada et al., 2011). Also, a flight simulation task was performed using X-Plane 10 simulation (KC-10 aircraft model) connected to force feedback sidestick control inceptors.

Experimental Procedure

The experimental session lasted approximately 4 hours. Subjects completed informed consent documentation. Subjects were briefed on the operation of the ROBD-2 and connected to physiological recording equipment. Subjects completed training sessions for each experiment task. Subjects sat quietly breathing room air while wearing mask to establish physiological baseline. Subjects performed each task three times under the following conditions: 1) breathing room air while wearing mask; 2) breathing sea level gas mixture through mask; and, 3) breathing 15,000 feet gas mixture through mask. Subjects recovered from hypoxia exposure by breathing 100% O_2 for 2 minutes following the 15,000 feet exposure. Subjects completed self-reported workload measure (NASA-Task Load Index, NASA-TLX) after each trial. After completing all trials, subjects were debriefed regarding the study purpose.

Dependent Measures

The experimental design was within-subjects, so all subjects experienced hypoxia while performing all of the experiment tasks. Self-reported hypoxia symptoms are recorded and will be examined to better understand the individual variability of hypoxia exposure. Self-reported workload (NASA-TLX) was examined to assess effect of hypoxia on subjective experience of each task. Performance on each of the tasks was assessed for errors of commission and omission. Raw physiological data were reduced to variables indicative of OFS to determine the effect of hypoxia on indices of OFS.

More advanced analysis techniques were employed to examine complex coupling of multiple body systems. Specifically, ECG multivariate respiration entropy was calculated by temporally syncing the respiration tidal volume signal with the ECG signal. The ECG signal was then used as the reference where a QRS detector was applied to determine the indice locations of the R wave (Pan & Tompkins, 1985). Utilizing the indices of the array from the ECG signal, we extract the tidal volume of the respiration at that specific time instance of when the R-Wave occurred. We then obtained the R-R interval from the ECG signal and the 'downsampled' tidal volume of the respiration signal to apply the multivariate entropy calculation to the two sequences (Costa et al., 2002). The configuration parameters of the multivariate entropy have been specified in the literature as follows: delay vector using a time delay vector of tau= (1,1), embedded dimension vector M = (2,2), and tolerance value (threshold) of r = 0.2 of the standard deviation (Richman & Moorman, 2000; Riedl et al., 2013).

The EEG data was analyzed using inspiration around the core idea proposed by von Tscharner (2000) and is similar to bandpass filtering and the concept of equalizers. The connection with wavelet theory is that the filter is constructed by rescaling a single basis function, $\hat{\varphi}(f) = e^{-\alpha(f-f_c)^2+-\beta\,(f-f_c)^4}$, using a special array of scales in the frequency domain with no imaginary components, where f represents the frequency, f_c is the center frequency of the wavelet, and α and β are tuning parameters that aid in maintaining an appropriate filter bank plateau value. These wavelets were then projected in the time domain using the Fast Fourier Transform. This allows us to obtain a complex wavelet design (real and imaginary components) in the time domain. Thus, when convolved with the EEG signal we produce a filtered signal intensity as a function of time. The filter bank design and optimization for this methodology is discussed in Napoli et al (2017). The signal is then smoothed using a Gaussian filter, providing 12 wavelet filters each with their own specified frequency bands tailored for EEG analysis. These frequency bands represent the typical delta, beta, theta, alpha and gamma bands. Each subject's EEG band intensities are z-scored prior to conducting statistical tests.

The alternative hypothesis is: human subjects experience cognitive impairments to a greater extent ($p < 0.05$) during ROBD-2 equivalent altitude = 15,000 ft (11.2% Oxygen) than during ROBD-2 equivalent altitude = sea level (21% Oxygen). ANOVA were used to test for main and interaction effects of subjective, behavioral, psychophysiological, and performance indices during sea level condition compared to hypoxia condition. Additional analyses of physiological and cortical responses were conducted.

Results

Statistical analyses of subjective workload ratings revealed significant difference between the sea level and 15,000 ft normobaric hypoxia conditions only during the flight simulation task:

- NASA-TLX Overall Workload: $t(52) = 1.8136$, $p = 0.0036$
- NASA-TLX Mental Effort: $t(52) = 1.1726$, $p = 0.0488$
- NASA-TLX Performance: $t(52) = 2.668$, $p = 0.0412$
- NASA-TLX Frustration: $t(52) = 2.189$, $p = 0.0154$

ANOVA statistical analyses of the EEG and task performance revealed no significant difference between the sea level and 15,000 ft normobaric hypoxia conditions.

However, the ECG-Respiration Multivariate Coupling revealed a significant difference between the sea level and 15,000 ft normobaric hypoxia conditions:

- $t(361) = 5.7053$, $p < 0.001$

The analysis demonstrated a significant decrease in the normalized power of wavelet W_6 (mid-level beta band: 15.19-18.37 Hz) across all but three sites (O_1, F_1, and C_2) during hypoxic trials.

Discussion

Analyses involving coupling across physiological systems and wavelet transforms of cortical activity revealed patterns that can discern between the simulated altitude conditions. Specifically, multivariate entropy of ECG/Respiration components were found to be significant predictors ($p< 0.02$) of hypoxia. Furthermore, in EEG, there was a significant decrease in mid-level beta (15.19-18.37 Hz) during the hypoxic condition in thirteen of sixteen sites across the scalp. Task performance was not appreciably impacted by the effect of 15,000 feet simulated altitude but self-reported indices of workload were found to be statistically significant due to hypoxia. Analyses of psychophysiological responses evince the potential of biomarkers for mild hypoxia onset.

The potential for identifying shifts in underlying cortical and physiological systems could serve as a means to identify the onset of deteriorated cognitive state. Enabling such assessment in future flightdecks could permit increasingly autonomous systems-supported operations. Augmenting human operator through assessment of cognitive impairment has the potential to further improve operator performance and mitigate human error in safety critical contexts. This study represents on-going work at NASA intending to add to the current knowledge of psychophysiologically-based input to automation to increase aviation safety.

References

Ahmed, M. U., & Mandic, D. P. (2011). Multivariate multiscale entropy: A tool for complexity analysis of multichannel data. *American Physical Society, Physical Review E, 84*(6).

Comstock, J. R., & Arnegard, R. J. (1991). The multi-attribute task battery for human operator workload and strategic behavior research. Hampton, VA: National Aeronautics and Space Administration.

Costa, M., Goldberger, A. L., & Peng, C. (2002). Multiscale entropy analysis of complex physiologic time series. Physical Review Letters, 89(6), 068102.

Dille, J. R. (2002). In the Beginning. In R. L. Dehart & J. R. Davis (Eds.), *Fundamentals of Aerospace Medicine*, 3rd ed, pp. 1-18. Philadelphia: Lippncott Williams & Wilkins.

FAA (2011). Next Generation Implementation Plan. Federal Aviation Administration.

Gradwell, D. P. (2006). Hypoxia and hyperventilation. In D. J. Rainford & D. P. Gradwell (Eds.), *Ernsting's Aviation Medicine*, 4th ed, pp. 51-54. Boca Raton, FL: CRC Press, Taylor & Francis Group.

Jasper, H. H. (1958). Report of the committee on methods in clinical examination in electroencephalography. *Electroencephalography*, 10, 370-375.

Kay, G. (1995) CogScreen Aeromedical Edition Professional Manual. Washington DC: CogScreen LLC.

Napoli, N. N., Demas, M. W., Stephens, C. L., Kennedy, K. D., & Barnes, L. E. (2017). An Adaptable Filter Bank Optimization Method for Evaluating EEG Activation Complexity Using Intensity Analysis. In *Proceedings of the International Conference on Biomedical and Health Informatics (BHI)*, Orlando, FL.

Pan, J., & Tompkins, W. J. (1985). A Real-Time QRS Detection Algorithm. *IEEE Trans. Biomedical Engineering*, *32*(3), 230–236.

Richman, J. S., & Moorman, J. R. (2000). Physiological time-series analysis using approximate entropy and sample entropy. American Journal of Physiology - Heart and Circulatory Physiology, 278(6), 2039-2049.

Riedl, M., Müller, A., & Wessel, N. (2013). Practical considerations of permutation entropy: A tutorial review. The European Physical Journal Special Topics, 222(2), 249-262.

Santiago-Espada, Y., Myer, R. R., Latorella, K. A., & Comstock, J. R. (2011). The Multi-Attribute Task Battery II (MATB-II) Software for Human Performance and Workload Research: A User's Guide. NASA, Langley Research Center. Hampton: NASA/TM-2011-217164.

Stephens, C.L., Scerbo, M.W., & Pope, A.T. (2011). Adaptive Automation for Mitigation of Hazardous States of Awareness. In Matthews, G., et al. (Eds.), The Handbook of Operator Fatigue. Farnham, Surrey, UK: Ashgate Publishing.

Von Tscharner, V. (2000). Intensity analysis in time-frequency space of surface myoelectric signals by wavelets of specified resolution. *Journal of Kinesiology and Electromyography*, *6*(10), 433–45.

Westermann, R., A., (2004) Hypoxia familiarisation training by the reduced oxygen breathing method. Aviation Medicine, 5, 11-15.

WHAT INFORMATION ABOUT CONSUMERS PREDICTS THEIR TRUST IN AUTOPILOTS?

Mattie Milner[1]
Rian Mehta[2]
Stephen Rice[1]
Scott Winter[2]
Matt Pierce[1]

[1]Embry Riddle Aeronautical University
Daytona Beach, Florida

[2]Florida Institute of Technology
Melbourne, Florida

Introduction

In order for pilots to carry passengers from one destination to another they must rely on their indicators and controls to take them there safely. This also includes their ability to use autopilot technology to help facilitate ease of travel and safety for not only pilots but passengers, as well. Autopilot technology is always advancing and now we are seeing that technology being incorporated into commercial aviation and UAS's doing tasks such as transporting goods to people's homes to performing military operations Despite this advancement in technology, passengers may feel nervous about pilots relying on autopilot technology for flight instead of flying manually. What types of passengers trust autopilots and find that autopilots are reliable?

Autopilots have been around for many years and the technology keeps improving. However some may argue pilots rely too much on autopilots which can be looked at as a handicap instead of flying with manual controls. Autopilots are a valuable resource in not only helping to aid pilots with flying safely and accurately but also aid in long duration flights. In *Plane and Pilot* (2012) Bill Cox talks about the usefulness and importance of autopilots. Pilots traveling for long duration flights become fatigued and may have to navigate through turbulent weather thus relying on autopilots for aid and accuracy in flying. Without the aid of autopilot, flying could be more difficult with higher degrees of human error. To begin utilizing autopilot technology we must first have trust in automation.

With more processes and technology becoming automated consumers may have mixed feelings and perceptions which could lead to a lack of trust in automation. In order for automation to be successful there needs to be a certain level of trust from consumers. Three studies conducted by Dzindolet, Peterson, Pomranky, Pierce and Beck (2003) looked at the relationship among automation trust, reliability, and resilience. Participants were shown slides of Fort Sill terrain and were asked to specify whether or not there was a camouflaged soldier while being assisted by an automated decision aid. Initially they found participants were trusting of the automated aid until it started making errors. An explanation of errors was needed to regain trust in the automation so they would know why the error occurred. This study helps to show how trust of automated systems can be swayed depending on the reliability of the system. Our study was focused on discovering what factors determined a person's trust in autopilot. Our hypothesis is as follows:

Ho: There will be no significant predictors of trust in autopilots when controlling for all other variables
Ha: There will be at least one significant predictor of trust in autopilots when controlling for all other variables

Methods

Participants

Eighty-nine (48 females) participants from the United States participated in this study. The study utilizes participants that were recruited via Amazon's ® Mechanical Turk ®, and were compensated for their completion of the survey. The mean age was 37.12 ($SD = 13.16$).

Procedure, Materials and Stimuli

First, the participants were asked to fill out a consent form and given instructions. Participants were given a hypothetical scenario about flying on a commercial flight from one major city to another. The participants were told that an autopilot would control the entire flight from takeoff to landing. The study utilized a previously validated trust scale adapted to fit the context of this research (Rice, Mehta, Winter & Oyman, 2015). Participants responded to the trust statements along a 5-point Likert-type scale from Strongly Disagree (-2) to Strongly Agree (+2) with a zero neutral option. A second aspect to the study was that participants were presented with another hypothetical situation. Participants were told that they had ordered a package from an online retailer, and that the package would be delivered via a drone (Unmanned Aerial Systems – UAS) operated by an autopilot. The same scale was used to rate participants' level of trust in the autopilot. The participants were then asked for demographic information, as well as a series of questions about personality traits, after which they were debriefed and dismissed.

Design

The study employs a correlational design using two stepwise regressions in order to create two regression equations in order to find significant predictors to autopilot trust. The two prediction equations being created refer to trust in autopilots as it relates to commercial air travel, and the use of UAS for package delivery. The factors being tested in this study as potential predictors of autopilot trust are: gender, age, political affiliation, education level, income, Frequency of air travel per year, trust in technology, number of high-tech devices owned, ratings of aviation technology encountered, general attitudes towards technology, general attitudes towards machine, knowledge about autopilots. The dependent variable is the participants' trust scores.

Results

In this study, a regression analysis was conducted of the dataset with respect to participants' trust in autopilots as it related to a commercial airline flight. The predictors being tested were gender, age, political affiliation, education level, income, Frequency of air travel per year, trust in technology, number of high-tech devices owned, ratings of aviation technology encountered, general attitudes towards technology, general attitudes towards machine, knowledge about autopilots. A backward stepwise regression was employed to eliminate statistically insignificant predictors. The resulting model included two significant predictors, general attitudes towards machines, and general attitudes towards technology out of the original twelve predictors. The regression equation created as a result of this analysis was:

$$Y = -0.09 + 0.29X_1 + 0.20\, X_2$$

where Y is predicted trust score trust in autopilots relating to commercial airline flights, and X_1 and X_2 are general attitudes towards machines, and general attitudes towards technology respectively. The model accounted for 29.40% (27.80% adjusted) of the variance in the criterion, $F(2,84) = 17.52$, $p < 0.05$.

Another similar regression analysis was conducted on participants' trust in autopilots as it relates to the use of UAS for package delivery. The predictors being tested were once again, gender, age, political affiliation, education level, income, Frequency of air travel per year, trust in technology, number of high-tech devices owned, ratings of aviation technology encountered, general attitudes towards technology, general attitudes towards machine, knowledge about autopilots. A backward stepwise regression was employed to eliminate statistically insignificant predictors. The resulting model included two significant predictors, trust in technology, and general attitudes towards machines out of the original twelve predictors. The regression equation created as a result of this analysis was:

$$Y = -0.60 + 0.01X_1 + 0.22\, X_2$$

where Y is predicted trust score trust in autopilots relating the use of UAS for package delivery., and X_1 and X_2 are trust in technology, and general attitudes towards machines respectively. The model accounted for 26.90% (25.10% adjusted) of the variance in the criterion, $F(2,84) = 15.43$, $p < 0.05$.

Discussion

As the field of aviation becomes increasingly automated, particularly around the topic of fully-automated commercial flights and UAS's, it is important to discuss the consumers' attitude toward the automation. The purpose of this study was to determine what factors were significant indicators of a person's trust in autopilots in the context of a commercial air travel and the use of UAS for package delivery. Twelve potential factors were considered; however, the stepwise regression model only identified two factors as being significant predictors: general attitudes toward technology and general attitudes towards machines. Presumably, participants who have had consistently reliable experiences with electronic devices, and who had a more positive attitude toward technology, were more trusting of autopilots. Majority of participants probably have a large amount of trust in their own electronic devices; therefore, they have more feelings of trust toward all electronic machines. Likewise, if they have a more positive attitude toward technology in general then they will be more likely to have a more positive attitude toward autopilots.

We hypothesized that at least one of the twelve factors would be a significant predictor of trust in autopilots. As predicted, general attitude toward technology and general attitude toward machines were significant predictors of trust in autopilot. During an interaction with automation, a person's trust is expected to be dynamic depending on their experience in the past and during the present. Social psychology literature has found that when lacking contradictory information, people tend to view each other, and unfamiliar or unknown things, as good (Cacioppo, Gardener, & Berntson, 1997; Dzindolet, Peterson, Pomranky, Pierce, & Beck, 2003). If people already have positive feelings toward machines and technology, then it may be possible that this positivity bias is extending to autopilots, as well.

Theoretical Contributions

Previous studies have shown that a person's willingness to use an automated device is moderated by the automation's reliability and the operator's trust in automation (Itoh, Abe, & Tanaka, 1999; Lee & Moray, 1992; Muir & Moray, 1996). In this context, "trust can be defined as the attitude that an agent will help achieve an individual's goals in a situation characterized by uncertainty and vulnerability" (Lee & See, 2004). When people trust the automated devices they already own, because there is high reliability, this translates to trust in the autopilots within commercial aviation and UAS's.

In addition, the level of reliability of the automated device plays an important role in determining the consumers' level of trust in the device (Cohen, Parasuraman, Freeman, 1998; Dzindolet et al., 2003; Parasuraman & Riley, 1997). The more reliable an automated device is, the more trust a consumer will have in that device. Likewise, consumers tend to have less trust in automated devices that are less reliable. In everyday life, people's experiences with automated devices tend to be fairly reliable; our computers don't crash every single day, phones reliably send texts and receive calls, air conditioning turns on and off as scheduled, etc. Therefore, people who experience high reliability with the automation devices they already own might judge autopilots in airplanes and UAS's as highly reliable, as well.

Applications

As society becomes increasingly automated, it is important to consider how consumers feel about fully automated technology, such as self-driving cars and fully autonomous airplanes. Companies will need to consider the best methods of encouraging trust between the user and the automation. Our study provides evidence of two factors that are significant predictors of trust: general attitudes toward machines and general attitudes toward technology. Companies should consider that automation that is highly reliable encourages consumers to have higher trust in the automation, and therefore use it more frequently. Future research should consider how to strengthen a person's attitude toward machines and technology. If this relationship can be strengthened, then it may be possible to influence the amount of trust a person has in automation and their willingness to use automated devices.

Limitations

One limitation of our study may be the use of convenience sampling via MTurk. MTurk has been shown to have similar reliability, gender, and ethnicity data composition as data that is collected in the lab (Johnson & Borden, 2012). However, since we are not in control of who participates, it is possible that our pool of participants did not contain a large amount of variability. Additionally, participants responded to the questionnaire using pre-determined answer choices. While this allows for everyone to have to same options, we may have missed information identifying potential predictors because participants were not allowed to write in their own answers.

Conclusions

Technology has allowed for several advances in automation and it is important to consider what factors predict consumers' trust, particularly in high-risk environments, such as autopilots for commercial aviation and UAS's. Previous research has shown that a person's willingness to use an automation device is moderated by the automation's reliability and the operator's trust in automation (Itoh, Abe, & Tanaka, 1999; Lee & Moray, 1992; Muir & Moray, 1996). Our study determined two factors that were significant predictors of consumers' trust in autopilots in commercial aviation and UAS's, general attitudes toward machines and general attitudes toward technology.

Participants were given a hypothetical scenario about flying on a commercial flight and responded to trust statements along a 5-point Likert-type scale from Strongly Disagree (-2) to Strongly Agree (+2) with a zero neutral option. In a second hypothetical situation, participants were told that they had ordered a package from an online retailer, and that the package would be delivered via a drone (UAS) operated by an autopilot. The same scale was used to rate participants' level of trust in the autopilot. A regression analysis was conducted of the dataset with respect to participants' trust in autopilots with twelve different factors being considered. The resulting model included two significant predictors, general attitudes towards machines, and general attitudes towards technology. Further research should be done to explore the relationship between general attitudes towards machines/ general attitudes towards technology and trust in autopilots.

References

Cacioppo, J., Gardener, W., Berntson, G. (1997). Beyond bipolar conceptualizations and measures: The case of attitudes and evaluative space. Personality and Social Psychology Review, 1(1), 3-25. doi: 10.1207/s15327957pspr0101_2

Cohen, M.S., Parasuraman, R., Freeman, J.T., 1998. Trust in decision aids: A model and its training implications. Proceedings of the 1998 Command and Control Research and Technology Symposium. CCRP, Washington, DC.

Cox, B. (2012). *Autopilots equal safety*. Braintree: Madavor Media LLC.

Dzindolet, M., Peterson, S., Pomranky, R., Pierce, L., & Beck, H. (2003). The role of trust in automation reliance. International Journal of Human-Computer Studies, 58(6), 697-718. doi:10.1016/S1071-5819(03)00038-7

Itoh, M., Abe, G., & Tanaka, K. (1999). Trust in and use of automation: Their dependence on occurence patterns of malfunctions. IEEE International Conference Proceedings. doi: 10.1109/ICSMC.1999.823316

Johnson, D., & Borden, L. (2012). Participants at your fingertips: Using Amazon's Mechanical Turk to increase student-faculty collaborative research. Teaching of Psychology, 39(4), 245-251. doi: 10.1177/0098628312456615

Lee, J., N. Moray (1992). Trust, control strategies and allocation of function in human machine systems, ergonomics, 35(10), 1243- 1270.

Lee, J., & See, K. (2004). Trust in automation: Designing for appropriate reliance. The Journal of the Human Factors and Ergonomics Society, 46(1). doi:10.1518/hfes.46.1.50_30392

Muir, B., Moray, N. (1996). Trust in Automation. Part 11. Experimental Studies of Trust and Human Intervention in a Process Control Simulation, Ergonomics, 39(3), 429-460.

Parasuraman, R., & Riley, V. (1997). Humans and automation: use, misuse, disuse, and abuse. The Journal of the Human Factors and Ergonomics Society, 39(2), 230-253. doi: 10.1518/001872097778543886

Rice, S., Mehta, R., Winter, S., & Oyman, K. (2015). A trustworthiness of commercial airline pilots (T-CAP) scale for American consumers. Journal of Aviation Technology and Engineering, 4(2), 55.

APPLICATION OF BIG DATA SYSTEMS TO AVIATION AND AEROSPACE FIELDS; PERTINENT HUMAN FACTORS CONSIDERATIONS

Chang-Geun Oh, Ph.D.
Kent State University,
Kent, Ohio 44242

The aviation and aerospace are typical areas that can apply big data systems due to their scales. This paper identifies aviation/aerospace areas that can utilize big data infrastructures to enhance their operational performances, and lightens human factors considerations related to the use of big data. The NextGen's network-centric infrastructure defines sharing a huge amount of aeronautics, flight, and weather data under the system wide information management program. Sensors installed on aircraft components extract huge numbers of aircraft health and operational status data. All professionals who work in the different aviation sectors require this shared situational awareness information for their own distinctive purposes, and big data systems will enable the effective use of the information. The improved prediction model by the big data analytics will improve aviation safety, reduce flight delays, and save the time and cost for maintenance. The pilot behavior research can adopt the naturalistic study method to supplement limitations of simulation test. The naturalistic flying study needs to consider collecting and analyzing data through big data systems. Human factors research questions naturally arise as aviation/aerospace fields apply big data systems pervasively.

Introduction

The aviation field has encountered a drastic growth of air traffic demand and required the effective management of aviation systems. It is going to be more difficult to ensure the safety for passenger and cargo. Recently the data acquisition collected from aviation infrastructures (satellite systems, ground stations, and airport radar systems) and sensors installed on aircraft became to be shared to most people who work in various aviation fields or customers who use airports or other aviation services. The volume of data extracted from these systems and sensors is too big to handle using the traditional computing capabilities with databases. For example, the average flight data collected during a current flight operation is up to 1000 gigabytes (Wholey, Deabler, & Whitfield, 2014). This "big data" is considered as one solution to enhance the aviation safety for the increased traffic volume and produce higher revenues for airlines.

The big data is referred to a huge volume of data that cannot be managed by the traditional data management paradigm (Schroeck, Shockley, Smart, Romero-Morales, & Tufano, 2012). The structured and unstructured data in non-unified format collected from various machines and sensors can be stored and utilized to discover new correlations or hidden information (Schroeck, Shockley, Smart, Romero-Morales, & Tufano, 2012). Many business sectors are interested in constituting big data infrastructures in their business environments as decision-making aids (Schroeck, Shockley, Smart, Romero-Morales, & Tufano, 2012). Different from the traditional data management, the big data systems employ separate application software components for data collection, data store, data curation, data use, data analytics, data update, and data transfer to next levels in an independent operating system. Tremendous efforts are required to design and test these systematic big data architectures in the ad-hoc manner. The big data analytics even enable users to utilize the hidden unstructured data that never actively used for any purpose. Based on the big data's four properties (volume, variety, velocity, and veracity; Schroeck, Shockley, Smart, Romero-Morales, & Tufano, 2012), users and operators can discover patterns, relationships, and insights that had not been easily identified with a limited volume of data. Developing a big data environment and applying the big data analytics for the aviation field can provide valuable novel information and insights to pilots, air traffic controllers, dispatchers, maintenance, and business leaders to improve the safety and operational performance. The federal aviation administration (FAA), industry, and research organizations became interested in the big data infrastructures.

This study identified three different aviation areas that need or already developed the big data systems for their subject matters and examined human factors considerations while dealing with the big data in each area. Following sections specified histories and plans of the big data application for (1) aviation infrastructure, (2) aircraft, and (3) operator. To highlight the human factors professionals' role in the big data environment, human factors questions for aircraft pilots, air traffic controllers, aviation dispatchers, and aircraft maintenance staffs were created in each field.

Aviation Fields Considering Big Data Application

Aviation Infrastructure

Next Generation Air Transportation System defines a concept of network centric infrastructure. Under the net-centric infrastructure, every aircraft that install automatic dependent surveillance-broadcast (ADS-B) system have the access to all the aeronautical/flight/weather data for their precision flight operations (JPDO, 2011).

This information-sharing program has been evolved from Aircraft Situation Display to Industry (ASDI) to system wide information management (SWIM) program. Feeding the ASDI data stream including aircraft in-flight location, flight plan, altitude, airspeed, destination, estimated time of arrival, designated identifier to all airliners and aviation organizations was initiated by the department of transportation (DOT) in 1990s (Ayhan, Pesce, Comitz, Sweet, Bliesner, & Gerberick, 2013). Many airline industries subscribed to this program to access the datasets for their businesses. The performance of this program stayed limited since the aircraft that want any data only connected to the data source remotely on demand that required complex procedures (Verma, 2016). The SWIM program is the modernized one that solved the point-to-point access problem. The SWIM employs a centralized common data platform of national airspace system (NAS) data that connects all data sources and users easily and rapidly (Verma, 2016). The FAA publishes all data stream in the SWIM so that all users with the FAA permission can have the access to whatever data they want (Verma, 2016). The data list expanded to a larger dataset adding airport operational status, weather information, status of special use airspace, and NAS restrictions. Stored in the cloud, the SWIM data is expected to increase the common situation awareness among all aviation communities during their operations since the ASDI was decommissioned at 2016 (Verma, 2016).

The dataset of SWIM is maintained best using the big data system since its volume is very big and it contains a mixture of structured and unstructured data. Users need to make a decision on which two heterogeneous data lists will be relevant to extract any valuable information. As conducting updated analytics with accumulated data, the insight becomes more accurate. Applying machine learning, automated flight management systems can recommend better alternative paths when encountering bad weather ahead of ownship based on the better prediction as machines themselves accumulate the information of rerouting recommendation for pilots (Akerkar, 2014). Ayhan, Pesce, Comitz, Sweet, Bliesner, and Gerberick (2013) demonstrated the actual vs. planned route based on the flight computer's big data information. Kasturi, Prasanna, Vinu, and Manivannan (2016) proposed an airline route profitability-optimization model based on big data analytics. This best recommended rerouting path are shared with air traffic controllers for shared situation awareness. The big data from the SWIM infrastructure is shared to everyone who want to know the flight information that affect the airplane delay or cancellation. Airline passengers and airport limousine services utilize the flight information using applications on their mobile devices (e.g. FlightAware).

Human Factors Challenges: human factors professionals may have these questions related to the utilization of big data systems for aviation infrastructure.

- How do we indicate predictive information or insights for any specific flight operation on the limited cockpit display screen?
- How can a pilot evaluate the information accuracy for their situation awareness? (recent but old information vs. near real-time information vs. real-time information)
- Which level of SWIM big data analytics information should be allowed to access and interact with for pilots and for air traffic controllers?
- Will the big data analytics reduce the workload for pilots and air traffic controllers?
- How does cockpit displays visualize multiple variables of information for pilots?
- How does air traffic control displays visualize multiple variables of information?

Aircraft

Recent aircraft install very high number of sensors on engines, avionics, or electrical components. Airbus A380-1000 model is expected to have 10,000 sensors in each wing; the number of sensors and the captured data using the sensors will further increase in the future (Marr, 2015). The purpose of installing these sensors on the aircraft parts is to monitor the aircraft health and extract status information during specific operational stages (Bellamy, 2014). This data enables the predictive maintenance – identifying what components are in bad conditions and repairing the components before they fail. Like the state-of-the-art automobile technology, aircraft also can monitor the fuel consumption in real-time. Accumulating the fuel consumption data in different operational stages, the smarter fueling decision can be made (Wholey, Deabler, & Whitfield, 2014). As well as the smarter operational performance prediction, this aircraft monitoring strategy may increase reliability and help accident investigations. Maintenance staffs can integrate the spare part-supply status data into the monitoring part status data to make a quick maintenance decision (Wholey, Deabler, & Whitfield, 2014). Identified component vulnerability results analyzed by the sensor data may also provide insights about the aircraft component design and development (Wholey, Deabler, & Whitfield, 2014). Since the quantity of updated data is a huge amount, the aircraft sensor data should be managed in big data systems. Many aerospace manufacturers developed big data architectures for diagnosis of their products (Chen et al., 2016).

Human Factors Challenges: human factors professionals can consider following questions related to the use of big data of aircraft sensors.

- How to design the interface of sensor data for technicians, engineers, and pilots?
- Is it required to integrate the sensor data to the SWIM infrastructure for the comprehensive management?
- How does a pilot maintain the SA of aircraft health even if the number of sensor increases?
- How to train engineers, safety managers and maintenance specialists to have the knowledge about big data for aircraft components?
- What are human factors considerations for precision maintenance based on the sensor data?

Operators (Pilots)

Like monitoring aircraft component statuses, operators' (pilots') behaviors can be monitored and the behavioral data can be collected to discover the potential human performance degrades or errors in specific operational stages. However, the environment of collecting human behavioral data is different from the aircraft condition data. Unlike the data from thermal, vibration, or pressure sensors, the sensor types to collect the human behavior are limited; video or audio sensors can be used, and the history of interaction with computer systems can be collected. To collect practical human behavioral data, it is important to make human operators comfortable while they are monitored to avoid the Hawthorne effect (i.e. the behavioral differences when participants are aware of being observed). Psychology fields defines this study methodology as the naturalistic study. The naturalistic study has been applied for the surface transportation. Virginia Tech Transportation Institute (VTTI) exploited "naturalistic driving study (NDS)" installing video sensors inside the car to monitor safety critical drivers' behaviors. The number of sensor-equipped car for the NDS was more than 100, and the period of time for data collection was several months to a year. The NDS experimenters have maintained separate storages for the data management and analysis.

The naturalistic study methodology can be applied to the aviation field for "naturalistic flying study (NFS)." Compared to car drivers, aircraft pilots have more list of safety critical task. Even the flight data including altitude, attitude, speed, and GPS signal should be recorded in line with the pilot behaviors. The NFS may have benefits to evaluate the pilot behaviors in the cockpit with multiple variables that was difficult to test in the simulated environment (Caponecchia, Wickens, Regan, Steckel, & Fitch, 2014). Researchers recently started the NFS. The collected data can apply the big data analytics to explore the hidden insights per specific stage of flight operation and pilot expertise level. To make the genuine big data system for the NFS, the experimenter should consider incorporate many external factors besides flight data, because the concept of big data for this matter is not merely an expansion of data volume of simulated study levels.

Human Factors Challenges: If the NFS passed their preliminary stages, several human factors research questions assuming more advanced testing environment may arise as follows.

- What is the privacy problem of videotaping pilot behaviors?
- Does the NFS validate the human-in-the-loop simulation test results for similar studies?
- Is it possible to integrate the NFS data into the SWIM infrastructure to create more comprehensive testing environment?

- Is it possible to integrate the NFS data into the aircraft sensor data to discover the relationship between aircraft sensor statuses and human behavior in specific operational stages?
- Is it possible to constitute real-time pilot behavior monitoring system in bigger aircraft?
- What is the security problem in implementing the NFS?
- Can the implications from the NFS with limited number of aircraft represent the larger pilot group in the same class?
- What kind of properties has been discovered while conducting the NFS compared with NDS?

Limitation of Big Data System for Aviation Applications

The FAA is interested in constituting the big data environment in air transportation system, but it has not been progressed as expected. There are some reasons for this. First, the big data system inherently demands connection with other dataset that is not directly related to the given dataset to create the hidden information. However, the investigation on which information should be discovered by connecting two information groups that are not directly related to each other, such as aircraft sensor data and meteorological data. Since connecting two datasets is a difficult task within a system, the obvious benefit by the connection should be found. Industries and aviation communities are still investigating the benefits and the current integration capability (Valeika, 2016).

Second, the data scientists often need to manipulate the dataset for analysis. However, the direct manipulation of scripting in the big data system is very difficult due to its scale (Fisher, DeLine, Czerwinski, & Drucker, 2012).

Third, the visualization techniques of analyzed information with high number of variables in the big data system needs to be studied. The visualization of huge statistically analyzed results may not fit in an average size screen and requires complex display techniques to understand (Fisher, DeLine, Czerwinski, & Drucker, 2012). Gorodov and Gubarev (2013) identified the problems of visualization in big data applications: visual noise, large image perception, information loss, higher performance requirements, and high rate of image change. This is also a human factors problem.

Fourth, large volume of data may not be always good. The provided big dataset should be evaluated if the dataset represents the larger group in many perspectives.

Fifth, the aviation fields generally require higher security level than other fields. Therefore, the higher security considerations should be applied when designing and developing a specific big data system for aviation. This could be a blocking factor to proceed human factors research activities.

Finally, any aircraft not equipping sensors will not reflect what happens in their components in the big data. Therefore, it is possible to have an inequality problem for representation of certain situation excluding the unequipped aircraft group (Wholey, Deabler, & Whitfield, 2014).

Conclusion

Employing big data systems to manage the data generated from the aviation infrastructure, aircraft sensors, and naturalistic flying study may provide benefits to discover hidden correlations and insights in all aviation sectors. Human factors professionals need to recognize challenges in these sectors including integrating two different datasets for the sake of users and comprehensible result visualizations when the big data systems are applied.

References

Akerkar, R. (2014). Analytics on big aviation data: turning data into insights. *International Journal of Computer Science and Applications, 11*, 116-127.

Ayhan, S., Pesce, J., Comitz, P., Sweet, D., Bliesner, S., & Gerberick, G. (2013, April). Predictive analytics with aviation big data. In *Integrated Communications, Navigation and Surveillance Conference (ICNS), 2013* (pp. 1-13). IEEE.

Bellamy, W., III. (2014, December 01). Avionics Big Data: Impacting All Segments of the Aviation Industry. Retrieved December 01, 2016, from http://www.aviationtoday.com/2014/12/01/avionics-big-data-impacting-all-segments-of-the-aviation-industry/ Avionics Magazine Column

Caponecchia, C., Wickens, C., Regan, M., Steckel, R., & Fitch, G. (2014, September). Naturalistic Flying Studies: Innovation in Aviation Safety Research (Discussion Panel). In *Proceedings of the Human Factors and Ergonomics Society Annual Meeting* (Vol. 58, No. 1, pp. 69-71). Sage CA: Los Angeles, CA: SAGE Publications.

Chen, J., Lyu, Z., Liu, Y., Huang, J., Zhang, G., Wang, J., & Chen, X. (2016, April). A Big Data Analysis and Application Platform for Civil Aircraft Health Management. In *Multimedia Big Data (BigMM), 2016 IEEE Second International Conference on* (pp. 404-409). IEEE.

Fisher, D., DeLine, R., Czerwinski, M., & Drucker, S. (2012). Interactions with big data analytics. *interactions, 19*(3), 50-59.

Gorodov, E. Y. E., & Gubarev, V. V. E. (2013). Analytical review of data visualization methods in application to big data. *Journal of Electrical and Computer Engineering, 2013*, 22.

JPDO (2011). Concept of Operations for the Next Generation Air Transportation System (Version 3.2). Joint Planning and Development Office, Washington, DC.

Kasturi, E., Devi, S. P., Kiran, S. V., & Manivannan, S. (2016). Airline Route Profitability Analysis and Optimization Using BIG DATA Analyticson Aviation Data Sets under Heuristic Techniques. *Procedia Computer Science, 87*, 86-92.

Marr, B. (2015, April 9). That's Data Science: Airbus Puts 10,000 Sensors in Every Single Wing! Retrieved November 30, 2016, from http://www.datasciencecentral.com/profiles/blogs/that-s-data-science-airbus-puts-10-000-sensors-in-every-single Data Science Central Blog Post

Schroeck, M., Shockley, R., Smart, J., Romero-Morales, D., & Tufano, P. (2012). Analytics: The real-world use of big data. *IBM Global Business Services*, 1-20.

Valeika, R. (2016, January 4). Opinion: 'Big Data' From Aircraft? Great. But For What? Retrieved December 2, 2016, from http://aviationweek.com/optimizing-engines-through-lifecycle/opinion-big-data-aircraft-great-what Aviation Week & Space Technology Editorial

Verma, A. (2016, April). SWIM consumer implementation. In *Integrated Communications Navigation and Surveillance (ICNS), 2016* (pp. 10B1-1). IEEE. JPDO, 2011

Wholey, T. J., Deabler, G., & Whitfield, M. M. (2014). *Commercial Aviation and Aerospace: Big Data Analytics for Advantage, Differentiation and Dollars* (Tech. No. GBW03316-USEN-00). Somers, NY: IBM Global Business Services.

"FINAL RESULTS OF MULTIMETHODOLOGY APPLICATION AT CIVILIAN AIR NAVIGATION COMPLEX ENVIRONMENTS"

CABRAL, Lisia Maria Espinola da Silva Pacheco
Airport Infrastructure Brazilian Organization (INFRAERO) - Retired
Rua Marques de Abrantes n°. 95 / 701, Flamengo, Rio de Janeiro, RJ, CEP 22230-060, Brazil, Phone (55) 21-999569228, E-mail - *lisiacabral@gmail.com*
ESTELLITA LINS, Marcos Pereira
Federal University of Rio de Janeiro (UFRJ), Alberto Luis Coimbra Institute of Post-graduation and Research in Engineering (COPPE), Production Engineering Program (PEP), Problem Structuring and Indicators Group for Modeling and Assessment (PSIGMA)
Av. Horácio Macedo, n°. 2.030, Fundão, Ilha do Governador, Rio de Janeiro, RJ, CEP 21.941-914, Brazil, Phone (55) 21- 997179771, E-mail - *marcos.estellita.lins@gmail.com*

The paper addresses the final results of a brazilian doctoral research developed at civilian Air Navigation environments (2011-2014), with partial results already presented at past ISAPs (2011-2015). The study adopted a qualitative, systemic and anticipatory approach to increase metacognition about Team Resource Management (TRM) Training abilities, focused to Threat and Error Management (TEM) practice, with the main purpose of reinforcing operational safety as a whole. It used Multimethodology, aiming at identifying, structuring, analyzing and monitoring problems upon participants´ different perspectives - operators and heads of distinct sectors. Multimethodology embraced four phases, yearly, covering multiple instruments and Theoretic Base, as Conceptual Map, System Thinking and Complexity. Some conclusions indicate: (i) organizational trend to reactive and bureaucratic cultures characterized by difficulties to deal with unexpected situations, not prescribed on standards, and to prioritize solutions to their possible effects that might be aggravated in the course of time; (ii) improvements of TRM behavior´s abilities - Communication, Situational Awareness, Stress and Health Management, Team Dynamics and Decision Making, derived of critical debates and perceptions of restrictive and positive aspects at work, promoted by iterations and interactions among a diverse scope of complex system´s segments, although this didn´t affect directly the update of TRM Training contents, from Error Management (EM) to TEM, towards predictive interventions; (iii) global understanding about a variety of operational realities with common safety purposes, helping to manage, without guilt, conflicts and paradoxes, although this didn´t seem to reach significant projections for future changes. The study suggests that Multimethodology may be adapted to other applications under validation.

This study was realized at civilian Air Navigation sets of the Airport Infrastructure Brazilian Organization (INFRAERO) and the partial results presented at past ISAPs can be found on the following articles:

A) "A Preliminary Analysis of Aeronautical Services in Air Navigation Activity" (CABRAL, MENDES et al, 2011) – This article described the structure of military and civilian Air Navigation services in Brazil, showing the importance to increment psychologists´ participation in contribution to safety, in reply to Human Factors requirements of International Civil Aviation Organization (ICAO) standards. It indicated some issues for discussion as demands to be implemented, among others: (i) intensification of proactive and predictive interventions to support aeronautical services in this area; and (ii) improvements on TRM upon TEM approach to improve operators´ interdisciplinary performance.

B) "Structuring, Analysing and Monitoring Problems and Decision Making Processes at Civil Air Navigation Sets of a Public Organization" (CABRAL & ESTELLITA LINS, 2013) – This article described the initial results of the 1st. Phase of Multimethodology applied to one of the Air Navigation sets (J) studied to stimmulate interactions and iteractions among workers and heads for better dealing with problems in the following services: (i) Air Traffic Control (Tower); (ii) Aeronautical Information System (AIS); (iii) Telecommunication Operation; and (iv) Meteorology. It emphasized the main purpose of identifying, structuring, analysing and monitoring problems at work, upon a collective perspective, derived of complex systems´ characteristics and reinforced by the operational safety and organizational cultures, as well as their negative reflexes, supported by some Theoretic Base, as follows: (i) Conceptual Map; (ii) System Thinking; (iii) Metagovernance; and (iv) Complexity Paradoxes.

C) "Contribution of Multimethodology to Human Factors in Air Navigation Systems" (CABRAL & ESTELLITA LINS, 2015) – This article described the results´ outline of all Air Navigation sets studied, mainly, involving Complexity Paradoxes analysis on Multimethodology and its different instruments´ application, aiming at reiforcing TRM / TEM abilities and better dealing with complex systems´ characteristics.

The present article intends to describe the final results of the same study addressed by the referred articles, taking one of the Air Navigation sets studied as a practical example of the complete application of Multimethodology (CABRAL & ESTELLITA LINS, 2015 APUD MINGERS, 2006).

1. Historical Background and Main Characteristics

In Brazil, the Air Space Control Department (DECEA) is a federal and military institution, subordinated to the Aeronautical Command (COMAER), which represents the aeronautical authority accountable to prescribe standards and fiscalize their application into military and civilian Air Navigation organizations, homologated by it to provide services in this area. DECEA standards (CABRAL & ESTELLITA LINS, 2015 APUD BRASIL 2005, 2008b, 2009b, 2012a, 2012b) are, mainly, based on COMAER standards (BRASIL, 1986; CABRAL & ESTELLITA LINS, 2015 APUD BRASIL 2007-2015) and ICAO standards (CABRAL & ESTELLITA LINS, 2015 APUD ICAO, 1998, 2000, 2002, 2003, 2008). Safety Management Manual (SMM) is one of the ICAO standards to be fulfilled by all countries´ members, aiming at increasing, continually, safety all over the world (CABRAL & ESTELLITA LINS, 2015 APUD ICAO, 2013, 2009, 2005a), which led Brazil to establish two standards, as follows: (i) the National Safety Program (Programa Nacional de Segurança Operacional - PNSO) to be fulfilled by the brazilian Aviation and Air Navigation aeronautical authorities; and (ii) the Safety Operational Managing System (Sistema de Gerenciamento da Segurança Operacional - SGSO) to be fulfilled by the Aviation and Air Navigation services providers (BRASIL, 2009a, 2009b, 2010).

The present study took place at some Air Navigation environments of INFRAERO, one of the organizations homologated by DECEA to provide the following services: (i) Air Traffic Control and Management on Tower and Approach (APP); (ii) Aeronautical Information System (AIS); (iii) Aeronautical Telecommunication Operation; and (iv) Meteorology (CABRAL & ESTELLITA LINS, 2015 APUD BRASIL 2010). It was realized in compliance to some ICAO Human Factors standards (CABRAL & ESTELLITA LINS, 2015 APUD ICAO, 1998, 2000, 2002, 2003, 2008), considering its important contribution to monitor aeronautical risks, as well as to decrease aeronautical incident and accident ocurrences, in reply to SGSO requirements.

Initially, the study contributed to the development of some specific Human Factors standards at INFRAERO to support: (i) TRM implementation (CABRAL & ESTELLITA LINS, 2015 APUD BRASIL, 2012d), started under appraise with DECEA from TRM Facilitators Training, homologated by it to enable TRM development by INFRAERO facilitators, as a formal organizational training for Air Navigation sets, submitted to continuous improvements, although not always observed, which represented one of the study demands; and (ii) psychologists´activities in Air Navigation sets (CABRAL & ESTELLITA LINS, 2015 APUD BRASIL, 2010b, 2012e), mainly, with the formalization of the Psychological Monitoring Program to be implemented by them, with the use of tests and interviews, to deal with Human Factors issues in operational safety practices, which represented an opportunity to proceed the study in parallel, complementing it.

2. Study Structure

2.1. Goals and Method

The Method characterizes a situated study, as investigative and interventionist (CABRAL 2015 APUD TRIPP, 2005), as well an active and ethnographic research (CABRAL 2015 APUD DE MATTOS, 2001), complementing the official Psychological Monitoring Program (CABRAL 2015 APUD BRASIL, 2010b, 2012b), implemented by psychologists of INFRAERO at six Air Navigation environments in the following services: (i) Air Traffic Control and Managing on Tower and Approach (APP); (ii) Aeronautical Information System (AIS); (iii) Aeronautical Telecommunication Operation; (iv) Meteorology; and (v) Airport Operation.

Considering there´s a trend to quantitative, reductionist, immediate, reactive and linear Human Factors´ approaches, raising difficulties to future foresee, which is proper of complex system´s activities, this study used a qualitative, systemic and antecipatory approach at the Air Navigation sets mentioned with the main purpose of reinforcing TRM abilities to identify, structure, analyze and monitor problems and decision making processes, upon different perspectives, for better dealing with Human Factors issues in operational safety practices, characterized by systemic complexity (CABRAL & ESTELLITA LINS, 2015 APUD ICAO, 2002; HOLLNAGEL, 2007; ESTELLITA LINS, ANTOUN NETO et al, 2010).

2.2. Theoretic Base and Methodology

The study chosed the following Theoretic Base to support it: (i) Human Factors Approaches comprising "Z" Theory (CABRAL & ESTELLITA LINS, 2015 APUD HOLLNAGEL, 2007) and Resilience Engineering (CABRAL & ESTELLITA LINS, 2015 APUD WOODS & COOK, 2002; WOODS, 2015); (ii) Metagovernance (CABRAL & ESTELLITA LINS, 2015 APUD JESSOP, 2002); (iii) Cultures embracing Operational Safety and Organizational ones (CABRAL & ESTELLITA LINS, 2015 APUD ICAO, 2013, 2009, 2005a); (iv) Soft Operational Research (Soft-OR) in Problems Structuring Methods (PSM) comprising Muitimethodology (CABRAL & ESTELLITA LINS, 2015 APUD MINGERS, 2006) and Conceptual Map (CABRAL & ESTELLITA LINS, 2015 APUD ESTELLITA LINS, ANTOUN NETO et al, 2010); (v) Complexity and Complex Systems covering System Thinking (CABRAL & ESTELLITA LINS, 2015 APUD GHARAJEDAGHI, 2011), Complexity Paradoxes (CABRAL & ESTELLITA LINS, 2015 APUD ESTELLITA LINS, 2011, 2014) and Organizational Metaphors (CABRAL & ESTELLITA LINS, 2015 APUD MORGAN, 1996, 1997, 2007 APUD SOUZA, 2009; KUNSCH, 2003 APUD MONTEIRO; 2011; ESTELLITA LINS, 2014); and (vi) Hierarchical Human Basic Necessities (CABRAL & ESTELLITA LINS, 2015 APUD COSTA, 1980, NEVES, 2009).

Multimethodology was the methodology chosed for this study and it was preceeded by a survey realized on investigation and safety visits, during the 1st. semester of 2011, to find possible demands that could justify its continuity. Multimethodology embraced a wide scope of instruments: (i) Group Dynamics, Video Summary and Speeches, emphasizing variability and emergency to achieve Problem´s Consciousness Goal; (ii) Group Exercises (Brainstorm Register, Symbol and Simulation Register, Debate and Oral Presentation), emphasizing interactions and perceptions to achieve Problems´Representation; and (iii) Conceptual Map, Report and Debriefing, emphasizing new paradigms and paradoxes to achieve Problems´Formalization. These instruments were applied yearly (from the 2nd. semester to 2011 to 2014), consisting of four phases, as continuous "interative loops" to promote successive interactions among the participants and to achive the goals mentioned of Problems´Consciousness, Representation and Formalization. Multimethodology was substantiated by the Theoretic Base mentioned, as showed in Figure 1, which will be commented later on the final results.

RELATION OF THEORETIC BASE TO MULTIMETODOLOGY, ITS INSTRUMENTS AND GOALS

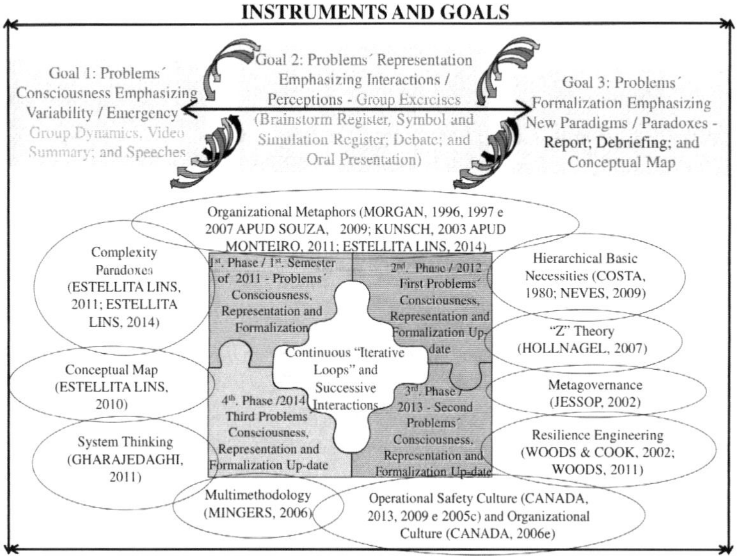

"Figure" 1. Relation of Theoretic Base to Multimethodology´s instruments and goals.

Each phase of Multimethodology ended up with Conceptual Map, either as a conceptual base and as an instrument, to "achieve the goal" of Problems´Formalization (CABRAL & ESTELLITA LINS, 2015 APUD ESTELLITA LINS, ANTOUN NETO et al, 2010). Debriefing was applied only on the 1st. Phase because of time limitations to join managers.

Table 1 indicates: (i) the six Air Navigation sets submitted to the study, referred by their first names´letter; and (ii) the participation rates on each set, from 2011 to 2014. One of the points to emphasize is that there were no negative impact derived of low participation rates. This article will comment the final results of one of these sets (G), which had the following participation rates: (i) 89,74% (1st. semester of 2011); (ii) 74,30% (2nd. semester of 2011); (iii) 16,66% (2012); (iv) 47,15% (2013); (vi) 44,11% (2014); and (vii) 51,28% (total).

Table 1.

Global Rates of Participants (CABRAL & ESTELLITA LINS, 2015 APUD MINGERS, 2006)

WHERE	2011 / 1st. Semester	2011 / 2nd. Semester	2012	2013	2014	TOTAL
→ (G)	89,74% (70 participants from the total of 78)	74,39% (61 participants from the total of 82)	16,66% (13 participants from the total of 78)	47,14% (33 participants from the total of 70)	44,11% (30 participants from the total of 68)	55,05% (207 participants from the total of 376)
(M)	69,47% (66 participants from the total of 95)	53,60% (52 participants from the total of 97)	30,52% (29 participants from the total of 95)	31,81% (35 participants from the total of 110)	35,29% (36 participants from the total of 102)	43,68% (218 participants from the total of 499)
(C)	91,66% (11 participants from the total of 12)	75% (9 participants from the total of 12)	100% (12 participants from the total of 12)	77% (10 participants from the total of 13)	84,61% (11 participants from the total of 13)	86,48% (53 participants from the total of 62)
(T)	100% (7 participants from the total of 7)	85,71% (6 participants from the total of 7)	75% (6 participants from the total of 8)	33% (3 participants from the total of 9)	100% (8 participants from the total of 8)	76,92% (30 participants from the total of 39)
(D)	60% (34 participants from the total of 57)	52,54% (31 participants from the total of 59)	33,33% (21 participants from the total of 63)	34,37% (22 participants from the total of 64)	48,39% (30 participants from the total of 62)	45,24% (138 participants from the total of 305)
(J)	79,54% (35 participants from the total of 44)	62,5% (30 participants from the total of 48)	42,55% (20 participants from the total of 47)	41,66% (20 participants from the total of 48)	55,55% (25 participants from the total of 45)	54,62% (130 participants from the total of 232)
TOTAL	76,10% (223 participants from the total of 293)	61,96% (189 participants from the total of 305)	33,33% (101 participants from the total of 303)	39,17% (123 participants from the total of 314)	46,97% (140 participants from the total of 298)	51,28% (776 participants from the total of 1513)

Table 2 shows this study participation at (G), yearly, during four periods, constituted by different phases, characterized by continuous "iterative loops" and successive interactions. It doesn´t indicate the study participation on the 1st. semester of 2011, already mentioned in Table 1, because this doesn´t consist of a Multimethodology phase or "iterative loop", but a period of previous survey to detect demands that would justify its implementation, from the 2nd. semester of 2011 to 2014. A point to emphasize in Table 2 is that Debriefing to managers about the problems detected on the 2nd. semester of 2011 was only realized on the 1st. Phase because of difficulties in time to join them, considering operational priorities. Table 2 also indicates that the study´s participants at (G) embraced the following operational functions: (i) Aeronautical Information Service´s Techinicians (AIS); (ii) Aeronautical Telecommunication Service´s Tecnicians (OEA); (iii) Meteorology Techinicians (PMET); (iv) Meteorology Professionals; (v) Air Navigation Specialists (ENA); Airport Operation Technicians (PSA); and leaderships (managers, coordinators and supervisors) (CABRAL & ESTELLITA LINS, 2015 APUD BRASIL, 2010a, 2010b). Apart to this, it´s necessary to explain that there were no participation of Air Traffic Controllers (PTA) of INFRAERO at (G), once this service over there is provided by military technicians of DECEA.

Table 2.
(G) Multimethodology Phases´Participation (CABRAL & ESTELLITA LINS, 2015 APUD MINGERS, 2006)

(G): "Iterative *Loops*"´ Participation in all Phases / 2011 to 2014 of Multimethodology - Problems´ Conscienciousness, Representation and Formalization													
PERIODS	*DEBRIEFINGS TO MANAGERS*	PARTICIPATION								GLOBAL PARTICIPATION			TOTAL
		AIS	OEA	PTA	PMET	MEG	ENA	PSA	Mana-ger	Presence	% Parti-cipation	Absence	
05 a 07, 09, 13 and 23.12.11	15.01.12	7	5	0	20	12		16	1	61	74,40%	21	82
07 to 21.12.12	*	4	2	0	2	1	0	4	0	13	16,66%	65	78
03 to 21.06.13	*	7	0	0	11	5	0	8	2	33	47,14%	37	70
25.03 to 07.04.14	*	7	1	0	9	3	1	6	3	30	44,11%	38	68
Total		25	8	0	42	21	1	34	6	137	45,98%	161	298

LEGEND: * No Debriefing
AIS - Aeronautical Information Service´s Professionals
ENA - Air Navigation Specialists
MEG - Meteorological Professionals
OEA - Aeronautical Telecommunication Service´s Professionals
PMET - Meteorological Technicians
PSA - Safety Airport Professionals
PTA - Air Traffic Controllers (Tower)

3. Final Results

The quantitative analysis of the study was realized after each Multimethodology´s Phase and is related to the Opinion Survey Questionnaire´s answers, which results pointed out to a prevalence of "satisfatory" compared to "over expected", "regular", "insufficient" and "not necessary" answers. Furthermore, the qualititive analysis was realized by integrating the results of all Multimethodology´s Phases and was divided into: (i) Opinion Survey Questionnaire related to the suggestions´ answers; (ii) Compatible Analysis related to the problems and situations detected; (iii) Theoretic Base Contribution related to the analysis of each concept´s effectiveness on supporting the goals of Problem´s Consciousness, Representation and Formalization "to be achieved" or presenting "difficulties to be achieved" or even "not achieved"; (iv) Global Demands related to the analysis of the problems plotted compared to the demands introduced by the previous survey (1st. semester of 2011), which justified the study, confirming them or not and verifying if there were any improvements; and (v) Other Considerations aside these analysis described. The final results of the qualitative analysis are extense and will not be described completely in this article, which is restricted to some points considered relevant.

Firstly, in the Opinion Survey Questionnaire, some relevant points to be emphasized as final results are: (i) importance of different functions and sectors´ participation, mainly during the 1st. Phase, when the heads´ participation showed to be decisive; (ii) demand for videos of real work situations (CABRAL & ESTELLITA LINS, 2015 APUD VIDAL & MÁSCULO, 2011) ; (iii) more time for problems´ debate and consciousness about work routines; and (iv) improvements in communication and interaction between workers and heads, as well as in integration among all systemic segments and levels as a whole, involving either human and organizational issues.

According to Table 3, Compatible Analysis classifications are, as follows: (i) Material and Organizational Problems involving external decisions, not depending on (G) iniciative, but with negative impacts on its operation; (ii) Human Group Problems involving internal conflicts and relationships as negative barriers to work contexts; (iii) Operational Problems inherent to service and related to standards as restrictions to operation; (iv) General Problems involving global aspects with indirectly reflexes to harm work routines; and (v) Positive Aspects, which, fortunately, were raised. Also, each of these classifications adopted different colours to distinguish the problems and situations plotted, as follows: (i) black to initial ones; (ii) blue to "new" ones; (iii) wine to "reincident" ones; and (iv) brown to the ones "on approval", which needed to be confirmed. Some main points to be emphasized are, among others: (i) although the high quantity of reincident problems, there were more Positive Aspects compared to all classifications of problems; (ii) there were more Material and Organizational Problems than other ones, with higher reincidence compared to them, probably because they refers to subjects which decisions depend on higher organizational levels outside (G); (iii) mostly Human Problems are affeccted by communication´s limitations, relationships´conflicts and a trend to find guilties; (iv) Operational Problems, in majority, need continuous standards´adaptation and up-date focused on daily activities´practice, prone to constant variability (internal and external), derived of complexity characteristics, proper of complex systems. Table 3 demonstrates, not all, but some examples derived of Compatible Analysis, based on the classifications described.

Table 3.
Study Compatible Analysis Examples

SOME EXAMPLES OF COMPATIBLE ANALYSIS´ RESULTS				
	Initial	New	Reincident	On Approval
Material and Organizational Problems (TOTAL = 15)	Insufficient training	-	Airport privatization with posterior consolidation	Slowness on material reposition
			Change of work shift for worst	Substitution of INFRAERO Air Traffic Management System (SGTAI) by Aircraft Services Messages Handling System (AMHS)
			Failures on internet	
Human / Group Problems (TOTAL = 13)	Different procedures for workers´ rest payment	Expectations about Meteorological Information Translator System (STIM) implementation	Centralization of informal routines on the same workers	Trend to work in homogeneous teams
	Trend to find guilties		Problems on workers´ vacations planning	Indication of higher valuation of Meteorological activity compared to others
Operational Problems (TOTAL = 8)	Low frequency of Operational Meetings	Burocratic routines at Meteorological Sector involving Meteorological Briefing	Over-use of taped phone at AIS Room deviating line to Meteorological Sector because of the high quantity of Flight Plans related to the number of workers	Deficiency on addressing messages
				Permanence at workplace during rest time
General Problems (TOTAL = 12)	Individualism, isolation	Emphasis on Selfdeception and Subject Indivisibility Compelxity Paradoxes	Unbalanced of Physiological Necessity related to sleep because of worker shift change	Alert to SGTAI not to accumulate message
	Dispersion, lack of interest, desmotivation		Emphasis on Isolated Parts, Information	
	Trend to follow standards and difficulties on improvisation in unexpected situations		Localization and Preservation involving Complexity Paradoxes	
	Do strictly what´s necessary		Mechanicist and Domination Metaphors	
Positive Aspects (TOTAL = 23)	Knowledge and cooperation among different functions to "armor" against error	Meeting-room	Improvements on communication and cooperation among different sectors	Improvements on problems´ feedback and planning
	Criation of new services procedures	Planning of work shift by each sector		Improvements on training
	Change of meteorological teams	Learning and Art of Analysis Metaphor		
	Common lunch-room and coffee place	Separate rest-rooms for men and women		
		More space of work rooms		

About Theoretic Base Contribution, there are some points to be emphasized, among others: (i) all goals of Problems´ Consciousness, Representation and Formalization were achieved related to Conceptual Map (CABRAL & ESTELLITA LINS, 2015 APUD ESTELLITA LINS, ANTOUN NETTO et al, 2010), which represented, at the same time, a conceptual base and a Multimethiodology instrument (CABRAL & ESTELLITA LINS, 2015 APUD MINGERS, 2006), used in all its Phases and functioning as a chain of connection among the others, as well as to System Thinking (CABRAL & ESTELLITA LINS, 2015 APUD GHARAJEDAGHI, 2011), once all instruments used promoted it, representing the main theoretic framework of this study; (ii) non of these goals related to Organizational Metaphors (CABRAL & ESTELLITA LINS, 2015 APUD MORGAN, 1996, 1997, 2007 APUD SOUZA, 2009; KUNSCH, 2003 APUD MONTEIRO; 2011; ESTELLITA LINS, 2014) were achieved, considering these weren´t used during the study phases, but only on its analysis, but, nevertheless, brought significant questions; (iii) the goal of Problems´ Formalization indicated "difficulties to be achieved" in the majority of conceptual base compared to the goals of Problems´ Consciousness and Representation, which includes Multimethodology itself, Cultures, Metagovernance and "Z" Theory, mainly, because of the negative impacts derived of reactive and bureaucratic cultures represented by Mechanicist Metaphor and hierarchical governance, disabling to make collective and antecipatory changes with future prospectives; and (iv) the majority of goals of Problems´ Consciousness and Representation were achieved, reinforcing the Theoretic Base Contribution to this study implementation, for instance, the ones related to Complex Paradoxes (CABRAL & ESTELLITA LINS, 2015 APUD ESTELLITA LINS, 2011, 2014), which, on the other hand, had difficulties to achieve the goal of Problem´s Formalization, for the trend to Localized Information, Preservation of Processes, Subject Indivisibility, Selfdeception and Unification (CABRAL & ESTELLITA LINS, 2015 APUD ESTELLITA LINS, 2011, 2014).

The majority of the Global Demands raised on previous survey (1st. semester of 2011) derived of the investigation visits weren´t contemplated, once the study focus was safety, except the one related to TRM, which indicated the need for some important improvements, such as: (i) participation of all segments, including heads; (ii) use of Multimethodology to reinforce it, as an organizational diagnosis to adequate its contents and framework; (iii) inclusion of improvisation and criativiy abilities to deal with internal and external variabilities, once the other abilities were improved with the study; (v) inclusion of TRM on formation courses, based on TEM model, once EM is still adopted. On the other hand, the study confirmed many demands derived of safety visits, some mentioned at the Compatible Analysis, which weren´t removed completely and, although there´s concerning about them, the improvements were temporary and limitaded to internal solutions, which were considered insufficient to safeguard their reocurrencies.

Finally, about General Considerations, there are some relevant points, as: (i) dual negative and positive meaning for some terms, as "adversities", "standard" and "technology"; (ii) application of Multimethodology in parallel to the Psychological Monitoring Program, limiting time for both; (iii) except on the 2nd. Phase, which didn´t have heads participation, but without significant damage on its results, there was a gradual increase either on their participation and their convergence to workers; (iv) Airport Operation Service´s workers represented the connection among others, indicating Learning and Art of Analysis Metaphor; (v) no Air Traffic Control and Management Service´s participation didn´t reduce the importance of the other services studied upon the whole system appreciation; and (vi) non-existence of civilian national standards for Air Navigation workers to support the problems plotted.

4. Conclusion

This study was realized into six Air Navigation sets and this article presented the final results of one set (G) studied. It chosed Multimethodology as the methodology to implement, based on a qualitative, systemic and antecipatory approach, which was applied after raising some demands derived of investigation and safety visits (1st. semester of 2011). Multimethodology consisted of four anually phases (from 2nd. semester of 2011 to 2014) and used different instruments to achieve the goals of Problems, Consciousness, Representation and Formalization. The final results embraced a quantitative analysis, indicating a prevalence of "satisfatory" answers, aa well as a qualitative analysis, both derived of the Opinion Survey Questionnaire. The qualitative analysis also comprised: (i) Compatible Analysis; (ii) Theoretic Base Contribution; (iii) Global Demands; and (iv) General Considerations. All goals of Conceptual Map and System Thinking were achieved, considering the Conceptual Map as the chain of connection among all other instruments, as well as System Thinking as the main theoretic framework of the study. These results indicated some limitations and benefits, such as: (i) promotion of interactions among the participants (workers and heads), as its main benefit, learning to develop criativity in face to work problems and positive situations, considering complex characteristics upon a systemic perspective, in complementation and reiforcement to TRM, as well as giving some suggestions to improve it; and (ii) reactive and bureaucratic cultures, as its main limitation, representing a negative barrier to collective and antecipatory agreements related to the problems plotted and future projections for their necessary changes, which indicates researches as positive to enable interactions´ intensification beyond internal contexts to outsider segments, sectors and organizations.

Acknowledgements

Thanks to: (i) INFRAERO colleagues which participated, directly and indirectly, of this study, as well as to heads for the opportunity and confidence to develop it; (ii) UFRJ, specially my doctoral advisor, for the technical and academic support on its implementation, as well as to colleagues for the continuous exchange of knowledge and difficulties during the doctoral course; (iii) my husband for the extreme helpfullness assistance and tolerance during the long period of the doctoral course; (iv) my parents for the educational support which contributed to my doctoral degreee; and (v) my friends for the motivation to face and overcome this challenge.

References

BRASIL. Brazilian Aeronautical Code (CBA), Lei nº 7.565, Presidência da República, 1986.

BRASIL. COMAER and ANAC, PSO-BR. Joint Act nº 764 / GC5 of COMAER and ANAC, 2009a.

BRASIL. ANAC. PSOE-ANAC, 2009b.

BRASIL. COMAER. PSOE-COMAER, 2010. Act of COMAER nº. 368, 2010 and COMAER Bulletin nº. 108, 2010.

CABRAL, L. M. E. S. P.; ESTELLITA LINS, M. P. Doctoral Thesis: "Use of Multimethodolgy to Study Civilian Air Navigation Enviornments". Federal University of Rio de Janeiro (UFRJ), Alberto Luis Coimbra Institute of Post-graduation and Research in Engineering (COPPE), Production Engineering Program (PEP), 2015. In: http://minerva.ufrj.br/F/RPGAF7SKKP4BRAEVHS2MTCT95HHV748GB7JMY87K2AECIGXT8C-69824?func=short-rank&action=RANK&W01=Lisia&W02=Maria&W03=Espinola&W04=da&W05=Silva

CABRAL, L. M. E. S. P.; MENDES, S. P. et al. International Symposium on Aviation Psychology (ISAP) article: "A Preliminary Analysis of Aeronautical Services in Air Navigation Activity". Dayton, OH, USA, 2011.

CABRAL, L. M. E. S. P. & ESTELLITA LINS, M. P. International Symposium on Aviation Psychology (ISAP) article: "Structuring, Analysing and Monitoring Problems and Decision Making Processes at Civil Air Navigation Sets of a Public Organization". Dayton, OH, USA, 2013.

CABRAL, L. M. E. S. P. & ESTELLITA LINS, M. P. International Symposium on Aviation Psychology (ISAP) article: "Contribution of Multimethodology to Human Factors in Air Navigation Systems". Dayton, OH, USA, 2015.

INTELLIGENT MULTI-UNMANNED VEHICLE PLANNER WITH ADAPTIVE COLLABORATIVE/CONTROL TECHNOLOGIES (IMPACT)

Mark Draper, Gloria Calhoun, Michael Hansen, Scott Douglass
Sarah Spriggs, Michael Patzek, Allen Rowe
AFRL/RHCI, Wright Patterson Air Force Base, Dayton OH
Dakota Evans
University of Dayton Research Institute, Dayton OH
Heath Ruff, Kyle Behymer, Michael Howard
Infoscitex Corporation, Dayton OH
George Bearden, Elizabeth Frost
Wright State Research Institute, Dayton OH

This symposium provides an overview of a research effort that integrated several autonomy advancements into a control station prototype to flexibly team a single human operator with heterogeneous unmanned vehicles. The autonomy related technologies optimize asset allocation, plan vehicle routes, recommend courses of action and provide a distributed support architecture featuring an extensible software framework. This effort also integrated these technologies with novel human-autonomy interfaces that allow operators to effectively manage UxV via high level "play" commands. Evaluation results indicate that the innovative approach supports operator-autonomy teaming for effective management of a dozen simulated vehicles performing base defense tasks.

Agility in tactical decision-making and mission management is a key attribute for enabling teams of heterogeneous unmanned vehicles (UxV) to successfully manage the "fog of war" with its inherently complex, ambiguous, and time-challenging conditions. This agility requires effective operator-autonomy teaming including the achievement of trusted, bi-directional collaboration and the flexible, high-level tasking required for team task sharing and decision superiority. A tri-service team has conducted research focused on instantiating an "Intelligent Multi-UxV Planner with Adaptive Collaborative/Control Technologies" (IMPACT) by combining flexible play calling for task delegation, bi-directional human-autonomy interaction, cooperative control algorithms, intelligent agent reasoning and autonomic technologies to enable command and control of cooperative multi-UxV missions (Figure 1). A command and control operator in IMPACT could task a total of 12 UxV (4 air, 4 ground, and 4 sea surface vehicles) in response to several unexpected events that arose during a base perimeter defense mission. This symposium will provide an overview of four key aspects of IMPACT that AFRL led: operator-autonomy interfaces, intelligent agent architecture, testbed framework and distributed architecture, and human-in-the-loop prototype evaluation.

Figure 1. IMPACT Control Station Prototype.

IMPACT: Interfaces for Operator-Autonomy Teaming

Gloria Calhoun, Heath Ruff, and Elizabeth Frost

IMPACT's displays and controls (Figure 2) feature video game inspired pictorial icons that present information in a concise, integrated manner to facilitate retrieval of the states/goals/progress for multiple systems and support direct perception and manipulation principles. Multi-modal controls (speech, touch, and mouse) augment a "playbook" delegation architecture and enable seamless transition between control states (from manual to fully autonomous). With this adaptable automation scheme, the operator retains authority and decision-making responsibilities that helps avoid "automation surprises" (Calhoun, Ruff, Behymer, & Frost, in press). By supporting a range of interactions, flexible operator-autonomy teamwork enables agility while responding to a dynamic mission environment. At one extreme, the operator can

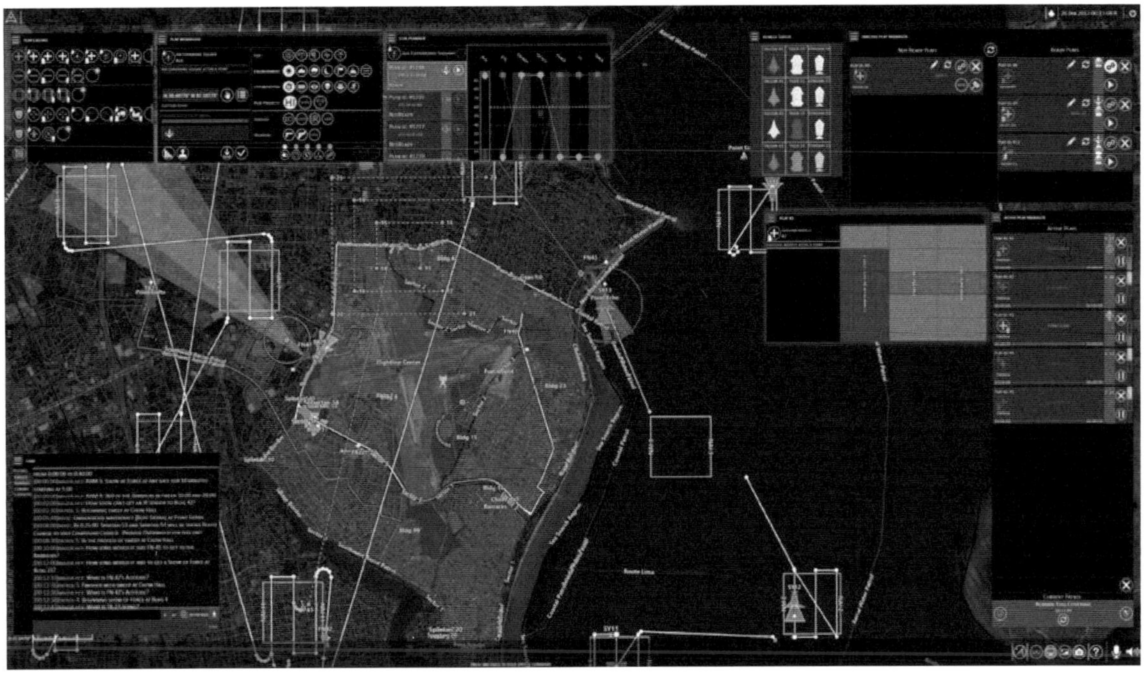

Figure 2. IMPACT Operator-Autonomy Play-based Teaming Interfaces.

manually control UxV movement or build plays from the ground up, specifying detailed parameters. At the other extreme, the operator can quickly task one or more UxV by only specifying play type and location with an intelligent agent determining all other parameters. For example, when an IMPACT operator calls a play to achieve air surveillance on a building, the intelligent agent recommends a UxV to use (based on estimated time enroute, fuel use, environmental conditions, etc.), a cooperative control algorithm provides the shortest route to get to the building (taking into account no-fly zones, etc.), and an autonomics framework monitors the play's ongoing status (e.g., alerting if the UxV won't arrive at the building on time). IMPACT's play calling interfaces also facilitate operator-agent communication on mission details key to optimize play parameters (e.g., target size and current visibility) as well as supporting operator/autonomy shared awareness (e.g., illustrated by a display showing the tradeoffs of multiple agent-generated courses of actions (COAs) across mission parameters). Play progress is depicted in a matrix display reflecting autonomics monitoring and a tabular interface aids play management (e.g., allocation of assets across plays). Additional detail on all the play-related interfaces is available (Calhoun, Ruff, Behymer, & Mersch, 2017).

IMPACT: Intelligent Agent Framework for Course of Action Generation

Dakota Evans, Michael Hansen, and Scott Douglass

An intelligent agent was developed using the Cognitively Enhanced Complex Event Processing (CECEP) framework, a complex event processing framework with extended procedural and domain knowledge aspects. Agents that use procedural knowledge were developed using a discrete finite state machine (FSM) representation called behavior models that include states and transitions between states that are guarded by patterns. A pattern language called Esper matches complex patterns for behavior model state transitions. The developed IMPACT agent has a set of patterns related to operator interactions for play calling. Behavior models can also produce behaviors (e.g., feedback for the operator or UxV play assignment). Agents that use domain knowledge were developed using cognitive domain ontologies (CDOs). A CDO is a rooted tree structure with features that are connected via relations. CDOs can be processed using the artificial intelligence process of constraint satisfaction to produce configurations, possible worlds, or courses of action (COAs). In IMPACT, CDOs were developed to capture the domain for UxV play calling and produce COAs for play to vehicle(s) assignment.

The IMPACT agent serves as a decision aid to a multi-UxV operator. The agent is integrated with a UxV route planner (UxAS), Fusion framework, plan monitoring service (Rainbow), and UxV simulator (AMASE). The operator's play calls are used as a starting point for generating COAs, with the play type (e.g., air point inspect), presets (e.g., cloudy, windy), and optimization criteria (e.g., time, fuel) forming the basis of domain knowledge used to constrain and rank possible COAs. Figure 3 describes IMPACT's play calling process.

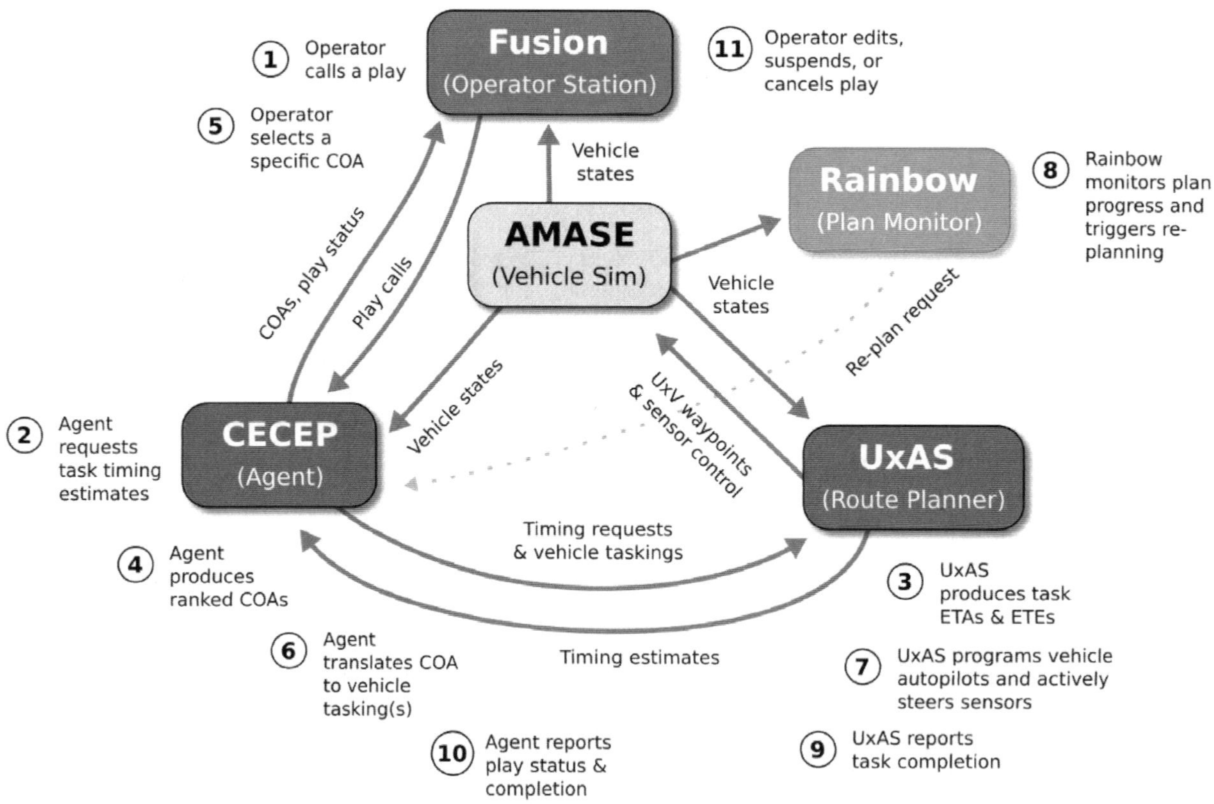

Figure 3. Play Calling Process in IMPACT.
(See list below for further information on each numbered item in the figure).

1. A play call commonly originates from the operator. However, the agent is capable of monitoring the vehicle locations and recommending opportunistic or serendipitous plays, such as inspecting the fuel dump by a vehicle already in the area.
2. The agent transforms play calls into lower level tasks for the UxAS route planner and requests task assignment utility from it (knowledge about task timings, fuel usage, and communications issues from the route planner).
3. The agent asserts the acquired knowledge to the play calling CDO domain representation.
4. The agent applies all constraints corresponding to the operator provided play details. The CDO is processed to produce all constraint compliant COAs. The agent uses an objective function to rank COAs and identify the Pareto optimal COA in a list and on a map presented to the operator (Hansen, Calhoun, Douglass, & Evans, 2016). A visual is also produced that allows the operator to compare COAs by solution utility. If no constraint compliant solutions exist the agent informs the operator.
5. The agent waits for operator acceptance, edit, or cancelation of a COA. Upon acceptance; the agent produces the vehicle action command to the UxAS to execute the COA.
6. The UxAS programs vehicle autopilots and actively steers sensors and sends other behaviors to the simulator.
7. The plan monitoring service monitors the active play and displays feedback to the operator if a constraint is violated (e.g., a UxV enters a restricted operating zone).
8. The agent waits for a task complete message from the UxAS.
9. The agent reports plan status to the operator to improve operator situation awareness.
10. The agent waits for operator to edit, suspend, or cancel the active play.

IMPACT: Fusion and the Distributed Architecture and Services

Sarah Spriggs, George Bearden, and Michael Howard

Fusion (Rowe, Spriggs, & Hooper, 2015) is a software framework that enables natural human interaction with flexible and adaptable automation. This is enabled by employing a distributed service oriented architecture that is composed of multiple disparate systems, unified representationally through negotiated communications protocols and physically through a common communications hub. The decentralization of the architecture enables logging, monitoring, and substitution of components with minimal effect on other components. Thus, several different systems can indirectly interact with one another through a publish/subscribe hub to provide a greater service to the user. All connected pieces communicate through a common messaging protocol to send and receive information. As a result, every component that connects to the hub has awareness of real time scenario and operator activity. Connected services developed for IMPACT include intelligent agent reasoning among disparate domain knowledge sources, autonomics monitoring services, intelligent aids to the operator, cooperative planners, and advanced simulation via instrumented, goal oriented operator interfaces. The distributed architecture along with an extensible software framework enables the system to be easily expanded for other human-automation research. For instance, modification of IMPACT is underway to support multiple stations whose operators share assets and potentially offload or gain tasks based on workload.

The Fusion architecture, as shown in Figure 4, includes the core (customizable) aspects that are common across applications as well as the features that support the IMPACT project. The Fusion test bed also displays the scenario environment, presents mission events that prompt UxV management tasks, provides a workspace for the operator to team with the autonomy in task completion, and records task performance measures. Other IMPACT specific components provide interfaces for calling and modifying plays, viewing agent generated candidate COAs, and presenting the results of an autonomics service monitoring play progress.

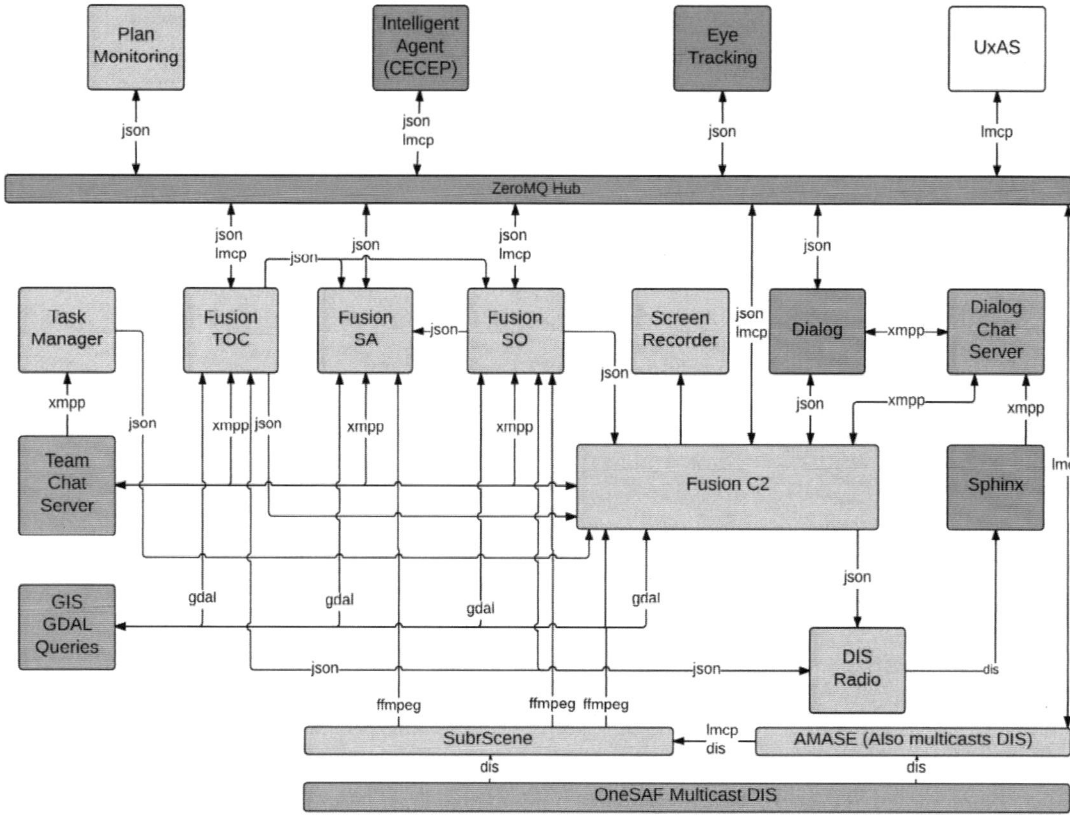

Figure 4. Illustration of Fusion and the Distributed Architecture and Services.

IMPACT: Operator-in-the-Loop Evaluation of Operator-Autonomy Teamwork
Kyle Behymer, Michael Patzek, and Allen Rowe

A high-fidelity human-in-the-loop simulation was used to compare the IMPACT prototype to a baseline system that represented the current state-of-the-art at the beginning of the effort. The baseline system included a subset of IMPACT's capabilities including the route planner and an associated interface. However, the baseline system lacked agent vehicle recommendation support, plan monitoring, and speech control. The experimental design was a 2 (Baseline, IMPACT) x 2 (low, high mission complexity) within-participant design with the order of conditions blocked by system (half of the participants used IMPACT first, the other half baseline) and counterbalanced across task complexity. Mission complexity was manipulated by varying the number and timing of tasks. Each of the eight participants familiar with base defense and/or unmanned vehicle operations performed four 60-minute base defense missions. Participants completed a variety of defense mission related tasks involving twelve simulated UxV. Participants' task performance was better on multiple mission performance metrics with the IMPACT system in comparison to the baseline system. Participants were also able to execute plays using significantly fewer mouse clicks with IMPACT as compared to baseline. The overall usability of each system was assessed using the System Usability Scale (SUS; Brooke, 1996). Participants rated IMPACT higher than baseline on all ten SUS items, and IMPACT's overall SUS score was significantly higher than baseline's overall SUS score. Participants also subjectively rated IMPACT significantly better than baseline in terms of its perceived value to future UxV operations as well as its ability to aid workload. In fact, every participant gave IMPACT the highest possible score for potential value, and all but one participant gave IMPACT the highest possible score for its ability to aid workload.

Way Ahead

The IMPACT project and its resulting control station prototype have enabled a deeper exploration into the critical issues that influence flexible and effective human-autonomy collaboration. Although the IMPACT evaluation demonstrated value in several aspects related to operator-autonomy teaming, several deficiencies were also identified and improvements are underway. These include novel methods enabling bi-directional communication and management of temporal constraints, more naturalistic dialogue and sketch interactions, and consideration of information uncertainty in decision-making tasks. Additionally, research is investigating the effects of increased decentralized replanning capability, real-time operator functional state assessment, and alternative team structures on overall human-autonomy teaming. The results will provide a much richer understanding of this area.

Acknowledgement

This research supported the ASD/R&E sponsored Autonomy Research Pilot Initiative (ARPI) "Realizing Autonomy via Intelligent Adaptive Hybrid Control."

References

Brooke, J. (1996). "SUS: A 'quick and dirty' usability scale." In P. Jordan, B. Thomas, I. McClelland, & B. Weerdmeester (Eds.), *Usability Evaluation in Industry.* Bristol, PA: Taylor and Francis Inc.

Calhoun, G. L., Ruff, H. A., Behymer, K. J., & Frost, E. M. (in press). Human-autonomy teaming interface design considerations for multi-unmanned vehicle control. *Theoretical Issues in Ergonomics Science.* (Special Issue: Human-Autonomy Teaming).

Calhoun, G. L., Ruff, H. A., Behymer, K. J., & Mersch, E. M. (2017). Operator-autonomy teaming interfaces to support multi-unmanned vehicle missions. In P. Savage-Knepshield & J. Chen (Eds.), *Advances in Human Factors in Robots and Unmanned Systems: Advances in Intelligent Systems and Computing 499* (pp. 113-126). Springer.

Hansen, M., Calhoun, G., Douglass, S., & Evans, D. (2016). Courses of action display for multi-unmanned vehicle control: A multi-disciplinary approach. *Artificial Intelligence and Human-Robot Interaction (AAAI) Fall Symposium Series.*

Rowe, A. J., Spriggs, S. E., & Hooper, D. J. (2015). Fusion: A framework for human interaction with flexible-adaptive automation across multiple unmanned systems. *International Symposium on Aviation Psychology.*

A DYNAMIC SYSTEM SIMULATION DEVICE TO DEFINE HUMAN-SYSTEM INTERFACE REQUIREMENTS FOR THE DASSAULT RAFALE

Daniel Hauret
Air Warfare Centre of the French Air Force
Mont-de-Marsan, FRANCE
Julien Donnot
Air Warfare Centre of the French Air Force
Mont-de-Marsan, FRANCE

Our study aimed to collect enhancement proposals of Rafale fighter aircraft human-system interface. Proposals had to be innovative and complied with the needs of information for pilots regarding Rafale future capabilities. We developed a methodology based on a device enabling the simulation of a dynamic system activity that is the Rafale integrated in its war environment. Creativity of front line pilots participating in this experimentation has been stimulated but constraint by the necessity of a useful production due to the risks associated to the modification of a fighter aircraft already operating since 2006. Each proposal has been analyzed and synthetized through the abstraction hierarchy model of Rasmussen (1986). Results showed that for prospective and retrospective fields, a specific tactical support built with models favored the expression of functional objective and that a board as a basic human-system interface favored the expression of physical functions. In the discussion, we supported the relevance of our methodology for the definition of human-system interface requirements in various dynamic systems.

Our study is part of an ergonomic intervention led for the French Navy and the French Air Force. We were tasked to define human system interface (HSI) specifications focused on the weapon delivery arena needed for the development of the future Dassault Rafale NG. In the field of HSI conception, we had to face a technological drift due to the multiple opportunities brought by the glasscockpit technology. We observed the well-known trend to orient conception of new HSI according to the offered technical capabilities more than by the user's needs. By example, more and more tactical screens are spreading up in cockpits whereas one of the essential needs for a fighter pilot is to keep his hands on the throttle and sticks (HOTAS). In the conception process, thinking might not be guided by the identification of the possible uses with a new technology which lead designers to provide the maximum of available information and create clutter, but by the search of the best technology responding to the user's operational need, in the field of action. The fact is that the observation of the fighter pilot activity is almost impossible because of the isolated location of the pilot in a supersonic single seat aircraft and the dynamic feature of a high-level risk environment.

Our approach aimed to place user's information needs as the ultimate objective of the ergonomic intervention (Hauret, Donnot & Van Belleghem, 2016). We decided to build a simulation device (Maline, 1994) with two objectives and one main constraint. Our device should permit to collect and/or create proposals in order to (1) simplify the current Rafale HSI, which takes place in the retrospective field of activity and in order to (2) integrate the new long-range air/air missile (i.e., METEOR), which is related to prospective field. We analyzed innovative proposals regarding the abstraction hierarchy (Rasmussen, 1986). We expected a higher level of abstraction in the prospective field because HSI designers need flexibility to develop technological solutions and a lower level of abstraction in the retrospective field when HSI designers intent to correct the current functions. The main constraint is to find a way to produce only useful proposals (Loup-Escande, Burkhardt & Richir, 2013) that are proposals matching a proved need.

Our approach consist in simulate the Rafale pilot activity to provoke the expression of needs to act. Thus, beyond the necessity of tangible supports (Barcellini, Van Belleghem & Daniellou, 2013) the request of fighter pilot as participants was unavoidable. However, fighter pilots are not HSI designers and need to be guided to produce proposals directly transposable in specifications.

Activity simulation on tangible support

The paradox of ergonomics in conception is to create before to use a product. How can we create a product if we do not know how we will use it and how will we use a product if we do not know what we can do with it ? (Theureau & Pinsky, 1984). We choose to simulate activity with tangible support based on models to avoid participants to call on prescribed uses. The tangible supports we created allow participants to be both actor and analyst. At each step of the simulation, the participant can either take on an allocentric view (in the mission environment) or an egocentric view (in the cockpit).

Building of the dynamic system simulation device

Preliminary analysis of fighter pilot activity

The prerequisite of the simulation of fighter pilots' activity is to collect sufficient data to know and understand tasks and skills of a Rafale pilot. During a week, we gathered knowledge by taking part in flight training briefing and debriefing in a Rafale squadron. By working with fighter pilots we understood that being creative is one of their core cognitive skills. This ability was a key feature for the success of our methodology.

Construction of Tactical support and HSI support

To reproduce a faithful environment of a Rafale mission we needed two main supports. Obviously a whiteboard (blank at start) was intended to reorganize the cockpit HSI but was not appropriated for simulating actions of the aircraft in a tactical environment. That is the reason why we built Rafale models destined to maneuver on a tactical map. These models lamp equipped and free to vary in altitude projected circles of light representative for each weapon domains. Our simulation device was composed by the combination of both supports for which one of the major points is to offer the opportunity of a static (step by step) simulation of a high risk dynamic system.

Simulation was guided by a three-part question. First, and at each step of the scenario, the pilot was asked to express his intent that is the aircraft status he wished to reach. Then, he described actions associated to this objective. Finally he listed required information by giving details about location, form and access of each mission and flight data.

Participants

Six French Air Force pilots and two French Navy pilots took part to the study. They got at least the pair leader qualification and claimed either an only Rafale flight experience or another combat aircraft proficiency.

Scenarios

Scenarios were created to be as closed as possible to real pilot activity and to integrate all the events related to the use of the new long-range missile. During an all-day session, first scenarios dedicated to handling and navigation were simple and became harder along the session with weapon management. Thus, the pilot progressively reconstructed his HSI and could focus on complex issues once the base of the HSI was redefined.

Procedure

The experimental setup was presented to the pilot. He was asked to realize specific mission just as he was in his real cockpit, which means we expected him to apply the same uses as he does in flight. Then, the mission was briefed by himself as a real mission. It was the time to reveal his own tactical schemas. The tactical support and the Rafale models were designed to permit him to realize the same aircraft actions than those required in the real environment (Figure 1). He was told to limit his highly trained ability to anticipate because he would progress step by step in the mission simulated on a static simulation device. Each step included decisions and actions realized during about one minute. The fact that simulation is static at each step favored a better understanding of the tactical situation and allowed him to deeply analyze and speak out his thoughts and actions to come. The main objective was to lead the pilot to identify the information needed in the HSI to act. Because the pilot is focused on his actions, available but not required current information should not be evoked and led us to a pure list of useful information.

Figure 1. The dynamic system simulation device with a pilot, an experimenter and an air traffic controller.

Data analyses

All the sessions were recorded with cameras. Two types of data were collected. Regarding both the retrospective and prospective fields, we recorded on one hand current necessary information displayed in the Rafale HSI and on the other hand, all the innovative proposals. Retrospective label was related to the evolution of existing functions in the aircraft. Prospective label concerned all the proposals related to the use of the METEOR or the use of a helmet mounted sight device.

These proposals were analyzed by a couple of experimenter and classified according to the Rasmussen's abstraction hierarchy (Rasmussen, 1986). Thanks to this classification, proposals of all the participants have been compared, ordered and synthetized in a unique integrative proposal.

We adapted levels of abstraction as followed from the first (concrete) to the fifth (abstract):

1- Graphic/auditory solution (e.g., the weapon load is displayed in a rear view of the aircraft).
2- HSI function (e.g., be warned of an alternative weapon shoot opportunity)
3- Avionic (e.g., calculation of weapon flight time)
4- Rules (e.g., switch in autonomous mode of the missile, namely pitbull mode).
5- Goal (e.g., simultaneous management of *air to air* and *air to ground* weapons).

Collected proposals have been synthetized in three lists of specifications. The first list, related to information in the head-up display (HUD) is already the subject of a specific test in a dynamic flight simulator. The second list presents the specification of a helmet mounted sight device and the third list, still in development, the specifications of the tactical display in head-down.

Results

For several reasons included confidentiality agreements, results presented in this paper are only related to the list of HUD specifications. Proposals were ranked depending on the first level induced by the pilot. During the session, pilots suggested creative ideas starting at a specific level of the abstraction hierarchy but they were guided by the experimenter to explore higher or lower levels of abstraction. Levels presented in the following figures are the first levels spontaneously addressed by the participant.

For both retrospective and prospective field, our results showed that higher levels of abstraction, appreciated by designers, were reached with the tactical map whereas lower levels were get through the whiteboard support (Figure 2).

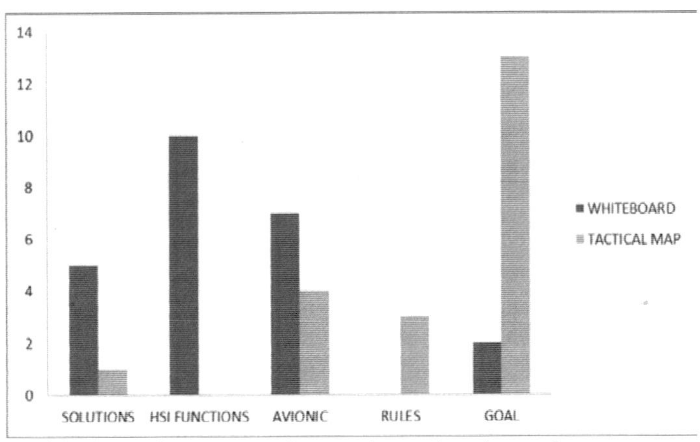

Figure 2. The whiteboard support favored production of concrete proposals whereas the tactical map support favored production of abstract proposals.

For the prospective field, our results confirmed the relevance of the tactical map to get abstract innovative proposals but it seemed interesting to underline that lower levels of abstraction are concerned by a few proposals with the whiteboard support (Figure 3a).

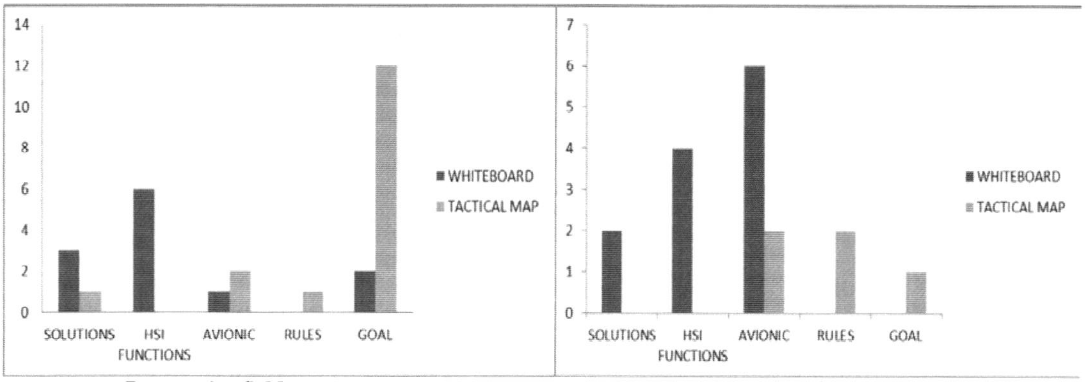

Prospective field ***Retrospective field***

Figure 3a and 3b. The tactical map support favored the production of prospective proposals at the highest level of abstraction. The whiteboard support favored production of retrospective proposals mainly at low levels of abstraction.

Concerning the retrospective field (Figure 3b), we confirmed that the whiteboard support brought more concrete proposals than the tactical map brought abstract proposals.

Discussion

We insist on the relevance of combining the two static supports to simulate the activity of a dynamic system. One of the strengths of our device is to allow the pilot to switch between the supports to be in the best conditions for revealing his needs of information. In this step by step approach, elaboration of the scenarios was crucial. An in-depth knowledge of fighter pilot activity is required preliminary to the simulation session because events occurring in scenarios will influence the pilot to be creative. The choice of front line pilots as participants improved the capacity of the simulation device to reveal useful needs which must not be confused with a user friendly feature. The resulting effect of soliciting representative front line pilots was to get a diversified sample of participants producing various innovative proposals. The use of abstraction hierarchy was justified and helpful to class, to regroup and to order all the proposals. Sometimes, two pilots suggested different proposals at a low level of abstraction but these same proposals were convergent at higher level of abstraction. Thus, we managed to produce integrative specifications. In fact, our lists of specifications, providing the identification of the appropriate level of abstraction, incorporated all the pilots' proposals.

In a near future, we will assess the relevance of our specifications for designing the future HSI in the Rafale program. We also consider reproducing our methodology and will apply it to other current functions of the aircraft such as the failure management or to prospective tactical concept such as handling remotely piloted aircrafts from a Rafale cockpit. In addition, we hope that the promotion of our methodology will create opportunities to investigate other complex jobs in aviation.

References

Barcellini, F., Van Belleghem, L. & Daniellou, F. Les projets de conception comme opportunité de développement des activités. In Falzon, P. (coord.) *Ergonomie constructive*. Presses Universitaires de France (2013), 191-206.

Daniellou, F. L'ergonomie dans la conduite de projets de conception de systèmes de travail. In P. Falzon (Ed.), *Ergonomie* Presses Universitaires de France (2004) (pp. 359-373).

Hauret, D., Donnot, J., & Van Belleghem, L. (2016). Intégration de la simulation de l'activité sur support tangible dans le processus de conception de l'interface homme-système du Rafale. Oral presentation, 15th conference Ergo'IA, Biarritz, France.

Loup-Escande, E., Burkhardt, J-.M., & Richir, S. (2013). Anticipating and Evaluating the usefulness of emerging technologies in ergonomic design : A review of usefulness in design. Le travail humain, *76*, 27-55.

Maline, J. Simuler le travail : une aide à la conduite de projet. Anact (1994).

Rasmussen, J. (1986). *Information processing and human-machine interaction.* New-York, NY : North Holland.

Theureau, J., & Pinsky, P. Paradoxe de l'ergonomie de conception et logiciel informatique. *Revue des Conditions de Travail* (1984), 9, 25-31.

AVIONICS TOUCH SCREEN IN TURBULENCE:
SIMULATION FOR DESIGN (PART 2: RESULTS)

Sylvain Hourlier & Xavier Servantie
Thales Avionics
Mérignac, France

Consumer market touch screens ubiquity has driven the avionics industry to launch in depth evaluations of touch screen for cockpit integration. This paper is a follow-up from ISAP 2015 paper where a methodology for turbulence simulation design was discussed. One of the challenges was to verify touch screen compatibility with in flight use under turbulent conditions, ranging from light to severe. The avionics industry recognized early on the need to alleviate such usability risk and the results of our evaluations enabled us to define recommendations for our HMI designs. Using our validated turbulent profiles, basic touch screen interaction performances were analyzed and this paper will focus on the results we gathered using our turbulence simulator.

Designing profiles for turbulence simulator

In our prior paper (Hourlier & Servantie, 2015), we presented the process that led to the design and validation of representative turbulence profiles and the selection of an hexapod as the best simulator for acceptable validity. In flight accelerometer (both linear and rotation) collections were performed to provide us with a baseline for choosing between possible simulation solutions. Given the 6 axis accelerometer profiles that were collected, a number of potential candidate simulation platforms were selected. They were reviewed in terms of performance and cost. A hexapod structure (figure 1) capable of reproducing those profiles with acceptable validity was selected. 6 simulated profiles were designed to mimic the "inflight" references. Tests were performed with pilots to validate the best profiles for each level of turbulence.

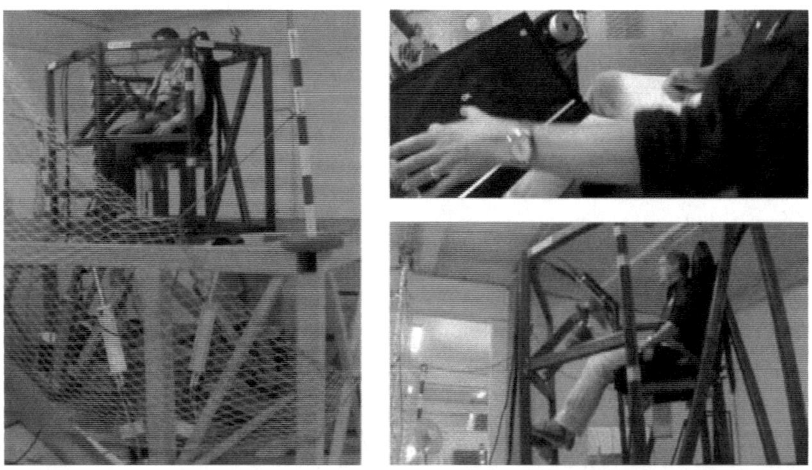

Figure 1: The Hexapod at ENSAM with the test bench on top

The selected profiles were then used to evaluate validate specific complex touch/gestures in light to severe turbulent conditions, using all the potential of touch interactions for novel cockpit Human Machine Interfaces. The result of these tests is presented here.

Using a turbulence simulator for interaction design

Once our selected turbulence profiles were validated by pilots, they were used for evaluation of technical solutions in the design of Avionics touch interactions. We needed first to assess the performance of basic interactions with regard to the 3 levels of turbulence relevant to the certification process: light, moderate and severe. This evaluation ran over a period of two weeks with 30 subjects in 2014. The results that are presented here are linked to our validated (Thales proprietary) levels of turbulence and should be considered as suggestions for design, as many other factors can influence touch interaction (existence of finger/palm rests or anchors to cite but the most obvious one).

Population and means

Population. 30 subjects performed this evaluation: 5 left handed, 25 Right handed; 4 women, 26 men; 6 aged 20—29, 11 aged 30-39, 8 aged 40-49, 5 aged 50-59; 7 men had more than 100h of piloting experience (5 with significant flight experience); 9 reported being sometimes sea sick or simulator sick.

Means & Method. The detailed account of the materiel used can be found in our prior paper (Hourlier & Servantie, 2015).

- The Hexapod (+/-2g, +/- 50cm Y,X,Z displacements and 3 axis angular acceleration), property of ENSAM Bordeaux was fitted with a specific "cage" replicating the conformation of the Thales AV2020 cockpit design.
- An in-house recording system collected all interactions with the touch screen (time stamps, screen XY localization).
- Videos using GoPro cameras were recorded: one filming the screen interactions, the other filming the subject. A wireless headset enabled communications between subjects and experimented. An emergency stop button was always accessible to the subject (but was never used)
- Four turbulence profiles (table 1) were preprogramed on the hexapod and could be played on demand: none, light, moderate and severe.

Table 1.
Turbulence profiles used for tests (acceleration in m/s^2)

Turbulence level	None	Light	Moderate	Severe
Maximum	-	2,29	5,52	8,11
Mean	-	0,65	1,53	2,60
Median	-	0,57	1,32	2,29

A typical run would comprise successive 4mn evaluations of basic interactions in successive turbulence profiles (no turbulence, light turbulence, moderate turbulence and severe turbulence). An individual session would last 1h30mn on average. A pause in the middle was added to accommodate the test subject, the experience being somewhat tiring.

Protocol. Subjects were asked to perform, at various levels of turbulence, simple tasks replicating basic interactions with Touch devices. These were: Press, Release, Double tap and Long press.

- Press & Release. A colored circle (Ø 7-12-15-18 mm) would appear on a black screen at random places along with a target (cross) at another random position. The task being to drag the circle to the cross and release on the center of as precisely as possible to make it disappear (speed and precision measurements collected).
- Double tap. A colored target circle (Ø 7-12-15-18-28 mm) would appear on a black screen at random places. The task being to double tap on its center as fast as possible to make it disappear (speed and precision measurements collected).
- Long press. A colored target circle (Ø 18mm) would appear on a black screen at random places. The task being to press it at least 2 seconds on its center to make it disappear (movement and precision measurements collected).

The objective of these trials being to identify size and time related recommendation for efficient touch interactions in turbulent conditions.

Results

All results presented here account for finger rest interactions (except for the few mentioned in table 2). Basic results are presented as an error rate outcome with regards to the analyzed variables.

For instance the figure 2 presents the error rate when pressing a target button in 3 conditions no, light or moderate turbulences. For example, if one considers 10% an acceptable error rate, the figure presents the size of the interacting zone radius 13.5mm for moderate turbulence (the zebra arrow) and 8mm for light turbulence (the dotted arrow). For example, if one wants to secure an interaction with a round button in moderate turbulence for an expected success rate of 90%, one should choose a 27mm diameter interaction zone.

To obtain our results, numerous trials were recorded. See table 2 for reference.

Discussion

The overall Gaussian shape of our data representation (figures 2 to 6) and their increasing logic with higher turbulence accredit the validity of our data and enable us to obtain explanatory mathematical transfer function from turbulence level to interaction error.

From the double tap spatial analysis we can recommend double tap effective zones and from the temporal analysis we can recommend on the delay before addressing a double tap as a single one and also recommend on the size of the zone to reduce the time delay.

From these results one can also analyze the involuntary finger movements (given a certain level of turbulence) and thus recommend a threshold before considering a movement as a drag. For instance such results could serve to differentiate between dragging a map and creating a marker on the map.

Finally, as analyzed for press interactions, (figure 7) the error rate can be more than 50% higher without finger rest in moderate turbulence level. Hence, in an aeronautical environment, FINGER REST is MANDATORY.

Table 2.
Occurrence collected during our basic tests

Interaction type	No Turbulence	Light Turbulence	Moderate Turbulence	Severe Turbulence
Press*	180/*514	482/*573	480/*367	342/*168
Release**	100	100	100	100
Double tap	307	306	186	121
Long press	182	369	322	248

*Without finger Rest for comparison (figure 7). **Protocol limited to 100 interactions for technical reasons.

Press analysis

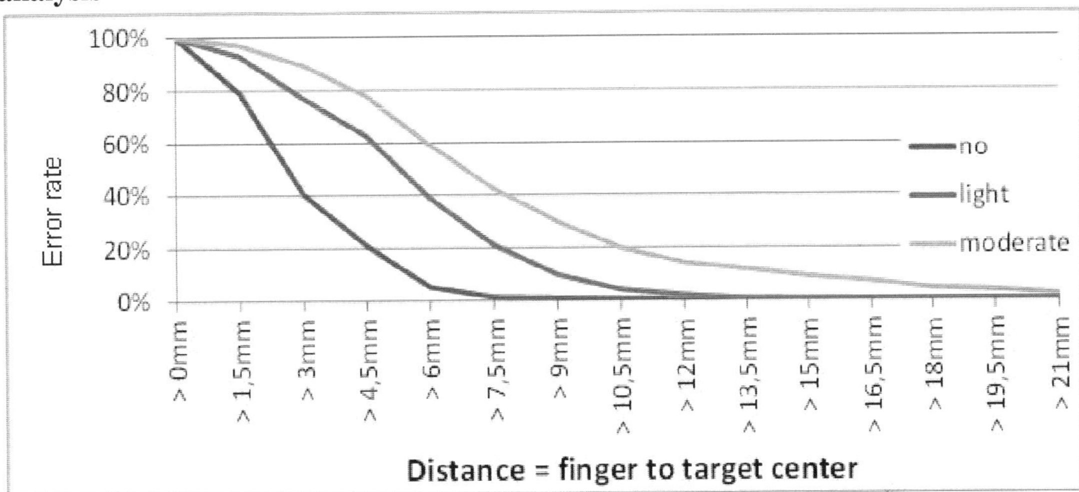

Figure 2: Error rate according to distance (touch to target in mm) and turbulence level for finger press

Release analysis

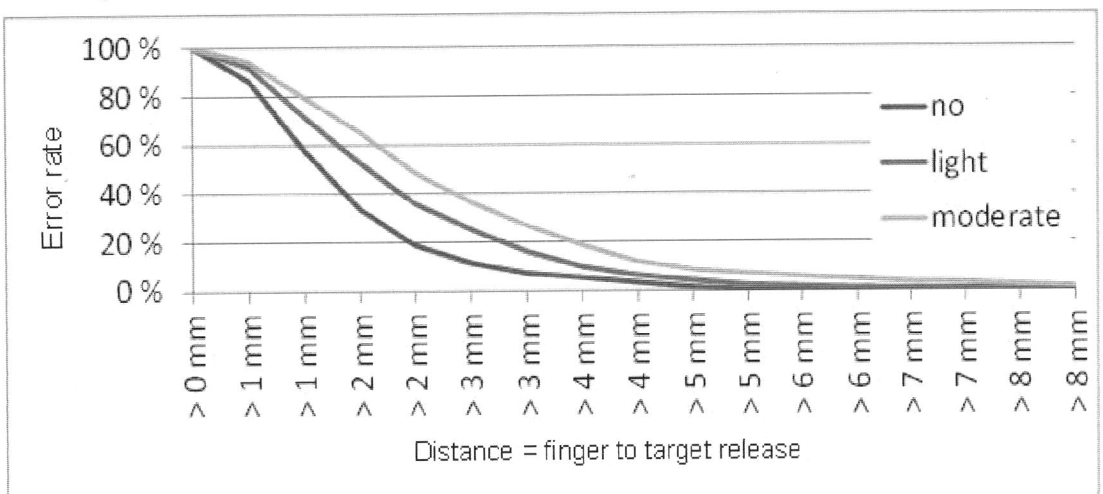

Figure 3: Error rate according to distance (release from target in mm) and turbulence level for finger release

Double tap analysis

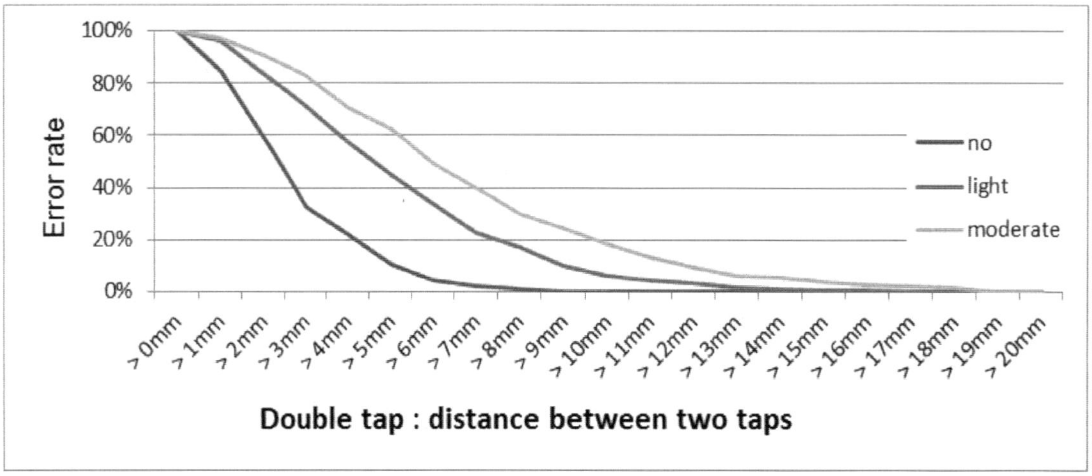

Figure 4: Error rate according to distance between taps (in mm) and turbulence level

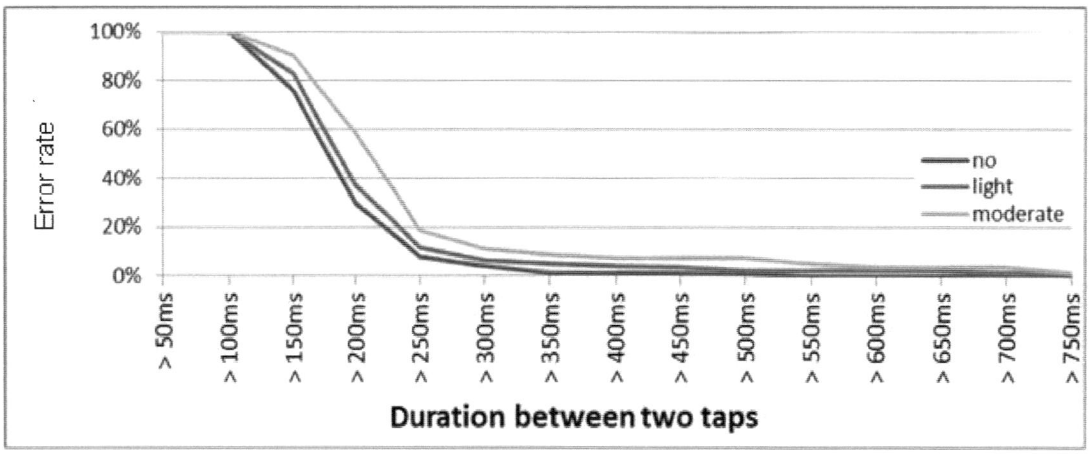

Figure 5: Error rate according to duration between taps (in ms) and turbulence level

Long press analysis

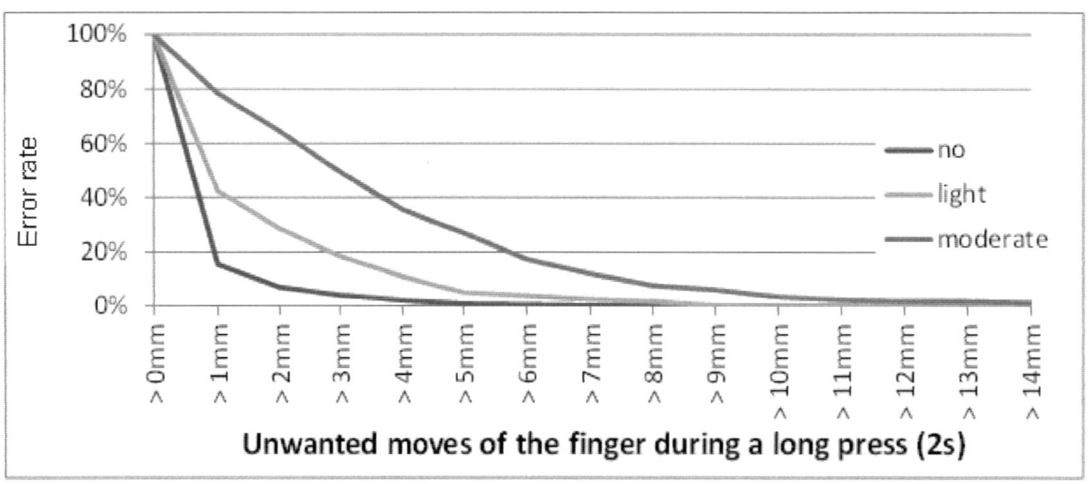

Figure 6: Error rate according to distance for a 2s long press and turbulence level

Error rate with and without finger rest (FR)

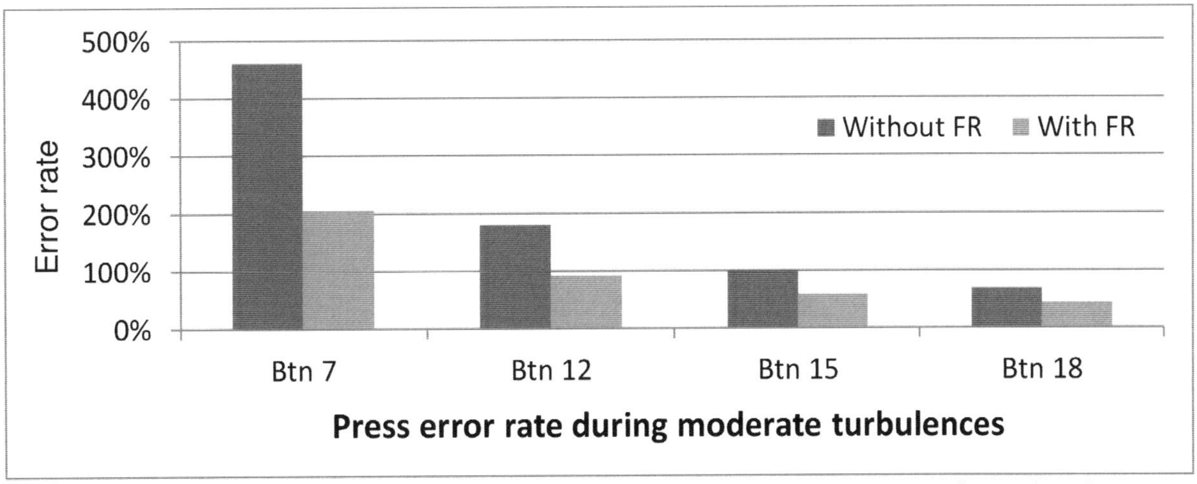

Figure 7: Error rate for various button sizes (in mm) with or without finger rest in moderate turbulence level.

References

Hourlier, S. & Servantie, X., (2015) Avionics touch screen in turbulence: simulation for design, proceedings of the 18th International Symposium on Aviation Psychology, May 4-7, Dayton Ohio

EFFECTS OF AN ECOLOGICAL INTERFACE ON FLIGHT TRAINING EFFECTIVENESS

Ronald Deerenberg, M. M. (René) van Paassen, Clark Borst, Max Mulder
Aerospace Engineering – Delft University of Technology
Delft, The Netherlands

For Ecological Interface Design (EID), the underlying constraints and properties of an operator's work domain are analysed and used as a basis for the design of the information displays, so that these may reveal these underlying mechanisms. Most evaluations for EID have been performed with expert or trained participants. However, it can be hypothesised that the effects of EID will also change the way tasks are learned by novices; since the EID designs support direct manipulation, and at the same time show the constraints in the work domain, a novice would be able to perform the task as a skill, employing the direct manipulation features of the interface, while at the same time learning the underlying constraints from the work domain. Our interest is the effect of an EID display on skill acquisition in a flying task. To this end we evaluated the EID display by (Amelink, Mulder, van Paassen, & Flach, 2005) in a study with novice pilots, learning flight path and speed control of a simulated aircraft. It was found that initial performance by the EID group was better than by a control group, the EID group also showed more consistent and homogeneous behavior. The EID display did not lead to increased workload, as measured with the Rating Scale for Mental Effort. Asymptotic performance levels for both groups were not significantly different.

Introduction

In order to reduce the time and cost involved with flight training, flight simulators are becoming more commonly used. Much effort is being put into understanding the contribution of motion feedback and visual cues towards increasing the effectiveness of simulator-based pilot training (Mulder, Pleijsant, van der Vaart, & van Wieringen, 2000; Pool, Harder, & van Paassen, 2016). When using a simulator for initial training, alternative interfaces that increase the instructional value may be considered. It can be argued that interfaces based on Ecological Interface Design (EID) may be applied to this. Even though EID is mostly considered to "facilitate human adaptivity and flexibility to cope with unforeseen events" (Borst, Flach, & Ellerbroek, 2015), the way in which users are supported might also aid in the learning process. The effect of EID interfaces on learning received limited attention, two notable exceptions are a longitudinal study using the DURESS II process control microworld (Christoffersen, Hunter, & Vicente, 1996, 1998), and a series of experiments in which the Oz display – not designed by the EID approach, but as a functional aviation display – is evaluated (Smith, 2007; Smith, Boehm-Davis, & Chong, 2004).

The objective of this research is to evaluate the training effectiveness of EID interfaces during a manual control task, by comparing the skill acquisition for task-naive subjects using conventional instrumentation to those training with the total energy-based perspecive flight path display (Amelink et al., 2005). During an approach, or any other situation in which altitude and airspeed changes are requested, energy management is the underlying principle for the coordination between throttle and elevator. The pilot tries to control two aircraft states (airspeed and altitude) by using two control inputs (throttle and elevator) but these inputs do not directly map onto the controlled aircraft states. During flight training, pilots are often taught to apply a control strategy in which the inputs do directly map onto the outputs in order to simplify the control task. The two variations of these simplified control strategies are the "throttle-to-path & elevator-to-speed" and "throttle-to-speed & elevator-to-path" strategies. However, these simplified strategies are sub-optimal and lead to adverse control couplings (Lambregts, 1983; Langewiesche, 1944).

The experiment investigates whether the visualization of the aircraft's underlying energy relations supports student pilots in learning the basics of flight. The intention is to evaluate the EID display as a training tool, so the experiment uses a quasi transfer-of-training set-up, in which participants' performance is tested with the conventional display.

Total Energy-Based Perspective Flight-Path Display

The total-energy display is based on a tunnel-in-the-sky display, which shows a three-dimensional guidance situation with respect to the trajectory to be followed: "It allows a direct spatial orientation of the aircraft's position, attitude and motion relative to a fixed landmark -the tunnel geometry- in the environment" (Mulder, 1999). This

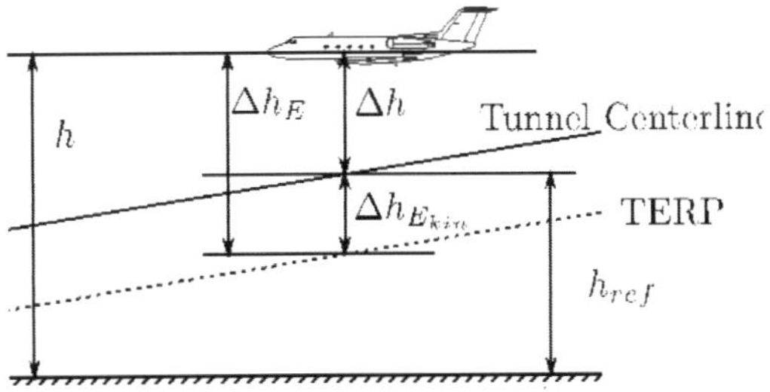

Figure 1: Definition of the Total Energy Reference Profile (TERP) (Amelink, Mulder, van Paassen, & Flach, 2005)

Figure 2: Human-Machine Interaction Laboratory (HMI-Lab) at Delft University of Technology

orientation and motion can also be interpreted as a visualization of potential (height) energy. In addition to the physical motion, the display also visualizes the energy inventory of the flight, by showing the Total Energy Reference Profile (TERP) and the Total Energy Angle (TEA). From the TERP and the tunnel visualizations, pilots can observe total energy, potential energy and kinetic energy deviations. Likewise, the TEA and the flight path angle show rates of these quantities (Figure 1).

Experiment

Goal of the experiment

The goal of the experiment is to evaluate the training effectiveness of the total energy-based perspective flight-path display during a manual flying task. In order to evaluate the training effectiveness, the skill acquisition of participants is compared to that of participants using a baseline tunnel-in-the-sky representation. Also, the natural progression of participants is the object of study, therefore the amount of feedback during training is limited as much as possible. This also means that task complexity had to be reduced, therefore the aircraft control was limited to purely longitudinal motion, effectively limiting flight training to climbs, descents, and straight-and-level flight.

Apparatus

The experiment was performed in the Human-Machine Interaction Laboratory (HMI-Lab) of the faculty of Aerospace Engineering at the Delft University of Technology (Figure 2). Subjects were able to control a six-degree of freedom, non-linear model of the Cessna Citation 500, by means of a right-handed electro-hydraulic side-stick and a throttle located to their left. The interface was presented by means of an 18-inch LCD monitor. The side-stick was configured such that only fore-aft movement was possible. Also, rudder control was disabled, flaps were kept to 15 ◦ and the landing gear remained retracted during the entire simulation. In terms of atmospheric models, zero wind was present but some turbulence was included. No outside visual was used in the experiment but engine sound was generated and the lights were dimmed during the experiment.

Participants

A total of 24 task-naive participants took part in the experiment, of which 20 male and 4 female. All participants indicated normal color vision and none of the participants had prior flying experience. All participants filled in the revised Study Preference Questionnaire (Jeske, Backhaus, & Rossnagel, 2014) before the experiment, in order to score participants based on the holist/serialist cognitive style division. The same questionnaire was completed by 110 pilots (RPL, PPL, and ATPL) in order to select participants such that both experimental groups were representative for a typical pilot population. In addition, participants were selected and divided over the two experimental groups in order to balance both groups as much as possible in terms of prior experience.

Control task

During the experiment, participants were asked to follow an altitude and airspeed profile along a longitudinal trajectory. A simulation run consists of a series of change in either: altitude, airspeed or a in knots combined change in altitude and airspeed. An example of this can be seen in Figure 3. The requested altitude profile

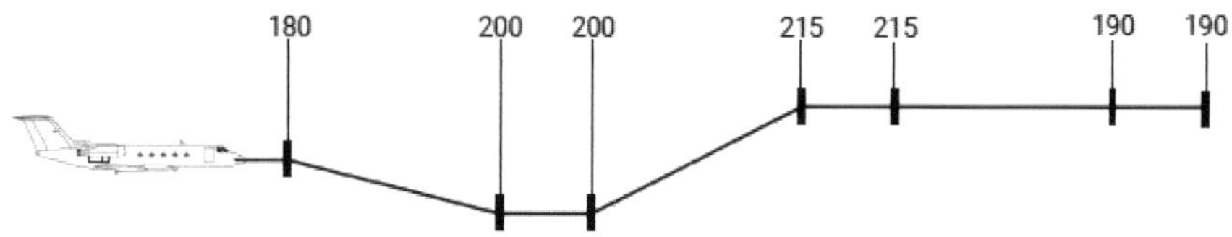

Figure 3: A section of the trajectory, indicating the requested altitude profile and the goal speeds in knots

is indicated by the tunnel-in-the-sky representation in both interfaces. The currently requested airspeed is indicated by the location of the speedbug and the current and following airspeed goals are also indicated numerically in the interfaces, both of which can be seen in Figure 4. Participants were told to control the aircraft through the tunnel and to follow the airspeed commands as accurately as possible. i.e., to minimize all occurring position and airspeed errors.

The simulation trajectory consisted of fifteen different changes in altitude and/or airspeed. After each requested change in either aircraft state, a section was always present where participants were requested to fly straight-and-level with a constant airspeed. Effectively doubling the amount of changes in aircraft state. This was done in order to allow participants some time to recover from any errors in the required aircraft state and thus to limit any error propagation through following sections of the trajectory. In order to avoid pattern recognition and boredom, the trajectory was split into three sections and the order was mixed according to Table 1. Between subjects there was no randomization performed, meaning that all subjects had the same order of trajectories as indicated in Table 1.

Experimental design

The experiment has a between-subjects design with a quasi-transfer-of-training manipulation. During this transfer-of-training experiment, there are two phases referred to as the training and transfer phase. The training phase consists of nine simulation runs carried out during the morning. In the afternoon, on the same day, participants completed another six simulation runs during the transfer phase, as can be seen in Figure 5. The control group, referred to as BASE, used the baseline tunnel-in-the-sky interface during the entire experiment. The other group, referred to as the EID group, used the total energy-based perspective flight-path interface in the training sessions and participants used the same baseline tunnel-in-the-sky interface. As each simulation run lasts approximately seven

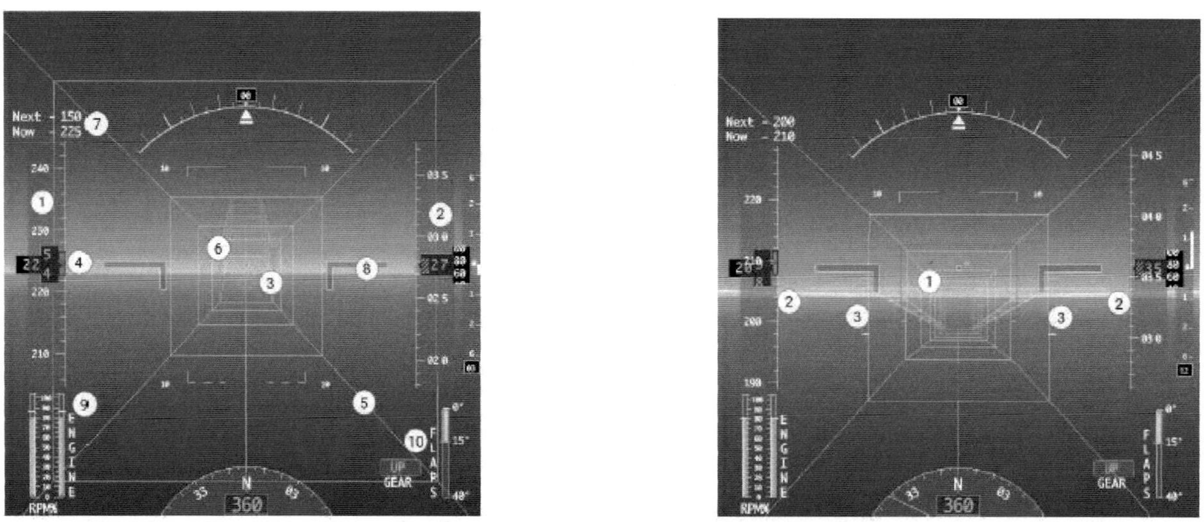

(a) **Baseline Interface:** 1) Airspeed Tape, 2) Altitude Tape, 3) Flight Path Vector, 4) Speedbug, 5) Tunnel, 6) Purple Goal Frame, 7) Goal Speeds, 8) Aircraft Symbol, 9) Tachometer, 10) Flap and Gear Indicator

(b) **Total Energy-Based Perspective Flight-Path Display:** 1) Energy Angle, 2) Total Energy Reference Profile, 3) Speedmarks ($\pm$ 2 knots)

Figure 4: The two interfaces that were used in the experiment, highlighting the various display elements.

minutes, a fifteen-minute break was scheduled after each block of three simulation runs, in order to avoid fatigue.

Each participant received an experimental briefing two days before the experiment, which also included a recap of the relevant theory of flight needed to complete the task. This included an explanation of the effect of the controls and a summary of the relevant material regarding straight-and-level flight, climbs, and descends according to the FAA's Instrument Flying Handbook (Anon., 2012). Before starting the experiment, there was room for questions regarding the theory of flight and the functioning of the interface, but no feedback was provided during the experiment except for explaining some display features.

Furthermore, the baseline group was explicitly told to use the throttle-to-speed, elevator-to-path control strategy. On the other hand, the EID group was explicitly told to use the throttle to control total energy and the elevator to control altitude. However, both groups received the same experimental briefing through which they were made aware of the different control strategies.

Table 1: Order of the sections A, B, and C of the trajectory for each simulation run

Run	Order	Run	Order
1	A-B-C	10	B-C-A
2	A-C-B	11	C-A-B
3	B-A-C	12	C-B-A
4	B-C-A	13	A-B-C
5	C-A-B	14	A-C-B
6	C-B-A	15	B-A-C
7	A-B-C		
8	A-C-B		
9	B-A-C		

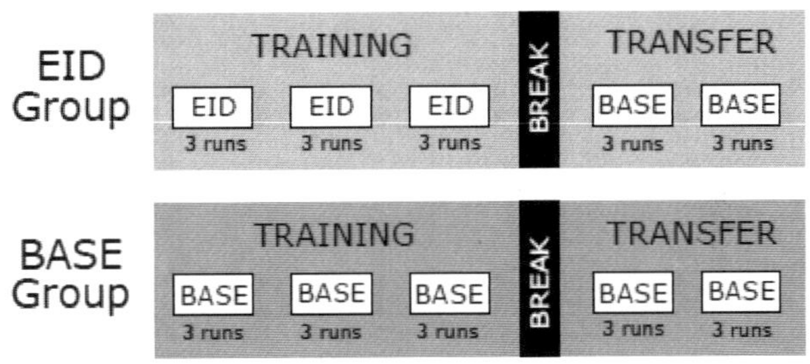

Figure 5: Experiment group definition, indicating which interfac was used by both groups during the experiment

Hypotheses

It was expected that the added energy information to the baseline tunnel-in-the-sky interface would increase mental effort during the training phase due to the apparent increase in display complexity as noted in previous evaluations of the EID display (Amelink et al., 2005). On the other hand it is expected that the EID group will perform initially better than the control group, however, with the expected amount of practice both groups might end at the same asymptotic level of performance, as measured in speed/altitude deviation. The EID group is also expected to have lower control activity, since the display information on energy and energy rate can resolve the cross-coupling between the control inputs and control target values.

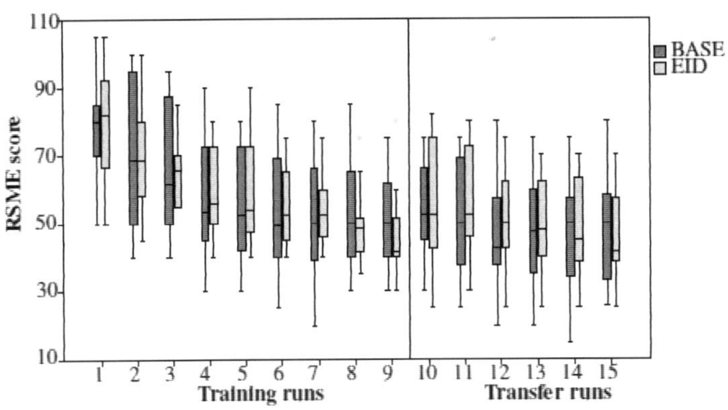

Figure 6: Boxplot of the experienced mental effort according to the Rating Scale Mental Effort, for each of the fifteen simulation runs for both the EID and baseline group. With simulation run 1-9 being the training phase, and 10-15 the transfer phase

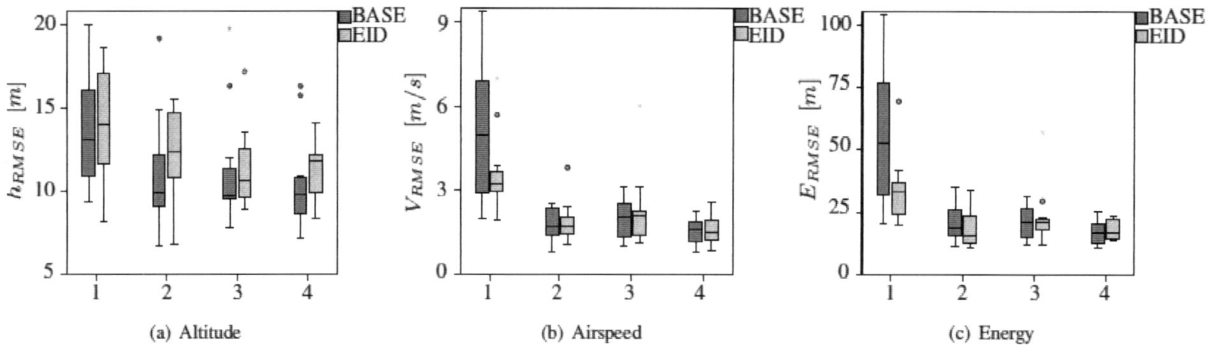

Figure 7: Boxplots of the performance measures, indicating: 1, the initial run; 2, the average scores for the last three runs of the training phase; 3, the first run after the transfer manipulation; and 4, the average scores for the last three runs of the training phase

Results and discussion

Mental Effort

The mental effort was measured with Zijlstra's RSME (Zijlstra, 1993). Previous evaluation with experienced pilots indicated a considerable increase in mental effort for the EID display. The results can be seen in Figure 6. A clear decrease over the runs is visible, but no significant difference between the two groups is present.between the final training run and the initial transfer run, a small increase in mental effort can be seen, however this was not statisticallly significant (Wilcoxon, $z = -1.735$; $p = 0.083$). There appears to be no difference in mental effort between the EID and baseline groups, indicating that the perceived complexity of the display as reported by expert pilots plays a smaller role in novices.

Performance

Values for the different performance measures are given in Figure 7. Overall, altitude tracking performance quickly reached proficient levels of performance and performance was adequate almost from the start of the experiment, resulting in only a small learning rate. The averaged performance in airspeed tracking of the two groups showed no significant differences (Wilcoxon, $z = -1.569$; $p = 0.117$), however the variance within the baseline group was significantly larger (Levene, $F = 6.004$; $p = 0.023$). Both groups reach the same level in airspeed tracking, even though the airspeed deviations are presented in a markedly different manner for the EID group.

The airspeed and total energy error both show larger errors for the first run only, however, asymptotic performance levels are quickly reached by both groups and by all participants. Regarding the total energy error, there is a significant difference between the two groups (Wilcoxon, $z = 1.961$; $p = 0.050$) and the associated variance of the EID group is significantly less than that of the Baseline group (Levene, $F = 8.385$; $p = 0.008$). When considering the average energy error for both groups in the final three training runs, or the final three evaluation runs, there were no significant differences. Thus both groups mastered energy control to a level that they were indistinguishable.

Conclusion

The flight-training effectiveness of an EID interface for learning a manual longitudinal flying task in a fixed-base simulator was evaluated by means of a between-subjects quasi-transfer-of-training experiment with 24 task-naive participants. Since the EID interface is considered for a training aid, the transfer is to a non-EID conventional interface.

Participants who trained with the EID interface showed better initial performance in terms of airspeed and total energy tracking, however these differences quickly disappeared. Also the variance between participants was significantly lower for the EID group. The usage of the EID interface also leads to an increase in control activity as evidenced by large elevator deflection rates and an increase in the number of throttle reversals. Contrary to expectation, since apparent display complexity is higher for the EID interface, there was no significant difference in terms of mental effort. The different interfaces did not result in a significant difference in performance once asymptotic performance levels were reached. There was no evidence of over-reliance on the energy cues by the EID group as there were no significant transfer effects. It appears that the functional information presented in the EID interface provide improved support during the initial phase of the training, without negative effects such as over-reliance on display features or increase in mental effort. In general, the effects of the EID display on performance were small, and only visible in the first few training runs. The evaluation did not include unanticipated situations for the participants, and thus did not test all aspects of the EID support. However, training benefits of EID displays might be larger for more complex systems that involve collaborative problem-solving and require higher order cognitive processes.

References

Amelink, M. H. J., Mulder, M., van Paassen, M. M., & Flach, J. M. (2005). Theoretical foundations for a total energy-based perspective flight-path display. The International Journal of Aviation Psychology , 15 (3), 205–231.

Anon. (2012). Instrument flying handbook . FAA-H-8083-15B. Federal Aviation Administration Flight Standards Service.

Borst, C., Flach, J. M., & Ellerbroek, J. (2015). Beyond ecological interface design: lessons from concerns and misconceptions. Human-Machine Systems, IEEE Transactions on , 45 (2), 164–175.

Christoffersen, K., Hunter, C., & Vicente, K. J. (1996). A longitudinal study of the effects of ecological interface design on skill acquisition. Human Factors: The Journal of the Human Factors and Ergonomics Society , 38 (3), 523–541.

Christoffersen, K., Hunter, C., & Vicente, K. J. (1998). A longitudinal study of the effects of ecological interface design on deep knowledge. International Journal of Human-Computer Studies , 48 (6), 729–762.

Jeske, D., Backhaus, J., & Rossnagel, C. S. (2014). Evaluation and revision of the study preference questionnaire: creating a user-friendly tool for nontraditional learners and learning environments. Learning and Individual Differences , 30 , 133–139.

Lambregts, A. (1983). Integrated system design for flight and propulsion control using total energy principles. In American institute of aeronautics and astronautics, aircraft design, systems and technology meeting, fort worth, tx (Vol. 17, 19).

Langewiesche, W. (1944). Stick and rudder : an explanation of the art of flying . New York: McGraw-Hill.

Mulder, M. (1999). Cybernetics of tunnel-in-the-sky displays (Doctoral dissertation, Delft University of Technology).

Mulder, M., Pleijsant, J. M., van der Vaart, J. L., & van Wieringen, P. (2000). The effects of pictorial detail on the timing of the landing flare: results of a visual simulation experiment. The International Journal of Aviation Psychology , 10 (3), 291–315.

Pool, D. M., Harder, G. J., & van Paassen, M. M. (2016). Effects of simulator motion feedback on training of skill-based control behavior. Journal of Guidance, Control, and Dynamics , 1–13.

Smith, C. (2007). Using displays to improve training: functional aviation displays improve novice piloting knowledge. In Proceedings of the human factors and ergonomics society annual meeting (Vol. 51, 25 , pp. 1583–1587). Sage Publications.

Smith, C., Boehm-Davis, D., & Chong, R. (2004). The effect of functional information in an avionics display. In Proceedings of the human factors and ergonomics society annual meeting (Vol. 48, 1 , pp. 1–5). SAGE Publications.

Zijlstra, F. R. H. (1993). Efficiency in work behaviour : a design approach for modern tools . Delft: Delft University Press.

INTERVIEWS OF GENERAL AVIATION PILOTS: AN INSIGHT TO AIRSPACE INFRINGEMENTS

Elena Psylou, Arnab Majumdar, Washington Ochieng
Imperial College
London, United Kingdom

This paper finds contributory factors to airspace infringements concerning the planning undertaken by general aviation pilots. Twenty seven recreational pilots who flew a light fixed-wing motor, glider and ultralight aircraft were interviewed using semi-structured interviews in Finland, Norway and United Kingdom. These countries experienced a major problem with the number of reported AIs. Interview transcripts were analysed using thematic analysis. The success of this study is attributed to the carefully design of both the questions of the interview and the sample that comprises the diverse general aviation sector. The newly found contributory factors are associated with a pilot's performance as well as airspace design features that can influence the pilot's flight route decision-making, e.g. wished flying altitude is higher than the lower boundary of controlled airspace in the capital of a country. The findings can aid the incident investigation and the development of mitigation actions of these incidents.

General aviation (GA) represents a unique group of airspace users that fly for a range of purposes using a diverse aircraft fleet that can sometimes be ill-equipped to fly in controlled airspace (Civil Aviation Authority United Kingdom, 2006; International Civil Aviation Organization, 2009). Typically, most GA pilots fly for recreation purposes at the weekends and when the weather conditions offer good visibility because most pilots fly under visual flight rules (VFR). GA pilots increasingly use digital devices to plan their flight pre- and in-flight. As with all such technologies, their use can improve as well as degrade a GA pilot's performance. Such influences can lead GA aircraft to fly into controlled and restricted airspace without receiving permission from the Air Traffic Controller (ATCO), who is responsible for managing the traffic in these areas. Such airspace infringement (AI) incidents can cause safety and other air traffic management problems, e.g. delays, with the worst case being a mid-air collision. On average, there are approximately 100 and 600 AIs every year involving GA in Norway and United Kingdom respectively (General Aviation Safety Committee, 2016).

This paper, therefore, aims to find contributory factors (CFs) of AIs involving GA flights and these CFs will relate to the flight planning undertaken by GA pilots. This paper is structured as follows. In the following section, the studies of AIs, conducted by European stakeholders, will be briefly disccused regarding the data, method and key findings, and the potential of findings pilot's related CFs in interviews will be discussed. Next, the participants, the interview design and the method to analyse transcripts used in this paper will be outlined. The CFs will be presented and discussed before concluding.

Literature review

During the past decade, two major studies of AIs in Europe were conducted in order to understand the underlying reasons behind the occurrence of AIs by two stakeholders (European Air Traffic Management, 2007a; European Air Traffic Management, 2007b; European Air Traffic Management, 2008; Safety Regulation Group, 2003). The CFs found in these studies are not exhaustive. There are generic factors, e.g. airspace design and flight planning, that indicate their importance with AIs; however, they are of limited use and further study is needed to distinguish these generic factors. There are also ill-defined factors and their poor definitions limits their use as well. Furthermore, the CFs do not comprise of factors related to the impact of technologies currently used by GA pilots on AIs as found in (O'Hare & Stenhouse, 2009; M. Wiggins, 2007). In general, factors related to a pilot's performance were also not found in the studies whilst such factors are found in the literature of aviation psychology, e.g. a pilot pursues the flight into adverse weather due to a past successful situation. (Molesworth, Wiggins, & O'Hare, 2006; M. W. Wiggins, Azar, Hawken, Loveday, & Newman, 2014).

Literature in decision-making used: questionnaires in which participants rated scenarios (Hunter, Martinussen, & Wiggins, 2003), simulated flights (Molesworth et al., 2006), scales (Hunter, 2005) and incident and accident data (M. W. Wiggins et al., 2014). It is evident that research questions that are broad or explorative, questionnaires are preferred over simulation studies. It is remarkable, though, that interviews were not commonly used in the literature given their evident success to address explorative research questions that knowledge in the field as in (Nascimento, 2014). Regarding the sample design used in the literature, the sample often consisted of GA pilots and commercial pilots whilst the results were presented for all the participants (Hunter et al., 2003; M. Wiggins, 2007). This aggregation might have prevented differences between these two types of pilots from becoming apparent. Hence, the design of the sample should account the diversity of GA sector. Last but not least, the validation method used in the literature was often not clearly stated whilst validation is essential. Validation can be conducted by a subject matter (Nascimento, 2014) and by a comparison with similar studies or data (Hunter et al., 2003; Hunter, 2005) .

In order to identify CFs of AIs related to the flight planning, interviews of GA pilots, who are the key contributors to AIs, will be conducted and the sample of the study will represent the diversity of GA sector. The sample, the interview design and the method of analysis of the interviews are presented in the following section.

Method

Interviews of recreational GA pilots were conducted in Finland, Norway and United Kingdom (UK) that possess a problem with AIs involving GA flights and their aviation stakeholders collect AI incident reports. Interviews were conducted between March and November in 2015 and their duration was between 45 and 70 minutes. A convenient time for the face-to-face interview was arranged at the participants' flying club or city of residence. Participants were found directly from flying clubs in the UK and through the airspace navigation service provider and national aviation authority in Finland and Norway.

Participants

Participants were selected based on four criteria as follows given analysis conducted of reported AIs in these countries. The geographical location of their flying base was a selection criterion in that approximately 80% of the participants used an aerodrome located in the region of the capital and subsequently 20% of the participants departed from other cities. The reason is that the safety analysis of reported AIs in these countries showed that most AIs located in the region of the capital of the countries whilst the airspace design might also relate to AIs. In order to ensure that the diversity of GA fleet is represented in the sample, even though most reported AIs occurred by fixed-wing motor aircraft, pilots of an ultralight and glider aircraft will also be interviews as follows: 80% of the participants flew a light fixed-wing motor aircraft, 10% flew an ultralight aircraft and 10% flew a glider aircraft.

In order to control the diverse activities, pilots must fly for recreational purposes and have a VFR-rating. Given that the flying hours of GA pilots can vary, the sample must consist of GA pilots who were recently issued their flying licence and have been flying for a long time. The flying activity in the last three months is also considered to account the inactive flying period in the winter. Pilots who fly cross-country flights will also be interviewed. Finally, participants must be fluent in English language as the interview will be conducted in this language. The involvement of the participants in an AI, their age, occupation and gender are not taken into account in the sample design.

The sample consisted of 27 GA recreational pilots as shown in Table 1. There were 20 pilots that flew a fixed-wing motor aircraft, three pilots that flew a glider and three pilots that flew an ultralight aircraft. Ultralight and glider pilots were difficult to find and thus, the minimum required number of these pilots was selected. In Norway, participants, who were based in the capital region were found only.

Table 1.
Design of the sample

Criterion	Fixed-wing motor aircraft	Glider aircraft	Ultralight aircraft	Total
Country				
Finland (Helsinki)	6	1	1	10
Finland (Southern Finland)	2			
Norway (Oslo)	5	1	1	7
United Kingdom (Greater London)	5	1	2	10
United Kingdom (South England)	2			
Total	20	3	4	27
Total flying hours*	505.5 (1310.7)	500 (822.7)	100 (35)	N/A
Flying hours in the last three months*	14.5 (24.3)	8 (24)	11 (1.5)	N/A
Number of pilots who also flew cross-country flights	17	1	0	N/A

Note. *median (standard deviation), N/A: not applicable

Interview design

A semi-structured interview was designed to address research questions beyond the research question of this paper. Participants were asked ten questions whose objectives were the description of the planning of a flight, the material and devices they use for planning and navigation and the pilots' involvement in AIs and other safety related incidents. For this paper, the responses concerning the description of the manner to which they decide the flight route for their desired destination including the difficulties they expect to experience will be used. The questions were open-ended and probe questions were asked, e.g. 'will the temperature affect your flight route decision?'

Analysis of Interview

The interview transcripts were analysed using the phenomenological method thematic analysis (Coyle & Lyons, 2007). The transcripts were coded and the codes were grouped to develop the themes and their sub-themes that will be the findings of the analysis. The analysis followed the guidelines for 'Publication of Qualitative Research Studies in Psychology and Related Fields' (Elliott, Fischer, & Rennie, 1999). For the analysis of the interviews, the qualitative data analysis and research software 'ATLAS.ti' was used. In particular, the analysis was conducted as follows.

Two randomly selected interview transcripts from each country were read so that the author became familiar with the content. For these transcripts, codes were created for meaningful text chunks. Once the coding was completed, codes were revised to remove duplicated codes, combine similar codes and then group the codes into meaningful categories. The revised list of codes was used to code the remaining transcripts and it was again revised at the end of this step. If the codes changed, the transcripts were coded again and the above process was repeated three times. Finally, the codes were grouped into themes and their sub-themes. Again, the themes were revised to remove duplicated sub-themes and combine similar sub-themes and themes.Whilst the aim of this paper is to present key CFs, the themes regarding the manner in which pilots plan the flight route and the features pilots consider were transformed to CFs. The participants' recall of AI incidents was also used to identify CFs.

The results were validated by a SME, who had ten years of expertise in aviation safety and interview analysis. The SME was provided with the themes at level 1 and 2 and was requested to assign the theme for 100 quotes. In the first stage, the description of the themes was not provided and the agreement was at 68%, which was below the minimum expected rate of agreement, i.e. 85%. In the second stage, the SME re-assigned the themes for each quote whilst the SME was provided with the description of the themes. At this stage, the agreement was 90%, and thus, the themes were successfully validated. The suggestions for re-naming two sub-themes were incorporated.

Results

A key finding of the interviews was the decision that pilots make to fly in uncontrolled airspace and near the boundary of controlled and restricted airspace (FB decision). In such a flight, the pilot can unintentionally infringe for a range of reasons (i.e. contributory factors), e.g. the pilot does not notice the minor change of the wind direction that succeeded to change the heading of the aircraft towards controlled airspace. This FB decision is influenced by a range of factors and these factors are also CFs of AIs. Such CFs can relate to the aircraft design, airspace design, airspace procedures, flight-route decision, communication skills of pilots, the pilot's personal factors and their risk management. For example, a *'pilots' wishing flying altitude is higher than the altitude of the lower boundary of controlled airspace'* and thus, the pilot flies as close as possible to the desired altitude. In such a situation, pilots who believe that the gliding distance is inadequate may fly very close to the boundary, e.g. 10ft below.

Other factors that can influence the FB decision can be the following. In the situation that the *'flight route passes through many controlled airspace areas'*, e.g. cross-country flights, the flight route is modified to pass through a fewer number of controlled airspace areas and subsequently the pilot will contact a fewer number of ATCOs given that communications can increase a pilot's workload. The *'pilot wishes to fly only in controlled airspace'* and thus, in areas that an entry to controlled airspace is less likely to be permitted, the pilot will make the FB decision. Another factor can be the *'ill-fitted ultralight and glider aircraft that cannot fly in controlled airspace'* and thus, these aircraft divert the route around controlled airspace; however, the diverted route is almost similar to the initial planned route.

The manner in which pilots plan their flight route pre-flight can also contribute to AIs and was found to be influenced by the technologies used by GA pilots as follows. Animated planning apps suggest a straight, direct flight route that might not be optimal for the aircraft, the area and the weather conditions. Pilots that do not change this suggested route might infringe, especially if *'the flight route is near controlled airspace'*. This CF is the *'unchallenged flight route that is suggested by the planning* app'. Due to the use of such planning apps, the *'pilot plans the flight route quickly'* and *'starts the planning closer to the time of departure' even just prior to take-off.* Both CFs can result in the situation whereby the *'pilot uses less number of landmarks'* and *'the pilot is inadequately prepared for the flight'* in that the pilot do not visualise the shape of the airspace, the local weather, e.g. wind of varied direction over mountainous area, and the potential traffic density in certain segments of the flights.

It was evident in the interviews that the pilots were confident of the accuracy of both the planning and navigation devices, e.g. animated apps and Global Positioning System receivers. Whilst the benefits of using these emerging technologies were stated by almost all the participants, their limitations and their potential to contribute to AIs were not clearly shared by all the participants. Such devices can run out of battery, freeze at any time and the positioning, especially that of tablets, might not be as accurate as the pilots believe. In the situation that the navigation device fails, the pilots have to switch to traditional navigation by comparing landmarks on the ground and the map. If the pilot did not find the landmarks that he/she flies over, the pilot would probably prioritise the tasks to identify the position of the aircraft and thus, this might lead to a loss of situational awareness.

Discussion

This study found factors related to planning that can contribute to airspace infringements involving GA flights. The CFs are detailed and thus, they overcome the limitation of the past studies of AIs that found generic CFs. The findings were validated by a SME. The newly found CFs are associated with a GA pilot's performance as well as airspace design features and these CFs can influence the pilot's flight route decision. For example, a low altitude of the lower boundary of controlled airspace in the capital of a country influences pilots to fly in uncontrolled airspace and near the boundary in order to maximise their gliding distance in the event of an engine-failure. The study also identified the impact of the planning apps that are increasingly used by pilots. In particular, pilots that use such apps might start the planning just prior to take-off and make more flight-route decisions in-flight due to access to information when airborne.

The key to this achievement was both the carefully designed semi-structured interviews of recreational GA pilots and of the sample. The enriched results were derived from the participants' description of the manner in which they typically decide the flight route. CFs, such as *'unchallenged flight route that is suggested by the planning app'* was found for the first time and this can be an example of the potential degradation of a GA pilot's performance due to the use of emerging technologies. Of equal importance, consideration of the diversity of the aircraft type and the flying base of the pilots succeeded in identifying their differences. A distinctive finding was that ultralight and glider pilots decided to fly in uncontrolled airspace due to the fact that the aircraft were ill-equpped to fly in controlled airspace. Another key finding is the impact of the heavily controlled airspace in the capitals of the countries in that pilots, who were based in these areas, e.g. London, where the uncontrolled airspace was narrow, and consequently planned the flight near controlled airspace.

The findings of this study shed light on the AI domain and thus, the findings can be used to develop a bespoke taxonomy of CFs of AIs. This taxonomy can aid the AI incident investigation and analysis as well. Such detailed findings, e.g. airspace design features, use of apps by the pilots, could also be used to develop mitigations actions of AIs involving GA flights.

Conclusions

This study successfully found contributory factors of AIs involving GA flights. The findings presented in this paper focused on the planning of pilots that is essential for completing a safe flight. For the purpose of this study, semi-structured interviews were conducted in Finland, Norway and UK and a sample of recreational GA pilots was carefully designed based on four criteria concerning operations and personal factors, e.g. country, aircraft type, city and flying hours. The findings can be used to develop a bespoke taxonomy of contributory factors and this taxonomy will comprise the diversity of GA operations, the environment the GA pilots fly and a GA pilot's performance. The findings can aid the incident investigator and support aviation stakeholders to design mitigation actions of AIs.

Acknowledgements

The authors would like to thank Lloyds Register Foundation who funded this research. The authors would also like to thank the participants of this study as well as Per Julius Helweg, Heli Koivu, Armin Zuger and Bettina Bajaj who organised the interviews in Finland, Norway and UK.

References

Civil Aviation Authority United Kingdom. (2006). *Strategic review of general aviation in the United Kingdom*. United Kingdom:

Coyle, A., & Lyons, E. (2007). *Analysing qualitative data in psychology* SAGE Publications.

Elliott, R., Fischer, C. T., & Rennie, D. L. (1999). Evolving guidelines for publication of qualitative research studies in psychology and related fields. *British Journal of Clinical Psychology, 38*(3), 215-229. doi:10.1348/014466599162782

European Air Traffic Management. (2007a). *Airspace infringement initiative risk analysis part I, Safety analysis of airspace infringements in Europe.* (No. 07/11/29-48).EUROCONTROL.

European Air Traffic Management. (2007b). *Airspace infringement risk analysis part II general aviation airspace infringement survey, analysis of pilot-reported causal factors and prevention measures.* (No. 08/01/07-01).EUROCONTROL.

European Air Traffic Management. (2008). *Airspace infringement risk analysis part III, case study Switzerland.* (No. 08/01/16-04).EUROCONTROL.

General Aviation Safety Committee. (2016). Fly on track. Retrieved from http://flyontrack.co.uk/statistics/

Hunter, D. (2005). Measurement of hazardous attitudes among pilots. *The International Journal of Aviation Psychology, 15*(1), 23.

Hunter, D., Martinussen, M., & Wiggins, M. (2003). Understanding how pilots make weather-related decisions. *The International Journal of Aviation Psychology, 13*(1), 73.

International Civil Aviation Organization. (2009). *Tenth session of the statistics division, review of the classification and definitions used for civil aviation activities.* (No. STA/10-WP/7).International Civil Aviation Organization.

Molesworth, B., Wiggins, M. W., & O'Hare, D. (2006). Improving pilots' risk assessment skills in low-flying operations: The role of feedback and experience. *Accident Analysis & Prevention, 38*(5), 954-960. doi:http://dx.doi.org/10.1016/j.aap.2006.03.006

Nascimento, F. (2014). *Hazard identification and risk analysis of night-time offshore helicopter operations*

O'Hare, D., & Stenhouse, N. (2009). Under the weather: An evaluation of different modes of presenting meteorological information for pilots. *Applied Ergonomics, 40*, 688.

Safety Regulation Group. (2003). *On track – A confidential airspace infringement project.* (No. 2003/2005).Civil Aviation Authority United Kingdom.

Wiggins, M. (2007). Perceived GPS use and weather-related decision strategies among pilots. *The International Journal of Aviation Psychology, 17*(3), 249.

Wiggins, M. W., Azar, D., Hawken, J., Loveday, T., & Newman, D. (2014). Cue-utilisation typologies and pilots' pre-flight and in-flight weather decision-making. *Safety Science, 65*(0), 118-124. doi:http://dx.doi.org/10.1016/j.ssci.2014.01.006

FLIGHT EXPERIENCE AND MENTAL REPRESENTATIONS OF SPACE

Mikayla Keller
The University of Western Ontario
London, Ontario, Canada

Jennifer E. Sutton
Brescia University College and The University of Western Ontario
London, Ontario, Canada

Spatial skills are critical for flight safety. The current study investigated whether increased flight experience as a pilot was associated with improved spatial skills, and in particular, the ability to form a mental representation of a novel virtual environment. Pilots completed small-scale spatial ability tasks, travelled along four routes in a virtual environment, and then completed two tests that assessed memory for the locations of landmarks in the environment. Pilots with more flight experience did not have more accurate mental representations of the environment than individuals with less flight experience. Increased flying experience was, however, linked to better performance on a perspective-taking test. Perspective taking has been proposed as central to navigation awareness during flight, and the current data suggest it improves with experience.

Successful navigation of an aircraft is a complex cognitive skill that demands pilots plan a route from point A to point B and be able to quickly plan alternate routes in the event of an emergency (Transport Canada, 2010). The foundation for such wayfinding is an ongoing understanding of the plane's current location relative to landmarks and other objects during flight. Gibb, Ercoline, and Scharff (2011) estimate that spatial disorientation, a situation in which a pilot mistakes the plane's location, motion, and/or attitude, accounts for 25-33% of all aviation accidents. In many cases, the pilot unknowingly makes this misjudgement and remains unaware of the mistake until it is too late. Gibb et al. also argue that mishaps attributed to spatial disorientation, which are underreported, have the highest fatality rate in comparison with other causes of crashes, indicating the critical importance of spatial cognition for flight safety.

It is well-established that spatial skills are not fixed abilities and can be improved through training in the laboratory (see Uttal et al., 2013 for a meta-analysis), but evidence that flight experience, in particular, can lead to improved spatial skills remains mixed (Dror, Kosslyn, & Waag, 1993; Sutton, Buset, & Keller, 2014). For instance, Dror et al. found that military pilots performed better than non-pilot controls on a mental rotation task but showed no difference on judgements of categorical spatial relations or mental image scanning. Furthermore, whether the mental rotation finding is attributable to spatial skills acquired in flying is unclear, as small-scale spatial abilities are only partially related to performance on large-scale navigation tasks (Hegarty, Montello, Richardson, Ishikawa, & Lovelace, 2006). On the other hand, Sutton et al. (2014) found that early-career civil aviation pilots formed more accurate cognitive maps of a novel virtual environment than non-pilot controls matched to the pilots on age and video game experience. These findings suggest that the spatial skills pilots acquire transfer to other, non-flight, navigation tasks and result in more accurate mental representations of the environment.

A cognitive map is a map-like mental representation of the configuration of landmarks in an area, often described as a mental "birds-eye-view" that is orientation independent (O'Keefe & Nadel, 1978; Tolman, 1948), and Sutton et al. (2014) suggested that multiple aspects of flight may improve cognitive mapping skills. For instance, the unique aerial views and/or the demands of constantly updating the aircraft's spatial position during flight may facilitate encoding the environment in memory as a cognitive map. In addition, assessments for pilot licensure include a requirement for candidates to assume a new heading and anticipate the locations of objects along the new path (Transport Canada, 2010), a skill greatly facilitated by having a mental map of the environment. Quick calculation of a detour using a map-like memory of an area requires an understanding of the current positions of objects relative to the plane and to each other, and the ability to mentally transform those relationships to correspond to a new heading, a process known as perspective taking (Thurstone, 1950). According to Aretz (1991), perspective taking is central to a pilot's navigation awareness during flight, and when combined with pilots' unique aerial viewpoint akin to a map, perspective-taking practice may lead to improved precision both in the storage and retrieval of cognitive maps in memory.

The current study extended the findings of Sutton et al. (2014) by investigating whether cognitive map accuracy and/or perspective taking improve with increasing hours of flight experience. Pilots explored a virtual environment, *Silcton*, via four separate routes and afterwards were tested for their ability to combine the routes into a single map of the environment. We predicted that pilots with more flight hours would form more accurate cognitive maps of Silcton than pilots who had fewer flight hours. In addition, we predicted that more hours would be associated with better memory of the routes travelled in Silcton. Because we hypothesized that perspective taking skills were a potential mechanism facilitating cognitive map encoding and retrieval in individuals with flight experience, a paper-and-pencil perspective-taking task was also administered in order to assess the association of perspective taking with flight hours and cognitive mapping skills. We expected that, as with the measures of Silcton, perspective taking would improve with increasing flight experience.

Method

Participants

Forty-two students (36 males, 6 females, mean age = 20.48) with at least one hour of flight experience were recruited from The University of Western Ontario. Participants were in the early stage of their careers with a varying number of flight hours ($M = 75.79$, $SD = 70.94$). Twenty-three participants (20 males, 3 females; M age = 20.48, $SD = 3.94$, range = 17 - 37) held a Private Pilot Licence or higher (e. g., Commercial Pilot Licence) and 19 had not yet obtained a licence (16 males, 3 females; M age = 20.47, $SD = 3.34$, range = 18 - 32). Some participants received $15 in compensation for participating in the study and others received course credit. Data for every participant (N = 42) are reported for all measures, except the same route and different route direction estimation tasks and the map building task, where N = 41 due to a technical error. The study was approved by the University of Western Ontario Non-Medical Research Ethics Board.

Materials and Procedure

After providing written informed consent, participants completed a demographic questionnaire where information on hours of flight experience, licences and ratings obtained, and GPS usage during flight was collected. Next, participants completed the Santa Barbara Sense of Direction scale (SBSOD; Hegarty, Richardson, Montello, Lovelace, & Subbiah, 2002), which assessed self-reported spatial abilities. After the SBSOD, participants completed the paper-and-pencil-based Spatial Orientation Test (SOT; Hegarty & Waller, 2004), a measure of perspective-taking ability where participants are required, while looking at a static array of objects on the page, to assume an imagined heading direction in the array and indicate the direction of another object in the array. Next, participants completed a spatial n-back test of spatial working memory. Note that of these tasks, only data from the demographic questionnaire and the SOT are presented in this paper.

After the small-scale spatial tasks, participants completed the virtual environment task using the Silcton environment (Weisberg, Schinazi, Newcombe, Shipley, & Epstein, 2013) presented on a 15.6" laptop (Samsung R525, Samsung Electronics, Suwon, South Korea) running Windows 7 64-bit with an AMD Phenom II Quad-Core N970 2.2 GHz Processor and an AMD Radeon HD 6600M Graphics card (Advanced Micro Devices, Sunnyvale, CA). First, participants familiarized themselves with the controls (arrow keys and mouse) and practiced moving around in Silcton. When participants were comfortable with the controls, they were instructed that they would be exploring four different routes (2 main routes and 2 connecting routes) through the same town used for practice. They were instructed to remember the names and the locations of eight buildings marked with blue diamonds in Silcton, as the tasks that followed would test their knowledge for these buildings. Participants travelled each route from start to finish, following red arrows marked on the path, and back to start. Participants were given an unlimited amount of time to complete the travel on each route.

Immediately after traversing all four routes, participants completed a direction estimation task based on memory for Silcton (Weisberg et al., 2013) where they were asked to place the eight target buildings around the perimeter of a circle in their appropriate directions from given heading directions. For instance, on one item, participants were instructed to imagine they were standing at Harris Hall facing the Batty House. From this heading, they positioned the eight buildings to indicate their directions relative to the imagined heading. This task provided measures of route and cognitive map accuracy, as participants were asked to estimate the directions of buildings on the same route (a measure of route knowledge) and buildings on different routes (a measure of map knowledge).

Participants completed a final map-building task (Weisberg et al., 2013) where they were shown a blank rectangle on the computer screen and were instructed to drag and drop bird's-eye images of the eight Silcton buildings into their appropriate locations.

Results

Means and standard deviations for all measures reported here are shown in Table 1. Paired *t* tests showed that pilots were more accurate (i.e., showed less error) at estimating directions between landmarks along the same route than across different routes, $t(40) = -8.28$, $p < .001$, and means for both estimation measures were significantly better than chance (90°), same route: $t(40) = -14.87$, $p < .001$; different routes: $t(40) = -4.28$, $p < .001$.

Table 1.
Means and Standard Deviations for Flight Hours, Spatial Orientation Test (SOT), and Silcton Measures.

	Flight hours	SOT	Silcton Direction Estimation Error		Silcton map building
			Landmarks on the same route	Landmarks on different routes	
M	75.79	21.58	64.72	81.13	.52
SD	70.94	0.07	10.88	13.27	.28

Note: SOT and direction estimation error measures are reported in absolute degrees. Accuracy on the Silcton map building task was scored using a bidimensional regression procedure resulting in an R^2 value with a potential range from 0 – 1.0.

Table 2 shows the results of two-tailed Pearson correlations. As expected, measures based on memory for Silcton were significantly correlated. Hours of flight experience was not significantly correlated with cognitive map accuracy of Silcton, as reflected in the measures of different-route direction estimation error and map building. Similarly, there was no correlation between flight experience and route knowledge on the same-route direction estimation task. Scatterplots and R^2 values for the associations between hours and same- and different-route direction estimation measures are shown in Figure 1 (panels A and C). In addition, experience was measured by dividing pilots into those holding at least a Private Pilot Licence (n = 23) or no licence (n = 19), and direction estimation error for both groups can be seen in Figures 1B and 1D. There was no significant difference between the groups on same-route direction estimation error, $t(17) = -1.75$, $p = .09$, nor on different-route estimation error, $t(17) = 0.11$, $p = .92$.

Table 2.
Pearson Correlations for Flight Hours, Spatial Orientation Test, and Silcton Measures.

	SOT	Direction estimation error: same route	Direction estimation error: different routes	Silcton map building
Flight hours	-.44**	.07	-.03	.09
SOT	-	-.10	-.04	-.07
Direction estimation error: same route	-	-	.46**	-.59***
Direction estimation error: different routes	-	-	-	-.49**

Note. * $p < .05$, ** $p < .01$, *** $p < .001$.

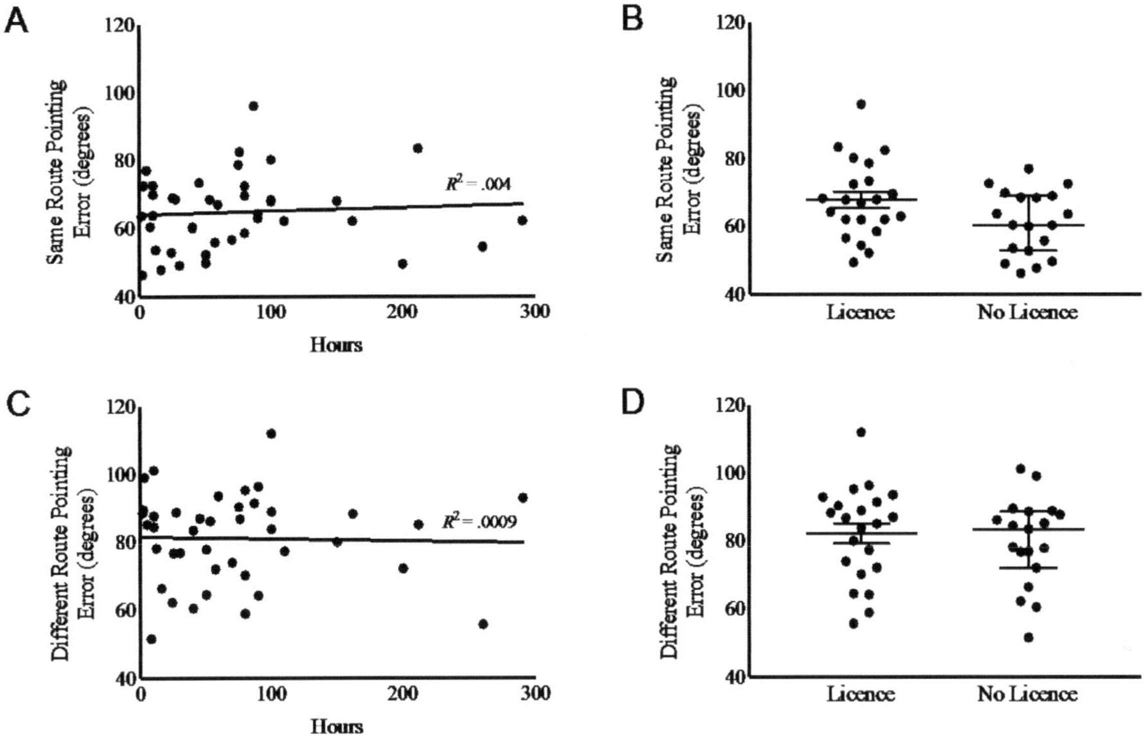

Figure 1. (A) Correlation between hours and error on same route direction estimation measure; (B) Means (center horizontal lines) and standard errors (outer horizontal lines) on the same route direction estimation measure for pilots holding at least a private pilot's licence versus those with no licence; (C) Correlation between hours and error on the different route direction estimation measure; (D) Means (center horizontal lines) and standard errors (outer horizontal lines) on the different route direction estimation measure for pilots holding at least a private pilot's licence versus those with no licence. Filled circles show individual scores.

Table 2 also shows that flight hours were significantly associated with performance on the SOT perspective-taking test, where participants with more flight experience showed lower error scores. Figure 2A shows a scatterplot of the relationship between flight hours and SOT error. A linear regression model showed that hours significantly predicted SOT error, $\beta = -.44$, $p = .004$, accounting for 19% of the variance in SOT performance, $R^2 = .19$, $F(1, 40) = 9.60$, $p = .004$. Figure 2B shows SOT error for pilots holding at least a private pilot's licence versus pilots with no licence. Licence holders showed significantly less error on the SOT than those without a licence, $t(18) = 2.99$, $p = .01$.

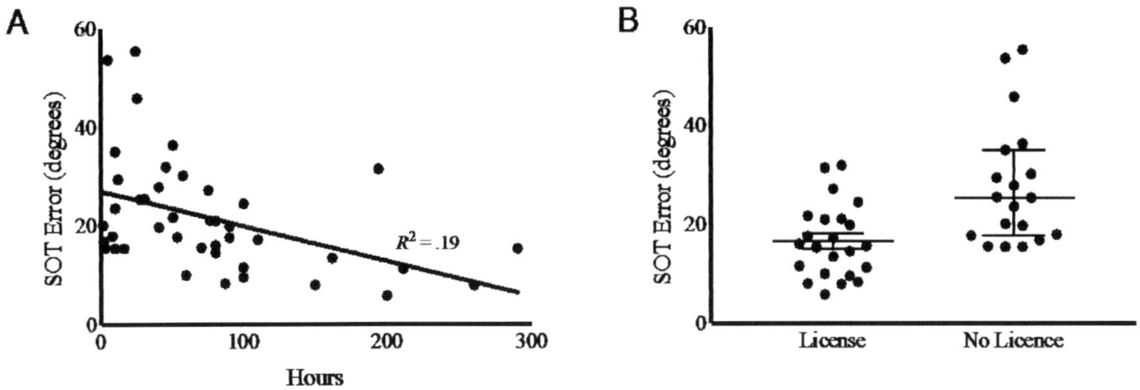

Figure 2. (A) Correlation between hours and SOT perspective-taking error measured in degrees. (B) Means (center

horizontal lines) and standard errors (outer horizontal lines) on the SOT for pilots holding at least a private pilot's licence versus those with no licence. Filled circles show individual scores.

Discussion

While we hypothesized that increasing hours of flight experience would be associated with better scores on all our assessments of spatial cognition, more flight hours were associated only with better small-scale perspective-taking ability and not the ability to form a cognitive map or route-based representation of a novel virtual environment. This pattern of findings was also evident when pilots' flying experience was categorized based on licensure status: those holding at least a Private Pilot Licence were more accurate on the perspective-taking task than others, but there was no difference in the virtual environment. Overall, pilots remembered specific routes more accurately than the overall map of Silcton, consistent with other research showing that only some individuals can integrate separate routes into a single mental map, both in real-world and virtual environments (Ishikawa & Montello, 2006; Weisberg et al., 2013; Weisberg & Newcombe, 2015).

The lack of an association between hours of flight experience and the Silcton direction estimation measures may be due to the difficult nature of these tasks that left little room for variation (i.e., a floor effect). In the direction estimation tasks used here, individuals must rely on memories formed during exploration of Silcton when estimating landmark directions. Error scores were closer to chance under these conditions than when participants are placed back in the virtual environment to make the estimations (e.g., see Weisberg et al., 2013). On the Silcton map-building task, hours again did not predict performance. Even though direct statistical comparisons are not possible, the pilots in our study appear to be slightly more accurate on map building ($M = .52$), compared to non-pilot samples tested with similar procedures by Weisberg et al. (2013) ($M = .48$) and Weisberg and Newcombe (2015) ($M = .47$). So, it could be speculated that pilots are marginally better than the general population on at least one Silcton task, even though performance within pilots does not vary according to hours of flight experience. Further research will be necessary to support this assertion.

The finding that more flight experience was associated with better perspective taking suggests that the skills pilots practice when flying generalize to this paper-and-pencil, non-flying task. Perspective taking involves an individual mentally transforming her heading and demonstrating accurate knowledge of the locations of objects relative to the new heading. Hegarty and Waller (2004) have asserted that perspective taking is distinct from mental rotation, another small-scale task in which the individual remains in a static orientation and imagines an object rotating around its own axis. Notably, Dror et al. (1993) found that pilots outperformed non-pilots on a mental rotation task, so it could be that both types of spatial mental transformation are improved with flight experience. We propose that perspective taking is actually the more critical ability in aviation, however, as updating the spatial position of the plane and surrounding landmarks is fundamental to maintaining navigation awareness (Aretz, 1991).

Our results suggest that better perspective taking can be acquired through increasing flight experience, although an alternative explanation is that individuals with better perspective-taking skills are more likely to progress in aviation, while those with weaker skills drop out and pursue other careers. A longitudinal design, where pilots are tested before flight training begins and then at specified intervals during training, is required to make stronger conclusions about the effect of flight on perspective taking. A similar design confirmed that structural changes in the hippocampus associated with driving a taxi were changes that occurred over the course of training rather than via attrition of those with weaker skills (Woollett & Maguire, 2011). Nonetheless, even in the absence of such longitudinal data, our findings here, coupled with our previous work (Sutton et al., 2014) point to better spatial abilities in pilots than non-pilots, and an experience-dependent effect on perspective taking.

Acknowledgements

Funding for this project was provided by the Natural Sciences and Engineering Research Council of Canada (NSERC) in the form of a Discovery Grant to JES.

References

Aretz, A. J. (1991). The design of electronic map displays. *Human Factors, 33*, 85-101.

Dror, I. E., Kosslyn, S. M., & Waag, W. L. (1993). Visual-spatial abilities of pilots. *Journal of Applied Psychology, 78*, 763-773.

Gibb, R., Ercoline, B., & Scharff, L. (2011). Spatial disorientation: Decades of pilot fatalities. *Aviation, Space, and Environmental Medicine, 82*, 1-8.

Hegarty, M., Montello, D. R., Richardson, A.E., Ishikawa, T., Lovelace, K. (2006). Spatial abilities at different scales: Individual differences in aptitude-test performance and spatial-layout learning. *Intelligence 34*, 151–176.

Hegarty, M., Richardson, A. E., Montello, D. R., Lovelace, K., & Subbiah, I. (2002). Development of a self-report measure of environmental spatial ability. *Intelligence, 30*(5), 425-447. doi: 10.1016/S0160-2896(02)00116-2

Hegarty, M., & Waller, D. (2004). A dissociation between mental rotation and perspective-taking spatial abilities. *Intelligence, 32*(2), 175-191. doi: 10.1016/j.intell.2003.12.001

Ishikawa, T., & Montello, D. R. (2006). Spatial knowledge acquisition from direct experience in the environment: Individual differences in the development of metric knowledge and the integration of separately learned places. *Cognitive Psychology, 52*(2), 93-129. doi: 10.1016/j.cogpsych.2005.08.003

O'Keefe, J., & Nadel, L. (1978). *The hippocampus as a cognitive map.* Oxford: Clarendon Press.

Sutton, J. E., Buset, M., & Keller, M. (2014). Navigation experience and mental representations of the environment: Do pilots build better cognitive maps? *PLoS ONE, 9*(3): e90058

Thurstone, L. L. (1950). *Some primary abilities in visual thinking (Rep. no. 59).* Chicago, IL: Psychometric Laboratory, University of Chicago.

Tolman, E. C. (1948). Cognitive maps in rats and men. *Psychological Review, 55*, 189-208. doi: 10.1037/h0061626

Transport Canada. (2010). *Flight test guide: Private pilot license – Aeroplane (TP 13723E).* Retrieved from http://www.tc.gc.ca/publicaVons/en/tp13723/pdf/hr/tp13723e.pdf

Uttal, D. H., Meadow, N. G., Tipton, E., Hand, L. L., Alden, A. R., Warren, C., & Newcombe, N. S. (2013). The malleability of spatial skills: A meta-analysis of training studies. *Psychological Bulletin, 139*, 352-402.

Weisberg, S. M., & Newcombe, N. S. (2015). How do (some) people make a cognitive map? Routes, places, and working memory. *Journal of Experimental Psychology: Learning, Memory, and Cognition, 42*, 768-785. doi: 10.1037/xlm0000200

Weisberg, S. M., Schinazi, V. R., Newcombe, N. S., Shipley, T. F., & Epstein, R. A. (2013). Variations in cognitive maps: Understanding individual differences in navigation. *Journal of Experimental Psychology: Learning, Memory, and Cognition, 40*(3), 669-682. doi: 10.1037/a0035261

Woollett, K., & Maguire, E. A. (2011). Acquiring "the Knowledge" of London's layout drives structural brain changes. *Current Biology, 21*, 2109-2114.

INDIVIDUAL PILOT FACTORS PREDICT DIVERSION MANAGEMENT DURING A SIMULATED CROSS- COUNTRY VFR FLIGHT

Kathleen Van Benthem, Ph.D.
ACE Lab, Carleton University
Ottawa, ON, Canada
Chris M. Herdman, Ph.D.
ACE Lab, Carleton University
Ottawa, ON, Canada

The risk of an accident during general aviation (GA) flight increases when pilots are required to make unexpected diversions. Specifically, a diversion may result in loss of situation awareness (SA). Loss of SA is associated with controlled flight into terrain, incorrect trajectory for orbiting or landing, or becoming lost en route. In the present study, 44 GA pilots (aged 41 to 74 years) flew a cross-country route in a Cessna 172 simulator and encountered an unexpected diversion to an alternate aerodrome. The outcome measure consisted of a diversion management score. Significant predictors of diversion management were pilot age and license, a measure of prospective memory in the cockpit, and response times from an executive cognitive function subtest of the CogScreen-AE. A model of performance derived from a "best subsets" linear modeling algorithm included pilot license, prospective memory, and executive function. Importantly, less skill in managing the diversion also predicted a greater likelihood of critical incidents during the cross-country flight. Understanding the role of pilot factors in identifying those most at risk when flying an unexpected diversion can better prepare pilots for these rare events, and inform customized learning opportunities during check rides and flight instruction.

General aviation continues to show higher rates of accidents per mile flown when compared to scheduled operations (AOPA, 2015). Thus, identifying high risk aspects of general aviation operations, and the factors associated with these risks is in the best interest of pilots and the public. Managing unplanned diversions, such as rerouting to an alternate aerodrome due to weather, relies on a sequence of cognitive factors, including rapid situation awareness updating and accurate and speedy decision-making, while safely navigating, communicating, and piloting the aircraft (Wright, 2013). Thus, pilot characteristics, which are known to predict situation awareness and decision-making, might also show associations with diversion management.

Situation awareness has received considerable attention in the aviation literature. Van Benthem, Herdman, Brown and Barr (2011) found that objective measures of situation awareness (knowledge of ownship and details and location of other aircraft) predicted the occurrence of critical incidents during simulated general aviation flight. Case analyses of actual accidents suggest that loss of situation awareness is associated with over 70% of pilot-caused general aviation accidents (Endsley, 1999). The construct of situation awareness has been described as a mechanistic model, and this model provides a framework for identifying predictors of situation awareness. Per Endsley (1988; 1995) situation awareness relies on three general cognitive mechanisms. The first is the perception and integration of stimuli into meaningful units of information. A second mechanism binds relevant information into a comprehensive model of the environment. The third process projects the current model into a likely future model of the environment. By this characterization, situation awareness is reliant on several cognitive functions that work in tandem to produce accurate and frequently updated representations of relevant aspects of the world. Situation awareness is responsive to top-down direction such as pilot attention and goals. At the same time, some aspects of situation awareness are affected by foundational cognitive factors such as working memory and processing speed, which support the production of situation awareness in a bottom-up fashion (Bolstad, 2001; Gugerty & Tirre, 2000; Gutzwiller & Clegg, 2013). Van Benthem et al. (2011) found that a cluster of pilot characteristics, including age, experience, perceptual-motor response times, and a situational judgement test for drivers predicted the second and third mechanisms of situation awareness (the current and future comprehensive model, as per Endsley's descriptions above). Perceptual motor speed and recent flight hours were the only two factors to predict situation awareness level one.

Decision-making during flight is also logically associated with outcomes of unplanned diversions, though few flight simulation studies have examined the predictors or outcomes of diversion-related decision-making. Along this line however, Goh and Wiegmann (2001) found that poor decisions to fly visual flight rules into instrument meteorological conditions were associated with an overconfidence in personal ability and an inaccurate diagnosis of visibility conditions. Causse, Dehais, Arexis, and Pastor (2011) examined the predictors of a landing decision task (due to wind factors on approach) and report that executive cognitive functions significantly predicted the landing decision. In the landing study, the wrong landing decision was associated with less accuracy in visual working memory updating and greater errors in detecting rule-shifts during the card sort task (Causse et al., 2011). Similarly, Kennedy, Taylor, Reade and Yesavage (2010) found that while flying simulated approaches older general aviation pilots showed a less conservative response bias in comparison to younger pilots, and were more likely to decide to land when visibility was poor. It appears that relevant predictors of decision-making during flight may be associated with individual pilot factors such as age, basic aviator competencies, executive cognitive abilities, and personality factors.

There appears to be considerable overlap between predictors of situation awareness and decision-making. This overlap also supports the notion that predictive models of unplanned diversion management will benefit from a range of factors that include cognitive functions, pilot characteristics, and aviator competencies. In the present study, general aviation pilots flew a cross-country route in a Cessna 172 simulator and encountered (and managed) an unexpected diversion to an alternate aerodrome. Considering that pilot personality and basic aviator competencies have been linked to situation awareness and decision-making we hypothesized that a broad range of predictors would be required to account for a reasonable amount of variance in diversion management scores. Using a "best subsets" technique for linear regression we compared simple to more complex models of diversion management.

Methods

The present study is part of an ongoing research agenda examining general aviation, aging and cognitive health. The sample was comprised of 44 volunteer pilots (all male) recruited from local flying clubs and schools. Inclusion criteria included age 40 years and older, having flown within the last year with a valid pilot's license and medical certification. Table 1 provides a description of the range of pilot age and experience. The study had approval from the university ethics review board, and all study participants provided informed consent after a description of the study activities was provided. Pilots attended two sessions: the first session was comprised of cognitive testing and practice flights in the simulator, and the second session consisted of three practice patterns followed by a cross-country route and diversion scenario.

Table 1.
Pilot Characteristics

	Age	Licence/Rating	Total Hours Flown	Total Years Licensed
Mean	54.80	2.455	556.3	12.83
Standard Deviation	9.065	1.044	1281	13.42
Minimum	41.00	1.000	1.000	1.000
Maximum	74.00	4.000	8000	50.00

Notes. License/Rating was based on a four-point scale where 1 = students, 2= visual flight rules (no additional ratings), 3 = visual flight rules with additional ratings, and 4 = instrument rated pilots, commercial pilots, and instructors.

Simulation Environment

The simulator structure was a converted Cessna 172 partial fuselage with a cockpit outfitted with instruments and controls specialized for flight simulation linked with Microsoft Flight Simulator X software (FSX) (Microsoft Game Systems, 2006). Projection graphics were produced by FSX "on the fly" and were not pre-rendered. Locations were geo specific in that they produced terrain modeled on actual aerodromes in Canada. The graphics architecture incorporated a broad-angle display system utilizing eight theater-quality 1080p projectors and a 14-foot tall, 180-degree curved screen to create a highly immersive visual environment. The data application computer logged the time and the pilot's location, airspeed, heading, bank, pitch, and altitude at one Hertz.

Figure 1. Cessna 172 simulator in situ with Broad-Angle Display System.

Flight Plan and Unexpected Diversion

Before entering the aircraft, pilots were briefed on a predetermined visual flight rules flight plan. Pilots were instructed to communicate with air traffic control or ground services as per the aerodrome. The weather experienced by the pilots was clear with no winds. The flight plan included a short leg from a large airport to a nearby general aviation aerodrome for two touch and gos. After departure from the small aerodrome pilots thought they were to follow a broad river to another large airport, where they were to complete their flight. After the final touch and go and departure from the aerodrome an unexpected instruction from ATC required pilots to divert from their plan and fly to an alternate aerodrome, and orbit at a prescribed altitude until further instructions were provided. A possible ground stop due to weather was the reason provided by ATC for the diversion. The cockpit was outfitted with visual flight rules navigation charts, a flight supplement document, and all non-electronic materials necessary for locating the new airfield. Pilots were expected to locate the alternate airfield on the map(s) provided and to determine an appropriate heading without assistance from ATC. Tasks also included changing radio frequencies as necessary. Throughout the flight, pilots heard other aircraft communicating with ATC or ground services. Listening to other pilot communication was the primary method of determining the location and intentions of other relevant aircraft.

Two unexpected pauses of the flight scenario occurred after the initial instructions from ATC to fly to the alternate airfield and provided the data for the diversion management metric. The diversion management score was comprised, in part, of key elements directly associated with the diversion and captured five minutes after the diversion message: speediness of response (0 or 1), acknowledgement of the alternate aerodrome (0 or 1), ability to locate the alternate aerodrome on a map (0 to 2), and accuracy in noting ownship on the map (0 to 2). An awareness of other key elements of the diversion were captured at a pause approximately 15 minutes after the diversion message (before the pilot returned to a final aerodrome as per ATC instruction), which included ownship, altitude, airspeed, and heading, and location, call sign, type, and altitude of other aircraft following similar ATC instructions (each element scored at 0 to 2 points). Pilots were also expected to make several radio calls while orbiting the alternate airfield (0-8). All elements of the diversion management score were equally weighted and summed to possible maximum score of 30. In sum, the diversion metric was based on the ability determine new flight plans in a speedy manner and maintain accurate situation awareness, while continuing to aviate, navigate, and communicate.

Prospective Memory

Pilots were expected to make radio calls at prescribed times during the scenario. Previous work in this flight simulation laboratory (Van Benthem, Herdman, Tolton & LeFevre, 2015) has found that pilot prospective memory for radio calls in the cockpit were sensitive to pilot experience, workload, age, and cognitive factors. Prospective memory for cockpit tasks have also been associated with critical incidents in the real world (Dismukes & Berman, 2010). Due to the particular sensitivity of prospective memory for infrequent radio calls in high workload situations (Van Benthem et al., 2015) only the calls for the mid-downwind position in pattern flight during higher traffic volume occasions were used to create the prospective memory metric in this analysis.

Critical Incidents

All critical events related to piloting behaviour were recorded during the flight simulation. Critical incidents included dangerous landings, incorrect response to ATC instructions, mis-dialing radio frequencies without detection etc. To avoid the inflation of a possible relationship, critical events occurring during the diversion management portion of the scenario were not counted in this performance metric.

Executive Cognitive Function

CogScreen-Aeromedical (AE) is a computerized cognitive screening tool designed to assess cognitive processes deemed relevant to the complex tasks of an aviator (Kay, 1995). CogScreen-AE measures attention, immediate and short-term memory, working memory, visual-perceptual functions, sequencing functions, logical problem solving, calculation skills, reaction time, and dual-task processing. CogScreen-AE testing was conducted using a Windows XP computer with Elo -Touch systems 2216 AccuTouch USB Touch monitor (Elo Touch Solutions). Eleven subtests of the CogScreen-AE were administered: Backward Digit Span, Math, Visual Sequence Comparison, Symbol Digit Coding, Matching to Sample, Manikin, Divided Attention, Auditory Sequence Comparison, Pathfinder, Shifting Attention, and Dual Task. The CogScreen-AE was administered to all the pilots in their first session. Only the Shifting Attention subtest was used in the present analysis because of its strong association with executive functions (Kay, 1995). In the Shifting Attention subtest participants determine and then update a repeatedly changing rule, which relates to the direction and colour of arrows and governs correct selection of arrow stimuli.

Results

A best subsets linear regression analysis was undertaken to determine the relative importance of each predictor. Despite the strong bivariate correlation found for age and the diversion score (see Table 2), age was not a significant predictor in the final model. The best combination of factors included pilot license, executive function, and prospective memory, r^2=.42. As shown in Table 2, the executive function factor (a response time metric) was strongly correlated with pilot age. Replacing executive function with age in the final model resulted in a drop of 11% of accounted variance, thus executive function was a more informative variable than age alone. In order of importance the variables were executive function, license, prospective memory, and age.

Table 2.
Correlations between Diversion Scores and Predictors

	Age	Licence	Executive Function	PM
Diversion Score	-0.457 **	0.336	-0.527 **	0.537 **
Age	—	-0.007	0.496 ***	-0.265
Licence		—	0.025	0.132
Executive Function			—	-0.426 *

* p < .05, ** p < .01, *** p < .001. *N*=34 due to random missing data. The relationship of executive function and diversion management is negative because the cognitive metric is based on participant response times.

A linear regression using Bayesian statistical analysis was also completed to confirm the order of importance of each variable, as the final linear regression results were quite different from the pattern of bivariate correlations. Bayes Factors (BF) also demonstrated that the combination of executive function, prospective memory, and pilot license best predicted diversion performance (total BF= 131.8). Although, when the factors were examined individually, age (BF= 6.9) was a stronger predictor than license (BF= 1.5).

Finally, the relationship of diversion management to critical incidents was examined using a Pearson correlation analysis. A significant negative relationship was shown, such that more a higher number of critical incidents were associated with lower diversion management scores, r= 0.343, p=.047.

Conclusion

The present findings suggest that pilots with poorer executive functions (perhaps associated with older age), lower levels of expertise, and difficulty with prospective memory in high workload situations may be at risk for poor outcomes from unplanned diversions. Low scores for diversion management were associated with a greater likelihood of critical incidents, suggesting that diversion management assessment may also provide an indication of general risk during flight. Corroboration for these results are found in a study of self-reported incidents and accidents: O'Hare (2006) found that pilots who had experienced critical incidents, in contrast to those pilots with no history of incidents or accidents, were also significantly more likely to have experienced weather-related diversions. Either choosing not to, or showing an inability to follow ATC instructions, and quickly locating ownship and alternate airfields on a well-known aviation chart may be a warning sign to any pilot who flies cross-country.

A key finding in the present work was that cognitive factors were shown to be more informative than pilot age and experience in relation to diversion management. This superiority of cognitive assessment over pilot age was also shown in similar work examining predictors of pilot deviations during pattern flight (Van Benthem & Herdman, 2016). Thus, pilot screening for cognitive factors, such as executive functions and prospective memory for cockpit tasks may be promising methods for reliable identification of at-risk pilots. Understanding the role of pilot factors in identifying those most at risk when flying an unexpected diversion can better prepare pilots for these rare events, and inform customized learning opportunities during check rides and flight instruction.

Acknowledgements

Research infrastructure was supported through the Canadian Foundation for Innovation and the Ontario Innovation Trust. We thank Anne Barr and Andrew Staples of the Carleton University, ACE Laboratory for their consultation and engineering expertise.

References

AOPA. (2016). *25th Joseph T. Nall Report:General Aviation Accidents in 2013*. Fredericksburg, MD: Author.

Bolstad, C. A. (2001). Situation Awareness: Does It Change with Age? In *Proceedings of the Human Factors and Ergonomics Society Annual Meeting* (Vol. 45, pp. 272–276). Retrieved from http://drivingassessment.uiowa.edu/drivingmetrics/Bolstad%20HFES%202001%20.pdf

Causse, M., Dehais, F., Arexis, M., & Pastor, J. (2011). Cognitive aging and flight performances in general aviation pilots. *Aging, Neuropsychology, and Cognition, 18*(5), 544–561.

Dekker, S. W. (2015). The danger of losing situation awareness. *Cognition, Technology & Work, 17*(2), 159–161.

Dismukes, R., & Berman, B. (2010). *Checklists and monitoring in the cockpit: Why crucial defenses sometimes fail.* NASA- A/TM—2010-216396. Retrieved from http://citeseerx.ist.psu.edu/viewdoc/download?doi=10.1.1.206.2615&rep=rep1&type=pdf

Endsley, M. R. (1988). Design and evaluation for situation awareness enhancement. In *Proceedings of the Human Factors and Ergonomics Society Annual Meeting* (Vol. 32, pp. 97–101). SAGE Publications. Retrieved from http://pro.sagepub.com/content/32/2/97.short

Endsley, M. R. (1995). Toward a theory of situation awareness in dynamic systems. *Human Factors: The Journal of the Human Factors and Ergonomics Society*, *37*(1), 32–64.

Endsley, M. R. (1999). Situation awareness in aviation systems. *Handbook of Aviation Human Factors*, 257–276.

Goh, J., & Wiegmann, D. (2001). An investigation of the factors that contribute to pilots' decisions to continue visual flight rules flight into adverse weather. In *Proceedings of the Human Factors and Ergonomics Society Annual Meeting* (Vol. 45, pp. 26–29). SAGE Publications Sage CA: Los Angeles, CA. Retrieved from http://journals.sagepub.com/doi/abs/10.1177/154193120104500205

Gugerty, L. J., & Tirre, W. C. (2000). Individual differences in situation awareness. *Situation Awareness Analysis and Measurement*, 249–276.

Gutzwiller, R. S., & Clegg, B. A. (2013). The Role of Working Memory in Levels of Situation Awareness. *Journal of Cognitive Engineering and Decision Making*, *7*(2), 141–154. https://doi.org/10.1177/1555343412451749

Kay, G. (1995). *CogScreen Aeromedical Edition Professional Manual* (Professional Manual). Florida: Psychological Assessment Resources, Inc.

Kennedy, Q., Taylor, J. L., Reade, G., & Yesavage, J. A. (2010). Age and Expertise Effects in Aviation Decision Making and Flight Control in a Flight Simulator. *Aviation, Space, and Environmental Medicine*, *81*(5), 489–497.

O'Hare, D. (2006). Cognitive functions and performance shaping factors in aviation accidents and incidents. *The International Journal of Aviation Psychology*, *16*(2), 145–156

Van Benthem, K., & Herdman, C. M. (2016). Cognitive Factors Mediate the Relation Between Age and Flight Path Maintenance in General Aviation. *Aviation Psychology and Applied Human Factors*

Van Benthem, K., Herdman, C. M., Brown, M., & Barr, A. (2011). The Relationship of age, experience and cognitive health to private pilot situation awareness performance. In *Proceedings of the 16th International Symposium on Aviation Psychology* (pp. 2–5). Retrieved from https://www.researchgate.net/profile/Matthew_Brown23/publication/264236484_The_Relationship_of_age_experience_and_cognitive_health_to_private_pilot_situation_awareness_performance/links/551aaa270cf2bb754076a917.pdf

Van Benthem, K., Herdman, C.M., Tolton, R. & LeFevre, J. (2015). Prospective memory failures: effects of cue salience, workload, and pilot individual differences. *Aerospace Medicine and Human Performance, 86*(4), 366-373.

Wright, C. (2013). *Diversions and Aeronautical Decision Making*. AOPA. Retrieved from https://blog.aopa.org/aopa/2013/05/08/diversions-and-aeronautical-decision-making/.

HUMAN-MACHINE ARTICULATION WORK: FUNCTIONAL DEPENDENCY DIALOGUE FOR HUMAN-MACHINE TEAMING

Clayton D. Rothwell
Infoscitex Corporation and Wright State University
Dayton, Ohio
Valerie L. Shalin
Wright State University
Dayton, Ohio

Articulation work is an overlooked requirement for successful human-machine teams. Articulation work captures the often hidden task management activities human-human teams regularly perform in response to functional dependencies amongst team members. While human-human teams demonstrate articulation work through language, human-machine teams currently do not. Aviation is replete with examples, from the superficially mundane adaptation inherent in the turnaround of commercial aircraft to the life-threatening misunderstanding in Turkish Airlines Flight 1951 and Asiana Flight 214. Current research in human-machine language-mediated interaction has failed to study tasks that are sufficiently complicated to require articulation work, resulting in a misleading optimism about the state of the art. More realistic scenarios in human machine teaming will promote attention to this fundamental limitation and motivate the development of analogous capability.

For several decades, initiatives such as crew resource management acknowledge team processes as critical to performance and safety. As both military and commercial aviation evolve in a stream of technological advances, human-machine teams are a new possibility and goal. Legacy frameworks for designing automation-based technologies are a natural starting point for human-machine teams. However assumptions of a static environment make these frameworks inherently rigid and brittle. In contrast, human-human teams are particularly fluid in the dynamic task management activities known as *articulation work,* typically accomplished through natural language dialogue. While natural language technology could enable human-machine articulation work, current technology assumes overly simplistic tasks and notions of cooperation, omitting articulation work as a requirement. This position paper argues that 1) articulation work is critical to the success of human-machine teams just as it is in human-human teams, 2) frameworks for designing human-machine teams do not allow for sufficient articulation work, 3) current language technology does not support articulation work, 4) requirements for articulation work using dialogue must drive new research and development.

The Importance of Articulation Work

Cooperative work in a dynamic environment necessitates articulation work, that is, task management activities aimed at functionally decomposing a task, negotiating goals, identifying dependencies, and divvying up who will do what and when. Articulation work establishes the team's varying *functional dependencies*, that is what team members are committed to doing and what other teammates are depending on them to do. Schmidt and Simone (1995) note that articulation work also serves to improve collaborations as team members:

> tacitly monitor each other; they perform their activities in ways that support coworkers' awareness and understanding of their work; they take each others' past, present and prospective activities into account in planning and conducting their own work (p. 17)

Articulation work obtained prominence in research areas focusing on teamwork and team performance, such as computer supported cooperative work (CSCW; Schmidt and Bannon, 1992; Malone and Crowston, 1990). Importantly, the articulation for a particular situation is not rigid, it's adapted as members leave and join the group, become fatigued over time, demonstrate competence or incompetence, learn new capabilities, or deplete certain resources. The hallmark of articulation is how it enables robust teamwork in the face of the moment-by-moment unexpected—uncertainty and environmental perturbations that cause a collision between the team's plans and execution. Teams discuss the nature of perturbations, their existing plans and commitments, and discuss whether or not the perturbations merit a change in approach from one or more members.

Consider what a team of airport and airline employees must face while performing turnarounds, the process of taking a plane that's just arrived, unloading and servicing it, and loading it again so it can take off again (Wales, O'Neill, Mirmalek, 2002). Articulation work allows a team to discuss perturbations as well as existing constraints and determine how to respond in concert, continuing the interdependent activities required to meet the goals. While the flight schedule determines resourcing staff and equipment, perturbations emerge in the form of flight delays, changes and cancellations due to weather and mechanical problems. When a flight is canceled and passengers are moved to another plane by operations personnel, bag handling personnel have to accommodate these changes. Turnarounds take longer or shorter than expected, spawning gate changes. Staff members report in ill or equipment breaks, such as fuel or catering trucks. On close inspection the plan *never* unfolds exactly as anticipated, but (hidden) articulation work fills in the gaps, including gaps in the technology, to create the misleading impression of (mostly) seamless integration.

Historical Human-Machine Frameworks

Legacy frameworks for designing automation-based technologies are a natural starting point for human-machine teams. Fitt's (1951) list men-are-better-at, machines-are-better-at (MABA-MABA) approach seeks to divide responsibility between humans and machines. It accomplishes this by identifying the (relatively) superior capabilities of humans and machines and then allocating tasks to whomever or whichever is the most proficient. Allocations suggested in the Fitt's list have been commonly understood as static assignments (see de Winter & Dodou 2011 for discussion). Later researchers identified the MABA-MABA approach as overly simplified and sought to have multiple levels of automation that can be isolated for different stages (e.g., Parasuraman, Sheridan, & Wickens, 2000). However, the levels of automation framework remains rigid. These use static task decompositions rather than allowing for articulation work. Dynamic function allocation, adaptive automation with machine-initiated changes, and adaptable automation with human-initiated changes (for a review see Scerbo, 1996), though they provide for change, are insufficient. These approaches have predefined the possible changes, the notifications of change, and the triggers for changes. Consistent with

Norman (1996), fine-grained articulation work is always necessary to generate novel team structures or distribution of responsibilities, provide flexibility in how to notify team members of change, and use triggers that can't be predicted or exhaustively programmed. Moreover, changes in automation status are poorly communicated to human teammates and therefore often missed, leading to mode confusion and automation surprise (Sarter, Woods & Billings, 1997). While Woods (1996) recognizes the expansive consequences of new automation on the distribution of responsibility, the molar time-scale of work practice adaptation neglects the moment-by-moment adaptation that teams require. Coactive design is a relatively new approach that seeks to detail human-machine interdependencies (Johnson, Bradshaw, Feltovich, Jonker, van Riemsdijk, & Sierhuis, 2014), however its breadth lacks specific guidance for dialogue system development.

Functional Dependencies in Aviation Accidents

Machines in use are currently inept at articulation work. Typically, expert humans resort to workarounds to distribute the functions amongst system members while managing the tasks supposedly distributed to the machine. Two similar commercial aviation accidents support this assertion: Turkish Airlines Flight 1951 and Asiana Flight 214. In each of these accidents, the humans were depending on the machine to perform a function, the machine was not aware the human was depending on it and the humans were not aware of the machine's tacit decline of responsibility. In both accidents, the critical function was to maintain thrust on approach through the autothrottle.

Turkish Airlines Flight 1951. On 25 February, 2009, Turkish Airlines Flight 1951 crashed during its approach to the Amsterdam Schiphol airport. The first officer as pilot flying was using Line Flying Under Supervision, which utilizes the autothrottle for airspeed control. Though the pilot flying and the crew were relying on the autothrottle to maintain the airspeed of the aircraft on the approach, the aircraft could not be informed of this functional dependency. Rather, the flight crew attempted to create the airspeed function through management of the autothrottle mode selections. The approach was higher than the glidepath, so a member of the flight crew selected the 'vertical speed' mode to increase the descent. After this change, the autothrottle entered RETARD mode, which was displayed on the left and right primary flight displays, and the autothrottle moved the thrust levers into the idle position. In contrast to crew assumptions, the autopilot would not maintain airspeed in this configuration. However, the machine's exclamation of RETARD to the flight crew was unspecific and did not communicate the breakdown of the expected function. There are two types of RETARD, one for flight level changes and one for flaring to land, and the primary flight display annunciation panel does not distinguish between the two (Silva and Hansman, 2015). The type of RETARD depended on the altitude information reaching the autopilot, with a threshold altitude of 27 feet. The autopilot believed the aircraft to be below 27 feet in altitude because the autopilot was receiving and using erroneous altitude data indicating the aircraft height at -8 feet, which disagreed with the altitude data presented to the pilot flying. The pilot's primary flight display showed a conflicting but correct altitude status, leading to confusion over the situation. Ultimately, the lack of clear functional dependency and the misunderstanding about the meaning of RETARD led to an unrecoverable stall and the aircraft crashed killing 9 and injuring 117 (Dutch Safety Board, 2010).

Asiana Flight 214. An accident of similar origin occurred when Asiana Flight 214 crashed on July 3rd, 2013 during approach to San Francisco International Airport. Asiana has an informal practice for visual approach of turning off both flight directors and then turning back on the pilot monitoring's flight director during the approach (National Transportation Safety Board, 2014). This practice results in the autothrottle entering speed mode and determines a distribution of functions: the autothrottle maintains airspeed and the pilot flying can focus on pitch and roll. In Flight 214, the aircraft was above the glidepath and needed to descend. The informal practice of toggling the flight director was followed 'loosely,' both flight directors were not off at the same time and therefore speed mode was not entered. The pilot flying moved the thrust levers and inadvertently caused the autothrottle to change to HOLD mode. The HOLD mode created a breakdown in the function of maintaining airspeed—HOLD deactivated automatic airspeed control. The burden of recognizing the mode change and the implications for the function being provided in the approach falls on the flight crew, but they did not note the change to HOLD mode. These events led the aircraft to descend below the glidepath at a high rate and collide with a sea wall, killing 3 and injuring 187 (National Transportation Safety Board, 2014).

Both cases hinge on the absence of human-machine articulation dialogue concerning the retention or abandonment of otherwise tacit commitments to act. These problems with managing functional dependencies are relatively simple when compared to the envisioned applications for human-machine teams.

Implications for Human-Machine Communication

Human teams routinely perform articulation work through dialogue. Research into human-machine communication and natural language dialogue has exerted a great deal of effort studying and improving clarification of utterances or lexical ambiguities, but not communication, clarification or negotiation of functional dependencies, which require a more complicated ontology including agent beliefs (e.g., Clancey, Sierhuis, Damer, & Brodsky, 2005). Many of the classic human-machine communication tasks do not provide opportunity for articulation work and therefore do not reveal these deficiencies.

Common application domains for natural language processing, such as shopping and navigation, are restricted and fail to address the ways in which these tasks vary. As the scope of artificial intelligence grows and natural language processing technologies become more integrated into work practices, their applications will not be limited to the subset of activities with overly simplified team processes. When a richer task is used, such as a collaborative problem-solving task similar to the board game Clue (Traum & Dillenbourg, 1996), the proportion of communication spent on articulation work is apparent. In particular, the authors noted a frequent topic was decomposition of who does what and when. Simple tasks lead to the impression that a simple ontology can work, whereas a moderately complex task (still quite simple in comparison to the wild) easily sets a high bar for a rich ontology comprising not just the task content but the possible conceptualizations and organizations of the cooperation. Research is needed to push dialogue-mediated tasks into more realistic scenarios that will require articulation work.

Future Research

The persuasive macro-level case for articulation capability does not provide concrete requirements for designers and software developers for building natural language interfaces. We specify three topics that should shape the research agenda.

Initiate team research utilizing task settings that require articulation. A primary goal of future research should be to have some aspect of the task that prompts articulation work. We envision articulation work to be prompted by an element of dynamic disparity in the task context that differs between partners and thus affects the rate of progress for one of the partners, which merits announcing to or discussing with the other partner. This could be due to a change in a sub-goal of the task, which requires discussion of a change in strategy, approach, sequence, or the like. This could also be due to introducing a problem that perturbs the existing strategy.

Formalizing a taxonomy of articulation work. A taxonomy is needed to enable diagnosis of the disconnects between human and machine teammates. Research programs from J. Allen, H. Clark, and J. Searle provide theoretical inspiration. A critical requirement is a conceptual distinction between real-time execution and planning activities (Shalin, 2005). It is the interaction between execution and planning, frequently initiated by a perturbation, that spawns articulation dialogues. While resources such as Aviation Safety Reporting System and specific accidents provide data, what is required is the conceptual framework to generalize the limitations, across instances in aviation and ideally, extending to other domains including laboratory tasks. Being accountable for providing a specific function is one facet of this taxonomy. Terminology grounding (e.g., which meaning of RETARD) and explicit task completion acknowledgement (as in toggling the flight director) is another.

Translate articulation requirements into functional machine analogues. Attempting to replicate human team members with machines is fraught with philosophical problems and an impractical near term goal at best. Nevertheless specific functions are well within technical reach without imbuing technology with human processes, e.g., for clarifying the grounding of terminology, confirming mode change communications are received, identifying that a functional dependency will not be upheld. These functions will drive the requirements for natural language interfaces.

Conclusion

In this position paper, we have argued that 1) articulation work is critical to the success of human-machine teams just as it is in human-human teams, 2) frameworks for designing human-machine teams do not allow for sufficient articulation work, 3) current language technology does not support articulation work, 4) requirements for articulation work using dialogue must drive new research and development. Human-machine teaming research to date has largely ignored the challenge of articulation work and cannot ignore it any longer.

Acknowledgements

This research was partially supported by the Air Force Research Laboratory's Human Interface Research Technologies Contract #: FA8650-14-D-6501. We thank Asaf Degani and Immanuel Barshi for helpful discussions early on in the development of this work.

References

Clancey, W.J, Sierhuis, M., Damer, B., Brodsky, B. (2005) Cognitive modeling of social behaviors. In R. Sun (Ed.), *Cognition and Multi-Agent Interaction: From Cognitive Modeling to Social Simulation*, (pp. 151-184). New York: Cambridge University Press.

Dutch Safety Board (2010). Crashed during approach, Boeing 737-800, near Amsterdam Schiphol Airport, 25 February 2009. *Dutch Safety Board Final Report.*

Fitts, P. M. (ed) (1951) Human engineering for an effective air navigation and traffic control system. National Research Council, Washington, DC

Johnson, M., Bradshaw, J. M., Feltovich, P. J., Jonker, C. M., Van Riemsdijk, M. B., & Sierhuis, M. (2014). Coactive design: Designing support for interdependence in joint activity. *Journal of Human-Robot Interaction*, *3*(1), 43-69.

Malone, T. W., & Crowston, K. (1990, September). What is coordination theory and how can it help design cooperative work systems?. In *Proceedings of the 1990 ACM conference on Computer-supported cooperative work* (pp. 357-370). ACM.

National Transportation Safety Board (2014). Crash of Asiana Flight 214 Accident Report Summary.

Norman, D. A. (1992). Coffee cups in the cockpit. In *Turn Signals are the Facial Expressions of Automobiles*, (pp. 154-174). Boulder, CO: Westview Press.

Parasuraman, R., Sheridan, T. B., & Wickens, C. D. (2000). A model for types and levels of human interaction with automation. *IEEE Transactions on systems, man, and cybernetics-Part A: Systems and Humans*, *30*(3), 286-297.

Sarter, N. Woods, D. & Billings, C. (1997). Automation surprises. In G. Salvendy (Ed.) Handbook of Human Factors & Ergonomics 2, (pp. 1926-1943). Hoboken, NJ: Wiley.

Scerbo, M. (1996). Theoretical perspectives on adaptive automation. In R. Parasuraman & M. Mouloua (Eds). *Automation and human performance: Theory and applications*., (pp. 37-63). Hillsdale, NJ: Lawrence Erlbaum.

Schmidt, K. and Bannon, L. (1992). Taking CSCW seriously: Supporting articulation work. *Computer Supported Cooperative Work*, 1(1), 7–40.

Schmidt, K. and Simone, C. (1995). Coordination mechanisms: An approach to CSCW systems design. In Navaro, L. (Ed). Demonstrator prototypes of computational mechanisms of interaction, (pp. 15-72). Lancaster, UK: Lancaster University.

Shalin, V. L. (2005). The roles of humans and computers in distributed planning for dynamic domains. *Cognition, Technology & Work*, *7*(3), 198-211.

Silva, S. S. and Hansman, R. J. (2015). Divergence Between Flight Crew Mental Model and Aircraft System State in Auto-Throttle Mode Confusion Accident and Incident Cases. *Journal of Cognitive Engineering and Decision Making*, 9(4), 312–328.

Traum, D., & Dillenbourg, P. (1996, August). Miscommunication in multi-modal collaboration. In *AAAI Workshop on Detecting, Repairing, And Preventing Human-Machine Miscommunication* (pp. 37-46).

Wales, R., O'Neill, J., & Mirmalek, Z. (2002). Ethnography, customers, and negotiated interactions at the airport. *IEEE Intelligent Systems*, *17*(5), 15-23.

de Winter, J. & Dodou, D. (2011). Why the Fitts list has persisted throughout the history of function allocation. *Cognition Technology & Work, 16(1), 1-11.*

Woods, D. D. (1996). Decomposing automation: Apparent simplicity, real complexity. In R. Parasuraman & M. Mouloua (Eds). *Automation and human performance: Theory and applications*, (pp. 3-17). Hillsdale, NJ: Lawrence Erlbaum.

FACILITATING COMMUNICATION FOR AVIATION TRAINING AND MAINTENANCE

Lorraine E. Holtaway
Kyle Jackson
Sarah M. Hubbard, Ph.D.
Purdue University
West Lafayette, IN

Maintenance errors are the primary cause of approximately 8% of commercial aircraft accidents worldwide. One factor that contributes to human errors is miscommunication. Clear communication is critical in aviation education and in aviation maintenance operations. A fundamental concept for clear communication is both the transmission and receipt of a common message. This research explores the miscommunication and misinterpretation of instructions used in maintenance training. Miscommunication may be due to ambiguity, use of jargon, and different individual interpretations and methods for standard practices. First, an example of a commonly misunderstood process is identified. Next, enhanced training tools are developed to reduce the likelihood of miscommunication. These enhanced training tools include detailed illustrations and the addition of descriptive text to provide more information, including additional physical characteristics and technical context. Finally, the proposed training aids are assessed in a controlled study to determine their effectiveness.

The Purdue University School of Aviation and Transportation Technology offers an Aviation Engineering Technology (AET) bachelor's degree that includes the opportunity for students to participate in a Federal Aviation Administration (FAA) 14 CFR Part 147 program. Students who complete the Part 147 program are eligible to for Airframe and Powerplant (A&P) certification, the designation for licensed aircraft mechanics. One course required to meet the Part 147 requirements is Advanced Aircraft Powerplants, which involves learning maintenance and overhaul procedures for reciprocating aircraft engines.

One of the laboratory activities the students perform in the Advanced Aircraft Powerplants course is manually adjusting the valve clearance, the amount of space between the valve stem and the rocker arm, on a Lycoming O-290 engine. The students are given a Service Instruction written by Lycoming as the procedure for completing the laboratory activity. One problem the students encounter is that the Lycoming Service Instruction was written for trained mechanics, not students. In previous years, the instructor would demonstrate the process to a small group of students, and they would only use the Service Instruction as a reference, not their primary source of instruction. As class sizes grow, instructors are not able to give as much individualized attention to the students. The students need to be able to learn more independently and the course materials need to communicate instructions clearly.

The research objective of this study is to assess the effectiveness of training tools that provide more technical context and descriptive illustrations. We hypothesize that enhanced training tools will reduce the likelihood of miscommunication so students will be better able to understand the instructions and learn independently. The objective of this paper is to describe the research and the methodology, and present preliminary results.

Literature Review

The Federal Aviation Administration (2014) states the importance of communication for aviation maintenance in *The Operator's Manual for Human Factors in Aviation Maintenance.* A research study in 2007 found that in eight percent of the commercial aircraft accidents from 1990 to 2006, the primary cause was maintenance. The leading factor for the FAA initiating Letters of Investigation (LOI) and taking administrative action on Aircraft Maintenance Technicians (AMTs) is failure to follow written procedures. Approximately 83% of maintenance Aviation Safety Reports (ASRs) from 2010 to 2013 were related to technical publications and other written company procedures. Training is a critical activity in the aviation industry, and it is identified as the top intervention for risk reduction (Federal Aviation Administration, 2014).

According to Chaparro and Groff (2002) in *Human Factors Survey of Aviation Technical Manuals*, an analysis of aircraft maintenance error causation ranked information as the highest contributing factor. Only a small number of the errors attributed to information were due to incorrect data, however, and more often the technicians did not refer to the information, misunderstood the information, or disregarded it in favor of an alternative method of performing the maintenance procedure. While this problem could be addressed through training or disciplinary action, it could also be a result of a problem with the usability of the technical documents (Chaparro & Groff, 2002).

The usability of aircraft manuals "includes how easy they are to use, how well they match the technician's representation of a task, how easy they are to read and interpret, and how useful the information is they contain" (Chaparro & Groff, 2002, p. 2). If the maintenance manuals contain misleading information, insufficient information, or unclear procedures, they can contribute to maintenance error. The work in *Human Factors Survey of Aviation Technical Manuals* researches the human factors issues in the development of aviation technical manuals and recommends improvements to the documents (Chaparro & Groff, 2002).

The results of the survey indicate that the documentation provided to maintenance technicians needs to contain accurate technical information and needs to be presented in a way that matches the way technicians actually do their job. A high percentage of survey responses were "disagree" or "strongly disagree" to the questions "the manual describes the best way to do a procedure" and "the manual writer understands the way I do maintenance." These responses show that manual usability is a common problem, and the potential consequences of these problems are the safety, speed, and cost of aircraft maintenance. The recommendations for addressing usability problems are increased feedback from the users, including an error reporting system, and controlling formatting consistency and reading level through standardization guidelines, including the ordering of procedural steps, the wording of procedures, the use of illustrations, and the level of detail (Chaparro & Groff, 2002).

A Design Aid for Improved Documentation in Aircraft Maintenance: A Precursor to Training provides background, research, and recommendation for writing documentation that reduces the likelihood of errors. The demand for error reduction in aviation maintenance is increasing, and the study evaluates a tool to help present complex work instructions in a way that will minimize error opportunities. The FAA Office of Aviation Medicine (FAA/AAM) has been funding research into human error, and one area studied was the information environment of the people performing inspection and maintenance activities. They found that much of the paperwork used to control hangar-floor activities did not follow good human factors practice. Aviation maintenance documentation is often used under non-optimal environmental conditions and with time stress, so any means of reducing errors, such as better workcard design, is cost-effective (Drur & Abdulkadir, 1997).

Drur and Abdulkadir (1997) undertook their study to provide a job aid for document writers to help them apply good human factors practices to their documents. They worked with an airline partner to examine existing workcards for specific problems using task observation, interviews with mechanics and inspectors, and survey data. Existing human factors research findings were also used to determine good practices. The research resulted in a Document Design Aid (DDA) that was arranged in steps, with sign-offs at each step, that could be used as a checklist to ensure that a document was well designed (Drur & Abdulkadir, 1997).

The researchers then evaluated the DDA for its usability and effectiveness. Usability was defined as a job aid being usable for its intended purpose by intended users, and effectiveness was defined as whether or not the intended users perform their job better with the job aid. A sample of intended users were assigned the task of modifying and existing engineering order (EO) to conform to the guidelines in the DDA. Usability was measured by user rating scales, and effectiveness was measured by comparing the changes made to the EO by each user to a master list of changes made by expert users. Users had sixty minutes to mark up the test EO, and they found and average of thirty-five percent of the changes suggested by the experts. The researchers considered this performance to be adequate because most of the major changes were found (Drur & Abdulkadir, 1997).

The literature review indicates that maintenance is a critical part of aviation safety, and that misuse or misunderstanding of documentation are leading causes of error. It also establishes the importance of measuring the usability and effectiveness of technical publications in order to decrease the occurrences of miscommunication and the resulting errors.

Method

The participants in this study were students enrolled in the Advanced Aircraft Powerplants class at Purdue University during the spring semester of 2017. The students in the class are working toward earning their FAA A&P certificate. Twenty-six students were enrolled at the time of the study.

The study was true experimental research with experimental and control groups. The participants were randomly assigned into two groups by drawing their name out of a hat. The students assigned to the control group were given the original set of instructions in the existing lab manual. The students in the experimental group were given a set of instructions enhanced with pictures and more description in addition to the original set of instructions. The participants were directed to perform the valve clearance laboratory activity.

The researchers observed the students as they performed the laboratory activity. The data collected was the date, the student's participant number, whether or not the student received the enhanced instructions, the number of attempts the student took to complete the activity, the amount of time taken to complete the activity, the student's perception of the quality of the instructions, and what questions the student asked during the project. The researchers also recorded comments on the types of errors the students made during the process. An attempt was defined as each time a student tightened the jam nut on the rocker arm. The valve clearance laboratory activity was considered complete when the student was able to set the distance between the valve stem rocker arm and demonstrate to the instructor that it was within the manufacturer's limitations. The student's perception of the quality of the instructions was measured on a scale of one to five.

Preliminary Results

Preliminary observations suggest that the primary metric to evaluate the quality of the instructions may be the number of questions the students ask while performing the laboratory activity. This metric reflects a student's ability to understand the instructions on their own, and predicts how much time the instructor would need to spend with individual students to provide support information needed to complete the lab. Fewer student questions indicate that a student has a better understanding of the process, presumably due to better instructions, which allow the student to learn more independently with less instructor involvement. The questions the students ask and the researchers' comments on the errors the students make also provide insight into the types of problems encountered during the learning process. The specific questions can be used to further improve the instructions.

The number of attempts the students take to complete the activity and the student perception of the quality of the instructions appear to be useful measures reflecting the quality of instruction. These measures appear to suggest differences between the current and proposed instructions based on preliminary results.

The amount of time students take to complete the activity, however, may not be a reliable metric to assess the quality of the instructions. A student using the current instructions written for experienced mechanics could become frustrated earlier in the process and ask the instructor for help sooner, which could allow them to complete the activity in less time than a student who works independently with the enhanced instructions. Alternatively, a student may find the enhanced instructions help them to work quickly through the activity while a student using the current instructions may take time to struggle to understand the process.

Further research is needed to fully determine the effectiveness of the enhanced instructions. Increasing the sample size in this study would reduce the impact of any outliers within the population. For example, some students have more mechanical experience in general, which would affect their performance. A larger sample size would encompass a broader range of abilities. In addition, some students did not follow the procedure correctly, but still managed to produce acceptable results. The researcher observing these students attributed their successful completion of the activity to luck, which would also have less of an effect in a larger sample size. The current study also does not address if the enhanced instructions enable the students to better retain the information they learned. This may be likely, since often visualization techniques are recommended as memory aides. Adding a secondary evaluation to assess how well the students remember the process after some time has passed (perhaps two weeks or a month) could lend another dimension to assess the effectiveness of the enhanced instructions.

References

Chaparro, A., & Groff, L. S. (2002). Human Factors Survey of Aviation Technical Manuals. *16th Human Factors in Aviation Maintenance Symposium.* Retrieved from https://pdfs.semanticscholar.org/e83d/24811b445cf1df6ab5b09a8a592a39350ec3.pdf

Drur, C. G., & Abdulkadir, S. (1997). A Design Aid for Improved Documentation in Aircraft Maintenance: A Precursor to Training. *Proceedings of the Human Factors and Ergonomics Society 41st Annual Meeting-1997*, (pp. 1158-1162). Retrieved from http://journals.sagepub.com/doi/abs/10.1177/107118139704100297

Federal Aviation Administration. (2014). *The Operator's Manual for Human Factors in Aviation Maintenance.* Washington, DC: Office of Aerospace Medicine. Retrieved from https://www.faa.gov/about/initiatives/maintenance_hf/library/documents/media/human_factors_maintenance/hf_ops_manual_2014.pdf

AVIATION ENGLISH INTELLIGIBILITY

Julia Trippe
Eric Pederson
University of Oregon
Eugene, Oregon

Aviation English (AE) is the "primary dialect" of international aviation. Demonstrably, AE and Conversational English (CE) are distinct varieties of English. Past research shows that AE is spoken more rapidly, with less inflectional variation and different rhythm than CE. Differences are strong enough that AE and CE may not be mutually intelligible. However, flight students are not trained in AE production and perception prior to flight training. This study examines the intelligibility relationship between AE and CE by comparing native English speaking non-pilots and native English speaking pilots responding to actual air traffic controller transmissions. A difference between these groups was predicted, given their comparative AE familiarity. However, the difference in AE intelligibility proved to be stronger than expected. Additionally, results from licensed pilots indicate that AE learning continues with flight experience, suggesting there may not be adequate training prior to reliance on AE in flight.

Aviation English is the mandatory language for pilots and air traffic controllers (ATC) at international airports, if they do not share a first language. Proficiency in Aviation English (AE) and conversational English (CE) are required by the International Civil Aviation Organisation (ICAO), yet little is known about how AE and CE interact in language learning and usage. These requirements as well as AE training conventions are based on the assumption that CE proficiency aids in AE proficiency, although this may not be the case (see discussion of "plain English" in Background section below). Indeed, past research shows that AE is different from CE in ways that may affect intelligibility (Trippe & Baese-Berk, submitted). The current study examines AE intelligibility differences between native English speaking pilots and native English speaking non-pilots. If AE is not intelligible to CE speakers without aviation experience, CE proficiency cannot be sufficient to predict AE proficiency. The goal of this research is to further establish the intelligibility relationship between AE and CE and influence development of effective AE training to improve international flight safety.

Aviation English Description

AE is a variety of radiotelephony developed to convey critical information between pilots and Air Traffic Control (ATC). Although AE includes both standard phraseology and "plain English", in this study the term Aviation English (AE) is used to denote standard phraseology and "plain English" is referred to as such. Ambiguity in AE is avoided by fixing a single meaning to each word and phrase. Words whose pronunciation may cause confusion are assigned distinct pronunciations. For example, AE require that *five* and *nine* be pronounced *fife* and *niner*. Additionally, word and phrase inventories of AE are restricted. Articles, prepositions and possessives are not used except to resolve ambiguity. Any ambiguous word is given a single meaning or substituted with another word. AE standard phrases use lexical topic identifiers and specific number expressions to signify aviation topics. For example, *wind three fife zero at one two,* or *turn right heading three fife zero* both use single digits to express direction, but each phrase has a lexical identifier denoting the aviation topic addressed (i.e. *wind* v. *heading*). Lexical and grammatical differences as well as environmental factors (i.e. multiple speakers, no face-to-face contact, signal static and reduced frequency range), lead to differences in the sound profiles of AE and CE. AE is faster than CE, with fewer intonational cues and a different rhythmic signature than CE (Trippe & Baese-Berk, submitted). These differences could make AE unintelligible to CE speakers.

Aviation English Regulation

High loss-of-life accidents caused in part by communication problems (Cookson, 2011) compelled ICAO to require AE proficiency in international airspace as of 2011. However, this requirement has yet to be thoroughly operationalized. While ICAO has published general proficiency-rating guidelines, there is no agreed upon standard protocol by which to attain or prove proficiency. Dozens of tests have been developed internationally and several are in use, although ICAO recognizes only one (English Language Proficiency for Aeronautical Communication). The new requirements also pertain to CE proficiency (ICAO, 2004), although the vast majority of pilot-ATC communication is in AE, which was designed to convey all typical transactions. When AE is not sufficient to convey messages, ICAO regulations stipulate the use of "plain English". Generally this caveat applies to unusual or emergency situations. Although the implementation of AE recognizes the need to keep communications succinct and unambiguous, it is impossible to control for these needs in "plain English", because the parameters of "plain English" are not defined. Native English speakers often speak quickly and colloquially during times of duress. Although such interactions usually aid in clarification of complex situations between native English speakers, these communications may not be comprehensible to non-native English speaking interlocutors (Kim & Elder, 2009). Additionally, second language English users have more difficulty conversing in CE under conditions of stress or high cognitive load that typically trigger "plain English" use in native speakers (Farris, Trofimovich, Segalowitz & Gatbonton, 2008). Further, the requirement to use "plain English" is confounded by the fact that there exists no consistent guidance as to what is meant by *plain* English. The regulatory intent is clear: this English variety should be readily understandable to one's interlocutor. Unfortunately, it is impossible to ascertain what level of English proficiency, or indeed what model of Standard English, one's interlocutor has. In fact, language experts recommend "plain English" be avoided as much as possible in aviation communications (Day, 2004: Moder, 2012). AE fluency reduces repetitions, delays, and misunderstandings.

As the international flying community becomes more diverse, pilots will operate in airspace and on crews with individuals from different language backgrounds, increasing the potential for misunderstanding and miscommunication (Kim & Billington, 2016). In this environment, it is critical to utilize AE standard phraseology, to reduce the potential for confusion as much as possible. Rather than relying on "plain English", consideration should be given to expanding AE so that unusual situations may be addressed using this clear and constrained format and lexicon.

Aviation English Testing and Training

Testing. Newly developed AE testing protocols differ greatly. However, a common element of AE tests is a face-to-face interview with a language evaluation specialist wherein the pilot must discuss unusual situations that may arise while flying, to determine if they have a working knowledge of aviation terminology and can convey ideas in CE. Interviews are typically conducted by English-language teaching specialists who are not aviation professionals nor fluent in AE. This type of testing does not evaluate AE speech used in most pilot-ATC interactions. In fact, listening and responding to actual ATC transmissions may not be included in the pilot's proficiency test, although this is the vast majority of pilot communications (Alderson, 2009). Additionally, when ATC speech is used in testing, it is created for that purpose and is often slower, without static, accents and multiple speakers that occur in actual transmissions. Therefore, passing an AE proficiency test does not guarantee a pilot's ability to fulfill their job requirements. In their study of non-native English speaker ATC oral proficiency, Moder and Halleck (2009) found that there was no consistent relationship between AE and CE scores. Additionally, Kim & Elder (2009) asserted that CE-focused testing protocols unfairly benefit native English speakers, who are assumed to be fluent in AE, but often do not comply with AE standard phraseology.

Training. The standard for AE training has long been that radiotelephony is learned simultaneously with flight training. It is assumed that pilots will learn through immersion: monitoring and interacting with ATC. Anecdotally, it is common knowledge that student pilots are as anxious about talking on the radio as they are about flying the plane. However, the AE immersion strategy has been adopted as the model for non-native speakers training in native English speaking countries, which is where a great deal of international commercial flight training takes place. Although many flight-training programs for non-native English speakers include language training, AE courses are designed by English language teaching experts in consultation with aviation professionals, focusing on face-to-face communication in CE with emphasis on aviation terminology. AE instructors are generally not fluent in AE. This learning environment does not reflect pilots' experience or needs. In actual flight conditions, pilots must interpret messages through static and reduced frequency range, without seeing their interlocutor. If AE is as different from CE as prior research indicates (Trippe & Baese-Berk, submitted), training in CE with non-AE speakers will not enhance AE skills as much as dedicated AE training will.

Aviation English Intelligibility

To further understand the intelligibility relationship between AE and CE, it must be determined if native CE users can understand AE and vice versa. The current study addresses the first of these proposals, examining the differences between native English speakers with and without AE experience, perceiving actual ATC transmissions. If AE is intelligible to CE users, then teaching and testing CE for aviators is practical. If AE is not intelligible to CE users, teaching and testing of CE for aviators may be a misuse of time and energy.

Method

Participants

Two groups of native English speaking participants were recruited for the study. The non-pilot population was made up of 26 (17 female) University of Oregon undergraduates, mean age 20.69 (SD = 3.03). The pilot population was made up of 23 licensed pilots (4 female) from Lane Aviation Academy and Hillsboro Aero Academy in Oregon, mean age 28.30 (SD = 7.77). The pilot group consisted of licensed pilots ranging in age from 19 to 55 (*median* = 26) with flight hours from 67 to 7000 (*median* = 350), including 4 to 2500 hours under Instrument Flight Rules (*median* = 56 hours).

Procedure

Participants performed three verbal repetition tasks, starting with a 15-minute verbal working memory task, followed by a five-minute Standard American English intelligibility task to establish CE competency. A 15-minute AE intelligibility task concluded each trial. Tasks were self-paced and computer-administered using Psychopy (Peirce, 2007) software. Participants completed language background questionnaires reporting other language and/or professional radio experience. Working memory (WM) was evaluated using the Word Auditory Recognition and Recall Measure (WARRM) (Smith, et al., 2016) which required participants to repeat Standard English monosyllabic audio stimuli with intervening unrelated cognitive tasks. WM was scored on a scale from 2 to 6 points, reflecting the number of words the participant was able to remember consistently. This score was then multiplied by 16.67 to make the highest possible score 100, to be comparable with percentage scores for the other tasks.

The second task was a CE intelligibility task in which participants repeated ten sentences from the Harvard Sentence recordings (Open Speech Repository, 2016) which were approximately fifth grade reading level, phonetically balanced for Standard American English, and from seven to ten words long. Responses were tape-recorded for later analysis. Score for the CE task was the percentage of words correctly reproduced of the 83 possible words in the ten CE sentences combined.

The third verbal repetition task was an AE intelligibility task in which participants repeated 84 ATC transmissions selected from the Air Traffic Control Complete corpus (Godfrey, 1994), based on number of topics and terminology. Since past studies indicate that subjects show a sharp decrease in navigational performance for transmissions with more than three propositions (Farris & Barshi, 2013), selected transmissions were limited to two topics (i.e. *[traffic no factor] [turn right heading two zero zero]*). Half of the selected ATC transmissions had one aviation topic and half had two. Equal numbers of transmissions were chosen from 22 (3 female) apparently native American English ATC. Responses were tape-recorded for later analysis. Stimuli were pseudo-randomized so that every dozen transmissions included an equal number of one- and two-topic tokens. AE task transcription was done by two trained lab technicians and the first author. Inter-coder reliability tests resulted in 98% agreement. Words were correct if they were in order relative to other words in the transmission (see Table 1).

Table 1.
Sample Points Awarded for Participant Response

Original transmission	TURN	RIGHT	...	HEADING	...	TWO	FOUR	ZERO	(6 words)
Response	...	right	turn	heading	zero	two	...	zero	
Points	0	1	0	1	0	1	0	1	4
Percentage									66.67

Results

Statistical Analysis

Verbal repetition task scores by group. Pilot group average for the CE task (M = 95.55, SD = 3.55) did not differ significantly from non-pilots' CE scores (M = 97.00, SD = 3.11) (t (44.12) = 1.52, p = 0.14). Nor did pilots' WM task scores (M = 77.30, SD = 13.60) from non-pilots' WM scores (M = 71.82, SD = 16.56) (t (46.77) = -1.27, p = 0.21). However, Average pilots' AE task scores (M = 87.97, SD = 18.22) were significantly higher than non-pilots' (M = 57.27, SD = 26.18) (t (46.69) = -15.81, $p < .001$). The only apparent learning effect in the data was for non-pilots showed a learning effect between the first to the second set of AE transmissions (see Table 2).

Table 2.
Mean Aviation English Percentage Correct Over Testing Period by Group

	Testing Period						
Group	AE1	AE2	AE3	AE4	AE5	AE6	AE7
Non-Pilots	46.44^{a}	51.28^{ab}	53.86^{b}	55.71^{b}	55.36^{b}	54.83^{ab}	55.60^{b}
Pilots	82.42^{c}	83.46^{c}	84.13^{c}	84.00^{c}	85.86^{c}	86.13^{c}	87.99^{c}

Note: Values with different superscripts are significantly different p < .05

Factors predicting AE intelligibility. A linear mixed effects regression was performed using the lme4 package (Bates, Maechler, Bolker, & Walker, 2015) in R (R Core Team, 2014) on all responses in the data. Fixed effects included CE score, WM score, group, radio experience, language experience, age, sex, number of words per transmission, number of topics per transmission and interactions with group for each fixed effect. Number of topics and number of words were collinear ($R^2 = 0.54$). However, given the significant interaction of number of topics by group, both number of words and number of topics were retained as factors in the model (see Table 3).

Table 3.
Linear Mixed Effects Analysis of AE Intelligibility Scores (Random effects: Subject, Transmission and Order)

Predictor	*Coefficient*	*Std. Error*	$\chi^2(1)$	*p-value*
Intercept	43.77	25.91		
CE Score	0.42	0.27	2.42	0.120
WM Score	0.11	0.06	3.21	0.073
Pilot Group	18.07	2.38	290.99	< .001***
Number of Words	-2.20	0.55	23.54	< .001***
Number of Topics	-10.74	3.54	1.66	0.197
PilotGroup*Words	-0.84	0.23	13.02	< .001***
PilotGroup*Topics	13.34	1.50	78.61	< .001***

*Note. Significance codes: .001 '***', .01 '**', .05 '*'. Non-pilot group coded as default treatment.*

Model fit determination using piecewiseSEM package in R (Lefcheck, 2015), gave a marginal (fixed effects) R^2 value of 0.46 and conditional (including random effects) R^2 value of 0.66. Regression results indicate that pilots had significantly higher AE intelligibility scores than non-pilots. Additionally, non-pilots' scores decreased with increases in number of words and number of topics, whereas pilots' scores decreased slightly more with number of words and increased with number of topics (see Figure 1). Model statistics indicated multicollinearity between CE ($R^2 = .12$) and WM ($R^2 = 0.11$). However, their addition to the model significantly increased model fit from $R^2 = 0.45$ to 0.46 ($\chi^2(2) = 7.71$, p = 0.021).

Flight experience effect on AE scores. A separate regression was done on pilot group AE scores, to determine flight experience effect on AE score (see Table 4).

Table 4.
Linear Mixed Effects Analysis of Pilot AE Intelligibility Scores (Random Effects: Subject and Transmission)

Predictor	*Coefficient*	*Std. Error*	$\chi^2(1)$	*p-value*
Intercept	104.13	3.25		
Number of Words	-2.74	0.28	95.30	< .001 ***
ln(IFR)	1.88	0.44	18.69	< .001 ***

*Note. Significance codes: .001 '***', .01 '**', .05 '*'*

The full model included the above factors in addition to total flight hours (TT) and Instrument Flight Rules hours (IFR). Substitution of ln(IFR) for IFR increased model fit by R^2 of .01, therefore it was included in the model. The model resulted in a marginal R^2 value of 0.27 and conditional R^2 value of 0.50. Pilots' AE scores were significantly predicted by number of words in the transmission and by flight experience.

Types of AE Errors

One randomly selected participant's responses from each group were preliminarily coded for descriptive analysis. Errors were coded as transpositions, substitutions, number-number substitutions, omissions, and readback omissions (reflecting standard AE terminology). About half of the transmission data consisted of repetitive phrases and 42.76% were numbers. Both pilots and non-pilots transposed, or produced wrong, numbers (see Table 5). Otherwise, observation of the data, as well as analysis of these two participants, indicates that pilots typically produced errors of omission, while non-pilots' errors were more often substitution. For example, responding to the transmission *Turn right heading two seven zero*, a pilot produced ____ *right two seven zero*, whereas a non-pilot responded: *Turn right <u>hitting</u> two seven zero.* Non-pilots' also included substituted numbers for non-number words.

Table 5.
Breakdown of AE Errors by One Participant From Each Group

Group	total errors	transposition	substitution	wrong number (substitution)	omission	readback omission
non-pilot	96	10 (10%)	42 (44%)	13(14%)	31 (32%)	na
pilot	76	4 (5%)	10 (13%)	11 (14%)	25 (33%)	26 (34%)

Discussion and Conclusion

Results of this study indicate that AE is not intelligible to non-pilot native English speakers beyond a low threshold (53%) and acoustic learning of AE with no feedback peaks early at a level far below ceiling (~ 55%). Examining the data, we can theorize as to why CE proficiency does not imply AE intelligibility. Firstly, regression results indicate that, whereas number of words in a transmission is the primary factor in determining AE difficulty for pilot and non-pilot groups, this effect was mitigated for pilots by number of aviation topics in the transmission. This finding is consistent with the observation that expert language-users chunk information to efficiently interpret language streams. Data examination also indicates that, since AE topic identifiers are frequent and predictable, they are rapidly produced and monotone, making them less intelligible to naïve listeners. Therefore, non-pilots substituted novel terms for topic identifiers (i.e. *<u>try to</u> maintain* for *climb and maintain*). On the other hand, pilot AE errors reflected patterns of standard pilot-ATC communication, in which pilots repeat only critical elements of transmissions. Therefore, although instructed to repeat the entire transmission, pilots often omitted words that could be implied, (i.e. *runway, heading, turn, left, right, of, and, to, the, at*).

The logarithmic relationship of pilot flight experience with AE scores suggests that the AE learning curve is steep for low-time pilots and shallows out with experience. During flight, a small percentage of time is in direct communication with ATC and a higher percentage of time is in passive exposure. A training program in which pilots are exposed to recorded ATC transmissions including periods of active response and periods of passive listening would expose students to both flight language experiences. This type of training protocol would enable pilots to dedicate their attention in a low-stress, focused, efficient language-learning environment, rather than struggling to allocate cognitive resources during flight training, allowing them to acquire AE proficiency in far less time than it takes in flight. Although native English speakers may not be able to learn AE without feedback, AE language itself is formulaic, employing a constrained lexicon and restricted phrase inventory. This makes AE ideal for teaching, particularly when taking into account the chunking methodology that lends itself to pilot comprehension. If focus in training is on topic identifiers, novices may quickly learn how to recognize these rapidly produced language chunks.

Conclusion

This study seeks to improve international pilot language training by showing the need for pilots to learn the language they use every day on the job. Previous studies have shown that AE's rhythm and intonation are different from CE. The current study shows that AE is scarcely intelligible to CE speakers. Therefore the assumption that CE proficiency enhances AE proficiency is in doubt. The most efficient way of teaching AE is to focus on the AE language that pilots actually hear: including static, fast speech, real accents and a reduced frequency range. Because of the emphasis on CE in training, pilots may not be getting enough AE training before relying on it in flight. A small amount of classroom and/or online training focusing on familiarization with the limited inventory of AE words and phrases, as well as exposure to the rhythm and intonation of real ATC transmissions could enable pilots to effectively and confidently communicate in AE as soon as they get off the ground.

References

Alderson, J. C. (2009). Air Safety, Language Assessment Policy, and Policy Implementation: The Case of Aviation English. *Annual Review of Applied Linguistics, 29*, 168-187. doi:10.1017/s0267190509090138

Bates, D., Maechler, M., Bolker, B., Walker, S. (2015). Fitting Linear Mixed-Effects Models Using lme4. Journal of Statistical Software, 67(1), 1-48. <doi:10.18637/jss.v067.i01>.

Cookson, S. (2011). Zagreb and Tenerife: Airline accidents involving linguistic factors. *Australian Review of Applied Linguistics, 32*(3).

Day, B. (2004). Heightened awareness of communication pitfalls can benefit safety. *ICAO Journal.*

Farris, M. C., & Barshi, I. (2013). *Misunderstandings in ATC Communication: Language, Cognition, and Experimental Methodology*: Ashgate Publishing, Ltd.

Farris, C., Trofimovich, P., Segalowitz, N., & Gatbonton, E. (2008). Air traffic communication in a second language: Implications of cognitive factors for training and assessment. *TESOL Quarterly, 42*(3), 397-410.

ICAO, (2004). *Manual on the implementation of ICAO language proficiency requirements.* (Doc 9835 AN/453). Retrieved from http://caa.gateway.bg/upload/docs/9835_1_ed.pdf)

Kim, H., & Billington, R. (2016). Pronunciation and Comprehension in English as a Lingua Franca Communication: Effect of L1 Influence in International Aviation Communication. *Applied Linguistics*, amv075.

Kim, H., & Elder, C. (2009). Understanding aviation English as a lingua franca: Perceptions of Korean aviation personnel. *Australian Review of Applied Linguistics, 32*(3), 23-23.

Lefcheck, Jonathan S. (2015) piecewiseSEM: Piecewise structural equation modeling in R for ecology, evolution, and systematics. Methods in Ecology and Evolution. 7(5): 573-579. doi:10.1111/2041-210X.12512

Moder, C. L. (2012). Aviation English *The Handbook of English for Specific Purposes* (pp. 227-242): John Wiley & Sons, Ltd.

Moder, C. L., & Halleck, G. B. (2009). Planes, politics and oral proficiency. *Australian Review of Applied Linguistics, 32*(3), 25-1.

Peirce, J. W. (2007). *PsychoPy-Psychophysics software in Python.* Journal of Neuroscience Methods, *162* (1-2):8-13

Open speech repository (2016). Retrieved from http://www.voiptroubleshooter.com/open_speech/american.html

Godfrey, J. J. (1994). *Air Traffic Control Complete*, (LDC94S14A). CD-ROM. Linguistic Data Consortium, Philadelphia, USA.

Smith, S. L., Pichora-Fuller, M. K., & Alexander, G. (2016). Development of the Word Auditory Recognition and Recall Measure: A Working Memory Test for Use in Rehabilitative Audiology. *Ear and Hearing*, *37*(6), e360-e376.

FEDERAL AVIATION ADMINISTRATION
FLIGHT DECK HUMAN FACTORS RESEARCH PROGRAM

Sheryl L. Chappell, Regina G. Bolinger, and Ashley C. Awwad
Federal Aviation Administration
Washington, DC

The next generation air transportation system (NextGen) is a comprehensive suite of state-of-the-art technologies and procedures that improves national airspace system (NAS) capacity and efficiency, while maintaining world-class safety. In order to realize these improvements, the roles and the systems of pilots and controllers are changing. Advanced technologies and new procedures make the information and the tasks more complex. The Federal Aviation Administration's (FAA) Flight Deck Human Factors Research Program examines flightcrew interaction with current and future technology and pilot performance of flight procedures. Human factors scientists across industry, government, and academia produce scientific and technical data-driven recommendations to support the FAA's development of regulatory standards, policies, and other guidance materials for aircraft manufacturers and operators' procedures, training, and equipage. A sample of the program's scope, methodology, findings, future needs, and challenges is described below.

In addition to providing the United States with air traffic control (ATC) services, the Federal Aviation Administration (FAA) is responsible for regulating U.S.-registered aircraft and their operation. The FAA recognizes the importance of human factors in both controlling air traffic and ensuring aircraft are built, maintained, and operated safely. In 1993, the FAA published the Human Factors Policy to establish the "policy and responsibilities for incorporating and coordinating human factors considerations in FAA programs, facilities, and activities to enhance aviation safety, capability, efficiency, and productivity" (FAA Order 9550.8). The order defines human factors as a "multidisciplinary effort to generate and compile information about human capabilities and limitations and apply that information to equipment, systems, facilities, procedures, jobs, environments, training, staffing, and personnel management for safe, comfortable, and effective human performance."

According to the FAA's definition, human factors research involves the scientific acquisition of information about human capabilities and limitations related to the following:

- Hardware
- Procedures
- Environments
- Errors
- Personnel management
- Software
- Jobs
- Training
- Situation awareness
- Decision support tools
- Facilities
- Organizations
- Staffing
- Workload
- Other performance implications in which the human is a component

The Human Factors Division in the Office of the Next Generation Air Transportation System (NextGen) has the responsibility of managing human factors aviation research for the agency. This paper provides a description of some of the research related to aircraft, pilots, and maintainers. The division also manages air traffic control human factors research, but that is not highlighted here. The Human Factors Flight Deck Research Program has the challenge to provide improved knowledge of the human-system interface and a reduction in accidents and incidents through enhanced aerospace vehicle, air traffic, and technical operations that adapt to, compensate for, and augment the performance of the human.

Human factors research provides the foundation for FAA guidelines, handbooks, orders, advisory circulars, technical standards orders, and regulations, which ensure the safety and efficiency of aircraft operations. This research also provides the aviation industry with information for use in designing and operating aircraft as well as training pilots and maintenance personnel. Sponsors from across the FAA determine research needs and the urgency. These are driven by operational safety trends and the timing of new aircraft and ATC system capabilities.

The Human Factors Division engages top human factors scientists in industry, government, and academia to conduct both short-term, sharply-focused and longer-term, comprehensive research. It is useful to think of the broad range of flight deck research—which currently exceeds 50 projects—as falling into two general categories: (1) the ability of the pilots and maintainers to perform their jobs safely, and (2) the design, operation, and maintenance of aircraft systems.

Pilots and Maintainers

To address the ability of pilots and maintainers to perform their tasks safely, the FAA is conducting studies on fatigue mitigation, pilot training and performance assessment, and maintenance risk-based decision making.

Fatigue Mitigation

Airlines are required to manage and mitigate pilot fatigue during day-to-day flight operations by developing and implementing fatigue management policies and procedures within their operations; providing fatigue awareness and education to improve alertness and reduce the potential for errors; and continuously assessing the performance of these policies and practices, revising them as necessary. Air carriers can also develop a fatigue risk management system, allowing them to safely conduct specific flight operations not found in the prescriptive regulations. The carriers submit an alternative method of compliance supported by sleep and wake-time data and simple task performance data during a series of flight duty periods, including layover and post-trip recovery, to assure safety of flight. The air carriers monitor the effects of circadian rhythm changes, adequacy of layover rest, and returning flight schedules. Following the data collection exemption flights, the FAA evaluates the data and only authorizes those schedules exceeding regulation table limits that demonstrate that pilots are alert and well rested during those flight operations.

The Human Factors Division also manages research in fatigue management for maintenance personnel. The FAA provides training materials to individuals and flight organizations to educate them on the hazards of—and mitigations for—maintainer fatigue.

Air Carrier Pilot Training and Evaluation

FAA air carrier policy makers, inspectors, and airline training departments constantly evaluate the performance of pilots and ask researchers how to make training more targeted for areas in need of improvement. Some areas currently under study include the following:

- Manual and cognitive skill degradation with increasing automation
- Crew resource management best training and evaluation practices
- Flight path monitoring
- Response to unexpected events, and
- Training on the increased complexity of instrument procedures and flight deck system automation.

The FAA uses data from these research programs to provide updates to advisory circulars and inspectors' handbooks, and the airlines use the data to improve their training curricula.

Maintenance Risk-Based Decision Making

As the industry and the FAA mature their risk-based decision-making capability, they measure human performance and take into account the assessment of risk. Safety management systems at the FAA as well as flight operators collect data from aircraft and air traffic operations. These collected data provide a rich source of human performance data.

One area this science is increasingly mature in is the maintenance of aircraft. FAA-funded human factors research products include the following:

- Maintenance-line operations safety assessment tools, a method to collect data during normal operations from those doing the work
- Fatigue risk management techniques, and
- Design principles for technical documentation.

These products are currently undergoing field-testing, the results of which will be used to underpin implementation guidance for FAA inspectors and maintenance operators.

Design and Operation of Aircraft

Research, performed under the management of the Flight Deck Human Factors Research Program on the design and safe operation of aircraft systems, covers most types of aircraft and flight operations, including unmanned aircraft systems (UAS), single-engine private pilot flying, rotorcraft operations, and air carrier operations. Studies include the following:

- Ability of pilots to taxi in poor visibility using enhanced vision displays
- Design of unmanned aircraft system control stations to provide flight information to the pilot on the ground
- Pilot's management of the aircraft's flight path
- Use of digital communications between pilots and controllers, and
- Information needed for time-based navigation.

The tools researchers use to investigate human performance can range from a tablet computer to a full-mission simulator with ATC and other traffic. Experimental scenarios are key to providing the proper level of context and workload. Dependent measures include:

- Response time
- Response accuracy
- Number of control inputs
- Flight control movement
- Course, altitude, and speed deviations, and
- Number and length of communications.

Other measures include subjective workload, preference ratings, and the discriminability of symbols and flight parameters.

As flight deck systems and procedures evolve, the FAA must address fundamental human factors issues. The FAA recognizes that the increased complexity of both the systems and the procedures introduces brittleness. Pilots are confronted with elaborate failure modes and a vast array of possible alerts. Not only how, but also where, to convey this information is an area of current study. To reduce the impact of system and procedure complexity, the FAA is also sponsoring research on the efficacy of displaying aggregated flight parameters, such as the aircraft's energy state.

New Flight Deck Systems

A Federal regulation (14 CFR 25.1302) requires new systems for transport category aircraft to be "designed so that qualified flightcrew members trained in their use can safely perform all of the tasks associated with the systems' and equipment's intended functions." The regulation requires controls and information to be clear and unambiguous and to enable the flightcrew to manage errors. FAA human factors research evaluates flightcrew use of new technologies for both displays and controls on the flight deck.

The addition of electronic flight bags, which provide updated charts, manuals, weather, and safety information to the flight deck, brings standard user interface human factors issues to be studied, e.g., managing multiple applications and verifying the integrity of the data. The shift in communications from verbal to digital is another area of study, as the technology becomes more available and advantageous. New international standards are being formed with the knowledge resulting from the FAA's datalink communications research program.

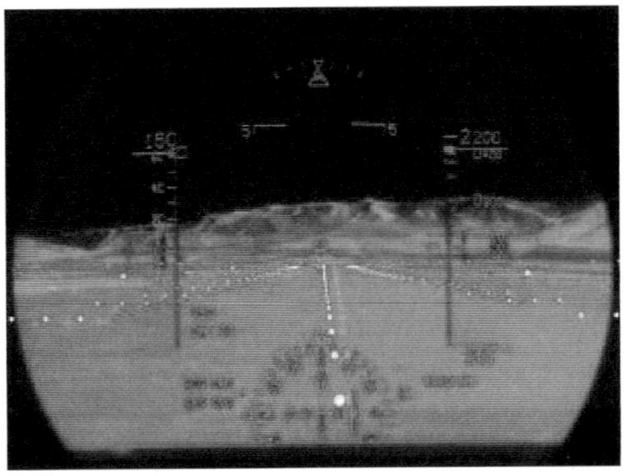

New vision systems that are available for use by the pilot, and which are the subject of FAA human factors research, include advanced vision systems permitting low-visibility taxi and takeoff operations. The FAA is also studying enhanced flight vision systems that allow landing at airports with reduced airport infrastructure. Other areas of research focus on determining minimum requirements with synthetic vision and combined vision systems.

Rotorcraft in near-to-ground operations have a significant number of incidents that involve striking obstructions or obstacles. Research is ongoing on display technologies to provide additional awareness of the presence of obstacles, especially head-mounted displays, which are a logical extension of natural-vision-referenced flight guidance. This research will provide guidance for both the certification and the operational approval for these new devices and will help to identify potential hazards associated with head-mounted systems. The results could be applied immediately to generate an advisory circular, with updates to relevant regulations to follow later.

Not only are visual flight deck displays being studied, other sensory modes are also explored. Presenting information aurally and tactually reduces the load on the visual information stream and the FAA is researching how to use these modes effectively on the noisy, vibrating flight deck. The FAA is also researching controls using other sensory modes. The speed and accuracy of touch, gaze, and voice interactions are being evaluated for control of flight deck systems.

New Flight Deck Procedures

The FAA's NextGen implementation is transforming the NAS in order to advance growth and increase safety while also reducing aviation's environmental impact. New ATC and flight deck procedures are enabling this transformation. These new procedures shift certain decision-making abilities from the controller to the pilot. Measuring the pilots' performance on the new procedures is an important part of the work managed by the Human Factors Division.

A NextGen capability, interval management time-based sequencing and spacing, will improve schedule predictability and system performance by maximizing throughput to use available system capacity and by reducing vectoring and holding, thereby improving fuel efficiency. A current study of this capability evaluates both controller and pilot performance. This research will identify the minimum information controllers and pilots need and will recommend procedures for successful implementation.

Unmanned Aircraft Systems

The Human Factors Division manages several research projects looking at the human performance of UAS pilots and of the air traffic controllers who are interacting with these new systems. Research is underway to determine what current flight deck standards apply to the design of the UAS controls station and how to substitute the information a pilot senses when in the aircraft. For example, the pilot is unable to physically see, and therefore avoid, other aircraft. Sensor systems providing data on the relative position of other aircraft as well as displays with alerting are necessary to provide this information in a meaningful way for pilots to remain well clear of other aircraft. Human factors research data on pilots' use of displays and alerting feed directly into industry standards.

When the datalink between the control station and the unmanned aircraft is unavailable, the aircraft will revert to a lost-link flight path. The air traffic controller responsible for separation of that aircraft with other traffic must know (1) that the loss of control has occurred, and (2) where the aircraft is going to go and when. Currently, most UAS operations take place in military airspace. This will not be true in the near future. Information and procedures are necessary for both the controller and the pilot to accommodate safe integration with the NAS. This requires a study of air traffic controller and UAS pilot performance using a high-fidelity simulation and realistic scenarios. Data collected from this research will result in modifications to controllers' displays and inform new procedures for controllers and pilots.

Summary

The Flight Deck Human Factors Research Program examines both flightcrew interaction with current and future technology and pilot performance of flight procedures. Research data are used to change or develop new avionics and air traffic procedures through regulations and guidance materials. FAA aircraft certification officials apply the findings of human factors research to the approval of aircraft systems. Other FAA personnel, such as air carrier principal operations inspectors and maintenance operations inspectors, incorporate research findings into their airline oversight. The airlines use these research findings to improve their pilot and maintainer procedures and training. Finally, aircraft manufacturers use data from the Flight Deck Human Factors Research Program to improve the functions as well as the displays and controls of their flight deck equipment. As FAA's NextGen technologies continue to evolve and enter into service in the NAS, flight deck human factors research will continue to play a vital role in increasing safety and improving the movement of aircraft through the National Airspace.

ALERTS ON THE NEXTGEN FLIGHT DECK

Angelia Sebok, Christopher D. Wickens, and Brett Walters
Alion Science and Technology
Boulder, Colorado

Karl Fennell
United Airlines
Denver, Colorado

The future generation of cockpit may substantially change the nature of displays, automation, and their implications for alerting systems. Flight deck automation and information systems imposed by NextGen will create system states that may need to be alerted. New concepts of providing continuous flight information can benefit pilot awareness. Ecological interface displays keep the operator aware of the system status and constraints, informing the pilot of emerging concerns before an alert is triggered. Continuous auditory and tactile displays support pilot awareness without requiring focused operator attention. Both approaches improve awareness of system state, but suffer poor operator acceptance. Automation can support pilots through carefully considered degrees of automation, through transparent automation design, and through adaptive automation that identifies when pilots fail to respond to alerts, and increases the salience of alerts or assumes control to implement the necessary actions. This paper summarizes a report that reviewed empirical research regarding these approaches.

Since the early days of aviation, there has been a need to improve alerts on the flight deck. Initially, the need was simply to provide alerts that would draw the pilots' attention to important status information (e.g., fuel level low). As experiences were gained and technologies improved, the alerting systems became increasingly sophisticated, warning pilots about proximity to ground or the nearby presence of other aircraft, or predicting the collision potential of surrounding traffic based on their trajectories. Aviation personnel have done a commendable job tracking incidents and accidents, and systematically evaluating them to identify lessons learned that enable continual improvements in alerting systems. Despite these efforts, challenges remain, and are likely to become even trickier in NextGen operations. The implementation and integration of new technologies will result in even more information on the flight deck (FAA, 2016), and additional automated systems may further increase complexity.

Some challenges with current-day systems include keeping the pilot aware of the status of the aircraft and automated systems, and appropriately applying alert suppression. For example, flight mode advisories (FMAs) inform the pilot of the current autoflight status and mode that result in changes in throttle or pitch control. The pilot uses this information to react to changes and correctly employ the automation. The FMAs are located at the top of the primary flight display and require directed visual focus and attention. During visual approaches, takeoffs, and other high workload situations, the pilot flying is primarily looking outside. Important changes on the FMA may easily be missed, possibly leading to lack of mode awareness. Improved salience or repositioning of the indicators appear to be needed.

Alerting systems on current flight decks attempt to support pilot performance by inhibiting nuisance information during critical phases of flight. For example, on many modern airplanes, caution alerts are inhibited above 80 knots during takeoff, with the intent of minimizing unnecessary information and helping the pilot make a go / reject decision. However, if a malfunction occurs, a message may still appear on the engine indicating and crew alerting system (EICAS). This message is presented without a caution light or sound, and the pilot is left to decide if the message warrants a rejected takeoff. Other information may also be presented, such as "high engine temperature" or "low oil pressure," displayed in red font on the EICAS. The result is that takeoffs are sometimes rejected when an actual takeoff would have been less risky. While the intent of inhibiting nuisance information makes sense, the current implementation seems to be clumsy and in need of improvement.

Dynamic situations, uncertain contexts, and fallible operators all contribute to challenges in the design of robust, appropriate and informative alerting systems. But there are also numerous opportunities for improved displays and intelligent automation to support performance. This paper, integrating many of the findings from a longer report (Wickens, Sebok et al., 2016) discusses potential future directions in flight deck alerting systems regarding ecological, predictive and multi-modal displays, and of automation; its more aggressive forms, its transparency and its adaptivity.

Alerts can fail to perform their intended function for several reasons. They may not be noticed, because of deficiencies in pre-failure monitoring, fatigue, cognitive tunneling, or insufficient alert salience or being located too far out of the normal field of view (Wickens, Sebok, McDermott & Walters, 2016). Alerts may be noticed, but not interpreted correctly. There may be too many alerts, confusing or misleading the pilot (Martensson & Singer, 1998).

Displays to Support Intuitive Monitoring

Displays can be designed to provide the operator with better awareness of the system and improved ability to predict undesirable future states than is provided by current-day displays and alerts. Such improvements should mitigate surprise and possible startle caused by the alert. Several techniques have been found to assist both the detection and subsequent diagnosis, above and beyond the alerts themselves. These displays provide a contextual background to support operator anticipation of an alert (improving detection) and diagnosis (understanding and prediction). Such displays depend upon **pre-attentive reference** (Woods, 1995), in which a perceptual cue (a sound or visual indication) provides information to the operator regarding the current state of the system. This can occur naturally, as part of the system operation, or it can be artificially added. Examples include the hum of an engine that changes in frequency as the throttle is applied or decreased. Changes to the cue, such as an engine that begins emitting a "knocking" noise and vibrating, can rapidly draw the operator's attention to a potential concern, without invoking the startle characteristics of an auditory alert (Rivera et al., 2014). These features support perception and understanding in a way that does not require effortful processing to realize that something is wrong or even perhaps to identify what is wrong (Woods, 1995). This approach to supporting pilot detection of non-normal events and hence supporting alert management has been investigated from several perspectives, including ecological interface design (EID), predictive displays, sonification, and tactification.

Visual Displays: Ecological, Configural and Predictive Displays

The goal of EID displays is to integrate data into intuitive graphics that present important information to the operator or pilot. This requires identifying the most important factors and parameters for performing the tasks, and putting that information together in a meaningful, readily understood graphical representation (Bennett & Flach, 2013; Muller, Manzey et al., 2015). These displays show not just the status of individual sensors, but current and predicted states. One example in aviation is related to energy management (Muller et al., 2015). Pilots think of flying tasks in terms of energy management, yet current displays do not directly support that concept. The following figure shows one example of an aviation EID energy management display.

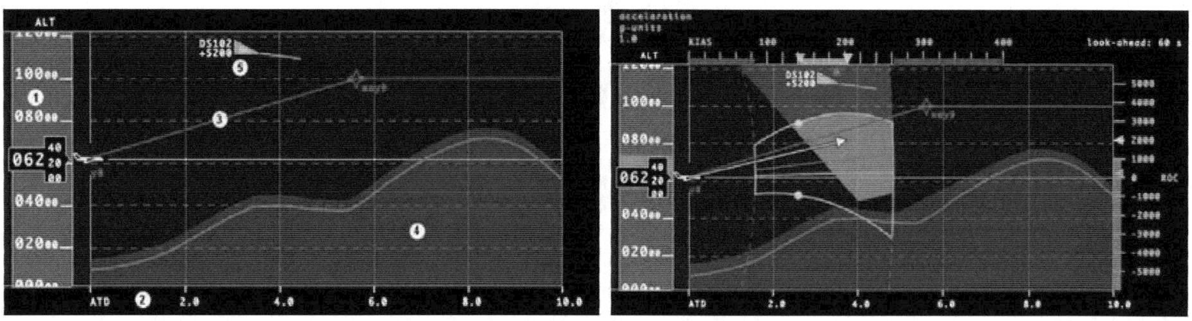

Figure 1: Vertical Situation Displays (from Borst et al., 2011). The left figure is a current-day display, and the right figure includes energy management information. Both show the ownship, a yellow aircraft at the left of the displays. The cyan outlined area in the right figure shows the potential future locations of the aircraft, given the minimum and maximum speed and climbing characteristics. The orange area above the brown "terrain" line shows the space in which the pilot can safely fly to avoid colliding with the other aircraft (green triangle) and the terrain.

The EID display allows the operator to monitor those key parameters with relatively low workload, and determine their proximity to danger boundaries that would trigger discrete alerts. Thus the EID provides operators with information regarding relationships in the system data and system constraints that are not normally presented on more conventional displays, or are presented in a less integrated fashion. In aviation, such a constraint might be the combinations of angle of attack, power and pitch that cause a stall (Wickens & Andre, 1990), or the combination of potential energy, sink rate, altitude and available thrust that creates unstable flight (Muller et al., 2016; Lambergts et al., 2008). By explicitly and graphically depicting the proximity of the current state of the system to these constraint boundaries, an EID can prepare the operator to detect a failure when the boundaries are crossed, to prevent that boundary crossing through proactive control, and to better diagnose the reasons why they are violated so that corrective actions can be applied appropriately. Thus by providing additional information, the EID should support operator monitoring, maintaining situation awareness (SA), detecting, diagnosis, decision making and fault management. Another benefit of EID displays is that, by integrating important system status information into a single display, the pilot can maintain awareness without needing to more widely distribute visual attention. However, the pilot will still need to seek information that is not included on the EID.

Two important components within many EID displays, used to help present the constraints and constraint boundaries, are configural object displays and predictive displays. A configural display presents multiple parameters graphically, so that their combined values form a shape or object. This object changes shape depending on the relative values of the parameters. Thus, the object can easily depict a departure from a normal state and its shape indicates the nature of the abnormal state. Successful examples include an octagon display indicating non normal conditions with a distortion of symmetry by the change in the location of one of the eight points (Beringer & Chrisman,1991); a rectangle display whose departure from the perfect symmetry of the square depicted the approach to stall, with deviations of appropriate airspeed and angle of attack (Wickens and Andre,1990), or the adjacent depiction of indicators of angle of attack and total energy angle, to signal the preservation, or departure from, minimum energy capabilities in vertical maneuvering (Muller et al., 2016).

Wickens, Sebok, Walters, et al. (2016) examined studies that compared performance with ecological displays against performance with conventional displays on the four aspects (monitoring, detection, diagnosis, fault management) of human processing of the non-normal events that trigger alerts. In many of these studies, the EID display condition presented more information than the traditional displays with which they are compared. Four aviation studies (Comans et al., 2014; Ellerbroek et al., 2013; Borst et al., 2011; and Rijneveld et al., 2010) examined non-normal events of the sort that might be alerted. All of these concerned traffic conflicts. Two of these studies (Ellerbroek et al., 2013; Comans et al., 2014) suggested a significant advantage of the EID concept, while the other two (Borst et al., 2011; Rijneveld et al., 2010) showed neither an advantage nor disadvantage. Of 11 aviation studies that were examined, 8 revealed superior performance in the EID conditions versus conventional conditions in some aspect of performance relevant to alert processing. The remaining 3 studies found no difference between conditions. These findings suggest the potential for EID displays to support more effective performance.

Predictive displays have long been known to increase control performance by inferring the future dynamic state, and hence allowing proactive control (Jensen, 1981). The advantages of predictive displays in flight path control are well documented (Wickens, 2003). These predictive displays include the "noodle" on the navigational display, or the predictive aircraft and 3D tunnel on synthetic vision system displays (Prinzel & Wickens, 2009). If aircraft state is trending toward a hazardous boundary (e.g., loss of separation, loss of sufficient potential energy or excessive temperature), the predictive algorithm can trigger the alert before the boundary is crossed. Yet often, as in the case of the traffic collision avoidance system (TCAS) alert, the only information displayed to the pilot is the discrete onset. There appears to be an advantage to also presenting the continuous predictive trend toward the boundary, so that a maneuver or control adjustment can be implemented prior to the time the alert would have occurred (to forestall the alert), or can be implemented more effectively after the alert, avoiding surprise. The benefits of such a continuous predictor have been validated for collision avoidance in cockpit displays (Alexander et al., 2005; Wickens, Gempler & Morphew, 2000) or engine parameters (Trujillo et al., 2008). Thus, much like ecological displays, a predictive display presents a broader context, which supports the pilot in predicting the future state of the aircraft. This, in turn, supports more expedient responses to the discrete alerts if they do occur, or more proactive control that will prevent the occurrence of an alert altogether.

Multi-modal Displays

The concept of multi-modal displays and alerts for the flight deck has received some recent attention (Lu et al., 2013). One approach has been to deliver alert indications in non-visual modalities, e.g., in auditory, tactile, or a redundant combination, typically redundant with a visual indication. Another approach is through the display of continuously changing parameters, such as the bearing of a potential traffic conflict, or the engine power through the tactile or auditory modality. These continuous displays are referred to as tactification and sonification respectively. Such an approach has the clear advantage of capitalizing on different perceptual resources than the visual channel which is predominately involved in flying, and thus allowing some degree of parallel processing (Wickens, 2008). A continuously changing auditory or tactile signal might also provide the same sort of pre-attentive reference and predictive information that was seen above to offer an advantage to proactive response to out of tolerance parameters. However there is one key limitation. While changing pitch or tactile intensity are effective for displaying the rate of change in a parameter (routine control), they are not as effective as vision for depicting the **absolute value** of the parameter, which is how alert boundaries or thresholds are characterized. In general, studies that investigate the use of sonification indicate that it is most effective when used in combination with a visual indication (Wickens, Sebok, Walters et al., 2016). Tactification approaches have also shown promise in terms of supporting situation awareness and early response to developing problems, particularly when used in combination with visual information presentation.

Both sonification and tactification however currently suffer poor operator acceptance (as reviewed in Wickens, Sebok, Walters et al., 2016). This can be due, in part, to a novelty effect, but it is also related to the inappropriateness for the particular environment. For example, sonification has been evaluated in simulated medical environments, where there are typically many patient monitoring systems that present auditory alerts, as well as verbal communications among the surgical team members. This noisy environment is a problem for effective sonification. Similarly, the flight deck currently has discrete auditory and voice alerts, and interpersonal communication. Sonification, in today's environment, simply adds another auditory signal to an already noisy operational context. It appears more likely that sonification and tactification would be used in remotely piloted aircraft, where there is a good deal more control over the pilot's environmental conditions.

Implications of Automation for Alerting Systems

Future forms of automation in NextGen and beyond have three direct implications for alerting systems. These are discussed in much greater detail in Wickens, Sebok, Walters et al., (2016) and summarized below.

Degree of Automation

The degree of automation (DOA) characterizes how aggressively and authoritatively automation assists the pilot's task (Parasuraman et al., 2000; Ferris et al., 2010; Sebok & Wickens, 2016). With respect to alerting systems, a low degree of automation may simply **inform** the pilot of the likely state, e.g., a low fuel alert, or the cause of the non-normal condition, such as the TCAS traffic alert. The alerting automation may more aggressively *recommend an action* (the TCAS resolution advisory), or even **implement** the action (the "pull up" function of the automatic ground collision avoidance system (Auto-GCAS) in the military F-16), representing the highest DOA. Empirical research is needed to establish the appropriateness of high degrees of automation because existing research has indicated that automated action advice or implementation, when based on uncertain inferences, may be quite problematic on the infrequent occasions when the inferences upon which the recommendations are made are incorrect (Sarter & Schroeder, 2001; Onnasch et al., 2014, Sebok & Wickens, 2016). Empirical research indicates that automation wrong is more problematic than automation gone failures, particularly when the automation provides a wrong (but plausible) diagnosis or recommends an incorrect course of action (Wickens, Clegg et al., 2015; Sauer, Chavaillaz & Wastell, 2015). As Onnasch et al., 2014 found, automation **can** potentially support operator performance during a failure, **if** the displays provide information needed to support SA.

Transparency of Automation

One technique for mitigating the costs of automation errors at higher DOA is to provide **transparency** (Sebok & Wickens, 2016) or **observability** (Ferris et al., 2010), sometimes in the form of a display to indicate what automation is doing (and why). This directly supports pilot SA. On the flight deck, such transparency can support mode awareness (Ferris et al., 2010), and the transparency offered by the traffic display of TCAS renders it easier to follow the advice of the resolution advisory. In air traffic control, Trapsilawati et al. (2017) have found that the transparency offered by a vertical situation display can offset imperfections of a conflict resolution aid. A number of studies investigating different approaches to enhancing transparency were found to provide better operator performance, both in routine and off-nominal conditions (Wickens, Sebok, Walters et al., 2016). The only potential drawback identified in these studies was that sometimes the more transparent automation drew the operator's attention and placed additional workload on the operator. Generally, though, these costs were mitigated by better performance in the case of automation failures or situations that were outside the realm of typical operations.

Adaptive Automation

It has been argued that automation should not necessarily always be present, but only be invoked in circumstances when it is needed because of high pilot workload (Dorneich, 2016; Kaber, 2013). This is considered **adaptive automation**. While adaptive automation has shown some promise in aviation systems, it has spawned another class of mode change alerting systems in the cockpit, namely alerting the pilot, as to when automation has taken control of the relevant aviation system (given that workload is assumed to be high), or when automation has returned control to the pilot. Failure of the pilot to be aware of the second of these mode changes can be particularly problematic. Another concern with adaptive automation is the logic and criteria used to determine when the pilot is overloaded and needs assistance. Techniques such as eyetracking, physiological parameters (heart rate, respiration), or time required to respond to requests for information are all used to predict when the operator needs assistance. If the automation incorrectly interprets that the pilot needs help, and offers assistance when it is not needed, that can be annoying to the pilot. Perhaps even worse is the condition where the pilot does need assistance, yet the automation does not detect or offer it. These problems can be addressed through the use of adaptable systems, or hybrid adaptive / adaptable systems that give the pilot control over when the automation is invoked.

Summary of Alerting Systems on the NextGen Flight Deck

A variety of display techniques can be, and have been (at least in experimental settings) used to support operator monitoring of a system, maintaining system awareness and responding to non-normal events. Some empirical evidence indicates that ecological displays and configural displays support rapid, accurate operator detection of non-normal conditions and accurate responses to these conditions. Predictive displays support operators in anticipating the future state, and avoiding alerts. In other modalities, empirical results are less conclusive than for visual displays, yet sonification and tactification can potentially provide techniques for supporting continual awareness of system state. Automation is expected to be more pervasive in the future flight deck, and can contribute to the detection of and response to alerts. Transparent automation systems can help pilots in assessing the appropriateness of diagnoses or recommended courses of action. Adaptive automation can assist pilots by increasing the salience of not-noticed yet critical alerts, or by deferring a low-priority alert that occurs during a high-workload situation. In summary, advanced, integrated visual displays or predictive displays; auditory and tactile alerts that provide continual system state information; and intelligent, transparent, and adaptive automation are potential techniques for supporting pilot performance on the NextGen flight deck.

Acknowledgments

This work was sponsored under contract 3003293921 with the University of Michigan under the FAA Cooperative Agreement 13-G-019. The opinions presented in this paper are those of the authors, and do not necessarily reflect those of UM or the FAA. The authors thank Nadine Sarter (UM PI), Ashley Awwad (FAA TPOC), Sherry Chappell, Michelle Yeh, Clark Davenport, Cathy Swider, and others at the FAA for their contributions to this work.

References

Alexander, A. L., Wickens, C. D. & Merwin, D. H. (2005). Perspective and coplanar cockpit displays of traffic information. *The International Journal of Aviation Psychology*, *15*(1), 1-21.

Bennett, K. B. & Flach, J. M. (2013). Configural and pictorial displays. Chapter 35 in J.D. Lee and A. Kirlik (Eds.) *The Oxford Handbook of Cognitive Engineering*. Oxford, England: Oxford University Press. 517-533.

Beringer, D. B., & Chrisman, S. E. (1991). Peripheral polar-graphic displays for signal/failure detection. *The International Journal of Aviation Psychology, 1*(2), 133-148.

Borst, C., Rijneveld, P., Mulder, M., & Van Paassen, M. M. (2011). Towards integrating traffic and terrain constraints into a vertical situation display. *ISAP Conf Proc,* 393-398. Dayton, OH: Wright State University.

Comans, J., Borst, C., van Paassen, M. M., & Mulder, M. (2014). Risk perception in ecological information systems. *ISAP Conference Proceedings,* 436-441. Dayton, OH: Wright State University.

Dorneich, M. (2016). Human performance risks and benefits of adaptive systems on the flight deck. *International Journal of Aviation Psychology, 26*(1-2), 15-35.
Ellerbroek, J., Brantegem, K. C., van Paassen, M. M., & Mulder, M. (2013). Design of a coplanar airborne separation display. *IEEE Transactions on Human-Machine Systems, 43*(3), 277-289.
Federal Aviation Administration (2016). *NextGen Implementation Plan 2016*. Washington, DC: Office of NextGen.
Ferris, T., Sarter, N. B. & Wickens, C. D. (2010). Cockpit automation: Still struggling to catch up. In E. Salas, & D. Maurino (Eds.), *Human factors in aviation* (2nd ed.). Amsterdam, The Netherlands: Elsevier.
Jensen, R.S. (1981) Prediction and quickening in perspective flight displays for curved landing approaches. *Human Factors. 23*(3), 355-363.
Kaber, D. B. (2013). Adaptive automation. In J. D. Lee & A. Kirlik (Eds.), *The Oxford handbook of cognitive engineering* (pp. 594-609). New York, NY: Oxford University Press.
Lambergts, T., Rademaker, R. & Theunissen, E. (2008). A new ecological primary flight display concept. In *2008 IEEE/AIAA 27th Digital Avionics Systems Conference* (pp. 4.A.1-1 - 4.A.1-20). IEEE.
Lu, S.A., Wickens, C.D., Prinet, J.C., et al. (2013). Supporting Interruption Management and Multimodal Interface Design. *Human Factors. 55*(4), 697-724.
Martensson, L., & Singer, G. (1998). *Warning systems in commercial aircraft – an analysis of existing systems.* Stockholm, Sweden: The Royal Institute of Technology.
Muller, S., Manzey, D., Schreiter, K. & Luckner, R. (2015). Implementing energy status in head-down cockpit displays: Impact of augmented energy information on pilot's performance. *HFES Proc., 59*(1), 926-930.
Muller, S., Schreiter, K., Manzey, D. & Luckner, R. (2016). nxControl instead of pitch-and-power. *CEAS Aeronautical Journal, 7*(1), 107-119.
Onnasch, L., Wickens, C. D., Li, H., & Manzey, D. (2014). Human Performance Consequences of Stages and Levels of Automation An Integrated Meta-Analysis. *Human Factors, 56*, 476-488.
Parasuraman, R., Sheridan, T. B., & Wickens, C. D. (2000). A model for types & levels of human interaction with automation. *IEEE Transactions on systems, man, & cybernetics-Part A: Systems and Humans, 30*, 286-297.
Prinzel, L. & Wickens, C.D (2009). Preface Special Issue. *Intl Journal of Aviation Psychology, 19*(2).
Rijneveld, P., Borst, C., Mulder, M. & van Paassen, M.M. (2010). Towards integrating traffic and terrain constraints into a vertical situation display. *AIAA Guidance, Navigation, and Control Conf*, AIAA 2010-8169, 1-20.
Rivera, J., Talone, A. B., Boesser, C. T., Jentsch, F. & Yeh, M. (2014). Startle and surprise on the flight deck: Similarities, differences, and prevalence. *HFES Proc., 58*(1), 1047-1051.
Sarter, N. B. & Schroeder, B. (2001). Supporting decision making and action selection under time pressure and uncertainty: The case of in-flight icing. *Human Factors, 43*(4), 573-583.
Sauer, J., Chavaillaz, A., & Wastell, D. (2015). Experience of automation failures in training: Effects on trust, automation bias, complacency and performance. *Ergonomics, 59*(6), 767-780.
Sebok, A. & Wickens, C.D. (2016) Implementing Lumberjacks and Black Swans into Model-Based Tools to Support Human-Automation Interaction. *Human Factors*. DOI: 10.1177/0018720816665201
Trapsilawati, F., Wickens, C., Chen, C.H., Qu, X. (2017). Transparency and Conflict Resolution Automation Reliability in Air Traffic Control. *ISAP Conference Proceedings*. Dayton, OH: Wright State University.
Trujillo, A., Bruneau, D., & Press, H. (2008). Predictive information: Status or alert information? In *2008 IEEE/AIAA 27th Digital Avionics Systems Conference* (pp. 4A4-1-4A4-9). St. Paul, MN: IEEE.
Wickens, C. D. & Andre, A. D. (1990). Proximity compatibility and information display: Effects of color, space, and objectness on information integration. *Human Factors, 32*(1), 61-77.
Wickens, C. D. (2003). Aviation displays. In P. Tsang, & M. Vidulich (Eds.), *Principles and practice of aviation psychology* (pp. 147-199). Mahwah, NJ: Lawrence Erlbaum Publishers.
Wickens, C. D. (2008). Multiple resources and mental workload. *Human Factors, 50*(3), 449-455.
Wickens, C. D., Clegg, B. A., Vieane, A. Z., & Sebok, A. L. (2015). Complacency and automation bias in the use of imperfect automation. *Human Factors, 57*(5), 728-739.
Wickens, C. D., Gempler, K. & Morphew, M. E. (2000). Workload and reliability of predictor displays in aircraft traffic avoidance. *Transportation Human Factors Journal, 2*(2), 99-126.
Wickens, C., Sebok, A., McDermott, P. & Walters, B. (2016) Field of View Issues on the Flight Deck. *HFES Proceedings, (60)*1, 56-60.
Wickens, C., Sebok, A., Walters, B., et al., 2016. *Alert Design Considerations.* Final Report submitted to the University of Michigan under PSA 3003293921 and to the FAA under Cooperative Agreement 13-G-019.
Woods, D. D. (1995). The alarm problem and directed attention in dynamic fault management. *Ergonomics, 38*(11), 2371-2393.

NEXTGEN HUMAN FACTORS ACCOMPLISHMENTS AND CHALLENGES IN THE FUTURE

Edmundo A. Sierra, Jr.
Federal Aviation Administration
Washington, D.C.

The Federal Aviation Administration (FAA), under the Next Generation Air Transportation System (NextGen) program, made significant progress in terms of infrastructure modernization of the National Airspace System (NAS). Development of infrastructure incorporated a broad scope of human performance considerations which this paper discusses at a high level. Since the FAA completed much of the infrastructure modernization, focus shifted to NextGen Transformation so that benefits can be realized. NextGen Transformation involves insuring that FAA systems are seamlessly integrated; that integrated pilot and controller procedures are in place; new roles and responsibilities are defined; and that stakeholders are informed, trained, and adapted to the NextGen technologies and procedures and are comfortable in their use. This paper concludes with an overview of challenges and opportunities the FAA is facing as it shifts its focus to NextGen Transformation - specifically as it pertains to human factors, safety, training, and the workforce of the future.

After a recent painful introduction I was required to provide as a student at the start of a class; the instructor asked, "Shouldn't NextGen be done?" I sighed, rubbed my forehead, and as I formulated my answer realized it was a fair question for a program the Federal Aviation Administration (FAA) called "one of the most ambitious infrastructure and modernization projects in U.S. history" (FAA, 2007c). I told the instructor that NextGen delivered a lot of the infrastructure for modernization and is focused on operational integration. That is, NextGen delivered the physical underlying parts of a system that was adapted to meet modern needs. I elaborated on that answer in this paper for the human factors practitioner and included emerging and far-term human system challenges NextGen is facing.

Next Generation Air Transportation System (NextGen)

To determine what NextGen has accomplished to date, I identified when it began and its objectives. In the United States, an authorization establishes a federal program (or agency). I determined that NextGen began within the law that reauthorized the legal operation of the Federal Aviation Administration in 2003. NextGen's goals were included in the authorization language.

NextGen Initiation

The 108th Congress endorsed the concept of NextGen in the Vision 100 – Century of Aviation Reauthorization Act, signed into law in December 2003. Congress found that the United States revolutionized the way people traveled by developing new technologies and aircraft. In addition, past investments in research and technology benefitted the economy and security of the United States. Congress saw continued leadership was needed, growth in aviation was necessary, and revitalization and coordination must begin. NextGen would encounter many challenges and there would be constraints by concerns related to safety, security, and environment.

NextGen Goals

The Vision 100 – Century of Aviation Reauthorization Act set goals for NextGen. I listed the seven goals from 117 STAT. 2584 in Table 1. I would overwhelm the reader if I organized human factors accomplishments by the goals under Subsection (c) of the reauthorization. I would also find it difficult to describe human factors accomplishments at a consistent level. Therefore, I organized accomplishments by major NextGen Investments to date for a subset of domains: Communications, Navigation, Surveillance, and Automation.

Even at a high level, my analysis identified many human factors accomplishments to date. For example; human factors made numerous contributions throughout phases of systems' research, concept development, engineering, development, and implementation. Human factors also contributed to overarching policies, standards, and guidance. Therefore, I identified the human components of systems and used their implementation as synonymous with human factors accomplishments. In the FAA, human factors must be included in planning, analysis, development, implementation, and in-service activities for systems (FAA, 2017a). The FAA verifies compliance with its policy throughout the lifecycle. For example, an in-service decision (ISD) authorizes deployment of a solution into the operational environment. A tool named the *in-service review* (ISR) *checklist* is used to identify and resolve readiness issues before the ISD. The checklist includes a human integration section with items verified by a human factors subject matter expert. The items include compliance with human factors policy, standards, and guidance; consideration of human performance; operational suitability; knowledge, skills, abilities; human error; functional design; and suitability of documentation. Systems the FAA implements meet FAA human factors goals.

Table 1.

The Next Generation Air Transportation System shall—.

Paragraph	
(1)	Improve the level of safety, security, efficiency, quality, and affordability of the National Airspace System and aviation services
(2)	Take advantage of data from emerging ground-based and space-based communications, navigation, and surveillance technologies
(3)	Integrate data streams from multiple agencies and sources to enable situational awareness and seamless global operations for all appropriate users of the system, including users responsible for civil aviation, homeland security, and national security.
(4)	Leverage investments in civil aviation, homeland security, and national security and build upon current air traffic management and infrastructure initiatives to meet system performance requirements for all system users
(5)	Be scalable to accommodate and encourage substantial growth in domestic and international transportation and anticipate and accommodate continuing technology upgrades and advances
(6)	Accommodate a wide range of aircraft operations, including airlines, air taxis, helicopters, general aviation, and unmanned aerial vehicles
(7)	Take into consideration, to the greatest extent practicable, design of airport approach and departure flight paths to reduce exposure of noise and emissions pollution on affected residents

Note. 117 STAT. 2584.

NextGen Built Foundational Infrastructure (2003-2016)

NextGen implemented the foundational infrastructure for Communications, Navigation, Surveillance, and Automation domains. Communications included digital communications between controllers and pilots and between FAA systems and facilities. Navigation included precise Global Positioning System (GPS) routes and procedures in all airspace domains. Surveillance provided high update rate surveillance used by controllers and pilots. Automation was found in every Air Traffic Control (ATC) domain and included decision support. I described the domains in more detail, included examples, and human factors accomplishments in this section.

Communications

Communications was comprised of elements that performed transmission or recording functions for voice and data communications within and external to the National Airspace System (NAS; FAA, 2017d). Communications supported connectivity between air-to-ground, NAS sub-systems and facilities, NAS and external systems and users. Air/Ground systems provided a wireless communication conduit between aircraft and other NAS users and systems.

Communication Exemplars. Three major programs were examples of Communications infrastructure: Data Communications, NAS Voice System, and System Wide Information Network. Data Communications (Data Comm) enabled controllers and pilots to communicate with digitally delivered messages in addition to voice. The NAS Voice System (NVS) allowed controllers and pilots to speak with each other without being limited by geography (versus radio). System Wide Information Management (SWIM) offered a single point of access to aviation data for controllers and operators including airlines, cargo carriers, business jet operators, and airports.

Human Factors Accomplishments. Data Communications were delivered at airport towers through Departure Clearance Tower Services (FAA, 2016b). These services allowed an Airport Traffic Control Tower (ATCT) controller to enter flight departure clearance instructions into a computer and push a button to electronically send the information to a flight deck. Flight crews viewed the instructions, pressed a button to confirm receipt, and pressed another button to enter the instructions into the flight management system. Preliminary qualitative benefits of Data Comm seen during trials included reduced communications time resulting in faster taxi outs, reduced delays, and reduced pilot and controller workload.

NAS Voice System will provide voice connectivity by linking incoming and outgoing communication lines to controller workstations (FAA, 2016b). The target users of NVS were air traffic controllers and pilots, including pilots of Unmanned Aircraft Systems (UAS). NVS successfully completed critical design review. The NVS program conducted Early User Involvement Events for Air Traffic and Technical Operations users. The program also held training manual conferences and technical manual conferences for the development of training and documentation. The program will complete NVS operational testing and evaluation, and key site testing in 2019.

System Wide Information Management implemented four key services: Interface Management Service, Messaging Management Service, Security Management Service, and Enterprise Service Management Service (FAA, 2017d). The target users for SWIM were air traffic controllers and operators including airlines, cargo carriers, business jet operators and airports (FAA, 2016a). The services supported three key domain areas and Community of Interest capabilities in the areas of Aeronautical Information Management, Weather, and Flight & Flow Management. Flight and Flow Management saw benefits when traffic managers used surface data to balance traffic demands with capacity demands across the NAS. Surface data also made it easy for Terminal Radar Approach Control (TRACON) controllers to identify departure congestion and anticipate changes that would impact their control of traffic.

Navigation

Navigation consisted of elements that provided visual and instrument based guidance to pilots during all phases of flight operations including airport surface navigation. Surveillance, which I included after this section, shared surface movement radar data with the Navigation element in order to aid pilots in navigating safely through airport surface, departure, and arrival operations (FAA, 2017d).

Navigation Exemplars. Three major programs were examples of Navigation: Performance Based Navigation (PBN), Metroplex PBN Procedures, and Wide Area Augmentation Systems (WAAS). PBN used satellites and onboard equipment for navigation procedures that are more precise and accurate and enabled aircraft to fly more directly from departure to arrival (FAA, 2017b). Metroplex PBN Procedures delivered large scale integrated PBN procedures in complex interdependent airspace. WAAS equipment and software augmented the Global Positioning System Standard Positioning Service.

Human Factors Accomplishments. Performance Based Navigation certified, published, and implemented procedures for Area Navigation (RNAV) Standard Instrument Departures (SID) and Standard Terminal Arrivals (STAR) at airports. Target users for PBN were controllers and pilots. RNAV SIDs and STARs increased predictability of repeatable flight paths and thereby enhaced safety and controller productivity. PBN also enabled En Route Automation to enhance the controller's ability to assign clearances to a pilot to operate on performance restricted routes.

Airspace congestion and limiting factors, such as environmental noise contraints, combined to reduce efficiency in Metroplexes. Study teams that included the FAA and aviation community analyzed the operational challenges of three Metroplexes using a consistent, repeatable approach. The FAA implemented their solutions, including PBN procedures, at Washington DC, North Texas, and Northern California Metroplexes (FAA, 2016b).

NextGen produced over 4,000 RNAV (GPS) Approaches for airports. The procedures were for WAAS localizer performance (LP) and localizer performance with vertical guidance (LPV). Pilots were able to fly approaches comparable to those of an Instrument Landing System (ILS) without the need for ILS's ground infrastructure. The capability also improved access for general aviation pilots who were able to file and fly to a greater number of airports during low visibility day or night. (FAA, 2017b)

Surveillance

Surveillance was comprised of elements that detected and reported the presence and location of aircraft and other targets in the air and on the airport surface movement areas. The data collected and created by Surveillance supported pilots, air traffic controllers, and other users via integration and data sharing within Automation (FAA, 2017d).

Surveillance Exemplar. Automatic Dependent Surveillance–Broadcast (ADS-B) is the successor to radar. ADS-B features the ability for an aircraft to broadcast its current location and other important information about the aircraft. The broadcast is received by ADS-B ground stations and by other aircraft. ADS-B uses GPS satellites to determine an aircraft's location, ground speed, and other data (FAA, 2017b). The surveillance coverage that ADS-B provides is nation-wide and NextGen also extended it to remote areas in Alaska and the Gulf of Mexico. ADS-B technology has also enabled the broadcast of non-coperative (unequipped) air traffic and weather information to be received by aircraft in flight without a service fee.

Human Factors Accomplishments. The FAA completed nationwide deployment of ADS B ground stations (FAA, 2017b). ADS-B was integrated into automation platforms at all en route air traffic control facilities and more than one-third of all terminal facilities. ADS-B traffic and weather broadcasts were also available nationwide. The target users for ADS-B were air traffic controllers; aircraft owners and pilots flying above 10,000 feet mean sea level, within Class C airspace, the airspace surrounding most major airports, or low altitude airspace along the Gulf of Mexico coastline; and airport surface vehicle operators (FAA, 2016a). Controllers used ADS-B to track aircraft during radar outages in controlled airspace. Airport Surface Detection Equipment–Model X, a ground-surveillance system, alerted controllers to potential runway and taxiway conflicts using ADS-B and other data sources. One ADS-B In capability gave pilots an audio alert to warn of other aircraft that might be a collision risk. Another ADS-B In capability allowed pilots to keep track of aircraft flying in front of them during a visual approach to a runway. General aviation pilots received current weather and airspace status information from the FAA's free FIS-B service.

Automation

ATC Automation provided air traffic control functions including ATC, flight service, traffic management, time management and information management (FAA, 2017d). It included seven sub-elements that supported air traffic controller operations and pilot situational awareness. Automation performed functions by receiving and processing data from the Surveillance, Navigation, and Weather systems. ATC Automation relied on Communications systems to send and receive both voice and data transmissions. As ATC Automation provided function to the controller workstation, Aircraft Automation Systems provided automation function to the aircraft.

Automation Exemplars. Five major programs were examples of Automation: En-Route Automation Modernization (ERAM), Terminal Automation Modernization and Replacement (TAMR), Terminal Flight Data Manager (TFDM); Collaborative Air Traffic Management (CATM), and Time Based Flow Management (TBFM). ERAM replaced aging Air Route Traffic Control Center (ARTCC) automation systems (FAA, 2017d). The TAMR program modernized the air traffic control systems that controllers used to control traffic approaching or leaving the United States' major airports. TFDM automated the flow of flight and other tower data between ATCT and other ATC domains, and provided decision support capabilities to improve airport surface traffic management. CATM coordinated flight and flow decision-making by flight planners and FAA traffic managers. TBFM leveraged the capabilities of the Traffic Management Advisor (TMA) decision-support tool system that was deployed to all contiguous United States ARTCCs.

Human Factors Accomplishments. En-Route Automation Modernization and its associated hardware, software and backups were the backbone of en route operations. Instead of 20 separate systems, the FAA has a single system and ERAM was designed to support the evolution to NextGen. The target users for ERAM were air traffic controllers at en route centers. ERAM accommodated increased air traffic flow and allowed air traffic controllers to handle traffic in greater geographic areas. ERAM processed flight and surveillance radar data, enabled controller-pilot communications, and generated display data to air traffic controllers (FAA, 2016a). ERAM enabled controllers to coordinate traffic beyond the boundaries of the airspace controlled by their center so they could more efficiently transition traffic from one airspace sector to another. ERAM automated traffic conflict alerts and minimum safe altitude warnings. ERAM added capabilities to allow controllers to separate aircraft with variable separation standards. ERAM increased flexibility in routing around congestion, bad weather, and other airspace restrictions.

The TAMR program upgraded multiple terminal ATC technologies into a single platform, the Standard Terminal Automation Replacement System (STARS; FAA, 2016a). Controllers used STARS to provide ATC services to pilots in the airspace immediately surrounding major airports. The target users for STARS were air traffic controllers at towers and TRACON facilities. STARS significantly improved flight plan processing with a four-dimensional trajectory (lateral, vertical, horizontal, and time) of every flight which improved a controller's situational awareness, decision making, and routing of aircraft.

The Terminal Flight Data Manager program implemented the Surface Visualization Tool (SVT) and Advanced Electronic Flight Strips (AEFS). SVT allowed TRACON controllers to spot departure congestion and anticipate changes. AEFS replaced paper flight strips and manual tracking of incoming and outgoing flights with an electronic flight data display (FAA, 2017b). AEFS is updated with a touch screen or mouse and a finger swipe sends the data to another station allowing controllers to stay engaged with traffic at all times.

Collaborative Air Traffic Management delivered Pre-Departure Reroutes and Airborne Rerouting to controllers (FAA, 2017d). Pre-departure reroutes enabled controllers to more quickly execute revised route clearances needed to accommodate changing weather. ERAM added the capability to receive amended reroutes pre-departure and provide controllers with updated flight data so they can monitor and react to non-compliance issues. Airborne Rerouting allowed a traffic manager to propose trajectory modifications to meet flow constraints for an airborne flight to the appropriate sector controller for action. Controllers may amend the intended trajectory for the flight, deliver the route clearance to the cockpit via voice, and the traffic manager may track the amended trajectory when considering further constraint adjustments for the flight.

Time Based Flow Management implemented Extended Metering, Groundbased Interval Management Spacing, and the Integrated Departure/Arrival Capability (FAA, 2016a). Target users were air traffic controllers and pilots. Extended Metering enabled metering to begin further from the airport so controllers can manage aircraft with minor speed adjustments. Groundbased Interval Management Spacing (GIMS) introduced automation support for en route controllers to sequence and schedule en route arrival flows at one or more meter points upstream from terminal arrival meter fixes such that the schedules were deconflicted at all meter points and fixes. GIMS also provided en route controllers with speed advisories to help deliver aircraft to meter points and fixes in accordance with the arrival flow schedule. Integrated Departure/Arrival Capability (IDAC) provided decision support capabilities for departure flows to controllers that automated the process monitoring departure demand and identification of departure slots, and deconflicted departure times between airports with traffic departing to common points in space. IDAC provided situational awareness of available departure times to air traffic control tower personnel so they could select and plan their operations to meet the times. TBFM also implemented Traffic Management Advisor's (TMA) Adjacent Center Metering Capability and the ability to use of RNAV Route Data to calculate trajectories used to conduct Time-Based Metering operations.

NextGen (2016-2020)

NextGen's mid-term is through 2020. NextGen will continue implementing parts of several key enabling technologies to realize additional operational improvements. Key technologies include data communications, digital voice switching, performance-based navigation, network-enabled information sharing, satellite-based surveillance, integration of weather into decision-making and collaborative air traffic management (FAA, 2015). NextGen will continue to meet human factors challenges throughout these systems' development lifecycle as well as those of integrating new entrants, Unmanned Aircraft Systems and Commercial Space Operations.

NextGen Transformation (Beyond 2020)

When NextGen was initiated, controllers provided air traffic services tactically based on the location of the aircraft and distance to other aircraft to ensure safe separation. The FAA will transition the NAS to more strategic time-based management. Air traffic will be controlled strategically based on what the location of the aircraft will be at designated times along its projected path, thereby ensuring safe separation.

Trajectory Based Operations (TBO) will enable time-based management. TBO will leverage the technologies and operational improvements made during the mid-term. The target users of TBO will be pilots, controllers, air traffic managers, airlines, and other NAS operators. For strategic planning, users will share four-dimensional information about the aircraft: lateral (latitude and longitude), vertical (altitude), and time. Users will have better knowledge of the estimated departure and the arrival time at waypoints along the route of the flight. Strategic planning will decrease the need for tactical intervention (FAA, 2016c). When it occurs, air and ground automation systems will quickly share clearances provided to the flight deck resulting in a consistent view of the four-dimensional trajectory across the NAS.

NextGen Transformation will face human factors challenges. Time-based management will require more reliance on automation. Seamless integration of automation platforms will be needed as users share information to make safe and efficient use of time-based services. User culture will need to transition from legacy operations: the transition will require procedural changes, training, and methods to achieve user acceptance. There will be new sources for safety hazards such as knowledge of and performance with automation reversionary modes, human automation interaction, and maintaining situational awareness in a system of systems.

Acknowledgements

I express my gratitude to Drs Paul Krois and Charles Buntin who assisted in critiquing this paper. I based this paper on a variation of related reports on NextGen progress. The views of the paper are my own and do not reflect the views of the Federal Aviation Administration.

References

Federal Aviation Administration. (2015). *National Airspace System Enterprise Architecture (NAS EA), Mid-Term Overview and Summary Information (AV-1), Version 4.0, January 30, 2015*. NAS-EA-AV-1-Mid-Term-v4.0-013015. Washington, DC.

Federal Aviation Administration. (2016a). NextGen Implementation Plan 2016. Retrieved from https://www.faa.gov/nextgen/media/NextGen_Implementation_Plan-2016.pdf

Federal Aviation Administration. (2016b). NextGen Priorities Joint Implementation Rolling Plan 2017-2019 Executive Report. Retrieved from https://www.faa.gov/nextgen/media/NG_Priorities_Joint_Implementation_Plan.pdf

Federal Aviation Administration. (2016c). The Future of the NAS. Retrieved from https://www.faa.gov/nextgen/media/futureOfTheNAS.pdf

Federal Aviation Administration. (2017a). Acquisition Management Policy. Retrieved from http://fast.faa.gov/docs/acquisitionManagementPolicy/AcquisitionManagementPolicy4.7.pdf

Federal Aviation Administration. (2017b). Delivering NextGen. Retrieved from https://www.faa.gov/nextgen/

Federal Aviation Administration. (2017c). FAA Statement [Press release]. Retrieved from https://www.faa.gov/news/press_releases/news_story.cfm?newsId=21435

Federal Aviation Administration. (2017d). NAS Systems Engineering Portal (SEP) [Data]. Retrieved from https://sep.faa.gov/

Vision 100—Century of Aviation Reauthorization Act, Pub. L. No. 108–176 (2003).

AN INVESTIGATION OF MINIMUM INFORMATION REQUIREMENTS FOR AN UNMANNED AIRCRAFT SYSTEM DETECT AND AVOID TRAFFIC DISPLAY

Kevin W. Williams
Civil Aerospace Medical Institute
Federal Aviation Administration
Oklahoma City, OK

This study was conducted to support the development of Minimum Operational Performance Specifications for UAS Detect and Avoid traffic displays being developed by RTCA Special Committee 228. The experiment tested four different display configurations. These were a baseline display, an indication of Closest Point of Approach (CPA), an avoidance area (blob) indication, and a banding display. Also manipulated in the study were two levels of pilot experience and two types of control interface. Analysis of the well clear violations showed a significant effect due to display type. Individual comparisons revealed that both the avoidance area and banding displays significantly decreased the likelihood of violating well clear relative to the baseline display. Performance with the CPA display was not significantly different from the baseline display.

One of the requirements for successfully integrating Unmanned Aircraft Systems (UASs) into the National Airspace System (NAS) is that UAS pilots be able to conform to Title 14 Code of Federal Regulations (14CFR) Part 91.113 which requires pilots to "see and avoid" other aircraft. Achieving this conformance requires research to assist in the development of technology that would allow UAS to detect other aircraft that the UAS pilot cannot see and to enable the UAS pilot and/or system to transmit maneuver commands to the unmanned aircraft (UA) so that it can avoid those other aircraft. As part of that effort, human factors research is required to determine what control station displays and controls are needed to support the UAS pilot in performing this traffic avoidance task.

Building primarily off previous work from the FAA (Rein, Friedman-Berg & Racine, 2013) and NASA (Fern, Rorie, Pack, Shively, & Draper, 2015; Rorie & Fern, 2015; Rorie, Fern & Shively, 2016; Santiago & Mueller, 2015), four traffic display formats were compared with regard to their effectiveness in assisting the pilot in remaining well clear from other aircraft. The first display format, based on the work of Rein et al., 2013, was considered a baseline format. The other three formats used the baseline display and added additional information to the display to see if there was a significant increase in the ability to remain well clear from other traffic. In addition to manipulating display format, the experiment tested two different types of control station pilot interfaces and two levels of pilot experience levels. For a complete description of the experimental design and results, the reader is directed to the FAA Technical Report by Williams, Caddigan, and Zingale (2017).

Method

Thirty-two pilots were recruited for the study. Sixteen of the pilots had UAS experience and the other 16 were instrument-rated manned aircraft pilots with no UAS experience. Two separate control stations were used for the study. The Predator Station pilot interface includes controls on the joystick but also accepts keyboard commands. For most flight commands, both the joystick and keyboard must be used. The ICOMC2 Station consists of a single screen. Interaction with the system is accomplished using a mouse and keyboard. Inputting flight commands can be accomplished either by typing values in certain locations on the screen or by clicking and dragging with the mouse. For both stations, a separate 19" monitor was used for the traffic display. Figure 1 shows the baseline traffic display depiction and symbology used for the other display configurations.

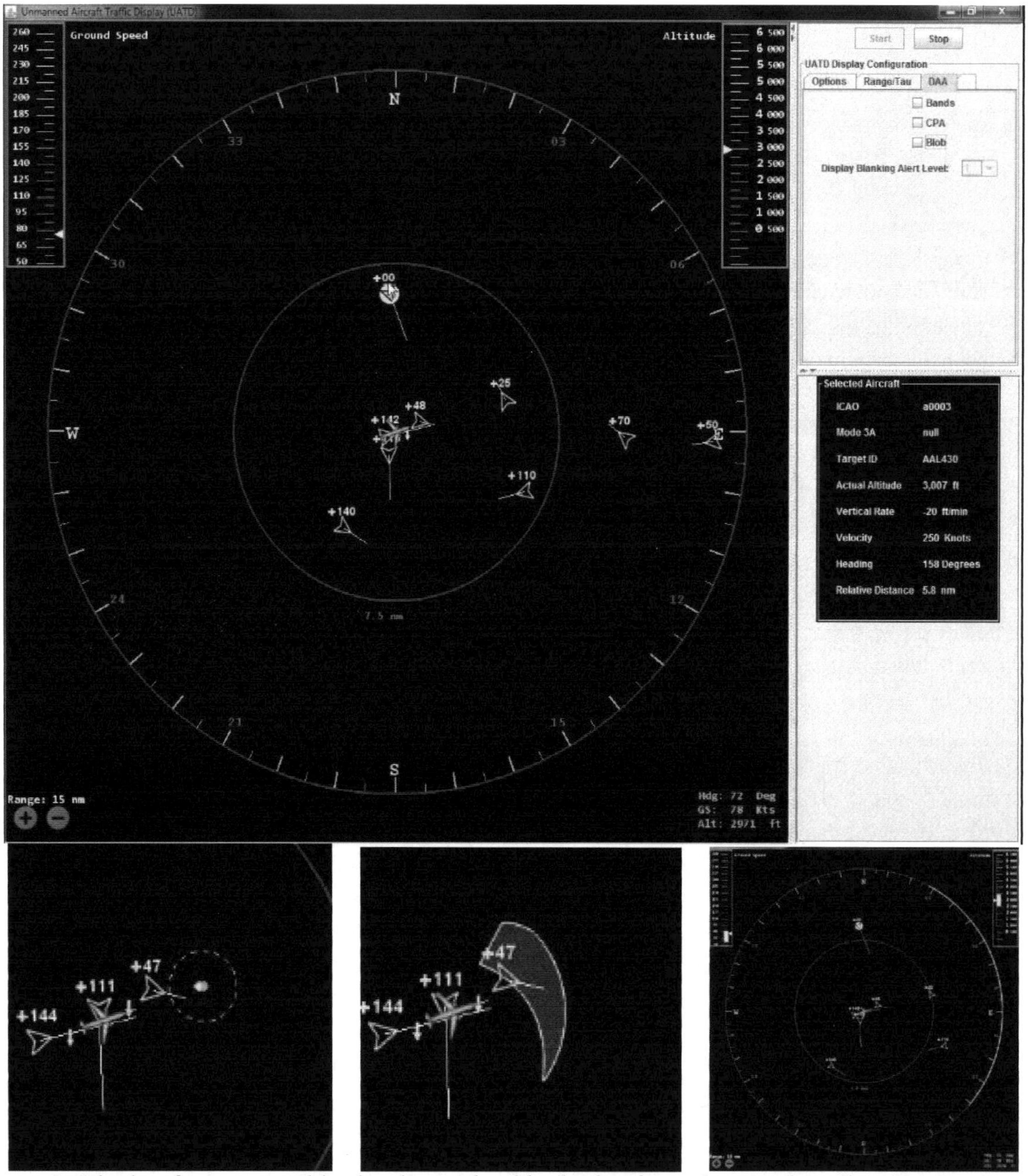

Figure 1. Display formats used in the the study. Clockwise from the top, baseline display, banding format, avoidance area (blob) format, and closest point of approach (CPA) format.

The alerting algorithms used for this study are collectively called DAIDALUS (Detect and Avoid Alerting Logic for Unmanned Systems) and were developed by NASA Langley Research Center personnel (Muñoz et al., 2015). The selection of timing parameters of the alerts, as well as the selection of traffic alert symbols and auditory alert messages was based on work accomplished by the RTCA SC-228 DAA working group. Figure 2 shows the visual and auditory alerts used in the study.

Preventive DAA Alert *"Traffic, Monitor"*	*Corrective DAA Alert* *"Traffic, Avoid"*	*DAA Warning Alert* *"Traffic, Manveuver Now* *Traffic, Maneuver Now"*

Figure 2. Visual and auditory alerts used in the study.

The lowest priority alert, the Preventive DAA Alert, did not require an action on the part of the pilot but was intended to draw attention to an aircraft that needed to be monitored. The other two alerts, the Corrective DAA Alert and the DAA Warning Alert both indicated that a loss of well clear would occur if both aircraft remained on their current courses. The main difference between the two was that the Corrective DAA Alert was intended to provide more time for the pilot to make a maneuver than the highest priority DAA Warning Alert. Participants were given instructions that, if they felt they had enough time to do so, they should contact air traffic control and request permission to deviate from their flight plan before performing the maneuver.

Eight different encounter geometries were used for the study (see Table 1). Variations in the scenarios were generated by altering the position of non-intruder "distractor" aircraft to create four versions of each encounter, thus resulting in 32 different scenarios. Each scenario contained 2-4 distractors, an intruder, and ownship.

Table 1. *Encounter geometries used in the study.*

Encounter	Horizontal Geometry	Vertical Geometry Ownship	Vertical Geometry Intruder
1	Head-on	Level	Level
2	Head-on	Descending	Level
3	Intruder Overtaking	Level	Level
4	Intruder Overtaking	Level	Climbing
5	Crossing	Level	Level
6	Crossing	Level	Level
7	Crossing	Descending	Level
8	Crossing	Level	Descending

Procedure

After arriving at the facility, the participant viewed an introductory briefing. They then read and signed an Informed Consent Statement and completed a background questionnaire. Next, the participant was given familiarization training on the appropriate UAS simulator.

Participants completed eight encounter scenarios for each traffic display configuration. Order of the display configurations was counterbalanced across participants. Before flying the encounter scenarios for a particular display configuration, participants completed one or two practice scenarios to ensure complete understanding of the display configuration being flown.

All traffic scenarios began with the UA already in the air. Each scenario assumed that the aircraft was following an instrument flight plan. Each scenario contained one traffic encounter, maneuver/s to avoid the traffic, and command/s to return to course. To increase the difficulty of the encounter, the traffic display did not display any traffic other than ownship until the occurrence of a traffic alert. This prevented the pilot from anticipating a potential avoidance maneuver before the alert. The scenario ended once the aircraft had started its return to course. Depending on the encounter and pilot responses, each scenario lasted from three to six minutes.

After the last scenario in each display configuration, the participant completed the Post-Display Questionnaire. After completing all four of the display configurations, the participant was given a post-study questionnaire. More complete details of the procedure and questionnaires can be found in Williams et al. (2017).

Results

Figure 3 presents the mean number of well clear violations as a factor of display type. Analysis of the well clear violations showed a significant effect due to display type, $F(3, 78) = 3.465$, $p = .02$. No other main effects or interactions were found in the analysis of well clear violations. Individual comparisons revealed that both the blob display, $t(31) = 3.66$, $p = .0005$, and banding display, $t(31) = 1.80$, $p = .04$, significantly decreased the likelihood of violating well clear relative to the baseline display. The CPA display was not significantly different from the baseline display, $t(31) = .61$, $p = .27$.

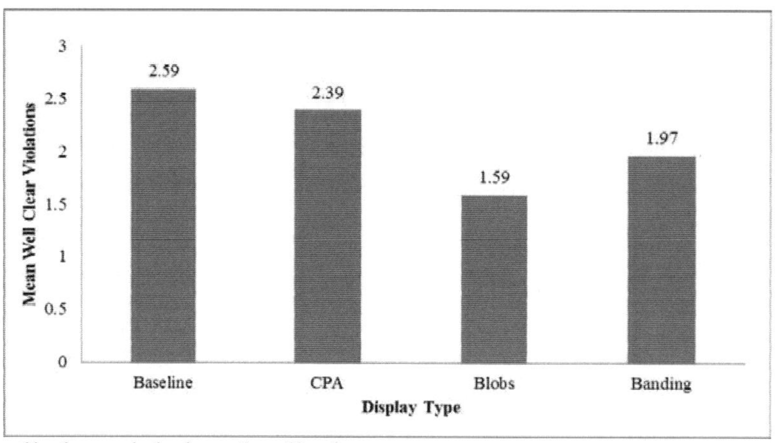

Figure 3. Mean well clear violations by display type.

Figure 4 presents the mean well clear violations across display types separated by pilot type.

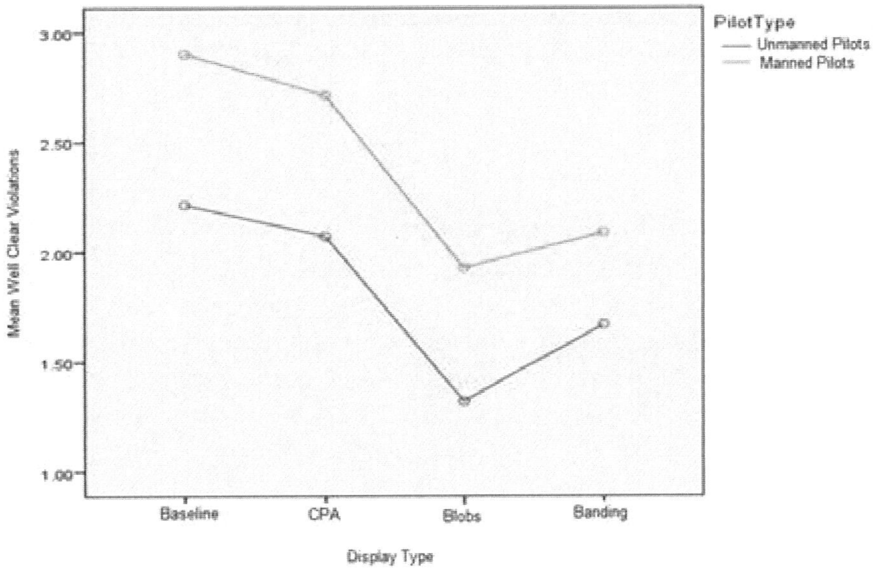

Figure 4. Mean well clear violations across display type by pilot type.

Looking Figure 4, the green (top) line is the mean well clear violations for manned aircraft pilots across display type and the blue line (bottom) is the mean well clear violations for the unmanned pilots across display type. Overall the pattern of well clear violations for both pilot types is nearly identical to the overall findings shown in Figure 3 with the baseline display having the most well clear violations, followed by the CPA display, banding display, and the blob display having the fewest number of well clear violations. While performance between UAS and manned pilots was not significantly different, $F(1,26) = 3.616$, $p = .068$, both pilot groups responded similarly across display configurations in regard to avoiding well clear violations.

Discussion

This study replicated the findings of other studies showing the benefits of suggestive maneuver guidance in the form of banding information, in addition to baseline information, for a UAS detect and avoid traffic display. Evidence for these benefits came from both objective and subjective measures. Objectively, use of the banding display resulted in significantly fewer well clear violations compared to the baseline information display. This effect was seen across a more varied population of pilots than have been looked at in previous studies as well as different control station interface designs than were used in previous studies. The pilot sample included both manned and unmanned pilots across a wide range of ages and flight experience levels. This gives strong support for the decision made by the RTCA SC-228 committee to require banding information as part of the minimum requirements.

In addition to the banding display, the study also found strong support for a different form of suggestive maneuver guidance implicitly provided in the avoidance area (blob) information. Objective measures of performance suggested that the blob display was as effective as the banding information. The relative success of the blob display raises a separate issue regarding traffic display information requirements. While the banding display contained an altitude band on the altitude tape instrument, the blob display only had suggestive guidance for a horizontal maneuver. The only information available for making a vertical avoidance maneuver was the same as was available on the baseline display, which consisted of relative altitude and vertical speed information located next to each traffic symbol.

That the blob display was as effectiveas the banding display, suggests that the vertical banding information as a form of suggestive guidance is not as useful as horizontal guidance. Further research on this issue is warranted.

Acknowledgements

I would like to acknowledge my coauthors on the technical report, Dr. Carolina Zingale with the FAA William J. Hughes Technical Center in Atlantic City, NJ and Dr. Eamon Caddigan with T.G. O'Brien & Associates, Inc. I also wish to express my thanks to all the personnel at the FAA William J. Hughes Technical Center who assisted in the running of this study. Unfortunately, there is not enough room to name everyone involved.

References

Fern, L., Rorie, R. C., Pack, J., Shively, R. J., & Draper, M. (2015). *An evaluation of detect and avoid (DAA) displays for unmanned aircraft systems: The effect of information level and display location on pilot performance.* Proceedings of the 15th Annual American Institute of Aeronautics and Astronautics (AIAA) Aviation, Technology, Integration and Operations Conference, Dallas, Texas, June 22 - 26, 2015.

Muñoz, C., Narkawicz, A., Hagen, G., Upchurch, J., Dutle, A., Consiglio, M., & Chamberlain, J. (2015). *DAIDALUS: Detect and avoid alerting logic for unmanned systems.* Proceedings of the 34th Digital Avionics Systems Conference (DASC 2015), Prague, Czech Republic, September 13-17, 2015.

Rein, J.R., Friedman-Berg, F. & Racine, N. (2013). *Information requirements for sense and avoid: High-fidelity human-in-the-loop simulation.* U.S. Department of Transportation, Federal Aviation Administration summary report for UAS Task 12-01a.

Rorie, R.C. & Fern, L. (2015). *The impact of integrated maneuver guidance information on UAS pilots performing the detect and avoid task.* Proceedings of the 59th Human Factors and Ergonomics Society Annual Meeting, Los Angeles, California, Oct 26-30, 2015.

Rorie, R.C., Fern. L., & Shively. R. J. (2016). *The impact of suggestive maneuver guidance on UAS pilots performing the detect and avoid function.* Manuscript in preparation. Proceedings of the 2016 AIAA SciTech Conference, San Diego, California, Jan 4–8, 2016.

Santiago, C., and Mueller, E.R. (2015). *Pilot evaluation of a UAS detect-and-avoid system's effectiveness in remaining well clear.* Eleventh USA/Europe Air Traffic Management Research and Development Seminar (ATM2015), Lisbon, Portugal.

Williams, K.W., Caddigan, E., & Zingale, C. (2017). *An investigation of minimum information requirements for an unmanned aircraft system detect and avoid traffic display*. U.S. Department of Transportation, Federal Aviation Administration, Office of Aerospace Medicine Technical Report #DOT/FAA/AM-17/14.

INCREASING ACCEPTANCE OF HAPTIC FEEDBACK IN UAV TELEOPERATIONS BY VISUALIZING FORCE FIELDS

Victor Ho, Clark Borst, René van Paassen, and Max Mulder

Faculty of Aerospace Engineering, Section Control and Simulation,
Delft University of Technology, 2629 HS Delft, The Netherlands

In tele-operating an UAV, human operators fully rely on cameras to control the vehicle from a distance. To increase operator situation awareness and reduce workload, haptic feedback on the control stick has been developed which acts as an automatic collision avoidance system. A virtual force field surrounding the moving vehicle interacts with obstacles surrounding it, yielding repulsive forces on the stick that lead the vehicle away from them. Albeit successful in significantly reducing the number of collisions, the haptic interface received low user acceptance ratings. Operators do not always fully understand the collision avoidance automation intentions, and they experience the haptic forces as intrusive. This paper discusses the development and testing of several visualizations of the underlying automation intentions, primarily the artificial force field. Results of a human-in-the-loop experiment show that these visualizations indeed led to higher user acceptance ratings, without affecting the operator's safety, performance and workload.

Operating UAVs can be a challenging task, especially beyond the operator's line of sight, where the drone is controlled by teleoperation. Locations with low visibility, e.g., due to the lack of light or because of obstructions like smoke, pose a threat to teleoperation since the on-board cameras and other electro-optical sensors cannot provide quality images. In addition, the teleoperator lacks multiple-sensory information of the surrounding environment (e.g., vehicle motion, vibrations, environment/vehicle sound and outside view) compared to pilots flying a manned aircraft. The information is usually provided by visual displays from on-board cameras and sensors which have limited resolution and Field of View (FOV) (Draper & Ruff, 2001).

To compensate for the lack of direct sensory input from the environment, a haptic interface has been developed for collision avoidance (Lam, Mulder & Van Paassen, 2007, 2008). A Haptic Collision Avoidance System (HCAS) uses an Artificial Force Field (AFF) to map environmental constraints to steering commands for avoiding collisions with objects.The resulting haptic feedback system provides information through the sense of force on the control device: a *shared* control system between human and automation. Research shows that these shared control systems are often not optimal, with low user acceptance ratings which are often caused by a lack of information of how the haptic system works, i.e., why are forces felt?, and where do these forces originate from? (Griffiths & Gillespie, 2005; Seppelt & Lee, 2007; Lam, Mulder & Van Paassen, 2007, 2008).

This paper presents two visual displays that can accompany the haptic feedback and that aim to provide some visual explanation of what the haptic feedback system's intentions are. Seppelt and Lee (2007) showed that these visualisations of the haptic feedback intentions can increase higher user acceptance ratings. The paper will first briefly discuss the principles of our haptic feedback systems, followed by the visualizations developed, and then describe the results of a first human-in-the-loop experiments to test our novel system.

Haptic Collision Avoidance System

Figure 1 illustrates the basic building blocks of the Haptic Collision Avoidance System developed by Lam, Mulder & Van Paassen (2007, 2008). The HCAS informs the operator if a certain control input will lead to a higher risk of an obstacle collision. To realize this function, the environment surrounding the UAV is evaluated by an obstacle detection algorithm. Detection is done by a Laser Imaging Detection And Ranging (LIDAR) sensor, which measures the object-UAV distance by analyzing the reflected light by the laser beam mounted below the UAV. The laser scans the environment in two dimensions, returning distance measurements at specific angle intervals. With this mapping, which resembles the visual control task of the pilot but extends it to all directions, the Artificial Force Field (AFF) computes the risk of collision. This risk is converted to a haptic force on the control stick, yielding a continuous haptic feedback that warns the operator of a potential collision.

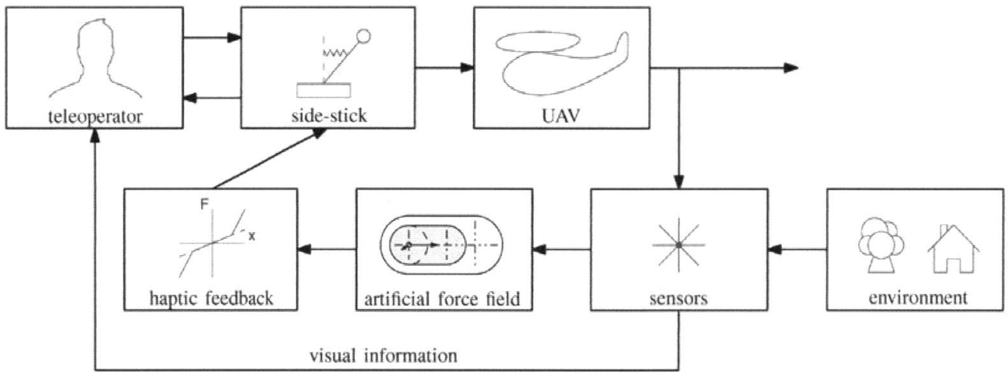

Figure 1: Schematic representation of haptic interface for UAV teleoperations.

The AFF used by Lam, Mulder & Van Paassen (2007, 2008) is programmed as a Parametric Risk Field (PRF), a "potential field" that extends outside the physical limits of the UAV being tele-operated and that shrinks and extends dependent on the direction of the UAV velocity. Figure 2 illustrates that all obstacles which fall into the potential field lead to repulsive forces; these are summed and averaged, yielding a single Final Avoidance Vector (FAV), with a direction and amplitude, equivalent to the force feedback on the operator's control manipulator.

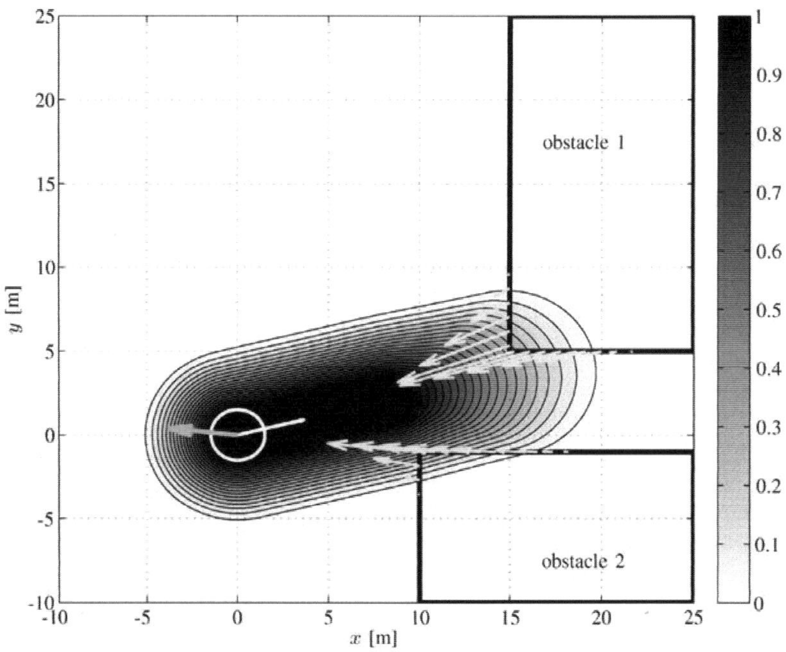

Figure 2 : Risk vectors (yellow, from the obstacles) and the FAV (blue vector extending from the UAV center).

Our previous research showed that our haptic feedback system significantly reduced the number of collisions and increased task performance, compared to a situation without haptic feedback. Subjective workload measured with NASA-TLX also increased with haptic feedback, especially the physical workload and subject frustration levels (Lam, Mulder & Van Paassen 2007, 2008). Subjects explained that at some moments the haptic feedback was 'too strong', and 'unpredictable', as subjects could not decipher what reasoning was underlying the HCAS feedback forces. In the next section we discuss two visualizations developed to mitigate this experience.

Haptic Feedback Visualizations

In our previous work, the UAV tele-operators had a (simulated) on-board forward-looking camera view display, as well as a two-dimensional 'navigation display', which presents a top-down bird's eye view of the situation, including the UAV, the obstacles, and a triangular shape which showed the field-of-view of the forward-looking camera. To avoid clutter on the three-dimensional camera image, and since our AFF is currently still a two-dimensional, horizontal force field, two visualizations were developed to be added on the navigation display.

The PRF Contour Risk Field (PRF-CRF) is our first visualization, Figure 3. It is almost a 1:1 visualization of the virtual force field of Figure 2 on the navigation display. However, to reduce clutter, all information within the red outer contour is deleted, and all risk vectors are reduced to colored dots, color-coding how much risk they represent. White dots mean low-risk and barely feelable haptic feedback, yellow dots mean medium risk with noticeable feedback, and red dots mean maximum risk corresponding with maximum force feedback. The FAV is a vector line attached to the UAV center, which changes its length and direction perfectly corresponding with the haptic feedback force put to the control manipulator.

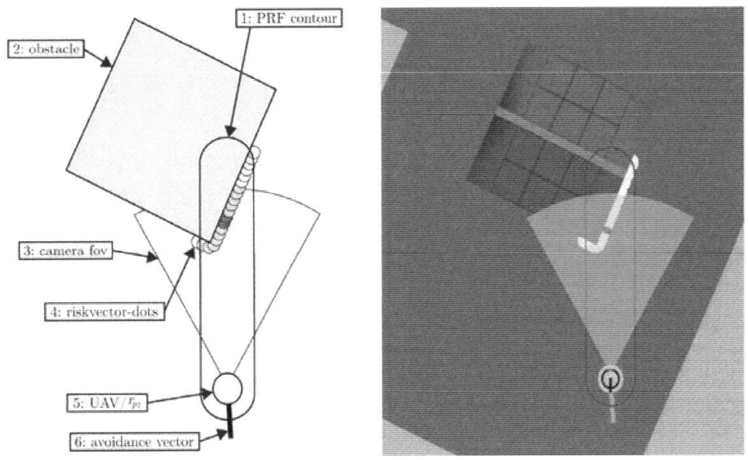

(a) PRF-CRF description (b) PRF-CRF during flight

Figure 3 : The PRF Contour Risk Field (PRF-CRF) visualization, developed to work together with the HCAS.

The Static Circular Risk Field (SCRF) is the second visualisation, Figure 4. In contrast to the first visualization, the SCRF does *not visually correspond to the HCAS algorithm.* Inspired by our previous research on supporting pilots in self-separation (Van Dam, Mulder & Van Paassen, 2008), only a circle is shown the size of which does not change (hence: static). Within the circle, white, yellow and red lines show the directions of risk vectors coming from obstacles with low/medium/high risk, respectively. In such a way, a 360 degree 'risk map' is shown within the circle, which we hypothesized to be easier to understand by teleoperators as compared to the PRF-CRF. Whereas the latter can shrink and extend rapidly, depending on the UAV's velocity and acceleration, possibly overlapping the obstacles, the SCRF only presents the risk map within the fixed static circle, without much overlap.

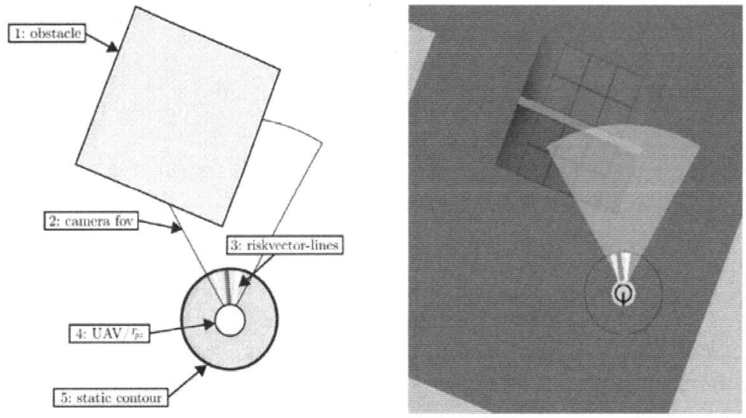

(a) SCRF description (b) SCRF during flight

Figure 4 : The Static Circular Risk Field (SCRF) visualization developed to work together with the HCAS.

Experiment

We performed an experiment to test the usefulness of our haptic interface with the two novel visualisations. Our main interests were whether the participants could better understand, predict, and because of that would better appreciate our HCAS. The setup of the experiment was very similar to the experiments reported by Lam, Mulder & Van Paassen (2007, 2008) and will be only briefly explained in this section; for more details see our earlier work.

The experiment had 12 participants (all males, right-handed, average age 25 years) and was conducted in the fixed-base human-machine interaction simulator. Subjects controlled a simplified UAV helicopter in the horizontal plane only (altitude was fixed) using a side-stick. Two displays were presented: (i) a simulated on-board camera view (60 degrees field of view) was projected on a wall 3 meters in front of the subjects as an outside visual display; (ii) a navigation display was presented on an 18 inch LCD screen in the simulator cockpit (i.e., head-down).

Subjects were instructed to fly the UAV from waypoint to waypoint, visualized with smoke plumes, in an obstacle-filled urban environment containing multiple buildings and artefacts. The obstacle course was fixed, and was clearly shown on the two visual displays. Subjects were instructed to fly the course as fast as possible (low priority), as closely as possible to the waypoints (medium priority), while avoiding any collisions (highest priority). When a collision did occur, the simulator was frozen for 10 seconds, i.e., inflicting a time penalty, while a loud beeping sound was heard. After the penalty, the UAV was repositioned to a fixed starting point corresponding to the 'subtask' where the collision occured. Each obstacle course was constructed by randomly 'connecting' 6 different sub-tasks, similar to Lam, Mulder & Van Paassen (2007, 2008).

The experiment had two independent variables: the six subtasks and the HCAS display configuration. The latter had three levels: No Visualization (NV), and the novel PRF-CRF and SCRF visualizations. Objective dependent measures were related to safety (number of collisions), performance (total elapsed time per run), control activity (stick rate), haptic activity (haptic forces). Subjective dependent measures related to workload (NASA TLX) and operator acceptance (using the Controller Acceptance Rating Scale (CARS), Lee et al., 2001). In addition, we asked our subjects to complete a small questionnaire (focusing on acceptance and preference) after the experiment.

Results

Contrary to our hypotheses, neither of the two visualizations led to any significant changes, that is, improvements in safety and task performance were not found. Regarding safety, the number of collisions was very low, 26 in 108 runs; it increased from 7 (NV) to 10 (PRF-CRF) and 9 (SCRF). Most collisions (15) occured in subtask 4 which required subjects to control the UAV through a long, narrow corridor. The mean risk vector magnitude slightly decreased with the two visualizations, and the average minimum distance to obstacles increased, both not significantly; there was no dependency between subtasks. Considering task performance, there were no significant differences in the total elapsed time and average UAV velocity. None of the control activity metrics, neither any of the haptic feedback force metrics, changed significantly.

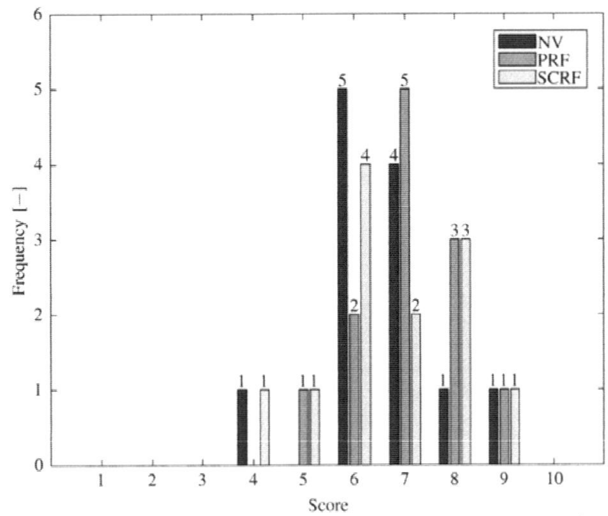

Figure 5 : Controller Acceptance Rating Scale (CARS) results (NV = No Visualization; PRF = PRF-CRF, and SCRF = Static Circle Risk Field).

When considering the subjective data, contrary to our hypothesis the subjective workload (TLX) did not reduce significantly with either of the two visualizations, which both led to slightly higher TLX frustration levels. Figure 5 shows that the CARS scores were slightly higher for the SCRF and PRF-CRF visualizations, as compared to the NV condition, an effect that was also not significant. The end-of-experiment questionnaire confirmed the CARS results, in that the PRF-CRF was preferred by most subjects in most conditions except for subtask 4 (see below). Subjects commented to use the PRF-CRF outer boundary to see when the haptic feedback would trigger and thus could make sharper turns, i.e., helping them to understandi the HCAS functioning and also improving the timing of their control actions.

Only in subtask 4 the SCRF was preferred. Here, while flying through the small corridor, the PRF-CRF visualization cluttered the screen with many risk vector dots, making it difficult to see how far the UAV was away from the walls. Occasionally, also a large drop in visual display update rate occurred with the PRF-CRF display because of drawing the many risk dots. Subjects commented also that in the PRF-CRF condition, showing the Final Avoidance Vector was not very useful; here, especially the outer contour mattered and was considered helpful.

The end-of-experiment questionnaire contained four questions where subjects had to rate their agreement to statements. When asked "*Did the visual feedback give you enough information about the workings of the haptic feedback?*", "*Have you felt any contradictions between the information received by the haptic feedback and the information shown on the display?*", the PRF-CRF scored significantly better than the SCRF, confirming our design aims with the former display which was to visualize 1:1 the virtual force field. When asked "*Did the visual feedback interfere with your flight performance compared to having no feedback?*", no differences between the two visualizations were found, and most subjects scored "low" on this interference. Finally, when asked "*Did you use the visual feedback to change your control strategy?*". The PRF-CRF scored significantly higher than the SCRF, and subjects reported that with the PRF-CRF they could fly closer to walls; yet, our other objective metrics did not provide any evidence that could possibly confirm this statement.

Conclusions

This research aimed to design and test two novel visualizations developed to obtain higher user acceptance ratings for a haptic feedback system in UAV tele-operation. Previous research showed that adding visualizations to haptic interfaces led to improved safety and operator performance. The human-in-the-loop experiment showed that our participants did not change their control strategies when either visualization was provided. Only marginal differences in objective dependent measures were found, and the new designs did neither significantly improve, nor deteriorate, the operator's safety, task performance and workload. Subjective data obtained through the CARS rating scale and the end-of-experiment questionnaire did show, however, that both visual displays were a welcome addition, as they provided more clarity of the internal functioning of the haptic feedback system, and helped increasing spatial awareness and timing of control actions. Differences were small, however, between conditions, and although the acceptance of the HCAS was increased, this increase was not significant. In our future work we will focus on using more subjects and developing experimental scenarios to better evaluate user acceptance and distringuish between the conditions with and without additional visualizations.

References

Draper, M.H., and Ruff, H.A. (2001). Multi-Sensory Displays and Visualization Techniques Supporting the Control of Unmanned Air Vehicle. *IEEE Int. Conf. Robot. Automation*: 1–6.

Lam, T.M., Mulder, M., and Van Paassen, M.M. (2007). Haptic Interface for UAV Collision Avoidance. *The International Journal of Aviation Psychology*, 17(2): 167–195.

Lam, T.M., Mulder, M., and Van Paassen, M.M. (2008). Haptic Feedback in UAV Tele-operation with Time Delay. *Journal of Guidance, Control & Dynamics*, 31(6): 1728–1739.

Griffiths, P.G., and Gillespie, R.B. (2005). Sharing Control Between Humans and Automation Using Haptic Interface: Primary and Secondary Task Performance Benefits. *Human Factors*, 47(3): 574–590.

Seppelt, B.D., and Lee, J.D. (2007). Making adaptive cruise control (ACC) limits visible. *Int. J. Hum. Comput. Studies*, 65(3): 192–205.

Van Dam, S.B.J. and Mulder, M. and Van Paassen, M.M. (2008). Ecological Interface Design of a Tactical Airborne Separation Assistance Tool. *IEEE Trans. on Systems, Man & Cybernetics, Part A*, 38(6): 1221–1233.

Lee, K. K., Kerns, K., Bone, R. and Nickelson, M. (2001). Development and Validation of the Controller Acceptance Rating Scale (CARS). *Proc. of the 4th USA/Europe Air Traffic Management R&D Seminar, Santa Fe (NM), December 7-13, paper 161*: 1-14.

AN INVESTIGATION INTO THE INFORMATION REQUIREMENTS FOR REMOTELY PILOTED AIRCRAFT CREW WHEN DEALING WITH CYBER THREATS

Dr. Kristen K. Liggett
Air Force Research Laboratory
Dayton, OH
Mr. Peter Venero
CAMO, LLC
Dayton OH
Dr. Gina Thomas
Air Force Research Laboratory
Dayton OH

Remotely piloted aircraft (RPA) crews of the future will encounter more than the traditional threats to their aircraft. In addition to air-to-air and surface-to-air missiles, future conflicts will most likely include cyber weapons. While cyber weapons can certainly cause physical damage to these aircraft, the potential also exists to turn the friendly RPA against their own forces. The goal of the Resilient and Assured UAS Systems and Operations (RAUSO) program is to develop a cyber security module (CSM) that will detect and defend RPAs from cyber attacks. In some cases, the CSM will need to act automatically to defeat the threat. In other cases, the threat can be isolated and dealt with at an appropriate time in the mission. As powerful as the CSM will be, it will not be able to determine an appropriate time to address an attack. In order to maintain the best mission performance, the CSM will have to "negotiate" with the human crew as to when the appropriate time is to address an attack. A study was conducted to determine the most effective way to present relevant information to RPA crews to inform them of cyber attacks, courses of action, and mission impacts for successful negotiation of actions with the CSM. Five two-person crews (pilot and sensor operator) executed simulated missions, and data were collected to determine mission performance degradation under two levels of cyber attack and how that degradation was impacted by the CSM. Alerting improved performance for both levels of attack.

Over the past 15 years, the use of remotely piloted aircraft (RPA) to conduct Air Force missions has increased exponentially, and this trend is expected to continue into the future (USAF, 2014). RPAs are used effectively for intelligence, surveillance, and reconnaissance (ISR), tracking targets, and delivering weapons in a variety of missions. One of the biggest advantages of using RPAs is that the RPA operators can conduct dangerous missions without jeopardizing their lives because operators in safe locations (typically the US) communicate with the vehicles in theatre via a satellite connection. The arrangement of vehicle, operator ground control station (GCS), and satellite represents nodes in a network and may be vulnerable to cyber compromises/attacks. Also, the vehicle itself contains a network of line replaceable units connected to a 1553 bus, also susceptible to cyber attack. One of the biggest challenges facing our Air Force today is making the avionics of air assets resilient to cyber threats (Gross, 2016; Skowronski, 2016). Members of the Resilient and Assured UAS Systems and Operations (RAUSO) team are trying to develop a cyber security module (CSM) that will detect and defend RPAs from cyber attacks. From a human factors perspective, the challenge is determining how a new technology capable of detecting cyber events should behave to ensure that the RPA crew can sustain mission performance. Integral to that challenge is ensuring the RPA crew understands how they can leverage the information made available by the technology.

In preparation for this study, the researchers participated in a series of knowledge acquisition activities with RPA operators from Springfield Air National Guard (SPANG) and Syracuse Air National Guard (SANG). The goals were to better understand the RPA missions and to determine how operators regarded the potential threat of cyber attacks (Dukes, Fox, Rigrish, Durkee, & Feeney, 2016). Specifically, the researchers were hoping to better understand how crews develop and maintain their mental model of the cyber/RPA space. Surprisingly, RPA crews don't fully understand the RPA system vulnerabilities to cyber attacks. The team also determined methods crews currently use to handle traditional anomalies in emergency situations in order to study how those methods might change or need to change in cyber situations that are expected to be detected by the CSM. One of the objectives of this study was to expose operators to cyber threats to determine if they would recognize them as cyber threats without being alerted. The second objective was to determine effects of applying traditional alerting procedures on operators' mission effectiveness under cyber attack.

Method

Simulation Environment

The study was conducted using the Vigilant Spirit Control System (VSCS), an interface testbed with a virtual simulation capability (Feitshans, Rowe, Davis, Holland, & Berger, 2008). Three components of the VSCS were used for this study: a pilot control station, a sensor operator control station, and a simulation component that allowed the researchers to create ecologically-valid and repeatable mission scenarios. The scenarios were created in the VSCS virtual world and then executed for data collection. VSCS currently has alerts and checklists embedded in their system, which is modeled after the MQ-9 Block 50. The alerts and checklists developed for the cyber attacks had the same look and feel as other alerts and checklists. Figure 1 shows the sensor operator control station with a cyber checklist activated.

Figure 1. VSCS Sensor Operator Control Station [tactical situation display on the left monitor and checklist, sensor feed, and chat windows on the right monitor].

Mission Scenario

The basic scenario used for this study was a variation of a scenario created by MITRE (Dinsmore et al., 2015). The task started with the RPA in a loiter pattern around a building in a compound where the planning of nefarious activities was believed to be taking place. The crew was instructed to observe the building for people exiting, follow anyone leaving the building, and report the locations of activities they observed while following the vehicle. In the scenario, a man left the building and drove directly to another building believed to be a weapons cache. Then, he drove the vehicle to a second location, exited the vehicle and displayed signs of digging (as if planting an improvised explosive device [IED]).

During scenarios that contained a cyber attack, attacks could present in one of two ways. The first was a loss of the sensor ball feed (a low-sophistication attack). The second was an intentional drift of the sensor's global positioning system (GPS) coordinates (a high-sophistication attack). The low-sophistication attack would occur at the time when the man exited the building and drove away in a red truck, preventing the sensor operator from observing this activity. The high-sophistication attack was less obvious and resulted in the passing of inaccurate coordinates for the location of a weapons cache and the IED implantation. For the scenarios that had the CSM activated, when a cyber attack occurred, an alert was presented on both the pilot and sensor operator's screens describing the type of attack that was detected along with a checklist with instructions of how to remedy the problem. These types of RPA cyber vulnerabilities are consistent with the results of the USAF Scientific Advisory Board Report on Operating Next-Generation Remotely Piloted Aircraft for Irregular Warfare (Scientific Advisory Board, 2010).

Participants

Participants were recruited from SPANG. Based on availability, eight members of the guard and two former RPA operators who served as subject matter experts participated in the study for a total of five two-person crews (pilot and sensor operator). The average RPA (MQ-1 Predator) flight hours for the pilots was 930 hours. The average experience of the sensor operators was 1572 total hours with an average of 1500 hours on the MQ-1.

Experimental Design

There were two independent variables in this study: two levels of CSM (active and not active) and two levels of sophistication of cyber attack (low-sophistication and high-sophistication) for a total of four conditions in a full-factorial within-subjects design. The dependent variable was a compiled measure of the crew's successful detection of a person leaving the building and accurately reporting the truck's location and the locations of the weapons cache and the IED implantation. In order to control for any learning effect, the initial location of the loiter pattern, the route the truck took to the second building, and the location of second building and the IED implantation was varied in the five scenario runs.

Procedure

Participants were greeted, and the purpose of the study was explained. They were given a short demographic questionnaire, a briefing about the details of the study, and were trained on VSCS. All crews performed an initial scenario with no cyber attacks to establish baseline performance. Then, all crews performed two missions with the CSM not present; one with a low-sophistication attack and one with a high-sophistication attack. Since the CSM was not present for these two conditions, the crew had no alert or checklists for resolving the problems. The order of these conditions was balanced across the five crews. After completing the missions without the CSM active, the crew performed two missions with a cyber threat present and the CSM active. Again (separately), the order of these two conditions was balanced across the five crews. Following data collection, crews were asked a series of questions to provide further insight into current operations and how cyber information should be presented in the future.

Results

The purposes of this initial study were to determine if there were practical effects of alerting crews to cyber intrusions and to ascertain an approximation of the size of expected effects for future research planning. It should be noted that there were no statistically significant effects found, which is not surprising due to the lack of statistical power resultant from the very low number of subjects (low N so effect size would have to be extremely large to show statistical significance).

When no cyber attack was present, crews, on average, were able to perform 95% of the tasks in the mission, but when a cyber attack was present without the CSM, crews completed only 25% of mission tasks. However, adding the CSM brought task completion back up to 83% (Figure 2). There appears to be a slight interaction between the presence of CSM and the type of attack (Figure 3). Crews were generally better at performing their tasks under drift than when the screen went blank regardless of whether or not CSM was present.

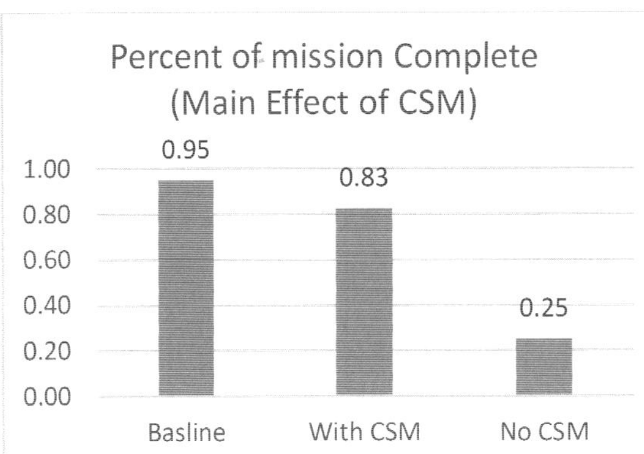

Figure 2. Effect of CSM.

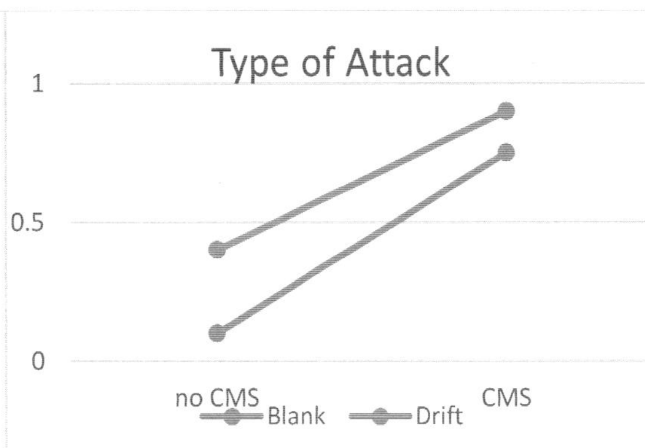

Figure 3. Use of CSM by Type of Cyber Attack.

Although the presence of the CSM improved performance in both attack conditions, CSM appeared to provide a bit more of a boost to performance in the blank condition, perhaps due to there being more room for improvement. Results of subsequent discussions with the crew members are addressed in the discussion.

Discussion

The effects of having the CSM active confirmed expectations that a system that would alert the crew of a cyber-based problem would result in better performance. The lower performance on the low-sophistication threat (sensor ball blanking) occurred because of a combination of the sensor ball setting and the rules of engagement that the crews were following. Crews were instructed to loiter over a building and watch for anyone leaving the building. The low-sophistication attack occurred when the person left the building, so when the sensor ball came back up, if the sensor operator had it zoomed in, they missed the person leaving the building. If they maintained their rules of engagement (stay loitering until you observe a person leaving) they would not zoom out and search for the red truck unless instructed to do so (by the customer in the real world or the experimenter in the study), so they missed the rest of the mission events. Note that accuracy was not 0% when the CSM was not active because some of the sensor operators had the sensor zoomed out to a level such that when the sensor feed came back on, they could still see the red truck leaving the compound and follow it. For the high-sophistication cyber threat (GPS coordinates drift), participants could complete the first two objectives (watch for a person and follow the truck) regardless of the attack but the coordinates they passed for the location of the weapons cache and the IED implantation would be incorrect when the CSM was not active. Therefore, any difference in performance (more improvement for low-sophistication attack) is probably due to having more room for improvement in the low-sophistication attack condition.

In terms of the crew-member feedback, much information was obtained regarding how the RPA crews go about dealing with emergencies and maintaining their mental models throughout these situations. Figure 4 shows the state space of possible alignments between the operator's mental model and the actual state of the system. Clearly the goal is to have the crews' mental model aligned with the current state of the world (shown in green). When there is a mismatch between the actual state and the operator's mental model (shown in pink), two things can result: 1) operators don't recognize a problem, or 2) they spend time on a problem that doesn't exist. Both of these situations lead to decreased mission effectiveness. When the operators think things are normal, they seem to be employing a more passive scan pattern, simply consuming key pieces of information that verify this situation. The crew does not seem to actively cross-check multiple sources of information against expected values to make sure they all indicate that the plane is, in fact, in a normal state. In other words, it is only when something unexpected and relatively obvious happens (i.e., an alert activates) that the crew detects the abnormal state and they subsequently seek out information that is not readily apparent. So one way in which alerting improves performance is that it provides that unexpected event causing the operator to move from the incorrect state of believing an abnormal system is normal to the correct state of believing the abnormal system is abnormal, minimizing the time spent in the incorrect state.

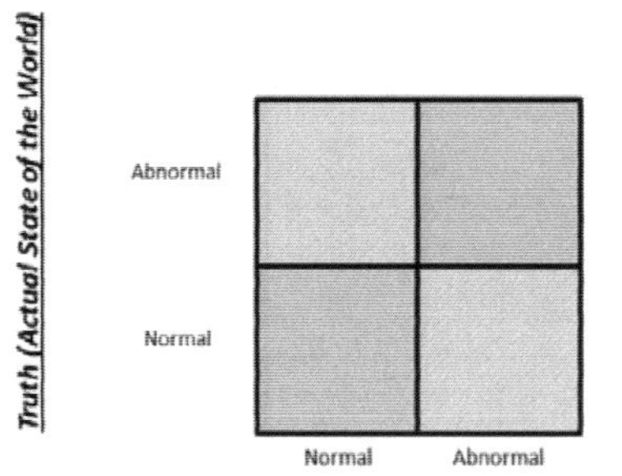

Figure 4. Actual State and Operator Mental Model Space.

Figure 5a expands the abnormal/abnormal state by showing this space with traditional mechanical and electrical failures. Figure 5b represents the current state of affairs in which a cyber attack is a possible cause of the abnormal state, but the operator's mental model about potential causes of abnormal states has not updated to include cyber attacks. In this case, when a cyber attack happens, the operator's understanding about the cause of the problem will always be incorrect. Figure 5c shows the possible conditions once operators are made aware of the potential for cyber attack. In this case, there is at least a chance that the operator will correctly identify the cause of the abnormality. In post-trial interviews, experimenters learned that some alerts cause multiple checklists to appear. When operators were asked how they decided which checklist to follow, they indicated that the one they chose was based on experience, but what they described was a process of looking at various information elements to determine which path to go down. This shows a shift in their procedure when an alert is present – from passive scanning of available information to actively seeking information relevant to the situation. The green boxes in figure 5c show the goal, which is to align the operator's mental model of possible abnormalities with the actual cause.

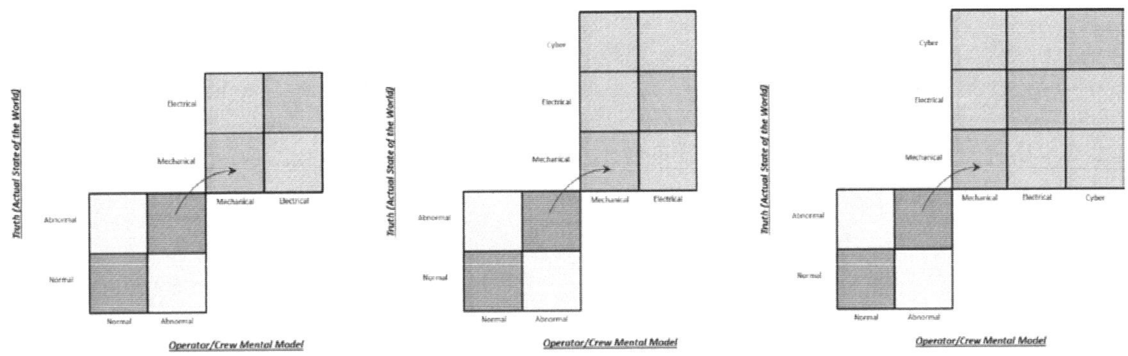

(a) (b) (c)

Figure 5. Actual State and Mental Model Abnormal Situations.

In the case of a cyber attack, the ability of the operator to shift from normal to abnormal relies on the CSM accurately detecting cyber events and providing e the additional information elements necessary for the operators to determine what course of action to take. These requirements for CSM are being fed back to the technology developers in the RAUSO program to ensure maximum benefits may be provided by the CSM.

Unfortunately, cyber events can occur for which providing an alert to the operator is not possible as some attacks are subtle and stealthy. How can the operator move quickly from normal/normal to abnormal/abnormal cyber when no alert is present? Recall from earlier in the discussion, the hypothesis is that operators in the normal/normal state are using a more passive reasoning method to maintain their awareness of the actual state of the system and, thus, are typically not actively cross-checking information against expected information for multiple sources but are instead fitting observed information into their definition of normal. This method will likely cause them to miss an unalerted cyber event that does not cause an obvious change in system behavior. Operating in this new environment requires imparting knowledge of various potential cyber threats and how those events would likely manifest themselves to the crew. The challenge will be in getting operators to change their reasoning approach. Another alternative is to design new interfaces that highlight the information elements necessary for crews to understand the situation in such a way that they can readily perceive mismatches in the case of cyber attack, allowing them to process the information in a way. The design of such interfaces is a challenge for future research, but raising awareness that cyber events are possible, describing how those could manifest themselves during a mission, and alerting operators to detectable cyber events is a good first step.

Conclusion

This study provides a significant first step in understanding how RPA operators need to receive information about cyber attacks in order to maintain mission effectiveness. Clearly, the best situation is to have a CSM that can detect the type of threat and provide information on how best to respond. Integrating cyber alerts and checklists into the standard format for mechanical and electrical alerts and checklists provides a sense familiarity for the operators when dealing with these new types of threats. However, in the future, new interfaces will need to be designed so operators can cross-check multiple sources of information quickly and efficiently so they can also detect and appropriately respond to cyber attacks that have gone undetected by the CSM.

Acknowledgements

We would like to acknowledge Dr. Steven King from ASD(R&E) for supporting the RAUSO program, Mr. Herb Mullens for providing programming support, and the Vigilant Spirit Control Station Team in 711 HPW/RHCI; specifically Greg Feitshans, Jason Davis, Jimmy Whelan, and Mark Squire.

References

Dinsmore, M., Jella, C., Lewis, M., Neal, D., Rush, J., Gay, C., Horowitz, B., Lau, N., and Leach, K. (2015). Developing cyber security CONOPS for remotely piloted aircraft (RPA) missions. *MITRE Technical Report 150032*, Bedford, MA: The MITRE Corporation.

Dukes, A. W., Fox, O. M., Rigrish, R. N., Durkee, K. T., and Feeney, J. J. (2016). *Resilient & Assured UAS Systems & Operations (RAUSO) Hybrid Cognitive Task Analysis.* (Air Force Research Laboratory Publication). Wright-Patterson AFB, OH: Author.

Feitshans, G. L., Rowe, A. J., Davis, J. E., Holland, M., & Berger, L. (2008). Vigilant spirit control station (VSCS) "The face of COUNTER". *AIAA Guidance, Navigation and Control Conference and Exhibit*, 1-12. doi: 10.2514/6.2008-6309

Gross, C. (2016). AFMC commander says cyber threats are real, need to get ahead of them. *USAF News and Information.* Washington DC: USAF. Retrieved from http://www.af.mil/News/ArticleDisplay/tabid/223/Article/951715/afmc-commander-says-cyber-threats-are-real-need-to-get-ahead-of-them.aspx

Scientific Advisory Board. (2010). *Operating Next-Generation Remotely Piloted Aircraft for Irregular Warfare.* SAB-TR-10-03, Washington DC: Author. Retrieved from https://publicintelligence.net/usaf-drones-in-irregular-warfare/

Skowronski, W. (2016). Vulnerability in cyberspace. Air Force Magazine, November/December, 52-53. Retrieved from http://www.airforcemag.com/MagazineArchive/Magazine Documents/2016/November 2016/1116cyber.pdf

United States Air Force. (2014). *RPA Vector: Vision and Enabling Concepts 2013-2038.* Washington DC: Author.

PEGASAS: WEATHER TECHNOLOGY IN THE COCKPIT

Ian Johnson
Federal Aviation Administration
Washington, DC

Geoff Whitehurst, Vladimir N. Risukhin, Lori J. Brown, and William Rantz
Western Michigan University
Kalamazoo, MI

Thomas K. Ferris, Trey Roady, and Carolina Rodriguez-Paras
Texas A&M University
College Station, TX

Kathryn Tippey
Vanderbilt University
Nashville, TN

Mel J Futrell
Glendale Community College
Glendale, CA

Research shows that a high percentage of weather-related General Aviation (GA) accidents can be attributed to pilots flying into Instrument Meteorological Conditions (IMC) without experience or appropriate certifications to safely operate beyond Visual Flight Rules (VFR). To make safety-critical decisions, pilots often use weather indication delivered on screens of portable electronic devices. This information often is obsolete with a latency up to 20 minutes. Web-based experiential education modules, using a flight simulation system for demonstration of this weather indication latency, can potentially mitigate this problem. Modules will be designed to provide pilots with the ability to "experience" different weather phenomena and will include tools to improve knowledge and skills for assessing deteriorating conditions and making effective decisions at imbedded decision points. This research also studied methods for delivering weather alert messages to pilots in-flight to support pilot reception of critical messages, examining effects of decision-making, workload, and situation awareness.

The Weather Technology in the Cockpit (WTIC) program is a Federal Aviation Administration (FAA) Next Generation Air Transportation System (NextGen) weather research program comprised of a portfolio of research projects in the Partnership to Enhance General Aviation Safety, Accessibility, and Sustainability (PEGASAS). The overarching goal of the FAA's WTIC Program is to identify, develop, verify, and validate a set of FAR Part 121, 135, and 91 Minimum Weather Service (MinWxSvc) recommendations to enhance pilot weather decision-making when faced with potentially hazardous weather conditions. The WTIC MinWxSvc will be associated with: the minimum cockpit meteorological information; the minimum performance standards and characteristics of the meteorological information; rendering guidance for the meteorological information; and enhanced meteorological information pilot and technology training.

The portfolio of projects will perform the research necessary to address the overarching WTIC research questions. Instrumental in gaining answers to these questions is research to identify and address weather-related "gaps" and "shortfalls" in; cockpit meteorological (MET) information, pilot training, pilot understanding and interpretation of MET information, technological and human factors issues associated with presenting cockpit MET information, and any operational, efficiency or safety risks associated with these gaps and shortfalls.

Information Technology Based Demonstration of Weather Indication Latency

Weather related accidents in GA often are caused by a pilot's inability to adequately perceive and assess actual meteorological conditions (Pearson, 2002; Aarons, 2014). While making in-flight weather relevant decisions, pilots of many GA light aircraft rely on information displayed on screens of various portable electronic devices capable of producing weather radar images (Pope, 2015). These images show weather situations that existed some time ago at the aircraft's current position. The time difference between the radar weather image and actual flight weather conditions may be as long as 20 minutes (Trescott, 2012; Zimmerman, 2013). The outdated weather information can prompt GA pilots' making safety-threatening decisions to continue flights into rapidly deteriorating weather conditions for which they or their aircraft are not certified.

Studies aimed at understanding and preventing the weather radar latency harmful influence on decisions made by GA pilots can benefit from researchers' ability to demonstrate the weather radar latency in simulated aircraft flight environment. This ability has been achieved by means of simultaneous utilization of commercially available software and hardware products in a weather information latency demonstrator (WILD). The WILD consists of two flight simulators that can be run simultaneously imitating the same flight scenario. While the aircraft current geographical position in both of the simulated flights is identical, a time difference (latency) can be set between two moments of the weather system development shown in each of the flight simulators at the geographical area corresponding to the aircraft actual position.

One of the flight simulators replicates a GA aircraft cockpit surroundings showing a visual picture of the weather environment seen through the aircraft windows. Another flight simulator generates the weather system status as it was some time ago. The information of this previously existed status of the same weather system is shown in the first flight simulator as an image on an electronic display imitating the weather radar information that GA pilots can see on screens of their portable electronic devices. The researcher can establish the time difference (latency) between two manners of the same weather system representation: the actual weather picture seen from the cockpit at the current moment of the flight, and the radar screen image showing the status of the same weather system that existed earlier at the current location of the aircraft.

In each of the two flight simulators, the Microsoft Flight Simulator X (Williams, 2006) or its successor Prepar3D (Lockheed Martin Corporation, 2015) software programs generate the aircraft cockpit environment and the flight progress. Active Sky (HiFi Simulation Technologies, 2015) software products run in each flight simulator generate correspondingly ether current or previously existed (delayed) weather conditions expressed in two appearances: as a visual picture seen from the aircraft cockpit, and as a weather radar image [Figure 1].

Figure 1. The weather radar image on the portable device screen shows a better weather situation that existed earlier at the aircraft current position.

To achieve realism of the weather relevant flight simulation, identical flight control inputs must be applied in both of the flight simulators. The WidevieW (Napolitano, 2015) software makes possible simultaneous control of the simulated aircraft flight path in both of the flight simulators by using only one set of flight controls (control wheel, rudder pedals, and engine power control) located in the flight simulator that imitates the aircraft cockpit.

Experiential Education - A New Approach to Helping Improve Judgement and Decision Making Skills of General Aviation Pilots in Adverse Weather

As previously stated, several gaps associated with VFR-IMC transitions, have been identified by the WTIC research team. These gaps include factors addressing pilot knowledge, skills and abilities (KSAs), where KSAs are defined as: Knowledge - Memorized facts/information - assessed by question/answer tests compared to a standard; Skills - Understand how to apply the knowledge - assessed by manipulating something compared to a standard; Abilities - The proficiency to take action at appropriate prompt(s) - assessed by timely application of appropriate knowledge and skills to novel prompts compared to a standard time range. The gaps identified by the WTIC team in these areas are:

Knowledge Gaps

1. *Lack of training (mainly due to little opportunity) for student pilots to fly in and experience different weather patterns and their associated visual and other cues.*
2. *GA pilots often do not understand the limitations of the technology in the cockpit.*

Skill Gaps

3. *There is a perceived gap in skills related to VFR-into-IMC decision-making.*
4. *Lack of Situational Awareness relating to VFR-into-IMC.*
5. *Retention of weather knowledge was identified as a gap.*

Ability Gap

6. *Lack of ability of pilots to correlate, interpret, and apply weather information related to VFR-into-IMC weather factors, specifically convection, icing, lowered ceilings, quickly emerging weather events, precipitation, or pilot-reported turbulence.*

Research conducted during a previous phase of the WTIC research project to address Gaps 4 and 5, showed only limited effects when using classroom-style, knowledge-focused education modules to affect pilot situation awareness and decision making in simulator tasks.

Other researchers (Wiggins & O'Hare, 2003; Ball, 2008; Knect, Ball & Lenz, 2010a; Knect, Ball & Lenz, 2010b; and Vincent, Blickensderfer, Thomas, Smith, and Lanicci, 2013) have also studied pilot awareness and decision-making, and much of their research has been studying training in the use of and understanding of weather products and weather information in aviation. One of their findings shows that pilots rarely receive any formal training on the use of weather-related equipment and tools, and often lack the skills to apply weather knowledge to effective decision-making.

Since increasing weather knowledge did not result in improving GA pilot's skill in applying their increased knowledge, the next step in trying to mitigate the higher-level gap, the "Ability" gap, is to develop a method of improving "skill" rather than "knowledge." Therefore, in this phase of the research, the team is developing several "Experiential Education" modules, which are designed to improve GA pilot's skill, gaps 3, 4 and 5, in applying weather knowledge rather than just increasing a GA pilot's knowledge/understanding of weather. An analogy would be to "show" someone how to do something rather than try to just try to "explain" how to do it.

The Experiential Education modules are being developed using the WILD simulator to provide video clips of specific flight environments. The flight environments will be based in different areas of the United States with a typical weather phenomena that may be associated with that specific area, e.g. Lake Effect Snow in Western Michigan. The modules will "fly" the pilot through weather phenomena such as decreasing visibility, developing convective clouds (thunderstorms), icing conditions, etc. The modules will start with a practice session on recognizing important aspects of the flight conditions in which they are "flying," such as estimating flight conditions (VFR, MVFR or IFR) based on their estimate of in-flight visibility, or identifying possible icing conditions, etc. Following the practice session, the pilots will "fly" a scenario which will include decision points, where the pilot will be asked, based on the in-flight conditions at that time, if the flight should continue or a diversion be initiate. Feedback will be provided, based on the decision made. These modules are designed to help improve a pilot's skill in applying weather knowledge to effective decision-making, and thus help mitigate the "Skill" and "Ability" gaps.

General Aviation Weather Alerting: Effectiveness of Display Characteristics in Supporting Weather-Related Decision Making

To employ effective decision making, pilots must first be aware of changes in conditions requiring an amendment to their actions (i.e decision points). Weather and visibility at takeoff can degrade quickly while in-flight, prompting pilots to rely on in-cockpit weather technologies to alert them to these changes, and provide some information necessary to modify their flight plan.

Methods

A research study was conducted in flight training devices at the FAA's William J. Hughes Technical Center and investigated two methods for receiving weather alert messages: via text messages embedded in a complex graphical map display, and via messages displayed on a smartwatch worn by participants. The text of the alerts followed the Lockheed Martin Flight Services Adverse Condition Alerting Service (LMFS ACAS) format. See Figure 2 below for an illustration of these two display conditions, as well as a representative flight training environment.

Figure 2. Flight training environment with weather alerts displayed as textboxes embedded in graphical display (right side, touch activated) and as text on a Pebble smartwatch worn by pilots on the left wrist.

Various measures of pilot decision-making accuracy and timing, as well as situational awareness, were collected to assess differences attributable to the display format. Additionally, vibrotactile cueing conditions (presented via the vibrating motor in the smartwatch) were examined to determine how nonvisual cues could support attention and interruption management during flight. While each participant completed two flight scenarios, one set in Alaska and one in New Mexico, each with a different display configuration (complex graphical and smartwatch), the vibrotactile cueing condition was handled as a between-subjects variable. Participants either received no vibratory cues (visual text only), received a single vibration pulse with each incoming message, or received "urgency-mapped" graded pulses, which would present pulses with higher frequency and duration when messages were more important and urgent.

Thirty-two pilots (3 female; ranging from age 20 to age 79 with an average age of 53 years) participated in the study. Each scenario began with the pilots mid-flight and with different types of degrading weather and visibility developing at the intended destination. A scenario was complete when pilots either verbalized their intent to divert or otherwise flew into IMC conditions. Completion of each scenario took approximately between 10 and 25 minutes. Scenarios involved three experimental alert messages of varied importance/urgency, and delivered when pilots were under varied task loads. Time and accuracy in responding to the alerts were coded by multiple observers and represent the main dependent variables of interest. Additionally, situation awareness probes in the forms of ATC-issued status checks were issued following the Situation Present Assessment Method (Durso, Dattel, Banbury, & Tremblay, 2004), and NASA-Task Load Index (TLX) surveys (Hart & Staveland, 1988) and functional near-infrared spectroscopy (fNIRS) were used to measure pilot workload while performing flight-related tasks.

Results

Due to unforeseen issues in data recording, some datasets were lost, leaving the data from 27 pilots for statistical analysis. Of these, few pilots (9 of 27, 33%) made the explicit decision to divert their aircraft as conditions deteriorated, and not a single pilot correctly diverted in both scenarios. For the measure of decision making accuracy, no significance was found for the main variables of interest: display configuration and vibrotactile cuing condition.

Response time to the alerts, however, was significantly ($p < 0.05$) affected by the vibrotactile cuing condition, showing considerably faster responses when a vibratory cue accompanied the delivery of the text-based alert message (see Figure 3, left side). Although there is an apparent trend showing faster responses for the text embedded in the graphical display of the Tablet, these differences did not reach significance.

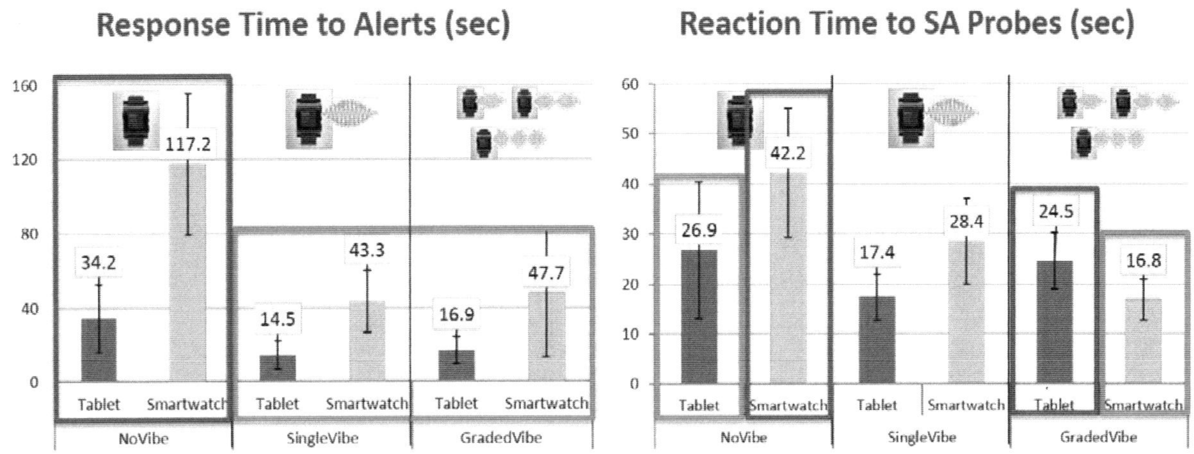

Figure 3. Left: Response time (in seconds) to alerts under each display configuration (Tablet or Smartwatch) and vibratory cuing condition (no vibration, single vibration, or graded vibration). Right: Reaction time to situation awareness (SA) probes. In each case, boxes around the data represent significant differences in the data.

The response time to situation awareness (SA) probes (status report requests from ATC) was also significantly affected by an interaction between display configuration and vibratory cuing condition (see Figure 2, right side). This effect shows that when no vibratory cues are presented, probe responses are faster (thus, SA levels are higher; Durso et al., 2004) with the Tablet-based presentation of alerts. With urgency-mapped graded vibratory cues, however, the Smartwatch conditions showed the higher levels of SA.

Workload measures showed no significant differences in NASA-TLX survey data, but a significant effect was found with the fNIRS data that suggested lower overall workloads when alerts were embedded in the graphical display (mean HbO_2 level of 0.21; max of 2.94) than when they were displayed on the smartwatch (mean HbO_2: 0.72; max: 3.34).

Conclusion

The results offer some answers for appropriate means of providing weather information to GA pilots and also new questions for future research. Messages displayed in existing onboard visual displays, such as GPS maps, show some significant benefits over exclusively watch-based messaging in terms of the time it takes to receive and process the messages. However, using vibrotactile cuing to accompany the arrival of new weather messages was more impactful, showing a substantial benefit over conditions that did not involve vibratory cuing. Building on this work and those of others that investigated vibrotactile cuing to support pilot awareness (e.g., Ahlstrom, Caddigan, Schulz, Ohneiser, Bastholm, & Dworsky, 2015; Sklar & Sarter, 1999), future research will seek to define more clearly effective urgency-mapped tactile encodings for common weather alerts and messages.

References

Aarons, R. N.(2014). Weather-related loss of control. *Aviation Week Network.* July 1, 2014. Retrieved from http://aviationweek.com/business-aviation/weather-related-loss-control

Ahlstrom, U., Caddigan, E., Schulz, K., Ohneiser, O., Bastholm, R., & Dworsky, M. (2015). The Effect of Weather State-Change Notifications on General Aviation Pilots' Behavior, Cognitive Engagement, and Weather Situation Awareness. Technical report DOT/FAA/TC-15/64, available at: http://elib.dlr.de/104081/1/full_text.pdf

Ball, J. (2008). The impact of training on general aviation pilots' ability to make strategic weather-related decisions (Rep. No. DOT/FAA/AM-08/3). *Washington, DC: Federal Aviation Administration. Retrieved from http://www.faa.gov/data_research/research/med_humanfacs/oamtechreports/2000s/media/20083.pdf*

Durso, F. T., Dattel, A. R., Banbury, S., & Tremblay, S. (2004). SPAM: The real-time assessment of SA. *A cognitive approach to situation awareness: Theory and Application, 1*, pp. 137-154.

Hart, S. G., & Staveland, L. E. (1988). Development of NASA-TLX (Task Load Index): Results of Empirical and Theoretical Research. In A. H. Peter & M. Najmedin (Eds.), *Advances in Psychology, Vol. 52*, pp. 139-183): North-Holland.

HiFi Simulation Technologies (2015). Active Sky Next user's guide. Apr 16, 2015. Retrieved from http://nimbus.hifitechinc.com/ASNInfo/ASN_User_Guide.pdf

Knecht, W.R., Ball, J., & Lenz, M. (2010a). Effects of video weather training products, Web-based preflight weather briefing, and local versus non-local pilots on general aviation pilot weather knowledge and flight behavior, Phase 1. *(Technical Report DOT/FAA/AM-10/1). Washington, DC: Federal Aviation Administration, Research and Development Service.*

Knecht, W.R., & Smith, J. (2010b). Effects of training school type and examiner type on general aviation flight safety. *(Technical Report). Washington, DC: Federal Aviation Administration, Research and Development Service.*

Lockheed Martin Corporation (2015). Prepar3D Product Overview. Retrieved from http://www.prepar3d.com/product-overview/

Napolitano, L. (2015). WidevieW. Retrieved from http://www.wideview.it/wideview.htm

Pearson, D. (2002). "VFR flight not recommended". A study of weather-related fatal aviation accidents. *NOAA/National Weather Service, Atlanta Center Weather Service Unit.* Retrieved from http://www.srh.noaa.gov/topics/attach/html/ssd02-18.htm

Pope, S. (2015). 8 Offbeat aviation apps. *Flying - magazine.* May 28, 2015. Retrieved from http://www.flyingmag.com/avionics-gear/pilot-supplies/8-offbeat-aviation-apps

Sklar, A. E., & Sarter, N. B. (1999). Good vibrations: Tactile feedback in support of attention allocation and human-automation coordination in event-driven domains. *Human factors, 41*(4), 543-552.

Trescott, M. (2012). NTSB Alert - In-cockpit NEXRAD weather age misleading and killing pilots. *Aviation Training, General Aviation Safety.* June 20, 2012. Retrieved from http://www.maxtrescott.com/max_trescott_on_general_a/2012/06/ntsb-alertin-cockpit-nexrad-weather-age-misleading-and-killing-pilots.html

Vincen M., Blickensderfer E., Thomas R., Smith M., and Lanicci J., 2013. In-Cockpit NEXRAD Products. *Training General Aviation Pilots, 57(1), p. 81-85.*

Wiggins M. W. & O'Hare D., (2003). Expert and novice pilot perception of static in-flight images of weather. *International Journal of Aviation Psychology, 13 (2003), pp. 173–187 http://dx.doi.org/10.1207/S15327108IJAP1302_05*

Williams, B. (2006). Microsoft Flight Simulator as a Training Aid: A Guide for Pilots, Instructors, and Virtual Aviators. *Aviation Supplies and Academics, Inc.*, Newcastle, Washington, 2006, 241 pp.

Zimmerman, J. (2013). Thunderstorm flying in the age of datalink weather. *Weather Geek.* September 8, 2013. Retrieved from http://airfactsjournal.com/2013/09/thunderstorm-flying-in-the-age-of-datalink-weather/

DOES THE PROJECTED-HAND ILLUSION HELP IN TELEOPERATION?

Lloyd Wajon[a], M. M. (René) van Paassen[a], David A. Abbink[b], Erwin Boer[b], Max Mulder[a]
Aerospace Engineering[a] & Mechanical Engineering[b] – Delft University of Technology
Delft, The Netherlands

A body illusion, commonly known in the form of the "Rubber Hand Illusion", is an illusion wherein visual inputs on an inanimate object and simultaneous tactile inputs on a part of the body lead to a situation where the inanimate object is identified as the body part. This study investigated the possibility of inducing a body illusion during a teleoperated reaching task, to see if this leads to increased telepresence and improved accuracy. Three conditions were presented in random order; the Direct Control (DC) condition, where the participant's hand is shown on the screen, the Projected Hand Illusion (PHI) condition, showing the slave device consisting of a 3D-printed hand designed to induce a body illusion, and the no Projected Hand Illusion (nPHI) condition, showing the slave device consisting of a 3D-printed object of appropriate shape but designed to not induce a body illusion. Reaching performance was interpreted in terms of position error, for which a significant difference was found between conditions PHI and nPHI. In the nPHI condition, participants kept more distance to the obstacle than in the PHI condition. Potential causes for this difference are an increased perception of risk due to a difference in visual perception, or subtle visual differences in between the two conditions.

Introduction

Teleoperation enables humans to interact with a remote or hostile environments, while preserving their capability of adapting to and coping with dynamic and unpredictable situations. The ideal for teleoperation is to reach optimal "telepresence" (Sheridan, 1989), so that tasks can be performed as if the human were physically in the remote environment. However, limitations in communication bandwidth, time delays and other restrictions make that in teleoperation, simple tasks can be made more challenging, as spatial awareness is degraded (Chen, Haas, Pillalamarri, & Jacobson, 2006). Solutions and improvements include the presentation of additional visual information (Azuma, 1997), aid from automation with shared control (Boessenkool, Abbink, Heemskerk, van der Helm, & Wildenbeest, 2013) or techniques to enhance teleoperation transparency (Okamura, 2013). This study investigates whether Body Illusions (BI) can play a role in improving teleoperation.

Normally, humans perceive the world and their body by a process called sensory processing. Two types of sensory processing are distinguished (Ernst & Bülthoff, 2004), sensory combination and sensory integration. In sensory combination, non-redundant information from different modalities regarding some environmental property are combined to provide a more robust estimate of the property in question. For instance, visual and haptic information about the shape of an object can be combined to provide a better estimate than information from either of the modalities alone could (Helbig & Ernst, 2007). Sensory integration integrates redundant signals from different sensory modalities such that "a coherent multisensory percept is formed" (Ernst & Bülthoff, 2004).

Sensory processing can be seen as a bottom-up process; an estimate of the body or environment is constructed solely from sensory information. However, this is complemented by a top-down process. Prior knowledge provides a logical framework to make sense of the incoming signals. Even though one's prior knowledge is built up over all the years of one's life, it is possible for "bottom-up perceptual mechanisms" to temporarily override it (Slater, Spanlang, Sanchez-Vives, & Blanke, 2010). Body illusions are one example of this phenomenon. During a BI, one experiences an illusory ownership over a fake limb, or even a complete fake body (e.g., (Botvinick & Cohen, 1998), (Ehrsson, 2007), (Slater et al., 2010)). For instance, in the original Rubber Hand Illusion, congruent visuotactile stimulation applied with a paintbrush on a fake hand and one's own unseen hand invokes the illusory perception that the fake hand is actually one's own hand (Botvinick & Cohen, 1998). It has also been proven that a visuomotor correlation (i.e. through initiating active movement and visually perceiving an identical and congruent movement in a fake limb or body) can induce a similar illusion (Dummer, Picot-Annand, Neal, & Moore, 2009), (Sanchez-Vives, Spanlang, Frisoli, Bergamasco, & Slater, 2010).

This study proposes the concept of applying the unique properties of body illusions (e.g., feeling of ownership over fake limbs or body) to the field of teleoperation, to try and enhance teleoperation performance by increasing spatial awareness. This concept is in accordance with Sheridan's supposition: "Identifying with remote arms, eyes or body, especially when there is geometric correspondence, would seem to have advantages", from

(Sheridan, 1989, p. 497). But, as Sheridan also stated: "However, it is not well understood why, or even whether, a feeling of presence enhances observing or acting, whether remotely or not", also from (Sheridan, 1989, p. 497).

To find an answer to this question, a simplified teleoperation set-up was created in which an added tactile stimulus was introduced to increase the induction of the body illusion. In different conditions, the performance of teleoperation with this set-up was evaluated.

Method

Participants

Sixteen participants, aged between 20 and 54, all male, participated in the experiment. One of the participants was left handed, but stated to normally control a computer mouse with the right hand, indicating that he would not have a problem performing the teleoperation task with the right hand.

Experimental set-up

The experimental set-up used a 3-DoF custom-built planar teleoperation system at the department of Mechanical Engineering of Delft University of Technology (Christiansson, 2007). It consists of a parallel, non-compliant, master and a serial, compliant slave. The master and slave were mounted on the base of two separate, custom-built wooden set-ups. These were each equipped with a webcam (Logitech HD Pro Webcam C920), a transparent acrylic plate holding three target obstacles and a cloth curtain. The webcams were used to provide a video-feed of the workspace on either the master or slave side, which was then displayed on a 17" monitor (HP 1740), on a stand behind the master set-up.

The obstacle plates both held three obstacles. The plates' position can be adjusted, to accommodate for different-sized hands and fingers. In the same way, the relative position between the manipulators and the obstacles of both sides can be properly adjusted to match the master and slave teleoperation set-ups. A cloth curtain was spanned across the width of the master-setup to provide tactile input to the participant's hand during the experiment. For visual consistency between both setups, the curtain is also present in the slave setup.

Paricipants were seated in front of the master setup, and placed their right hand on the computer mouse-like manipulator attached to the master end of the teleoperator. Their hand was never directly visible; depending on the experiment condition, either the webcam image of the hand and master side, or the webcam image of the slave side was shown (Fig. 4).

Three different attachments were created with a 3D printer, these were: (1) A mouse-like manipulator for the master side, with a ramp for the finger of the participant's hand; (2) A realistic hand, with the index finger extended and mounted on top of the same mouse-like manipulator. A nitrile glove is wrapped around the hand, to obscure its plastic nature and increase the visual similarity between the attachment and the participants' own hand, for this participants wore an identical glove during the experiment. This attachment was used for the slave side; (3) A second mouse-like shape for the slave side, with a rod attached to the ramp to have the same effective dimensions as the manipulator, but made to not resemble the participant's own hand. For added distinction, a plastic tie-wrap was tied around the rod. The different attachements as fixed to the manipulator are shown in Fig. 1.

Task description

The task environment contained three obstacles which needed to be avoided, suspended slightly above the participant's hand, with the obstacles' heads at about the same height as the participants' fingertips. The front target is the base (or B), the latter targets are called L and R, for left and right obstacle. See Fig. 3. Four targets were defined relative to the L and R obstacles, two lateral targets (located left of the obstacles) and two longitudinal targets in front of the obstacles. The targets are strictly virtual and they are not visible on the video-feed. The targets were numbered from 1 to 4 for analysis purposes. During the experiment, targets were indicated on the upper part of the screen, above the webcam view, and participants were asked to quickly move from the base position to the target, without hitting or touching any of the obstacles.

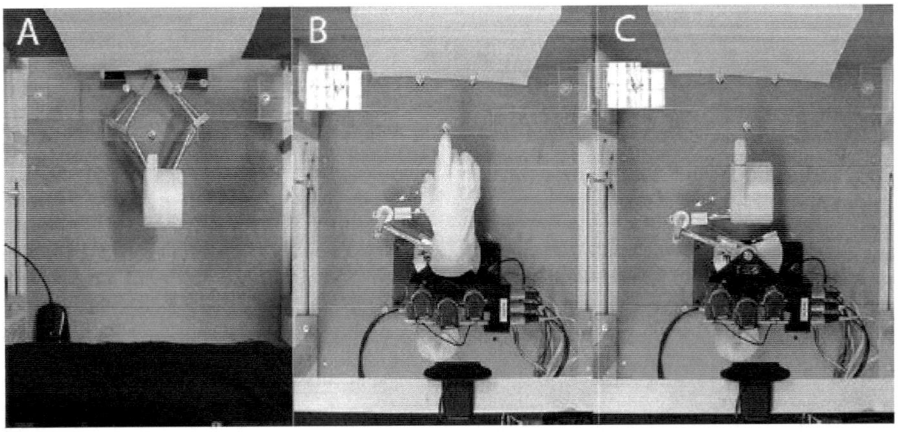

Figure 1: All three attachments, as mounted on the setup. A: attachment for the participants to hold, mounted on the master-side; B: attachment with the realistic hand, mounted on the slave-side; C: attachment with the finger-like boom, mounted on the slave side

Experiment Design

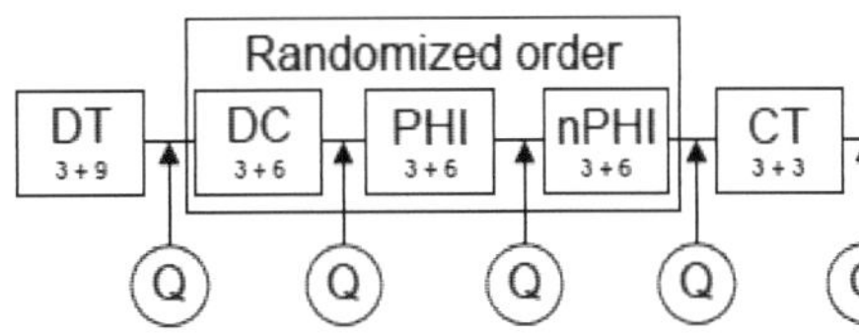

Figure 2: Schematic overview of the order of conditions. The numbers below the abbreviated condition names indicate the training and measurement sets, respectively. The Q's represent the moments of filling out the questionnaire.

The experiment started with a training consisting of three parts. During the first part subjects received on-screen feedback on the registered position relative to the target and were asked to touch the L or R obstacle before moving to the target. In the second part of the training, the obstacle was not to be touched, but feedback was still provided. The third training set had no feedback out the questionnaire. Then 9 measurement sets (of 4 movements each) were performed (DT). Subjects were shown the master setup and their own hand on-screen. This, in combination with the movement, and the sensation from the curtain should induce the Projected Hand illusion.

The measurement conditions, Direct Control (DC), with view of the master side and the subject's hand, Projected Hand Illusion (PHI), with view of the slave side and the fake hand, and non-Projected Hand Illusion (nPHI), with view of the slave side and fake finger, were offered in randomized order, each of these consisted of 3 training sets and 6 measurement sets. The experiment was closed off with a control condition (CT), equal to the DT condition, to check for learning during the experiment. A questionnaire on the strength of the body illusion was taken after each condition (Fig. 2).

Hypotheses and metrics

A 20 item questionnaire on the body illusion, adapted from (Graham, Martin-Iverson, Holmes, & Waters, 2014; Longo, Schüür, Kammers, Tsakiris, & Haggard, 2008), was answered after each condition. From this questionnaire, the question that most seemed to define the occurence of the body illusion, "It felt like . . . the wooden hand was my hand." was used to divide the participants into two groups. The 5 participants who scored higher on this question for the PHI condition compared to the nPHI condition were grouped in a qualifying group (Q), the remaining 11 participants were assigned to the non qualifyng (nQ) group. It was hypothesized that the sensation of ownership of the artificial hand would affect performance on the task, and that would make performance in the DC and PHI conditions similar, and different from the nPHI condition, while for the nQ group the performance in the PHI and nPHI condition is expected to be similar. Data were recorded on the slave and master positions, and performance on reaching the target position was calculated. Data on timing proved to be not consistent enough for further analysis.

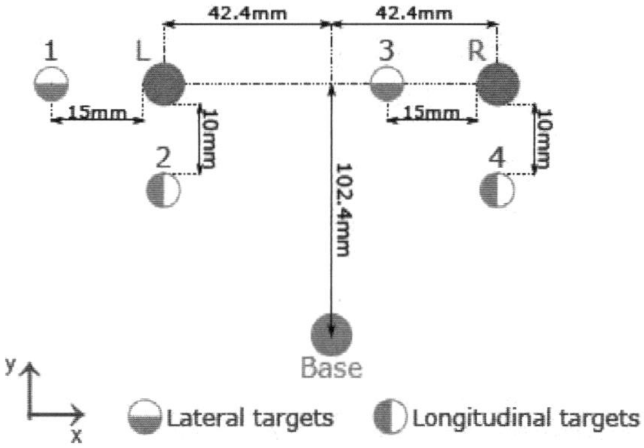

Figure 3: Schematic lay-out of the 3 obstacles and the 4 targets as seen from the top, including dimensions and target numbering. The diameter of the obstacles is 6.5mm. Note: figure not to scale.

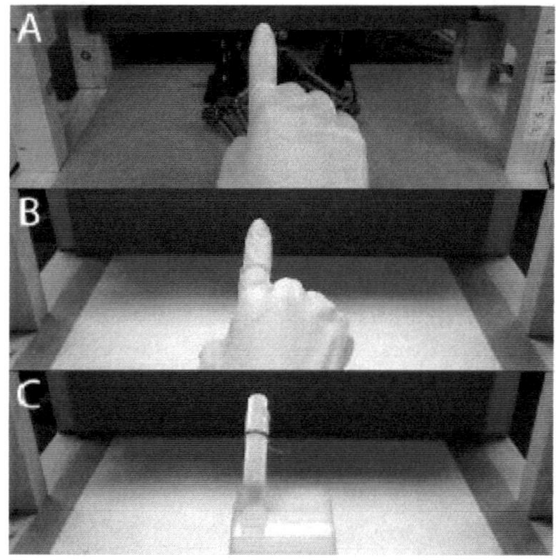

Figure 4: Screenshots (cropped) showing what participants saw on the monitor during the experiment fc the different conditions. A: DT condition, DC condition and CT condition (own hand, master side); B: PHI condition (fake hand, slave side); C: nPHI condition (fake finger, slave side).

Results

On the basis of the questionnaire only 5 participants were grouped in the Q-group. However all but one of the participants in the Q-group judged the experience of the BI as "neutral" or "positive" in condition nPHI as well. As condition nPHI was designed not to induce the Projected Hand Illusion, this is an unexpected result. This was not limited to the Q group, most participants actually did not experience a real difference in BI between conditions PHI and nPHI, based on their responses and the used rules for assessment (BI is present when score is 0 or higher); of the 16 total participants, 11 participants filled out a score of 0 or higher in both PHI and nPHI, and 2 participants filled out a score lower than 0 for both conditions. This means that only 3 participants noticed an actual difference in BI between the two conditions; only 1 participant experienced the BI only in the PHI condition (in accordance to the design of the experiment), while 2 other participants experienced the BI only in the nPHI condition (the opposite of what was designed to happen).

Fig. 5 shows the deviation between the position of reached and intended target in y-direction (longitudinal direction from the participant's point of view), for all trials of the participants in the qualifying group (Q-group, n=5). The four targets and the three conditions of interest (i.e. DC, PHI, nPHI) are depicted separately. The positive direction is defined as being directed away from the participant. Each participant completed each target 6 times during each condition, except for one participant who completed each target 5 times in the DC condition due to an error during measurements.

A one-way analysis of variance (ANOVA) was performed over all three conditions for each target separately. Therefore, a Bonferroni correction was applied, reducing the alpha level to 0.0125 instead of 0.05. A significant difference was found for target 2 (**F**(2, 84) = 19.36, **p** < 0.001) and target 4 (**F**(2, 84) = 11.61, **p** < 0.001). Subsequently, dependent t-tests were performed over conditions PHI and nPHI for target 2 and target 4, confirming significant differences in performance between these conditions.

Similarly, Fig. 6 depicts the deviation between the position of reached and intended target in y-direction, but now for all trials of the participants in the non-qualifying group (nQ-group, n=11). Again, the four targets and the three conditions of interest (i.e. DC, PHI, nPHI) are depicted separately, and the positive direction is defined as being directed away from the participant. Each participant completed each target 6 times during each condition.

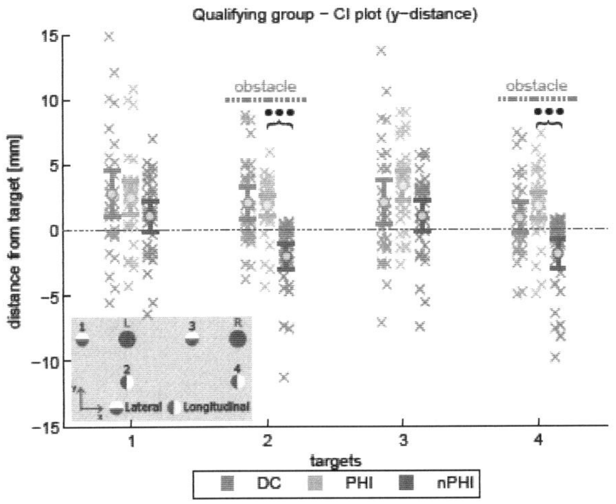

Figure 5: Distance in y-direction between reach and intended target, for the Q-group (n=5). Each cross represents the result of one trial, while the circles show the mean of all trials and the error bars indicate the 95% confidence intervals. Significance is denoted by "•", "••" and "• • •", representing $p \leq 0.05$, $p \leq 0.01$ and $p \leq 0.001$, respectively.

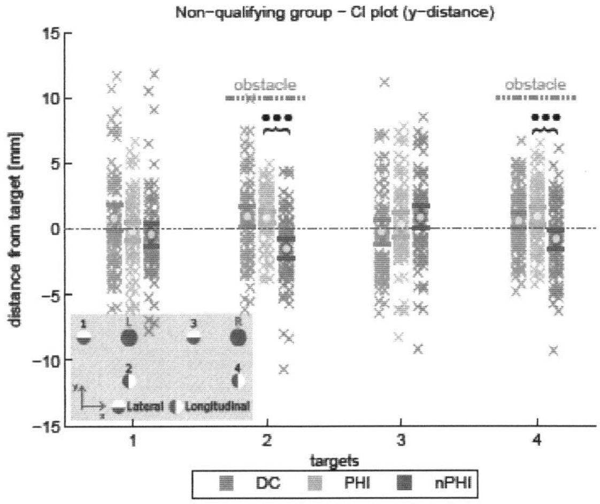

Figure 6: Distance in y-direction between reach and intended target, for the nQ-group (n=11). Each cross represents the result of one trial, while the circles show the mean of all trials and the error bars indicate the 95% confidence intervals. Significance is denoted by "•", "••" and "• • •", representing $p \leq 0.05$, $p \leq 0.01$ and $p \leq 0.001$, respectively.

An ANOVA over all three conditions for each target separately (again, a Bonferroni correction was applied, such that _ = 0:0125) showed that there is a significant difference for target 2 (F(2, 195) = 15.18, p < 0.001) and target 4 (F(2, 195) = 7.47, p < 0.001). Post-hoc dependent t-tests again showed the differences in between conditions PHI and nPHI.

Data for lateral direction were also analysed, and showed a significant difference between the PHI and nPHI conditions for targets 2 and 3, with small offsets to the left and to the right respectively (these data are not presented in this paper).

Discussion

Differences between performance in the PHI and nPHI conditions were found for two of the targets. However, the differences do not directly indicate an effect of the BI on teleoperation performance, since both the Q- and nQ- groups showed a similar effect. It can be argued that the manipulation – nPHI versus PHI condition – was ineffective in controlling the BI. The questionnaire results indicate that in many cases the judgment of BI was independent of the type of slave device shown, possibly due to a lack of sensitivity in the questionnaire, or possibly a gloved hand versus the more abstract computer mouse-like shape did not provide enough differentiation in sense of ownership of the controlled and visible device.

Indeed, the majority of participants verbally reported that after a while, controlling the finger-like boom felt natural and similar to controlling the fake hand or their own hand – even though they clearly noticed and reported that the finger-like boom did not look like a hand. Thus, despite clear visual discrepancies, participants may have accepted the fake finger as being their own, see again Fig. 4 for a visual comparison of the conditions as seen by the participants.

The remaining persistent differences between the nPHI and PHI conditions might also be attributable to other factors. To avoid systematic errors in the movement, the master and slave set-up were calibrated as accurately as possible to match up the distances to the target positions. For instance, a difference in end-point of the "fingertip" of the attachments used in conditions PHI and nPHI can easily cause a bias in the y-direction, and despite the calibration, these differences cannot be completely excluded. Also, as the attachments of conditions PHI and nPHI show quite some geometrical differences, making visual references provided by the attachments different. This difference can also influence the distance estimation by the participants. In addition, the attachments are not symmetrical and thus the left and right side of the attachments can provide differing visual references, which can specifically influence spatial estimations laterally.

Conclusion

The effect of Body Illusions (BI) on tele-operated reaching performance under three different conditions: Direct Control (DC, showing the own hand on the master side, gloved and visible on-screen), the Projected Hand Illusion (PHI, with 3D printed gloved hand visible on the slave side) and the no-Projected Hand Illusion (nPHI, with a mouse-like shape on the slave side). Based on differences in subjective responses to a question on how the virtual hand was experienced for the nPHI and PHI conditions, participants were assigned to a qualifying group or a non qualifying group, that experienced less or no body illusion sensation. Errors between target and final position after reaching were used as performance measure.

Consistent longitudinal differences in movement to the target position were found for the two positions where the finger longitudinally approached the target (complicated by limited depth perception), for the nPHI and PHI conditions. No significant differences were found between the DC and PHI condition. The results could not be unequivocally attributed to BI effects. Participants did indicate in the majority of the cases (80%) "ownership" of the shown device, both in the nPHI and PHI conditions, suggesting that the effect of a body illusion might already be present in situations where the manipulated device cannot be mistaken for a body part.

References

Azuma, R. T. (1997). A Survey of Augmented Reality. Presence, 6(4), 355–385.

Boessenkool, H., Abbink, D. A., Heemskerk, C. J., van der Helm, F. C., & Wildenbeest, J. G. (2013, January). A task-specific analysis of the benefit of haptic shared control during telemanipulation. IEEE Transactions on Haptics, 6(1), 2–12.

Botvinick, M. & Cohen, J. (1998, February). Rubber hands 'feel' touch that eyes see. Nature, 391(6669), 756.

Chen, J. Y. C., Haas, E. C., Pillalamarri, K., & Jacobson, C. N. (2006). Human-Robot Interface: Issues in Operator Performance, Interface Design, and Technologies. (July).

Christiansson, G. A. V. (2007). Hard Master, Soft Slave Haptic Teleoperation.

Dummer, T., Picot-Annand, A., Neal, T., & Moore, C. (2009). Movement and the rubber hand illusion. Perception, 38(2), 271–280.

Ehrsson, H. H. (2007, August). The experimental induction of out-of-body experiences. Science (New York, N.Y.) 317(5841), 1048.

Ernst, M. O. & Bülthoff, H. H. (2004, April). Merging the senses into a robust percept. Trends in cognitive sciences, 8(4), 162–9.

Graham, K. T., Martin-Iverson, M. T., Holmes, N. P., & Waters, F. A. (2014, August). The projected hand illusion: component structure in a community sample and association with demographics, cognition, and psychotic-like experiences. Attention, perception & psychophysics.

Helbig, H. B. & Ernst, M. O. (2007, June). Optimal integration of shape information from vision and touch. Experimental brain research. Experimentelle Hirnforschung. Expérimentation cérébrale, 179(4), 595–606.

Longo, M. R., Schüür, F., Kammers, M. P. M., Tsakiris, M., & Haggard, P. (2008, June). What is embodiment? A psychometric approach. Cognition, 107(3), 978–98.

Okamura, A. (2013). Teleoperation. Stanford: Stanford University.

Sanchez-Vives, M. V., Spanlang, B., Frisoli, A., Bergamasco, M., & Slater, M. (2010, January). Virtual hand illusion induced by visuomotor correlations. PloS one, 5(4), 1–6.

Sheridan, T. (1989). Telerobotics. Automatica, 25(4), 487–507.

Slater, M., Spanlang, B., Sanchez-Vives, M. V., & Blanke, O. (2010, January). First person experience of body transfer in virtual reality. PloS one , 5 (5), 9.

INVESTIGATING PILOT'S DECISION MAKING WHEN FACING AN UNSTABILIZED APPROACH: AN EYE-TRACKING STUDY

Maxime Reynal [1]
Frank Rister [2]
Sébastien Scannella [1]
Christopher Wickens [3]
Frédéric Dehais [1]
[1] *ISAE-SUPAERO, Université Fédérale de Toulouse, France.*
[2] *Truestream Aerospace GmbH, Hamburg, Germany.*
[3] *Department of Psychology, Colorado State University, Boulder CO, USA.*

Unstabilized approach has been identified to be a major causal factor of approach-and-landing accidents (e.g. off-runway touchdowns, hard landing, tail-strikes, etc). We conducted an experiment in order to analyze pilots' performance during such approaches. Ten type-rated, commercial pilots flew each in a B737 full-flight simulator during an unstabilized approach at Hamburg airport. The Pilot Flyings' (PF) eye gazes were collected. The results revealed that half of the pilots persisted in an erroneous landing decision. These latter pilots had higher dwell time on the attitude indicator/flight director whereas the group of pilots who perfomed the go-around exhibited more fixations on the navigation display prior to their final decision. These findings indicate that the decision whether to land or to go-around is taken considerably long before the respective task is executed, and that the use of heuristics impair pilot's performance.

Introduction

Unstabilized approaches have been identified to be a major causal factor of approach-and-landing accidents. Poor pilot performance in aircraft handling, system control or crew resource management during approach and landing reveal that, from the years 2001 to 2010, 49% of all fatal accidents worldwide occurred during the initial approach, final approach, or in the landing phase (Boeing, 2010). To approach and land safely, pilots are required to follow approach profiles fulfilling predetermined stabilization criteria based on flight parameters defined by the authorities, such as vertical speed, airspeed, or landing configuration in relation to the height above ground (ICAO, 2006; Airbus, 2006). If the criteria for stabilized approach are not met at the stabilization height (e.g. 1000 feet), a go-around is mandatory. However, continuation of an unstabilized approach has been found to be a causal factor in 40% of all approach. Combined with a system philosophy based on a master (human)-slave (machine) relation (Tessier & Dehais, 2012), today's flight deck automation has a significant negative impact on this demanding flight phase. Mode confusion or improper system state awareness significantly contribute to approach destabilization (Sarter & Woods, 1995). The aerodynamic characteristics of modern aircraft wings aggravate the competing physical interplay of altitude loss, deceleration, (vertical) speed restrictions and airplane configuration. This always creates added complexity in the pilot's decision-making process leading to high workload situation (Dehais et al., 2012) and thus promoting perseveration to land at all cost (Causse et al., 2013; Curtis & Smith, 2013).

One promising approach to better understand pilot's performance during unstabilized approach is to measure their eye movements. For instance, several accidents analyses pointed out that poor monitoring was a contributive factor involved in many accidents during the approach (Spangler & Park, 2010; National Transportation Safety Board, 2014a; National Transportation Safety Board, 2014b; Dutch Safety Board, 2010; Civil Aviation Authority, 2013). Little is known on how pilots actually supervise the flight deck during critical phases such as the approach. Interestingly enough, several studies revealed the suitability of the eye tracking technique for understanding attentional vulnerabilities of pilots (Dehais, Causse, & Pastor, 2008; Huettig, 1999; Kasarskis et al., 2001; Mumaw, Sarter, & Wickens, 2001; Sarter, Mumaw, & Wickens, 2007; Wickens, et al., 2008; Dehais et al., 2015).

In order to better understand why trained pilots may fail to adequately monitor flight parameters, the DGAC/DSAC [1] initiated the Pilot Vision project that aimed at analyzing eye tracking data collected during more

[1] Directorate for Civil Aviation Safety (DSAC), a service of the French Directorate General for Civil Aviation (DGAC).

than 100 approaches by ISAE-SUPAERO. A first study was conducted over 32 stabilized approaches and revealed that both Pilot Monitoring (PM) and Pilot Flying (PF) exhibited a high percentage of dwell time out of the window during the short final. For PMs this was to the detriment of the monitoring of the speed indicator (Reynal, Colineaux, Vernay, & Dehais, 2016). In the present study, we focused our analyses on an eye tracking dataset collected with 10 pilots facing unstabilized approach at the Hamburg airport. The approach was segregated in two major sequences to investigate pilots' decision making. The first sequence started from around 3000 feet to 2500 feet (*Approach Initiation* sequence), and the second one (*Decision* sequence) started from 2500 feet point until the Missed Approach Point (MAPt). The 2500 feet FL was somewhat arbitrary, but meant to initiate a time during which the instability of the approach should begin to be noticeable. As some pilots continued with the unstable approach beyond decisoin height, our intention was to compare the ocular scanning strategy between those who did, and those who did not. These data were then compared with another eye tracking dataset collected during a stabilized approach to identify potential different ocular strategy.

Material and Method

Participants

Ten airline Captains, including one female, volunteered to partipate to the experiment. They all endorsed the role of Pilot Flying. Their mean age was 44.45 years old (SD = 17.91; min = 23; max = 71) with a mean flight experience of 11372.73 hours (SD = 11899; min = 1500; max = 33000). On Boeing 737 NG, this group had a mean flight experience of 4402.55 hours (SD = 3558.54; min = 64; max = 9000). A confederate pilot was involved as PM to play a particular role during the flight, but their data are not included in the analyzed group as they are part of the experimental protocol.

Flight simulator and scenarios

The experiments were conducted on a Boeing 737 NG full-flight simulator of the CAE 600 series. It has a hydrostatic motion system with six degrees of freedom (6DOF), a Rockwell Suprawide Vision System and is certified as Level D/Zero Flight Time—this means that the simulator reflects the aircraft so realistically that operator training can be accomplished without the necessity to do further training on the real aircraft before the trainee pilot is allowed to fly with passengers on board.

The scenario consisted of an approach to Hamburg and began approximatively 50NM south of the field at Flight Level (FL) 150 (15,000 feet). The tailwind component of the descent profile was stronger and the cloud layer with freezing conditions was thicker than forecast. The latter required the use of the engines' Thermal Anti-Ice system (TAI). The resulting, higher bleed demand drives the Full Authority Digital Engine Control (FADEC) system to schedule higher idle thrust in order to ensure sufficient air flow within the engines. Tailwind and high idle thrust had an impact on the descent profile such that the aircraft was approximatively 3,000 feet high on path. Immediately after the start of the scenario, a runway change was announced that shortened the distance to touchdown by 25NM. Altogether, the aircraft ended up being approximatively 10,000 feet high on path. The crews had full auto-flight system function available (autopilots, flight director, and auto-thrust).

Eye-tracker and Areas of Interest

Eye tracking data were collected with a Pertech eye-tracker (0.25° – 0.5° of accuracy). Head movements were corrected by an alignment of three infra-red emitters to map participants' fixations on an image of reference. The 10 following areas of interest (AOI) were created (see Figure 1): 1) Heading (HDG), 2) Attitude Indicator (AI), 3) Airspeed (Speed), 4) Flight Mode Anounciator (FMA), 5) External view (Ext.), 6) Flight Control Unit (FCU), 7) landing gears panel (Gears), 8) Engine-Indicating and Crew-Alerting System (EICAS), 9) Navigation Display (ND), 10) Altitude indicator (Alt.), and the two subsidiaries 11) No AOI, which is for all what is being viewed but does not correspond to any AOI, and 12) Uncaptured, which includes all the data that was not captured by the device (i.e. this is not an AOI but a non-captured quantity of data).

Eye-tracking and data analysis

Each approach was segregated into two sequences (see Figure 2), namely *Approach Initiation* (from the beginning of the recording to 2500 feet), and *Decision* (from 2500 feet to the decision to land or to go-around). As the temporal milestones that define the *Decision* sequence do not vary from one subject to another, this study is mainly focusing on this *Decision* sequence.

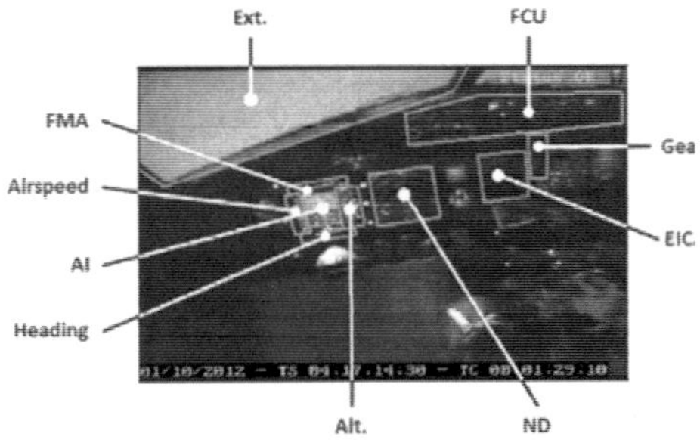

Figure 1. The different Areas of Interest (AOIs) in a Boeing 737 NG cockpit as they were drawn for the experiment.

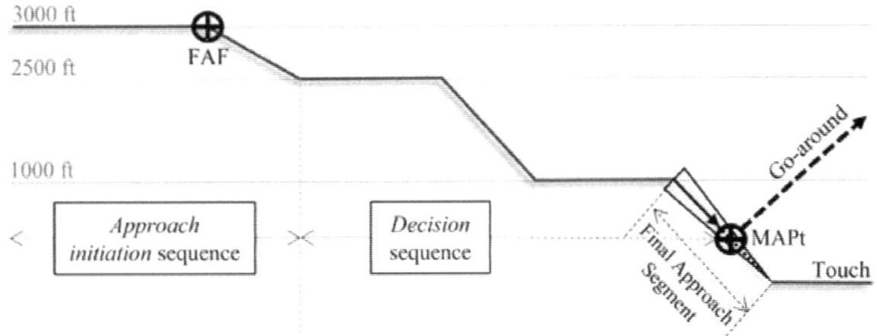

Figure 2. Schematization of a standard approach profile (FAF stands for Final Approach Fix; MAPt stands for Missed Approach Point, which is the point were the go-around procedure is effectively started after a go-around decision).

Behavioral and eye-tracking results

Our behavioral results disclosed that five pilots decided to make a go-around (GA group) and the five other persisted in an erroneous landing decision (Landing group). Therefore, we focused our eye tracking study on the comparison between the GA group and the Landing group to identify potential different ocular strategies that would characterize them.

The descriptive results for the study of pilots' ocular behavior (in terms of dwell time percentages per AOI during the *Decision* sequence on the D3CoS data) are shown on Figure 3. While 10 different AOIs were measured in the data analysis (see Figure 1), we focused our analysis on the seven that received the majority of scans. Therefore, only Airspeed, Attitude Indicator, Altitude, ND and External view AOIs were taken into account. The average percentage of dwell times on the different AOIs and for *Approach Initiation* and *Decision* sequences were plotted to reveal the differences in each AOI, between the cases of go-arounds and landings. As these descriptive results suggested, it appears that pilots who continued landing ganced at the AI more (M = 27 vs M = 11; t = -1.64) and at the ND less (M = 9.6 vs M = 23; t = 1.49) compared to those who correctly executed the go-around.

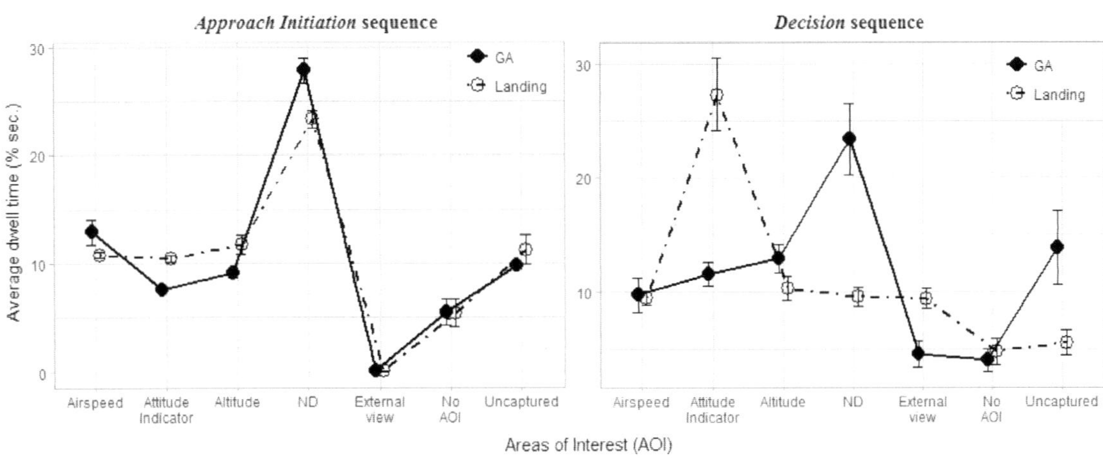

Figure 3. Average dwell time percentages centered standard error bars (in seconds; *y* axis) per AOI (*x* axis), for *Approach Initiation* sequence (on the left), and *Decision* sequence (on the right) during Hambourg approach (breakthrough in clouds), for GA (closed circles) and Landing (open circles) groups.

In order to identify ocular behaviors specific to unstabilized approach, we compared these eye tracking data with a previous one collected during a stabilized approach at Saint-Exupéry Lyon airport in Boeing 777 and Airbus A330 full-flight simulators (please report to Reynal, Colineaux, Vernay, & Dehais, 2016). Though the scenario and aircrafts were different from the ones that we used in this experiment, glide characteristics, level of automation and user interfaces were similar, thus allowing us for such comparisons. As the number of pilots were not similar (n = 8), we randomly removed 2 pilots from our unstabilized approach dataset. The descriptive results are shown on Figure 4. As in the previous graphs, the set of AOIs have been reduced. We conducted a second 7 x 2 ANOVA (AOIs [Indicator, Altitude, ND, External view, No AOI, Uncaptured] x Approach types [Stabilized, Unstabilized]), with AOIs implemented as within factor and Approach types implemented as between factor. This analysis disclosed a significant Approach types x AOIs interaction [$F(1, 14) = 4.52, p < .001, \eta^2_p = .18$]. Tukey's HSD post-hoc analysis revealed that the pilots who faced a stabilized approach glanced more at the External view ($p < .05$) and the Attitude Indicator ($p < .05$) than the ones who experienced unstabilized approach. The data also reveal a trend toward less scanning on te ND for the stablized group.

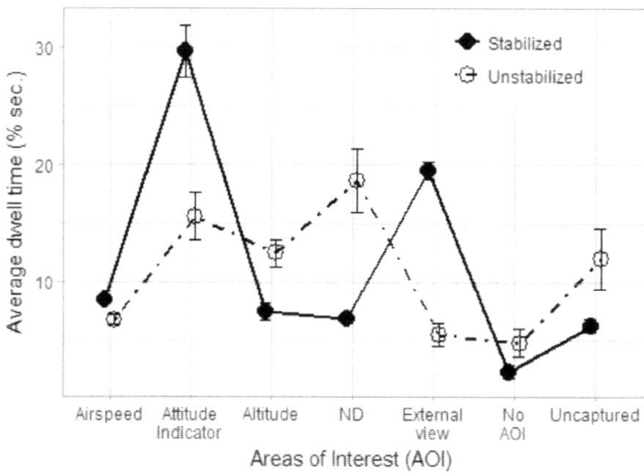

Figure 4. Average dwell time percentages centered on standard error bars (in seconds; y axis) per AOI (x axis), starting from 2500 feets until the decision to go-around or to land (i.e. *Decision* sequence), for stabilized approaches (closed circles) and unstabilized ones (open circles).

Discussion

The objective of this paper was to better understand pilot's ocular and behavioral performance when facing an unstabilized approach. To the authors' knowledge, this study was the first to measure the PF's eye movements during such kind of approach. The scenario was designed in such a way that the aircraft would never become stabilized at the 1000 feet-gate. Our behavioral results, shown that half of the crew persisted in an erroneous landing decision while the other half decided to go-around. Interestingly enough, these qualitative eye tracking findings disclosed that these two groups of pilots exhibited different ocular behaviors. This was particularly true during the *Decision* sequence as pilots from Landing group focused more on the Attitude Indicator whilst pilots from GA group glanced more at the ND (see Figure 3). The Attitude Indicator displayed the flight director behavior thus indicating the flightpath for landing whereas the ND provided information of the current position of the aircraft and of future trajectory (i.e. missed approach segment). This finding may indicate that pilots from the Landing group summoned up all their cognitive resources on supervising the landing trajectory to the detriment of the monitoring of other parameters related to alternative strategies (i.e. go-around). On the contrary, pilots from the GA group seemed to have a better management and anticipation of alternative strategies.It is important to note that the two groups did not differ in their scanning during the *Approach Initiation* sequence, indicating that there was nothing maladaptive about the landing group's overall visual performance. The difference between the two groups therefore seems to lie in the ability to notice specific cues regarding the instabilitiy, or in the decision criterion (e.g., acceptance of risk).

In the experiment, the flight profile changed as dynamical as the environmental conditions. Thus, the decision whether and how to re-stabilize the aircraft had to be constantly challenged and re-taken by the pilots. There are navigational rules that enable pilots to compute the vertical path of a trajectory. However, the closer the final approach, the least the cognitive resources are available to correctly calculate and follow such algorithms (Lacko, Osterloh, & Dehais, 2013). Instead, algorithms are replaced by heuristics, being built upon experience and recency, which make corrective actions to path divergence less trustworthy (Wickens, 2003). The eye-tracking results imply that the decision whether to land or to go-around is taken considerably long before the respective task is executed, and that heuristics aggravate the perseveration being observed in this study. This is well in line with statistical data, which show that still 97 percent of all unstabilized approaches end up with a decision to land (Curtis & Smith, 2013) thus exhibiting "perseveration" behavior (Causse et al., 2009; 2013, Dehais et al, 2012). Eventually the comparison of these eye tracking data with a previous ocular dataset collected during stabilized approach seemed to support that unstabilized approach also impacted pilots' gaze behavior. Hence, we finally compared our "unstabilized" dataset with a previous one collected during stabilized approaches. The statistical findings indicated that stabilized and unstabilized approaches induced different gaze patterns. Indeed, stabilized approaches makes pilots more confident to land and thus allow them to more focus on PFD and also integrate visual cues for a transition from instrument flight to the (always) visual landing manoeuvre. This is in line with training recommendations in the Flight Crew Training Manuals (FCTMs). However, it is interesting to note that pilots from the Landing group, in our unstabilized scenario, qualitatively exhibited similar pattern, especially on the PFD. Thus, traces of perseveration can be identified in that group as this gaze behavior doesn't justify otherwise.

We believe that these analyses demonstrate the potential of eye tracking studies to analyze PF's eye movements during critical phases such as unstabilized approach. These first descriptive results, with other (Reynal, Colineaux, Vernay, & Dehais, 2016), show that there is a need to establish standards on visual scanning pattern with regards to eye tracking to be consistent with SOPs during landing. These eye tracking results support recent recommendation by the Federal Aviation Administration (FAA), stated that "by March 2019, air carriers must include specific training pertaining to improve monitoring". However, one has to consider our study has several limitations that need to be considered as our sample was composed of only 10 pilots in simulated conditions, and the accuracy of eye tracking techniques still remains a challenge, especially in ecological conditions.

Acknowledgements

This research was conducted within the European project D3CoS (Designing Dynamic Distributed Cooperative Human-Machine Systems) and was funded by ARTEMIS-JU and national authorities under grant agreement no. 269336 and by DSAC/DGAC. We would like to thank Yvanne Colineaux (DSAC/DGAC) and André Vernay (DSAC/D3CoS) and all the pilots who took part in this experiment.

References

Airbus. (2006). Flight Operations Briefing Notes – Approach Techniques: Flying Stabilized Approaches. *Part IX, Ch. 1 – Stabilized Approach – Parameters.*

Boeing. (2010). Statistical Summary of Commercial Jet Airplane Accidents. *Retained from the Internet under: http://www.boeing.com/news/techissues/pdf/statsum.pdf.*

Causse, M., Dehais, F., Péran, P., Démonet, J. F., Sabatini, U., & Pastor, J. (2009). Monetary incentive provokes hazardous landing decision making by enhancing the activity of" emotional" neural pathways. *NeuroImage, 47*, S117.

Causse, M., Dehais, F., Péran, P., Sabatini, U., & Pastor, J. (2013). The effects of emotion on pilot decision-making: A neuroergonomic approach to aviation safety. *Transportation research part C: emerging technologies, 33*, 272-281.

Causse, M., Péran, P., Dehais, F., Caravasso, C. F., Zeffiro, T., Sabatini, U., & Pastor, J. (2013). Affective decision making under uncertainty during a plausible aviation task: An fMRI study. *NeuroImage, 71*, 19-29.

Civil Aviation Authority, L. o. (2013). Monitoring Matters: Guidance on the Development of Pilot Monitoring Skills.

Curtis, B., & Smith, M. (2013). Why are Go Around Policies Ineffective. *Missisauga, Canada: Presage Group.*

Dehais, F., Cauchard, F., Rister, F., Cao, Y., Lacko, I., Mikulu, F., Helmke, F., Osterloh, J. P. (2012). Evaluation of an Approach Stabilization Advisory system in a B737 full flight simulator.

Dehais, F., Causse, M., & Pastor, J. (2008). Embedded eye tracker in a real aircraft: new perspectives on pilot/aircraft interaction monitoring. *Proceedings from The 3rd International Conference on Research in Air Transportation. Fairfax, USA: Federal Aviation Administration.*

Dehais, F., Peysakhovich, V., Scannella, S., Fongue, J., & Gateau, T. (2015). Automation surprise in aviation: Real-time solutions. *Proceedings of the 33rd Annual ACM Conference on Human Factors in Computing Systems*, (pp. 2525-2534).

Dehais, F., Tessier, C., Christophe, L., & Reuzeau, F. (2010). The perseveration syndrome in the pilot's activity: guidelines and cognitive countermeasures. Dans *Human Error, Safety and Systems Development* (pp. 68-80). Springer.

Dutch Safety Board (2010). Crashed during approach, Boeing 737-800, near Amsterdam Schiphol Airport, 25 February 2009. Final report of Dutch Safety Board.

Flight Safety Foundation (2014). A Practical Guide for Improving Flight Path Monitoring.

Huettig, G. A. (s.d.). G., & Tautz, A.(1999). Mode awareness in a modern glass cockpit: Attention allocation to mode information. *Proceedings of the 10th International Symposium on Aviation Psychology.*

ICAO. (2006). Procedures for Air Navigation Services – Aircraft Operations (PANS OPS). *Part IX, Ch. 1 – Stabilized Approach – Parameters.*

Kasarskis, P., Stehwien, J., Hickox, J., Aretz, A., & Wickens, C. (2001). Comparison of expert and novice scan behaviors during VFR flight. *Proceedings of the 11th International Symposium on Aviation Psychology*, (pp. 1-6).

Lacko, I., Moravek, Z., Rister, F., Osterloh, J. P., Luedtke, A. & Dehais, F. (2013). Safety Aspects of Human-Machine Interaction in Cockpits during the Approach Phase of Flight. *MOSATT.*

Mumaw, R. J., Sarter, N., & Wickens, C. D. (2001). Analysis of pilots' monitoring and performance on an automated flight deck. *11th International Symposium on Aviation Psychology, Columbus, OH.*

National Transportation Safety Board (2014a). Descent Below Visual Glidepath and Impact With Seawall, Asiana Airlines Flight 214 Boeing 777-200ER, HL7742, San Fransisco, California, July 6, 2013.

National Transportation Safety Board (2014b). Crash During a Nighttime Nonprecision Instrument Approach to Landing UPS Flight 1354 Airbus A300 - 600, N155UP Birmingham, Alabama August 14, 2013.

Parasuraman, R., Molloy, R., & Singh, I. L. (1993). Performance consequences of automation-induced'complacency'. *The International Journal of Aviation Psychology, 3*, 1-23.

Reynal, M., Colineaux, Y., Vernay, A., & Dehais, F. (2016). Pilot Flying vs. Pilot Monitoring During the Approach Phase: an Eye-Tracking Study. *HCI-Aero 2016, International Conference on Human-Computer Interaction in Aerospace.*

Sarter, N. B., & Woods, D. D. (1995). How in the world did we ever get into that mode? Mode error and awareness in supervisory control. *Human factors, 37*, 5-19.

Sarter, N. B., Mumaw, R. J., & Wickens, C. D. (2007). Pilots' monitoring strategies and performance on automated flight decks: An empirical study combining behavioral and eye-tracking data. *Human Factors: The Journal of the Human Factors and Ergonomics Society, 49*, 347-357.

Smith, J. M., Jamieson, D. W., & Curtis, W. F. (2012). Why are go-around policies ineffective? The psychology of decision making during unstable approaches. *65th Annual FSF International Air Safety Seminar Santiago, Chile.*

Spangler, C., & Park, A. (2010). Loss of Control on Approach Colgan Air, Inc., Operating as Continental Connection Flight 3407 Bombardier DHC-8-400, N200WQ Clarence Center, New York February 12, 2009. *ACM SIGGRAPH 2010 Dailies*, (p. 7).

Tessier, C., & Dehais, F. (2012). Authority Management and Conflict Solving in Human-Machine Systems. *AerospaceLab*, (4), p-1.

Wickens, C. D. (2003). Pilot actions and tasks: Selections, execution, and control. *Principles and practice of aviation psychology*, 239-263.

Wickens, C. D., McCarley, J. S., Alexander, A. L., Thomas, L. C., Ambinder, M., & Zheng, S. (2008). Attention-situation awareness (A-SA) model of pilot error. *Human performance modeling in aviation*, 213-239.

INCOMPLETE KNOWLEDGE OF RESULTS AND THE MANIPULATION OF RESPONSE BIAS

Matthew J. Davis
CDO Technologies, Inc.
Dayton, Ohio

In signal detection theory, an optimal observer exploits all available information to achieve the desired goal of a particular decision strategy (Green & Swets, 1966). Detection experiments often provide the observer with complete knowledge of results (CKR) in order to ensure best possible performance for the task. If optimal behavior is indeed dependent upon CKR, then a degradation of that information should also reduce the likelihood of achieving optimal response bias. A single-interval auditory detection experiment was conducted to measure changes in response bias in the presence of *incomplete knowledge of results* (IKR) (i.e. feedback for some combination of true/false detections and true/false rejections) (Davis, 2015). The results were compared with the theoretical "optimal" bias level for the task. Statistical tests revealed significant differences between complete and incomplete feedback conditions. These results are consistent with the hypothesis that IKR can significantly degrade an observer's ability to achieve optimal response bias.

In the aviation industry, pilots and air traffic controllers are often presented with scenarios where it is important to detect nuanced changes between stimuli, such as detecting auditory alarms in loud environments, or correctly determining the distance between aircraft. In these instances, correct discrimination of the stimuli is improved with experience, and experience is coupled with knowledge of one's performance. The more information that is received about the outcome of a particular decision, the more that knowledge can be used to influence future decisions. Decisions about ambiguous stimuli can be described using signal detection theory (SDT), where decision outcomes are defined in terms of *sensitivity* and *response bias*. An observer who frequently detects an ambigious stimulus is considered to have a high degree of *sensitivity*. An observer who frequently responds with one decision over another (e.g. *"yes, there's a problem"* vs. *"there's no problem"*) is described as having a high degree of *response bias*. The definition of what is biased depends almost exclusively on the decision strategy being implimented, such as "maximize the proportion of correct responses", "maximize a weighted combination of hits and correct rejections", "maximize expected value", and the "Neyman-Pearson objective" (Green & Swets, 1966, pp. 20–26; Macmillan & Creelman, 2005). The ultimate goal of any decision maker is to not only obtain the highest degree of sensitivity possible, but also to obtain the optimal ratio of responses as dictated by an appropriate decision strategy.

Knowledge of results (KR) is known to be an important aid in the optimization of response bias (Green & Swets, 1966, p. 395; Macmillan & Creelman, 2005, p. 130). Many experimental tasks that require the detection or discrimination of ambigious stimuli utilize complete knowledge of results (CKR), where feedback is provided for every possible response. The real world, however, is more complex and often provides very little useful feedback information from which to optimize responses. Feedback that is not presented for every response-type is known as incomplete knowledge of results (IKR), and is comprised of continuous trial-by-trial feedback, but only for some combination of true/false detections or true/false rejections of the stimulus (Figure 1), (Davis, 2015). As more information is expected to increase one's ability reach an optimal response bias, incomplete feedback information may degrade the ability to respond optimally. Understanding the influence of incomplete feedback on response bias is imporant in understanding how humans utilize decisions strategies with incomplete information.

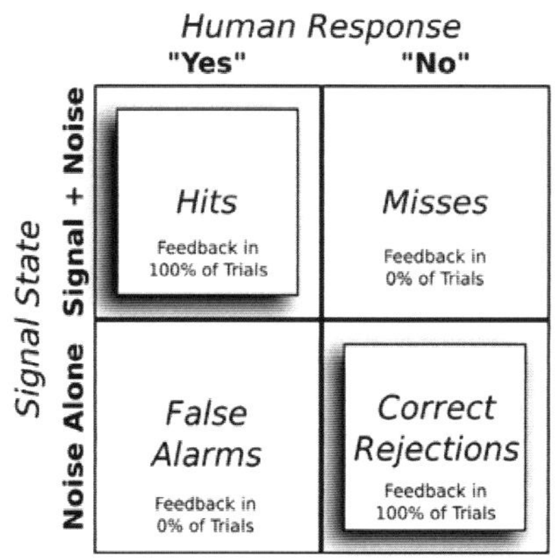

Figure 1. An example of incomplete knowledge of results. In this case, feedback is provided for hits and correct rejections, but not for misses and false alarms.

Background

Traditionally, feedback has been used in signal detection experiments as a means of stabilizing performance in sensitivity or, to a lesser extent, response bias (Green & Swets, 1966, p. 395; Macmillan & Creelman, 2005, p. 130). While feedback for every response may appear to be the most logical method of providing KR, early detection literature utilized different types of feedback and with varying nomenclature (Kaess & Zeaman, 1960; Wiener, 1963). Other studies have examined the effect of limiting feedback to a predetermined proportion of trials known as *partial knowledge of results* (Lurie & Swaminathan, 2009; McCormack, Binding, & McElheran, 1963; Szalma, Parsons, Warm, & Dember, 2000). Szalma et al. (2006) studied the effects of optimism and pessimism on stress states, and provided feedback for certain response-types and witheld them for other responses. In that study, the term "knowledge of results format" was used, though the term "incomplete knowledge of results" was proposed by Davis (2015) as a more accurate description. In each of these cases, feedback was designed to improve or at least modify behavior, though the effects on response bias were examined in only a few cases.

Methods

Procedure

An auditory detection experiment was conducted to examine the effects of IKR using a 1 kHz tone and a white noise masker. Participants were first presented with a practice task designed to increase familiarization with the single-interval paradigm and signal/noise characteristics. The next task was designed to measure a masked signal threshold (d' ≈ 1.3) for use in the IKR experiment by utilizing the single-interval adjustment matrix (Kaernbach, 1990). The final task contained conditions that manipulated IKR and used the individualized thresholds from the SIAM procedure to present the signal and noise stimuli at a constant SNR in a single-interval yes-no paradigm (Green & Swets, 1966). Each participant completed 10 conditions, and each condition contained 10 blocks of 50 trials. Subjects were presented with the stimulus (either "signal+noise" or "noise alone") and were asked to indicate if the target signal was present in the noise. In response to the question, subjects could select either "yes" or "no" by clicking the appropriate button on a graphic user interface with a computer mouse. Feedback of some type was provided for every trial, but only for the response types that were specified by the condition [e.g. some combination of hits (H), misses (M), false alarms (FA), and/or correct rejections (CR)]. Each condition consisted of feedback for (1) no response types, (2) H, (3) M, (4), FA, (5) CR, (6) H+M, (7) H+FA, (8) H+CR, (9) H+M+FA, and (10) H+M+FA+CR. Conditions were completed in random order with the exception of the first condition (no feedback), which was always completed first as a baseline condition; and the last condition (complete feedback), which was always completed last, to prevent the complete set of feedback from influencing other conditions.

Two main features of IKR were examined in this study: *quantity* and *implicitness*. The question of IKR *quantity* refers to the amount of response-types that receive feedback. IKR *implicitness* refers to the possibility of inferring KR from missing feedback. The features of each type may be important for explaining individual results of subjects as feedback is increased in the number of response-types across conditions. Subjects were asked to utilize the decision theory that *maximizes the proprotion of correct answers* (Green & Swets, 1966, pp. 20–26). Subjects were not told the *a priori* signal probability of the signal, and were thus unaware that the optimal decision strategy would require an 50% split in "yes" and "no" responses. The optimal decision criterion for this decision strategy and signal probability was $c = 0$, where $c = -1/2[z(H) + z(F)]$ (Macmillan & Creelman, 2005, equation 2.1).

Subjects

Participants consisted of 5 male and 5 female adults (ages 18-32). Hearing thresholds were tested at 125 Hz, 250 Hz, 500 Hz, 1000 Hz, 2000 Hz, 4000 Hz, and 8000 Hz. Acceptable thresholds were defined as < 15 dB HL loss at these frequencies. All subjects were part of an in-house, part-time subject pool. All subjects volunteered for the study and were given the option to leave at any time, and for any reason without penalty to their standing in the subject pool. All subjects participated through the completion of the study.

Stimuli

The target stimulus consisted of a 20 ms, 1 kHz sinusoidal signal that was present in exactly 50% of the randomized trials. The masking stimulus consisted of 500 ms of white noise and was present in every trial. Both the signal and the noise employed a cosine onset/offset ramp to the first and last 10ms of the stimuli to unintended artifacts. The distribution of trials with "signal+ noise" vs. "noise alone" was randomized. The center of the target signal (when present) always coincided with the center of the noise, so that the noise was always the first and last stimulus to be heard. The rms level of the noise was 60 dB SPL, and the average presentation level of the signal and the noise combined was no more than 60.3 dB SPL.

Results

IKR Quantity

Since the *magnitude* of response bias is of primary interest for the question of IKR *quantity*, the data were organized by the absolute value of the decision criterion, *c*. Negative values of *c* indicate bias toward "yes", and positive values of *c* indicate a bias toward "no". These data were modelled using individual exponential functions per subject; $f = ae^{bx}$ (Figure 2). Of the ten subjects who participated in this study, 60% demonstrated statistically significant *negative* slopes ($p < .05$); the remaining 40% yielded flat data with no significant slopes (Table 1). None of the subjects produced statistically significant *positive* slopes as the amount of feedback was increased across conditions.

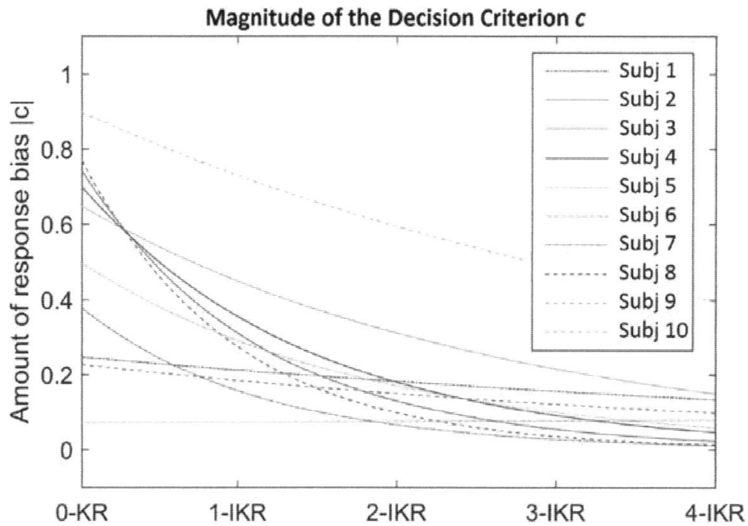

Figure 2. Analysis of inividual response biases for each subject across conditions with different quantities of feedback.

Table 1.
Exponential model coefficients of individual and group response bias.

Subject	*a*	pVal	*b*	pVal
1	.247	.013	-.153	.371
2	.741	.000	-.870	.013
3	.647	.000	-.367	.005
4	.699	.000	-.678	.002
5	.071	.105	.025	.910
6	.496	.000	-.535	.010
7	.377	.002	-.874	.064
8	.768	.000	-1.029	.019
9	.227	.001	-.208	.106
10	.896	.000	-.205	.009
Group	.72	.000	-.39	.009

Note. Results are significant at the $p < .05$ value.

IKR Implicitness

If subjects are able to use *implicit* feedback information to optimize their responses, then it is expected that the three implicit conditions being examined in this study (H+M, H+CR, and H+M+FA) would have an optimal decision criterion (c = 0). Across all subjects, a total of 23% of all *implicit* IKR conditions contained means in the range the optimal decision criterion. However, 70% of the conditions (across all subjects) contained means in the range of the complete feedback (CKR) condition. A total of 50% of individual subjects demonstrated similarity with $c = 0$, and 90% of all subjects contained bias similar to the CKR condition in at least one of the three implicit conditions.

Discussion

The primary purpose of this study was to better understand the degredation of optimal response bias as feedback information is provided at various levels of incompleteness. The results of IKR *quantity* demonstrate that conditions of different amounts of feedback can be modelled individually using a negative exponential curve. The data can be split into two types of behavior: subjects who become more optimal with more feedback information, and subjects who maintain near-optimal behavior from the beginning. It is important to note that even though some subjects did not yield a signficant negative exponential slope approaching $c = 0$, subjects did not significantly *increase* their bias as the number of feedback response types were increased. These data are consistent with the hypothesis that the type of feedback, not just the proprtion of feedback trials, is important for optimizing response bias for a given decision strategy.

The results of the *implicit* feedback conditions suggest a surprising inability of subjects to utilize missing feedback information to achieve optimal bias. Many subjects who did demonstrate optimal bias in these conditions were also relatively unbiased in every condition. One possible explanation for this behavior stems from the definition of optimal bias. In reality, the subjects had two tasks: (1) discover the optimal ratio of responses with limited information, and (2) optimize their responses with the aforementioned ratio. These two tasks, while similar, are not the same. It is entirely possible that the participants failed to properly estimate the optimal bias while also using the missing feedback to optimize responses to their own imperfect internal representation of the optimal strategy.

Conclusion

The results of this study reveal the importance of feedback in the attainment of optimal response biases for the decision strategy that maximizes the proportion of correct responses. As the number of response types associated with feedback increases, the probability that humans will respond optimally also increases. Additionally, it was expected that participants would be able to utilize the implicit feedback conditions to further optimize their responses. Instead, bias for the implicit conditions in most subjects contained greater similarity to the individual bias levels for the complete feedback condition than for the optimal bias level, (which were often not equal). These results suggest that humans are imperfect estimators of optimal response bias, though in general this imperfection is consistent with their internal representation of the optimal decision strategy. These results provide important insights into the decision making processes of humans, and reveal that the type of feedback information that is witheld is nearly as important as feedback that is accessible.

Acknowledgements

This research is based in part on the author's doctoral dissertation conducted at The Ohio State University. Financial support was provided by a research fellowship from the Dayton Area Graduate Studies Institute (DAGSI) and the Air Force Research Laboratory (711th Human Performance Wing).

References

Davis, M. J. (2015). *The Effects of Incomplete Knowledge of Results on Response Bias in an Auditory Detection Task* (Doctoral Dissertation, The Ohio State University).

Green, D. M., & Swets, J. A. (1966). *Signal Detection Theory*. New York: Wiley.

Kaernbach, C. (1990). A single-interval adjustment matrx (SIAM) procedure for unbiased adaptive testing. *Journal of the Acoustical Society of America*, *88*(6), 2645–2655.

Kaess, W., & Zeaman, D. (1960). Positive and Negative Knowledge of Results on a Pressey-Type Punchboard. *Journal of Experimental Psychology*, *60*(1), 12–17.

Lurie, N. H., & Swaminathan, J. M. (2009). Is timely information always better? The effect of feedback frequency on decision making. *Organizational Behavior and Human Decision Processes*, *108*, 315–329.

Macmillan, N. A., & Creelman, C. D. (2005). *Detection Theory: A User's Guide* (2nd ed.). New York: Lawrence Erlbahm Associates.

McCormack, P. D., Binding, F. R. S., & McElheran, W. G. (1963). Effects on Reaction Time of Partial Knowledge of Results of Performance. *Perceptual and Motor Skills*, *17*, 279–281.

Szalma, J. L., Hancock, P. A., Dember, W. N., & Warm, J. S. (2006). Training for vigilance : The effect of knowledge of results format and dispositional optimism and pessimism on performance and stress. *British Journal of Psychology*, *97*, 115–135.

Szalma, J. L., Parsons, K. S., Warm, J. S., & Dember, W. N. (2000). Continuous vs. Partial Knowledge of Results in Training for Vigilance. *Proceedings of the Human Factors and Ergonomics Society Annual Meeting*, *44*(21), 386–389.

Wiener, E. L. (1963). Knowledge of Results and Signal Rate in Monitoring: A Transfer of Training Approach. *Journal of Applied Psychology*, *47*(3), 214–222.

THE IN-FLIGHT AFFECTIVE BRAIN: DECISION MAKING UNDER UNCERTAINTY AND SAFETY IMPLICATIONS IN AVIATION

Mariateresa Sestito, Institut des Sciences Cognitives Marc Jeannerod CNRS UMR 5229, Bron, France
Current address: Department of Psychology, Wright State University, Dayton, OH, United States

Francesca Benuzzi, Department of Biomedical Sciences, Metabolism, and Neuroscience, University of Modena and Reggio Emilia, Italy

Seongmin Park, Institut des Sciences Cognitives Marc Jeannerod, CNRS UMR 5229, Bron, France

Pierre Wydoodt, Center for cognitive Neurosciences, CNRS UMR 5229, Bron, France

Ermanno Giombelli, Department of Neuroscience, University Hospital of Parma, Italy

Umberto Scoditti, Department of Neuroscience, University Hospital of Parma, Italy

Giacomo Pavesi, Department of Neurosurgery, University of Modena and Reggio Emilia, Italy

Paolo Frigio Nichelli, Department of Biomedical Sciences, Metabolism, and Neuroscience, University of Modena and Reggio Emilia, Italy

Jean-Claude Dreher, Institut des Sciences Cognitives Marc Jeannerod, CNRS UMR 5229, Bron, France

Understanding the relation between motivation and pilot impulsive decision-making is extremely important in the context of aviation human factors. One way to operationalize motivation is by presenting participants with different reinforces, either primary (e.g. food rewards) or secondary (e.g. financial incentives, arguably playing a crucial role in in-flight decision-making). To establish the role that different reward systems play in decision-making, we tested the extent to which distinct brain regions are sensitive to the reinforcement content. Combining a neuro-economics approach with a voxel-based lesion method, we found that distinct regions within orbito-frontal cortex (OFC) are differentially involved in impulsivity decisions based on the reinforcement type. In contrast, lesions in dorsolateral prefrontal cortex (DLPFC) were not associated with such decisions. These results suggest a distinction between reward types at the neural level, and thus emphasize the importance of investigating how different reinforces can affect flight activity especially in pilot impulsive, risk-taking behavior.

Most flight accidents occur during arrival, even though this flight phase represents only 4% of overall flight time (Boeing, 2005). One potential reason for this is pilot risky decision-making. In 2,000 cases of approaches under thunderstorm conditions studied, two aircrews out of three kept on landing in spite of adverse meteorological conditions, instead of going-around to perform a new attempt to land more securely or to divert to another airport (Rhoda & Pawlak 1999). Many factors might account for the difficulty for pilots to revise their flight plan given adverse weather conditions (Goh & Wiegmann 2002). The decision to go-around might incur a broad range of strong negative emotional consequences, as a go-around decision increases uncertainty and level of stress. Moreover, a go-around has important financial consequences for the airlines, resulting in airlines emphasizing fuel economy and getting passengers to their destinations rather than diverting the flight (Orasanu, 2001). Thus, the airlines might inadvertently send implicit messages to their pilots and these blurred messages might create conflicting motives, which can affect pilots' risk assessments and the course of action they choose. Together, all these emotional pressures could negatively impact pilots' rational decision-making.

As in the aviation context, everyday decisions are often made in the presence of risk and uncertainty. Risk refers to multiple possible outcomes, both positive and negative, that could occur with well-defined or estimable probabilities (Stearns 2000). In the field of neuro-economics, 'risk' refers to monitoring potential monetary outcomes (Tom et al. 2007). Studies exploring the effects of risk on cognition in the form of monetary reward or punishment highlighted that financial incentives can bias working memory and object recognition (Taylor et al. 2004). As reward/punishment manipulation may interfere both with cognition and emotion, one might expect similar effect in pilots placed in a conflict situation between systems of punishment (extra fuel consumption, fatigue caused by a second landing attempt etc.) and reward (bring passengers without delay).

The current study aims to explore the mechanisms underlying conflict in reward and its impact on decision-making. We adopted a voxel-based lesion symptom mapping (VLSM) approach (Bates et al. 2003) to investigate the relationship between the amount of brain tissue damage and the extent to which decision-making is affected by different types of reinforcement. Prior studies of the neural correlates of risk in the context of financial decision-making suggest that estimating monetary gains and losses involves activity in multiple brain areas (e.g. Huettel et al. 2006) including different regions within prefrontal cortex, particularly orbitofrontal cortex (OFC) and dorsolateral prefrontal cortex (DLPFC). DPLFC is involved in higher cognitive processes, such as reasoning, whereas OFC is involved in emotional processes, primarily modulating the anticipation of regret linked with financial loss (Coricelli et al. 2005). Further, neuroimaging studies found dissociation in the type of rewards activating specific sub-regions of OFC: erotic stimuli preferentially activated the posterior portion of the OFC whereas monetary gains activated the anterior lateral part of OFC (Sescousse et al., 2010; 2013). Given that posterior OFC is phylogenetically and ontogenetically older than anterior OFC, this might reflect differential processing of rewards based on their evolutionary significance.

We aimed to establish the extent to which distinct brain regions within the OFC show a differential sensitivity to the content of reinforcement. In this lesion-defined approach, behavioral performance is accounted by the amount of missing voxels derived from various groups of patients with lesions varying in location and extent (Bates et al. 2003). This may reveal whether the absence of a given brain region (e.g. OFC) can explain variation in behavioral performance (e.g. impulsivity). Notably, the VLSM method does not require patients to be grouped by either lesion site or behavioral cutoff, but instead makes use of continuous behavioral and lesion information. We selected patients with brain lesions located in the OFC cortex (OFg), patients with brain lesions located outside the frontal cortex (non-OFg), and healthy participants (CnTg). All participants were tested with the Delay Discounting (DD) paradigm that allows measuring impulsivity in decision-making by asking participants to choose between either a fixed amount of a reward that could be received immediately or a greater amount of reward that could be received after a specific delay. We administered the DD paradigm with two types of rewards: primary (food) and secondary (monetary). Based on the notion of a reward-based functional dissociation within the OFC, our hypothesis was that a higher amount of brain damage (in terms of missing voxels) involving the anterior OFC would result in a more impulsive behavior in the money-related task, whereas a greater damage involving the posterior OFC would result in a more impulsive behavior in the food-related task.

Materials and Methods

Participants

Participants comprised of 37 individuals: 11 patients with lesions involving the OFC (OFg; mean age 59.27 DS ± 10.71; mean education in years 9.45 DS ± 4.23; 7 females) located either in the anterior (involving BA 10; mean missing voxels = 1,020 N = 5) or the posterior part (involving BA 11 and 13; mean missing voxels = 1,126 N = 6), 9 patients with lesions located outside the frontal cortex (non-OFCg; mean age 50.22 DS ± 11.39; mean education in years 10.78; DS ± 3.93; 6 females) and 17 healthy controls (CnTg; mean age 53.12 DS ± 13.61; mean education in years 13.11 DS ± 5.48; 8 females). The three groups did not significantly differed for age, education and lesion extension (all $p_s > 0.05$).

Experimental paradigm and procedure

Participants were requested to choose between a fixed amount of a reward that could be received immediately and a greater amount of reward that could be received after a specific delay. The nature of the reward changed across sessions: one session assessed DD for money, whereas the other one assessed DD for food. Delays used for the money session were based on Peters & Büchel (2009) – i.e. 6 hours, 1 day, 7 days, 30 days, 90 days, 180 days. Since food reward, by its nature, cannot be delayed over long period of time, we used shorter intervals for the food condition than the money condition. Based on a pilot study (unpublished), the following delays were adopted: 30 min, 90 min, 3 hours, 10 hours, 1 day, 7 days. Participants were told that the compensation for their participation would be based on a raffle performed at the end of the study, in which a trial would have been chosen at random and delivered to them. Participants were tested individually. Two behavioral experimental sessions - one with primary (food) and one with secondary (money) rewards were performed. The session order was counterbalanced across subjects. The total duration of the experimental session was about 30 minutes.

Data analysis

Behavioral data. For each task, the rate at which the subjective value (SV) of a reward decays with delay (TD rate) was assessed through the discounting parameter (*k*) (Rachlin et al., 1991; Green and Myerson, 2004). The hyperbolic function $SV = 1/(1+kD)$, where SV = subjective value (expressed as a fraction of the delayed amount), and D = delay, was fit to the data to determine the *k* constant of the best fitting TD function, using a nonlinear, least-squares algorithm. The larger the value of *k*, the steeper the discounting function, the more participants are inclined to choose small-immediate rewards over larger-delayed rewards. The hyperbolic *K* constants were normally distributed after log-transformation (Kolmogorov-Smirnov $d < 0.19$, $p > 0.1$ in all cases) therefore allowing the use of parametric statistical tests.

Lesion localization and quantification. To identify patients' lesion, a high-resolution T1-weighted anatomical image (TR, 9.9 ms; TE, 4.6 ms; 170 sagittal slices; voxel size, 1x1x1) was acquired with a Philips Intera system at 3 T. In order to automatically identify patients' lesions avoiding to trace them manually, MRI scans acquired from 100 healthy participants (not included in the actual CnTg) divided in two subgroups based on gender (male subgroup: N = 50; mean age = 34.16; mean education = 15.96; female subgroup: N = 50; mean age = 42.6; mean education = 15.16) were used. To control for gender effect, each patient was compared with the appropriate gender subgroup. Lesion data from 20 patients belonging to the OFg (N = 11) and the non-OFg (N = 9) were analyzed. Matlab 7.1 (Mathworks Inc., MA, USA) and SPM12 (Wellcome Trust Centre for Neuroimaging, London, UK) software were used for *pre-processing*. This included the following steps: segmentation, template creation, normalization, modulation and smoothing. The number of voxels lost was calculated using xjView (http://www.alivelearn.net/xjview) for the following Broadmann areas (BA 10, BA 11, BA 13, BA 46, BA 47).

Statistical analysis. A hierarchical regression analysis (forward stepping) was then conducted separately for food and money conditions in order to determine the variance explained in the dependent variables (the log-transformed K constants for food and money tasks) with the missing voxels computed in BA 10, BA 11, BA 13, BA 46, BA 47 and sex as predictors. In order to correct for lesion extension, each BA value reflected the percentage of missing voxels computed out of the total missing voxels due to the lesion. For all performed analysis, $p < .05$ was considered to be statistically significant.

Results

Our hypothesis was that a brain damage predominantly located in the anterior OFC would result in a selective impairment in performing the money-related task, while a brain damage predominantly located in the posterior OFC would result in a selective impairment in performing the food-related task. To this end, missing voxels due to brain lesions located in different Regions of Interest (ROIs) within OFC and DLPFC were computed in order to see whether a greater amount of tissue damage involving the anterior/posterior part of OFC correlates with a more impulsive behavior in the DD task for money and food, respectively.

For the DD task with food as reward, the hierarchical regression analysis showed that the K constant was explained by three ROI predictors: missing voxels in BA 11 ($t = 2.27$ $\beta = 0.50$ $p < 0.03$), BA 46 ($t = -2.75$ $\beta = -0.53$ $p < 0.01$) and BA 13 ($t = 0.47$ $\beta = 0.09$ $p > 0.6$) with overall regression model ($F_{(3, 33)} = 3.52$, $p < 0.03$, $R = 0.49$, $R^2 = 0.24$) accounting for 24.25% of the variance. To estimate the independent contribution of each BA above and beyond the variance accounted for by the other ones, semi-partial correlation coefficients were calculated for each predictor. Only the correlations between the K constant and the missing voxels in BA 11 and BA 46 were significant (see Table 1). Note, though, that whereas the BA 11 was positively correlated, BA 46 was negatively correlated (see discussion).

Table 1.
Semi-partial correlations between predictors and K constants obtained for the Delay Discounting task with food as reward.

Predictors	R	R^2	p
BA 10	0.01	0.24	0.948
BA 11	**0.34**	**0.53**	**0.030***
BA 13	0.07	0.29	0.638
BA 46	**- 0.42**	**0.38**	**0.010***
BA 47	0.04	0.58	0.787

For the DD task with money as reward, the hierarchical regression analysis demonstrated that the K constant was explained by three predictors: missing voxels in BA 10 ($t = 2.31$ $\beta = 0.42$ $p < 0.03$), BA 46 ($t = - 2.25$ $\beta = - 0.52$ $p < 0.04$) and BA 47 ($t = 2.46$ $\beta = 0.51$ $p < 0.02$), with the overall regression model accounting for 29.94% of the variance ($F_{(5, 30)} = 2.56$, $p < 0.05$, $R = 0.55$, $R^2 = 0.30$). Semi-partial correlation coefficients for each predictor were calculated, and only the correlations between the K constant and the missing voxels in BA 10, BA 46 and BA 47 were significant (see Table 2). Note, though, that whereas the BA 10 and BA 47 were positively correlated, BA 46 was negatively correlated (see discussion).

Table 2.
Semipartial correlations between predictors and K constants obtained for the Delay Discounting task with money as reward.

Predictors	R	R^2	p
BA 10	**0.35**	**0.29**	**0.028***
BA 11	-0.02	0.66	0.909
BA 13	0.17	0.12	0.281
BA 46	**-0.34**	**0.57**	**0.032***
BA 47	**0.38**	**0.46**	**0.020***

Discussion

This study reports preliminary findings pointing towards a distinctive involvement of different anterior/posterior portions of OFC when performing a decision-making task with primary and secondary rewards. The more the brain damage involves BA 11, the higher the impulsivity towards the food. In contrast, the more the brain damage overlaps BA 10 (and BA 47), the higher the impulsivity towards money. Moreover, the impulsivity showed in the DD task for both food and money was negatively correlated with damage in DLPFC (BA 46, Petrides & Pandya 1999). The current results support the idea that the brain has distinct systems for different reward types, having a direct impact on impulsive decision-making behavior.

Critically, the two types of rewards considered here (money and food) have significant evolutionary differences, which are putatively paralleled at the cerebral level. While food can be considered as primary reward because it has an innate value and satisfy biological needs, money is a secondary reward which appeared recently in human history and whose abstract value needs to be learned by association with primary reinforces. Similarly, this distinction is reflected both phylogenetically and ontogenetically in the brain. The anterior part of the OFC is especially well developed in humans relatively to other non-human primates, and cytoarchitectonically, it is characterized by a granular cell layer, which is thought to be more recent than the agranular and dysgranular layers characteristic of posterior and medial parts of OFC (Wise 2008). Our current results are therefore in line with the idea that motivational factors (in the form of primary/secondary rewards) impact on decision-making is manifested in the brain as a frontal postero-anterior axis of complexity (Kringelbach & Rolls 2004; Sescousse et al. 2013).

The relation between impulsive motivation and decision-making described here is especially relevant in the context of aviation human factors, as it is well established that motivational factors play a crucial role in in-flight decision-making. These factors might include different incentives, such as financial or food rewards. Thus, a parallel could be drawn with the current results and previous studies that considered a different involvement of frontal regions when participants make in-flight decisions under financial incentives and uncertainty. Emotional, motivational factors have been shown to jeopardize decision-making and cognitive functioning in piloting tasks (Dehais et al. 2003). Notably, emotion has a fundamental role in rational decision-making, especially in risk assessment of situations with high uncertainty (Damasio 1994). A series of studies using a simplified landing task in a simulated flight environment have shown that financial incentive biases decision-making towards a more risky and hazardous landing behavior (Causse et al. 2010; 2013). Critically, such risky behavior was associated with activity in cortical reward circuits including OFC. In contrast, a decision to land in an emotionally neutral condition (i.e. no financial incentive) resulted in enhanced activity in DLPFC. Overall, these findings showed that a shift occurs from *cold reasoning* (rationally driven, underpinned by DLPFC) to *hot reasoning* (emotion-driven, subserved by OFC) under financial incentive. This suggests that pilot erroneous trend to land could be explained by a perturbation of the decision-making process due to negative emotional consequences associated with the go-around. According to these studies, our results show that decision-making under reward influence (either primary or secondary) crucially relied on OFC integrity, while negatively correlated with damage in DLPFC.

In conclusion, the neuro-economics approach herein adopted coupled with the neuropsychology lesion method suggests a neural distinction between reward types, thus emphasizing the importance of investigating in future ad-hoc designed studies, how reinforces can affect flight activity in accounting for pilot impulsive behavior. Further investigations in this direction would greatly contribute to the study of human factors in aviation, ultimately improving flight safety.

Acknowledgements

This work was supported by a Postdoctoral fellowship from Fyssen Fondation to M.S. The authors acknowledge Drs. Annette Puzzolante, Stefano Meletti, Paola Castellini for their help in patients' selection. We thank Assaf Harel for valuable comments and insights. Address correspondence to Mariateresa Sestito at mariateresa.sestito@wright.edu.

References

Bates, E., Wilson, S. M., Saygin, A. P., Dick, F., Sereno, M. I., Knight, R. T., & Dronkers, N. F. (2003). Voxel-based lesion–symptom mapping. *Nature neuroscience, 6*, 448-450. doi:10.1038/nn1050

Boeing (2005). Summary of Commercial Jet Airplane Accidents Worldwide Operations 1959–2004.

Causse, M., Dehais, F., Pastor, J., Péran, P., & Sabatini, U. (2010). Economic issues provokes hazardous landing decision-making by enhancing the activity of "emotional" neural pathways. In *Proceedings of the 3 rd International Conference on Research in Air Transporation.*

Causse, M., Dehais, F., Péran, P., Sabatini, U., & Pastor, J. (2013). The effects of emotion on pilot decision-making: A neuroergonomic approach to aviation safety. *Transportation research part C: emerging technologies, 33*, 272-281. doi: 10.1016/j.trc.2012.04.005

Coricelli, G., Critchley, H. D., Joffily, M., O'Doherty, J. P., Sirigu, A., & Dolan, R. J. (2005). Regret and its avoidance: a neuro- imaging study of choice behavior. *Nature Neuroscience, 8*, 1255–1262. doi:10.1038/nn1514

Damasio, A. (1994). *Descartes' error: Emotion, reason, and the human brain.* New York: Grosset/Putnam.

Dehais, F., Tessier, C., & Chaudron, L. (2003). GHOST: Experimenting conflicts coutermeasures in the pilot's activity. Paper presented at the IJCAI, Acapulco, Mexique.

Elliott, R., Newman, J., Longe, O., & Deakin, J. (2003). Differential response patterns in the striatum and orbitofrontal cortex to financial reward in humans: a parametric functional magnetic resonance imaging study. *Journal of Neuroscience, 23*, 303.

Goh, J., & Wiegmann, D. (2002). Human factors analysis of accidents involving visual flight rules flight into adverse weather. *Aviation, space, and environmental medicine, 73*, 817-822.

Green, L., & Myerson, J. (2004) A discounting framework for choice with delayed and probabilistic rewards. *Psychological Bulletin, 130*, 769 –792. doi: 10.1037/0033-2909.130.5.769

Huettel, S.A., Stowe, C.J., Gordon, E.M., Warner, B.T., & Platt, M.L. (2006) Neural signatures of economic preferences for risk and ambiguity. *Neuron, 49*, 765–775. doi: 10.1016/j.neuron.2006.01.024

Kringelbach, M. L., & Rolls, E. T. (2004). The functional neuroanatomy of the human orbitofrontal cortex: evidence from neuroimaging and neuropsychology. *Progress in neurobiology, 72*, 341-372. doi: 10.1016/j.pneurobio.2004.03.006

Mitchell, R., & Phillips, L. (2007). The psychological, neurochemical and functional neuroanatomical mediators of the effects of positive and negative mood on executive functions. *Neuropsychologia, 45*, 617–629. doi: 10.1016/j.neuropsychologia.2006.06.030

Orasanu, J., Ames, N., Martin, L., & Davison, J. (2001). Factors in aviation accidents: Decision errors. In E. Salas & G. A. Klein (Eds.), *Linking expertise and naturalistic decision making* (pp. 209–225). Mahwah, NJ: Lawrence Erlbaum Associates.

Peters, J., & Büchel, C. (2009). Overlapping and distinct neural systems code for subjective value during intertemporal and risky decision making. *Journal of Neuroscience, 29*, 15727-15734.

Petrides, M., & Pandya, D. N. (1999). Dorsolateral prefrontal cortex: comparative cytoarchitectonic analysis in the human and the macaque brain and corticocortical connection patternns. *European Journal of Neuroscience, 11*, 1011-1036. doi: 10.1046/j.1460-9568.1999.00518.x

Rachlin, H., Raineri, A., & Cross, D. (1991). Subjective probability and delay. *Journal of the experimental analysis of behavior, 55*, 233-244. doi: 10.1901/jeab.1991.55-233

Rhoda, D., & Pawlak, M. (1999). An assessment of thunderstorm penetrations and deviations by commercial aircraft in the terminal area. Massachusetts Institute of Technology, Lincoln Laboratory, Project Report NASA/A-2, 3.

Sescousse, G., Caldú, X., Segura, B., & Dreher, J. C. (2013). Processing of primary and secondary rewards: a quantitative meta-analysis and review of human functional neuroimaging studies. *Neuroscience & Biobehavioral Reviews, 37*, 681-696. doi: 10.1016/j.neubiorev.2013.02.002

Sescousse, G., Redouté, J., & Dreher, J. C. (2010). The architecture of reward value coding in the human orbitofrontal cortex. *Journal of Neuroscience, 30*, 13095-13104. doi: 10.1523/JNEUROSCI.3501-10.2010

Stearns, S. C. (2000). Daniel Bernoulli (1738): evolution and economics under risk. *Journal of biosciences, 25*, 221-228. doi: 10.1007/BF02703928

Taylor, S., Welsh, R., Wager, T., Luan Phan, K., Fitzgerald, K., & Gehring, W. (2004). A functional neuroimaging study of motivation and executive function. *Neuroimage, 21,* 1045–1054. doi: 10.1016/j.neuroimage.2003.10.032

Tom, S. M., Fox, C. R., Trepel, C., & Poldrack, R. A. (2007). The neural basis of loss aversion in decision-making under risk. *Science, 315*, 515-518. doi: 10.1126/science.1134239

Wise, S. P. (2008). Forward frontal fields: phylogeny and fundamental function. *Trends in neurosciences, 31*, 599-608. doi: 10.1016/j.tins.2008.08.008

HUMAN DECISION MAKING AND THREAT-AWARENESS RESPONSE DURING EMERGENCY AIRCRAFT EVACUATIONS

Sarah Hubbard and Tim Ropp
Purdue University, West Lafayette, IN

Aircraft emergencies requiring evacuation present unique safety challenges to both crew and passengers due to the confined space and the speed at which fire, extreme heat and smoke propagate. In this scenario, where a one or two second delay can literally determine survivability, rapid evacuation is paramount. Although evacuation capability is demonstrated through required and controlled drills for aircraft certification, during a real emergency human factors affect passenger decision making, in some cases resulting in the decision to retrieve personal items during actual emergency evacuations. This may pose a significant threat to post-accident survivability. This research evaluates evacuation decision making and the associated impact on passenger exit flow, during emergency evacuation scenarios. This paper provides an update on a controlled field study using a functional CRJ-100 50-seat aircraft to explore the contributing factors affecting passenger threat awareness and decision making during aircraft evacuations.

Unpredictable or adverse passenger decision making or lack of adherence to instructions during emergency aircraft evacuations have long been identified as contributors to delayed egress and fatalities from otherwise survivable crashes (Muir & Marrison, 1990). Even a few seconds delay can result in additional and needless fatalities. While aircraft design plays a key role, human behavior and decision making can also dramatically impact exit flows and potential survivability; this has been demonstrated even in very recent aircraft incidents (McGee, 2016). A National Transportation Safety Board (NTSB) study in 2000 found that 50% of people involved in an actual emergency evacuation reported having attempted to retrieve their carry-on bag and were the most frequently cited obstruction to evacuation (NTSB, 2000). Accidents as recent as the Delta MD-88 runway excursion in 2015 involved post-accident evacuations that were hampered by hesitancy to begin evacuation by crew members and passengers (Aerossurance, 2016; NTSB, 2016). Survivor reports of passengers attempting to retrieve carry-on bags were documented for the Asiana 777 runway crash in San Francisco, the British Airways 777 engine failure in Las Vegas, and the Emirates 777 fire in Dubai; these demonstrate that baggage retrieval and hesitancy remain a threat (RAS, 2016). Clearly, passenger decision making continues to play a significant role in survivability of modern aircraft, and could contribute to fatalities in what might otherwise be a survivable aircraft accident. This exploratory study evaluates passenger emergency evacuation scenarios using a 50 passenger jet, in an effort to gain a better understanding of the passenger human factors and to identify any appropriate measures to mitigate associated potential threats.

Background

As part of a rigid certification process, regulatory requirements spell out precise configuration and testing requirements for evacuation timelines for large aircraft certification. For example, Federal Aviation Administration (FAA) regulations require aircraft with seating capacity over 44 passengers be designed for complete evacuation through half of the available exits in 90 seconds. (U.S.C.F.R.-FAA (a), 2016) including formulas for total exit time calculations (FAA (b), 2012 p. 19) and required manufacturing design, g-force load carrying capabilities and safety technologies like fire suppression (FAA (c), 2010).

Findings from one NTSB safety study on emergency evacuations indicated that although exit row passengers may be pre-screened and briefed regarding the use of emergency exits, many passengers reported that they did not actually comply with instructions to read and understand emergency exit instruction cards (NTSB, 2000). Additionally, human intuition and perception of the way things operate (right or wrong) and ease of use can be strong determinants of behavior in an aircraft evacuation.

Aircraft design engineers report that evacuation speed can be improved through the use of automatic evacuation systems, doors that are easy to operate, and seats that will stay secure and do not block the aisles (Rosenkrans, 2014). To reduce confusion reported in accident investigations, Boeing redesigned a series of its Type III overwing exit doors to hinge up and out after door handle activation; this design eliminates the issue of where to place removable exit "plug" type doors after they have been opened. This is critical since cabin flight crew members may not always be available to give instructions (NTSB, 2000).

Aircraft cabin design changes and new technologies have improved post-crash survivability and egress capabilities and have allowed airlines to meet prescribed requirements (FAA(c), 2010) As a result, the percentage of survivable accidents is increasing (Rosenkrans, 2014). However, timely aircraft evacuation also depends upon passengers leaving all items on board. Unfortunately, passengers may make inappropriate decisions that can potentially compromise their own safety, or the safety of other passengers. Passenger decisions to bring personal items with them when they evacuate or to retrieve carry-on items before exiting slow down evacuation and block the egress path for other passengers (NTSB, 2000). Examples of this are shown in Figure 1. Trying to retrieve carry-on baggage has been cited as a contributing factor for injuries and fatalities in recent crashes (McGee, 2016). Personal items and carry-on luggage may also potentially damage the aircraft evacuation slide, or cause injury to passengers.

(Gold, 2015)

Figure 1. Social media response to passengers who evacuate with luggage in London

While bringing items when evacuating is against FAA rules and many passengers consider it selfish, other passengers acknowledge they would bring items, with justifications ranging from medical need (medicine in a carry-on), to business need (computer and paperwork) or personal inconvenience (loss of personal items). Some passengers put their passport, wallet and cell phone in their pockets for takeoff and landing, so they can evacuate unimpeded but still be assured possession of their essentials (Gold, 2015). Other passengers suggest that if people want to bring carry-on items, they should be the last to evacuate (Gold, 2015); however, enforcing this framework would be difficult if not impossible. One pilot suggests that carry-on retrieval not only delays evacuation, but also increases exertion and causes greater oxygen use, when oxygen may be in limited supply during an emergency (Gold, 2015). A flight attendant notes that during a planned evacuation, passengers are asked not only to leave all items behind, but also remove high heels and eye glasses, which could potentially damage the slide and prevent others from evacuating (Gold, 2015).

Purpose

Although FAA provides a tightly regulated framework for assessing the evacuation time for an aircraft, and federal regulations advise that passengers leave items behind and follow the directions of a crew member, behavior observed in recent emergencies suggests that many passengers will bring items with them when they evacuate an aircraft in an emergency situation. The purpose of this research is to document human decision making during a controlled field test of an aircraft evacuation, including both qualitative and quantitative findings. One associated research hypothesis is that the average time to evacuate an aircraft will be longer if passengers are carrying personal items or carry-on items; the correlating null hypothesis is that evacuation times will not be longer if passengers are carrying items during evacuation. The purpose of this paper is to describe the framework and methodology, as well as present preliminary findings from pilot tests.

Methodology

This research is a controlled field study using a 50-seat Canadair Regional Jet (CRJ) 100 model manufactured in 1995. The aircraft is a single-aisle, twin engine transport category aircraft with a 3 flight crew configuration (2 flight and 1 cabin crew member). The lower deck height of this aircraft eliminates the need for emergency exit slides, however, as shown in Figure 2, in preliminary tests stairs were used to facilitate passenger exit off the wing leading edge to ensure safety, which is permitted during test certification per the regulatory guidance in Advisory Circular AC25.803 1A (FAA, 2012). ARFF responders were on-site to assist, and the evacuation activity was leveraged to support ARFF aircraft familiarity training. One fire fighter was stationed on the wing to enhance safe evacuation. This aircraft has four passenger area exits:

- Forward left passenger entry door (Type 1) that opens down to the ground incorporating stairs for main cabin and flight crew entry and exit.
- Forward right galley service door (Type 1) that can be used for emergency exit.
- Left side overwing removable plug (Type III) exit door.
- Right side overwing removable plug (Type III) exit door.

In preliminary tests, only the left side entry and overwing exit doors were used for consistency with FAA regulatory certification requirements to use only one half of the aircraft's doors; this also simplifies video documentation of the evacuation. Cameras were also mounted in the cabin, as shown in Figure 3. The flight deck also has an emergency escape door at the top of the fuselage, however, for safety, this was not incorporated in the initial testing. Aircraft interior seating chart is shown n Figure 4, and external dimensions of the aircraft in Figure 5.

The aircraft is functional but non-flying. Seating configuration is two seats abreast, single class configuration. All systems operate including electrical, hydraulic, pneumatic, engines, APU, interior overhead, placard and Emergency Exit Lighting. The test location was the aircraft ramp area used for parking and working on the jet as a live learning and research laboratory.

A convenience sample was used and participants were recruited from personal contacts of the researchers. Although demographic information was not collected, the sample was predominantly college students with an interest in aviation, since the study was conducted at the Purdue Airport, which serves the Purdue Aviation Technology program. The subject pool included anyone over 18 without regard to gender, ethnicity, or health status, although participants were required to be physically mobile and able to safety enter and exit the aircraft. The study was approved by the University Institutional Review Board (IRB) and the purpose of the research was explained to the participants, who signed a consent form prior to participating. After boarding the aircraft, stowing any personal items and carry-on items (some of

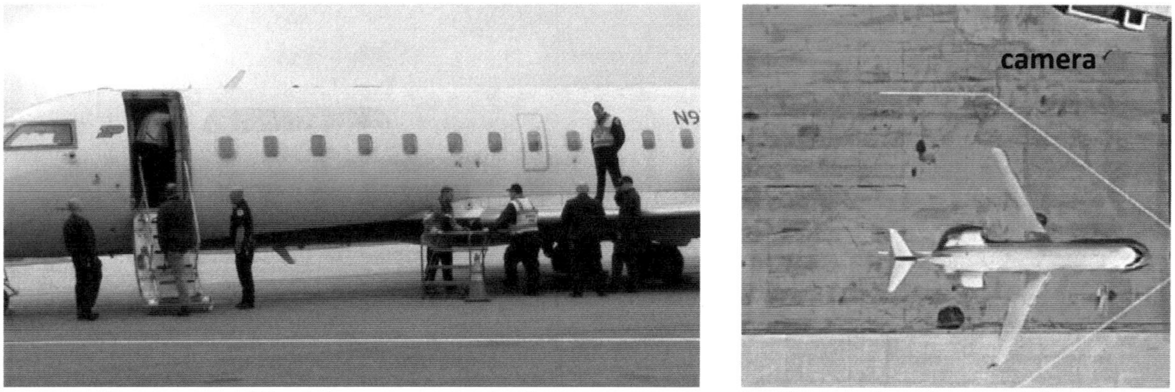

a. Passengers utilized the two port exits for egress   b. Camera location provided view of both port exits

Figure 2. Aircraft evacuation through port exits

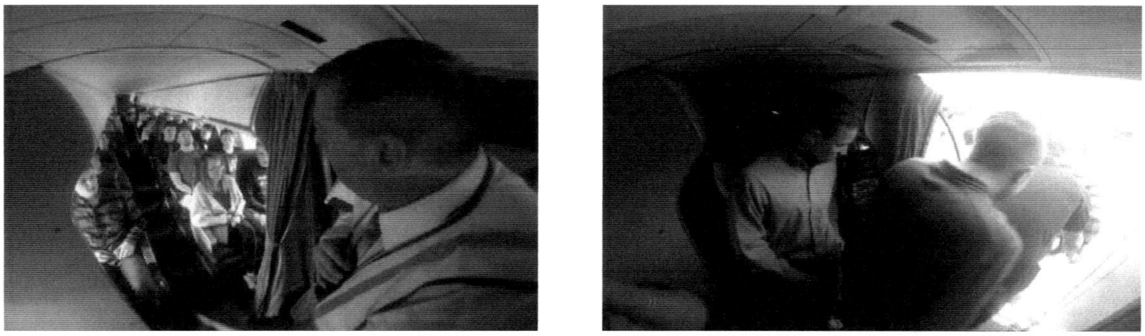

Figure 3. Camera mounted in the galley provided limited field of view in the cabin

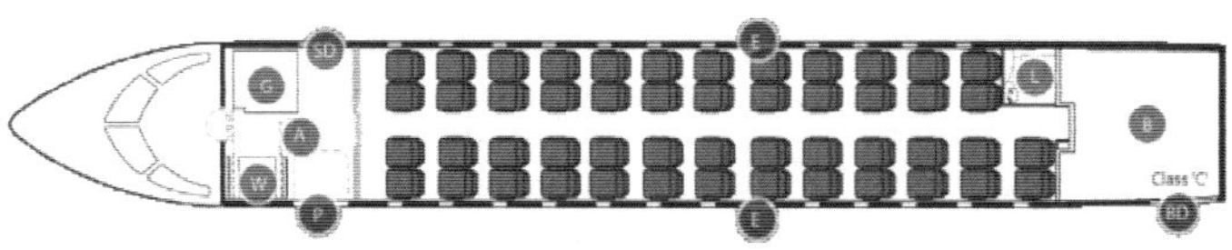

Figure 4. CRJ-100 interior seating chart (Bombardier)

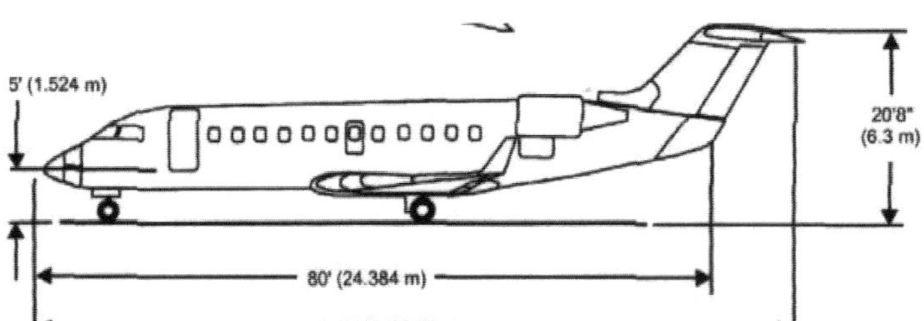

Figure 5. Exterior dimensions

which were provided by the researchers), participants sat in available seats of their choice in the aircraft. Researchers provided the flight safety briefing per the federal regulations for the CRJ aircraft, and participants were instructed to fasten seatbelts. Participants exited the aircraft via the left hand over wing emergency exit and left main entry door upon a verbal signal from a researcher on board the aircraft. Participants opened the exit and main doors using only the aircraft's actual safety information card without the assistance of researchers. Participants seated in row 8 (exit row) removed the exit door. In the first evacuation, they were instructed to place the door on their seat as they exited. Actual procedures would instruct that the door be thrown out of the aircraft, however, in the interest of safety and to protect the aircraft, in the remaining evacuation tests the door was handed to an ARFF or research personnel through the exit opening. After exiting, participants were directed to an emergency assembly area by ARFF personnel. Participants re-boarded the aircraft for another trial run but were instructed to sit in a different section of the aircraft so they would potentially exit from a different door. Trial runs included: passengers only (no items), passengers with carry-on items from below the seat in front of them or the overhead bin, and a mix of passengers with and without items.

Results

Pilot trials were held with two groups of less than ten passengers each, and with one group of 50 passengers. This allowed testing of the protocol and collection of preliminary test configuration data. The procedure generally went very well, and evacuation times were all within the 90 seconds required for a live drill. One limitation of these trials is the composition of the sample, namely able bodied aviation students who are very familiar with airplanes and are generally physically fit. The ease of evacuation within the required timeframe also suggests that the issue of retrieval of items may be more critical on larger aircraft.

Preliminary trials indicate that one potential issue is the placement of the exit door. When the door was placed in the cabin, it was an obstacle to egress (see Figure 5). Another possible challenge is the full utilization of all exits. Participants tended to use the exit closest to them, even if there was no line for the far exit. This makes sense, because the person at the front of the line wants to exit the aircraft as soon as possible, which would always be the exit right next to them. If the queue is near the window exit, it would be optimal to have half the people come to the front to exit, which flight crews are instructed to direct if able, but this is not optimal or intuitive in terms of fastest evacuation for the person at the front of the line. This imbalance would be presumably be exacerbated if there were smoke or other line of sight obstructions, or if a coordinating flight crew member were incapacitated or absent. One final observation is that, in some cases, the wing exit door was removed faster than the front door was opened. This could cause potential problems (due to an even longer ling for the wing exit) if people from both the front half and back half of the aircraft try to exit from through the wing exit.

One limitation of this study was that the stakes were low. In a real evacuation the consequences of evacuating quickly would be significant, and the stress of the situation would affect passenger decision making, which is hard to simulate in a controlled field study in which participant safety is a primary concern.

a. Exit door in cabin can be an obstacle to egress

b. Passengers evacuated through rear door only even after queue at front door cleared

Figure 5. Sample observations during preliminary trials

Recommendations and Conclusions

The human decision making element may be a more challenging factor to control in an emergency situation, based on observation of both historic data as well as very recent large aircraft accident events. Ironically the human element remains the greatest constraint in a NextGen system that has made substantial progress in terms of aircraft technologies, materials, and accident survivability. There is no question that timely and safe passenger evacuation is a critical component affecting safety and survivability following an aircraft incident or accident. Recent airline accident evacuation highlight the ongoing need to better educate passengers and develop more innovative intervention techniques to manage the passenger human factor and threat response during post-accident evacuation.

This paper presents a methodology as well as preliminary findings. These findings suggest that there is value to continued research in this area. Additional research includes additional data collection, the use of a larger aircraft, the use of a more diverse sample of passengers, and increasing the complexity of the evacuation scenario for a more accurate simulation of an actual emergency evacuation. Given the importance of this topic for passenger safety, additional research is clearly warranted to expand the understanding of the factors affecting safe and timely evacuation, and to develop educational and promotional information for passengers, as well as more detailed information regarding the cost, in terms of time as well as safety consequences, associated with passengers evacuating with items.

References

Aerossurance (2016). Delta MD-88 accident at La Guardia 5 March 2015. Posted September 14, 2016 in: Accidents & Incidents, Airfields / Heliports / Helidecks, Fixed Wing, Human Factors / Performance, Regulation, Safety Management. Retrieved: http://aerossurance.com/safety-management/runway-excursion-evacuation/

Federal Aviation Administration (a), (2016). Carry on baggage tips. Retrieved 03 October 2016 from: http://www.faa.gov/passengers/prepare_fly/baggage/

Federal Aviation Administration (b), (2012). Advisory Circular (AC) 25.803-1A. Emergency Evacuation Demonstrations. 03/12/12.

Federal Aviation Administration (c), (2010). FAA FactSheet Report – 2010. Improvements to Aircraft (Accident) Survivability. FA3017-01-0710. Retrieved online Dec 28, 2016: www.faa.gov/passengers/fly_safe/safety_improvements/media/FAA-FactSheet-Aircraft-Survivability.pdf

Gold, Ashely. The fury over 'hand luggage'plane evacuees. BBC Magazine. BBC News, Washington. 10 September 2015. Available online at http://www.bbc.com/news/magazine-34191035

McGee, B. (2016). USA Today: Emergency evacuation? Leave your laptop! Sept. 7, 2016.

Muir, H.C & Marrison, C. (1990). Human factors in cabin safety. Flight Safety Foundation Cabin Crew Safety. Vol.25 No.2 March/April 1990. Pg.2

National Transportation Safety Board (NTSB), (2000). Safety Study: Emergency evacuation of commercial airplanes. PB2000-917002 NTSB/SS-00/01.

National Transportation Safety Board (NTSB), 2016. Runway excursion during landing. Accident Report: NTSB/AAR-16/02 PB2016-104166. Delta Air Lines Flight 1086. Boeing MD-88 N909DL. New York, New York. March 5, 2015.

Rosenkrans, W. Flight Safety Foundation (2014). Aero Safety World 2014. Survival factors: NTSB hearing on Asiana Airlines Flight 214 expands insights into occupant protection, escape and rescue. March 17, 2014. Retrieved 03 Jan 2017: https://flightsafety.org/asw-article/survival-factors/

Royal Aeronautical Society (2016). Lives before luggage. Retrieved: https://www.aerosociety.com/news/lives-before-luggage/

U.S.C.F.R-FAA (a) (2016). Code of Federal Regulations (CFR), Part 121.291(a) Demonstration of emergency evacuation procedures. Retrieved online 03 Oct. 2016: Http://ww.faa.gov

U.S.C.F.R.-FAA (b) (2016). Code of Federal Regulations (CFR), Part 121.291 Appendix D. Criteria for Demonstration of Emergency Evacuation Procedures Under §121.291. Http://ww.faa.gov

TASK COMPLEXITY AND TIME PRESSURE AFFECT AIR TRAFFIC CONTROLLER'S PERFORMANCE AND WORKLOAD

Qianru Yang
Andrew R. Dattel
Dahai Liu
Embry-Riddle Aeronautical University
Daytona Beach, Florida

The purpose of this study was to investigate the effects of task complexity (TC) and time pressure (TP) on air traffic controller's (ATC) performance and mental workload. Sixteen students enrolled in an aviation college completed four scenarios which were a subset of the ATCPrep software for the AT-SAT. Fifteen performance variables were measured (e.g., conflict resolution). Additionally, NASA-TLX was used to test participants' mental workload. As expected, for most of the performance variables, high TC and high TP resulted in the lowest participant performance. For the three performance variables, TP had a differential effect on TC. Participants experienced the greatest mental workload when TC and TP were the highest. Although higher TC and higher TP was shown to have the greatest impact on performance variables, participants seemed to handle TC better than TP in several situations. When developing new technology, greater consideration should be given to TC rather than TP.

Introduction

As air traffic continues to grow, the associated demands for ATCs increase as well. ATCs play a very important role in the air traffic management system because they direct aircraft both on the ground and within the airspace. Controllers must prevent collisions, organize the flow of air traffic, and offer information to pilots. Their tasks become more cognitively demanding as traffic increases, which could compromise their performance and air traffic safety (Trapsilawati, Qu, Wickens, & Chen, 2015). Many factors have been found to have effects on ATCs' performance and mental workload, including weather, task complexity (TC), ATCs' fatigue, and time pressure (TP) (Edwards, Sharples, Wilson, & Kirwan, 2012).

ATC Task Complexity

The construct of complexity has been a largely augured matter in the ATC domain (Djokic, Lorenz, & Fricke, 2010). Many factors influencing ATCs' cognitive complexity have been indicated, such as organizational procedures, traffic environment, and display complexity (Marchitto, Di Stasi, & Cañas, 2012). ATCs play a necessary role in the safety and fluidity of the airspace as they prevent collisions, organize the flow of air traffic, and offer information to pilots. To complete their complex tasks, they use radar display to observe aircraft. When multiple aircraft show up on the display screen simultaneously, it requires greater visual attention and more working memory. This will put a high demand on mental workload for ATCs (Kaber, Perry, Segall, & Sheik-Nainar, 2007). Therefore, when all these factors are combined, it increases TC, which may influence ATCs' mental workload and their safety.

With the development of technology, aircraft and ground facilities are constantly improving and enhancing their reliability (Trapsilawati et al., 2015). However, the rate of aircraft or related equipment failure has been decreasing gradually. On the contrary, the rate of human error associated with unsafe acts has risen dramatically. (Trapsilawati et al., 2015). Mental workload assessment seems to be a recurrent problem in ATC field (Philippe, Christian, André, Sylvie, & Evelyne, 2004). Many factors can have an effect on the workload of ATCs, such as individual differences, working or living environment, TC, TP, salaries, attitude, motivation, or fatigue (Costa, 1996). ATC errors can lead to catastrophic consequences; therefore, it is important to study ATCs' performance and mental workload.

ATC Workload

An increase in the number and types of tasks is not necessarily a synonym for workload, but it also depends on individual differences, such as age, life styles, life events, work experience, and behavioral characteristics, such as mood and sleeping habits (Costa, 1996). Air traffic volume is continuing to increase worldwide. If air traffic management organizations want to meet future demands safely, they will be required to pay attention to controller's workload (Loft, Sanderson, Neal, & Mooij, 2007). Physiological measures have been used to study issues related to the effects of long-term stress on ATCs' health (Brookings, Wilson, & Swain, 1996). There are three factors that greatly affect ATC workload: time-based factors, task intensity-based factors, and operator's psychophysiological functional state (Philippe et al., 2004). High mental workload can also affect air safety due to its negative effect on ATC performance and limits traffic-handling capacities (Boag, Neal, Loft, & Halford, 2006).

Di Stasi, Marchitto, Antolí, Baccino, and Cañas (2010) found that combining different task complexities could be useful in creating different mental workload levels. The authors used an eye tracker and found that saccadic peak velocity was sensitive to variations in mental workload. During the same year, other researchers found that subjective workload hinges not only on the complexity of ATC, but also on the communication load of the ATC (Djokic et al., 2010). In addition, Fothergill and Neal (2008) used an ATC simulator and found that controllers were less likely to select the optimal solution under a high workload than under a low workload when the optimal solution was difficult to calculate. They also discovered that controllers were likely to select the optimal solution under both levels of workload when the optimal solution was easy to calculate. The following null hypotheses were tested in this study:

H_{01}: TC does not have a significant effect on an ATC's performance.
H_{02}: TP does not have a significant effect on an ATC's performance.
H_{03}: The interaction of TC and TP do not have a significant effect on an ATC's performance.
H_{04}: TC does not have a significant effect on an ATC's workload.
H_{05}: TP does not have a significant effect on an ATC's workload.
H_{06}: The interaction of TC and TP do not have a significant effect on an ATC's workload.

Method

Participants

Sixteen students at a private university in the southeastern United States were recruited. They were interested in the topic of this study. Gender and age were not factors considered. The grades (i.e., freshman, sophomore, and junior) will also not be considered for the participants.

Materials

AT-SAT software. AT-SAT is a pre-employment screening for Federal Aviation Administration ATC applicants. This software has seven cognitive tests: Air Traffic Scenarios Test, Dials Test, Analogy Test, Letter Factory Test, Angles Test, Scan Test, and Applied Mathematics Test. The Air Traffic Scenarios Test (ATST) was used in this study. In this subtest, participants should control traffic in interactive, dynamic, low-fidelity simulations of air traffic situations requiring prioritization (Dattel & King, 2010). Different scenarios can be set in the ATST. Participants handle aircraft to land at airports or go to exits efficiently. Finally, there were 15 categories scores (i.e., dependent variables), which were calculated by the software to reflect participant's performance.

NASA-TLX. In addition to the objective measures by AT-SAT of ATC's performance, their mental workload was measured by using the NASA-TLX. It is the most commonly utilized tool to measure workload (Noyes & Bruneau, 2007). The TLX is a scale with six subscales that are scored from 0 to 100. The six subscales include mental demand, physical demand, temporal demand, performance, effort, and frustration. NASA-TLX combines subscale ratings, which are weighted according to participant's subjective importance to subjects for a research (Cao, Chintamani, Pandya, & Ellis, 2009). When using NASA-TLX, participants should select two subscales of those six subscales first. These subscales are what participants find to be most relevant to the situation. Then they identify scores about these two subscales.

Procedure

Upon arrival, participants were first briefed about the purposes and procedures of the study and presented the informed consent form to review and sign. After signing the informed consent form, participants were trained how to use the AT-SAT software. The training lasted 10 minutes, which included practice trials. The test trials consisted of four scenarios, which are shown in Table 1. During the test, all aircraft in the AT-SAT software were instructed either to land at "airports" or go to "sector exits." Low task complexity scenarios started with five aircraft; high task complexity scenarios started with 10 aircraft. In the low time pressure scenario, the airplanes were moving at a slow rate; in the high time pressure scenario, the airplanes were moving at a fast rate.

Table 1
Four Scenarios

Independent Variables	Low Task Complexity	High Task Complexity
Low Time Pressure	Low Task Complexity and Low Time Pressure	High Task Complexity and Low Time Pressure
High Time Pressure	Low Task Complexity and High Time Pressure	High Task Complexity and High Time Pressure

The study was a 2 x 2 within-subjects design. The order of four scenarios were counterbalanced using a Latin Square Design. Each scenario lasted 10 minutes. After a participant finished one scenario, he or she completed NASA-TLX, then had a 5-minute break.

Results

AT-SAT results. AT-SAT provided the following 15 performance variables. Fifteen two-way within-subject ANOVAs were run in SPSS with the alpha-level set at .05. Table 2 shows the descriptive statistics of these 15 categories scores. Table 3 shows the results of these two variables and their interaction on ATCs' performance for 15 categories.

Table 2
Descriptive Statistics of Total Scores

	N	Minimum	Maximum	Mean	Std. Deviation
Dis Ned	64	72.10	100.00	99.45	3.54
Ti Ned	64	49.60	100.00	96.89	8.79
Conflicts	64	.00	100.00	73.26	31.88
Collisions	64	46.20	100.00	98.80	6.77
Pro Airs	64	98.40	100.00	99.83	.25
Pro Airp	64	77.80	100.00	98.82	4.36
Exit Airs	64	57.10	100.00	94.10	10.17
Exit Spd	64	57.10	100.00	96.10	8.69
Exit Alt	64	57.10	100.00	92.90	10.48
Land Des	64	.00	100.00	86.78	24.96
Land Head	64	.00	100.00	76.53	24.19
Land Spd	64	.00	100.00	81.85	23.22
Land Alt	64	.00	100.00	85.90	24.59
Set Dif	64	66.70	77.80	72.25	5.59
Total Result	64	41.60	102.60	71.64	21.85

Note. Dis Ned = Distance Needed, Ti Ned = Time Needed, Pro Airs = Prohibited Airspace Border Crossings, Pro Airp = Prohibited Airport Border Crossings, Exit Airs = Exiting the Airspace Correct Destination, Exit Spd = Exiting the Correct Speed, Exit Alt = Exiting the Correct Altitude, Land Des = Landing at Airports Correct Destination, Land Head = Landing at Airports Correct Headings, Land Spd = Landing at Airports Speed, Land Alt = Landing at Airports Correct Altitude, Set Dif = Set up Difficulty.

Table 3
Significance of Independent Variables and Performance

	Task Complexity			Time Pressure			Task Complexity and Time Pressure		
Per Va	*F*	*p*	Rejected or Retained	*F*	*p*	Rejected or Retained	*F*	*p*	Rejected or Retained
Dis Ned	$F(1, 15) = .70,$	.42	Retained	$F(1, 15) = .70$	.42	Retained	$F(1, 15) = 1.59$	.23	Retained
Ti Ned	$F(1, 15) = .08$	.78	Retained	$F(1, 15) = 4.99$	< .05	Rejected*	$F(1, 15) = .89$	.36	Retained
Conflicts	$F(1, 15) = 49.36$	< .05	Rejected*	$F(1, 15) = 92.78$	< .05	Rejected*	$F(1, 15) = 27.09$	< .05	Rejected*
Collisions	$F(1, 15) = 1.42$	.25	Retained	$F(1, 15) = 1.65$	.22	Retained	$F(1, 15) = 1.64$	.22	Retained
Pro Airs	$F(1, 15) = .002$	.97	Retained	$F(1, 15) = 1.31$	.27	Retained	$F(1, 15) = 2.74$	.12	Retained
Pro Airp	$F(1, 15) = 1.47$	.25	Retained	$F(1, 15) = 1.48$	.24	Retained	$F(1, 15) = .13$	.73	Retained
Exit Airs	$F(1, 15) = 9.58$	< .05	Rejected*	$F(1, 15) = 16.59$	< .05	Rejected*	$F(1, 15) = 36.48$	< .05	Rejected*
Exit Spd	$F(1, 15) = 25.54$	< .05	Rejected*	$F(1, 15) = 30.67$	< .05	Rejected*	$F(1, 15) = 25.54$	< .05	Rejected*
Exit Alt	$F(1, 15) = 6.67$	< .05	Rejected*	$F(1, 15) = 14.12$	< .05	Rejected*	$F(1, 15) = 19.14$	< .05	Rejected*
Land Des	$F(1, 15) = 22.75$	< .05	Rejected*	$F(1, 15) = 20.55$	< .05	Rejected*	$F(1, 15) = 18.03$	< .05	Rejected*
Land Head	$F(1, 15) = 6.85$	< .05	Rejected*	$F(1, 15) = 3.09$	.10	Retained	$F(1, 15) = 1.27$	.28	Retained
Land Spd	$F(1, 15) = 8.31$	< .05	Rejected*	$F(1, 15) = 3.79$	.07	Retained	$F(1, 15) = 6.88$	< .05	Rejected*
Land Alt	$F(1, 15) = 18.10$	< .05	Rejected*	$F(1, 15) = 15.48$	< .05	Rejected*	$F(1, 15) = 13.86$	< .05	Rejected*
Set Dif	$F(1, 15) = 1$	.33	Retained	$F(1, 15) = 225$	< .05	Rejected*			
To Re	$F(1, 15) = 179.23$	< .05	Rejected*	$F(1, 15) = 2.31$	.15	Retained	$F(1, 15) = 38.43$	< .05	Rejected*

Note. Per Va = Performance Variables, Dis Ned = Distance Needed, Ti Ned = Time Needed, Pro Airs = Prohibited Airspace Border Crossings, Pro Airp = Prohibited Airport Border Crossings, Exit Airs = Exiting the Airspace Correct Destination, Exit Spd = Exiting the Correct Speed, Exit Alt = Exiting the Correct Altitude, Land Des = Landing at Airports Correct Destination, Land Head = Landing at Airports Correct Headings, Land Spd = Landing at Airports Speed, Land Alt = Landing at Airports Correct Altitude, Set Dif = Set up Difficulty, To Re = Total Result.

NASA-TLX results. There are six subscales in NASA-TLX. The NASA-TLX provides two results for participants' mental workload. One of the six subscales were the most relevant to workload. The other one of the results was mean value of overall workload. Table 4 shows the description of overall workload.

Table 4
Description of Overall Workload

	Valid	Missing	Mean	Median	SD	Min	Max
LT_LTP	16	0	89.13	85	49.86	30	175
LT_HTP	16	0	133.31	147.50	43.02	60	188
HT_LTP	16	0	120.88	132.50	46.29	30	175
HT_HTP	16	0	164.81	172.50	33.89	95	200

Note. SD= Std. Deviation, Min = Minimum, Max = Maximum, LT_LTP = Low Task Complexity and Low Time Pressure, LT_HTP = Low Task Complexity and High Time Pressure, HT_LTP = High Task Complexity and Low Time Pressure, HT_HTP = High Task Complexity and High Time Pressure.

A two-way within-subject ANOVA was run in SPSS with the alpha level set at .05. The results that were analyzed the overall score. Therefore, the results of overall workload were: $F(1, 15) = 14.72, p < .05$ for TC; $F(1, 15) = 45.86, p < .05$ for TP.

Discussion and Conclusion

Task complexity and time pressure affect performance.

There are 15 categories for AT-SAT to reflect ATC performance. The data yielded some intriguing findings. Results show that distance, airspace border, and airport border were not affected by TC and TP. Second, when TC is higher and TP is lower, ATCs had better results of Exiting the Airspace Correct Destination than when TP is higher; therefore, the level of TP had a greater negative impact on performance than TC. For the "Total Result" variable, when TP is lower and TC at the same level, ATCs performed better. In addition, when TC was higher, regardless of TP level, ATCs had better performance. This indicates that an increased number of aircraft yields greater workload, high TC may promote better performance.

TC and TP affect workload.

Results showed that mental workload occurred more frequently than any of the other five subscales (i.e., physical demand, temporal demand, performance, effort, and frustration). This means that mental workload is the most relevant subscale of ATCs workload. As expected, high TC or high TP had a greater effect on ATCs' workload than low TC or TP respectively.

Future research should consider these suggestions for improvement. First, conducting this experiment utilizing trained ATCs may yield more reliable results. Second, scenarios may have different levels of difficulty, such as low, medium, and high. Moreover, higher TC is not necessarily bad in all situations. Therefore, when doing selecting and training of ATC in the future, TC might be considered more than TP.

Reference

Boag, C., Neal, A., Loft, S., & Halford, G. S. (2006). An analysis of relational complexity in an air traffic control conflict detection task. *Ergonomics, 49*, 1508-1526. doi: 10.1080/00140130600779744

Brookings, J. B., Wilson, G. F., & Swain, C. R. (1996). Psychophysiological responses to changes in workload during simulated air traffic control. *Biological Psychology, 42*, 361-377. doi: 10.1016/0301-0511(95)05167-8

Cao, A., Chintamani, K. K., Pandya, A. K., & Ellis, R. D. (2009). NASA TLX: Software for assessing subjective mental workload. *Behavior Research Methods, 41,* 113-117. doi: 10.3758/BRM.41.1.113

Costa, G. (1996). Occupational stress and stress prevention in air traffic control. International Labour Organization.

Dattel, A. R., & King, R. E. (2010). The impact of reweighting the AT-SAT subtests on group score differences. *The International Journal of Aviation Psychology, 20,* 111-123. doi: 10.1080/10508411003617748

Djokic, J., Lorenz, B., & Fricke, H. (2010). Air traffic control complexity as workload driver. *Transportation Research Part C: Emerging Technologies, 18*, 930-936. doi: 10.1016/j.trc.2010.03.005

Di Stasi, L. L., Marchitto, M., Antolí, A., Baccino, T., & Cañas, J. J. (2010). Approximation of on-line mental workload index in ATC simulated multitasks. *Journal of Air Transport Management, 16*, 330-333. doi: 10.1016/j.jairtraman.2010.02.004

Edwards, T., Sharples, S., Wilson, J. R., & Kirwan, B. (2012). Factor interaction influences on human performance in air traffic control: The need for a multifactorial model. *Work, 41*, 159-166. doi: 10.3233/WOR-2012-0151-159

Fothergill, S., & Neal, A. (2008). The effect of workload on conflict decision making strategies in air traffic control. *Proceedings of the Human Factors and Ergonomics Society Annual Meeting, 52*, 39-43. doi:10.1177/154193120805200110

Kaber, D. B., Perry, C. M., Segall, N., & Sheik-Nainar, M. A. (2007). Workload state classification with automation during simulated air traffic control. *The International Journal of Aviation Psychology, 17*, 371-390. doi: 10.1080/10508410701527860

Loft, S., Sanderson, P., Neal, A., & Mooij, M. (2007). Modeling and predicting mental workload in en route air traffic control: Critical review and broader implications. *Human Factors: The Journal of the Human Factors and Ergonomics Society, 49*, 376-399. doi: 10.1518/001872007X197017

Marchitto, M., Di Stasi, L. L., & Cañas, J. J. (2012). Ocular movements under taskload manipulations: Influence of geometry on saccades in air traffic control simulated tasks. *Human Factors and Ergonomics in Manufacturing & Service Industries, 22*, 407-419. doi: 10.1002/hfm.20320

Noyes, J. M., & Bruneau, D. P. J. (2007). A self-analysis of the NASA-TLX workload measure. *Ergonomics, 50,* 514-519. doi: 10.1080/00140130701235232

Philippe, A., Christian, C., André, D., Sylvie, A., & Evelyne, V. (2004). Mental workload in air traffic control: An index constructed from field tests. *Aviation, Space, and Environmental Medicine, 75*, 333-341.

Trapsilawati, F., Qu, X., Wickens, C. D., & Chen, C. (2015). Human factors assessment of conflict resolution aid reliability and time pressure in future air traffic control. *Ergonomics, 58*, 897-908. doi: 10.1080/00140139.2014.997301

USING PAIR-WISE RANKINGS IN THE ASSESSMENT OF ADAPTIVE AIDING

Christina Gruenwald
Oak Ridge Institute for Science and Education
Belcamp, Maryland
Matt Middendorf
Middendorf Scientific Services
Medway, Ohio
Chelsey Credlebaugh
Ball Aerospace and Technologies Corp.
Fairborn, Ohio
Jonathan Mead
Oak Ridge Institute for Science and Education
Belcamp, Maryland
Scott Galster
Air Force Research Laboratory
Wright-Patterson AFB, Ohio

In remotely piloted aircraft (RPA) operations, operator cognitive workload is an important concern. High workload could result in performance decrements and operational mishaps. In research, physiological data can be used by models to assess the operator's cognitive state. When a model detects the onset of cognitive overload, assistance could then be provided to the operator to help mitigate the overload in some form of augmentation. However, it is imperative that the assessment is accurate and completed in a timely manner. The accuracy of a workload assessment model and augmentation application can be evaluated using a psychometrically determined scale of man/machine conditions. Both the operator and machine can be in various conditions at any point in time. In three prior studies, eighteen participants were asked to perform pair-wise rankings of sixteen conditions to generate the rankings. These rankings will be used to evaluate the accuracy of the workload assessment model in future research.

Operator cognitive workload is an important concern in remotely piloted aircraft (RPA) operations. RPA use is increasing for missions in hostile environments, or those considered too dangerous for manned aircraft (U.S. Department of Defense, 2011). This places more importance on monitoring the cognitive state of the RPA operators. When task demands are high and cognitive overload occurs, operator performance can suffer and lead to mission failure (Young & Stanton, 2002).

The Sense-Assess-Augment (SAA) paradigm was developed for use in research studying operator cognitive state (Galster & Johnson, 2013). In general, the paradigm serves to *sense* physiological measures, *assess* the state of the operator, and if necessary, provide tools to *augment* the operator's performance.

The most complex aspect of the SAA paradigm is within the assessment stage. Models must be able to accurately assess cognitive state using physiological measures. If the assessment is inaccurate, the augmentation strategy applied may not be beneficial. In the current research, the SAA paradigm is applied to simulated RPA operations in order to prevent detrimental errors that could occur during a mission as a result of cognitive overload.

Adaptive augmentation is best suited for RPA operations to mitigate high workload. The Yerkes-Dodson Law states that workload must fall within a middle range (not too low and not too high) in order to achieve optimal performance (Cohen, 2011). Augmentation must be adaptive for two reasons. First, if augmentation is not necessary for the operator's current cognitive state, the resources necessary for the augmentation would be unnecessarily tied up. Secondly, if the operator's cognitive state is already in the middle range of workload, we do not want to provide an augmentation that will allow the operator to fall into a cognitive underload state. A low workload state has been shown to be just as detrimental to performance as having high task demands (Desmond & Hoyes, 1996).

In past research, artificial neural network (ANN) models have been used in an attempt to model cognitive workload (Hoepf et al., 2016). These models can be used to trigger augmentation and close the SAA loop. This report focuses on the development of a methodology that can be applied to studies that use an assessment model to determine if augmentation is being triggered at the correct time. Further, augmentation should only be provided when it is truly needed and not when the workload level is already manageable. Previous research has used control groups such as yoked or random augmentation to determine if augmentation is being applied correctly (e.g., Bailey, Scerbo, Freeman, Mikulka, & Scott, 2006). However, in field operations, having experimental trials is not possible and there will still be a need for ensuring accurate cognitive state assessment. The development of the pair-wise rankings methodology and how it can be applied to future research is reported.

Method

Participants

A total of 18 individuals from three previous studies participated in the pair-wise rankings evaluation after completion of the studies. Eleven of the participants were male and seven were female. The average age of the participants was 21.3. All participants read and signed an informed consent document prior to participating. Study procedures were reviewed and approved by the Air Force Research Laboratory (AFRL) Institutional Review Board.

Task

Prior to the pair-wise rankings, each participant completed one of three studies that each consisted of surveillance, tracking, and communication tasks. The main differences between the three studies were the physiological measures being collected and modeling techniques used to estimate workload. Therefore, the experimental tasks and manipulations were consistent across these studies. All studies consisted of two 2 x 2 within-subjects designs.

For the surveillance task, participants were instructed to search a market place to find four high value targets (HVTs). The HVTs carried a sniper rifle whereas all other pedestrians in the market place were distractors: not holding anything, holding a handgun, or holding a shovel. Once the HVT was found, participants pressed the F key and tracked the target (i.e., kept the target on screen) until he went under a tent. Participants would then proceed to search and find the next HVT. Only one HVT was present in the scenario at a time. Once an HVT was found, participants started accumulating points. Therefore, the sooner the participant found the HVT, the higher performance score they received.

There were two experimental manipulations, each consisting of two levels for the surveillance task: distractors and fuzz. Distractors could be low or high, meaning there were either 16 or 48 other entities walking around the market place during the trial. The fuzz manipulation could either be off or on. When fuzz was off, the camera feed was clear and it was fairly easy to identify the HVT. However, when fuzz was on, the camera feed was degraded.

For the tracking task, participants were instructed to follow HVTs that were traveling by motorcycle. The two manipulations for the tracking task were number of HVTs and route the HVT was traveling. Participants would either track one or two HVTs. Two HVTs required continuous clicking back and forth on two camera feeds. The HVT(s) would either travel a country route, which consisted of a paved, straight road, or a city route in which the HVT(s) turned often and could be occluded by buildings. Performance for the tracking task was based on keeping the HVT(s) in the camera feed. In addition, more points were awarded for keeping the HVT(s) closer to the center of the feed, compared to the edges.

In conjunction with the surveillance and tracking tasks, participants had a secondary task to perform. Evenly distributed throughout the trial, mental math questions were asked over the headset. These questions consisted of altitude change, travel time, and speed change questions relative to the RPA. For example, if the RPA was traveling at 40 knots, a question might ask how long it would take to arrive at a location 100 nautical miles away with a headwind of 15 knots.

Man/Machine Conditions

Throughout the experiment, participants experienced different man/machine conditions. Both the man and machine could be in various conditions depending on the experimental manipulations and the specific point of time within a trial. There are 16 possible combinations of various man and machine conditions for both the surveillance and tracking tasks as seen in Table 1 and Table 2.

For surveillance, the machine could be in a condition of fuzz on or off and distractors high or low (experimental manipulations). The man could be in a condition of either looking for an HVT or tracking the HVT until it goes under a tent. The man could also be in a condition of answering a communication question or not having a question to answer. Although the machine condition was consistent throughout each trial, the man condition could change within a trial.

Similarly, for tracking, the machine could be in a condition of tracking one or two HVTs, along a country or city route (experimental manipulations). The man could be in a condition of either successfully tracking the HVT or searching for a lost HVT. Also, the man could be answering or not answering a communication question. Identical to the surveillance task, the machine condition was constant throughout each trial, but the man condition varied.

Table 1.
Possible Man/Machine Conditions for the Surveillance Task.

Condition	HVT	Question	Fuzz	Distractors
A	Looking	Yes	On	High
B	Looking	Yes	On	Low
C	Looking	Yes	Off	High
D	Looking	Yes	Off	Low
E	Looking	No	On	High
F	Looking	No	On	Low
G	Looking	No	Off	High
H	Looking	No	Off	Low
I	Tracking	Yes	On	High
J	Tracking	Yes	On	Low
K	Tracking	Yes	Off	High
L	Tracking	Yes	Off	Low
M	Tracking	No	On	High
N	Tracking	No	On	Low
O	Tracking	No	Off	High
P	Tracking	No	Off	Low

Table 2.
Possible Man/Machine Conditions for the Tracking Task.

Condition	Tracking	Question	Route	Targets
A	Lost	Yes	City	2
B	Lost	Yes	City	1
C	Lost	Yes	Country	2
D	Lost	Yes	Country	1
E	Lost	No	City	2
F	Lost	No	City	1
G	Lost	No	Country	2
H	Lost	No	Country	1
I	Tracking	Yes	City	2
J	Tracking	Yes	City	1
K	Tracking	Yes	Country	2
L	Tracking	Yes	Country	1
M	Tracking	No	City	2
N	Tracking	No	City	1
O	Tracking	No	Country	2
P	Tracking	No	Country	1

Procedure

After completing each experiment, participants performed the pair-wise comparison survey. Each of the 16 man/machine conditions were compared to each other. Participants simply selected which of two conditions had higher workload. There were 120 comparisons for each task.

The number of times a condition was selected as the higher workload condition was summed for each participant. This number was then averaged across participants producing an overall ranking of the man/machine conditions.

Results

The results from the pair-wise comparisons are reported in Table 3 for the surveillance task. It is clear that the largest factor to affect workload was whether the participant was looking for the HVT or had already found the HVT and was tracking it. This is evident by "Looking" for the HVT being in the top seven rankings. The lowest workload man/machine condition was tracking the HVT, with no question being asked, fuzz off, and distractors low, as anticipated. In contrast, the highest workload man/machine condition was looking for the HVT, answering a question, fuzz on, and distractors high.

Table 4 shows the results from the pair-wise comparisons for the tracking task. It was rated that when a target was lost, a question was being asked, and they were tracking two targets in the city (Condition A) was the highest level of workload. Conversely, Condition P had the lowest level of workload.

Table 3.
Surveillance Man/Machine Rankings.

Condition	Rank	Average	HVT	Question	Fuzz	Distractors
A	16	14.9	Looking	Yes	On	High
E	15	13.3	Looking	No	On	High
B	14	12.3	Looking	Yes	On	Low
C	13	11.7	Looking	Yes	Off	High
F	12	10.0	Looking	No	On	Low
G	11	9.7	Looking	No	Off	High
D	10	8.3	Looking	Yes	Off	Low
I	9	8.3	Tracking	Yes	On	High
H	8	6.5	Looking	No	Off	Low
J	7	5.9	Tracking	Yes	On	Low
K	6	5.7	Tracking	Yes	Off	High
M	5	5.5	Tracking	No	On	High
L	4	2.8	Tracking	Yes	Off	Low
N	3	2.6	Tracking	No	On	Low
O	2	2.3	Tracking	No	Off	High
P	1	0.1	Tracking	No	Off	Low

Table 4.
Tracking Man/Machine Rankings.

Condition	Rank	Average	Tracking	Question	Route	Targets
A	16	14.8	Lost	Yes	City	2
E	15	13.4	Lost	No	City	2
C	14	12.3	Lost	Yes	Country	2
I	13	10.9	Tracking	Yes	City	2
G	12	10.3	Lost	No	Country	2
M	11	9.2	Tracking	No	City	2
B	10	9.0	Lost	Yes	City	1
K	9	8.1	Tracking	Yes	Country	2
F	8	7.6	Lost	No	City	1
D	7	6.3	Lost	Yes	Country	1
O	6	5.6	Tracking	No	Country	2
J	5	4.4	Tracking	Yes	City	1
H	4	4.1	Lost	No	Country	1
N	3	2.3	Tracking	No	City	1
L	2	1.7	Tracking	Yes	Country	1
P	1	0.1	Tracking	No	Country	1

Discussion

The pair-wise ranking methodology gives insight into how the man/machine conditions compare to each other. For the surveillance task, it was determined that looking for the HVT usually ranked as higher workload than tracking the HVT after it was already found. Similarly, for the tracking task, tracking two targets was harder than tracking one target in most conditions.

This methodology can be applied to future experiments to evaluate the accuracy of the workload assessment model and the activation of augmentation. This can be accomplished by taking the average of the man/machine condition when the augmentation was triggered and comparing it to when performance decrements occurred or if the man/machine ranking was in the upper portion of the rankings. This information can then be used to evaluate if ANN models are activating the augmentation at the correct times. The augmentation should be provided when the man/machine condition was in a condition that was rated as having higher workload, although the point in which augmentation is needed can vary due to individual differences in performance. In conclusion, the pair-wise ranking methodology will be applied to ongoing and future studies, in addition to other statistical analysis to ensure modeling and adaptive augmentation is working properly.

Acknowledgements

The authors would like to thank Paul Middendorf, Kevin Durkee, Avi Hiriyanna, and Noah Depriest for assistance in data collection and software support.

This research was supported in part by an appointment to the Student Research Participation Program at the U.S. Air Force Research Laboratory (USAFRL) administered by the Oak Ridge Institute for Science and Education through an interagency agreement between the U.S. Department of Energy and USAFRL. The views expressed in this report are solely those of the authors and do not necessarily reflect the views of the employers or granting organizations.

References

Bailey, N. R., Scerbo, M. W., Freeman, F. G., Mikulka, P. J., & Scott, L. A. (2006). Comparison of a brain-based adaptive system and a manual adaptable system for invoking automation. *Human Factors, 48*, 693-709. doi: 10.1518/001872006779166280

Cohen, R. A. (2011). Yerkes–Dodson Law. In *Encyclopedia of clinical neuropsychology* (pp. 2737-2738). Springer New York. doi: 10.1007/978-0-387-79948-3_1340

Desmond, P. A., & Hoyes, T. W. (1996). Workload variation, intrinsic risk and utility in a simulated air traffic control task: evidence for compensatory effects. *Safety Science, 22*(1), 87-101. doi: 10.1016/0925-7535(96)00008-2

Hoepf, M., Middendorf, M., & Mead, J., Gruenwald, C., Credlebaugh, C., Middendorf, P. & Galster, S. (2016). Evaluation of Physiologically-Based Artificial Neural Network Models to Detect Operator Workload in Remotely Piloted Aircraft Operations (Report No. AFRL-RH-WP-TR-2016-0075) Wright-Patterson Air Force Base: Air Force Research Laboratory, Human Effectiveness Directorate.

Galster, S. M., & Johnson, E. M. (2013). Sense-assess-augment: A taxonomy for human effectiveness (Report No. AFRL-RH-WP-TM-2013-0002). Wright-Patterson Air Force Base: Air Force Research Laboratory, Human Effectiveness Directorate.

U.S. Department of Defense. (2011). Unmanned systems integrated roadmap FY2011-2036 (Reference No. 11-S-3613). Washington, DC: Department of Defense.

Young, M. S., & Stanton, N. A. (2002). Attention and automation: New perspectives on mental underload and performance. *Theoretical Issues in Ergonomics Science, 3*, 178-194. doi: 10.1080/14639220210123789

AN EEG DATA INVESTIGATION USING ONLY ARTIFACTS

Chelsey Credlebaugh
Ball Aerospace and Technologies Corp.
Fairborn, Ohio
Paul Middendorf
Oak Ridge Institute for Science and Education
Belcamp, Maryland
Matt Middendorf
Middendorf Scientific Services
Medway, Ohio
Scott Galster
Air Force Research Laboratory
Wright Patterson AFB, Ohio

For decades, it has been reported that the electroencephalogram (EEG) is a positive indicator of mental workload. However, EEG signals are easily affected by artifacts. An artifact mediation approach, called artifact separation, was developed to enable the consumer of the EEG data to decide how to handle artifacts. The current investigation uses only data contaminated by artifacts and discards the artifact free data. This was done to solve a problem associated with data collection. Specifically, in an experiment, EEG electrode leads for T3 and Fz were swapped where they were connected to the signal acquisition hardware. To facilitate analysis of the data, it was essential to determine when the swap occurred. This was accomplished using only EEG data that were contaminated by blinks. Power associated with a blink is lower at site T3 than Fz. The artifact separation technique supported this investigation to determine when the swap occurred.

The reliable assessment of mental workload is important due to the effect increased workload can have on human operator performance. One potential solution to offset the risk of operator overload is to monitor workload in real-time and provide assistance before performance decrements occur (Hankins & Wilson, 1998). One challenge in the study of cognitive workload is the problem of how to effectively measure it (Gevins & Smith, 2003). Tsang and Wilson (1997) classified workload measurements into three general categories, which include: performance, subjective evaluation, and physiological measures, including electroencephalography (EEG) and electrooculography (EOG). In the current line of research, EEG data were used for this purpose. However, in this paper, EEG is being used for a different purpose. Specifically, solving a problem when electrode leads were inadvertently swapped.

The Electroencephalogram

Electroencephalography (EEG) is a noninvasive sensing technique that uses electrodes placed on the scalp to measure brain activity (Credlebaugh, Middendorf, Hoepf, & Galster, 2015). The locations of these sites are based on the International 10-20 system (Jasper, 1958). Researchers have reported the sensitivity of EEG to changes in mental workload (Gevins & Smith, 2003). These researchers found that the spectral peaks in the delta band (1-3 Hz) and theta band (4-7 Hz) increase in power during high workload tasks. In contrast, multiple studies have shown that power decreases in the alpha band (8-12 Hz) during high workload (Dussault, Jouanin, & Guezennec, 2004; Prinzel, Parasuraman, Freeman & Scerbo, 2003; Wilson, 2002).

Although EEG has often been used as a measure of cognitive workload, it has some limitations that must be considered. EEG signals are easily corrupted by a number of artifacts. That is, in addition to the brain's electrical activity recorded at the scalp, the EEG signal can include contaminating potentials from rapid eye movements (saccades) and blinks (Gevins & Smith, 2003). A handful of existing artifact mediation techniques are widely used, including artifact avoidance, rejection and removal. In many cases, artifacts will eventually be accounted for during data processing and analysis. The existing artifact mediation techniques can facilitate the analysis of artifact-free data. The work presented here is unique because the artifact separation approach allowed only data contaminated by artifacts to be analyzed.

The Electrooculogram

The Electrooculogram (EOG) is a measure of electrical signals associated with eye activity, including blinks and saccades. Typical blink measures include: amplitude, duration, and frequency. It has been reported that

Distribution A: Approved for public release. 88ABW Cleared 02/10/2016; 88ABW-2016-0540.

when faced with increased cognitive workload; individuals will blink with reduced duration and frequency (Recarte, Perez, Conchillo, & Nunes, 2008). Typical saccade measures include: amplitude, velocity, and length. Wang and Zhou (2013) reported that the peak saccade velocity will increase as workload increases.

Among EOG artifacts, blinks cause the largest distortions in the EEG, mainly because when the eyelid covers the cornea during a blink, it acts as a "sliding electrode" that effectively short-circuits the positively charged cornea to the skin of the eyelid (Picton et al., 2000). This result causes a large potential difference that travels posteriorly across the scalp. The voltage spike creates an EEG artifact that is most prominent in the frontal electrodes and attenuates the further back it travels (Barry & Jones, 1965).

Artifact Mediation Approaches

Considering the effects of artifacts on the EEG signal, a great deal of research has been directed towards artifact mediation (Gevins & Smith, 2003). Common methods of dealing with artifacts include: artifact avoidance, artifact rejection, and artifact removal. The artifact avoidance method attempts to avoid artifacts all together by instructing the participants to not blink. Artifact avoidance has the advantage of being the least computationally demanding, since it is assumed that no artifact is present in the signal (Fatourechi, Bashashati, Ward, & Birch, 2007). Having the inability to control eye movements gives this approach a disadvantage.

Artifact rejection refers to the process of rejecting the data affected by artifacts (Fatourechi et al., 2007). Artifact rejection can be done manually or automatically. During the manual rejection method, data is visually checked by an expert and the contaminated EEG data are removed from the analysis (Fatourechi et al., 2007). Automatic rejection discards segments that are contaminated automatically using the EOG signals or by using EEG signals contaminated with artifacts (Gratton, 1998). Automatic artifact rejection approaches are less labor intensive than manual approaches but still suffer from the loss of valuable data.

Artifact removal is the process of reducing the impact of the artifact on the EEG signal. This may be thought of as an attempt to 'fix' the signal in the time domain. Common methods for artifact removal include: linear filtering, linear combination, regression, blind source separation, and principle component analysis (Gotman, Skuce, Thompson, Gloor, Ives & Ray, 1973; Croft & Barry, 2000). EOG artifacts primarily affect the low frequency bands during EEG analysis. The removal of artifacts in these low frequency bands will also result in the removal of the underlying EEG signals, resulting in the loss of data (de Beer, van de Velde, & Cluitmans, 1995).

A new technique for artifact mediation, known as artifact separation, was recently developed (Credlebaugh et al., 2015). This technique relies on blink and saccade detection algorithms using EOG data. EEG data is typically analyzed using time domain windows. If a blink or saccade occurs during a window, the spectral results are flagged as contaminated. Having the spectral results flagged as containing an artifact, means that the consumer of the data has the freedom to decide how to use the artifact flags during data analysis. This paper will focus on the artifact separation technique and how it was used to resolve an unusual issue associated with data collection.

One could reasonably argue that artifact separation is the same thing as automatic artifact rejection. One difference is artifact rejection is typically done in the time domain, and the artifact separation approach is applied during data analysis.

Problem Description

In this paper three studies are discussed. They are referred to as Study 1, Study 2, and Study 3; with the main focus of the paper on Study 3. In Study 3, physiological measures (EEG & EOG) were collected and explored as indicators of cognitive workload.

In the course of conducting this experiment, it was discovered that the EEG channels Fz and T3 were swapped on the signal acquisition hardware. The exact date when the electrodes were inadvertently swapped was unknown. Realizing the serious implications due to the mislabeled data, the date when the swap occurred was needed so that the EEG data could be properly processed. The artifact separation technique was used to solve this problem.

Methods

Participants

There were a total of 13 participants in Study 3, with 6 males and 7 females. The age of participants ranged from 19-25 (M=21.8). Participants were recruited from a local mid-western university. They read and signed the informed consent document before participating and were compensated for their time. All study procedures were reviewed and approved by the Air Force Research Laboratory Institutional Review Board.

Task

Each trial consisted of two separate primary tasks and one secondary task. Trials were presented to the participants as a simulated remotely piloted aircraft (RPA) mission. Each trial started with a surveillance task and then transitioned into a tracking task. The same secondary task was present during both primary tasks. The secondary task was a communications task in which the participants were asked cognitively demanding questions. These tasks were implemented using a RPA simulator called Vigilant Spirit. This software was produced by the Air Force Research Laboratory Supervisory Control Interfaces Branch (RHCI).

For the surveillance task, participants were required to search a market for high value targets (HVTs). The number of distracters (non-HVTs walking around in the market) and the visibility of the camera served as experimental manipulations to affect workload. During some conditions, an automation feature was implemented to help the participants find the HVT. When the HVT was within the sensor footprint, a tone would play in the headset. The participant would then simply need to examine the entities within the footprint rather than search other areas of the market.

In the tracking task, participants were instructed to track one or two HVT(s) using RPAs. This was accomplished by continuously clicking in each video feed while the HVT(s) traveled by motorcycle. Dependent upon the condition, the HVT would either take a route through the city or country. Half of the trials consisted of tracking one HVT and the other half consisted of tracking two HVTs. Similar to the surveillance task, an automation feature was incorporated that would help the participant track one HVT. In this situation, an experimenter would take over tracking of one HVT.

Procedure

Participants were brought into the laboratory for two training sessions and eight data collection sessions. For training, participants were asked to read through a PowerPoint presentation briefing them on task instructions. The researchers then provided training on each individual task (surveillance and tracking), followed by eight practice trials. On data collection days, participants were equipped with physiological sensors which included EEG and EOG. Participants then completed four trials per day, for a total of 32 trials.

Apparatus and Measures

Seven channels of EEG data were recorded during this study which included: F7, Fz, F8, T3, T4, Pz and O2. The frequency ranges of the seven bands of EEG were delta (1-3 Hz), theta (4-7 Hz), alpha (8-12 Hz), beta (13-30 Hz), gamma 1 (31-40 Hz), gamma 2 (41-57 Hz), and gamma 3 (63-100 Hz). The EOG data were acquired using four electrodes. Two were placed above and below the left eye and the other two laterally to the outer canthus of the eyes. Mastoids were used as reference and ground points. Electrode impedances were below 5kΩ for EEG and 20kΩ for EOG. The EEG and EOG data were sampled at 480 Hz using the Cleveland Medical Devices BioRadio 150. This device has hardware high pass filters with break frequencies of 0.5 Hz.

Analysis Approach

EEG signal processing. The raw EEG data were split into two-second windows and filtered using a 4th order Butterworth band pass filter with pass bands set as described earlier. A Hanning window was applied and a power spectral analysis was performed. The resulting power in each window was then averaged. The two-second time domain windows had a 50% overlap, thus yielding one average power measure every second for each frequency band and site. This produced a total of 49 measures per second (7 frequency bands at 7 sites).

Blink detection algorithm. The blink detection algorithm uses vertical EOG to identify blinks in real-time. The main features computed for each blink are its amplitude and duration. After two or more blinks are found, blink rate can be computed. See Epling et al., 2015 for a detailed explanation of the blink detection algorithm.

Saccade detection algorithm. A saccade detection algorithm was used to process EOG data and detect saccades. The algorithm uses both vertical EOG and horizontal EOG, and reports saccades in magnitude and angle (polar coordinates). See Middendorf et al., 2015 for a detailed description of the polar saccade detection algorithm.

Propagation of a blink

When determining whether to use blinks or saccades at sites Fz or T3 to solve the swapping problem, literature was consulted. Picton et al. (2000) states, ocular potentials can be recorded at some distance from the eyes and can thus distort recordings of the EEG. Blink potentials are mainly produced by the downward movement of the upper eyelid over the cornea (Matsuo, Peters, & Reilly, 1975; Antervo, Hari, Katilla, Ryhanen, & Seppanen, 1985). The EOG contamination is at its highest in the electrodes near the eyes and decreases with increasing distance away from the eyes (Picton et al., 2000). Picton et al. (2000) reported blink potentials are significantly larger than the saccade potentials at Fz. The data from Picton et al. (2000) also suggests that Fz will record more low frequency power during a blink than T3. This investigation uses power in the delta frequency band.

Results

The artifact separation technique discussed earlier was used to determine the delta power due to blinks at each EEG site, using data from the surveillance task for one participant. The artifact flags were used to look only at EEG spectral data that coincided with blink artifacts. This data was used to generate a graph that represents the average power due to blinks at each EEG site in the delta band (Figure 1).

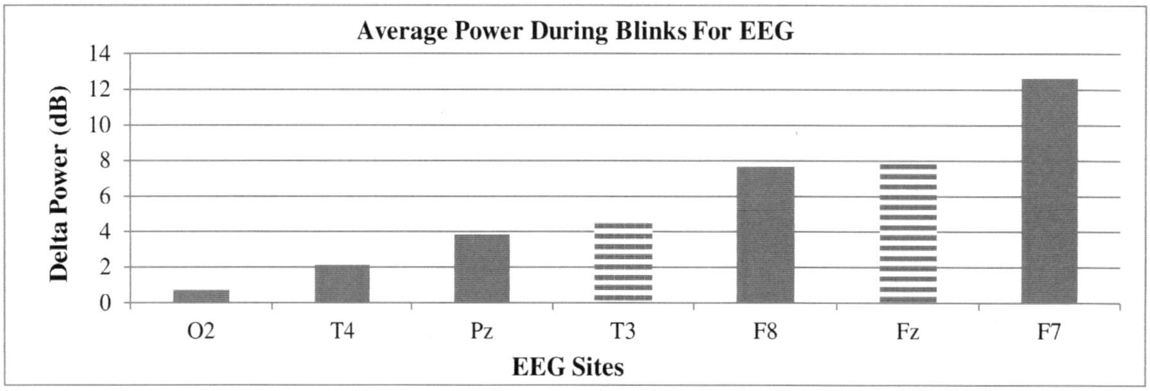

The delta band is used in the graph because it shows the greatest power due to blinks compared to other frequency bands (Gevins & Smith, 2003). The difference between sites is clear; Fz shows a greater power when a blink occurs than T3. This knowledge allows the signals to be differentiated from each other based solely on a characteristic of the data.

Graphs were created showing just these two sites for every trial of participant seven. This is the participant that was running when the problem was discovered and the electrodes were swapped back to the proper configuration. This occurred just prior to trial 25. Figure 2 shows the average power in the delta band at sites Fz and T3. The average power is computed using only blink contaminated data. This data is for participant seven for all trials from the surveillance task. A trend is easily seen in the data; the two sites clearly show a different response to a blink. For the first 24 trials T3 shows higher power than Fz, however the last eight show the opposite. The data from the last eight trials show the expected behavior, and were collected after the electrodes were corrected. This means the data from Fz and T3 were mislabeled for the first 24 trials for this participant.

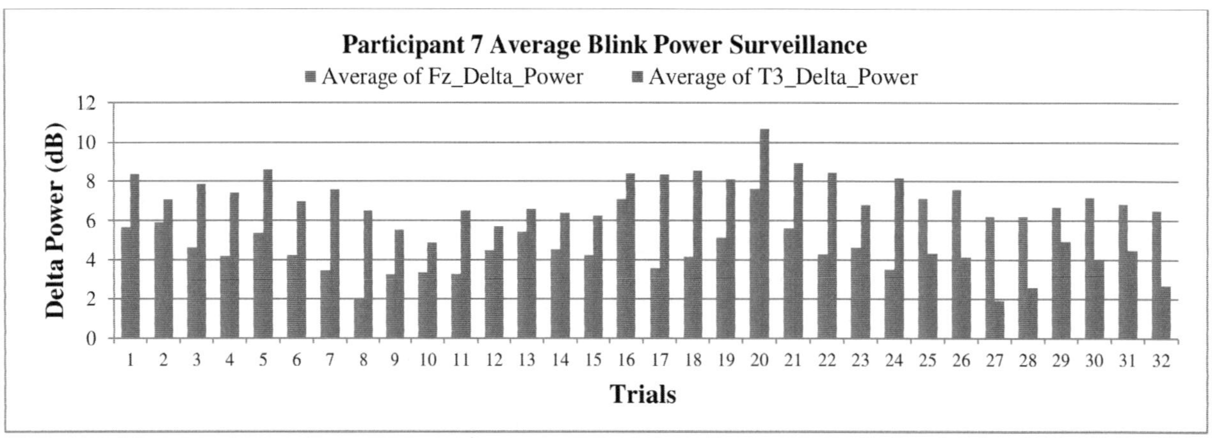

Figure 2. Average power in the delta band at Fz and T3, during a blink. Note that the average power at T3 and Fz changed on trials 25-32. This is when the electrodes were swapped back to the correct locations.

This technique was then applied to the data for every participant. Graphs were used to isolate when the electrode leads were plugged in to the wrong locations. This was accomplished by observing the power at Fz and T3 over the course of all trials for each participant. A timeline of the study, during which the problem was corrected, was developed using these graphs (Figure 3). This was possible because these graphs allowed us to see when T3 and Fz did not fit the expected behavior. This timeline shows the data being mislabeled from the beginning of the study. Therefore, previous studies had to be examined to determine the date when the swap occurred.

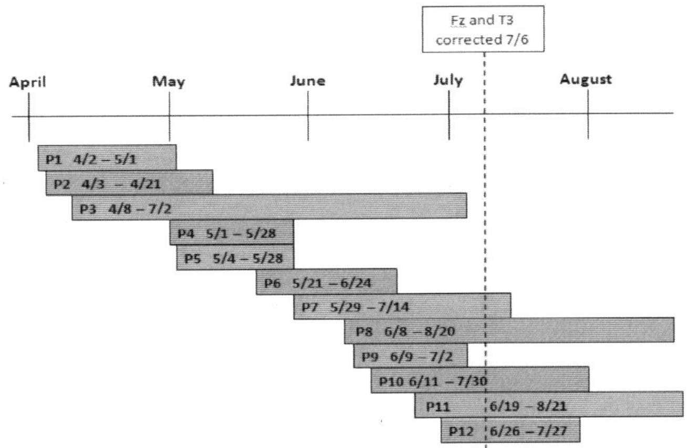

Figure 3. Timeline of when the participants started and completed the current study. Note that the abbreviation P1 indicates participant one, P2 indicates participant two, etc.

Data from two previous studies (Study 1 & Study 2) were processed with this technique and a larger timeline was determined (Figure 4). This figure shows the three studies conducted in the laboratory. The time when the electrodes were initially swapped was determined to fall between Study 1 and Study 2, as shown by the dotted line labeled "Fz and T3 leads swapped." Now that the date of the swap has been found, the EEG spectral data can be easily corrected using software.

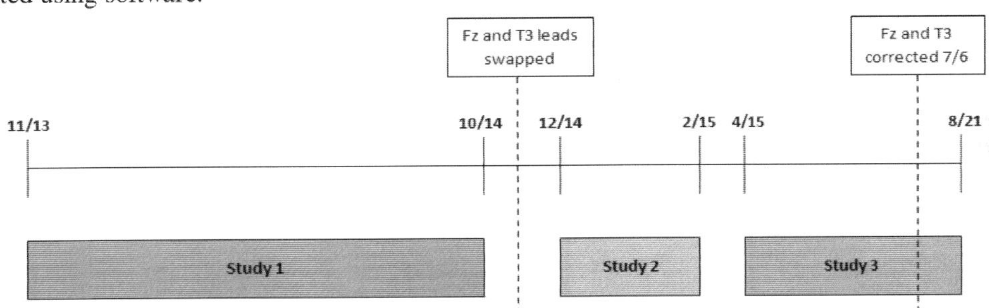

Conclusion

The artifact separation technique is a powerful tool for data analysis. An important feature of the technique is that, it does not attempt to 'fix' the signal in the time domain and the technique allows the user of the data to decide what to do with the artifacts. In this case, contaminated data was flagged and later analyzed to determine when the electrode leads were initially swapped. The date when the electrodes were swapped was obtained and data was later reprocessed. The fact that the electrodes were swapped for part of the study had no negative repercussions.

Acknowledgements

The authors wish to express gratitude to Michael Hoepf and Jonathan Mead for assistance with data analysis and preparation of this manuscript. The authors would also like to thank Christina Gruenwald for discovering the problem. The views expressed in this report are solely those of the authors and do not necessarily reflect the views of their organizations.

References

Antervo, A., Hari, R., Katilla, T., Ryhanen, T., & Seppanen, M. (1985). Magnetic fields produced by eye blinking. *Electroencephalography and Clinical Neurophysiology, 61*, 247-253. doi: 10.1016/0013-4694(85)91091-0

Barry, W., & Jones, G. M. (1965). Influence of eye lid movement upon electro-oculographic recording of vertical eye movements. *Aerospace Medicine, 36*, 855-858.

Credlebaugh, C., Middendorf, M., Hoepf, M., & Galster, S. (2015). EEG data analysis using artifact separation. *Proceedings of the International Symposium on Aviation Psychology, 18*, 434-439.

Croft, R. J., & Barry, R. J. (2000) Removal of ocular artifact from the EEG: A review. *Clinical Neurophysiology, 30*, 5–19. doi: 10.1016/S0987-7053(00)00055-1

de Beer, N. A., van de Velde, M., & Cluitmans, P. J. (1995). Clinical evaluation of a method for automatic detection and removal of artifacts in auditory evoked potential monitoring. *Journal of Clinical Monitoring, 11*, 381-391. doi: 10.1007/BF01616744

Dussault, C., Jouanin, J. C., & Guézennec, C. Y. (2004). EEG and ECG changes during selected flight sequences. *Aviation, Space, and Environmental Medicine, 75*, 889-897.

Epling, S., Middendorf, M., Hoepf, M., Gruenwald, C., Stork, L., & Galster, S. (2015). The electrooculogram and a new blink detection algorithm. *Proceedings of the International Symposium on Aviation Psychology, 18*, 512-517.

Fatourechi, M., Bashashati, A., Ward, R. K., & Birch, G. E. (2007). EMG and EOG artifacts in brain computer interface systems: A survey. *Clinical Neurophysiology, 118*, 480-494. doi: 10.1016/j.clinph.2006.10.019

Gevins, A.S., & Smith, M.E. (2003). Neurophysiological measures of cognitive workload during human-computer interaction. *Theoretical Issues in Ergonomics Science,* 4, 113-131. doi: 10.1080/14639220210159717

Gotman, J., Skuce, D. R., Thompson, C.J., Gloor, P., Ives, J. R., Ray, W. F. (1973). Clinical applications of spectral analysis and extraction of features from electroencephalograms with slow waves in adult patients. *Electroencephalography and Clinical Neurophysiology, 35*, 225-235. doi: 10.1016/0013-4694(73)90233-2

Gratton, G. (1998). Dealing with artifacts: The EOG contamination of event-related brain potential. *Behavior Research Methods, Instruments & Computers, 30*, 44-53. doi: 10.3758/BF03209415

Hankins, T. C., & Wilson, G. F. (1998). A comparison of heart rate, eye activity, EEG and subjective measures of pilot mental workload during flight. *Aviation, Space, and Environmental Medicine, 69*, 360-367.

Jasper, H. (1958). Report of the committee on methods of clinical examination in electroencephalography. *Electroencephalography and Clinical Neurophysiology, 10*, 370-375. doi: 10.1016/0013-4694(58)90053-1

Matsuo, F., Peters, J. F., & Reilly, E. L. (1975). Electrical phenomena associated with movements of the eyelid. *Electroencephalography and Clinical Neurophysiology, 38*, 507-511. doi: 10.1016/0013-4694(75)90191-1

Middendorf, M., Gruenwald, C., Stork, L., Epling, S., Hoepf, M., & Galster, S. (2015). Saccade detection using polar coordinates – A new algorithm. *Proceedings of the International Symposium on Aviation Psychology, 18*, 518-523.

Picton, T., Roon, P., Armilio, M., Berg, P., Ille, N., & Scherg, M. (2000). The correction of ocular artifacts: A topographic perspective. *Clinical Neurophysiology, 111*, 53-65. doi: 10.1016/S1388-2457(99)00227-8

Prinzel, L. J., Parasuraman, R., Freeman, F. G., Scerbo, M. W., Mikulka, P. J., & Pope, A. T. (2003). *Three experiments examining the use of electroencephalogram, event-related potentials, and heart-rate variability for real-time human-centered adaptive automation design* (Tech. Rep. No. 2003-212 442). Hampton, VA: NASA Langley Research Center.

Recarte, M., Pérez, E., Conchillo, A., & Nunes, L. (2008). Mental workload and visual impairment: Differences between pupil, blink and subjective rating. *The Spanish Journal of Psychology, 2*, 374-385. doi: 10.1017/S1138741600004406

Tsang, P., & Wilson, G.F. (1997). Mental workload. In: G. Salvendy - (Ed.), *Handbook of Human Factors and Ergonomics.* Baltimore, MD: John Wiley.

Wang, Y., & Zhou, J. (2013). Literature review on physiological measures of cognitive workload. Machine learning research group – NICTA.

Wilson, G. F. (2002). An analysis of mental workload in pilots during flight using multiple psychophysiological measures. *The International Journal of Aviation Psychology, 12*, 3-18. doi: 10.1207/S15327108IJAP1201_2

INVESTIGATING FACIAL ELECTROMYOGRAPHY AS AN INDICATOR OF COGNITIVE WORKLOAD

Jonathan Mead
Oak Ridge Institute for Science and Education
Matt Middendorf
Middendorf Scientific Services
Christina Gruenwald
Oak Ridge Institute for Science and Education
Chelsey Credlebaugh
Ball Aerospace & Technologies
Scott Galster
Air Force Research Laboratory
Wright-Patterson AFB, Ohio

Facial electromyography (fEMG) is an electromyographic measurement technique that has primarily been used as a tool for measuring affect, but previous experiments suggest that it also has the potential to help quantify cognitive workload. In the current study, two task-irrelevant facial muscles, corrugator supercilli and lateral frontalis, were monitored in real-time to determine whether they were sensitive to workload changes in a remotely piloted aircraft (RPA) task environment. Real-time signal processing techniques were applied to derive the median amplitude and zero-crossing rate from windowed fEMG data. Statistical analysis of these features determined that both muscles were sensitive to variations in specific workload manipulations. This research suggests that real-time fEMG features extracted from the aforementioned muscles possess the potential to serve as, or contribute to, an index of cognitive workload. Future work aims to refine fEMG data collection techniques to produce a more responsive and representative measure suitable for workload assessment.

The ability to remain vigilant for extended periods of time is incredibly crucial to many positions in the aerospace domain. Pilots, sensor operators, and air traffic controllers, for example, must maintain high levels of situational awareness to ensure optimal safety and performance. Cognitive workload is an important factor in determining an operator's ability to perform at the level required to prevent hazardous consequences (Young & Stanton, 2002). Cognitive overload and underload can both induce performance decrements, while a moderate level of cognitive arousal facilitates an ideal performance capacity (Cohen, 2011).

In order to ease the vigilance burden on aerospace operators and to help them maintain ideal performance, the Sense-Assess-Augment (SAA) framework was developed to identify and alleviate cognitive workload imbalance across a wide range of task environments (Galster & Johnson, 2013). Because changes in cognitive workload have been shown to be correlated with a variety of physiological events, this framework can be applied to sense an array of physiological measures produced by an aerospace operator, incorporate those measures into a model that can assess the operator's cognitive state, and then augment the operator's performance to lessen performance abatement induced by cognitive overload or underload (Wilson & Russell, 2007; Hoepf, Middendorf, Epling, & Galster, 2015; Hoepf et al., 2016). In order for the SAA-based workload modeling approach to function across a wide range of task environments it is crucial that an extensive suite of physiological measures are incorporated as inputs to the model. The nature of the task being performed by the operator likely defines the usefulness of each type of physiological measure (cortical, cardiac, etc.) being used to assess workload (Hoepf et al., 2016). For example, during mental calculation type tasks it was found that cortical measures associated well with workload, while cardiac measures were sensitive to workload during flight-based tasks that primarily demanded the use of instruments, and ocular measures were related to workload in flight-based tasks that were very visually dependent (Hankins & Wilson, 1998).

Many psychophysiological scientists and engineers are researching the correlation between various physiological measures and cognitive workload in an attempt to further advance the ability to model an individual's cognitive state in real-time. One of the most recent physiological signals to be explored as a potential indicator of cognitive workload has been facial electromyography (fEMG). fEMG is an electromyographic (EMG) measurement technique that describes muscle activity by sensing and magnifying the minute electrical impulses that are generated

DISTRIBUTION STATEMENT A. Approved for public release: distribution is unlimited. 88ABW Cleared 02/09/2017; 88ABW-2017-0545.

by facial muscle fibers when they contract. In earlier studies, researchers recorded positive yet inconsistent results concluding the relationship between cognitive effort and EMG amplitude in task-irrelevant limb muscles. However, in 1993 van Boxtel and Jessurun determined that EMG amplitude of the lateral frontalis and corrugator supercilii muscles provided a sensitive index to the degree of cognitive effort exerted by a human participant (with amplitude increasing with cognitive effort; see Figure 1 for muscle locations). The scientists suggested that task-irrelevant activity of the facial muscles originates in the medial interneurons of the portion of the brainstem in contact with the facial cranial nerve and the limbic system. Somatic and limbic activity (stimulated by increasing levels of cognitive workload) are known to have a diffuse effect on the excitability of motor neurons throughout the brainstem and spinal cord. Thus, van Boxtel and Jessurun hypothesized that somatic and limbic influences congregating around the interneurons of the facial nerve could induce involuntary, spontaneous (task-irrelevant) activity within the facial musculature. A follow-up study added further support for a related hypothesis that EMG activity in specific facial muscles are related to the mobilization of non-specific energetic resources required by the body in order to maintain vigilance while compensating for increasing levels of cognitive workload (Waterink & van Boxtel, 1994). More recently, researchers concluded that corrugator based fEMG was effective in detecting confusion, and suggested that fEMG could be an effective addition to a sensor suite designed to monitor the cognitive state of operators in a variety of human-machine systems (Durso, Geldbach, & Corballis, 2012).

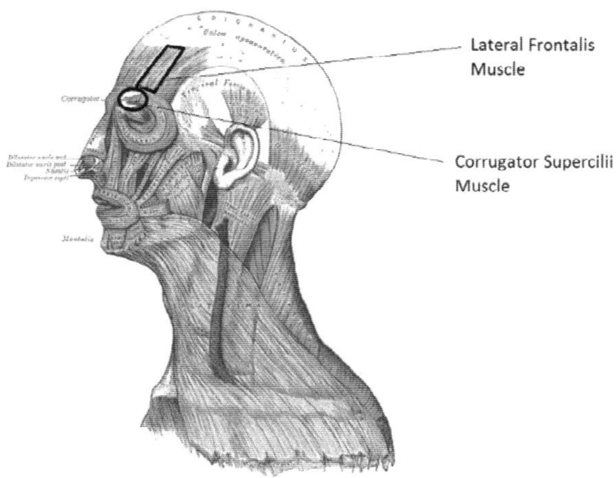

Figure 1. Anatomical diagram depicting the locations of the corrugator supercilii and lateral frontalis muscles

The aim of the current research is to investigate whether real-time periodic fEMG measures are sensitive to changes in cognitive workload, and thus, are suitable for use as inputs to a cognitive state model. The corrugator supercilii and lateral frontalis muscles (hereby referred to as corrugator and frontalis) were selected for this fEMG investigation as these muscles are task-irrelevant in a majority of aerospace operator positions, and have been most often associated with cognitive workload in prior studies (van Boxtel & Jessurun, 1993; Waterink & van Boxtel, 1994; Veldhuizen, van Boxtel, & Waterink, 1998; Durso, Geldbach, & Corballis, 2012). The real-time fEMG signals will be sampled and windowed in order to extract two periodic measures: median tonic amplitude (tension) and zero-crossing rate. Based on results found in prior research, it is hypothesized that as cognitive workload is increased, the median tonic amplitude of both the corrugator and frontalis muscles will increase while the zero-crossing rate of the same muscles will decrease. Zero-crossing rate is known to decrease with muscular tension and fatigue (Kilbom, Hägg, & Käll, 1992).

Method

Electrode Placement and Raw Data Acquisition

fEMG data were recorded using a Cleveland Medical Devices BioRadio 150. Two differential channels were utilized, including a corrugator and frontalis channel. In order to record the fEMG signals, two small 1 cm diameter Ag/AgCl cup electrodes were placed on the left corrugator and frontalis muscles (four electrodes total) following the placement instructions detailed in The Guidelines for Human Electromyographic Research (Fridlund & Cacioppo, 1986). Only signals from the muscles on the left side of the face were recorded because additional sensors involved in the experiment did not allow for proper placement on the right side. It has also been suggested that the power of fEMG activity on the left side of the face is stronger than that of the right (Zhou & Hu, 2006). An additional electrode was placed on the left mastoid to serve as a ground for the fEMG signals. Before placement, each electrode skin site was prepared with the use of an alcohol pad and abrasive gel. Each electrode was prepared by covering the electrode with grass electrode paste. A small gauze square was placed on the convex side of the electrode. The electrode was then placed on the skin, gauze side up, and held in place for a few seconds until the grass paste began to dry. Finally, each electrode was secured with the use of medical tape to ensure the electrode did not move or fall off throughout the data collection session. The initial fEMG electrode impedances were measured to be at or below 20 kΩ, while the ground electrode was measured to be at or below 5 kΩ. Data were sampled at 960 Hz, and subjected to a first order analog band pass filter with an input bandwidth of 0.5 - 250 Hz. The sampled data were transmitted wirelessly to a computer for processing and recording.

Biosignal Processing

The fEMG data were processed in real-time using two second windows with 50% overlap, thus yielding measures once per second. The windowed data was then filtered using a second order Butterworth high pass filter with a cutoff frequency of 25 Hz to attenuate eye blink artifacts while still maintaining a vast majority of EMG frequency power (Konrad, 2006). The filtered signal data was processed to count the number of zero-crossings. The number of zero-crossings is divided by the window length in seconds to compute the zero-crossing rate. The filtered data was full-wave rectified (i.e., absolute value) to prepare it for further processing to compute the normalized median amplitude. The rectified data was squared and convolved with a vector that contains the inverse of the window size. A square root of the convolved data was taken to perform root mean square smoothing. The median of the resulting data was found and normalized using the median amplitude calculated during baseline data collection. The baseline data collection was held prior to formal data collection each experimental session. The participant was instructed to sit still and watch a monitor displaying scenes of the task environment for three minutes while fEMG data was collected. The median of the middle 120 seconds of data was used for the normalization calculation. Zero-crossing rate and the normalized median amplitude were written to a file for data collection and analysis.

Participants

A total of ten individuals recruited from the Midwest region participated in this study. Eight participants were male and two were female. Age ranged from 18-33, with a mean of 21.9. Participants were screened for motor, perceptual, cognitive, and heart conditions. Similarly, if participants were taking any neurological medications or medications that caused drowsiness they were excluded from the study. The participants stated they were comfortable operating a computer, reading, hearing and comprehending verbal commands, and learning complex computer tasks. The participants were fluent in English and had normal or corrected-to-normal eyesight with no color blindness. They read and signed the informed consent document before participating and were compensated for their time. All study procedures were reviewed and approved by the Air Force Research Laboratory (AFRL) Institutional Review Board.

Task Descriptions and Experimental Design

In this experiment, trials alternated between the two primary tasks (surveillance or tracking) and both included a secondary communications task. Both primary tasks were implemented using a remotely piloted aircraft (RPA) simulator called "Vigilant Spirit." This software was produced by the AFRL Supervisory Control & Cognition Branch (RHCI). The secondary task was created using the Multi-Modal Communications (MMC) tool. This software was created by the AFRL Battlespace Acoustics Branch (RHCB). Each trial was presented as a simulated RPA mission. There were 48 scenarios (24 surveillance and 24 tracking) that each participant experienced once over the course of six data collection days. On any given data collection day, participants experienced eight total trials (four surveillance and four tracking trials). Each surveillance trial lasted four minutes and each tracking trial lasted four and a half minutes. Conditions were counterbalanced within task type, even though the tracking task always followed the surveillance task. As described below, this experiment can be viewed as two separate tasks, each having a 2 x 2 x 2 full factorial design.

The surveillance task required the participants to search a marketplace to find four high value targets (HVTs). Each HVT walked out from under a tent, walked around the marketplace, and went back under a different tent for one minute intervals. These four HVTs never appeared at the same time. Experimental manipulations included the presence or absence of sensor fuzz (on vs. off), and the number of distractors (other people walking around; 16 vs. 48). The tracking task required participants to track HVT(s) traveling by motorcycle(s). Participants were instructed to track the HVT(s) by continuously clicking back and forth in each video feed. Dependent upon the condition, the HVT on the motorcycle would either take a route through the city or country (city being harder, i.e., more frequent turns and occlusion behind buildings), or have to track multiple HVTs at once. Half of the tracking trials consisted of tracking one HVT, while the other half consisted of tracking two HVTs. The third manipulation for both primary tasks was a secondary communications task. This consisted of answering a variety of operationally relevant cognitively challenging mental math questions. Questions were asked verbally over a headset and transcriptions were displayed. These questions were evenly distributed throughout each trial, but dependent upon the condition, the quantity could change from two questions being asked to four, per trial. Prior to the start of the study, it was confirmed via visual inspection that oral communication produced no artifacts in either the corrugator or frontalis fEMG signals.

Subjective Workload

Self-reported workload assessments were obtained using the NASA-Task Load Index (TLX), a multidimensional measure that assesses perceived workload (Hart & Staveland, 1988). The NASA-TLX consists of six subscales that measure mental demand, physical demand, temporal demand, performance, effort, and frustration. On a scale from zero to one hundred, workload can be determined by averaging across these six-subscales. At the end of each trial, participants were asked to complete this survey, self-reporting their subjective workload. In past experiments, the NASA-TLX was administered to assure the independent variable workload levels were properly portrayed (Hoepf et al., 2016). For example, while completing the surveillance task, it is easier to find the HVT when the sensor fuzz is absent and the number of distractors is low. Likewise in tracking, it is easier to track one HVT traveling along the country route. When participants reported their subjective workload using the NASA-TLX, the workload condition levels were validated.

Procedure

Participants were brought into the laboratory for two days of task training and six days of data collection. For training, participants were asked to read through a PowerPoint presentation briefing them on specific task instructions, followed by completing part-task training for both primary tasks and the secondary task. Upon completion of the part-task training, participants had to fulfill eight comprehensive practice trials (four surveillance and four tracking). On data collection days, participants were equipped with physiological measurement devices, including fEMG electrodes. Participants then completed eight trials per day (four surveillance and four tracking), for a total of 48 trials. At the end of each trial, the participant completed the NASA-TLX. A structured debriefing was conducted at the end of the sixth data collection day.

Results

Prior to analysis, data were evaluated and removed if signal cancellation occurred resulting from improper electrode placement (more information on this can be found in the discussion). Only significant results are reported. For the surveillance task, a 2 x 2 x 2 (communications x fuzz x distractors) ANOVA was performed. As seen in Figure 2, there was a significant main effect of the distractor manipulation on corrugator median amplitude, $F(1, 7) = 5.94$, $p < 0.05$. Corrugator median amplitude was higher when distractors were high ($M = 117.58$, $SE = 5.47$) than when distractors were low ($M = 110.55$, $SE = 5.83$). Similarly, there was a significant main effect of the distractor manipulation on corrugator zero-crossing rate, $F(1, 7) = 13.77$, $p < 0.01$ (see Figure 3). Corrugator zero-crossing rate was higher when distractors were low ($M = 427.32$, $SE = 26.24$) than when distractors were high ($M = 405.43$, $SE = 24.79$).

For the tracking task, a 2 x 2 x 2 (communications x route x targets) ANOVA was performed. There was a significant main effect of the communications manipulation on frontalis median amplitude, $F(1, 6) = 12.22$, $p < 0.05$. Frontalis median amplitude was lower when only 2 communication questions were asked ($M = 114.73$, $SE = 6.61$) than when 4 communication questions were asked ($M = 119.65$, $SE = 6.03$) as shown in Figure 4.

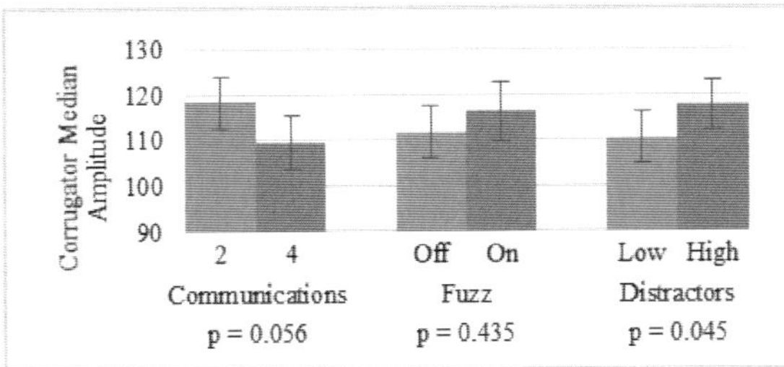

Figure 2. Main effects of the surveillance task manipulations on corrugator median amplitude.

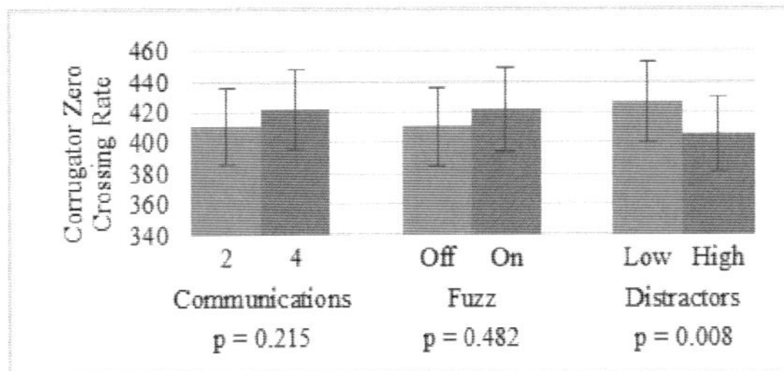

Figure 3. Main effects of the surveillance task manipulations on corrugator zero-crossing rate.

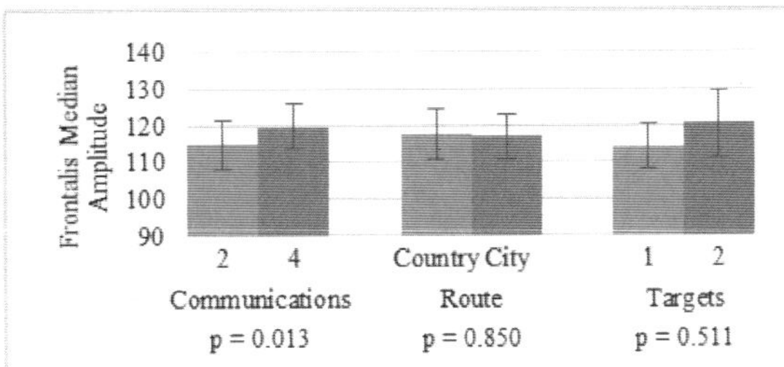

Figure 4. Main effects of the tracking task manipulations on frontalis median amplitude.

Discussion

While the study produced three significant results, they were each in the expected direction. In the surveillance task, corrugator median amplitude increased and zero-crossing rate decreased as the number of marketplace distractors was increased. The higher number of distractors makes the task more difficult so it would make sense that the corrugator amplitude would increase, suggesting that a higher level of muscular tension was produced by a higher cognitive workload. The significantly lower zero-crossing rate during the higher number of distractors adds further support for the hypothesis that corrugator tension increases with increases in cognitive demand. Frontalis median amplitude also increased significantly during the tracking task when more communication questions were present. Additional communication questions produced more cognitively challenging tracking trials. Although there was not an abundance of significant results, this data suggests that real-time fEMG measures may correlate with varying levels of cognitive workload.

This experiment contained a few limitations that potentially attenuated the scope of the results. It is suggested that miniature surface electrodes with 0.25 cm Ag/AgCl detection surfaces and 0.5 cm or 1 cm housings with an inter-electrode distance of 1 cm be used for fEMG measurement (Fridlund & Cacioppo, 1986). However, only electrodes with 1 cm detection surfaces were accessible in the current experiment. Due to the use of less than ideal electrode size, muscular crosstalk may have occurred and contaminated the data with noise. The larger electrodes also made site application difficult – occasionally, electrodes were placed too close together, causing the signal to cancel itself out. This data was flagged and removed from the results prior to analysis. In the future, similar experiments should be conducted with the use of smaller, more appropriately sized electrodes. An improved electrode placement technique is also recommended to ensure accurate site placement with consistent inter-electrode separation distances across experimental sessions. This will most likely produce a more responsive and representative measure suitable for workload modeling. A larger sample size across various task environments is also necessary to determine whether fEMG measures may serve as robust cognitive workload correlates.

Conclusions

The research produced by this study suggests that fEMG measures sourced from the corrugator and frontalis muscles may have the potential to serve as indicators of cognitive workload, and more importantly, as inputs to cognitive state models. Further experiments, which involve a larger number of participants, different task environments, and improved raw signal acquisition capabilities are necessary to endorse this theory and prove its reproducibility. Additionally, further developments in sensor engineering are essential to successfully employ this technology, with an off-body sensing capability being the ultimate goal.

Acknowledgements

The authors would like to thank Paul Middendorf, Michael Hoepf, and Adam Strang for their support, as well as Chuck Goodyear for his help with statistical analysis. This research was supported in part by an appointment to the Student Research Participation Program at the U.S. Air Force Research Laboratory (USAFRL) administered by the Oak Ridge Institute for Science and Education through an interagency agreement between the U.S. Department of Energy and USAFRL. The views expressed in this report are solely those of the authors and do not necessarily reflect the views of the employers or granting organizations.

References

Cohen, R. A. (2011). Yerkes–Dodson Law. *In* Encyclopedia of clinical neuropsychology (pp. 2737-2738). Springer New York. doi: 10.1007/978-0-387-79948-3_1340

Durso, F. T., Geldbach, K. M., & Corballis, P. (2012). Detecting confusion using facial electromyography. *Human Factors: The Journal of the Human Factors and Ergonomics Society*, *54*(1), 60-69. doi: 10.1177/0018720811428450

Fridlund, A. J., & Cacioppo, J. T. (1986). Guidelines for human electromyographic research. *Psychophysiology*, *23*(5), 567-589. doi: 10.1111/j.1469-8986.1986.tb00676.x

Galster, S. M., & Johnson, E. M. (2013). Sense-assess-augment: A taxonomy for human effectiveness (Report No. AFRL-RH-WP-TM-2013-0002). Wright-Patterson Air Force Base: Air Force Research Laboratory, Human Effectiveness Directorate.

Hankins, T. C., & Wilson, G. F. (1998). A comparison of heart rate, eye activity, EEG and subjective measures of pilot mental workload during flight. *Aviation, Space, and Environmental Medicine*, *69*, 360-367.

Hart, S. G., & Staveland, L. E. (1988). Development of NASA-TLX (Task Load Index): Results of empirical and theoretical research. *Advances in Psychology, 52*, 139-183. doi: 10.1016/S0166-4115(08)62386-9

Hoepf, M., Middendorf, M., Epling, S. & Galster, S. (2015). Physiological indicators of workload in a remotely piloted aircraft simulation. In *Proceedings of the Eighteenth International Symposium on Aviation Psychology* (pp. 428-433). Dayton, OH: Wright State University.

Hoepf, M., Middendorf, M., Mead, J., Gruenwald, C., Credlebaugh, C., Middendorf, P. & Galster, S. (2016). Evaluation of Physiologically – Based Artificial Neural Network Models to Detect Operator Workload in Remotely Piloted Aircraft Operations (Report No. AFRL-RH-WP-TR-2016-0075). Wright-Patterson Air Force Base: Air Force Research Laboratory, Airman Systems Directorate.

Kilbom, Å., Hägg, G. M., & Käll, C. (1992). One-handed load carrying—cardiovascular, muscular and subjective indices of endurance and fatigue. *European journal of applied physiology and occupational physiology*, *65*(1), 52-58. doi: 10.1007/BF01466274

Konrad, P. (2006). The ABC of EMG - A practical introduction to kinesiological electromyography (1.4th ed.). Scottsdale, Arizona: Noraxon U.S.A, Inc.

van Boxtel, A., & Jessurun, M. (1993). Amplitude and bilateral coherency of facial and jaw-elevator EMG activity as an index of effort during a two-choice serial reaction task. *Psychophysiology*, *30*, 589-604. doi: 10.1111/j.1469-8986.1993.tb02085.x

Veldhuizen, I. J. T., van Boxtel, A., & Waterink, W. (1998). Tonic facial EMG activity during sustained information processing: Effects of task load and achievement motivation. *Journal of Psychophysiology*, *12*, 188-189.

Waterink, W., & van Boxtel, A. (1994). Facial and jaw-elevator EMG activity in relation to changes in performance level during a sustained information processing task. *Biological Psychology*, *37*, 183-198. doi: 10.1016/0301-0511(94)90001-9

Young, M. S., & Stanton, N. A. (2002). Attention and automation: New perspectives on mental underload and performance. *Theoretical issues in ergonomics science*, *3*(2), 178-194. doi: 10.1080/14639220210123789

Zhou, R., & Hu, S. (2006). Study of posed emotion in facial EMG asymmetry. *Perceptual and Motor Skills, 102*, 430-434. doi: 10.2466/pms.102.2.430-434

EMPIRICALLY EVALUATING REPRESENTATIONAL AIDS FOR TARGET TRACKING AND SENSOR MANAGEMENT

Taleri Hammack
Department of Human Factors Psychology, Wright State University
Dayton, Ohio
Jared Neely, Terry Stanard, Jason Roll
711th Human Performance Wing, Air Force Research Laboratory
Dayton, Ohio
John Flach
Department of Human Factors Psychology, Wright State University
Dayton, Ohio

Today, security officers at military and civilian installations are often required to track people and vehicles (targets) moving in a remote space using a distributed array of stationary security cameras. A pervasive tracking challenge is maintaining view of the target as it moves through the restricted fields of view of different cameras. The current research explores how different display designs indicating camera fields of view impact the operator's situation awareness of the next best camera to continue viewing a moving target. Three different interface displays (Full North-Up Map, Peripheral Display, and Track-Up Mini-Map) were evaluated over four experimental conditions. While having all display types available was most preferred by participants, the Peripheral Display provided better situation awareness as indicated by a statistically significant increased ability to pick the best camera to continue following the target. This was an encouraging finding since the Peripheral Display was designed to complement the video feed information while preserving spatial relationship information resembling a map-like display.

Between 2012 and 2016, the number of HD CCTV units shipped worldwide increased 140x from 0.2 million to 28 million units (Cropley, 2016). With the expanding deployment of surveillance camera technologies, there is a persistent need for a multi-sensor management interface (MSMI) that will support an operator managing a distributed array of cameras. The MSMI is critical for the success of a wide variety of surveillance scenarios within military (e.g., base perimeter defense) and civilian settings (e.g., train stations, airports, shopping centers).

In high priority target tracking tasks (such as a military base defense scenario) the need to maintain *constant* visual of a moving target is not an uncommon performance standard. Using a network of CCTV cameras placed throughout an urban environment, Roll, Stanard, Ayala, and Bowman (2016) conducted a simulated target tracking task, where the user's objective was to maintain visual of a walking pedestrian (the target) in the video feed of at least one camera at all times. Results of this experiment revealed that participants maintained view of the target 72% of the time on average, falling short of ideal. These results inspired the current research.

The current research focused on a target tracking mission using an array of grounded cameras/sensors distributed throughout a virtual urban environment. Participants were tasked with identifying the next-best camera to maintain visual of a walking pedestrian as he moved beyond the field of view of the current camera. To improve the MSMI user's spatial awareness of the targets movement in the 3D environment, two new displays were developed with information about nearby camera locations and their current and possible fields of view. To evaluate whether use of the two new displays in the interface could improve tracking performance, participants had to retrieve information from the displays to maintain visual of the target, including information to decide the next camera to select and what direction it should be turned.

DISTRIBUTION STATEMENT A. Approved for public release: Distribution is unlimited. 88ABW Cleared 03/10/2017; 88ABW-2017-0956

Experimental Interface Displays

Full North-Up Map Display

All conditions provided the participant with a video feed from one camera (the right-most window seen in Figure 1) and the Full North-Up Map (seen on the left in Figure 1). The Full North-Up Map was a rectangular, north-up oriented map with symbology representing terrain (streets, buildings, parks), all cameras and their current and possible fields of view, and the target's initial starting location as indicated by the yellow dot.

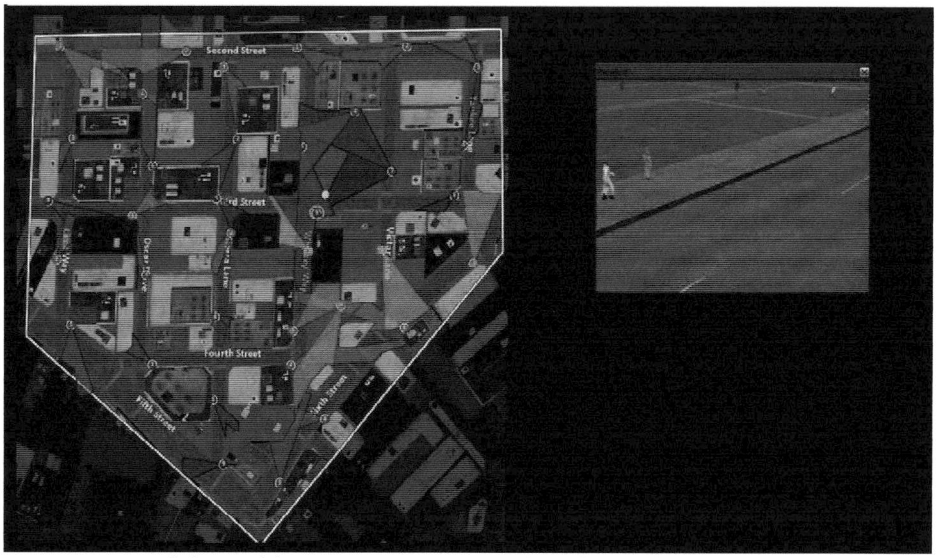

Figure 1. Screen capture of the multi-sensor management interface (MSMI) providing the Full North-Up Map display (left) and the video feed of a single camera (right).

In the video feed shown in Figure 1, if the pedestrian wearing a white robe continued walking up the sidewalk (towards the upper right corner), he would eventually move outside the possible field of view of the current camera. To maintain sight of this target, the user must determine what other camera to select and which direction to turn it to bring the target pedestrian back into view. To make these decisions the user must mentally project the 3D view of the environment seen in the camera feed into the top-down 2D map. These spatial transformations are both cognitively effortful and time consuming, since there is a mental reset time for the viewer to establish the context for the new scene in each display. This difficulty integrating data across successive displays is indicative of a MSMI with low "visual momentum" (Woods, 1984). To increase visual momentum and the ease of making camera selection and turning decisions, a "peripheral display" was designed.

Peripheral Display

Two conditions included the Peripheral Display (see Figure 2), which wrapped icons of the nearby cameras peripherally around the camera video feed. The location of each camera icon around the video feed corresponded to the approximate direction each camera would be located in the environment, relative to the scenery in the video feed. For example, the camera icon labeled "4," seen located in the upper left-hand corner around the video feed, indicates that this camera is located forward and leftward of the current camera view. Also surrounding the video feed was a colored border that indicated what part of the current camera view the nearby cameras could also see. A dashed line connected each camera icon to a border section, and overlapping border sections were darker colored (see Figure 2). This possible field of view was also indicated on the camera icon by the white arc. The yellow dash on the camera icon referred to the relative direction the camera was pointing. The colored border contained information somewhat redundant to the camera icons, providing an alternative representation of the fields of view of nearby cameras with respect to the current field of view in the video feed.

Track-Up Mini-Map Display

In addition to the Full North-Up Map and a camera video feed, two conditions also provided participants with the Track-Up Mini-Map (see Figure 2). The Track-Up Mini-Map was a smaller, circular map oriented so that the current field of view of the selected camera (providing the video feed) was centered and pointed upward. The Track-Up Mini-Map contained the same symbology for streets, buildings, nearby cameras and their current and possible fields of view, and the initial starting location of the target (indicated by the yellow dot). Surrounding the Track-Up Mini-Map were up to four icons of nearby cameras which were outside the boundary of the mini-map. These camera icons did not include the conical shapes indicating the current field of view. These camera icons surrounding the Track-Up Mini-Map were included to give the user knowledge about additional nearby cameras while minimizing interface clutter.

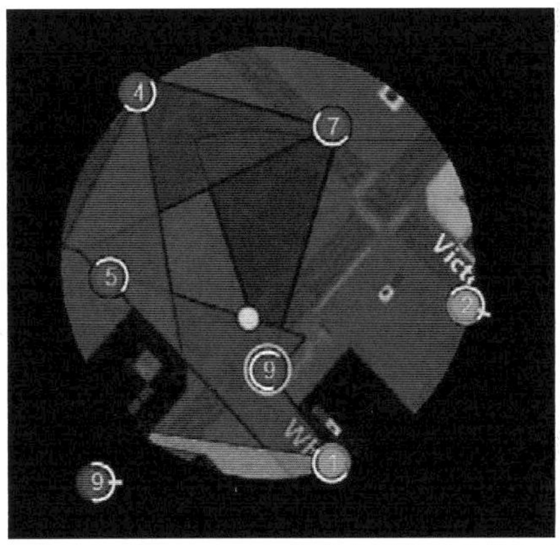

Figure 2. Screen capture of the Peripheral Display (left) and a screen capture of the Track-Up Mini-Map display (right). In the actual interface the Track-Up Mini-Map display was about two-thirds of the size of the Peripheral Display.

Method

Participants

A total of thirteen volunteer Wright-Patterson Air Force Base employees (8 males, 5 females) between the ages of 22 - 47 ($M = 27.38$, $SD = 6.59$) participated in this study. All participants reported normal/normal to corrected vision and normal color vision.

Experimental Design

The three display types were evaluated over four experimental conditions, and trials were blocked by interface configuration. Conditions 1-4 provided the video feed of a single camera showing a walking pedestrian (the target) and the Full North-Up Map. The Full North-Up Map was the only display provided in Condition 1. In Condition 2, the Peripheral Display was provided to participants in addition to the Full North-Up Map. For Condition 3 however, the Track-Up Mini-Map display was provided instead of the Peripheral Display. In Condition 4, all displays were made available (the Full North-Up Map, the

Peripheral Display, and the Track-Up Mini-Map). All participants completed the block of trials with each interface configuration (Conditions 1-4), with block order counterbalanced across participants. Each block contained 28 experimental trials, with the same randomized order of trials used for each participant. Each participant answered a total of 308 experimental questions.

Experimental Task

In each trial participants were presented an 8 - 18 second ($M = 13.81$, $SD = 2.59$) video clip of a walking pedestrian (the target) as viewed by one camera. Each trial required participants to answer two different multiple choice questions by retrieving information from the display(s) provided, and a third question based on the participants individualized use of the different available displays in that condition. The questions were: (1) "In order to maintain view of the target, what would be the next best camera to switch to?" (2) "In order to maintain view of the target, which way would the next best camera you selected need to be turned?" and (3) "Which display(s) did you use the *most* to answer the two previous questions?" The questions were presented sequentially so that once Question #1 was answered Question #2 would appear, and then once Question #2 was answered Question #3 would appear. Question #3 was not given to participants in the Condition 1 block of trials since there was only the Full North-Up Map display available.

Once all three questions were answered participants were presented a screen asking them if they would like to continue to the next trial; this was done to ensure participants were not rushed into the proceeding trial before they were ready. The target's physical appearance (gender, body type, clothing) did not change throughout the experiment. Feedback was not provided during the experimental trials, but a hard copy of the rules for selecting the next best camera to maintain view of the target (Question #1) was available to all participants throughout the experiment. The general rule was to select the closest camera the target would next approach if he continued walking in the same direction.

Procedure

Upon arrival, participants read and signed the informed consent document, filled out a short demographics questionnaire, and were given an overview of the study. In the overview, participants were presented with introductory training slides specifying the goals of the research, the nature of the task, and the requirements to successfully complete the upcoming trials, including the rules for selecting the "next best camera" (Question #1). Participants were then trained on the individual displays available in the interface configuration block they were going to receive next. After this training, participants were given 8 practice trails that they could repeat until they felt confident in their ability to retrieve the necessary information from the given interface display(s).

Participants were given a Post-Block Questionnaire after each condition, specific to the display(s) they just experienced. Questions included their perceived speed and accuracy, their ability to retrieve the necessary information, and their thoughts about possible display modifications. After all four blocks were completed (and the respective questionnaires) a Post-Experiment Questionnaire was administered. The Post-Experiment Questionnaire had participants compare the different display types, indicate their preferences, and provide any additional feedback or recommendations. Total session time for each participant was approximately 1.5 hours.

Results & Discussion

Data was collapsed across participants and analyzed for each condition. Performance data (response accuracy and time) and questionnaire responses were analyzed with a repeated measures Analysis of Variance (ANOVA) model with a Greenhouse-Geisser correction.

Accuracy

The results showed a significant difference in accuracy on Question #1 ("In order to maintain view of the target, what would be the next best camera to switch to?") across conditions ($F(2.13, 25.64) = 4.29, p < .05$), but not a significant difference for Question #2 accuracy ("In order to maintain view of the target, which way would the next best camera you selected need to be turned?"). Post hoc Bonferroni t-test results indicate that accuracy on Question #1 was significantly higher when the Peripheral Display was present with the Full North-Up Map than when only the Full North-Up Map was provided ($p = .013$). Accuracy on Question #1 was also significantly higher when the Peripheral Display + Full North-Up Map were provided than when the Track-Up Mini Map + Full North-Up Map were provided ($p = .028$). Finally, when all display types were made available (Full North-Up Map + Peripheral Display + Track-Up Mini-Map), Question #1 accuracy was marginally significantly higher than when only the Full North-Up Map was available ($p = .064$).

Response Time

In order to better reflect the time required to retrieve information to answer the questions correctly, response time was calculated from the time each question was presented until the participant selected their response. This enabled response times for both Question #1 and Question #2 to be recorded separately. There was not a significant difference in average response time on Question #1 across conditions, but there was a significant difference in average response time for Question #2 ($F(2.02, 24.32) = 3.34, p = .051$). Post hoc Bonferroni t-test results indicated that response times were significantly faster on Question #2 with the Full North-Up Map (only) than with the Peripheral Display + Full North-Up Map ($p = .042$). Average response time on Question #2 was also marginally faster with the Full North-Up Map (only) than with the Track-Up Mini-Map + Full North-Up Map ($p = .062$). Interestingly, average response time on Question #2 was not significantly different between the Full North-Up Map (only) and when all three displays were provided.

Subjective Data

The final Post-Experiment Questionnaire asked participants to rank the four different display configurations (i.e., experimental conditions) on: Ability to identify the next best camera, ability to identify the direction to turn the next best camera, predicted ability if they were tracking a target in real-time, and predicted ability if they were target tracking in real-time *and* had to track multiple targets. After collapsing the data across these four dimensions, the results showed a significant difference in condition preference ($F(1.88, 94.03) = 31.58, p < .01$). Post hoc Bonferroni t-test results indicate that the highest ranked, and thus most preferred option was when all display types were available (Condition 4). The next most preferred option was when the Peripheral Display was available (Condition 2), followed by having the Track-Up Mini-Map available (Condition 3), and then least preferred by participants was only having the Full North-Up Map (Condition 1). The only pairwise comparison that was not statistically significant at $p < .05$ was the preference for displays in Condition 3 over Condition 1. When averaging across the four ranked dimensions on the Post-Experiment Questionnaire, 9 of the 13 participants most preferred having all displays available.

Results from analyzing Question #3 responses revealed that people did not refer to the Full North-Up Map even half as often when they were given the Track-Up Mini-Map as they did when they were given the Peripheral Display (21.16% vs 55.53% respectively). This is a particularly interesting finding because accuracy performance was significantly higher when people were given the Peripheral Display compared to the Track-Up Mini-Map. These results suggest that the Peripheral Display, although useful, did not have all the necessary information to answer Questions #1 and #2. The need to include more information in the Peripheral Display, specifically information regarding the proximity of nearby cameras, was reiterated in several of the Post-Experiment Questionnaire comments made by participants regarding possible display improvements.

One limitation to the current study was the use of strict rules for choosing the one correct "next best camera" to continue viewing the target (Question #1). In the real world, the rules for selecting another camera are dynamic and depend on the context and goals of the operator doing the tracking. For example, the operator may seek a camera providing a close-in view of the target so details are visible, or the operator may instead seek a camera providing the longest view time of a moving target. Although the rules chosen for this experiment were found to be used by operators to track targets in a previous study (Roll, Stanard, Ayala, & Bowman, 2016), they are not necessarily the only criteria used by operators for selection of a camera to switch to.

Conclusion

Results from the subjective data revealed that participants most preferred having all display types available. This is not an unexpected finding since the layout of this interface allows participants to use each of the three displays independently or in combination, so having all displays available to them allows for the largest range of resources to answer each question. An encouraging finding was that situation awareness was increased when the Peripheral Display was made available, as indicated by a statistically significant increased ability to pick the best camera to continue following the target.

Future testing of the Peripheral Display should be explored in a more dynamic environment, with the operator tracking a target in real-time for extended periods of time. Furthermore, operators could be tasked with tracking multiple targets simultaneously, since this is a realistic scenario in real world applications such as in military base defense events. The utility of including camera proximity information in the Peripheral Display (e.g., icon size changes with camera distance, with further cameras having smaller icons) should be explored in future iterations of design. Including this information in the Peripheral Display could greatly reduce response times by reducing the need to consult a second display, namely a map. Furthermore, the camera icons wrapping around the Peripheral Display would support direct manipulation if operators could directly preview and/or switch to the desired camera video feed just by clicking on the icon. These design implementations tested in a real-time tracking task, would help verify that the Peripheral Display supports situation awareness by enabling faster, more accurate decision making when operators switch camera perspectives in order to maintain visual of a moving target.

Acknowledgments

The authors sincerely thank Elizabeth Frost and Antonio Ayala for their assistance in the experiment development process and to Mike Bowman for software development.

References

Cropley, J. (2016, February). *Top video surveillance trends for 2016.* Retrieved from IHS Markit Technology: https://technology.ihs.com/520142/top-video-surveillance-trends-for-2016

Roll, J., Stanard, T., Ayala, A., & Bowman, M. (2016). Enabling human-automation collaborative target tracking using an array of sensors. *Ground/Air Multi-Sensor Interoperability, Integration, and Networking for Persistent ISR VII, Proceedings of the SPIE, 9831. Baltimore, MD: SPIE.*

Woods, D. (1984). Visual momentum: A concept to improve the cognitive coupling of person and computer. *International Journal of Man-Machine Studies*, *21*(3), 229-244.

ABLE FLIGHT AT PURDUE

Wesley Major
Raymart Tinio
Sarah Hubbard
Purdue University
West Lafayette, Indiana

This paper discusses the human factors considerations associated with Able Flight at Purdue, a program that provides flight training to individuals with disabilities. The program requires a tailored approach to training due to the varied needs specific to each individual. Aircraft procedures and airfield operations are standard and all FAA regulations are followed, however, the way individuals with limited dexterity, limited hearing, or limited speech interact with the aviation system needs to be creatively approached with an open mind. The critical thinking used to address individual needs provides an excellent demonstration of problem solving that reflects human factors considerations, based on the SHELL model, reflecting the Software, Hardware, Environment and Liveware in the aviation system. Human factors considerations extend beyond flight preparation and flight, and encompass nontraditional methods of learning, in which flight instructors adapt traditional and standard techniques to provide effective and individualized training techniques. Training this unique population provides many benefits, including promoting diversity in the aviation industry and broadening the teaching skills of flight instructors.

Many developed nations have long recognized the benefits of social inclusion (The Charity Commission, 2001; United Nations, 2007; World Bank, 2013). The World Bank (2013) defines social inclusion as "both an outcome and a process of improving the terms on which people take part in society." Social inclusion is the opposite of social exclusion, and social exclusion can be a result of many factors, such as but not limited to age, sexuality, race, religion, mental illness, or physical disability. Social inclusion seeks to empower marginalized individuals or groups through integration of all societal members, with goals to improve peace, development, and human rights. Inclusion can have benefits that extend beyond the individual, fostering a culture of inclusion that may support a positive workplace culture, increased workplace satisfaction, and even increased innovation (Moon, Todd, Morton, & Ivey, 2012; Pearce & Randel, 2004).

There are many organizations and programs that exist with a mission to promote social inclusion. This paper describes one, Able Flight, including basic information and a discussion of the human factors components that are necessary to the program's success. Able Flight is a nonprofit organization that provides aviation related scholarships to individuals with a disability. Able Flight offers four scholarships:

- **Full Flight Training Scholarships** for people who wish to earn a Sport Pilot certificate,
- **Return to Flight Scholarships** for people who have become disabled after already having earned a pilot's certificate, and now wish to return to flying under the Sport Pilot Rule,
- **Flight Training Challenge Scholarships** for people who would benefit from dual instruction only, and have no current plans to seek a Sport Pilot certificate,
- **Career Training Scholarships** for people who wish to train to earn an FAA-issued Repairman Certificate (Light Sport Aircraft) with Maintenance Rating, or an FAA Dispatcher License, or to defray academic expenses while training for an aviation career (Able Flight, 2016b).

Able Flight has partnered with a number of organizations to deliver these scholarships. Purdue University is one of those partners, and fulfills the training for the Full Flight Training Scholarship. This partnership has benefited both Able Flight and Purdue.

Literature Review

Due to the technical nature of flight, Able Flight incorporates many aspects of Science, Technology, Engineering, and Mathematics (STEM) education. Broadening participation in STEM educational programs has been a goal of many programs, and the inclusion of underrepresented populations in STEM programs provides many benefits. The National Science Foundation has emphasized the importance on the development of a diverse STEM workforce in the United States (National Science Foundation, 2000, 2004). This same message was repeated by the National Science Board in 2010 through their report, Preparing the Next Generation of STEM Innovators (National Science Board, 2010).

A number of STEM oriented programs targeted to individuals with disabilities have been successful. The Experiential Learning for Veterans in Assistive Technology and Engineering (ELeVATE) program provides service members with vocational rehabilitation goals and results indicated that participants were more confident and had improved self-efficacy upon completion (Goldberg, Cooper, Milleville, Barry, & Schein, 2015). One major Midwestern university found that Student Learning Communities that provided knowledge, skills, and abilities curricula for students with disabilities were successful in engaging this unrepresented group to peruse STEM degrees (Izzo, Murray, Priest, & Mcarrell, 2011). Although there are numerous other examples, the Research Initiative for Science Excellence (RISE) program provides an excellent example of the potential benefits. RISE is a training program that seeks to further prepare and support minority students obtaining STEM degrees through research experience and mentorship (Schultz et al., 2011). Programs that target minority groups have the opportunity to positively impact underrepresented individuals by providing resources, education, and guidance that otherwise might not be available to them because of lack of inclusion. A secondary benefit of these programs comes through the direct interaction between more traditional students and minority groups, which results in a change in perceptions about the capabilities of people with a disability and their ability to positively influence society.

Able Flight at Purdue

The Able Flight program began a decade ago, and the first Full Flight Training Scholarship was awarded in 2007. For a short period of time, training was conducted at a variety of locations, including a personal hangar in Oshkosh, WI. In 2010, Able Flight partnered with Purdue University, which allowed the program to expand and provide a more robust training experience.

The Able Flight at Purdue program takes place every summer after Purdue University's spring semester and graduation ceremonies are over; the program typically starts in mid-May and finishes by early July. During this time, residential facilities are available and Able Flight participants live in one the university's newly constructed and fully accessible residence halls. Each participant has their own private bath connected to their room, which provides independence and privacy. Scholarship recipients also receive access to campus dining courts throughout their flight training. Since the basic needs of food and housing taken care of, Able Flight program participants can focus their time, energy, and attention on getting the most out of the Able Flight program.

Able Flight at Purdue requires a great deal of work and commitment on the part of the student. Many hours of studying and many hours in the plane are needed to become a confident, competent, and safe pilot. As recognized by Able Fight's mission statement, "individuals with a disability are presented with a unique way to challenge themselves through flight training, and by doing so, gain greater self-confidence and self-reliance" (Able Flight, 2016b). While an important goal for a Full Flight Training Scholarship recipient is an FAA Light Sport Pilot certificate, the program's impact is much greater than merely being a pathway to certification. The impact includes accomplishment and empowerment for participants, benefits associated with experiencing a fully inclusive environment for participants, and secondary benefits of broadening the perspective of participating CFI's, and positively influencing the culture of aviation through the association of Able Flight students with the larger aviation community. Able Flight at Purdue has realized a 100% completion rate, with 36 graduates as of 2016. It takes a dedicated team to ensure such a successful program, and it also requires application of human factors principals to modify traditional training methods to accommodate the unique characteristics of each Able Flight participant.

A Human Factors Approach to Able Flight

The most widely known framework for human factors in aviation may be the SHELL Model, which was developed in 1972 (Edwards, 1972) and subsequently endorsed by ICAO (International Civil Aviation Organization, 1989). The SHELL Model defines human factors in terms of the Software, Hardware, Environment and Liveware components, which work together, as centered around the Liveware (humans) as illustrated in Figure 1.

Figure 1. Diagram of SHELL components, with Liveware (human) in the center, surrounded by the Environment, and interacting with the Hardware and the Software and other Liveware (humans) (Hawkins, 1987).

In this case, the Able Flight program and its components are assessed from the perspective that the Able Flight student is the liveware in the center of the diagram. The CFI, Air Traffic Controller (ATC) and other pilots are the liveware at the bottom of the diagram. The hardware at the top of the diagram includes the aircraft, and adaptive physical aids that the pilot may use in training or flight. The software at the left of the diagram includes the rules, checklists and procedures, which are always based on traditional FAA flight protocols, but may include enhancements or modifications to reflect the need of the individual Able Flight student. The environment, at the right in Figure 1, includes the Able Flight system, including the natural environment as well as the social and economic factors.

The liveware considerations for Able Flight include the physical characteristics of the flight student, as well as the CFI. Since light sport aircraft are used, matching the size of the instructor to the size of the student may be important for weight and balance considerations, both to optimize aircraft performance and ensure that a reasonable fuel supply is available without exceeding the aircraft maximum weight limitations. Interaction with ATC is another liveware consideration, especially for deaf pilots. Deaf pilots rely on light gun signals when operating in a controlled airfield, and ensuring proper procedures and resources are available for both pilots and ATC are important. Purdue University upgraded their signal lamp to accommodate these operations.

Liveware considerations also reflect input characteristics, reflecting the way that pilots can receive information. This can affect the interaction between liveware and hardware. For example, it is appropriate for a deaf pilot to have a plane that allows the CFI to sit next to, rather than behind. The interaction between liveware and hardware also suggests that some aircraft may be better suited to some pilots, with examples as shown in Table 1.

Table 1.
Matching Able Flight Pilot Liveware with Aircraft Hardware to Ensure Compatibility

Aircraft	Aircraft Characteristics	Pilot Characteristics
Flight Design CTLS	Side by side pilot and co-pilot	Deaf or hearing impaired
	Longer cockpit	Tall pilot
Ercoupe 415C	Hand control for rudder	Pilot with limited use of legs
Sky Arrow L600	Finger brakes	Pilot with limited use of legs
	Ease of access to seat	Landing gear does not prevent wheelchair users from getting close

The general guidelines shown in Table 1 may be useful for conceptual discussion, however, it is important to note that individual characteristics and capabilities are more important than general guidelines, as evidenced by

Able Flight Pilot Jessica Cox, who demonstrates that a lack of arms does not restrict her from using the standard operating controls in the Ercoupe 415C, as shown in Figure 2.

Figure 2. Jessica Cox was born without arms and flies an Ercoupe 415C (Able Flight, 2016a).

Liveware considerations can also affect the interaction between the liveware and software, which includes the rules and procedures. For an Able Flight pilot in a wheelchair, the pre-flight checklist must be approached with a method to allow the pilot to instruct someone else to check the oil levels, rather than check the oil level themselves.

Liveware can affect the interaction between the liveware and the environment. In some cases, Able Flight pilots are more susceptible to heat due to the inability for individuals with a spinal cord injury to control their body temperature. To accommodate this consideration, training schedules reduced flights during the hottest part of the afternoon, and shifted training times to the early morning and evening when temperatures were lower. In addition to adapting to the environment, the Able Flight program results in changes to the environment, since the experience of Able Flight often changes the attitudes of CFIs and other aviation students and professionals who interact with Able Flight pilots. This change in attitude can foster a change in culture, which is one component of the environment.

Benefits

Problem solving and adaptive learning environments, which can be explained using the SHELL human factors approach, has facilitated the successful implementation and expansion of Able Flight at Purdue. This has resulted in documented success, both in terms of FAA certificates obtained, and the impact on the wider community. Able Flight participants have changed the attitudes of CFIs and other aviation professionals, and have contributed to a culture shift in the aviation community. Marketing and media for the Able Flight program has been extensive, with features on local television (Sullivan, 2015), in local newspapers (Flores, 2015; Higgins, 2015), on the internet, and even on national programs such as the Big Ten network (Tolley, 2016). This media attention fosters a broader impact, by changing how individuals and society perceive people with disabilities, their capabilities what they can accomplish, and how they can positively contribute to society.

The benefits to participants provide compelling evidence that the program impact and program success extends beyond successful completion of the program itself.

- Randy Green was born without hands or feet, but earned his Airline Transport Pilot (ATP) rating, the highest pilot certificate FAA recognizes. Randy received training through a Career Training Scholarship from Able Flight and is now flying professionally.

- Kevin Crombie received an Able Flight Full Flight Training Scholarship in 2011. After completing Able Flight at Purdue, Kevin enrolled in Purdue University's undergraduate aviation program. Since graduation, Kevin joined the aviation work force as an employee in the Commercial Space sector at the Federal Aviation Administration in 2015. Kevin purchased a Piper Cherokee 180 and continues to fly.
- Raymart Tinio is Deaf but dreamed to fly since he was a teenager. This dream became reality and was made possible through an Able Flight Full Flight Training Scholarship in 2015. Raymart is now an aviation graduate student at Purdue University; one of Raymart's goals is to improve communications between deaf pilots and air traffic controllers.
- John Robinson is a quadriplegic who received an Able Flight Full Flight Training Scholarship in 2015. After his successful completion of the program, he founded AV84all, a 501(c)(3) public charity with a mission to provide aviation for all and allow pilots with disabilities to fly (Robinson, 2016).
- Wesley Major was paralyzed in a motorcycle accident prior to participating in Able Flight. As a result of his positive experience in 2012, Wesley enrolled in the graduate program and is now a Ph.D. student at Purdue. Wesley's graduate research focuses on improving the airline transportation experience for disabled passengers. He has been a volunteer in the Able Flight program at Purdue since his Able Flight graduation, providing administrative support and serving as a mentor, as well as recruiting program participants, interviewing candidates, and providing media outreach.

Able Flight is a life changing program, and through the application of flight training, it allows individuals with a disability to expand their opportunities and challenge societal norms by demonstrating their ability to fly. During a luncheon among Able Flight staff, student pilots, and sponsors, Professor Bernard Wulle presented an important phase that changes how people think, "Employers should not see what people with disabilities can't do, but see what they can do" (Wulle, 2015).

Conclusion

The Purdue Able Flight program has been a dramatic success and has had a positive impact on both the participants, as well as the larger aviation community. The mainstream media generated by Able Flight at Purdue has provided an opportunity to positively impact the public's perception of people with disabilities, by focusing on their capabilities and abilities to positively contribute to society. The human factors associated with the Able Flight program are explained in the framework of the SHELL Model, which provides a context for the modifications and interaction between the Able Flight pilot as liveware, and the software, hardware and environment. Able Flight at Purdue is a successful program that can be duplicated elsewhere, using the human factors model explained in this article.

References

Able Flight. (2016a). Meet the Able Flight pilots. Retrieved February 2, 2017, from http://ableflight.org/meet-the-scholarship-winners

Able Flight. (2016b). Scholarships. Retrieved January 24, 2017, from http://ableflight.org/scholarships

Edwards, E. (1972). Man and machine: Systems for safety. In *Proceedings of British Airline Pilots Associations Technical Symposium* (pp. 21–36). British Airline Pilots Association.

Flores, T. (2015). Able Flight at Purdue on path to expand. Retrieved February 1, 2017, from http://www.jconline.com/story/news/2015/06/19/able-flight-purdue-path-expand/28996309/

Goldberg, M., Cooper, R., Milleville, M., Barry, A., & Schein, M. L. (2015). Ensuring success for veterans with disabilities in STEM degree programs: Recommendations from a workshop and case study of an evidence-based transition program. *Journal of STEM Education, 16*(1), 16–24.

Hawkins, F. H. (1987). *Human factors in flight.* Brookfiled, VT: Gower Publishing Company.

Higgins, K. (2015). Disabled from waist down, pilot enables scholarship recipients. Retrieved February 1, 2017, from http://www.purdueexponent.org/features/article_f7d56221-25e6-59c9-bbef-5e019105b51b.html

International Civil Aviation Organization. (1989). *Fundamental Human Factors concepts. Circular 216-AN/131 Human Factors Digest No. 1.*

Izzo, M. V., Murray, A., Priest, S., & Mcarrell, B. (2011). Using Student Learning Communities to recruit STEM students with disabilities. *Journal of Postsecondary Education and Disability, 24*(4), 301–316.

Moon, N. W., Todd, R. L., Morton, D. L., & Ivey, E. (2012). *Findings from research and practice for middle grades through university education.* Atlanta, GA: Center for Assistive Technology and Environmental Access, Georgia Institute of Technology.

National Science Board. (2010). Preparing the next generation of STEM innovators: Identifying and developing our nation's human capital.

National Science Foundation. (2000). *Land of plenty: Diversity as America's competitive edge in science, engineering and technology.* Retrieved from https://www.nsf.gov/pubs/2000/cawmset0409/cawmset_0409.pdf

National Science Foundation. (2004). *Women, minorities, and persons with disabilities in science and engineering (No. NSF-04-317).*

Pearce, J. T., & Randel, A. E. (2004). Expectations of organizational mobility, workplace social inclusion, and employee job performance. *Journal of Organizational Behavior, 25*(1), 81–98.

Robinson, J. (2016). AV84All. Retrieved February 1, 2017, from http://www.av84all.org/home.html

Schultz, P. W., Hernandez, P. R., Woodcock, A., Estrada, M., Chance, R. C., Aguilar, M., & Serpe, R. T. (2011). Patching the pipeline: Reducing educational disparities in the sciences through minority training programs. *Educational Evaluation and Policy Analysis, 33*(1), 95–114. https://doi.org/10.3102/0162373710392371

Sullivan, K. (2015). Purdue teaches people with disabilities how to fly; plans for expansion. Retrieved February 1, 2017, from http://wlfi.com/2015/06/12/purdue-teaches-people-with-disabilities-how-to-fly-plans-for-expansion/

The Charity Commission. (2001). *The promotion of social inclusion.*

Tolley, J. (2016). Pilots with disabilities soar via Able Flight program at Purdue. Retrieved February 1, 2017, from http://btn.com/2016/10/16/pilots-with-disabilities-soar-via-able-flight-program-at-purdue-btn-livebig/

United Nations. (2007). *Participatory dialogue: Towards a stable, safe and just society for all.* Retrieved from http://www.un.org/esa/socdev/publications/prtcptry_dlg(full_version).pdf

World Bank. (2013). Social inclusion. Retrieved October 6, 2016, from http://www.worldbank.org/en/topic/socialdevelopment/brief/social-inclusion

Wulle, B. W. (2015). Expanding industry's understanding the skills people with disabilities can bring to the workplace. Presentation presented at the 2015 Able Flight Luncheon, West Lafayette, IN.

Advancement in Pedagogical Foundations: Developing Language Proficiency for Student Success

Stacey L. McIntire
Juan R. Merkt
Jacqueline R. Luedtke
Timothy B. Holt
Brent D. Bowen
Hunter Watson

College of Aviation
Embry-Riddle Aeronautical University
Prescott, AZ 86301

Abstract

Interruptions in flight training and a corresponding increase in costs appear to be prevalent among universities with aviation training programs. Students in these programs have to manage both demanding academics and flight training. Additionally, international students, for whom English is not their primary language, have the added disadvantage of learning complex aviation concepts in English. In order to maximize retention in collegiate flight programs, an experimental aviation English course has been designed to help frontload aviation vocabulary and take a proactive approach to teaching language skills that are essential in flight training. This as a case study model includes the learning goals and objectives for this course. The primary intent is to develop an applied knowledge of international radiotelephony alphabet and numerals, basic flight fundamentals and maneuvers, airport operations, national airspace operations, and emergency procedures as a means to developing English language proficiency. Students will know how to comprehend basic air traffic controller (ATC)-pilot transmissions, basic runway navigational directions given by ATC, and communicate effectively with clear, understandable speech, and at an appropriate tempo. Other important goals and objectives for this course include students understanding the role of ATC and the importance of clear communication by conducting an interview and providing a written summary of the interview. Furthermore, students will learn the value in safety by recognizing situational awareness through the means of analyzing both fatal and non-fatal incidents, why these incidents occurred and how they could have been avoided.

Background

Aviation

Aviation industry professionals are continuously seeking new and innovative ways to improve safety among pilots. Human factors, more specifically communication, has been a primary topic of discussion for over 20 years in the International Civil Aviation Organization. In 1998, ICAO diligently began taking steps toward establishing language proficiency requirements (LPRs). In 2004, ICAO published the first edition of Document 9835 (ICAO, 2010), which offers a brief, yet comprehensive, overview of language acquisition by identifying and defining essential concepts in the field of language teaching for members of the aviation field who are unfamiliar with or have no previous experience in the language acquisition field. Additionally, ICAO acknowledged that while language-learning schools have strict mandates, guidelines and regulations, aviation English training schools are not monitored and regulated through an accreditation process for their language teaching. To circumvent potential setbacks and keep safety in the forefront, ICAO published Circular 323 (ICAO, 2009), a guideline for aviation English training design, delivery, trainer profiles and trainer training. One essential component identified in both publications is that aviation English trainers must first have a background in language teaching and second a willingness to learn and be exposed to the technical foundations of aviation (ICAO, 2009). Soon after Circular 323 was established, ICAO published their second edition of Document 9835. This document mandates that each state's civil aviation authority implement and enforce the LPRs. In our case, the Federal Aviation Administration published an advisory circular (60-28A) in 2013 that very briefly outlines the requirements for ensuring aviators meet basic English LPRs before being issued an Airmen Certificate under part 61, 63, or 65. What brought these publications into motion? Three accidents, over 800 peoples' lives lost; all with one common factor - insufficient English language proficiency (ICAO, 2010).

Language

The adage, "aviate, navigate, communicate" is well known in the aviation industry. The term 'communicate' is a fundamental component in the popular saying and holds a tremendous amount of weight. In order to communicate, one needs language. According to Mathews (2012), language is complex and is often a subtle, yet significant factor in some aviation accidents. ICAO identifies what language is not. It is not a set of grammar rules, vocabulary, or simple pronunciation of sounds. Language is a set of complex interactions with a number of skills and abilities that work together to enable communication (ICAO, 2010). To develop a language, there are two processes, language learning and language acquisition. Language learning is a conscious process that involves learning language forms, whereas language acquisition is an unconscious method of learning, much like an infant learning a language. While language learning has its advantages, learners will lack the ability to improve in spontaneous situations. Emery (2016) notes in his article, featured in *Changing Perspectives on Aviation English Training*, emergencies depend on a wider range of proficiency. This is why ICAO promotes a student-centered learning approach, in that activities are planned to engage all learners' styles and students given ample time to be immersed in language learning activities. Just as identifying language as a factor in aviation accidents is difficult, so is learning or acquiring a language. Students come to flight programs at various levels of proficiency. Acknowledging these differences and setting realistic expectations is essential when designing an aviation English program.

Statement of the Challenge

Prior studies indicate that student flight-training delays and corresponding increased costs are widespread among universities with aviation programs (Bryan and Thuemmel, 1996). Identification of students prior to failure in professional flight courses was conceived from an eighteen-month study of training trends in the Embry-Riddle Aeronautical University, Prescott, Arizona Department of Aeronautical Science. Through this study of flight training trends and assessment of how the College of Aviation could better assist students in ensuring the success of the flight training requirements, it was realized that, in order to maximize student retention, students at risk must be identified early, prior to training failure or financial distress. In evaluating this data it became apparent that international students, for whom English was not their primary language, had some difficulty mastering aviation concepts and communication while completing their initial phase of flight training.

Innovative Approach

To address this problem, an aviation English course was developed to assist non-native English speakers in mastering aviation-specific content and communications. The learning goals and objectives for this course are to have good knowledge of international radiotelephony alphabet and numerals, basic flight fundamentals and maneuvers, airport operations, national airspace operations, and emergency procedures. Students learn how to comprehend basic air traffic controller (ATC)-pilot transmissions, basic runway navigational directions given by ATC, and communicate effectively with clear, understandable speech, and at an appropriate tempo. Other important goals and objectives for this course include students understanding the role of ATC and the importance of clear communication by conducting an interview and providing a written summary of the interview. Students learn the value in safety by recognizing situational awareness through the means of analyzing both fatal and non-fatal incidents, why these incidents occurred and how they could have been avoided. This course follows the three models of pedagogy all of which are supported and encouraged by ICAO in Circular 323 and Document 9835.

Model 1: Content-Based Instruction

The first is a Content-Based Instructional (CBI) model, which allows for students to make strong connections to academic content while learning and mastering the language skills necessary to be successful in their flight training (Stoller, 2008). According to Stoller (2002), in CBI the focus of content objectives and language objectives can shift along a continuum to meet the needs of the students. Stoller also asserts in her plenary address at the TESOL conference (2002) that, as content is learned, students will then improve their language skills. For this course, a great portion of the class is structured on content rich objectives and discussions, which are centered on topics that the students are learning concurrently in their Private Pilot Ground School (AS 121) course. The students' Private Ground School course provides the fundamental aviation knowledge and technical foundation they will need in their foundational aeronautical coursework. As a complementary course, it provides additional language support through the use of content with the use of cooperative learning strategies, such as jigsaw reading, that promote student-to-student interaction (Johnson, 1998). Jigsaw reading is a strategy where students are put into small groups and then assigned a short reading. They then become "experts" of their content and teach their section to the other members of the class. Materials used for this portion of the course include but are not limited to the Pilots Handbook of Aeronautical Knowledge, Airport Facility Directory, the Aeronautical Information Manual, and other relevant industry approved publications.

Model 2: Language Acquisition Strategies

The second pedagogy model is English for Specific Purposes (ESP), which incorporates language acquisition strategies through the use of special topics, such as aviation (Master, 1997). The focus on this portion of the course is primarily on language objectives rather than content objectives. Language forms and grammatical structures that are relevant to the field of aviation are explicitly taught, and as a result communicative skills are strengthened (Hutchinson, 1987). Students are provided additional vocabulary and language support in areas, such as phraseology and pronunciation, so that students can master critical concepts and better prepare students for their initial phases of flight training. Pronunciation is a key aspect of this course and will facilitate the students in understanding, being understood, and building self-confidence in a communicative environment (Goodwin, 2001). The goal is for students to obtain functional intelligibility where errors do not interfere with the message and are not distracting; native-like fluency is not a goal for this course. Along with pronunciation and speaking fluency, listening comprehension is another key aspect of this course. Listening skills are supported through the use of authentic discourse in both routine and non-routine aviation situations. Both bottom-up and top-down processing strategies are used to support student learning. Richard and Burns define bottom-up processing as sounds, words, and phrases, whereas top-down processing is focused on meaning (2011). An example of top-down approach would be to provide students with an authentic audio transmission, such as a pilot communicating with ATC, and allow students to listen for meaning. After students had several attempts to listen to the audio transmission, students are then supplied with a written script to deconstruct the transmission on a word level, a bottom-up approach. In an airplane, there are no scripts for pilots to use, but by providing structured practice within the classroom, students are given the confidence they need to apply their skills in flight. In teaching skill-related content, a gradual release model is used, which is a form of scaffolded instruction (Pearson & Gallagher, 1983). In this type of model there is a release of responsibility from teacher to student; in other words, it transfers learning from a teacher-centered focus to a student-centered focus (see Figure 1).

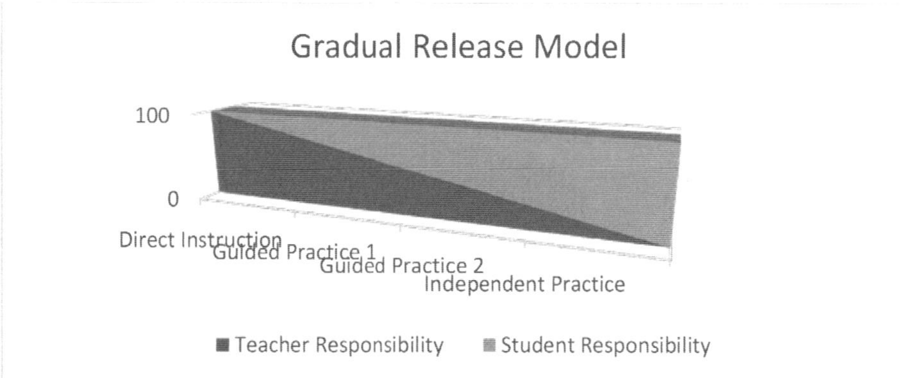

Figure 1

Model 3: Computer-Assisted Language Learning

The final model is the use of Computer-assisted Language Learning (CALL) where instruction is aided by digital learnings strategies, such as classroom capture technology, to improve and build language skills (Hubbard, 2009). Egbert (2005) asserts that CALL is focused on language development, not technology. In other words, technology is used as a learning tool. For the use of this course, a classroom capture technology is used to record lectures in the Private Pilot Ground School course. Students work at different paces, so by capturing vital lectures students are able to review and watch portions of the lecture multiple times. As a result, student anxiety level will more than likely decrease and result in increased learning (Egbert, 2005). Other technology, such as *TCS Learning System* ™ (TCS), a guided simulation program, and content specific videos created by Embry-Riddle Aeronautical University are used to help support student learning.

Benefits and Lessons Learned

Each semester students enter into aviation English with varying language proficiency levels, as well as a wide range of aviation knowledge. For this reason, keeping a certain amount of fluidity within the areas of content is essential. In the fall semester, for instance, students needed additional support in learning the E6B, an aviation flight calculator. They felt they were struggling more so than their native English speaking classmates in ground school. In order to meet their learning needs, the weather unit, which they felt confident in, was shortened so that class time could be utilized. A gradual release model of instruction was used in teaching the E6B. This was found to be an exceptionally useful strategy in teaching second language learners and students are encouraged throughout the course to self-reflect on their learning and offer areas in which they feel they need additional support.

Challenges

An unexpected challenge in designing this course was trying to frontload, or pre-teach, vocabulary before students were taught the same topics in ground school. Frontloading is found to be far more beneficial than re-teaching strategies. Typically this is done in a single lesson or unit, but this strategy is used to frontload content across multiple courses. The idea is to familiarize students with essential vocabulary within the aviation English class before learning about those concepts in the ground school course. The pace of the ground school course is extremely fast due to the large amount of content that needs to be taught and thus poses a difficulty in structuring the content in the aviation English course. Postponing ground school instruction is not always a viable option for students, especially those who have stronger language proficiency skills. Fortunately, students with stronger language proficiency skills are able to adapt with ease. Students with weaker language proficiency skills need more support. Our next steps will be to further consider how students with lower language proficiency skills can continue to be supported throughout their flight training.

Successes

A variety of methods were used to gather data on the effectiveness and success of the aviation English course, such as instructor observations, end of course evaluations, class assessment data, and a follow-up survey via a web-based survey collector. Overall, as noted the support that students received through their ground school training was highly successful. Students felt comfortable to request specific content to review, ask questions via our classroom "parking lot" and frequent the instructor's office hours for additional support. Often students would communicate their successes in exams or flights to the instructor and felt they were successful because of the support they received. Another success was the activities and methods of teaching that were used in designing this course. Students noted in a voluntary follow-up survey that the learning activities they preferred the most were classroom discussions, guest speakers, field trips to the local ATC tower and flight line, as well as using computer technologies to support learning. Lecture style instruction was favored less. Of the 18 students that were invited to partake in a voluntary follow-up survey, 15 respondents completed the survey. All 15 indicated that they would recommend this course to a friend. All of those flying (14) noted that the course supported them in their flight training. Due to the course's overall success and benefits gained for our international non-native English-speaking students, this course is in the process of becoming a permanent course within the Aeronautical Science program.

References

Barbieri, Brian Joseph. "Aviation English: History and pedagogy." *Journal of Teaching English for Specific and Academic Purposes* 2.4 (2015): 615-623.

Bedard, R. (2014). Private pilot course. *True course simulations.* www.tcsims.com. Web. 16 Sept. 2016.

Borowska, A. & Enright, A. (Eds.) (2016). *Studi@ Naukowe 29: Changing perspectives on Aviation English training.* Warsow: Uniwersytet Warszawski.

Bryan, J. L. and Thuemmel, W. L. (1996). An analysis of student programmatic delays in postsecondary flight training programs: A national study. Paper presented at the American Vocational Association Convention (Cincinnati, OH, December 5-8, 1996).

Egbert, J. (2005). *CALL essentials: Principals and practices in CALL classrooms.* TESOL.

Federal Aviation Administration. (1997). *Advisory circular, English language skill standards required by 14 CFR parts 61, 63, and 65, AC No: 60-28, September 23, 1997.* U.S. Department of Transportation.

Goodwin, J. (2001). Teaching pronunciation. Teaching English as a second or foreign language, 3rd Ed. Heinle. 117-131.

Hubbard, P. (Ed.) (2009). *Computer assisted language learning: Critical concepts in linguistics*, Volumes I-IV. London & New York: Routledge.

Hutchinson, T., and Waters, A. (1987). English for Specific Purposes *A learning-centered approach.* UK: Cambridge.

ICAO. (2009). *Guidelines for aviation English training programmes* (Cir 323-AN/185). Montréal, Canada: International Civil Aviation Organization.

ICAO. (2010). *Manual on the implementation of the language proficiency requirements* (Doc 9835-AN/453) (2nd ed.). Montréal, Canada: International Civil Aviation Organization.

Jenkins, Jennifer, Alessia Cogo, and Martin Dewey. "Review of developments in research into English as a lingua franca." *Language teaching* 44.03 (2011): 281-315.

Johnson, D.W., Johnson, R.T., & Smith, K.A. (1998). *Active learning: Cooperation in the college classroom.* Edina, MN: Interaction Book Company.

Master, P. (1997). Content-based instruction vs. ESP. *TESOL Matters*, 7(6), 10.

Mathews, E. (2012). Language gap. First of two parts. *AeroSafety World*, 22-27.

Molesworth, Brett RC, and Dominique Estival. "Miscommunication in general aviation: The influence of external factors on communication errors." *Safety science* 73 (2015): 73-79.

Pearson, P. D., & Gallagher, M. C. (1983). The instruction of reading comprehension. *Contemporary Educational Psychology, 8,* 317–344.

Richards, J.C., and Burns, A. (2011). Listening skills. *Tips for teaching listening: A practical approach.* White Plains, NY:Pearson Education. 17-34.

Richards, J.C., and Burns, A. (2011). Second language listeningp. *Tips for teaching listening: A practical approach.* White Plains, NY:Pearson Education. 1-16

Stoller, F.L. (2002). *Content-Based Instruction: A shell for language teaching or a framework for strategic language and content learning?* Article slightly modified from author's plenary address at the TESOL convention in Salt Lake City, Utah. Web. 23 Sept. 2016. http://carla.umn.edu/cobaltt/modules/strategies/stoller2002/stoller.pdf

Stoller, F. L. (2008). Content-based instruction. In *Encyclopedia of Language and Education* (pp.1163-1174). Springer. Web. 16 Sept. 2016. DOI 10.1007/978-0-387-30424-3_89

Williams, G. J. (2016, March). Solutions for improving the safety of aviation communication: *An investigation of pilots' and air traffic control officers' opinions on aviation English.* (Master thesis). University of Leicester, U.K.

COGNITIVE ERGONOMICS AND KNOWLEDGE MANAGEMENT: A NEW MODEL OF ADDRESSING AVIATION ISSUES

Ronald John Lofaro
Embry-Riddle Aeronautical University/Worldwide
Orange Beach, AL

The field of cognitive/knowledge engineering (CE/KE) has expanded to encompass a subset called knowledge management (KM). The main difference between KM and KE is that the knowledge manager establishes the direction the process should take, whereas the knowledge engineer develops the means to accomplish that direction. Cognitive Ergonomics, also a discipline within CE/KE, deals with decision-making, skilled performance(s) and training with a focus on the fit between human cognitive abilities/limitations and selected task(s)... all areas vital to aviation. This paper will briefly show the components and processes of a new model for decision-making using all these disciplines and Pareto analysis in a blended Delphi, beginning with a small group delphi (SGDP) addning in estimate-talk-estimate (ETE) techniques and ending in a real time Delphi (RTD). This paper will outline how this new paradigm can be used on a current aviation issue; showing a blueprint/framework for the actual process of the paradigm.

In 2016, one finds an expansion in the the field called cognitive/knowledge engineering (CE/KE) has occured and is still in progress. KE was defined in 1983 by Edward Feigenbaum and Pamela McCorduck as follows: "KE is an engineering discipline that involves integrating knowledge into computer systems in order to solve complex problems normally requiring a high level of human expertise." There is a new emphasis on a related discipline: knowledge management (KM). Knowledge management (KM) has been defined as "...the practice of selectively applying knowledge from previous experiences of decision making activities with the express purpose of improving the organization's effectiveness." (Jannex, 2014). The main difference between KM and KE seems to be that the (knowledge) manager establishes the direction the process should take, where the (knowledge) engineer develops the means to accomplish that direction.

There is also a somewhat new emphasis in the KE/KM fields on ontology, a term that comes from philosophy. A KE/KM ontology compartmentalizes the variables needed for some set of computations and establishes the relationships between them; thus, an ontology is used to limit complexity and to organize and structure information. It is then a practical application of philosophical ontology, with a taxonomy. Applications are AI, information science and technology, decision-making and much more. The author is here attempting to show how a Small Group Delphi Paradigm (SGDP)/Estimate-Talk-Estimate (ETE) amalgam with a modified Real Time Delphi (RTD) could be used on some specific problems, as well as adding in the use of a mini-Pareto as a start point. All done in the hopes that such efforts might be seen and be of some interest, even be of some help in decision-making. [Note; the author has authored/co-authored some 10 articles and and 2 book chapters on decion-making in the unforgiving aeronatical environment; space truly precludes a listing here and especially in the **referecences** section]. The article will present a "how to" for using this modification on a specific aviation problem.

The Delphi and its Characteristics

The Delphi method was developed by Project RAND during the 1950-1960s. Delphi techniques, a subset of CE/KE, have become common methodologies for eliciting analyses, expert opinions and evaluations on a variety of topics. Meister (1985) noted "The (Delphi) methodology is by no means fixed…[it] is still evolving and being researched." This is as true now as it was when Meister stated it. In point of fact, with the leaps in communication methods and related computer technology, this is even more true today. The rationale for this statement is actually two-fold. The first being that in the last 10 to 15 years, there has been quantum leaps in computer memory/power as well as communication technology that uses desk-top computer, even iphone technology. These leaps and advances seem now to occur almost daily. Concomitantly, Delphi techniques have recently begun to look at and attempt to take full advantage of these advances.

The following key characteristics of the Delphi method help the participants focus on the issues at hand and, what separates Delphi from other methodologies:

1. Anonymity of the participants.
2. Structuring of information flow.
3. Regular feedback.
4. Role of the facilitator.

The person coordinating the Delphi method can be known as a facilitator or leader, and facilitates the responses of their panel of subject matter experts (SMEs).

What has been presented above is the standard description and rationale for the Delphi process..

From Small Group Delphi Paradigm to Knowledge Management

This article was over 30 years in the making. It begins with a specific problem and task: to develop for the U.S. Army Aviation Command an aviator candidate selection test, later termed Multitrack, that also showed which of the current rotorcraft (the U.S. Army uses this term, not helicopter as do other U.S. military services.) would be the optimum operational aircraft placement for the candidate upon completion of initial training. At this point, the author, who was experienced in team training and group function/dynamics, decided to develop a modification to Delphi processes wherein the abilities approach of Fleishmann was used with face-to-face, small groups. Highly experienced and with high performance evaluations Army aviators were brought in from all Army posts world-wide, functioned as subject matter experts (SME's). In the ability requirements approach of *Taxonomies of Human Performance,* Fleischmann and Quaintance (1984, revised 2000): the ability requirements theory/approach is away of describing and classifying human tasks. In this approach, tasks are described, contrasted, and compared to the abilities required of the individual performing a specific task. Once a set of tasks is identified, a human performance taxonomy can be developed from it. Taxonomy, as used here, denotes a system that classifies and describes human tasks according to a particular focus, such as the abilities seen as essential to a specific task. Thus, as it was termed, the SGDP (Lofaro, 1992a), took the Delphi process in another direction by modifying it via merger with elements of group dynamics in order to have interactive (face-to-face) Delphi workshops

The SGDP then accomplished what had never been done before. It delivered an operational, computerized set of tests with scoring algorithms that not only selected the optimal aviator candidates for initial pilot traing but also showed which of the four then-existing rotorcraft types these candidates should be placed for transition training in upon completion of initial training.... Tthe U.S. Army Aviation Command not only used Multitrack but, there was a very high predictive validity as to both selection and placement. It now seems that the SGDP was a KM effort.

While the definition of measures of KM success did not exist in 1986 (and not even throughout the ensuing use of the SGDP into 2003), nor did the Jennex and Olfman KM Success Model (Jennex, 2013), it would seem, that in part, these measures were somewhat met by the Mutitrack test/scoring algorithms resulting from the SGDP. Examples: A KM Success Model measure of KM success is system quality. In the SGDP the SME's created/produced knowledge that was stored, able to be retrieved and was applied. That knowledge probably could not have been otherwise captured as the SME's may not have been available later, the need for their specific expertise had not, nor would be asked for again. But Army Aviation needed it exactly then. Knowledge quality was achieved as the knowledge was shown to be useful as to correctness and inclusion. The service quality was seen in the performance impact of the using (making operational) Multitrack and subsequent U.S. Army Aviation Command satisfaction.

The Small Group Delphi Paradigm: Past Uses

From 1987 through 2003, the SGDP was adapted and used in a broad spectrum of tasks, from ATCS selection (Gibb and Lofaro, 1993); to managerial core competencies(Lofaro,1998); to flightcrew performance evaluation (Lofaro, and Smith, K.M., 2007); to useability testing (Lofaro & Maliko-Abraham, 2002; Maliko-Abraham & Lofaro, 2001, 2003); to work on a mission performance model for flight crew resource management integration and evaluation. (Lofaro, 1992b); to selection and evaluation of aircarriery baggage screeners (Lofaro, Gibb and Garland, 1994; Gibb and Lofaro, 1994). It must be noted that the successful use of SGDP techniques by others indicate that the techiques are not dependent on who administers them; rather, they not only are effective across venues but also with different personnel directing/facilitating the SGDP workshops. These successful adaptations, modifications, implementions, across many venues, showed that the SGDP is both transferable and generalizable and possesses external validity. The SGDP can be used for any project that requires that a set of SMEs be used to identify, evaluate and criticality rate tasks (an enhanced task analysis); to identify core needs/skills; to recommend modifications to equipment, procedures and training. Finally, the SGDP can be used to sharpen, modify and revise existing methodologies.

The Small Group Delphi Paradigm: 2016 Technological Modifications

Some twenty-five years after the SGDP was devised, used and appeared in multiple publications, it has been re-discovered, as it were, and is now recognized as an acceptable CE/KE method. The use of face-to-face groups in a Delphi is now a fact and is termed Estimate-Talk-Estimate (ETE). New communication capabilities and technologies seem to have driven the development of what are generically referred to as a mini-Delphi or ETE, with many variants. There is also the Real Time Delphi (RTD) that maintains anonymity but uses computer linking for a high level of instantaneity (Gordon, T. J., & Pease, A., 2006). While the SGDP structure and processes are still relevant, they can be and are in real need of some level of integration with ETE/RTD techniques and current advances in computer and communication abilities and techniques. The author's belief is that the result of an integration would be a revised Delphi that will produce the same level, if not a higher level, of accurate information, decision-making guidance and products in the aviation arena, or many other venues. In modifying the SGDP for use with today's technology and advances in KE/KM, any problems to be investigated would require, as a first step, the building a model of a knowledge domain, defining the terms inside that domain and the relationships among them...developing an ontology. We will return to this later.

A Model Of A Blended SGDP/ETE/RTD...And Beyond

The core SGDP structures to be maintained are: the careful selection of a limited number of SMEs (however, there can be many small(er) SME groups functioning at one time), the use of an extensive read-ahead package for the SMEs, the use of some facilitation and group dynamics instruction, combined with some type face-to-face sessions. One example of a difference in an ETE or RTD (a computer-based Delphi with anonymity) versus either a tradtional or SGDP Delphi: the iteration structure used in the tradtional or SGDP Delphi, which is divided into three or more discrete rounds, can be replaced by a process of continuous (roundless) interaction, enabling SMEs to change their evaluations at any time. A new ETE/RTD/SGDP model would be computer-based. The reader is referred to the work of Turoff and Hiltz (1996) on computer-based Delphis.

Integration of the SGDP with a ETE/RTD (*sans* anonymity) approach can be achieved thusly: all participants can be logged on simultaneously, each participant can briefly state their name and credentials, the group dynamics instruction can be done by the facititator to all simultaneously (aside: it would seem that a linked network of all SMEs is possible and even *de rigueur).* The SGDP face-to-face group meeings would now be done on-line. This will allow for instanteous feedback by any SME during a session, as well as discussions). The iteration structure used in SGDP, which is divided into as many discrete rounds as needed for consensus, can be replaced by a process of continuous (roundless) on-line interaction, enabling SMEs to change their evaluations at any time and give a rationale with ensuing

discussion in real-time. Finally, the statistical group(s) response(s) can be updated in real-time and shown whenever a SME, or a group, provides a new evaluation. It is clear that "face-to-face" discussion will be virtual. This is, to the author, a real and significant loss. But, the speed, multiple iterations, real-time, access to a large number of SMEs and other aspects to be gained cannot be ignored. Another possible modification is a multi-tiered SGDP/ETE/RTD where the use of two or more groups working differnt problems can be convened and given objectives based on their expertise. As these groups come to consensus on their objectives, these new data can be integrated, built into a new re-ahead package and made available to a new set or participants with new or prior SME's.
A second modification would be use of only specific elements of a Pareto Analyis as the first step after the group dynamics instruction. This is because such a step would identify problems, then, sharpen (focus in on the ones that are amenable to resolution) while, at the same time, providing an ordering of which should be worked on via criticality ratings. The elements of the Pareto Analysis would be the first three steps in such an analytic technique. Step 1: Identify and list problems to be examined. Or, if and only if one problem/issue exists, a break-out of the sub-problems could be done. Step 2: identify causal factors inherent in each problem. Step 3: Score (in this case, criticality rate) the problems, resulting is a somewhat rough ontology. The Pareto, done in the context of a SGDP/ETE/RTD could also allow for many possible solutions to be evaluated by many types of SMEs with differing areas of expertise but in areas germane to the problem. This would result in a winnowing down of courses of action to those that were realiostic and possible of success. The advantages of using a Pareto, core SGDP structure and ETE/RTD (all computer/Internet driven) are that they are all content-area neutral and, in a real sense, generic in application.

A Specific Aviation Issue/Problem

This paper will only deal with one current commercial aviation issue (as the author has 26 years of experience in aviation; United States Air Force, Army Aviation Command, and the Federal Aviation Administration (FAA), that choice seemed simple). There is a significant amount of controversery over upset training; training pilots to recovery the airplane from unusual (unstable; dangerous; rarely encountered) attitudes thereby not having potentially fatal accidents. The Air France Flight 447 accident, for example: A series of errors by pilots and a failure to react effectively led to the crash of Flight 447. The plane went into a sustained stall, signaled by a warning message and strong buffeting of the aircraft. Despite these persistent symptoms, the crew never understood they were in a stall situation and, therefore, never undertook correct recovery maneuvers. In other words, a high altitude, high speed stall was a situation the crew was unable to recognize much less had ever trained for. The Colgan Air crash outside of Buffalo, NY is another fatal accident where the U.S. National Transportation Safety Board (NTSB) determined that the probable cause of this accident was the Captain's inappropriate response to the activation of the stick "shaker" (the airplane's main control device actually shakes/vibrates in the pilot's hand if the plane is in danger of stalling/spinning) which led to an aerodynamic stall from which the airplane did not recover. To further cloud the issue, the American Airlines upset prevention and recovery training (UPRT) ground school with flight simulator (FS) training, called advance aircraft manuevering program (AAMP), was seen by the NTSB as possibly a contributing factor in the American Airlines flight 587 crash in 2001 (Croft, John. 2014a).
The FAA, while not yet issuing an Advisory Circular (AC) or a Federal Aviation Regulation (FAR) has issued a document called Airline Upset Recovery Training Aid, version 2. The current issues seem to be use of a full motion FS, that will be part of an expected FAA pilot training rule by 2018 (Croft, John. 2014b), versus in-aircraft training (or some combination) and swept wing jet aircraft specialized training. It is believed that a modified SGDP/ETE/RTD, using senior pilots as SME's can point a way to types/procedures and applicability for upset training.

This SGDP/ECE/RTD would proceed thusly: A multi-tiered effort, beginning with the Pareto described above to identify 2 or 3 issues that most need solutions (a quasi-ontology) from the issues cited above: use of a full motion FS and/or in-aircraft training and/or swept wing jet aircraft specialized training.

Next, as always , multi-tiered, identify the sub-issues involved. Here, as with the remainder of the SGDP/ETE/RTD, the data needed for all tiers/sub-tiers would come come from SGDP/ETE/RTD's consisting of 5 to 9 (senior) pilots from as many aircarriers as possible. Possibly, each interested carrier initially could do such efforts. In this way, each carrier would have results based on their mission and objectives for possible upset recovery training. Since the multiple sub-tiers would only be dealing, at first, with one specific arena, the results could be used grist for another round(s) where consensus, via criticality ratings and discussion, is worked on. This multi-tiered approach can be used for all the issues listed above-simultaneously or sequentially. In a second, or as an extension of the first SGDP/ETE/RTD, the SME pilots can identify and sort these upset recovery maneuvers into taxonomies for those which are primary and necessary for what can be called an air carrier's "basic training" for aircrew, then for transition training and for recurrent training. Again, in a tiered series of SGDP/ETE/RTDs.
The use of a FS (What type? What recovery techniques can be taught in FS and which require in-plane training? And for which planes?). What are the performance standards for each manuever? Answers to these all would be results from such an effort. It may be possible that the large amounts of data being developed ("soft data" as they are the results of opinions and/or criticality ratings) could be handled by data & knowledge engineering (DKE) methods. Finally, these SGDP/ETE/RTD techiques can be used for many other existing problems; in aviation, flight crew rest/fatigue immediately comes to mind.

FUTURE DIRECTIONS

It must be strongly noted that the article is not a report on the results of prior experimentation, rather it is an attempt to meld a modified Delpi procedure with today's KE efforts and today's technology. It is an attempt to indicate the structure of this new, modified Delpi. More importantly, it is also a call for research efforts to try the new procedures and validate (or not) them. Would that the author were still in prior positions as a government agency researcher/manager; the author could have attempted these efforts. This is not the case today. This article provides a rough template for future research.

References

Croft, John. (2014a). Unambigous Upset. Aviation Week & Space Technology, Vol.176, No.20, 52.

Croft, John. (2014b). Unambigous Upset. Aviation Week & Space Technology, Vol.176, No.20, 53.

Feigenbaum, Edward A, McCorduck, Pamela (1983). The Fifth Generation (1st ed.). Reading, MA: Addison-Wesley.

Fleishman, Edwin A. and Quaintance, Marilyn K (reissued 2000). Taxonomies of human performance: The description of human tasks. Potomac, Maryland. Management Research Institute, Inc.

Gibb, G.D., Lofaro, R.J. et al. (1993). Computer-based Assessment of Air Traffic Controller Abilities: Experimental Selection Tasks in *Proceedings of 1993 Aerospace Medical Association Annual Symposium*. Alexandria, VA..

Gibb, G.M., Lofaro, R.J. (1994). Airport Security Screener and Checkpoint Security Supervisor Training Workshop. DOT/FAA/CT-94/109. Springfield, VA: National Technical Information Service.

Gordon, T. J., & Pease, A. (2006). RT Delphi: An Efficient, "Round-less," Almost Real Time Delphi Method in *Journal of Technological Forecasting and Social Change,73(4),* Amsterdam , NX., Elsevier B.V.

Jennex, Murray (2013). Knowledge Management Success in an Engineering Firm in *IEEE Engineering Management Reviews, volume 2, issue 2.* Victoria, Australia. IEEE Technology and Engineering Management Society.

Maliko-Abraham, Helene and Lofaro, R.J.(2003). Usability Testing: Lessons Learned and Methodology in *Proceedings* of the 2003 International Military Testing Association (IMTA) Conference.

Maliko-Abraham, Helene and Lofaro, R.J.(2001b). Air Traffic Delays: New ATC Equipment Coupled With New ATC Candidate Selection Methods May Be One Answer in *Proceedings* of the 2001 National Air Traffic Controllers Association (NATCA).

Lofaro, R.J., Gibb, G.M., and Garland, D.(1994). A Protocol for Selecting Airline Passenger Baggage Screeners. DOT/FAA/CT-94/110. Springfield, VA: National Technical Information Service.

Lofaro, R.J. (1992a). A Small Group Delphi Paradigm in *The Human Factors Society Bulletin, 35 (2),* 1-4.

Lofaro, R.J. (1992b). Workshop on Integrated Crew Resource Management (CRM). DOT/FAA/RD-95/5. Springfield, VA: National Technical Information Service.

Lofaro, R.J., (1998). Identifying and Developing Managerial and Technical Competencies: Differing Methods and Differing Results in *Proceedings of 42nd annual Human Factors and Ergonomics Society.*Santa Monica, CA.

Lofaro, R.J. and Maliko-Abraham, H. (2002). Techniques for Developing High Predictive Validity Selection and Screening Test Batteries in *Proceedings* of the 2002 World Aviation Congress; Phoenix, AZ.

Meister, David (1985). *Behavioral Analysis and Measuremen.* New York, NY. Wiley Press.

Turoff, Murray; Hiltz, Roxanne Starr (1996). Computer-based Delphi processes in Michael Adler, Erio Ziglio (eds.). *Gazing Into the Oracle: The Delphi Method and Its Application to Social Policy and Public Health,* 56-88. London, England: Jessica Kingsley Publishers

HOW AUTOMATION MAY TRANSFORM THE WAYS IN WHICH CREW MANAGE PEAK WORKLOAD AND INCAPACITATION

Nick McDonald
Alison Kay
Rabea Morrison
Margaret Ryan
Centre for Innovative Human Systems, Trinity College Dublin, Ireland

A 'Crew Manage the Operation' concept was developed as a unifying framework to analyse the combined role of diverse technologies in supporting the management of peak workload and incapacitation. Multiple-crew configurations mean that many of the same technologies are supporting quite contrasting instantiation of crew roles. While Proactive and Immediate modes support the independence of the crew in their cockpit, both reactive modes of workload management pose questions about the information flow between the cockpit and ground control, and in turn about the level of support or, in the final mode, effective intervention that could be provided from the ground. These issues are not just about the flow of information but about responsibility and accountability. Thus the technologies are neither determinate of the way they are operated nor are they bounded by cockpit environment; therefore a profound discussion about crew roles and the philosophy of automation is required.

New technologies on the flight deck are transforming the nature of automation to provide novel solutions to core issues concerning human performance such as the management of peak workload and pilot incapacitation. ACROSS was a large integrated European project established with the goal of designing technology systems that alleviate crew workload in current two-pilot operations to improve operational safety (ACROSS, 2016). Three main objectives were set out to guide the design, development and testing of new cockpit solutions: Objective 1 - Addressing peak workload situations; Objective 2 - Addressing reduced crew operations; and Objective 3: Identifying open issues for possible single-pilot operations. This paper simply outlines the development of concepts concerning workload and automation through the project workprogramme. Other papers will describe in more detail the methodologies used and present the analyses of results of the various studies involved.

The Project comprised more than 30 partners spread across Europe, involving multiple organisations, nationalities and large number of diverse technologies being developed in parallel and tested and evaluated in a range of separate test beds. The technology work packages were organised around the classic pillars: Aviate, Navigate, Communicate, Manage Systems, with additional technology workpackages for Crew Monitoring and Crew Incapacitation. Human Factors was one of several transverse workpackages designed to provide a coherent integrated approach across the project. While the project was driven by the development of technologies, its core philosophy always emphasised Human Factors as core to delivering its operational objectives.

In the absence of a physical integration of technologies at project level, the Human Factors team had to deliver a conceptual framework to address the technology scope for each flight function as well as their overall integrated assessment. The objective was to deliver an integrated concept of the crew and their activity with technology at three levels: 1) the operational process, 2) crew tasks, and 3) Human-Machine-Interface. the challenge was to manage the integration of human factors throughout the project ensuring the achievement of operationally valid solutions.

The Initial Workshop

The first Human Factors workshop to address the global operational level was attended Technology development leads, HF experts and the operational representatives in the project. A core purpose was to ask Technology development leaders to link the contribution of their technology to the crew maintenance of the Situational Awareness (SA) bubble. SA refers to the perception of the elements in the environment within a volume of time and space, the comprehension of their meaning and the projection of their status in the near future (Endsley, 1988). The term "Situational Awareness bubble" is used commonly to emphasise both the time and space aspects of SA. Technology leaders were asked to conceptualise the situational awareness bubble in terms of the situational dependencies that need to be managed by crew and the discussion centred around the following questions:

> How are the operational dependencies organised in time and sequence (e.g. flight phases)? How do they relate to each other? How should they be organised and prioritised? How should multiple dimensions be represented in the HMI? How to balance demand against capacity? How to transfer authority when capacity is limited?

The answers to these questions linking operational dependencies to technological functions led to the following general observations:

> 1) The crew should be at the centre of the flight operations process and pilots should always be the ones to make the final decision on the flight-deck. 2) There is a clear difference between managing a flight (gate-to-gate) and managing the flight operations process in terms of both system and stakeholder input and output required. 3) There is a need for an integrated crew support function on the ACROSS flight deck. 4) The technology interfaces and functions, considered individually, could not deliver a solution to crew maintaining the situational awareness bubble. 5) The crew should receive decision support to manage the tasks of Aviate, Navigate, Communicate, Manage Systems and Crew Monitoring. This decision support should be in the form of prioritised recommendations for action. 6) It would be advantageous if the recommendations could give an idea of what potential consequences of those actions would be.

The general conclusion of the integration meeting was that the technologies, on their own, cannot deliver an autonomous progressive automation concept, moving from one level of automation to another according to the level of workload, or an entirely reliable and valid incapacitation decision leading to automatic recovery of the aircraft. This also implied that the normative model of workload expressed as a function of demand over capacity (e.g. Fuller, 2000) would not be adequate to address the core function of crew as 'managers of the operation', including, of course, being active managers of workload and incapacity.

The Relationship between Technology and Operations

This conclusions begged the question: what is the relationship between advanced technology and operational performance? Generating an answer to this question was stimulated by a radical analysis of cognitive HMI by Hourlier and his colleagues which formed the basis of the Thalis ODICIS cockpit (Hourlier & Suhir, 2014; Lini et al., 2012; Lini, Favier, Servantie, Vallespir, & Hourlier, 2011). In this framework there are four cognitive HMI principles that enable a rupture between classical fly-by-wire technologies and the next generation of cognitively enhanced cockpits. Briefly, these principles suggested the following questions about the ACROSS technologies:

Schematise: what schematic representations can be supported by technologies that are critical to the management of workload by crew?

Anticipate: what advanced information can technologies support/provide that support proactive management of workload?

Delegate: What functions can be delegated (to other crew or technology) through technologies to support the crew in managing workload?

Routinise: In what way do technologies support management of workload to render them simple, intuitive, easily learned and reducing overall cognitive training effort?

These cognitive HMI principles tie into the cognitive behavioural cycle of operations that deals not only with separate technologies but with how these enable the management of concurrent tasks, taking into account both the past and projections for the future. These principles highlight what the technology can do for the crew. This then helped to focus the question: what can the crew do with the technologies?

The ACROSS Workload Management Concept

This approach led to the formulation of a generic workload management concept, as outlined in Table 1.

Table 1. ACROSS Workload Management concept

Workload Management Concept	Definition
Proactive Workload Management	Managing workload using timelines and other schematisations. This enables anticipation, which in turn enables planning and allocating resources along the timeline. This enables crew to spread the anticipated workload better and also to be more prepared and more capable of absorbing unexpected spikes in workload.
Immediate Workload Management	In ACROSS immediate workload management is achieved through the use of automation, which reduces demand and together with enhanced decision support reduces crew workload in the here-and-now.
Reactive Workload Management	Managing workload by reacting to events/situations after they have happened. The main focus of reactive workload management in ACROSS is the Crew Monitoring System, which can detect pilot incapacitation and suggest mitigations.

In the final phase of the project this workload concept was tested in a series of workshops involving operational and human factors experts following a set of operational scenarios deploying the technologies enbaling a focussed discussion of the operational aspects of the technologies under the different configurations envisaged by the project (full crew; one crew member incapacitated; both crew incapacitated).

The following provides a short schematic summary of the core relationships of each mode of managing workload, based on the analyses of the workshops with key stakeholders. The purpose of this exercise was to consolidate a model of how the ACROSS technologies could support the management of peak workload and pilot incapacitation, and to point to some basic issues that need further examination.

Proactive Workload Management does not stand on its own, it is a precursor to the other modes. It provides a barrier to peak workload through advanced information in relation to alternate airports and their characteristics, the weather, frequency changes, overall system status, amongst other things. This enables crew to, for example, plan and select alternates or to adjust the mission in the light of system status. It reinforces current crew roles and could have a strong input to crew briefing. The outcome is to spread workload more evenly and to ensure crew a both prepared for things that are foreseen, and ready to tackle unforeseen demands.

Immediate Workload Management operates through reducing demand in a number of ways: providing specific information to assist decision-making, actuation of decisions by use of automation, new HMI design which co-locates system status and action actuation, and interaction design which supports a check process. These barriers to escalating workload in turn support a number of mechanisms: making a decision (selecting nearest airport, for example), deploying automated systems, going through fault identification and rectification sequences that manage the systems' status. Again these mechanisms reinforce both crew roles. The outcome is easier and less demanding decisions, delegating tasks to automation, including automated checks on system status. The objective is an optimal division of labour in a highly demanding situation.

Reactive Workload Management in Single Crew Incapacitation involves the key barriers of monitoring information about the crew and about the aircraft operational status. These are also accompanied by all the other technologies we have just described in the previous two paragraphs – both information and actuation. The monitoring information enables on-going monitoring and decision-making, but the other technologies are crucial in reducing the demand on the remaining pilot in continuing the operation (for example in automated go-around). The focus here is on consolidating both crew roles in one pilot, though some of the experts referred to some automated functions as a potential co-pilot. A big question that arose in the workshop discussions concerned the sharing of information with ATC and Flight Operations Control. The outcome here is reasserting effective control over the operation, with some questions about the type of support that could be received from the ground.

Reactive Workload Management when all crew are incapacitated involves the critical barriers of constant crew monitoring, together with the availability of emergency support systems for the aircraft controlling all the automation options that can return the aircraft safely to the ground. The mechanisms centers around a clear signal that initiates transfer of control to the emergency support systems. This then potentiates functions like automated navigation and landing. The Ground Station is in strategic control. Again the question was raised about the level of sharing cockpit information with the Ground Station and ATC. The outcome is a safe landing. An issue that arose concerned the responsibility and capability of the ground support to deal effectively with any malfunction of the emergency support systems.

Conclusions

Two things stand out from this analysis:

Firstly, the Proactive and Immediate Workload management modes both reinforce the current crew configuration and their roles which make sense in terms of optimizing all resources in managing a demanding situation. On the other hand crew incapacitation involves transforming those roles – consolidating in one crew or transferring to an automated pilot and ground control. Thus many of the same technologies are supporting quite contrasting instantiation of crew roles.

Secondly, while the Proactive and Immediate modes support the independence of the crew in their cockpit, both reactive modes pose questions about the information flow between the cockpit and ground control, and in turn about the level of support or, in the final mode, effective intervention, that could be provided from the ground. These issues are not just about the flow of information but about responsibility and accountability.

The 'Crew manage the operation' concept has provided a unifying framework through which it is possible to see the combined role of these diverse technologies in a crew-centric way. It has enabled the exploration of the role of advanced technologies, and most particularly the 'rupture' in technology development that is typified by the ACROSS technologies which bring both rich meaningful content and dramatically expanded connectivity.

However that rupture demonstrates the need to rethink the philosophy of automation. In referring to Tarnovsky (2002), the question is raised whether or not this is still an authoritative comment on automation philosophy in the light of the cockpit technologies represented in ACROSS. Simplistic thinking along the lines of classic automation is no longer sufficient – it is not just a question of replacing one function after another with an automated system until one gets rid of a crew-member. The implications of technological step changes must be addressed in depth.

Trust in technology is core to trust in the future system. Some confidence was placed in this during the workshops – with use, people will learn to trust the new technology and exploit the functional benefits it brings. However single crew operation is for many a threatening concept viewed with great suspicion if not hostility. Crew monitoring raised questions about how future systems would "handle" data associated with crew and fatigue. There must be clear guidelines for data protection whilst ensuring safe practice for crew. The dynamics of system transformation by new technology is not a deterministic process and there are real choices to be made about the relative role of people and technologies. These choices need to be made in an informed way.

All of this will require new procedures and new training to capture reinforce the optimal relationships between new technology and operations and how to cope when the technology fails.

A new paradigm for system integration has emerged with new flight deck technologies together with the SESAR and Next-Gen mega-projects; however this has yet has to generate a new operational paradigm. The issues can no longer be resolved on the flight deck or at the controller's workstation - they increasingly concern the relationship between flight deck and ground control (both airline and ATC). It is not just the technology interface with the human that is important but the connectivity to the rest of the system. Because of this complexity the driving operational concept should be clearly embedded in a rich understanding of operational reality. This involves a capability for 'System Design for Operations' (McDonald, Morrison and Grommes, 2007). The development of new technologies pushes us further to consider not just the transformation of roles at local level where new automation can enhance human functioning as well as supplanting it; it forces us to consider how relationships are transformed across the system and it puts clearly on the agenda the requirements for effective and accountable governance of the next generation of operational systems

Acknowledgements

The authors appreciate the financial support for this research by the EC through the ACROSS project (Contract No ACP2-GA-2012-3141501).

References

ACROSS (2016).ACROSS website. Accessed 20th March, 2016, http://www.across-fp7.eu/

Endsley, M. R. (1988). *Situation awareness global assessment technique (SAGAT).* In Proceedings of the National Aerospace and Electronics Conference. NAECON (pp. 789-795).

Fuller, R. (2000). The task-capability interface model of the driving process. *Recherche - Transports - Sécurité, 66*, 47-57. doi:10.1016/s0761-8980(00)90006-2

Hourlier, S., & Suhir, E. (2014). Designing with Consideration of the human factor: Changing the paradigm for higher safety. Paper presented at the IEEE Aerospace Conference, Yellowstone Conference Centre, Montana.

Lini, S., Bey, C., Hourlier, S., Vallespir, B., Johnston, A., & Favier, P.-A. (2012). Anticipation in aeronautics: Exploring pathways towards a contextualizd aggregative model based on existing concepts. Paper presented at the Human Factors: A view fro an integrative perspective, HFES Europe Chapter Conference, Toulouse, France.

Lini, S., Favier, P.-A., Servantie, X., Vallespir, B., & Hourlier, S. (2011). Developing ASAP (Anticipation Support for Aeronautical Planning). Paper presented at the 16th International Symposium on Aviation Psychology, Dayton, OH

McDonald, N., Morrison, R., Groomes, P. (2007). *Knowledge space model – Human Factors and Operability.* Paper presented at the first CEAS European Air and Space Conference Century Perspectives. 10-13 September, 2007. Berlin, Germany.

Tarnowski, E. (2002): Cockpit Automation Philosophy. Paper presented at the RTO HFM Symposium on "The Role of Humans in Intelligent and Automated Systems", held in Warsaw, Poland, 7-9 October 2002, and published in RTO-MP-088. NATO/OTAN

UNDERSTANDING PILOTS' EXPLANATIONS OF AUTOMATION SURPRISES

Robert Mauro
Decision Research & University of Oregon
Eugene, Oregon
Loukia Loukopoulou
San Jose State University Foundation
Zurich, Switzerland
Julia Trippe
Decision Research & University of Oregon
Eugene, Oregon

Automation surprise may result from inadequate or mistaken "mental models" of the automation (Sarter and Woods, 1995). To study pilots' mental models of their automation, 202 airline pilots were asked to explain five different events involving unexpected behaviors of aircraft automation. Pilots' abilities to correctly explain the behavior of the automation differed systematically across the scenarios. The number of complete and correct responses varied from 19% to 86%, depending on the scenario. As the complexity of the automation increased, understanding decreased. Performance on the scenarios was not related to flight experience, automation experience, or source of automation training. But pilots' conceptions of automation were related to performance on the scenarios. Implications for training are discussed.

In normal operations, the automated flight system of the modern airliner can control nearly all functions required for flight. This automation has greatly reduced problems due to pilot fatigue and other human frailties and has allowed more consistent and precise navigation and performance. However, automation has given rise to new problems caused by faulty interactions between the pilot and the autoflight system (AFS). This class of problems has been variously termed lack of mode awareness (Javaux & De Keyser, 1998), mode confusion (Degani, Shafto, & Kirlik, 1999), and automation surprise (Woods, Sarter, & Billings, 1997, Woods & Sarter (2000), Burki-Cohen, 2010). In these cases, the flight crew expects the automation to command one behavior and is surprised when it commands another. Automation surprise may result from: undetected failures in aircraft sensors or other systems, problematic interfaces that do not provide adequate information about the status of the aircraft (Feary et al., 1998; Norman, 1990, Degani et al., 1999), and inadequate or mistaken pilot "mental models" of the AFS (Sarter & Woods, 1995).

Automation surprises are a nuisance and a source of inefficiency in normal operations. Pilots repeatedly complain that flight management systems misbehave. Although they rarely malfunction, these systems *appear* on occasion to unilaterally decide to "drop" fixes, void altitude restrictions, or change modes of operation. In turn, these events set the stage for other errors and create problems for other aircraft and air traffic as controllers. Indeed, automation surprises have contributed to a number of airline accidents and subsequent loss of life (e.g., Reveley, Briggs, Evans, Sandifer, & Jones, 2010; Sherry & Mauro, 2014). To prevent or mitigate the effects of these "automation surprises" one must first understand why they occur. One must ask why the behavior of the autoflight system was not expected. Based on their training and experience, pilots build an understanding -- a "mental model" -- of how the AFS functions. If pilots' mental models are not completely accurate, situations may arise in which their expectations of how the AFS will behave will depart from reality.

In the research reported here, we sought to examine discrepancies between pilots' mental models of the AFS and its actual functioning by asking pilots to explain events in which a properly functioning AFS surprised the pilots. This technique of asking individuals to explain events is used to assess knowledge in education and training (Lee, Liu, & Linn, 2011) and to assess expert knowledge and decision-making (Hoffman & Lintern, 2006). By comparing the situations which yield complete and accurate explanations to those that do not, one can obtain insight into what aspects of pilots' mental models may be inaccurate and develop interventions to correct those problems.

Method

Airline pilots completed a questionnaire that asked them to explain several different events involving unexpected behaviors of aircraft automation. These events were selected in part based on prior research on pilot reports of automation surprise events (Trippe & Mauro, 2015). Instructions for responding to the scenarios were as follows:

Each of the following scenarios may or may not apply to the type of aircraft that you are presently flying. Whether or not these scenarios apply to your present aircraft, please tell us what you think caused the event described... For each of these scenarios, list as many different possible explanations as you can.

The scenarios were:

- <u>Takeoff Default</u>: *Facing a short runway with a hill ahead, the pilot entered a heading 10 degrees off of the runway heading into the heading window on the Mode Control Panel (MCP). Immediately after takeoff, the pilot engaged the autopilot and then immediately pressed the "HDG SEL" (Heading Select) button on the MCP to turn away from the terrain. However, the aircraft did not turn but continued to climb straight ahead. Why?*

- <u>Altitude Capture</u>: *Air Traffic Control (ATC) issued a clearance to a new altitude. The pilot set the altitude in the MCP "ALTITUDE" window and the aircraft began to climb. As the FMA on the PFD began flashing ACQ (Acquire), ATC issued a new altitude below the altitude that the aircraft was climbing through at that moment. The pilot set the new clearance altitude in the "ALTITUDE" window, but the aircraft failed to level off at any altitude and continued the climb. Why?*

- <u>VNAV</u>: *The pilot properly programmed the descent as required in the STAR and engaged the FMS. However, the aircraft failed to make the crossing restrictions. Why?*

- <u>Runway Change</u>: *While on the transition from the en route segment to the approach, ATC changed the expected runway. The pilot acknowledged the change and entered the change correctly into the FMC. However, the aircraft continued on its heading and failed to join the approach to the new runway. Why?*

- <u>Localizer Intercept</u>: *ATC provided a sharp turn to intercept the localizer. The pilot entered the heading into the MCP and armed the localizer. However, the aircraft flew through the localizer and proceeded to attempt to recapture the course from the other side. Why?*

To make the questionnaire a manageable length, two forms of the questionnaire were used. Both included the "Takeoff Default" scenario. The first questionnaire also included the "VNAV" and "Localizer Intercept" scenarios whereas the second questionnaire included the "Altitude Capture" and "Runway Change" scenarios. The questionnaires also gathered background information on flight experience (e.g., flight time, years of experience, rank, and automation experience), automation training, and perceptions of aircraft systems (navigation, electrical, FMS, crew, engine and hydraulics). The two questionnaires were distributed randomly to an equal number of pilots in each of ten discussion sessions. Thirteen to 44 pilots participated in each session.

Coding Procedure

Pilot responses to the scenarios were categorized as "Complete" if they provided an explanation that described a cause and effect relationship between the conditions given in the scenario and the described actions of the autoflight system (AFS) that could have produced the behavior described in the scenario. Responses were categorized as "Correct" if they gave an accurate cause but did not describe how that cause led the AFS to produce the described behavior. A response was categorized as a "Solution" if it suggested a fix for the situation, but did not describe why it occurred. A response was categorized as a "Non-answer" if it ignored the conditions stipulated in the scenario. Responses were categorized as "Unknown" if they were not interpretable. Finally, responses were categorized as "Wrong" if they gave an implausible cause for the automation behavior, indicated that the pilot didn't know what could have caused the behavior, or gave no response.

Subjects

Two-hundred-and-two airline pilots completed the questionnaire. Twenty-four questionnaires had too much unclear or incomplete data to be used in the analyses. One hundred and seventy-eight questionnaires were used. Of these, 97 were Commanders (CMD) and 81 were First Officers (FO). Experience ranged from 205 to 20,500 hours and 0.2 to 42 years. Usable responses came equally from the two different questionnaire Groups (89 each).

Results

There were 81 First Officers (FO) and 97 Commanders (CMD) in the sample. As expected, First Officers reported fewer total flight hours ($t(176)=16.66$, $p<.001$) and fewer years of experience ($t(175)=13.44$, $p<.001$). On average, First Officers had been flying for 7.36 years (SD 7.43); Commanders had been flying for 22.6 years (SD 7.55). First Officers reported an average of 2418 flight hours (2793); Commanders reported having an average of 10,854 flight hours (SD 3776). Both First Officers and Commanders reported approximately the same recent flight time (FO: 271 hours, SD 98; CMD: 258 hours, SD 84; $t(176)=0.886$, n.s.). The pilots reported having substantial experience with flight in automated aircraft. Commanders reported having spent 88.5% of their flight time in automated aircraft. First Officers reported having spent 83.7% of their total flight time in automated aircraft, (difference by rank: $F(1,176)=3.91$, $p=.049$). Based on their reported hours, this means that on the average Commanders reported having 9,435 hours of flight experience in automated aircraft, whereas First Officers reported having 2,080 hours of flight experience in automated aircraft.

Sources of Automation Knowledge

In general, pilots reported knowing a "moderate" or "large" amount about aircraft automation (Mean 3.30, SD 0.828). Only 4.6% (8) of the pilots reported knowing only a "small" or "very small" amount about automation. When asked where they learned what they know about automation, the pilots reported learning most about automation from initial and recurrent airline training, materials provided by the airline, and their own experience. However, there were some differences between First Officers and Commanders. First Officers reported learning significantly more than Commanders from primary training (Means: CMD: 1.22, FO: 1.68; $F(1,172)=14.60$, $p<.001$), commercial training (Means: CMD: 1.95, FO: 2.47; $F(1,172)=11.01$, $p=.001$), initial airline training (Means: CMD: 3.02, FO: 3.69; $F(1,175)=28.99$, $p<.001$), and other pilots (Means: CMD 2.48, FO: 2.97; $F(1,166)=10.75$, $p=.001$). This may indicate a shift in the content of early training as advanced automation is becoming more prevalent in training aircraft.

Conceptions of Automation

Pilots were asked how reliable, predictable, complex, and understandable (to themselves and to pilots in general) they perceived the aircraft FMS and other aircraft systems to be. In general, pilots demonstrated a strong linear trend in their perceptions of aircraft systems (Linear trend $F(1,165)=94.32$, $p<.001$, $\varepsilon^2=.364$). Hydraulic systems were perceived to be the most reliable, predictable, and understandable, followed by engine systems, electrical systems, navigation systems, and Flight Management Systems. The crew was perceived to be the least reliable, predictable, and understandable. Overall, Hydraulic and engine systems were perceived to be the least complex, but pilots varied substantially in how complex they perceived the other systems to be. In general, Commanders perceived the aircraft systems to be more reliable ($\lambda=.887$, $F(6,169)=3.603$, $p=.002$) and understandable ($\lambda=.014$, $F(6,166)=1918.17$, $p<.001$) than did First Officers. In regards to the FMS, Commanders perceived the FMS to be more reliable (Means: CMD: 5.48, FO: 5.07; $F(1,176)=7.99$, $p=.005$), predictable (Means: CMD: 5.49, FO: 5.18; $F(1,174)=4.77$, $p=.03$), and understandable (Means: CMD: 5.52, FO: 5.20; $F(1,175)=4.17$, $p=.043$) than did First Officers.

Performance on Scenarios

Though encouraged to provide as many explanations for the scenarios as possible, 85% of the pilots provided two or fewer explanations per scenario (Mean 1.5, SD .67). No matter how many explanations they produced, the pilots tended to produce explanations for individual scenarios that were either consistently reasonable or not (see Table 1). More pilots were able to generate "complete" or "correct" responses for the Localizer scenario than for the MCP or Hill scenario. In turn, more pilots were better able to generate acceptable responses for the MCP and Hill scenarios than for the Runway and VNAV scenarios. Performance on the scenarios was not related to flight experience (Rank, years flying, flight hours, recent flight hours $F(4,168)=0.58$, n.s.), automation experience ($F(1,167)=3.26$, $p=.07$)[1], or source of automation training ($F(3,164)=1.61$, n.s.).

Table 1.						
Explanation Quality by Scenario: Percent of Pilots Providing "Correct" Explanations						
Appropriate Explanations		Scenario				
		Localizer	Takeoff Default	Altitude Capture	Runway	VNAV
All	Count	74_b	86_a	44_a	26_c	19_c
	Percent	83.1%	48.3%	49.4%	29.2%	21.3%
None	Count	8_b	57_a	37_a	52_c	54_c
	Percent	9.0%	32.0%	41.6%	58.4%	60.7%
Some	Count	7_b	35_a	$8_{b,c}$	$11_{a,b,c}$	$16_{a,c}$
	Percent	7.9%	19.7%	9.0%	12.4%	18.0%
Total	Count	89	178	89	89	89
	Percent	100.0%	100.0%	100.0%	100.0%	100.0%
	Note: Within each row, cells with different subscripts are significantly different $p<.05$.					

Pilots' conceptions of automation were related to performance on the scenarios (Criterion: total number of responses scored complete or correct; $R^2=.293$, $F(6,166)=11.463$, $p<.001$). Pilots who perceived the FMS as more predictable produced better explanations of the events described in the scenarios ($b=.298$, $t(166)=2.249$, $p=.026$). Given that pilots tended to produce explanations for individual scenarios that were either all "complete or correct" or all not "complete or correct," the set of explanations provided by each pilot for each scenario were categorized as either all "complete or correct" or not all "complete or correct". Using this classification, the effect of pilots' conceptions of the FMS on performance on the individual scenarios was examined using logistic regression. Perceived complexity and predictability of the FMS predicted performance on the VNAV scenario. Taking into account differences in the number of explanations given, pilots who perceived the FMS to be more complex ($b=0.568$, Wald $X^2(1)=4.364$, $p=.037$) and more predictable ($b=1.18$, Wald $X^2(1)=4.896$, $p=.027$) were more likely to provide complete or correct explanations. Perceived predictability also predicted performance on the takeoff scenario ($b=0.590$, Wald $X^2(1)=5.823$, $p=.016$). No other statistically significant relations between pilots' conceptions of automation and performance on individual scenarios were observed.

Discussion

Pilots' abilities to correctly explain the behavior of the FMS differed systematically across the scenarios. As the complexity of the automation increased, understanding decreased. The scenarios may be arranged in order of increasing complexity. The Localizer Intercept scenario is the least complex and it generated the greatest number of "complete" and "correct" responses. For the AFS to intercept the localizer, lateral control coupled to the approach track is required. The logic utilized by the AFS is relatively clear to the pilot and the issues are the same for the automated flight system as they are for a pilot manually executing this maneuver. If a pilot flying manually approaches the localizer course at too great of an intercept angle, too close to the ground station, and at too great a ground speed, the aircraft will overshoot the localizer course unless the pilot deviates from standard procedures by starting the turn early or using a steeper than standard bank angle. Hence, it is easy for the pilot to understand that the AFS must confront the same issues, but may not be able to alter the rules that it was programmed to obey.

The Takeoff Default and Altitude Capture scenarios generated the next greatest number of "complete" and "correct" responses. In both of these cases, understanding the scenario requires retrieving from memory knowledge

---

[1] Reported automation experience was *negatively* related to overall performance on the scenarios, but this effect was not statistically significant at conventional levels.

of AFS logic for which there is no clear manual flight analog. The pilot in the Takeoff Default scenario wants to make an early turn to avoid a potential obstacle. In manual flight there is no limitation on making turns after takeoff. Airline procedures may recommend against making steep banking turns close to the ground, but ATC will routinely instruct aircraft to make early turns and fly offset departures to improve traffic flow. In the Altitude Capture scenario, the aircraft fails to respond to the pilot's command to change altitude because the AFS has transitioned into an "altitude capture" mode and is not responding to additional inputs. There is no manual flight analog for this behavior. In addition, both scenarios involve limitations that are not under the control of the flight crew. Many aircraft have a knob on the MCP that allows the flight crew to change bank limitations as required in the localizer scenario. However, in most aircraft, the altitude at which takeoff track limitations are released is neither annunciated nor adjustable. Sometimes the altitude at which an aircraft transitions to an altitude capture mode is annunciated. But because this altitude varies with the energy state of the aircraft, it is not consistent across flights and it cannot be set by the flight crew.

Pilots provided the fewest number of "complete" and "correct" responses to the Runway Change and VNAV scenarios. Both of these scenarios require knowledge of the operation of the Flight Management Computer. In the previous scenarios, targets and modes could be set directly using the MCP. In these scenarios, the pilots must interact with the AFS through the CDU. To understand the behavior of the AFS in these conditions requires that the pilot have and retrieve knowledge of the manner in which data are stored and used by the FMC and the way in which the FMC uses these data to make complex flight plan calculations. Furthermore, the results of these data manipulations depend on the specific position and energy state of the aircraft relative to the desired targets. For example, in the VNAV scenario the ability of the aircraft to make the desired crossing restriction depends on the position of the aircraft relative to the fix, the aircraft's altitude, ground speed, and energy state. The calculations performed by the FMC to determine whether the restriction will be met are hidden from the pilot. Sometimes a negative result is annunciated, but this is not always the case. To understand the problem in the Runway Change scenario, the pilot must know that approach procedures are typically built off of runways, and hence a change in runway may result in a discontinuity in the flight plan, and that when this occurs the FMC may not resolve the discontinuity and instead command the aircraft to continue on the last good heading. Furthermore, the problem will not occur if the change is made prior to a fix common to both approaches. Thus, pilots may perceive the FMC as pernicious, sometimes succeeding in make a successful runway change, sometimes not. In both of these scenarios, the data used is hidden from the pilot, making it difficult for a pilot to discern the nature of the problem from the information easily available in the cockpit.

Pilot performance on the scenarios was clearly related to the level of automation implicated in each scenario. However, on every scenario except for the Localizer Intercept scenario there was considerable variation between pilots. No measure of flight experience, automation training, or automation experience could explain this variation. There may be factors not ascertained in this study that could explain this variation. However, this result suggests that current automation training is not sufficient to ensure that pilots consistently develop a deep understanding of aircraft automation.

Manufacturers of automated systems have often suggested that automation can simplify pilots' tasks while improving precision and efficiency. However, pilots and researchers have repeatedly noted that while aviation automation has improved the efficiency and precision of operations, it has not reduced complexity. Indeed, automation may have increased the complexity of the pilot's job. Pilots often plead for manufacturers to make the automation simpler. There may be modifications to interfaces that would help simplify pilots' tasks. However, the complexity of the automation itself cannot be substantially reduced. It must be complex because the operational environment is complex and dynamic and the automation has been tasked with operating the aircraft in this environment with minimal pilot intervention.

One response to training complex automation has been to reduce the perceived complexity by limiting the training to the mechanics of executing particular procedures and limiting pilot discretion to the execution of only these procedures. However, this strategy limits pilots' understanding of the automation. When conditions arise that do not correspond to those covered by the trained procedures, the actions of the automation may surprise the pilot. Without a deeper understanding of how the automation operates, pilots cannot be expected to reliably generate explanations of the automation's behavior or to deal with it efficiently. Methods for automation education need to be developed that can help pilots build a *functional understanding* of their automation that allows them to anticipate automation actions and not simply respond with a small set of canned procedures. For pilots to construct adequate mental models of their automation, they do not need to know the intricacies of the underlying engineering, but they must understand how the system interacts with the environment – how it obtains information, what it controls, and what targets it is trying to achieve. Hence, for each automation mode, pilots must be trained to understand: 1) what is being controlled (e.g., pitch, thrust), 2) what data about the current state of the aircraft is being used (e.g., altitude from the Captain's radio altimeter, lateral position from GPS (Global Positioning System)), 3) what targets are being pursued (e.g., altitude from the MCP, speed from the FMC), and 4) what actions will be taken when the targets are achieved or fail to be achieved (e.g., proceed on heading, revert to programmed flight plan). Without this deep understanding, pilots will continue to be surprised by the automation.

References

Burki-Cohen, J. (2010). Technical challenges of upset recovery training: Simulating the element of surprise. *Proceedings of AIAA Guidance, Navigation, & Control Conference*: Toronto, Ontario, Canada, http://dx.doi.org/10.2514/6.2010-8008.

Degani, A., Shafto, M., & Kirlik, A. (1999). Modes in human-machine Systems: Constructs, representation, and classification. *International Journal of Aviation Psychology*, *9,* 125-138.

Feary, M., McCrobie, D., Alkin, M., Sherry, L., Polson, P., Palmer, E., & McQuinn, N. (1998). Aiding vertical guidance understanding. *NASA Technical Memorandum NASA/TM- 1998-112217*, Ames Research Center, Moffett Field, CMD.

Hoffman, R. & Lintern, G. (2006). Eliciting and representing the knowledge of experts. In K. Ericsson, N. Charness, P. Feltovich, & R. Hoffman (Eds.) *Cambridge Handbook of Expertise & Expert Performance*. (pp. 203-222). New York: Cambridge University Press.

Javaux, D., & De Keyser, V. (1998). The cognitive complexity of pilot-mode interaction: A possible explanation of Sarter and Woods' classical result. In *Proceeding of the International Conference on Human-Computer Interaction in Aeronautics* (pp. 49-54). Montreal, Quebec: Ecole Polytechnique de Montreal.

Lee, H., Liu, O., & Linn, M. (2011). Validating measurement of knowledge integration in science using multiple-choice and explanation items. *Applied Measurement in Education*, *24*, 115-136.

Norman, D. A. (1990). The 'problem' with automation: inappropriate feedback and interaction, not 'over-automation.' *Philosophical Transactions of the Royal Society B: Biological Sciences*, *327*(1241), 585-593.

Reveley, M., Briggs, J., Evans, J., Sandifer, C., & Jones, S. (2010). Causal factors and adverse conditions of aviation accidents and incidents related to integrated resilient aircraft control. *NASA Technical Memorandum NASA/TM-2010-216261.*

Sarter, N. B., & Woods, D. D. (1995). How in the world did I ever get into that mode? Mode error and awareness in supervisory control. *Human Factors: The Journal of the Human Factors and Ergonomics Society*, *37*(1), 5-19.

Sarter, N. B., Woods, D. D., & Billings, C. E. (1997). Automation surprises. *Handbook of Human Factors and Ergonomics*, *2*, 1926-1943.

Sherry, L., & Mauro, R. (2014). Controlled Flight Into Stall (CFIS): Functional complexity failures and automation surprises. *Proceedings of the Integrated Communications Navigation and Surveillance Conference,* D1-1. IEEE.

Trippe, J. & Mauro, R. (2015). Understanding Automation Surprise: Analysis of ASRS Reports. Proceedings of the 18th International Symposium on Aviation Psychology, Dayton, Ohio.

Woods, D. & Sarter, N. (2000). Learning from automation surprises and "going sour" accidents. In Sarter, N. & Amalberti, R. (Eds.) *Cognitive Engineering in the Aviation Domain*. LEA: Mahwah, NJ.

TRANSPARENCY AND CONFLICT RESOLUTION AUTOMATION RELIABILITY IN AIR TRFFIC CONTROL

Fitri Trapsilawati
Nanyang Technological University
Singapore
Christopher Wickens
Alion Science and Technology
Boulder, Colorado
Chun-Hsien Chen
Nanyang Technological University
Singapore
Xingda Qu
Shenzhen University
Shenzhen, China

This paper investigates the automation reliability and the transparency in automation conflict resolution advisories for air traffic control. Four general effects: those of traffic load, those of expertise and, those of imperfect automation and its mitigation by automation transparency in the context of the lumberjack analogy were examined. The results showed that the two automation functions, the conflict resolution advisor (CRA) and the vertical situation display (VSD) offer benefits for both novice and professional controllers's performance and increased situation awareness across traffic loads, even when the former is of imperfect reliability.

Automation Conflict Avoidance Aids

The next generation airspace procedures will be coupled with a wealth of new technology and automation tools, in order to accommodate the anticipated 2-fold growth in traffic density (IATA, 2016). One such automation tool of particular interest to our research is the air traffic control conflict resolution aid (CRA), and it is its evaluation that we report here. In the following, we briefly examine conflict avoidance automation tools for ground (ATC) operations. We then describe some of the general principles of human interaction with imperfect automation before presenting a synopsis of three experiments that have evaluated the imperfect CRA.

Conflict avoidance operations can benefit from support for two different predictive automation tools: conflict detection, and conflict resolution aids. On the ground, the air traffic controller's conflict detection tasks are well supported by the automation conflict alert (CA) system (Wickens, Rice, et al. 2009). However the operational controller is not currently supported by the tool corresponding to the airborn Traffic Alert and Collision Avoidance System (TCAS). That is, not supported by an automation-ground based conflict *resolution* advisor, although prototypes have been evaluated (Prevot, et al. 2012; SESAR, 2016).

One feature of all such conflict avoidance automation tools, an inevitable consequence of their functionality of predicting the future which is inherently uncertain (Herdener, et al. 2016) is that they are *imperfectly reliable*, prone especially to generate false alarms (or nuisance alarms), and more so at longer look-ahead times (Dixon & Wickens, 2007). However the TCAS false alarm rate appears generally to be low enough (reliability high enough) so as to still offer considerable benefits; and the same has been found for the CA for controllers (Wickens et al., 2009). In these cases the automation error rate is at a level below a threshold of around 25%, above which assistance is no longer proffered (Wickens & Dixon, 2007).

However it remains uncertain the extent to which an imperfect CRA will offer assistance relative to unaided conflict resolution, because such empirical research does not appear to have been conducted outside our laboratory. Instead the general R&D evaluations have implemented automated resolution advisories that will always increase separation, relative to the trajectory of the uncorrected aircraft (i.e.;, 100% relliablity). Yet because of the extreme complexity and density of the future airspace, it is likely that some such "automation errors" could occur. The pilot, receiving advice from the CRA-assisted controller may receive three categories of such errors: advice to maneuver in a manner that clearly decreases the anticipated minimum separation, advice to maneuver in a different direction or axis than one preferred by the pilot from the standpoint of energy management, fuel consumption or passenger comfort, and advice that avoids an immediate conflict, but now places the aircraft on a trajectory toward a new one. Our implementation of the imperfect CRA employs the third of these categories.

The Lumberjack Analogy

We can place the imperfect conflict avoidance aids (both detection and resolution) within the context of the automation stages & levels taxonomy initially applied to air traffic control automation by Parasuraman, Sheridan and Wickens (2000), and subsequently supported by strong empirical evidence from a meta-analysis carried out be Onnasch Wickens, et al., (2014), who coined the term "degree of automation" (DOA) defining automation that did "more cognitive work" relative to the human operator who is supported by that automation. While the full taxonomy is more complex than space allows here (See Sebok & Wickens, 2016), within the current conflict avoidance framework, we specify two degrees of automation. At a lower degree is conflict detection, a diagnostic or **situation assessment aid** that advises "what is". At a higher degree is conflict resolution, a **decision aid** that advises "what to do". This increase in DOA from SA support to decision support was predicted (Parasuraman, Sheridan, & Wickens, 2000), found (Onnasch et al., 2014) and modeled (Sebok & Wickens, 2016) to (a) *improve* nominal performance and reduce workload when automation functioned correctly, but (b) lead to *greater* problematic, and sometimes catastrophic consequences on the infrequent occasions when automation failed (or failed to operate as expected by the human supervisor). In terms of the lumberjack analogy: "the higher the tree, the harder it falls".

One of the key features revealed by the meta-analysis carried out by Onnasch et al (2014) is that the greater problematic response to automation failures, with the higher degree of imperfect automation, was paralleled by a loss of situation awareness in those circumstances. This loss is triggered by being more "out of the loop" in decision automation which enables automation to select or advise actions, compared to SA-support automation that still forces the operator to actively choose actions. Such active choices better implant the state of the system in the operators's memory, i.e., an increaese of situationn, via a phenomenon known in memory theory as the "generation effect" (Slamecka & Graf, 1989; Hopkin, 1995).

The final piece in our puzzle and basis of our current predictions, is that, if SA is lacking with decision support automation, it can be restored by effective automation **transparency**, or displays that provide more graphic information about the current state of the environment from which automation draws its action recommendations (Bizantz & Seeong, 2008; Mercado et al., 2016). Thus our argument in the current project is that, to the extent that controller-CRA interaction is hindered by the occurrence of occasional imperfections or automation errors (a prediction we expect to confirm), this problematic response can be mitigated by a display supporting controller situation awareness. What then should this display be? In the typical ATC console, the controller is well supported in lateral awareness by the "radar display" or plan view display (PVD). But less so in vertical (altitude) awareness because most information about altitude and relative altitude is depicted in symbolic digital data tags, a less than ideal way of conveying trend information about the relative altitude of multiple aircraft. Hence our transparency mitigation was designed to provide controllers with a *vertical situation display* (VSD), a concept receiving substantial research in the flight deck CDTI (e.g., Battiste & Johnsons, 2002; Thomas & Wickens, 2008), but less so in ATC (SESAR, 2013). In particular, to our knowledge, no research has been carried out joining the two automation concepts of the VSD and the CRA, let alone in circumstances in which the CRA is imperfect. Our program of research does this.

In the three experiments described below, we first show that the CRA can assist resolution performance, and can even do so when it is imperfect, relative to fully manual performance. We do this with modest traffic load (experiment 1; Trapsilawati et al., 2015) and then with much higher traffic load (experiment 2; Trapsilawati, et al., 2016) evaluating the greater dependence on the CRA in the latter conditions. Because both of these experiments are published, we only describe them briefly here. Then in experiment 3, we evaluate the possible mitigation effectiveness of transparency provided for some participants by the VSD to support the human response to automation failures, within the framework of the lumberjack analogy. Because we do not examine conflict detection aids here, our tree is always high (decision aiding); we document its fall, but also show that we can lessen the impact of the fall with the VSD.

Methods

All three experiments involved the same general simulation and methods, described in some detail in Trapsilawati et al., 2015, 2016, and only briefly here. Participants, either students within the Aeronautical and Aerospace programs, or professional controllers viewed the TRACON display in an NLR ATC simulator (NARSIM). They were responsible for moving traffic through the sector, and avoiding loss of separation (conflict avoidance). During a typical 1 hour session, 5 conflicts would be imposed, at unpredictable times, leading to an LOS if not control action was taken. This action was implemented by a voice input (e.g., "change heading to 030") and carried out by a pseudopilot, where the changed trajectory would be then visible on the display.

The four experimental sessions differed from each other in terms of the automation support offered by a CRA. In this regard, the CRA was either absent (manual performance only) present and fully reliable, or present and "imperfect" such that one of the advisories directed an aircraft to change trajectory and avoid an immediate conflict, but in the process, created a predicted conflict with a second aircraft. The latter predicted conflict did NOT trigger advice from the CRA. As such erroneous advice occurred in one trial out of 5 in the imperfect automation block, the overall CRA reliability could be said to be 80%; although prior to the first time a failure was observed in the imperfect session block, the controller would experience it as having 100% reliability, since no failures were imposed during the training blocks. This first failure will be particularly relevant to our evaluation of support for the lumberjack analogy. Participants were free to comply with or ignore the advice of the CRA if they felt that an alternative maneuver was preferable. During each session, participants were periodically probed with a SPAM situation awareness question regarding the current status of the airspace (Durso & Dattel, 2004). The latency to respond to the ready probe assessed overall workload (OWL), and the accuracy measure of the probe response assessed SA.

In all experiments, a generic TRACON space was employed. In Experiment 1, employing 12 controllers who were primarily students, traffic density was 30 aircraft per hour In Experiment 2, employing 24 participants, again primarily students, traffic density was increased to either 60 or 90 (between groups) to simulate the projected growth of airspace congestion that would benefit more from automation assistance. In Experiment 3, employing exclusively professional controllers, in which the VSD was imposed, traffic density was set at a constant level of 60 aircraft. In the following we refer to students as "novices" and to professional controllers as "experts".

Results

Figure 1 shows, on the X axis all three experiments juxtaposed, with the three automation conditions along the X axis defining the shape of each line. The relative scale of each of these three dependent variables (performance, top; situation awareness, middle; OWL, bottom), is arbitrary as each has been transformed so that they show minimum overlap within the figure. The important factors are the shape of the profiles of each 3-point line, and the relative position of the three profiles across experiments. These relative positions are connected by dashed lines. The following general observations can be made:

Differences, due to Traffic Load, between Experiments 1 and 2

Experiment 2, with its higher traffic load shows an overall reduction of performance compared to Experiment 1 ($p= 0.02$). However, the reduction of SA ($p= 0.12$) and the increased workload ($p= 0.63$) were not significant. Experiment 2, with greater traffic load shows OWL to be greater in the manual condition than with automation. Stated in other terms, in Experiment 2, with its higher traffic load, in fact, there is a greater benefit of CRA automation to reducing workload, whether the CRA was reliable or not, and the CRA automation in Experiment 2 actually **restores** workload to a level equivalent to that of the lower traffic load in experiment 1, as indicated by the significant interaction between the automation condition and experiment/traffic load ($p= 0.04$).

Differences in Profiles between Experiments 1&2 (Novices) and 3 (Experts)

To allow for direct comparison between novices and experts, we did the analysis between Experiment 2 with medium traffic condition (novice participants) vs Experiment 3 without the VSD condition (expert participants) where the air traffic loads were similar. We found that overall performance was not significantly different between novices and experts ($p= 0.36$). However, the interaction effect was significant ($p= 0.03$), showing much better performance of experts than novices in the manual condition. Novices' overall SA was marginally higher than that of the experts ($p= 0.08$). However no difference of SA was found across automation conditions for either novices or experts (p= 0.20). The experts' workload is considerably lower than novices although the trend was not significant (p= 0.14).

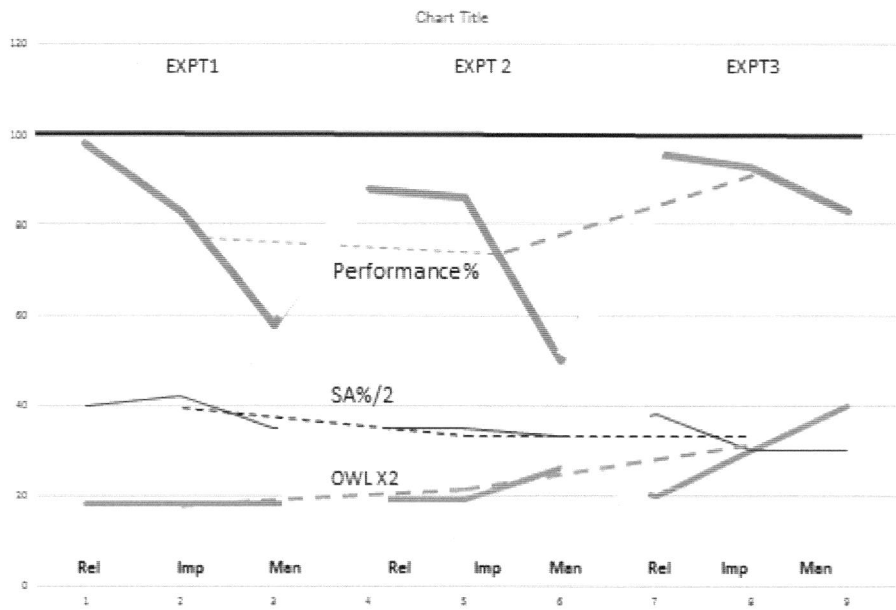

Figure 1 Results of the Three Experiments. The three color coded lines of different width within each experiment define each of the three critical independent variables; Performance (% resolved conflicts) at the top, Situation awareness (% correct) in the middle, and Objective workload (OWL: ready response latency) on the bottom.

The Examination of Lumberjack Analogy based on Data of Experiment 3

The presence of the VSD slightly improved conflict resolution performance (from 89% to 94%, $F(1, 18) = 1.35$, p= 0.26), substantially increased situation awareness (from 59% to 73%; $F(1, 16) = 4.13$, p= 0.059) and significantly lowered workload (from 7.78s to 5.38s; $F(1, 14) = 8.57$, p= 0.01). The VSD was found to have equivalent effects across all three automation conditions (i.e., no interaction between automation condition and display).

On the first failure trial, for the block in which the CRA was unreliable; performance accuracy was compared with all correct trials, in which automation was functioning correctly. Here the accuracy for the four combinations of automation accuracy and VSD support is shown in Table 1. Examining these data, we observed what could be interpreted as a significant interaction effect, in that a test of proportions revealed a substantial significant decrement of the 25% reduction without the VSD (Z= 2.36, p= 0.02), but no significant effect (Z= 1.08, p= 0.28) of the 7.5% decrement when the controller was supported by the VSD.

Table 1.
The First Failure Analysis.

Display Conditions	Automation Correct Trials	Automation Failure Trial
VSD	97.50%	90.00%
No VSD	95.00%	70.00%

Discussion

Overall the results allowed us to examine three general effects: those of traffic load, those of expertise and, most critically, those of imperfect automation in the context of the lumberjack analogy.

Workload/Traffic Load

In comparing experiment 1 (low density=30) with Experiment 2 (high density= 60 or 90), both using primarily novice controllers, we found that increasing density produced a decrease in performance, a trend of loss in situation awareness and only a very slight increase in objective workload. We might argue that, when these novice controllers confronted the high traffic density, their performance went over the "red line" of workload (Grier, 2008), which could not be rated higher (they were "maxed out"; and hence could give no more resources), even as the gap between resources demanded and those supplied increased, hence lowering performance. At maximum capacity in Experiment 2, the novices also diminished any resources available for maintaining SA. Hence there was a trend of SA degradation.

Controller Expertise

In comparing the overall results of Experiment 2 with those of Experiment 3, both at medium-high workload/traffic load levels, the most obvious difference is the increase in performance of the experts (Experiment 3), particularly when controlling manually (without CRA automation assistance). This is not surprising. Experts generally are better performers. This increase was attained with no change in workload, but with a marginal loss in SA, an effect that is somewhat surprising.

The Lumberjack Analogy

To examine the lumberjack analogy, we focus attention at greater depth only on the performance of the experts in Experiment 3, as this performance is most generalizable to the real world of air traffic control and only hee can we examine the mitigating effoects of the VSD. Here we find, as with the first two experiments, a benefit of automation, although this benefit was reduced, given the higher baseline level of manual performance of the experts in Experiment 3. Somewhat unexpectedly, we also found an increase in situation awareness with automation, contrary to the standard "folk lore" of automation (Sebok & Wickens, 2016) in which automation is assumed to produce an out-of-the-loop unfamiliarity syndrome, breeding complacency, dependency and "the automation bias" (Mosier & Skitka, 1996), and mediated by a **loss,** not a gain of situation awareness. In accounting for this departure from our expectations, we assume that, unlike some other cases, our expert professional controllers invested any resources saved by the CRA decision aid, into deeper processing the raw data from the display.

Insofar as the lumberjack analogy itself is concerned, we have partial support for its expression. On the one hand, experts did not perform significantly worse overall with imperfect (80% reliable), than with perfect automation blocks, even though there was a non-significant trend in that direction. On the other hand, on the single (and first) failure trial, they did perform worse, with a detection rate, when unsupported by the VSD that dropped from 95% (on the correct trials) to 70%. We also found that this problematic failure cost was mitigated by the automation transparency provided by the VSD relative to the control group. The former showed only a small (7.5%) non-significant loss of performance on the failure trial, while the latter showed a large loss of 25%. Finally, we ask if this failure recovery difference between the two groups was mediated by a difference in situation awareness. Here the interpretation is again ambiguous. On the one hand the VSD **did** substantially improve SA. But on the other hand, such an improvement was equally manifest on both manual trials and on those supported by perfect automaton. Hence we cannot infer that the differential performance improvement was associated with a differential increase in SA.

The ambivalence of theoretical interpretation notwithstanding, we can conclude with certainty that the two aspects of technology examined here, the automation of the CRA, and the SA support of the VSD are both of benefits to professional controllers, even when the former is of imperfect reliability.

Acknowledgements

This work was supported by grant No. [ATMRI:2014-R5-CHEN] from Civil Aviation Authority of Singapore (CAAS) and Air Traffic Management Research Institute (ATMRI), Nanyang Technological University (NTU), Singapore. We would like to thank Mr. Erik-Jan Hartlieb from Netherland Aerospace Lab for the technical assistance. The research presented in this paper was based on Fitri Trapsilawati's doctoral dissertation at the NTU,

Singapore. Any opinions, findings, conclussions, recommendations expressed in this paper are those of authors and do not reflect the views of the employers or granting organization.

References

Battiste, V., & Johnson, N. H. (2002). An operation evaluation of ADS-B and CDTI during airport surface and final approach operations. In *Proceedings of the Human Factors and Ergonomics Society 46th Annual Meeting* (pp. 36-40). Baltimore, MA: Human Factors and Ergonomics Society.

Dixon, S. R., Wickens, C. D., & McCarley, J. S. (2007). On the independence of compliance and reliance: Are automation false alarms worse than misses? *Human Factors, 49*, 564-572.

Durso, Frank T, Dattel, Andrew R, Banbury, S, & Tremblay, Sebastien. (2004). SPAM: The real-time assessment of SA. In S. Banburry & S. Tremblay (Eds.), *A cognitive approach to situation awareness: Theory and application* (Vol. 1, pp. 137-154). Vermont: Ashgate Publishing.

Grier, R. (2008). The redline of workload: Theory, research and design. A panel. In *Proceedings of the Human Factors and Ergonomics Society 52nd Annual Meeting* (pp. 54-58). Santa Monica, CA: Human Society.

Herdener, N., Wickens, C. D., Clegg, B. A., & Smith, C. A. P. (2016). Overconfidence in projecting uncertain spatial trajectories. *Human Factors, 58*, 899-914.

Hopkin, V.D., (1995) *The Human Factors of Air Traffic Control*. London: Taylor & Francis.

IATA. (2016). IATA Forecasts Passenger Demand to Double Over 20 Years: International Air Transport Association. Retrieved from http://www.iata.org/pressroom/pr/Pages/2016-10-18-02.aspx

Mercado, J. E., Rupp, M. A., Chen, J. Y., Barnes, M. J., Barber, D., & Procci, K. (2016). Intelligent Agent Transparency in Human-Agent Teaming for Multi-UxV Management. *Human Factors, 58*, 401-415.

Mosier, K. L, & Skitka, L. J. (1996). Human decision makers and automated decision aids: Made for each other. *Automation and human performance: Theory and applications*, 201-220.

Onnasch, L., Wickens, C. D., Li, H., & Manzey, D. (2014). Human Performance Consequences of Stages and Levels of Automation An Integrated Meta-Analysis. *Human Factors, 56*, 476-488.

Parasuraman, R., Sheridan, T. B., & Wickens, C. D. (2000). A model for types & levels of human interaction with automation. *IEEE Transactions on systems, man, & cybernetics-Part A: Systems and Humans, 30*, 286-297.

Prevot, T., Homola, J. R., Martin, L. H., Mercer, J. S., & Cabrall, C. D. (2012). Toward Automated Air Traffic Control—Investigating a Fundamental Paradigm Shift in Human/Systems Interaction. *International Journal of Human-Computer Interaction, 28*, 77-98.

Sebok, A., & Wickens, C. D. (2016). Implementing Lumberjacks and Black Swans Into Model-Based Tools to Support Human–Automation Interaction. *Human Factors*, 59, doi: 10.1177/0018720816665201.

SESAR. 2013. Final Project Report on the concept and benefits for improving TP using AOC data: Improved Airline Flight Plan Information into ATC Trajectory Prediction (TP) Tool. SESARJU Report. Retrieved from http://www.sesarju.eu/sites/default/files/ solutions/3_AOC_Data_for_TP_Final_Project_Report.pdf?issuusl=ignore.

SESAR. (2016). Automated support for conflict detection, resolution support information and conformance. ESSIP Plan Edition 2016. Retrieved from https://www.eurocontrol.in t/sites/default/files/content/documents/official-documents/reports/atc12-1.pdf

Slamecka, N. J., & Graf, P. (1978). The generation effect: delineation of a phenomenon. *Journal of experimental Psychology: Human learning and Memory, 4*, 592.

Trapsilawati, F., Qu, X, Wickens, C. D., & Chen, C.-H. (2015). Human factors assessment of conflict resolution aid reliability and time pressure in future air traffic control. *Ergonomics, 58*, 897-908.

Trapsilawati, F., Wickens, C. D., Qu, X., & Chen, C. H. (2016). Benefits of imperfect conflict resolution advisory aids for future air traffic control. *Human factors, 58*, 1007-1019.

Thomas L.C., & Wickens, C.D. (2006). Display dimensionality, conflict geometry, and time pressure effects on conflict detection and resolution performance using cockpit displays of traffic information. *International Journal of Aviation Psychology, 16*(3), 321-342.

Wickens, C. D, & Dixon, S. R. (2007). The benefits of imperfect diagnostic automation: A synthesis of the literature. *Theoretical Issues in Ergonomics Science, 8*, 201-212.

Wickens, C. D., Rice, S., Keller, D., Hutchins, S., Hughes, J., & Clayton, K. (2009). False alerts in air traffic control conflict alerting system: Is there a "cry wolf" effect? *Human Factors, 51*, 446-462.

INTELLIGENT SYSTEMS APPLIED TO WORKLOAD MONITORING AND TASK ALLOCATION

Scott Grigsby, Jacob Crossman, Rich Frederikson, Ben Purman, and Dylan Schmorrow
Soar Technology, Inc.
Ann Arbor, MI

We have implemented an integrated system, called Computational Situation Awareness (CSA), into a prototype control station to coordinate dynamic task sharing and allow UxV operators to be quickly brought up to speed with sufficient situation awareness to effectively understand and handle new tasking. Within CSA we have implemented a real-time workload model for estimating workload across a user's visual, tactile, and cognitive resources. This workload estimate enables the system to recognize when tasks should be handed off and to whom they should be handed. By tracking user tasks, user workload, and system state, the system builds an understanding of the team's effectiveness and current capabilities and tracks a task's progress within the team. This understanding allows the system to determine which team member can best perform each task and to determine the information each operator will need to obtain or maintain SA without unduly increasing workload.

Team coordination during unmanned systems (UxS) operations is relatively simple when an intelligence, surveillance, reconnaissance (ISR) mission is proceeding normally; however, when something goes wrong (e.g., vehicle control issue) or an interesting event occurs (e.g., new contact or threat), individual human team members become bogged down and can quickly be overwhelmed (Salas, Rosen, Burke, Nicholson, & Howse, 2007). **Dynamic task sharing** – the ability to quickly re-assign tasks and responsibilities - vastly improves team coordination. Task assignment can be traded off between the UV autonomy and multiple operators based on mission context.

Two elements are required for effective task sharing and handoff: **workload assessment** and **situation awareness** management. Workload assessment enables the system to recognize when tasks should be handed off, and to whom they should be handed (Breslow, Gartenburg, McCurry, & Trafton, 2014) (Solovey, Zec, Garcia-Perez, Reimer, & Mehler, 2014). SoarTech has developed a workload model for assessing real-time workload across a user's visual, tactile, and cognitive resources and has applied this model to the problem of automated offloading of tasks from one user to another.

Approach

The basic architecture for workload estimation is illustrated in Figure 1. Events and state data (most importantly, tasks) are pulled from the internal operating picture (IOP). For each event and task, the system looks up the workload model for that event and reads the expected instantaneous workload value for that event at the given time. The basic workload assessment capability uses a 3-dimensional model to estimate workload in the cognitive, visual, and manual dimensions (these three dimensions are taken from Wickens (2008), but here the auditory mode is not incorporated to simplify the model for initial implementation).

The equation used to compute the overall workload is as follows:

$$W(t) = \sum_{i=1}^{N} max_{k=1,2,3}\{w_k(t-t_i)\} + g(t) \quad (1)$$

$$g(t) = 0.1 \times 1.25^{N} \quad (2)$$

Here $W(t)$ is the estimated workload computed as the accumulation of the maximum of the component (e.g., cognitive, visual, manual) workloads for N events/tasks. Component workloads are computed as functions of time, where each t_i is the start time for the ith event or task. The $g(t)$ component approximates the non-linear effects caused by multiple simultaneous active events (attention switching). Its constant values were tuned during laboratory tests such that workload values approximated to 1.0 when a user appeared to be overloaded. g(t) presents some problems as it scales exponentially over the whole range of N, which is not realistic and can overestimate the workload for high Ns. It works reasonably well for $N < 15$, but provides excessive estimates. A future version will likely replace the exponential with a sigmoid function to model a diminishing effect as N increases to large numbers.

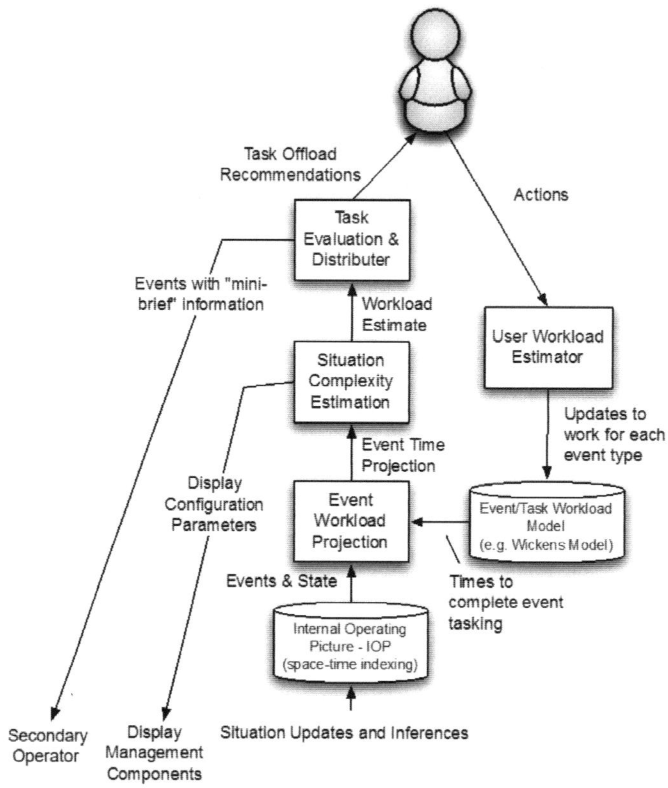

Figure 1. Workload estimation architecture.

This model is similar in concept to the Wickens' (2008) model, but differs in that it scales to an arbitrary number of tasks and events and accounts for change in workload over time such as temporal delays, workload ramp up, and workload ramp down. Our model is also designed to be more precise in that it seeks to derive a number that can be compared to the user's full load level (e.g., 1.0 = fully loaded).

The most important elements of the workload calculation are the workload models associated with each event and task. Figure 2 illustrates the workload model concept.

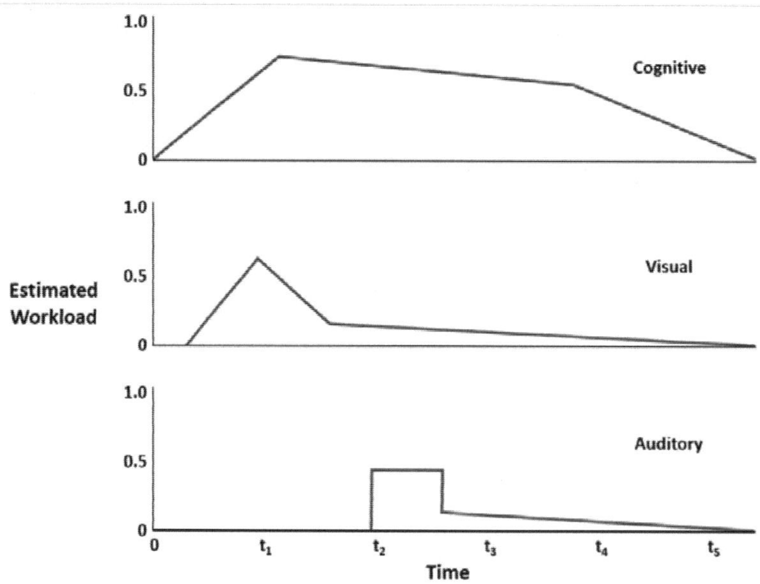

Figure 2. Sample time courses of workload estimates for a single event/task.

For each event, we define three curves, where the x-axis is time and the y-axis values are the estimated workload along the given dimension, at the given time. Each of these curves is defined with the start of the event defined as x = 0. For our model, we defined these curves as piecewise linear functions with four linear regions: (1) the delay time (time to start the effect), (2) the ramp up time (time over which the effect rises to its maximum value, (3) the peak duration, and (4) the ramp down time. Workload profiles were pre-defined for all events and all tasks within the scenario.

The expected instantaneous workload values are aggregated, using Equation 1 (abaove), to form an overall estimated workload. This value is fed into the task evaluation and distribution function where decisions on task distribution are made. When the workload is estimated to be above a single operator's capacity to execute effectively, events and tasks are offloaded to another operator. However, the decision when to offload is not straightforward. We found in our discussions with users during the evaluation that there are several ways tasks can be divided within a team.

1. They can be broken up purely based on their **timing** (e.g., round robin distribution, or secondary user gets all tasks after the main user becomes overloaded).
2. They can be broken up based on **load balancing** (e.g., the choice of which user gets which tasks is made so as to minimize user workload differences).
3. They can be made based on **geography** (e.g., each user takes care of a geographic segment)
4. They can be made based on **asset ownership** (e.g., each user gets events and tasks associated with the assets that he/she owns).

In practice, during our evaluation, operators used more than one of these approaches when they were given a choice.

We only had time and resources to select one offloading method for our initial implementation. We selected a variant of (2) for its simplicity. The variation was that the algorithm tries to load users one at a time – meaning it will not balance the load until at least one user is fully allocated. The idea behind this strategy is that it allows the secondary operator(s) to focus on other tasks while the primary operator is able to do the task alone. The algorithm is as follows:

```
1. Assign all unassigned events to the user with the highest workload (this is the primary user).
2. If this user's workload estimate is < 1.2 (combined) then stop, otherwise --
3. Find the maximum workload event assigned to the high workload user that has occurred in the last 30
   seconds. (i.e. offload the event that is the biggest component of the workload.
4. Assign this event to the lower workload user if (event.workload < 0.9 ( high_user.workload -
   low_user.workload ))
```

We note a couple of key points regarding this algorithm. First, events/tasks are not offloaded until the workload level reaches 1.2. This allows for the lack of precision in the workload model numbers. They are not intended to be correct to a single decimal place. Second, it moves tasks to other users from largest to smallest. The idea is to take the major focus tasks away from the primary user, so the primary user can continue to maintain global SA. In the future, task allocation will incorporate user role and other factors to best determine which tasks should be distributed. It may be that a task needs to stay with an already high-workload operator simply because it would lead to the best opportunity for mission success. These aspects would need to be incorporated into the world model of the task allocation system.

Evaluation

The workload and offloading evaluation was part of a bigger evaluation study looking at operator effectiveness and situation awareness within a multi-UxS scenario. The scenario had both a Primary and a Secondary operator handling the various tasking. The evaluation was executed as a comparative study using a single control (or base) case and a single evaluation case. Both cases used the same scenario design and simulation and the same core user interface (RaptorX). Communication between operators occurred only via a chat window. The independent variable for the study and the difference between the two cases was presence or absence of the workload estimation and offloading capabilities (which were presented to the user as extensions of the Raptor X capability).

Overall, the operators were responsible for monitoring activity at a virtual port. They were given initially 4, and as the mission progressed as many as 6 UAVs to execute the mission. These assets could be used to obtain tracks of vessels and ground vehicles in and around the port. Each track was associated with a set of property data (e.g., location, size, cargo), the completeness of which is dependent on the quality of the sensor reading on the vehicle.

To achieve the mission objectives the user could do two primary things: command UAVs and inspect/update information about tracks. The user could also execute supporting actions that help make decisions and commands easier. In addition to the main tasks, various situations can arise that require operator attention (communication failure, low fuel, navigation failure, air space breach, etc.). In many cases the operator has freedom in the details of the action taken. During the course of the scenario, tasks would be offloaded automatically per the workload and task allocation algorithms but also the Primary operator could manually offload tasks to the secondary operator when desired.

Results

An interesting aspect of the evaluation was that when using the system, users did not consciously notice that automated task offloading was occurring nor did they subjectively report decreased workload. Nevertheless, it influenced how they operated and allowed them to perform more efficiently due to how it affected asset and event ordering and display in the user interface. For example, while objective measures of mission performance (time to identify targets, target tracking, etc.) showed moderate improvement, we found that the primary operator's interactions with the system was significantly reduced when task allocation was implemented from an average of 375 down to 240 ($p<0.05$, 2-tailed T-test) and, at the same time, the secondary operator's interactions stayed roughly the same (215 vs 210) (Figure 3). Furthermore, while the user's subjective workload remained virtually the same, the system's estimated workload for that user, based on our tasking estimation model, showed that the overall estimated workload of the primary operator decreases when the system is used (Figure 4).

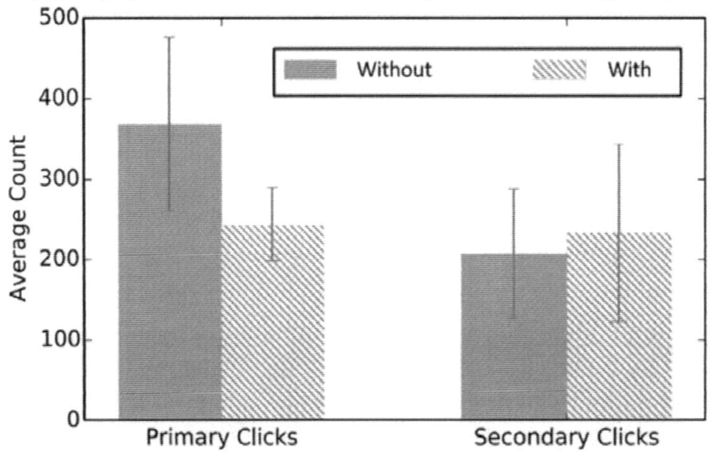

Figure 3. Average user interactions within the scenario.

So, while the specific workload numbers may or may not accurately reflect the user's subjective workload, because workload is estimated from the set of active tasks and events, this indicates that the primary user has fewer events high in his/her queue and that the offloading is actually functioning. As we see, the estimated workload when not using the system centers around 1.0, meaning that our workload calculations estimate that the primary user is almost always fully or nearly fully loaded. However, when using the system, the workload centers around about 0.6 indicating that the systems computes the operator's workload has reduced by about 0.4. This shows that events and tasks are being offloaded even if the effect is not easy to assess.

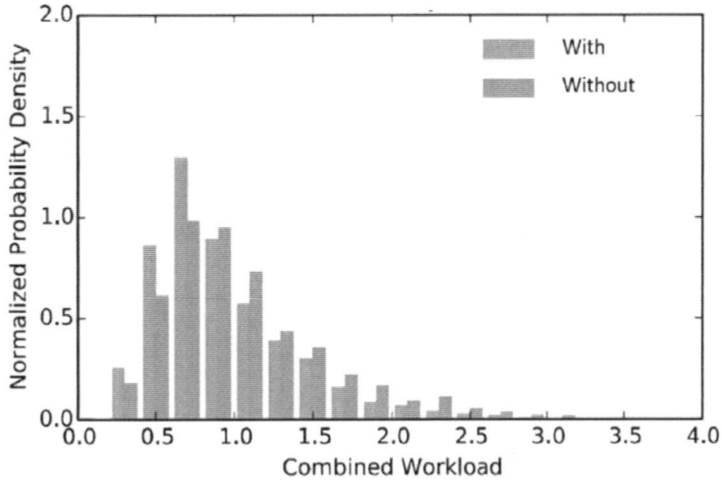

Figure 4. Distribution of Primary Operator Workload

While these results are encouraging, our workload system requires model tuning to be accurate/effective and during this evaluation, we did not have the time to go through multiple tuning/adjustment phases. Thus, our overall results reflect a first pass at modeling workload that we believe would improve significantly if further work were done, given the data collected. For example, the workload functions (*W*(t) in Equation 1) could be adjusted, at least for the manual component, based on the data we collected in the evaluation. Because we did not know how users would like to use the offloading capability, we structured the evaluation to allow the user significant freedom on how this was done. The users were allowed to offload their own events and assets if they wanted. Both users were allowed to task any assets they wanted to (leading, in some cases, to conflicts that had to be resolved via chat).

Furthermore, discussions after the evaluation led us to conclude that the offload method/algorithm should potentially be selectable by the user. That is, the system should implement multiple algorithms that the user can select among based on the mission and mission state.

Discussion and Future Directions

Our most important finding from our system evaluation is that the overall system appears to significantly reduce the manual interaction and intra-team interaction required for ISR missions using multiple UAVs. We hypothesize this effect is caused by (1) a reduction in the scan/check process required to maintain situation awareness and (2) better targeted tasking (few tasks issued, better sensor coverage). Though performance was improved and throughput was increased, subjectively, users did not report lower workload. This suggests that users were trading one type of work for another. In this case, we believe that the task allocation system reduced the user's manual workload allowing the primary user to focus more attention on cognitive and visual aspects of the problem. Our hypothesis is supported by the fact that operators using the enhancements were able to execute the most challenging parts of the mission, ground target tracking, more often and more effectively.

Despite the operational improvements, it was clear from our analysis that workload estimation and offloading requires additional refinement, especially to the workload model. This could be accomplished by using the evaluation data as a basis for model tuning/learning. For example, we could model each workload function as a probability distribution and learn the parameters of these distributions using user activity.

In addition, our current workload models are based only on *a priori* estimation of task workload and the combinatorial workload associated with multiple tasks. New learning models would be most effective if combined with physiological sensors that provided objective real-time workload measures such as heart rate variability (HRV) and pupillometry. For this approach, we would seek to leverage the considerable research in operational neuroscience (Schmorrow, Estabrooke, & Grootjen, 2009) and neuroergonomics (Parasuraman, 2003) which has shown the potential of real-time assessment of workload and situational awareness from behavioral and neurophysiological sensing. This physiological data is becoming surprisingly easy to measure. Several existing commodity fitness trackers are comfortable to wear for long durations while providing accurate measures of heart rate, galvanic skin response, respiration rate, and body temperature. In addition, other measures, such as eye tracking tools, are also becoming much more cost effective. This type of data holds promise to enhance real-time workload estimation by providing excellent data for offline training of workload models and real-time closed-loop feedback about user workload. This will improve the *a priori* task estimates over the long term and enable the estimation models to adapt to the specific user and operating conditions.

Acknowledgements

This research was sponsored by a Small Business Innovation Research (SBIR) award through NAVAIR PMA (contract #N68335-14-C-0015). The views and conclusions contained in this document are those of the authors and should not be interpreted as representing the official policies, either expressed or implied of the U.S. Navy or U.S. Government. The U.S. Government is authorized to reproduce and distribute reprints for Government purposes notwithstanding any copyright notation herein.

References

Breslow, L., Gartenburg, D., McCurry, J.M., & Trafton, J.G. (2014). Dynamic operator overload: A model for predicting workload during supervisory control. *IEEE Transactions on Human-Machine Systems 44*(1), 30-40.

Parasuraman, R. (2003). Neuroergonomics: Research and Practice. *Theoretical Issues in Ergonomics Science 4*(1-2), 5-20.

Salas, E., Rosen, M., Burke, C.S., Nicholson, D., & Howse, W. (2007). Markers for enhancing team cognition in complex environments: The power of team performance diagnosis. *Aviation, Space, and Environmental Medicine 71*(1 supple), B77-B85.

Schmorrow, D.D., Estabrooke, I.V., & Grootjen, M. (2009). *Foundations of Augmented Cognition. Neuroergonomics and Operational Neuroscience 5th International Conference.* Berlin Heidelberg: Springer-Verlag.

Solovey, E., Zec, M., Garcia-Perez, E.A., Reimer, B., & Mehler, B. (2014). Classifying driver workload using physiological and driving performance data: Two field studies. *Proceedings of the SIGCHI Conference on Human Factors in Computing Systems*. Toronto, Canada: ACM.

Wickens, C.D. (2008). Multiple Resources and Mental Workload. *Human Factors 50*(3), 449-455.

THE OCULOMETER TRAINING TAPE TECHNIQUE: THE REVIVAL

Carolina Diaz-Piedra
Mind, Brain, and Behavior Research Center; University of Granada
Granada (Spain)
Andres Catena
Mind, Brain, and Behavior Research Center; University of Granada
Granada (Spain)
Luis J. Fuentes
Department of Basic Psychology and Methodology; University of Murcia
Murcia (Spain)
Leandro L. Di Stasi
Joint Center University of Granada - Spanish Army Training and Doctrine Command
Granada (Spain)

The oculometer training tape technique (OT^3) aims to enhance aviation training by 1) allowing the flight instructors to provide real-time feedback, 2) improving debriefing sessions by playing back the trainee pilot's scan behavior, and 3) editing didactic videos based on the scan behavior of experienced pilots.

Despite the original positive evaluations of its usefulness, the OT^3 has failed to gain traction in aviation training programs. This is probably due to the technical difficulties as well as the intrusiveness/bulkiness of the equipment needed. Modern non-intrusive eye trackers, integrated with a forward facing scene-camera, can record pilots' eye movements and, simultaneously, capture what the pilot sees.

Here, we describe the implementation of an updated OT^3 and its potential benefits to the aviation training programs of the Spanish Armed Forces.

Since the pioneering studies of Jones and colleagues with aircraft pilots (Jones, Milton, & Fitts, 1949), eye movement recording techniques have represented one of the most reliable tools to improve aircraft instruments/panels design (e.g. Gainer & Obermayer, 1964) and to study pilots' biobehavioral states (Di Stasi, McCamy, et al., 2016). As flying is a complex perceptual task that requires, not only conceptual knowledge, but the skills to visually search for relevant information, eye tracking technology may be also used to enhance pilot training (Diaz-Piedra, Rieiro, et al., 2016). In the 70's and the 80's, NASA and US Armed Forces researchers (e.g. Barnes, 1970) already developed applications of the eye tracking technology consisting of using the pilot's scan behavior as an instructional aid (Figure 1): the oculometer training tape technique (OT^3) (Spady, Jones, Coates, & Kirby, 1982). The OT^3 objectively measures the eye positions while pilots are performing flight tasks. It provides information on the pilot's scan behavior (both direction of the gaze and fixation time). With this information, the OT^3 aims to enhance aviation training (Wetzel, Anderson, & Barelka, 1998) by 1) allowing the flight instructors to provide real-time feedback about the observed trainee pilot's scan behavior, 2) improving debriefing sessions by playing back the trainee pilot's scan behavior, and 3) editing didactic videos based on the scan behavior of experienced pilots.

Despite the original positive evaluations about its usefulness (Dennis H. Jones, Coates, & Kirby, 1982; Spady et al., 1982; Wetzel et al., 1998), the OT^3 has failed to gain traction in aviation training programs. This is probably due to the technical difficulties of recording eye movements, and the intrusiveness/bulkiness of the equipment needed (Di Stasi, McCamy, et al., 2016). In recent years, user-friendly, commercial, and portable eye trackers– e.g. located on lightweight eyeglass frames – have overcome many of these barriers. These non-intrusive devices, integrated with a forward facing scene-camera, can record pilots' eye movements and, simultaneously, capture what the pilot sees. Here, we describe the implementation of an updated OT^3 into two aviation training programs of the Spanish Armed Forces.

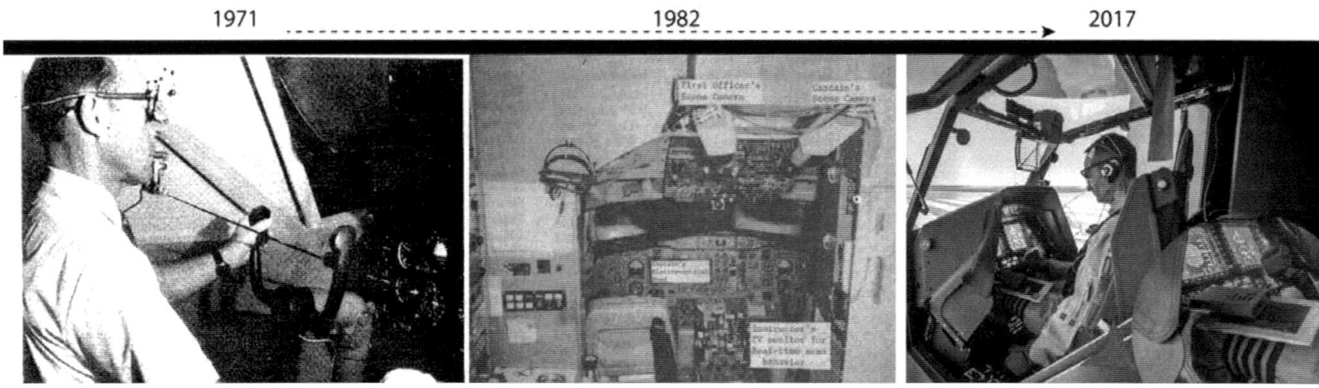

Figure 1. Advances in eye tracking technology in aviation settings. From left to right: **1971)** Eye Point-of-Regard System mounted in an eyeglass frame (Source: Weir & Klein, 1971). **1982)** Honeywell electro-optic head (Source: Harris, Glover, & Spady, 1986). **2017)** Tobii Pro Glasses 2.0 and its recorder unit (attached to the pilot's kneeboard) (Source: Diaz-Piedra, Catena, et al., 2016).

Our Experience

We recorded flight sessions of flight instructors and trainee pilots from the Spanish Army Airmobile Force (First Attack Helicopter Battalion I – BHELA I, Almagro, Ciudad Real) and the Spanish Air Force (78th Wing, Helicopter School, Armilla, Spain), while they performed simulated flight tasks. To record eye movements, we used the Tobii Pro Glasses 2.0 (Tobii AB, Sweden), a portable eye tracker, worn as normal glasses (in this case, comfortably under the helmet, see Figure 2). We performed the calibration procedure inside the aircraft/simulators (Airbus Helicopter Tiger and Sikorsky S-76). Gaze data and the first-person perspective video could be viewed in real time on a tablet computer. All data and videos were stored on a SD card for later replay (see below).

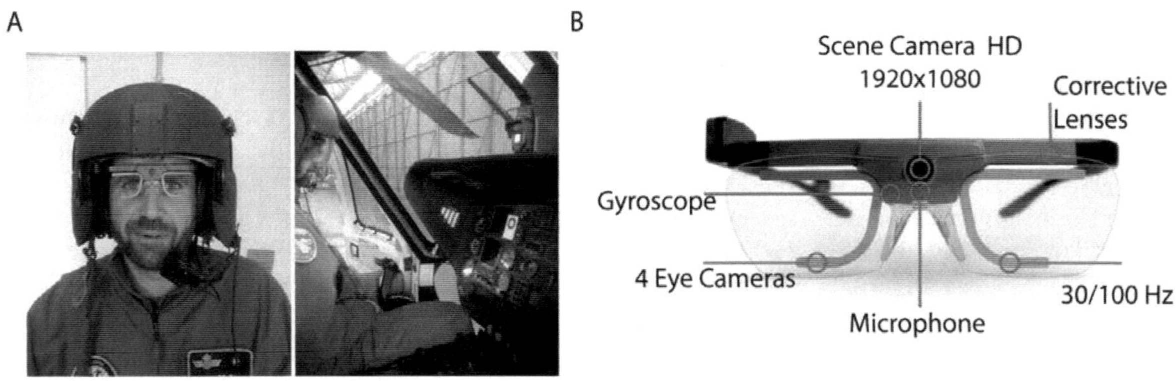

Figure 2. A) Left. A flight instructor from the Spanish Air Force Helicopter School (78th Air Base Wing) (Armilla, Spain), wearing his helicopter flight helmet and the eye tracker. Right. Calibration procedure inside a Sikorsky S-76 helicopter using a single point calibration. B) Tobii Pro Glasses 2.0 and its technical details.

Real-time Feedback

Visual scan patterns might be guided to improve the performance of flight tasks (Wetzel et al., 1998). However, in order to provide the best feedback to the trainee pilot (in the right form and at the right time), the flight instructor needs to know how the trainee pilot allocates his/her (visual) attention.

Eye tracking technology provides an objective measure of where the pilot is looking at. Furthermore, a wireless live view function allows the flight instructor to monitor the trainee's eye movement behavior online and to provide real-time feedback (Figure 3).

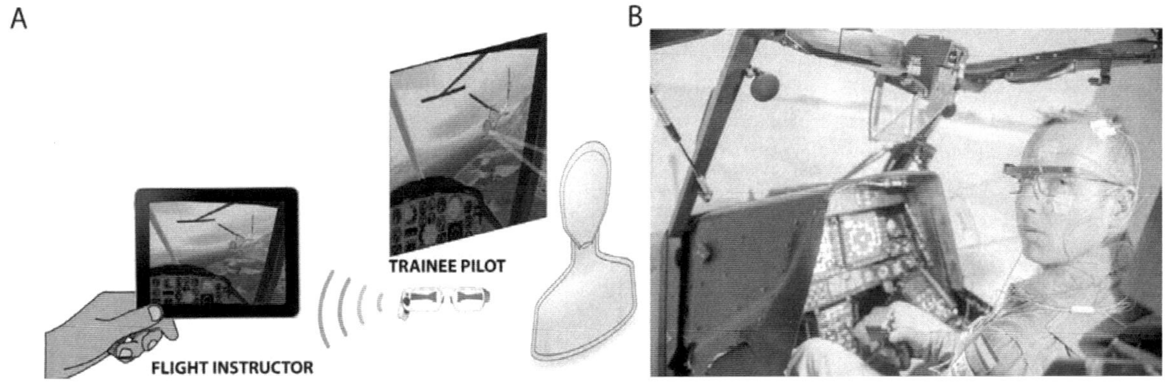

Figure 3. OT^3: *real-time feedback.* A) Sketch of our OT^3. Gaze fixations are illustrated by red circles, and circle diameter indicates fixation duration. During the flight simulation, trainee pilot's eye movements are continuously monitored by the flight instructor using a wireless system so that he/she can correct trainee pilot's scanning behavior in real time. B) A pilot member of the First Attack Helicopter Battalion I – BHELA I (Spanish Army Airmobile Force, Almagro, Ciudad Real, Spain) wearing the eye tracker inside the Airbus Helicopter Tiger simulator.

Debriefing Sessions

Another application of the OT^3 aims to improve debriefing sessions. Flight instructors might play back trainee pilots' flight videos and, consequently, show them their visual scan patterns (Harris et al., 1986), and where they focused during the flight session (pilot's eye movements are superimposed onto the recorded video). In this way, flight instructors can easily point out examples of missed cues or other events that could compromise flight safety.

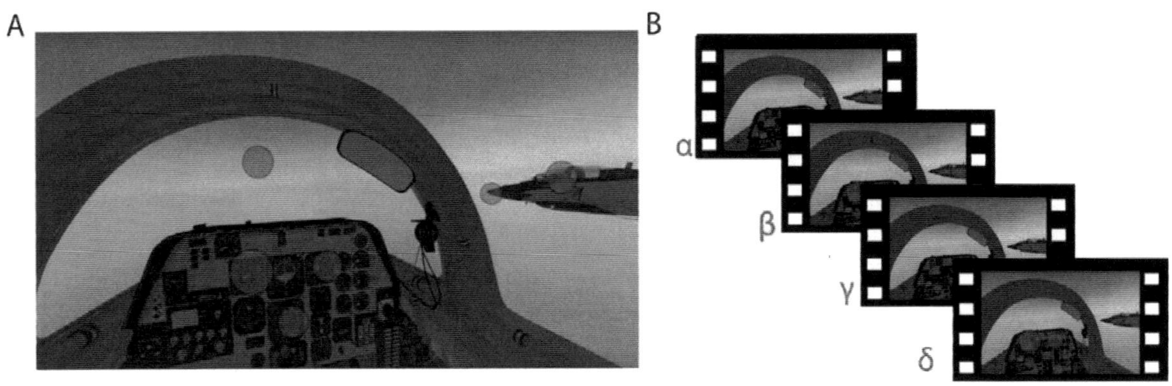

Figure 4. OT^3: *augmented videos.* A sketch of our OT^3 showing different frames of an augmented video. These videos contain the trainee pilot's eye movements superimposed onto the first-person perspective video captured by the integrated scene camera. Also, the communications between the flight instructor and the trainee pilot are recorded. Flight instructors can use these videos to point out accurate and mistaken maneuvers in debriefing sessions.

Expert Guidance

When accomplishing complex visual tasks, experts possess sophisticated visual observation skills which enable them to find relevant features of a visual stimulus with irrelevant features and to interpret these observations (Jarodzka, Scheiter, Gerjets, & van Gog, 2010). Therefore, even though Jones and colleagues already noted that flying training based on individualized feedback would be more helpful (Jones, Coates, & Kirby, 1983), using visual observations of experienced and successful task performers (for example, standardized videos) might also improve instruction by cueing (Gog & Jarodzka, 2013).

For our OT^3, we recruited expert pilots (flight instructors) to perform, in a didactic manner (i.e. avoiding knowledge-based shortcuts), several simulated flight maneuvers (mostly abnormal/emergency flight conditions) during the flight while their eye movements were recorded. Then, we created augmented videos superimposing the pilot's eye movements onto the recorded flight video and the verbal explanation about how he/she was performing the flight tasks (Figure 5). Trainee pilots could watch these augmented videos as part of their aviation training.

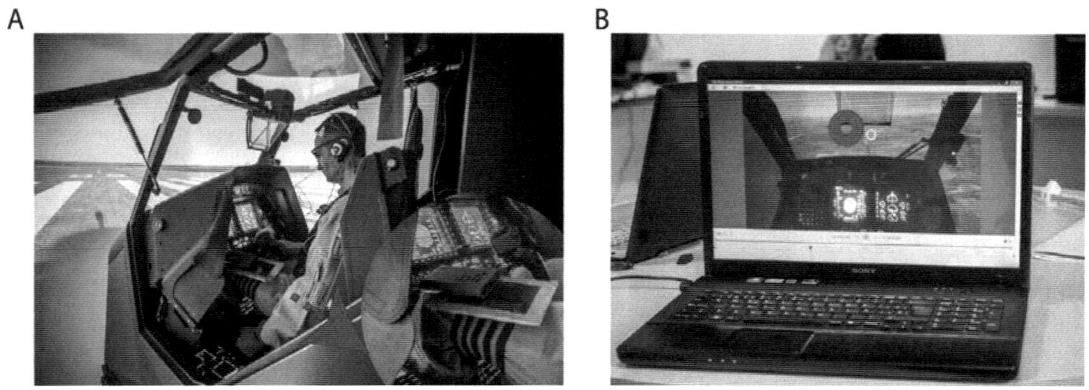

Figure 5. OT^3: *expert augmented videos.* A) A flight instructor from the First Attack Helicopter Battalion I (Spanish Army Airmobile Force, Almagro, Ciudad Real, Spain), performing a standard pre-flight checklist procedure while his eye movements are recorded. B) Example of video editing (cyclopean view [red circle] superimposed onto a frame of the scene) using the software Tobii Pro Glasses 2.0 Controller.

Final remarks

Flying an aircraft is a highly demanding cognitive task where performance heavily relies on visual search and the interpretation of visual information. Training is a key element to acquiring effective scanning patterns that allow safe interactions with the aircraft. During the training of scanning strategies, flight instructors often face the dilemma of knowing when and how to provide the best feedback to the trainee pilot (Sullivan, Yang, Day, & Kennedy, 2011). Modern eye tracking technology applied to flight training might offer a real opportunity to learn, as it provides valuable, objective, and real-time information for both the flight instructor and the trainee pilot. Furthermore, this information can be used to create augmented videos.

The expected outcome of our application will be to enhance flight safety, decrease in-flight errors, and optimize performance by developing complementary educational materials. Finally, the proposed OT^3 will also have applications across a wide range of disciplines in and outside of the aviation industry (e.g. medicine [Di Stasi, Diaz-Piedra, et al., 2016]).

Acknowledgements

This study was funded by the CEMIX UGR-MADOC (Project PINS 2014-16 to CDP & LLDS). We thank the Tobii Group for the Eye Track Awards Stipend awarded to LLDS. The funding organization and the sponsor had no role in the design or conduct of this research.

LLDS is supported by the Ramón y Cajal fellowship program (RYC-2015-17483). The authors have no conflicts of interest to declare.

We thank Colonel Alberto Cherino from the Spanish Army TRADOC, and Commander Juan Suarez from the Spanish Air Force for their assistance and help during the data collection. We thank Jose M. Morales, MSc and Hector Rieiro, MSc (Mind, Brain, and Behavior Research Center, University of Granada) for his help in data processing. We thank Michael B McCamy (NeuroMentix, Newtown, CT, USA) for his helpful discussions and comments on earlier versions of this paper.

References

Barnes, J. A. (1970). *Tactical utility helicopter information transfer study* (No. Technical Memorandum 7-70). Aberdeen Proving Ground, Maryland, US: U. S. Army Aberdeen Research & Development Center.

Di Stasi, L. L., Diaz-Piedra, C., Rieiro, H., Sanchez Carrion, J. M., Martin Berrido, M., Olivares, G., & Catena, A. (2016). Gaze entropy reflects surgical task load. *Surgical Endoscopy*, *30*(11), 5034–5043.

Di Stasi, L. L., McCamy, M. B., Martinez-Conde, S., Gayles, E., Hoare, C., Foster, M., … Macknik, S. L. (2016). Effects of long and short simulated flights on the saccadic eye movement velocity of aviators. *Physiology & Behavior*, *153*, 91–96. https://doi.org/10.1016/j.physbeh.2015.10.024

Diaz-Piedra, C., Catena, A., Fuentes, L. J., Cherino, A., Sáenz de Santa María, I., Suarez, J., & Di Stasi, L. L. (2016). Herramientas innovadoras para la mejora de los procesos de enseñanza en entorno militar basadas en los movimientos oculares: Proyecto EYES. In J. Serna, P. Sanchez-Andrada, & I. Alvarez (Eds.), *Actas: IV Congreso Nacional de i+d en Defensa y Seguridad, DESEi+d 2016* (pp. 1313–1320). Santiago de la Ribera, Murcia, Spain: Centro Universitario de la Defensa de San Javier.

Diaz-Piedra, C., Rieiro, H., Suarez, J., Rios-Tejada, F., Catena, A., & Di Stasi, L. L. (2016). Fatigue in the military: Towards a fatigue detection test based on the saccadic velocity. *Physiological Measurement*, *37*(9), 62–75. https://doi.org/10.1088/0967-3334/37/9/N62

Gainer, C. A., & Obermayer, R. W. (1964). Pilot Eye Fixations While Flying Selected Maneuvers Using Two Instrument Panels. *Human Factors*, *6*(5), 485–501. https://doi.org/10.1177/001872086400600507

Gog, T. van, & Jarodzka, H. (2013). Eye Tracking as a Tool to Study and Enhance Cognitive and Metacognitive Processes in Computer-Based Learning Environments. In R. Azevedo & V. Aleven (Eds.), *International Handbook of Metacognition and Learning Technologies* (pp. 143–156). Springer New York. Retrieved from http://link.springer.com/chapter/10.1007/978-1-4419-5546-3_10

Harris, R. L., Glover, B. J., & Spady, A. A. (1986). *Analytical techniques of pilot scanning behavior and their application* (No. NASA TP-2525). Hampton, VA, USA: NASA Langley Research Center.

Jarodzka, H., Scheiter, K., Gerjets, P., & van Gog, T. (2010). In the eyes of the beholder: How experts and novices interpret dynamic stimuli. *Learning and Instruction*, *20*(2), 146–154. https://doi.org/10.1016/j.learninstruc.2009.02.019

Jones, D. H., Coates, G. D., & Kirby, R. H. (1982). *The effectiveness of incorporating a real-time oculometer system in a commercial flight training program* (No. NASA-CR-169465). Hampton, VA, USA: NASA Langley Research Center.

Jones, D. H., Coates, G. D., & Kirby, R. H. (1983). *The effectiveness of an oculometer training tape on pilot and copilot trainees in a commercial flight training program.* Retrieved from https://ntrs.nasa.gov/search.jsp?R=19830012282

Jones, R. E., Milton, J. L., & Fitts, P. M. (1949). *Eye fixations of aircraft pilots, I. a review of prior eye-movement studies and a description of a technique for recording the frequency, duration and sequences of eye-fixations during instrument flight* (No. USAF-TR-5837). Wright-Patterson Air Force Base.

Spady, A. A., Jones, D. H., Coates, G. D., & Kirby, R. H. (1982). The Effectiveness of Using Real-Time Eye Scanning Information for Pilot Training. *Proceedings of the Human Factors Society Annual Meeting*, *26*(11), 1014–1017. https://doi.org/10.1177/154193128202601124

Sullivan, J., Yang, J. H., Day, M., & Kennedy, Q. (2011). Training simulation for helicopter navigation by characterizing visual scan patterns. *Aviation, Space, and Environmental Medicine*, *82*(9), 871–878.

Weir, D. H., & Klein, R. H. (1971). Measurement and Analysis of Pilot Scanning Behavior during Simulated Instrument Approaches. *Journal of Aircraft*, *8*(11), 897–904. https://doi.org/10.2514/3.59187

Wetzel, P. A., Anderson, G. M., & Barelka, B. A. (1998). Instructor Use of Eye Position Based Feedback for Pilot Training. *Proceedings of the Human Factors and Ergonomics Society Annual Meeting*, *42*(20), 1388–1392. https://doi.org/10.1177/154193129804202005

EFFICIENCY OF A SITUATION AWARENESS TRAINING MODULE IN INITIAL PILOT TRAINING

Gayraud, David
ENAC, University of Toulouse
Paris, France

Matton, Nadine
ENAC & CLLE-LTC, University of Toulouse
Toulouse, France

Tricot, André
CLLE-LTC, University of Toulouse
Toulouse, France

We designed a specific SA training module and tested whether it would enhance the SA competency of pilot trainees during initial training. Twelve pilot trainees took part to the empirical phase of the study. They were pseudo-randomly assigned to two conditions, the experimental group (*n*=6) who received the SA training module and the control group (*n*=6) who did not receive any specific SA training. All pilot trainees were assessed during a flight simulator session in order to evaluate objectively their levels of SA (SAGAT, Situation Awareness Global Assessment Technique, Endsley, 1995). Results highlighted that the SA levels were globally high and homogeneous for the experimental group (range of percentage of maximum SAGAT score= [79.3% ; 87.5%]), whereas the control group's scores varied more widely (SAGAT range = [57.5% ; 87.5%]). Moreover, a qualitative analysis revealed specific strategies used by those pilot trainees who had highest levels of SA.

Introduction

"Being aware of what is happening around you and understanding what information means to you now and in the future" has been named *situation awareness* (Endsley & Jones, 2012, p.13). Situation awareness (SA) has emerged since the 1980s as an important construct in human factors and applied ergonomics and is still the focus of research studies in various domains (e.g., Cordon, Mestre, & Walliser, 2017 in seafaring ; Afkari, Bednarik, Mäkelä, & Eivazi, 2016 in surgery ; Lu, Coster, & de Winter, 2017 in driving). The most widely cited definition is the one of Endlsey (1995b): "the perception of the elements in the environment within a volume of time and space, the comprehension of their meaning, and the projection of their status in the near future". In the field of aviation, SA has been recognized as a critical component. Indeed, an investigation of flight accidents between 1989 and 1992 (Endsley, 1995a) revealed that among the 17 accidents mainly attributed to human error, SA was the most prevalent factor in 15 of them (i.e., 88%). Another study published by the Flight Safety Foundation (Khatwa & Roelen, 1998/1999) focused on 156 CFIT accidents between 1988 and

1994. This study highlighted that 45% of accidents in which flight crew errors occurred involved a SA error.

Given the importance of SA, several researchers have tested the development of training approaches for improving it. For instance, Strater et al. (2004) tested a PC-based tool to improve SA of naval cadets. They developed two modules, one module which teached time management and task prioritization ("SA Planner") and another which focused on general strategies aimed at improving SA ("SA Trainer"). In order to assess the efficacy of their "Infantery SA trainer" (ISAT), they compared trained and untrained cadets on SA assessment and performance. They obtained mixed results and could only conclude on "tentative indications of training effects" (p. 671). Bolstad, Endsley, Costello, and Howell (2010) tested the efficacy of six different computer-based training modules (checklist completion, air traffic control comprehension, psychomotor skills, attention sharing and contingency planning) on general aviation pilots. They concluded that "no evidence was found to show that improvements in the basic and cognitive skills trained by the modules translated to improved flight skills performance". For airline pilots, a training programme for SA has been developed and tested by a european consortium (Hoermann, 2003) in the ESSAI (Enhanced Safety through Situation Awareness Integration in training) project. Eight crews received the ESSAI training and a control group of eight crews did not. Results revealed SA improvement and better performances during a simulator session for the trained group.

To our knowledge, no air training organization has designed a training program oriented towards SA, which would strengthen this competency during the ab-initio pilot training. The present study aimed at developing and testing a specific SA training module that would improve SA of ab-initio students during their pilot initial training.

Method

Participants

Twelve student pilots of the ENAC (Ecole Nationale de l'Aviation Civile, France), the a french national civil flight training organization, were recruited to perform the experiment. They had all completed a 18-week module of visual flight rules (VFR) training and were following completing a single engine instrument flight rating (IFR) module. They were pseudo randomly assigned to two conditions. Indeed, they were located at two different flight training centers, Carcassonne (*n*=8) and Grenoble (*n*=4). Student pilots from Carcassonne were at week #22 (131,25 flight hours on average) and those from Grenoble were at week #30 (194,25 flight hours on average) of the training syllabus. In order to control the flight training experience variable, each experimental group was composed of two student pilots from Carcassonne and four student pilots from Grenoble. Otherwise, pilot students were randomly assigned to each experimental condition. All participants were volunteers and provided written consent.

SAGAT questionnaire

A SAGAT questionnaire was designed in order to assess the level of situation awareness of the participants at three moments of the simulated flight. The questionnaire consisted of eight items assessing the three levels of situation awareness defined by Endsley (three items of level 1, three items of level 2 and two items of level 3).The accuracy and relevance of each item was assessed by the first author of the present paper, also flight instructor, on a five-points scale. Thus, SAGAT scores could range from 0 to 40.

Procedure

Participants of the control condition performed a simulated flight on a certified FNPT-II flight simulator of the Socata TB20 aircraft. After 30 min of explanations and briefing, participants performed a one-hour simulated flight. The scenario started in flight at flight level 110 over the Geneva Lake heading to Lyon St Exupery. The flight preparation proposed two alternate airports with good weather conditions, Saint Etienne and Grenoble. Weather at destination was sufficient enough for a non-precision approach (cloud base 400ft). No technical failures were scheduled during this simulator session but several and continous environmental changes like wind changes, airport constraints. At three times the simulation was freezed and the participant answered to the SAGAT questionnaire in order to evaluate their assessment of the new environment. At the end of the simulation, participants were debriefed during 20 min. Participants of the control group did not receive any specific training on SA (the concept was only introduced during their initial theoritical training).

Participants of the experimental condition received a specific training module the day before they performed the same simulated flight as the control group. This training module consisted of five hours of training: theoretical presentation of SA and related concepts (e.g., mental workload, mental schemas,...), discussions about case studies based on real incidents and familiarization with a self SA-assessment tool assessing subjectively the three levels of SA (this tool will not be detailed here because it will be presented in another paper). Participants of the experimental condition performed the same simulated flight as the control group. However, during the simulated flight, they also had to fill the self SA assessment tool before each SAGAT questionnaire.

Results

SAGAT scores

Given the small sample size of each group, we performed no statistical test to compare the two groups (see Table 1 for summary statistics). However, a qualitative analysis of the data highlighted that the SAGAT scores of the experimental group were globally higher than those of the control group (see Figure 1). More precisely, SAGAT scores of the experimental group were high and more homogeneous (all between 79.3% and 88.3%) than those of the control group (between 57.5% and 87.5%). Interestingly, the lowest SAGAT scores (less than 60%) were only found in the control group.

Table 1.
Descriptive statistics of SAGAT scores for each experimental group (EXP=experimental; CTL=control).

Group	*n*	*mean (%)*	*sd*	*min (%)*	*max (%)*
CTL	6	29.1 (72.7%)	5.4	23.0 (57.5%)	35.0 (87.5%)
EXP	6	34.1 (85.3%)	1.3	31.7 (79.3%)	35.3 (88.3%)

Note. Maximum score of SAGAT was 40. Numbers in parentheses represent percentages of this maximum score.

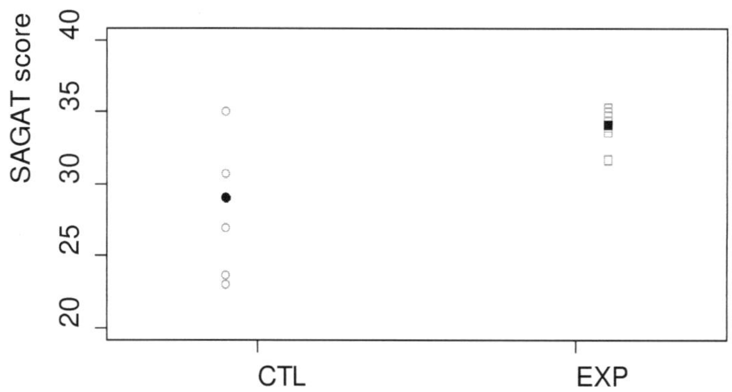

Figure 1. Means (black filled) and raw data (not filled) of total SAGAT scores for each group (control and experimental). SAGAT scores could range from 0 to 40.

Qualitative analysis

A qualitative analysis of the decisions made by each participant of each group revealed that all pilot students of the experimental group (6/6) made safer decisions based on a relevant information collection and comprehension. On the contrary, in the control group, two pilot students out of six uncorrectly assessed different airport weather conditions and had a wrong environment understanding leading for example to too steep approaches. Interestingly, these two "poorer" pilot students were at the week#22 stage of their training. Pilots of the control group at week#30 performed qualitatively as good as those of the experimental group.

Discussion

The present study aimed at testing the efficacy of a SA training module designed for pilot students during basic flight training on light aircraft. An experimental group who completed the 5-hours specific SA training module was compared to a control group who did not follow this training module. The two groups were compared on the basis of a simulated flight with assessment of their SA level through SAGAT queries. Results highlighted that pilot students of the experimental group had globally higher and more homogeneous SAGAT scores than those of the control group. A qualitative analysis of their flight performances suggested that experimental group pilots were better than the control group 'at understanding a constant evolving situation and adapting strategies in accordance to it.

However, several limitations of this study need to be adressed. Firstly, a larger sample size would be needed to confirm these results and to allow generalisation. Secondly, the SA training module seemed to improve flight performance only for pilot students at an early stage of the IFR training module (week#22). Clearly, one can assume that some pilot students are more able to generate high levels of SA without any specific training. For instance, Endlsey and Bolstad (1994) found that the best fighter pilots had SA scores that were 10 times better than those with the lowest SA and they did not get any specific SA training. However, the question is whether pilot students who would have difficulties in improving their SA on their own would benefit from a specific SA training module. Thirdly, the level of difficulty used for the flight scenario may have a large impact on the results of studies aimed at testing the efficacy of a SA training module. Indeed, maybe the flight scenario used in this study was too easy for pilot students at a more advanced stage of the IFR training module (week#30), and no differences between experimental groups could be observed.

References

Afkari, H., Bednarik, R., Mäkelä, S., & Eivazi, S. (2016). Mechanisms for maintaining situation awareness in the micro-neurosurgical operating room. *International Journal of Human-Computer Studies, 95*, 1–14. doi:10.1016/j.ijhcs.2016.05.004

Bolstad, C. A., Endsley, M. R., Costello, A. M., & Howell, C. D. (2010). Evaluation of Computer-Based Situation Awareness Training for General Aviation Pilots. The *International Journal of Aviation Psychology, 20(3),* 269–294. doi:10.1080/10508414.2010.487013

Cordon, J. R., Mestre, J. M., & Walliser, J. (2017). Human factors in seafaring: The role of situation awareness. *Safety Science, 93*, 256–265. doi:10.1016/j.ssci.2016.12.018

Endsley, M. R. (1995a). A taxonomy of situation awareness errors. In R. Fuller, N. Jonhston, & N. McDonald (Eds), *Human factors in aviation operations* (pp. 287-292). Aldershot, England: Avebury Aviation, Ashgate Publishing Ltd.

Endsley, M. R. (1995b). Toward a theory of situation awareness in dynamic systems. *Human Factors, 37*, 32-64.

Endsley, M. R., & Bolstad, C. A. (1994). Individual Differences in Pilot Situation Awareness. *International Journal of Aviation Psychology, 4(3),* 241–264. doi:10.1207/s15327108ijap0403_3

Endsley, M. R., & Garland, D. J. (2000). Pilot Situation Awareness Training in General Aviation. *Proceedings of the Human Factors and Ergonomics Society Annual Meeting, 44(11),* 357–360. doi:10.1177/154193120004401107

Endsley, M. R. & Jones, D. (2012). *Designing for situation awareness: An approach to user-centered design* (second ed.). Boca Raton: CRC press.

Hoermann, H.-J., Soll, H., Banbury, S., Dudfield, H., Blokzijl, C., Bieder, C., Paries, J., Lehmann, O., Field, P., Lodge, M., Aymeric, B., Colautti, A., Polo, L. (2003). ESSAI Enhanced Safety through Situation Awareness Integration in training: a validated recurrent training programme for situation awareness and threat management (ESSAI/NLR/WPR/WP6/2.0).

Khatwa, R, & Roelen, A.L.C. (1998, November/1999, March). An analysis of Controlled-flight-into-terrain accidents of commercial operators, 1988 through 1994. *Flight safety digest*, pp. 166-212.

Lu, Z., Coster, X., & de Winter, J. (2017). How much time do drivers need to obtain situation awareness? A laboratory-based study of automated driving. *Applied Ergonomics, 60,* 293–304. doi:10.1016/j.apergo.2016.12.003

Strater, L. D., Reynolds, J. P., Faulkner, L. A., Birch, D. K., Hyatt, J., Swetnam, S., & Endsley, M. R. (2004). PC-Based Training To Improve Infantry Situation Awareness. *Proceedings of the Human Factors and Ergonomics Society Annual Meeting, 48(3),* 668–672. doi:10.1177/154193120404800384

A PEDAGOGICAL APPROACH TO TEACH AVIATION STUDENTS HOW TO CONDUCT SITUATION AWARENESS RESEARCH

Andrew R. Dattel
Andrey Babin
Tianhua Li
Ziyi Dong
Stephanie G. Fussell
Qianru Yang
Embry-Riddle Aeronautical University
Daytona Beach, FL

Situation awareness (SA) has been investigated in the aviation industry for decades and recently has become more prominent in other industries. For example, healthcare has experienced tremendous growth in SA research and training. Despite agreement among researchers that SA is important for performance and safety in complex domains, less agreement exists for defining and measuring SA. Certain industries (e.g., aviation, healthcare, process control operations) often have specific methods on how to approach SA. These approaches of introducing employees and students to SA in a specific context may inadvertently limit their full appreciation and understanding of this construct.

To address this issue, graduate students with a wide variety of aviation interests were enrolled in a course titled, "Situation Awareness in Aviation/Aerospace." Common theories and applications of SA were discussed throughout the course. To increase their domain-general understanding of SA, students completed both aviation and non-aviation assignments. An example of the latter included giving individual presentations of National Transportation Safety Board (NTSB) reports of non-aviation accidents. Also, as a group project, students collected SA data from human participants in a non-aviation dynamic environment. Students responded favorably to these activities and reported that learning about SA in other environments gave them a better understanding and appreciation of SA. This paper discusses the steps taken to teach SA, the outcome measures of the class, and the unique perspective that students had of learning about SA outside of aviation. Feedback from students highlights the benefits of cross-domain pedagogical approaches to teaching SA.

The traditional college classroom is changing. Instead of only presenting a traditional one direction lecture approach, many college professors seem to be offering additional pedagogical approaches such as active learning, engaged learning, and learning via application to many domains (Barkley, 2010; Lee, 2004). Graduate students enrolled in a summer term *Situation Awareness and Performance* seminar in an aviation program studied the theories of situation awareness, but learned and applied situation awareness in both an aviation and non-aviation context. This seminar where students were more active in the pedagogical approach, as well as developing a respect for situation awareness outside of their comfort zone (i.e., aviation) proved to be not only a rewarding experience, but an effective learning experience as well.

Situation Awareness

Those in aviation may experience nostalgia when they remember the aircraft hangar talks when the term situation awareness (SA) became commonly accepted. Pilots would sit around and try to explain accidents, incidents, and mishaps as a loss of "situational awareness." Mechanics and air traffic controllers also participated in these lively discussions. From these domain-developed discussions, SA soon became a construct worthy of scientific research, inasmuch that millions of dollars in grants have been expended for the study of SA. Although many in the aviation community may feel like they "own" SA, it has been overwhelming applied to other dynamic environments such as healthcare, process control industries, and surface transportation (Gugerty, 1997; Lau, Jamieson, & Skraaning, 2016; Stubbins, Chaboyer, & McMurray, 2012). The multi-domain application of SA has several common denominators, but also contains many different applications such as ways to measure SA and how to train and teach SA.

Endsley (1995) created the most common cited definition of SA, which is a construct that operates on 3 levels: perception, understanding, and anticipation of future situations. Durso, Rawson, and Girotto (2007) offer a similar definition absent the hierarchical structure, which is defined as how one understands relevant information in a rapidly changing environment. SA is an important capability to have when operating in a dynamic environment. SA has been shown to be a distinct construct beyond other underlying mechanisms such as attention, memory, etc. (Durso, Bleckley, & Dattel, 2006). Good SA typically leads to better decision making, better understanding, and better performance (Endsley & Jones, 2012).

Class Design and Content

This first-time offered class for the graduate program drew a relatively large class roster. Enrolled students had a variety of aviation interests (e.g., pilots, management, education technology). During the first week of class, the instructor showed the film, *Premium Rush*, starring Joseph Gordon-Levitt. This action-packed drama follows the trials and tribulations of a New York City bicycle messenger for a day – quite a great way to start off a class! The film portrays the expedience of navigating a bicycle through a vastly changing dynamic environment by displaying the protagonist's processing of information to maintain a high level of SA. This movie quickly gave a good perspective of what SA is—now the challenge for the students was to understand the theory, apply the concept, and measure the construct. In another class period a guest lecturer, who has substantial experience in SA in industry and government, discussed real world experiences to the delight of the captured audience.

As is typical in a class seminar, much of the class period was devoted to discussing articles — in this case about 50%. Seminal, theoretical, and practical articles in SA were reviewed. In addition, articles about training, designing, and measuring SA were discussed. These articles included applications to aviation and non-aviation domains. Because of the students' varied backgrounds, topics could develop from an individual student's experience, leading to unexpected, but relevant discussions.

One of the first assignments for the class was to explain to a friend what SA was, and ask the friend to give examples of where he or she had witnessed good or bad SA. The student was then to report back to the class what the friend described. This exercise helped the classmate understand SA by articulating the definition, and then self evaluated their own way of describing SA in layman's terms.

Research Proposal

The most heavily weighted assignment was for to students to write a research proposal in regards to SA in aviation. For many students in the class, the research proposal was their first experience thinking and formulating ideas about research. Some of the proposals that students wrote for the class became the basis for their Master's thesis.

NTSB reports

Another assignment that received positive reviews required each student to present to the class findings an accident report from the National Transportation Safety Board (NTSB). Students made these presentation throughout the term. A requirement for the NTSB report was that the primary or secondary cause of the accident had to be attributed to poor SA. Additionally, the report had to be non-aviation related. It was obvious that each student presenter had thoroughly researched the report and delivered it well. However, it seems the most rewarding aspect were the comments and discussion that were generated in the class.

Class project.

Throughout the semester, the students worked on a class project. The requirements for the project was that an SA research study had to be conducted that involved human participants. The class had to design the research project, develop the stimuli, collect the data, analyze the data, and present the findings in a discussion and conclusion format. Approval from the Institutional Review Board (IRB) was required. Although it was required that the students conduct the study adhering to accepted research methods, the final reporting of the study was allowed to be more lenient. Initially, the class project was to involve SA in aviation; however, the students decided to conduct the study in a non-aviation dynamic environment. In retrospect, conducting the study in a non-aviation environment may have been facilitated a better understanding and appreciation of SA in the aviation environment. The following sections describe the students design, findings, and applications of the SA project.

Student Design of the Research Project

Part of the class coursework included a group project, which was a research study designed and performed by the students. This research project not only helped to foster students' critical thinking skills, active learning, natural inquiry, but also gave students an appreciatiation and better understanding of conducting SA research. Several topics were considered for the research project, including observing and interviewing air traffic controllers at a local Tower/TRACON, and observing and videotaping racecar drivers practicing on a racetrack. However, it was determined by the students and the professor that pilots, air traffic controllers (ATC), and racecar drivers are professionally trained maintain good SA at all times. The class concluded it would be beneficial to measure SA of the general public in a non-aviation dynamic environment.

Why EPCOT? Situation awareness has been associated with the field of aviation for decades (Salas, Prince, Baker, & Shrestha (1995). To increase their understanding of SA in a public, dynamic environment, aviation students measured the SA of patrons and employees at Disney's EPCOT theme park. This environment was chosen for a variety of reasons:

1. The number of patrons and employees available to survey. EPCOT provided the students with a large sampling pool due to the large volume of park patrons and the large number of staff.
2. The park provided the students with a wide variety of patrons and employees to survey, although demographics were not considered as part of this survey. The diversity of participants assures the undistinguished data collection without any directive trends (e.g. the data is not correct if the participants are mainly consisting of middle age).
3. The abundant and realistic SA-related activities. Students studied three types of SA in EPCOT: spatial awareness, time awareness, and general awareness
4. Emergency situation potential. The public environment of the theme park allowed the students to question participants on potential emergency situations.
5. Good signage and visibility. EPCOT is filled with maps, signs, and other directional information. The researchers hypothesized that the visual cues at EPCOT would increase patron SA.
6. Hospitality. The atmosphere if Disney is famous for being open, friendly, and willing to host student researchers.
7. The location and transportation. EPCOT provided the students with an easily-accessible dynamic environment for study that was outside of the field of aviation.

Of these, the potential pool of participants in terms of size and diversity were the most important reasons for location determination.

Formulation of measurements

During one class lesson, the students collectively filled out an application form for Institutional Review Board (IRB) approval and prepared a plan for interviews. Students discussed SA measurement methods and questions to ask patrons and employees. Students formulated two groups of questions, separated for patrons and for park employees.

Students began preparations in the a few weeks leading up to the trip to EPCOT. The purpose of the study was to interview park patrons and employees. The students focused questions on patrons' awareness of their surroundings, theme park arrangement, and time awareness. Questions for patrons were mainly focused on their navigation skills (i.e., where the nearest exit was, where their car was parked, and where the nearest restroom was). Park employees were interviewed about their knowledge of safety measures in the event of an emergency. Questions for staff members were related to their actions in case of an emergency. Some questions for patrons and staff also were related to the awareness of current time, such as park closure time, or, for staff, time until their next break.

To measure SA, students recorded and later compared participant response time for each question. If applicable, some answers (e.g., "Where is the nearest restroom?") were checked for accuracy (Yes/No).

Results

Patrons

Fifty patrons participated in the study, and each of them answered 12 questions. The patrons' situation awareness was measured by analyzing confidence ratings and response times as opposed to the number of questions correctly answered by the patron. The questions correctly answered were used to identify valid data. For example, the response time that a participant took to point out the location on a map correctly was analyzed; conversely, if a participant did not point out the correct location, the response time was abandoned. There were seven yes-or-no questions. The descriptive statistics are summarized in Table 1.

Table 1. Responses to Specific Questions.

	N	Number of Yes	Percentage of Yes (%)
Meet-up Location	50	13	26.0
Point Location	11	11	100.0
Exit Direction	48	41	85.4
Park Front	48	42	87.5
Nearest Restroom	48	39	81.3
Point on Map	48	38	79.2
Drive to Park	43	24	55.8
Know Where Parked	24	21	87.5
Current Time	40	29	72.5
Park Close Time	47	28	59.6
Have Water	45	37	82.2

Note. Point Location = Point to the Direction of Meet-up Location, Point on Map = Point out Current Location on a Map

Confidence rating. Two of the 12 questions measured patron SA using confidence ratings. They were Car Location and Arrival Time. The scales of confidence ratings were from 1 to 5. One indicated the participant was not confident at all in their given answer, and 5 indicated the participant was very confident in their given answer.

Twenty-one participants of the 50 surveyed rated their confidence in their knowledge of Car Location. Of the 21 participants who rated their confidence, 18 participants, or 85.7%, indicated a 5 or very confident. The other three participants selected 3, or neutral/unsure.

For the question concerning Arrival Time, 48 participants of the 50 responded. The numbers of participants for each rating level are summarized in Table 2.

Table 2. Confidence Rating for Arrival Time.

Rating Scale	Number	Percentage (%)
1	0	0.0
2	1	2.1
3	5	10.4
4	17	35.4
5	25	52.1

Employees

There were 15 theme park employees who participated in this portion of the study, and each of them answered six questions. The employee's SA was measured by analyzing response times to questions as opposed to the number of questions correctly answered.

Parallels to aviation

When analyzing the data, the yes-or-no questions from the survey were divided into three categories, including spatial awareness, time awareness, and planning and preparation. Spatial awareness consists of pointing to locations on a park map, direction to the closest exit, where the front of the park was, and the nearest restroom. Theme park patrons have to be aware of the spatial information so that they are able to enjoy the park by not getting lost, and to evacuate in a short time if necessary. Time awareness contains included questions about park closure time and current time. Time awareness is important because patrons need to plan the rest of their stay regarding current time and park closure time. Planning and preparation, which includes meet-up location, knowing where car was parked, and having fluids for hydration is another important factor. The results showed that the patrons' spatial awareness was much better than the time awareness and the planning and preparation. Good spatial awareness amount the partrons may be been due to repeat customers. However, repeat visits did not help them with the time awareness and planning and preparation. The mean percentage of yes for each category is shown in Figure 1.

Figure 1. Mean percentage of each category

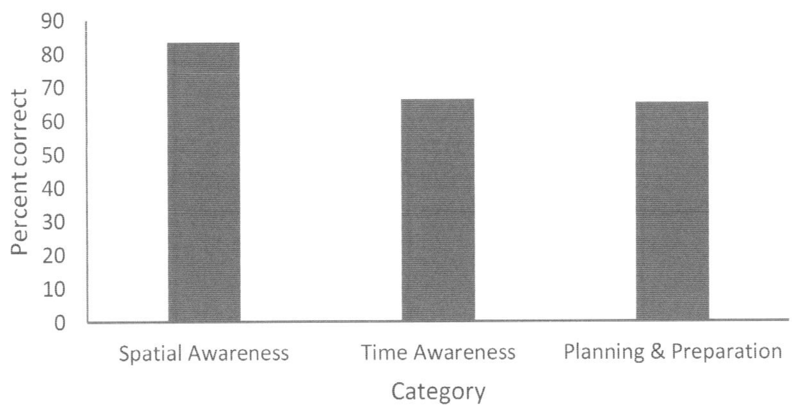

The main purpose of the study was to measure SA in a dynamic environment in a field not related to aviation. A local entertainment park was chosen for the location of the study. Although on the surface the environments are dissimilar, parallels between a theme-park environment and aviation environment were evident and gave the aviation students the tools to understand and measure SA. Many of the questions asked of the park patrons can be related to aviation. For example, knowing how to immediately egress an airplane or airport terminal can be the matter of life or death. Other planning skills, such as a meet-up location in a busy amusement park is similar to having a contingency plan or alternative flight plan route.

An explanation of why some patrons did not have a meet-up location may be because people have become more reliant on cellphones and prefer to call members of the group in the event of separation rather than determine a meeting point in advance. However, relying on cell phones call in case of separation from party members is poor planning because cell phones batteries can die, or reception could be unavailable. This is in contrast to a pilot who loses communication because there is a defined protocol to follow in case such failure occurs.

Discussion

Because this 3 credit class was conducted during the summer, the semester was condensed to a 6 ½ week period. However, the variety of assignments, emphasis on theory, application, active learning, and teamwork seemed to be very successful. The course was quite comprehensive, but students seemed to enjoy putting in the work effort. Below are selected comments from the end of course evaluation.

- "Being able to conduct a survey for a class project was very helpful in understanding how research is conducted. The process of completing the IRB, developing the project questions, and working together to complete the project was a lot of fun and informative. The 10-minute presentations about non-aviation related accidents was very useful and provided another way to interpret the levels of SA and apply them to non-aviation scenarios. At Embry-Riddle, we often focus on aviation everything, but this gave us a different understanding of the SA tool."

- "The detailed research papers allowed understanding in advance theories and concepts. The field work eased having experience on gathering data used to understand the construct of SA."
- "Group discussion and reading articles were an excellent source of knowledge. Overall, the class was very fun and made the time pass much faster than normal, in a good way."
- ".....the movie about SA is also a good example to learn the course content"
- "Examples of SA in real life and its applications in both aviation and other industries, e.g., medicine, driving, etc...... Also, accident reviews helped a lot. I liked the variety of readings (theoretical/empirical) that we went through during the course."
- "The course, I thought I would be an expert in because I am a pilot, and I thought I knew everything about Situation Awareness. Well, I was wrong, after the first class session I realized that I only knew the basics of SA. This course has given me a more thorough understanding of Situation Awareness and I can truly say this course has opened my eyes, it has changed the way I think about the situations while flying and I am confident that there has been an improvement in my personal Situation Awareness. This course was well involved, this course gave me exposure to experiments, this course opened my mind and perhaps changed my life."

Conclusion

There are three markers that appear to make this course design highly successful. The first marker is the end of semester evaluations. Student evaluation of the course across all questions was a 3.87 out of 4.0. However, it is the comments noted above that were most encouraging. The second marker is the reputation among students that the class was a positive learning experience as well was as an enjoyable experience. The third marker is that the maximum enrollment was reached soon after the course was offered for the second time. The instructor feels that this design of the class was highly successful, and looks forward to teaching it again.

References

Barkley, E. F. (2010). *Student Engagement Techniques: A Handbook for College Faculty*. San Francisco, CA: John Wiley & Sons.

Durso, F. T., Bleckley, M. K., & Dattel, A. R. (2006). Does situation awareness add to the predictive validity of cognitive tests? *Human Factors, 48*, 721-733.

Durso, F. T., Rawson, K. A., & Girotto, S. (2007).Comprehension and situation awareness. In F.T. Durso, R.S. Nickerson, S.T. Dumais, S. Lewandowsky, & T. J. Perfect (Eds.), *Handbook of applied cognition (2nd ed.)*, (pp. 163-193). Hoboken, NJ: John Wiley & Sons.

Endsley, M.R. (1995). Towards a theory of situation awareness in dynamic environments, *Human Factors, 27*, 32-64.

Endsley, M. R., & Jones, D. G. (2012). *Designing for Situation Awareness: An Approach to User-Centered Design,* Boca Raton, FL: CRC Press.

Gugerty, L. J. (1997). Situation awareness during driving: Explicit and implicit knowledge in dynamic spatial memory, *Journal of Experimental Psychology: Applied, 3*, 42-66.

Lau, N., Jamieson, G. A., & Skraaning, G. (2016). Situation awareness acquired from monitoring process plants-The Process Overview concept and measure. *Ergonomics, 59*, 976-988.

Lee, V. S. (2004). *Teaching & Learning Through Inquiry: A guidebook for Institutions & Instructors*. Sterling, VA: Stylus Publishing.

Salas, E., Prince, C., Baker, D. P., & Shrestha, L. (1995). Situation awareness in team performance: Implications for measurement and training. *Human Factors, 37*, 123-136.

Stubbings, L., Chaboyer, W., & McMurray, A. (2012). Nurses' use of situation awareness in decision-making: An integrative review, *Journal of Advanced Nursing, 68*, 1443-1453.

EFFECTS OF COUPLING AN OPERATIONAL PHILOSOPHY AND A CORPORATE CULTURE IN A SOCIOTECHNICAL SYSTEM

Oleksandr Petrenko, Oleg Petrenko
Center of Innovative Technologies for Human Factors Control
Kyiv, Ukraine

The article is considered the content, functions and displays of operating philosophy. Suggested the additional model for the analysis of areas of possible incompatibilities in the structure of human factors in view of the "operating philosophy" genesis and manifestations. The results of an empirical study of the interrelationship of the professional worldview on human factor issues and judgments on the desirable socio-cultural characteristics of ATS units among begining air traffic controllers are presented.

The human factor is distinguished by the complexity of its structure and linkages of elements. This makes it difficult to estimate and correct the effects of the ongoing evolution of technology and equipment from the viewpoint of human role and capacities.

In the general view, a human factor is commonly understood as a whole set of effects related to the contact of a human with information, tools, tasks and rules of activities, as well as a physical and social environment that have an impact on performance. However, the simple enumeration of the elements included in the classical model of the human factor, as well as the lines of their contact, does not reflect the whole complexity of the real picture. Everything that a person deals with, he fills with meanings that express his worldview and are caused by involvement in society. It seems obvious to us that these meanings impose an imprint on the functioning of any socio-technical system. A person in sociotechnical system manifests his professional worldview both at the level of the fundamental philosophy of activity and at the level of interpretation of the meaning of various rules. In addition, the organizational life in which people manifest themselves at the personal level leaves an imprint on any professional action in any workplace, even if we do not see obvious relationships.

It should also be noted that aviation technologies are spreading globally and thus are used in different cultures. We can expect that the philosophy of the same professional activity in different cultures would be different. Another point is that the professional worldview of technology developers and users often does not coincide.

This caused our interest in the professional representations of people who obtained aviation education in the field of operator activity, about the fundamental ideas of organizing and implementing their activities, and also the interrelationships between these ideas.

The Development of Ideas about Human Factor

The evolution of human factor concepts reflected the desire for exhaustive coverage of its important aspects. The transition from the Edwards model to the Hawkins model highlighted an attention to such factors as teamwork, communication, leadership, social norms (Hawkins, 1987).

Socio-cultural aspect found an explicit expression in a model of the human factor SCHELL, where component C is defined as "organizational and national cultures that influence the interaction" (SMS for Aviation – a Practical Guide, Civil Aviation Safety Authority, Australian Government, 2012).

We analyzed complex and contradictory processes manifestations of the socio-cultural factor in the post-soviet space in the context of regulation of joint activity in the composition of the flight crew and practices of crew training by programs for learning effective interaction. The analysis of crew's training programs prevalent on the post-soviet space testified that some of them directly contradict the modern concept of CRM, but consistent with the inherited corporate culture and individual professional worldviews (Petrenko, 2010).

In spheres where sophisticated equipment is operated, an important component of the human factor is the specific entity that we call operational philosophy. Under this concept we understand the basic ideas and requirements for the content and organization of the process of people activity as a part of socio-technical systems, as well as ways to ensure reliability of human component and the whole system.

One of the functions of operating philosophy is to ensure the unity of design solutions machine component and specified operational rules. But the fundamental role of operational philosophy is a conceptual agreement between all the components of the human factor.

Inconsistency of prescribed operational philosophy and socio-cultural features of the organization of the operator can pose a very real danger. Operation philosophy has non-contradictorily fit into the corporate culture and space of individual worldviews.

In our view, the operating philosophy is based on an understanding of the capabilities and vulnerabilities of human and, thus, on a notion situations and risks that could be handled through human potential, as well as on a notion of the risks and difficulties that may arise due to limitations of a human (Fig. 1). It is important that this concept should be applied to both individual and team activity.

Based on the above, we see possible entered in consideration of the human factor aspects such as "requirements for a human" and "operating philosophy", correlating them with the possibilities of human and socio-cultural features of team and organization (Fig. 2).

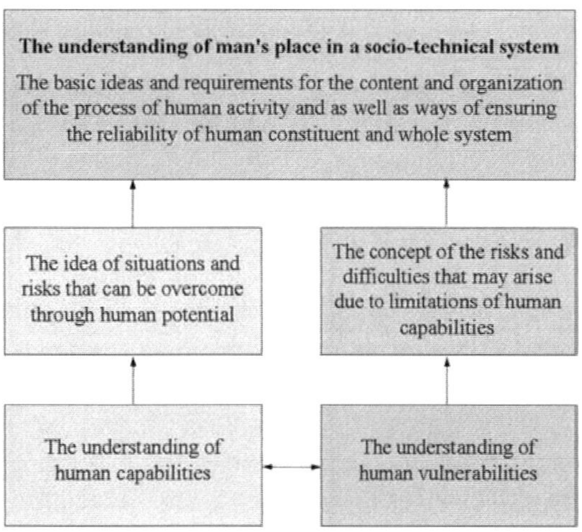

Figure 1. Genesis of operating philosophy

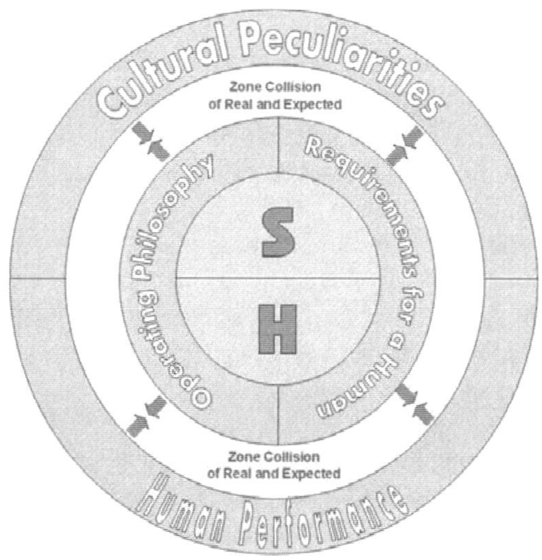

Figure 2. Areas of dangerous incompatibilities manifestation (Petrenko, 2015)

This model directs attention to the "zone of collision of a real and expected". It identifies four key areas for analysis:

- accordance between human capabilities and requirements for human;
- accordance between requirements for human and cultural peculiarities of social space;
- accordance between operation philosophy and human capabilities;
- accordance between operation philosophy and cultural peculiarities of social space.

Empirical Studies of Individual Representations of ATC-Beginners Regarding Safety Philosophy of Air Traffic Management and Organizational Culture of ATM Units

Methodology

The aim of this study is to establish interrelation patterns between individual ideas about the philosophy of safe work of an air traffic controller and the appropriate organizational culture traits for ATM units. The study is based on the idea of dependence of an organizational culture from a professional worldview of a personnel.

As responders were chosen novice air traffic controllers, as a carriers of ideas which have not yet been distorted by the organizational traditions and prejudices.

Using questionnaires with a set of scales respondents evaluated the significance of the assertions related to professional worldview (their list is given in Table 1), and reflect their ideas about the desired features of an organizational culture (Table 2). Stimulus material forms were drafted by expert interviews with open questions, and based on existing typologies of organizational cultures.

The stimulus material of the questionnaires was created based on the expert survey with open questions and on the existing classifications of organizational cultures.

To assess the degree of correlation r-Pearson coefficient was used.

Table 1.
Professional worldview statements offered to respondents for evaluation

1	Presence of emotional problems of the staff poses a threat
2	Individual capabilities of information processing require evaluation and consideration
3	A psychological climate in a team may endanger the safety of air traffic
4	It is necessary to constantly maintain readiness for difficult situations
5	Loss of motivation poses a danger
6	Personal attributes of air traffic controllers could be dangerous
7	The system must be protected from failures of an individual
8	Quality of a working activity is influenced by an objectivity of its evaluation
9	Quality of a working activity is influenced by a timeliness and adequacy of response to shortcomings
10	Quality of a working activity is influenced by a perfection of a rules
11	Quality of a working activity is influenced by a workplace ergonomics
12	Quality of a working activity is influenced by an automation
13	Quality of a working activity is influenced by a regular training and skills transferring
14	Quality of a working activity is influenced by a practice of staff motivation
15	Quality of a working activity is influenced by an appropriate health and well-being of staff
16	The positive psychological atmosphere improves the quality of a working activity

Table 2.
Organizational culture characteristics offered to respondents for evaluation

17	Control - Trust
18	Openness and communication - Psychological distance
19	Initiative – Diligence
20	Tough demands from management - Goodwill and tact of management
21	Team spirit, mutual assistance and support - Focusing in own responsibilities, autonomy
22	Enthusiastic work - Working without superfluous emotions
23	Focus on the organizational stability - Focus on the organizational development
24	Adherence to principles from the management
25	Permanent analysis of performance
26	Attention to detail
27	Criticism and analysis of shortcomings
28	The presence of the development strategy and the ability to see the perspective by the management

Table 2.

29	Interest in innovation
30	Commitment to corporate traditions and their strengthening
31	Attention to the moral aspects of life of an organization
32	Caring for people
33	Attention to the people's opinion
34	Participation of employees in making important organizational decisions
35	Team cohesion

Result and Discussion

We analyzed the relationship between the assertions of the professional worldview. The presence of branching statistical relationships (Fig. 3) leads to the conclusion that novice air traffic controllers in general have a quite complete and complex picture of professional views on basics of safe operation and risks related to human factor.

Interrelations between the characteristics of organizational culture, represented in individual perceptions of the respondents (Fig. 4) allow to see the presence of several fairly distinct groups of social and cultural features.

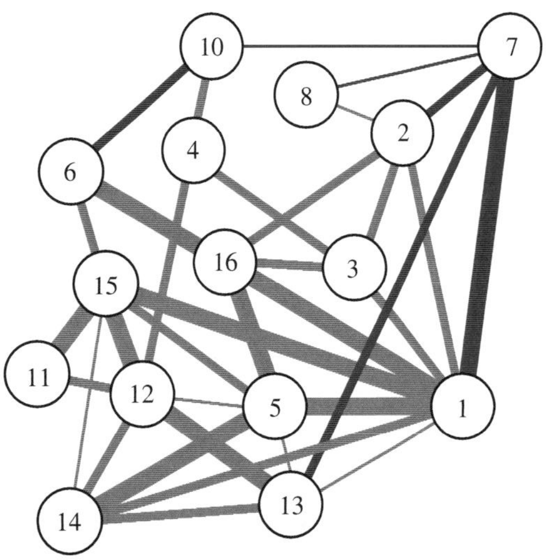

Figure 3. Pleiade of correlations between components of individual representations about the philosophy of safety air traffic management

Table 3.
Correlations (Fig. 3)

Couple	r	p	Couple	r	p
1 - 2	0.439	< 0.05	5 - 14	0.520	< 0.01
1 - 3	0.480	< 0.05	5 - 15	0.496	< 0.05
1 - 5	0.635	< 0.01	5 - 16	0.547	< 0.01
1 - 7	-0.512	< 0.01	6 - 10	-0.416	< 0.05
1 - 13	0.351	< 0.1	6 - 15	0.465	< 0.05
1 - 14	0.500	< 0.05	6 - 16	0.509	< 0.01
1 - 15	0.572	< 0.01	7 - 8	-0.337	< 0.1
1 - 16	0.509	< 0.01	7 - 10	-0.349	< 0.1
2 - 3	0.408	< 0.05	7 - 13	-0.474	< 0.05
2 - 7	-0.484	< 0.05	11 - 12	0.423	< 0.05
2 - 8	0.348	< 0.1	11 - 15	0.517	< 0.01
2 - 16	0.404	< 0.05	12 - 13	0.580	< 0.01
3 - 4	0.409	< 0.05	12 - 14	0.458	< 0.05
3 - 16	0.481	< 0.05	12 - 15	0.559	< 0.01
4 - 10	0.417	< 0.05	13 - 14	0.434	< 0.05
4 - 12	0.437	< 0.05	13 - 15	0.404	< 0.05
5 - 12	0.348	< 0.1	14 - 15	0.370	< 0.1
5 - 13	0.358	< 0.1			

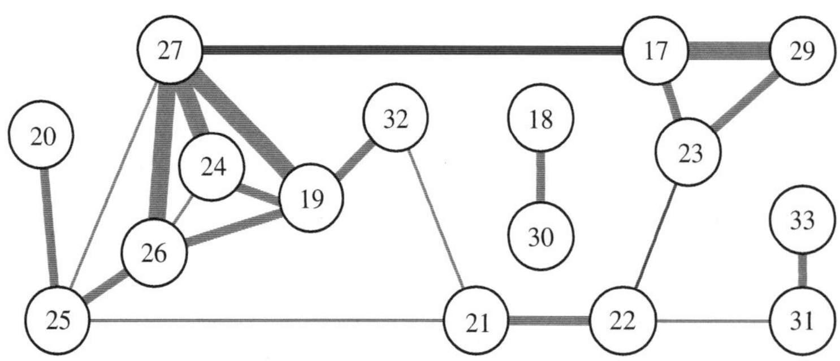

Figure 4. Pleiade of correlations between components of individual representations about organizational culture of ATM units

Table 4.
Correlations (Fig. 4)

Couple	r	p
17 - 23	0.486	< 0.05
17 - 27	-0.398	< 0.05
17 - 29	0.517	< 0.01
18 - 30	0.455	< 0.05
19 - 24	0.400	< 0.05
19 - 26	0.413	< 0.05
19 - 27	0.549	< 0.01

Couple	r	p
19 - 32	0.425	< 0.05
20 - 25	-0.457	< 0.05
21 - 22	0.458	< 0.05
21 - 25	0.356	< 0.1
21 - 32	0.347	< 0.1
22 - 23	-0.383	< 0.1
22 - 31	0.371	< 0.1

Couple	r	p
23 - 29	0.485	< 0.05
24 - 26	0.392	< 0.1
24 - 27	0.619	< 0.01
25 - 26	0.410	< 0.05
25 - 27	0.341	< 0.1
26 - 27	0.750	< 0.01
31 - 33	0.401	< 0.05

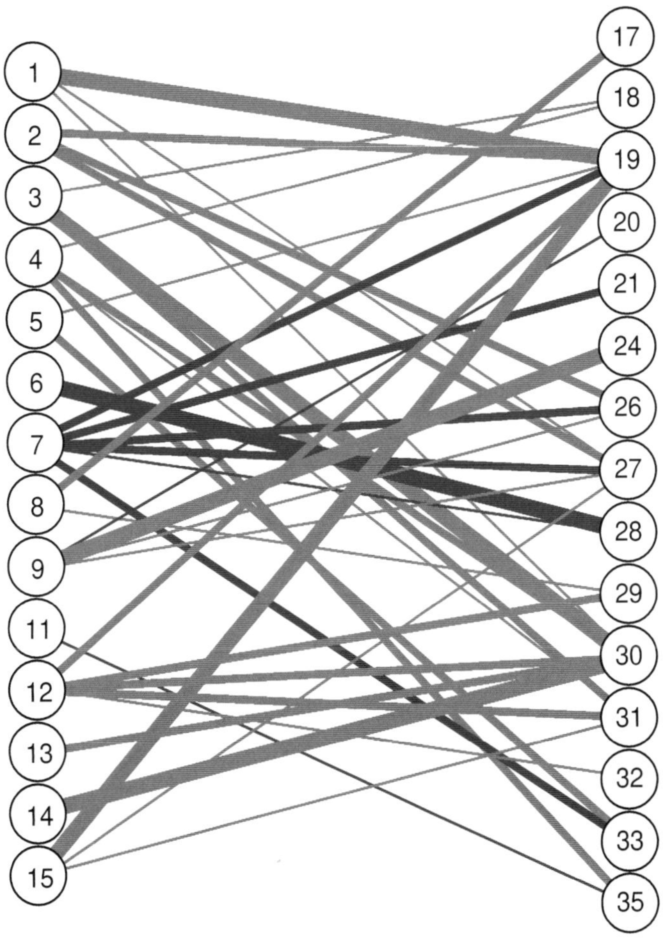

Figure 5. Statistical relationships between individual representations on approaches to safety of air traffic control and individual representations on optimal organizational culture of ATM system

Table 5.
Correlations (Fig. 5)

Couple	r	p
1 - 19	0.529	< 0.01
1 - 27	0.355	< 0.1
1 - 30	0.388	< 0.1
2 - 19	0.478	< 0.05
2 - 26	0.411	< 0.05
2 - 27	0.416	< 0.05
3 - 18	0.340	< 0.1
3 - 30	0.538	< 0.01
3 - 31	0.449	< 0.05
4 - 18	0.384	< 0.1
4 - 30	0.477	< 0.05
4 - 31	0.358	< 0.1
4 - 35	0.398	< 0.05
5 - 19	0.399	< 0.05
5 - 33	0.405	< 0.05
6 - 28	-0.512	< 0.1
7 - 19	-0.465	< 0.05
7 - 21	-0.410	< 0.05
7 - 26	-0.413	< 0.05
7 - 27	-0.451	< 0.05
7 - 28	-0.353	< 0.1
7 - 33	-0.398	< 0.05
8 - 17	0.404	< 0.05
8 - 29	0.349	< 0.1
9 - 20	-0.387	< 0.1
9 - 24	0.557	< 0.01
9 - 26	0.339	< 0.1
9 - 27	0.384	< 0.1
11 - 35	-0.394	< 0.1
12 - 19	0.397	< 0.05
12 - 29	0.439	< 0.05
12 - 30	0.467	< 0.05
12 - 31	0.450	< 0.05
12 - 32	0.373	< 0.1
13 - 30	0.426	< 0.05
14 - 30	0.517	< 0.01
15 - 19	0.555	< 0.01
15 - 27	0.369	< 0.1
15 - 31	0.345	< 0.1

The links 26-27-19, 17-23-29 seem quite understandable, as well as 31-33 and 21-22 (Fig. 4). But the link 18-30 appears to be paradoxical. One of the possible interpretations of this link may be in some formal attitude to the corporate values. This requires additional research.

Statistical relations between professional worldview on human factor in the air traffic management system and individual view on optimal organizational culture in the ATM system (Fig. 5) show sensitivity of the organizational life to the personnel qualification regarding human factors. But at the same time these relations demonstrate possible risk of appearance of incompetent actions due to deformation of organizational life. Positive effects of combination of certain organizational culture with a certain operating philosophy are possible as well as negative ones.

Through the efforts of the aviation community operator personnel gets knowledge on human factors at the appropriate level in any part of the world, but the practice of organizational management is more dependent on organizational and national cultures, which creates the likelihood of situations in which the required expertise simply would not be realized.

References

Cameron, K.S., Quinn, R.E. (1999). Diagnosing and Changing Organizational Culture. – Addison-Wesley Longman, Inc.

Hawkins, F.H. (1987). Human factors in flight (2nd Ed.). – Ashgate (Aldershot, UK).

Hofstede, G.H. (1991) Cultures and Organizations: Software of the Mind. – New York: McGraw-Hill.

Johnston, N. (1993) CRM: Cross-cultural perspectives. In: Earl L. Wiener, Barbara G. Kanki, Robert L. Helmreich (Eds.); with a foreword by John K. Lauber, *Cockpit Resource Management.* – London, Academic Press (pp. 367-395).

Petrenko, O.V. (2010). Psychological Aspects of Flight-Crew Resource Management. In: Actual Problems of Psychology: Scientific Research Journal, Vol. 10, Issue 17 – Kyiv, Ukraine (pp. 346-355).

Petrenko, O.V. (2015). Actual aspects of the modern model of human factor. *Proceedings of the XII International Conference "Avia-2015"*, National Aviation University, Kyiv, Ukraine (pp. 9.1-9.5).

SMS for Aviation – a Practical Guide. Civil Aviation Safety Authority, Australian Government. Retrieved from: https://www.casa.gov.au/sites/g/files/net351/f/_assets/main/sms/download/2012-sms-book6-human-factors.pdf

Trompenaars, F., Hampden-Turner, C. (1997). Riding the Waves of Culture.

SELF-DECEPTION AND IMPRESSION MANAGEMENT IN COMMERCIAL PILOTS: AN UNDERREPORTED AND POTENTIAL CONFOUND IN AVIATION RESEARCH

Tabitha T. Black
Eastern Washington University
Cheney, WA
Charalambos C. Cleanthous
Eastern Washington University
Cheney, WA
Ryan Sain
PsyCore
Spokane, WA
Steven M. Vera
Eastern Washington University
Cheney, WA

A concern when administering questionnaires is whether the participant is providing information that is accurate. The *Balanced Inventory of Desirable Responding* (BIDR) was used to assess commercial pilots' socially desirable responding resulting in two profiles: Impression management (IM; faking bad) and self-deceptive enhancement (SDE; faking good). These pilots' profiles were compared to the *Aviation Safety Locus of Control* (ASLOC) scale, used to measure external (ASLOC-E) or internal (ASLOC-I) orientation, and the *Crew Resource Management Training Survey* (CRMTS) developed from the Federal Aviation Administration's guidelines for CRM. The results from the SDE indicated that over a fourth of the participants responded in a socially desirable manner. Significant differences were also found between those scoring high on the IM subscale versus those scoring in the normal range of the CRMTS subscales.

The well documented issue of socially desirable responding continues to present a self-report validity concern to behavioral science researchers (van de Mortel, 2008). If uncontrolled, it may confound the validity of research results (Nederhof, 1985). Socially desirable responding occurs when a participant's response bias results in answering survey questions that present the participant in a favorable light. This bias is a function of the test behavior of the subject (King, Bruner, & Hensel, 1991) and not necessarily always indicative of malicious intent. SDR can result in self-deception, in which the participant believes the presentation to be true about oneself (Nederhof, 1985; Paulhus, 1991). Alternatively, the response bias may result in the attempt to present oneself as worse off than what one is, this is known as impression management (Paulhus, 1991). Several factors contribute to SDR including the test setting, participant motives, and the participant's expectation of repurcussions of responses (King, Bruner, & Hensel, 1991). Both qualitative (Stodel, 2015) and quantitative research efforts have attempted to identify SDR biased responding. One of the early attempts at quantifying the construct through the use of a questionnaire, that had acceptable validity and reliability was the Crowne-Marlow scale (Crowne & Marlowe, 1960). A variety of methods have been devised to control SDR, yet none are efficacious in controlling the response bias for specific settings or intended populations (Hunter & Stewart, 2009; Nederhof, 1985).

There is a dearth of research regarding SDR and commercial pilots, but the research that does exist strongly suggests that there is a need for this measure when assessing pilot responses (Butcher & Han, 1995). Pilots, through the process of aviation training and testing, become well adapted to positive self-presentation for the purposes of career advancement (Butcher & Han, 1995). They spend their career in regular training and testing for the purpose of maintaining proficiency in their work demands. This training consists of both written and oral tests, and concurrent validity of maneuver performance, also known as practical tests, of maneuvers while under the supervision of an examining authority. Interweaved into the fabric of crew resource management (CRM) training, exists a cultivation of personal confidence, assertiveness and authority as pilot in command. SDR factors, setting, personal motives, and expectancies (King, 1991) when applied in aviation, may or may not be amplified (Galic, Jerneic, & Kovacic, 2012). Moreover, individuals who have a motive to present themselves in superlative manner, such as commercial pilots, consistently produce higher profiles on defensiveness indexes (Butcher & Han, 1995). The concern over pilot defensiveness has led to the development of a scale specifically applicable to commercial pilots (Butcher & Han, 1995).

Directly relevant to SDR, is CRM as the in-cockpit activities directly impact the outcome of any given flight. A co-pilot, for example, who does not speak up when one should, due to one's self-preservation concern, a potential SDR issue, directly impacts the outcome of the flight. CRM training in the USA has been in effect for approximately two decades (Helmreich, Merritt, & Wilhelm, 1999). Overall, crewmembers find CRM training to be relevant and useful, and the aviation industry within the US has overwhelmingly endorsed the program (Helmreich & Wilhelm, 1991). The foundation of the concept and the challenges it aims to address have been as a result of workshops and meetings initiated by airlines and aerospace authorities for the purpose of aviation safety for the last 40 years. The evolution of CRM has been a reactive analysis initiated by various aviation forums (Helmreich, Merritt, & Wilhelm, 1999). Although generally accepted, one continuing question in CRM is its effectiveness and validity (Salas, 2006). One approach to investigating CRM's effectiveness is through aviation psychological research, that is, observation of CRM performance and surveying pilots' opinions and attitudes. CRM method performance and inquiry of the pilot. In order to evaluate the research, one must understand the degree to which participant bias affects the resultant data.

An area in which pilots, overall, have excelled is locus of control, that is, they exhibit an internal locus of control indicating that they are responsible and capable of dealing with events (Hunter, 2002; Skinner, 2011). In the aviation environment safety is paramount; the perceived locus of control of an event is important to predicting the outcome of emergency situations (Hunter, 2002). The research indicates that pilots who are at greater risk of aircraft accidents can be identified in advance (Hunter, 2002). The aviation safety locus of control scale (ASLOC) has been useful in this regard (Hunter, 2002). Given the concern with SDR, however, the question that arises is the extent to which a pilot's SDE or IM impact the degree of perceived control, and, CRM. Given the concern with SDR and the dearth of research in the area, the question that arises is whether those scoring high on either the SDE or IM also demonstrate an external ASLOC and, consequently, making them poorer managers of emergency situations.

The present study sought to examine whether pilots who endorsed socially desirable items had significantly different profiles than those who did not. If differences were identified, if those differences led to a lower endorsement of various crew resource management criteria as promoted by USA Federal Aviation Administration guidelines (Federal Aviation Administration, 2004) or an external locus of control profile.

Methods

Participants

With the permission from the site administrators, a link to the Crew Resource Management Study was posted on the Flights Above the Pacific Northwest Facebook page (https://www.facebook.com/groups/FLightsAboveThePNW/) and Airline Pilot Central Forums (http://www.airlinepilotforums.com/). Participation of the survey was restricted to USA/FAA commercial rated pilots, employed as an active pilot within the last 10 years. This restriction ensured that all participants would have completed a formalized CRM training per FAA regulation AC 120-51E (Federal Aviation Administation, 2004). The results indicated that the distribution by participant gender matched the current ratio of employed commercial pilots within the US: 65 total participants: 58 males, 5 females, 2 gender non-response (Bureau of Labor Statistics, 2016).

Measures

The *Balanced Inventory of Desirable Responding* (BIDR) was developed to measure two dimensions of SDR (Paulhus, 1991) that were absent in prior measures. The BIDR is comprised of 40 7-point Likert-type scale items. Paulhus (1999) reported convergent validity for SDE subscale with several other scales including, among others, repressive styles, defense mechanisms, and ways of coping. Convergent validity for the IM subscale was reported Eysenck's Lie scale and the MMPI Lie scale. The internal reliability resulted in Cronbach's alpha for IM=.84 and SDE=.75. Convergent validity for the SDE subscale was reported with several other scales including, among others, repressive styles, defense mechanisms, and ways of coping. The first subscale is referred to as impression management (IM) and is a bias that reflects a person's attempt to present oneself in an unrealistically positive manner; it is also referred to as faking good. The self-deceptive enhancement (SDE) index, the second subscale, measures the behavioral response tendency of a to answer items in a manner that portrays oneself in a positive light.

The *Aviation Safety Locus of Control* (ASLOC) is unique in this domain of scales as it was designed specifically to measure safety issues within an aviation environment; two subscales are produced from the 20 items: external (ASLOC-E) and internal safety locus (ASLOC-I) (Hunter, 2002). The items are presented on a 5 point Likert-type scale (*strongly agree* to *strongly disagree*). Hunter (2002) reported that the two subscales of the ASLOC exhibited acceptable internal consistency and were negatively correlated ($r = -0.419$, $p < 0.001$). Construct validity was reported by comparing the combined ASLOC scores with the resignation score from the Hazardous Attitudes Inventory.

A demographic survey was designed to provide a description of the participants including: gender, post-secondary education and total flight time. The *Crew Resource Management Training Survey* (CRMTS) was developed (Black) from the FAA guidelines (Federal Aviation Administration, 2004) for CRM training to assess the participants and opinions and self-reported use of CRM training procedures. The CRMTS was comprised of seven subscales including: pilot in command (PIC), communication (COM), management of a flight team (MFT), time management (TM), fatigue (FTG), stress (STR), and aeronautical decision making (ADM). The eighty items were made up of three response styles based on content: 5 point Likert-type scale (*strongly agree* to *strongly disagree*), true-false, and multiple choice items.

Results

There were 68 anonymous participants. Three were dropped from the analysis as their data was too incomplete to use. Of the 65 remaining participants, 5 were female and 58 males. These numbers match the current female-male ratio of pilots in the US. The small number of female participants makes it impossible to conduct adequate statistical analyses and will only be used to indicate trends. In terms of flight time, 8 (7 male, 1 female) pilots had between 1000-2000 hours, 16 (13 male, 3 female) between 2000-4000 hours, 17 (16 male, 1 female) between 5000-10,000 hours, and 22 (all male) had over 10,000 hours. The participants' educational level, inclusive of either completed or earned degree, 8 had an AA or AS, 33 a BA or BS, 21 with an MA or MS, and 2 with a doctorate.

The overall results showed that 48 pilots scored in the normal range on the BIDR, 13 scored high on IM and 4 on SDE. Table 1 shows the results from a correlational analysis of the relationship between the BIDR and ASLOC scales. Table 1 shows the significant correlations between the BIDR, ASLOC, and CRM. The only correlation that was significant between the BIDR and ASLOC was between the SDE and ASLOC-E that are positively correlated. There was a significant negative correlation between the ASLOC-I and time management. The ASLOC-E was also positively correlated with time management, and negatively correlated with pilot-in-command.

Table 1.

Pearson Product Moment Correlation: BIDR and ASLOC Subscales

	N	r	Sig.
BIDR (SDE) x ASLOC-E	64	.25	.047
ASLOC-I x CRM--Time Management	64	-.31	.012
ASLOC-E x CRM--Time Management	64	.34	.006
ASLOC-E x CRM--Pilot-in-Command	63	-.35	.005

The comparison of the BIDR IM and SDE and CRM subscales indicated several important results. Significant differences were found on four of the seven subscales and the overall CRM score. Comparisons across all groups were made using the Kruskal-Wallis independents samples test. Overall CRM, ADM, FTG, MFT, and COM subscales were significant at the .003, .006, .002, .029, and .009 levels, respectively. All follow-up comparisons using a Mann-Whitney were not significant after a Bonferroni correction or were invalid. Figure 1 shows that where there were significant differences, the normal group endorsed more CRM items than either the SDE or the IM group.

The overall CRM score indicated that the normal and the SDE group endorsed an average of 49% of the items, whereas the IM group endorsed 33%. On the ADM subscale, the normal group endorsed 77% and the SDE and IM group both endorsed 68% of the items. On the FTG subscale, the normal group endorsed 93% whereas the SDE group endorsed 90% and the IM group 69% of the items; this was the only subscale on which the SDE group endorsed more items than the IM group. The MFT subscale indicated that the normal group endorsed 37%, the SDE 29%, and the IM group endorsed 23% of the items. Finally, the COM results indicated that the normal group endorsed 38%, the SDE group 28%, and the IM 22% of the items.

Figure 1.

Comparison of BIDR on CRM Endorsement

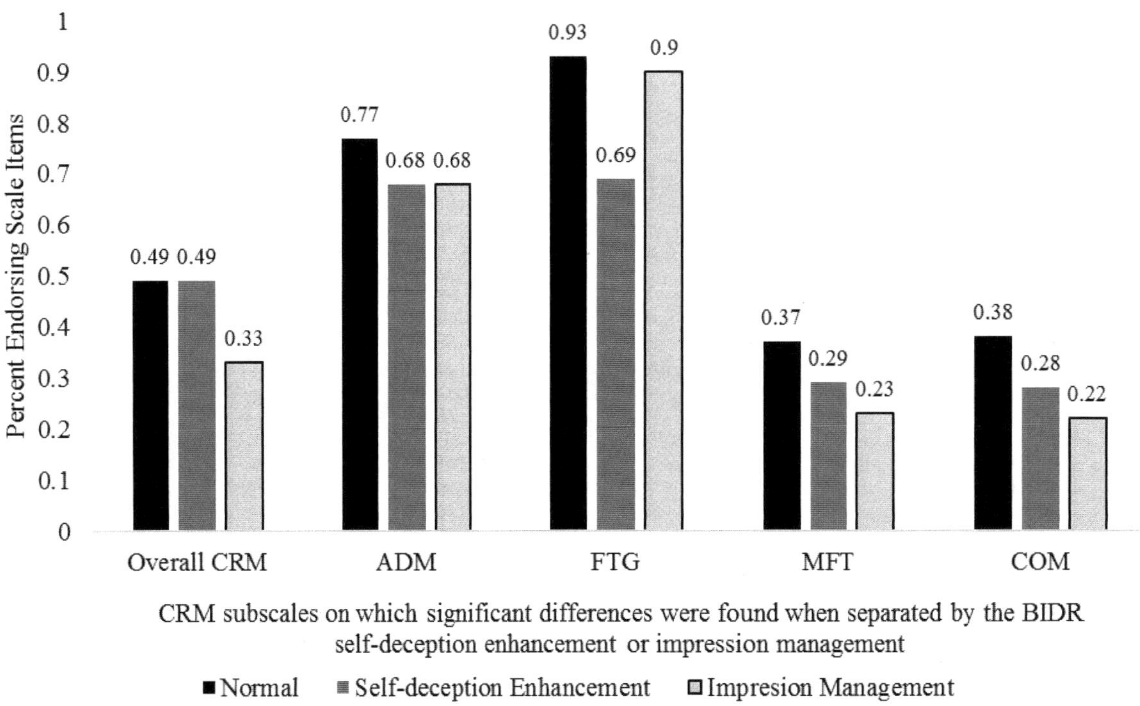

Discussion

The results speak to the necessity of using a scale that assesses socially desirable responding in aviation research. Although a variety of procedures have been developed to counteract socially desirable responding, all leave much room for improvement. In the present case, rather than excluding the socially desirable responders, their results made up the comparison group allowing us to detect important differences when compared to the typical responders.

The positive correlation between SDE and ASLOC-E, although relatively small, does suggest that these two factors are influencing one another. However, given that it is a weak relationship, it is possible that extreme scores are impacting this relationship. Further examination of this relationship is warranted, especially as the correlation is between an external locus of control and someone who is faking good. In the ASLOC-E case, it is someone who may be attributing to factors out of one's control, not the best strategy in the cockpit, and someone who is exaggerating one's strengths. The latter is dangerous as it may be masking one's limitations (King, 1991). Two correlations in opposite directions were found depending on whether the relationship was between ASLOC-I or ASLOC-E and TM. One group is indicating that TM is a problem (ASLOC-E) and the other (ASLOC-I) is suggesting that time management is not such a big problem. Further investigation of how these attitudes are impacting one's in-cockpit and work-related behavior is important as time is inescapable. The final significant correlation was between ASLOC-E and PIC. The results indicate that those scoring high on ASLOC-E had a diminishing view of the PIC's role. Again, given the nature of the cockpit and the necessity of working as a team, further exploration of this issue is warranted.

The other analyses were concerned with SDR and CRM. Examining the overall CRM endorsement, two conclusions can be reached. First, the overall endorsement of CRM practices is under 50%. Second, the IM group endorses CRM practices at 33%. The overall results suggest that there is much room for improvement irrespective of ones SDE or IM score. However, a finer grained analysis indicates that there is considerable variation in the items endorsed by all groups. Such a distribution argues for an interpretation at the individual scale level. By examining the items that are endorsed or not endorsed, would allow for further refinement of both the scale and more targeted CRM interventions. In other words, training would be targeted for those weak in the area of CRM. Based on that assessment of the individual subscales, one would have various groups to target, that is, the groups would be made up of the subscales on which one was weak, potentially, that could be all seven subscales. An important use of the overall score could be to determine weaknesses in an individual's understanding of CRM. Having identified such an understanding a more targeted intervention could be developed. All this can take place before the pilot, for obvious CRM reasons, is allowed in the cockpit. Moreover, such a targeted intervention could be used as a continual assessment of the impact of CRM training.

There were two limitations to the present study. The first limitation was the development of the CRMTS. It is a rationally developed survey based on the criteria established by the FAA. It is crucially important that such a survey be developed with the appropriate psychometric properties. The second limitation concerned the participant sample. In particular, the concern is with the low number of female participants. Even though the percentage of participants matched the USA commercial pilot rates, the low number made it impossible to draw any statistically meaningful conclusions. Obviously, it behooves researchers to pursue this matter with some urgency as the females displayed a higher rate of SDR than did their male counterparts.

A strength of the present research was to use a computerized version of SDR that research indicates is the best, current, method for decreasing SDR responding. It is possible that the current rate of 27%, bad as that may be, is lower than would have been the case had paper-and-pencil assessments been used. Given the weakness and strengths of the present study, the conclusion that SDR impacts what commercial pilots are endorsing about CRM training and practices cannot be overstated. Incorporating measures to assess and control SDR responding in commercial pilots is warranted.

References

Bureau of Labor Statistics (2016). *Labor Force Statistics from the Current Population Survey.* Transportation and material moving—aircraft pilot and flight engineers. Retrieved on February 20, 2017, from: https://www.bls.gov/cps/cpsaat11

Butcher, J. N., & Han, K. (1995). Development of an MMPI-2 scale to assess the presentation of self in a superlative manner: The S Scale. In J. N. Butcher, C. D. Spielberger, J. N. Butcher, C. D. Spielberger (Eds.), *Advances in personality assessment* (vol. 10, pp. 25-50). Hillsdale, NJ, US: Lawrence Erlbaum Associates, Inc.

Crowne, D. P., & Marlowe, D. (1960). A new social scale of social desirability independent of psychopathology. *Journal of Consulting Psychology, 24*, 349-354.

Flights Above The Pacific Northwest (2016). https://www.facebook.com/groups/FLightsAboveThePNW/

Federal Aviation Administration. (2004). Advisory Circular System 120-51E (2004) – Crew resource management training. Retrieved on August 4, 2016 from http://www.faa.gov/regulations_policies/advisory_circulars/index.cfm/go/document.information/documentid/22879

Galic, Z., Jerneic, Z., & Kovacic, M. P. (2012). Do applicants fake their personality questionnaire responses and how successful are their attempts? A case of military pilot cadet selection. *International Journal of Selection and Assessment*, 20(2), 229-242.

Helmreich, R. L., & Wilhelm, J. A., (1991). Outcomes of crew resource management training. *International Journal of Aviation Psychology, 1*(4), 287-300.

Helmreich, R. L., Merritt, A. C., & Wilhelm, J.A. (1999). The evolution of crew resource management training in commercial aviation. *International Journal of Aviation Psychology*, 9(1), 19-32.

Hunter, D. R. (2002). Development of an aviation safety locus of control scale. *Aviation, Space, and Environmental Medicine,* 73(12), 1184-1188.

Hunter, D. R., Stewart, J. E., (2009). Locus of control risk orientation, and decision making among U.S. army aviators. Technical Report 1260. Personnel, Performance and Training Technology Army Project Number 622785A790.

King, M. F., Bruner, G. C., & Hensel, P. J. A., (1991). Social desirability bias: A neglected aspect of validity testing. *Psychology and Marketing 2000, 17*(2), 79-103. Retrieved January 15, 2015 from https://www.researchgate.net/profile/Gordon_Bruner/publication/230034577

van de Mortel, T. F. (2008). Faking it: Social desirability response bias in self-report research. *Australian Journal of Advanced Nursing, 25*(4), 40-48.

Nederhof, A. J. (1985). Methods of coping with social desirability bias: A review. *European Journal of Social Psychology, 15*(3), 263-280.

Paulhus, D. L. (1991). Measurement and control of response bias. In J. P. Robinson, P. R. Shaver, & L. S. Wrightsman (Eds.), *Measures of personality and social psychological attitudes* (pp. 17-59). New York: Academic Press.

Salas, E (2006). Does crew resource management work? An update, an extension and some critical needs. *Human Factors and Ergonomics Society, 48*(2) 392-412.

Skinner, E. A., & Zimmer-Gembeck, M. J. (2011). Perceived control and the development of coping. In S. Folkman, S. Folkman (Eds.), *The Oxford handbook of stress, health, and coping* (pp. 35-59). New York, NY, US: Oxford University Press.

Stodel, M. (2015). But what will people think? Getting beyond social desirability bias by increasing cognitive load. *International Journal of Market Research, 57*(2), 313-323. DOI:10.25/IJMR-2015-024

ADAPTIVE AUTOMATION BASED ON AIR TRAFFIC CONTROLLER DECISION-MAKING

M. IJtsma, C. Borst, G.A. Mercado-Velasco, M. Mulder, M.M. van Paassen
Delft University of Technology
Delft, the Netherlands

Through smart scheduling and triggering of automation support, adaptive automation has the potential to balance air traffic controller workload. The challenge in the design of adaptive automation systems is to decide how and when the automation should provide support. This paper describes the design of a novel mechanism for adaptively invoking automation support. Whereas most adaptive automation support systems are reactive in that they invoke automation support *after* controller workload has increased, the aim of the designed mechanism is to proactively trigger automation support *prior* to workload increases. To do this, the mechanism assesses the quality of air traffic controller's decisions. The designed adaptive automation system has been tested in a human-in-the-loop experiment. Results indicate that the adaptive support helps to increase efficiency and safety as compared to manual control. However, lower triggering thresholds (resulting in more frequent automation intervention) increased the frustration level of participants (as measured with NASA TLX) and decreased acceptance of the support.

Currently, one of the main limiting factors toward increasing the airspace capacity is the workload of the Air Traffic Controller (ATCo) (Tobaruela et al., 2014). In managing ATCo workload, the concept of adaptive automation has the potential to balance workload between underload and overload. Contrary to static automation, adaptive automation does not operate at a single Level of Automation (LOA), but adapts itself by dynamically switching between multiple LOAs during operations. By smartly trading control between automation and the human controller, higher LOA support can be provided during times of high workload, while lower LOA support or manual control can be offered during low workload conditions.

Two main challenges in the design of adaptive automation systems are to determine *how* and *when* the automation should intervene. Firstly, a wide range of variables can be considered to trigger switches between LOAs. Secondly, whatever the type of triggering variable, different thresholds can be set at which to raise or lower LOAs. The goal here is to find a combination between the triggering variable and its threshold that yields the best possible timing of automation support (Parasuraman, Bahri, Deaton, Morrison, & Barnes, 1992).

Earlier studies have looked at a multitude of metrics for invoking automation support, for example the number of aircraft in an ATC sector (Hilburn, Jorna, Byrne, & Parasuraman, 1997) or electroencephalograms (EEGs) of the operator (Freeman, Mikulka, Scerbo, Prinzel, & Clouatre, 2000). Although these studies have shown promising results, there is room for developing "smarter" mechanisms that can adapt the automation support more reliably and intelligently. Instead of reacting to increased workload conditions, as most existing mechanisms do, novel triggering mechanisms can be designed that focus on preventing higher workload through the use of predictive measures. With these predictive measures, automation support can be invoked *prior* to the actual workload increase.

This study investigates a new combination of triggering variable and thresholds that can potentially prevent workload: decision-quality and the number of allowed "bad" decisions. Several studies have observed that a remarkable amount of traffic complexity is a direct result of the controller's wrong decisions and suboptimal control actions. For example, a study on operational data of Short-Term Conflict Alert (STCA) found that for one out of two STCAs, the controller implemented a resolution that triggered additional STCAs (Lillo et al. 2009). Similarly, a study on sector complexity found that in 120 out of 400 experiment runs participants had introduced one or more additional conflicts as a result of suboptimal control actions (Rahman, Borst, Mulder, & van Paassen, 2015). Detecting and preventing such "self-induced" complexity is expected to reduce workload considerably.

This study explores the use of adaptive automation to improve the quality of the controller's decisions and control actions. It is hypothesized that through the promotion of good decision-making, the adaptive automation system can help to balance the controller's workload. The study includes the design of a decision-based adaptive automation system for Air Traffic Control (ATC) and a subsequent experiment to test the effects of three different triggering thresholds – in terms of allowed "bad" decisions – on the workload, performance and situation awareness of ATCos.

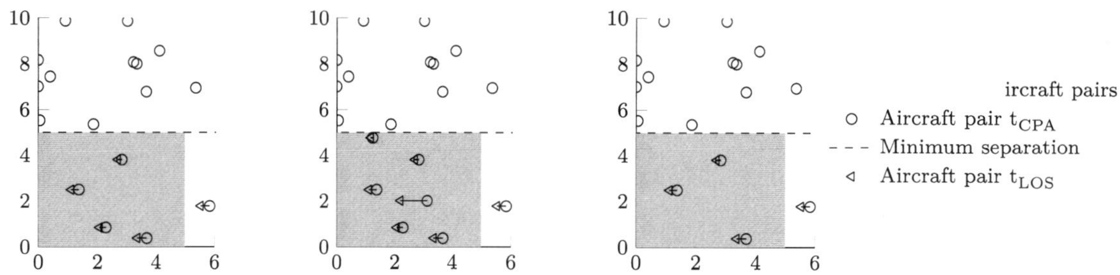

(a) (b) (c)

Figure 1. Three visualizations of time to CPA and separation at CPA. Subfigures (a) the initial condition, (b) a bad decision (increase in number of conflicts) and (c) a good decision (decrease in the number of conflicts).

Adaptive Automation Design

The designed adaptive automation system assesses decision-quality based on an increase or decrease of the number of conflicting aircraft pairs in the airspace. For each aircraft pair in the controller's sector, the projected separation at Closest Point of Approach (CPA) is an indication of whether the two aircraft will lose separation when no action is taken. For conflicting aircraft pairs, the time to Loss of Separation (LOS) indicates the urgency of the conflict. For the adaptive automation system, an aircraft pair is a conflicting pair when their projected separation at CPA is less than 5 NM and the time to CPA is less than 5 minutes.

Figure 1 visualizes the projected separation at CPA and the time to LOS (or time to CPA for non-conflicting pairs) for all aircraft pairs in a particular sector using the Separation Monitor (Irfan, Bull, Clinch, & Pember, 2012). Aircraft pairs within the red shaded area are in conflict. When taking actions to resolve a conflict, the controller's resolution should move the concerned aircraft pair outside this critical area. Additionally, other aircraft pairs should not be moved inside the critical area as a consequence of the controller's action. In this latter case, the controller has induced secondary conflicts, which will then result in self-induced workload because these conflicts also need to be resolved. Thus, a "good" decision should reduce the number of aircraft pairs in the critical area of the separation monitor. Vice versa, a decision and subsequent control action that increases the number of aircraft pairs in the critical area can be considered a "bad" decision.

With the *how* of triggering adaptive automation support being the occurence of bad decisions, one needs to decide what the appropriate threshold is for switching to a higher level of support. This will determine *when* the support is provided: should additional support be provided as soon as one bad decision is being made or should the system be more lenient and only intervene when multiple bad decisions are being made. To explore the effect of this triggering threshold on the system performance, we tested three different thresholds for the adaptive automation in a human-in-the-loop experiment that is described in the next sections.

Although adaptive automation can employ any number of LOAs, we limited the adaptive automation aid here to two LOAs to reduce the risk of over-complicating the design and thereby confounding our experiment. Two LOAs were defined, a low and a high level:

- The low LOA is manual control with short-term conflict warnings and alerts (STCAs). Visual STCA warnings are provided 130 seconds prior to LOS events. Aural and visual STCA alerts are provided 60 seconds prior to LOS.
- At the high LOA, additionally to the STCA warnings and alerts, the automation provides advisories for resolving conflicts between aircraft. The resolution advisories are provided on a management-by-exception basis, i.e., the controllers have a fifteen-second timespan to accept or reject an advisory, after that the advisory is implemented automatically.

The algorithm for automated resolution advisories has been designed to follow ATCo best practices for three types of conflict geometries: overtaking, crossing and reciprocal. In order to limit the scope of the automation algorithm design, the experiment focused on the use of pure heading changes to resolve conflicts: participants could not give altitude or speed commands.

Table 1.

Independent Variables

Condition	Description	Hypothesis
AA1	Triggering support after any single bad decision	Early intervention, lowest workload, highest efficiency & safety
AA2	Triggering support after two bad decisions	Intermediate intervention, lower workload, higher efficiency & safety
AA3	Triggering support after three bad decisions	Late intervention, medium workload, medium efficiency & safety
MAN	Baseline: manual control without automation support.	Highest workload, lowest efficiency & safety

Note. One run was performed with full automation, for which no participants were required. This full automation run was used as a baseline performance condition, in which the automation solution is regarded as the optimal solution. In the following, this condition is referred to as condition AUTO.

Experiment Design

An experiment was conducted to study the effects of different triggering thresholds for invoking adaptive automation on controller workload, automation acceptance, efficiency and safety. In addition, the experiment is a test case for the effectiveness of the decision-based triggering mechanism in balancing workload and preventing self-induced conflicts.

Participants and task. Eighteen participants (2 female, 16 male, average age 27 years) were selected, consisting of students and staff members of the Faculty of Aerospace Engineering. All participants had some experience with ATC, through participating in courses and earlier experiments related to ATC.. Participants could interactively control aircraft using the mouse and keyboard and thus no radiotelephony was needed. Commands were implemented automatically, corresponding to a situation in which a data link is available to communicate vector clearances to aircraft. The task of the participants was to vector (the experiment was limited to heading clearances only, speed and altitude was fixed) aircraft to their designated sector exit waypoints, whilst preventing and resolving any conflicts between aircraft.

Independent and control variables. As the independent variable, the triggering threshold for invoking the adaptive automation was varied. Table 1 lists the different conditions. Control variables include the duration that automation support is active after it is triggered (during pre-experiment testing it was found that an automation duration of 30 seconds provided appropriate support), the expiration time of resolution advisories (set at 15 seconds as determined during pre-experiment testing), and the traffic scenario (to prevent the participants from recognizing conflict geometries from earlier experiment runs, the airspace was between experiment runs).

Procedure. The experiment consisted of the pre-experiment briefing, a training phase and a measurement phase. Breaks were held between the training and measurement phases and halfway the measurement runs. The training phase consisted of eight training runs, which allowed the participants to become familiar with the working of the simulator and the automation support. Here, the adaptive automation system was used as a training tool: when a participant could manage the traffic in these last few runs without triggering the automation support (i.e., making bad decisions), it was an indication that a baseline performance was met and that the participant was sufficiently trained. Finally, the measurement phase consisted of four runs, one for each experiment condition. A Latin square design was used to randomize the conditions and prevent carry-over effects in the measurements.

Scenario. A single traffic scenario was used for the measurement runs. This scenario consisted of three traffic flows, between which conflicts emanated at two intersection points. The intersections were chosen such that in order to prevent self-induced conflicts, any action to resolve conflicts at the first intersection required careful consideration of the traffic at the second intersection. Runs were 15 minutes long and the simulation was run twice as fast as real-time, to make sure participant stayed engaged with the task. At the start the scenario, there was a fade-in period of two minutes, to allow the participant to become familiar with the traffic situation while the traffic density gradually built up. There were nine conflicting aircraft pairs precoded in the traffic scenario.

Dependent measures. During the experiment runs, participants were asked for Instantaneous Self-Assessment (ISA) ratings of their subjective workload once every minute. A metric for airspace complexity, which has been shown to correlate with ATCo workload (d'Engelbronner, Borst, Ellerbroek, van Paassen, & Mulder, 2015), was used to gain more objective insight in the participants' workload. This metric consists of the relative number of heading and speed commands that will create a conflict, averaged over all aircraft in the airspace sector, and provides an indication of the average "solution-space" that is available to the air traffic controller; the higher the number, the smaller the solution-space is and the more likely it is that the participant is experiencing a high workload. Furthermore, performance indicators such as number of conflicts, number of control actions (for this measure the AUTO condition was used as baseline) and number of Short Term Conflict Alerts (STCA) were recorded during the runs. After each run, participants filled out Controller Acceptance Rating Scales (CARS), a NASA-TLX workload form and a short questionnaire.

Results and Discussion

Performance. Figure 2 shows the total number of aircraft pairs in the critical area of the separation monitor, defined by $t_{CPA} < 5$ and $d_{CPA} < 5$. With higher thresholds, there are notably fewer conflicts, which confirms that the automation algorithm has worked as intended. A repeated-measures ANOVA indicated that the difference between conditions is significant ($F(3,45) = 14.478, p < 0.001$). A post-hoc test revealed that the significance is found between conditions AA1 and AA3, conditions AA1 and MAN and conditions AA2 and MAN.

The minimum required number of control actions for the scenario was sixteen (as solved in the AUTO condition). None of the participants was able to solve the scenarios with this minimum number of control actions, requiring on average nine more control actions. The data did show a clear reduction in the number of implemented control actions with lower triggering thresholds, shown in Figure 3. A repeated-measures ANOVA indicated that this reduction is significant ($F(3, 45) = 8.091, p < 0.01$). A post-hoc test revealed that this effect is found between conditions AA1 and AA3 and conditions AA1 and MAN. Although the number of implemented control actions reduced with lower triggering thresholds, the total number of actions (which also includes accepting and rejecting advisories) appeared to be constant or even slightly increasing. A repeated-measures ANOVA indicated that there was no significant difference between the conditions ($F(3, 45) = 0.825, p = 0.487$).

Table 2 shows the number of STCA alerts that were encountered during the experiment runs. The mean ranks clearly show that there were fewer STCA alerts with lower triggering thresholds. Indeed, a Friedman test indicated that there was a significant difference between the conditions ($\chi2(3) = 15.593, p = 0.001$). Post hoc analysis with a Wilcoxon sign-ranked test applying a Bonferroni correction indicated that this difference can be found between AA1 and the three other conditions: AA2 ($Z = -2.646, p = 0.008$), AA3 ($Z = -2.887, p = 0.004$) and MAN ($Z = -2.673, p = 0.004$); but not between other condition pairs.

Complexity and workload. The means of the ISA rating Z-scores over each experiment run are shown in Figure 3. With stricter automation it seems that the means of the Z-scores are reduced, which indicates a slight decrease in workload with lower triggering thresholds. A repeated-measures ANOVA indicated that this reduction is not significant ($F(2, 30) = 0.642, p = 0.592$). The total scores of the NASA-TLX ratings are shown in Figure 4. The data do not show a clear pattern as a result of the experiment manipulation, which contrasts with the ISA ratings. A Friedman test indicated that there was no significant difference between the conditions ($\chi2(3) = 4.208, p = 0.240$). A breakdown of the various components of the NASA-TLX score revealed that the stricter automation conditions have more workload originating from frustration, both in weighting as well as in rating.

From the questionnaire results, it appeared that frustration mainly originated from occasions in which the automation's advice interfered with the participant's own plans. This resulted in occasional "fights" between

Table 2

Medians and mean rank of the number of STCA alerts

Condition	Median (interquartile range)	Mean rank
AA1	0 (0 to 2)	1.72
AA2	2 (2 to 2)	2.56
AA3	2 (2 to 2)	2.84
MAN	2 (2 to 2)	2.88

Table 3

Medians and mean rank of CARS ratings.

Condition	Median (interquartile range)	Mean rank
AA1	7.5 (6.25 to 8)	1.84
AA2	7 (7 to 9)	2.00
AA3	8 (7 to 9)	2.16

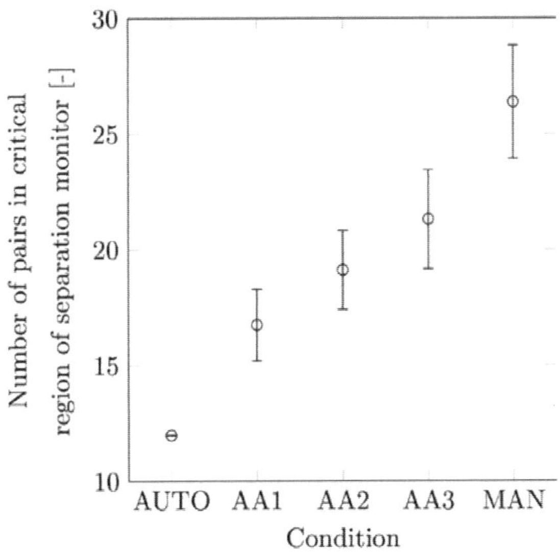

Figure 2. Total number of aircraft pairs in conflict, over duration of experiment run.

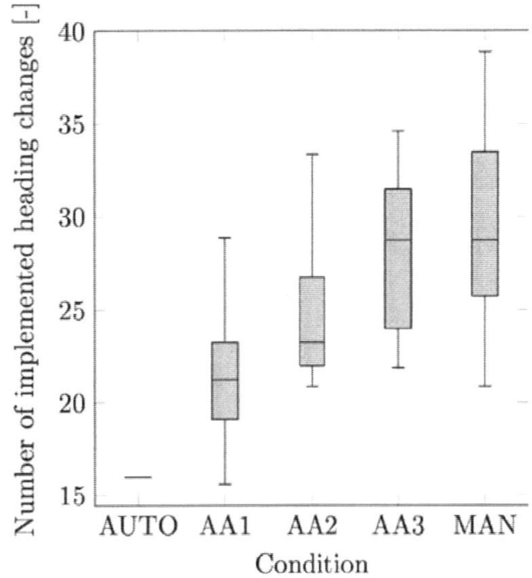

Figure 3. Number of implemented heading changes.

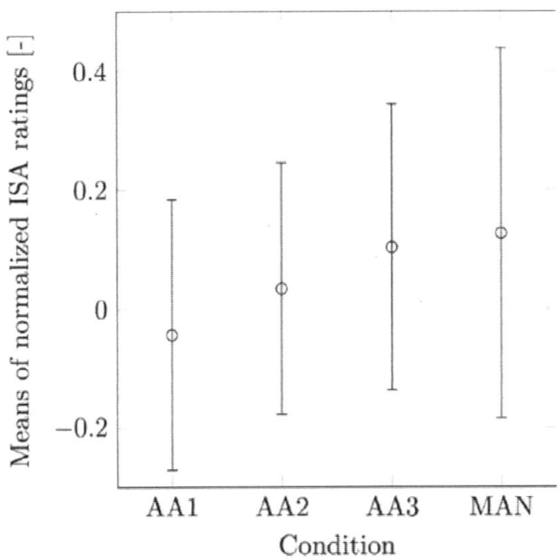

Figure 4. Means of normalized ISA ratings.

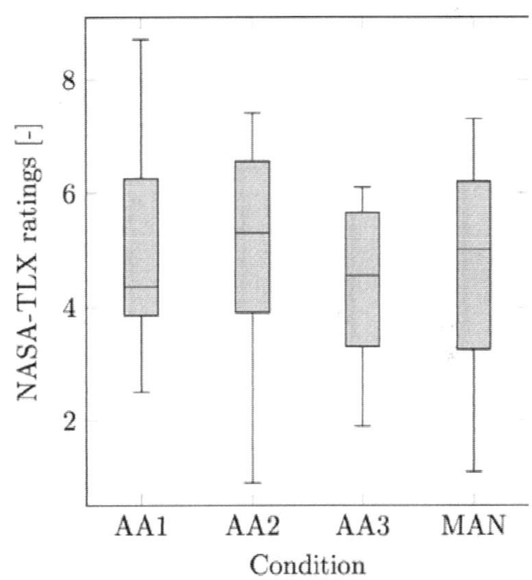

Figure 5. Total NASA-TLX ratings.

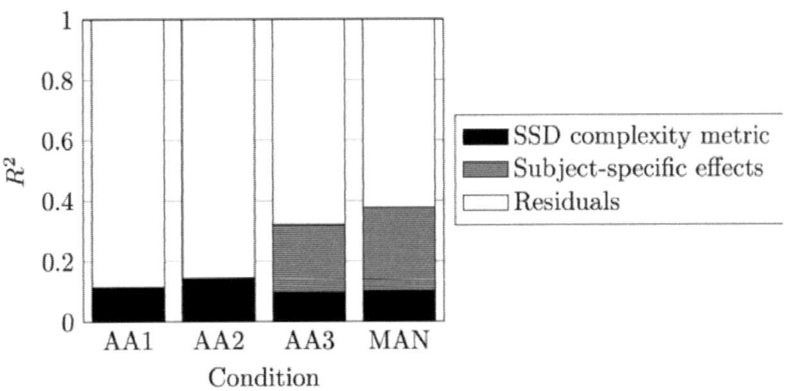

Figure 5. Explained variance of the linear mixed-effects model of airspace complexity

automation and participant. In these situations – even though sometimes automation proposed the better solution – the participant had a more elaborate plan than the automation, in that the participant had thought two or three control actions 'ahead'. In other words, possible workload gains were nullified by the automation's support missing the *intent* information of the human operator.

The complexity metric showed slight variations between conditions, but none were significant. In particular, condition AA3 showed a slightly higher and more variable complexity. A linear mixed-effects model with an intercept and random subject-specific effect was applied to test the correlation between ISA-ratings Z-scores and the complexity metric. Figure 5 shows the amount of variance in ISA-ratings Z-scores that can be explained by complexity and subject-specific effects. Approximately 10% of variance is explained by complexity, which is unaffected by the triggering threshold. For conditions AA3 and MAN, subject-specific effects start to play a more dominant role in the ISA-ratings indicating larger deviations in participants' performance.

Automation acceptance. Table 3 shows the medians and mean ranks of the CARS ratings. With lower triggering thresholds, fewer participants rated the automation at an eight or higher (corresponding to answering the question "is the system satisfactory without improvement" with YES). Instead, participants rated six or below more frequently. The mean ranks decrease with stricter automation, indicating a decrease in automation acceptance with lower thresholds. However, a Friedman test indicated that this difference is not significant ($\chi2(2) = 1.111$, $p = 0.574$).

Conclusions

The results from the experiment indicate that the designed adaptive automation system was effective in improving performance. With stricter automation, there were fewer self-induced conflicts and the overall system was more efficient and safe. However, subjective workload ratings indicated that with stricter automation, frustration of the controllers increased. This meant that although workload reduced on other aspects because of the automation support, workload reductions were nullified by higher frustration of participants. It was observed that automation with lower triggering thresholds reduced the acceptance of the automation support.

In conclusion, finding an optimal combination of triggering variable and threshold for adaptive automation, means that a trade-off must be made between performance, safety and workload. Whereas this research aimed to provide more insight in the design space for triggering mechanisms, future research can further explore the possibilities and effects of different interfaces, triggering variables and thresholds, and LOAs for adaptive automation.

Acknowledgements

The authors would like to thank the participants that volunteered to participate in this study.

References

D'Engelbronner, J.G., Borst, C., Ellerbroek, J., van Paassen, M.M. & Mulder, M. (2015). Solution-Space-Based Analysis of Dynamic Air Traffic Controller Workload. *Journal of Aircraft, 52*(4), 1146-1160.

Hilburn, B., Jorna, P.G., Byrne, E.A. & Parasuraman, R. (1997). The Effect of Adaptive Air Traffic Control (ATC) Decision Aiding on Controller Mental Workload. In Mouloua, M. & Koonce, J. M., (Eds), *Human automation interaction: Research and practice* (pp. 84–91). Mahwah, NJ: Lawrence Erlbaum Associates.

Irfan, M., Bull, M.J., Clinch, A.T. & Pember, S.J. (2012). Air Traffic Control.

Freeman, F.G., Mikulka, P.J., Scerbo, M.W., Prinzel, L.J. & Clouatre, K. (2000). Evaluation of a psychophysiologically controlled adaptive automation system, using performance on a tracking task. *Applied Psychophysiology and Biofeedback, 25*(2), 103–115.

Lillo, F., Pozzi, S., Tedeschi, A., Ferrara, G., Matrella, G., Lieutaud, F., Lucat, B. & Licu, A. (2009). Coupling and Complexity of Interaction of STCA Networks. In *Proceedings on the EUROCONTROL Conference 8th Innovative Research Workshop & Exhibition* (pp. 1–8). Bretigny-sur-Orge, France: EUROCONTROL.

Parasuraman, R., Bahri, T., Deaton, J.E., Morrison, J.G. & Barnes, M. (1992). Theory and Design of Adaptive Automation in Aviation Systems. *Naval Air Development Center, Washington, DC, Tech. Rep.*

Rahman, S.M.A., Borst, C., Mulder, M. & van Paassen, M.M. (2015). Solution Space Diagram in Conflict Detection Scenarios. In *Jurnal Teknologi (Sciences & Engineering), 75*(8), 53–58.

Tobaruela, G., Schuster, W., Majumdar, A., Ochieng, W.Y., Martinez, L., & Hendrickx, P. (2014). A Method to Estimate Air Traffic Controller Mental Workload Based on Traffic Clearances. *Journal of Air Transport Management, 39,* 59–71.

EXPERIMENTAL EVALUATION OF A SCALABLE AUTONOMY CONCEPT FOR COGNITIVE AGENTS ABOARD RECONNAISSANCE UAVS

Georg Rudnick
Axel Schulte
Universität der Bundeswehr München (UBM)
Institute of Flight Systems (LRT- 13), 85577 Neubiberg, Germany
{georg.rudnick, axel.schulte}@unibw.de

The concept of scalable autonomy with a guided access to lower automation levels allows human UAV pilots to select the level of autonomy and enables the onboard automation to understand the pilot's intent and offer support. To evaluate this concept, we conducted flight and simulation experiments with German military personnel performing reconnaissance missions with a small UAV. We compared three configurations. The high autonomy configuration completely prevents access to low-level functions. The naive approach configuration allows unrestricted access to all lower-level functions. The guided access configuration restricts access by forcing the pilot to communicate his intent first by entering mission tasks. As dependent variables, we measured mission performance and workload by the achievement of objectives, questionnaires (e.g. NASA-TLX) and secondary task performance evaluation. The low-level access configurations improved mission performance significantly, while keeping the workload on a normal level. The subjective workload was even slightly reduced.

Current UAV deployment, especially in military applications, seldom includes a known and predictable environment. Be it weather, enemy actions or just the inconsistency in the plans of own forces, change and unforeseen events are given in every battlefield situation. Humans have learned to live with that unpredictability. At the same time higher degrees in UAV automation are employed to increase efficiency, as for example the ratio pilots to controlled UAVs can be inversed from larger than one to a fraction (Schulte & Meitinger 2010) with their help. The downside of this approach is that higher automation functions also have to be able to handle change and unforeseen circumstances on the battlefield. This is only possible to a limited degree. In other words, there will always be situations for which an automation function is either not designed, performing less than optimally or just wrongly implemented. These are the cases were manned-unmanned teaming can show its strength, as a human pilot can improvise, act on a tactical level or supplement the flaws of automation in order to achieve a better mission performance. Thus, the team is stronger than its parts. For a pilot to have a chance to interact with the UAV in this way, direct access to at least some automation functions must be granted. If only the top level is accessible, a human is not able to fill in the previously mentioned shortcomings. If only low-level functions are used, the benefits of having higher degrees of automation are lost, resulting in more workload and lower mission performance. Therefore, a concept of scalable autonomy is necessary in order to adapt to most of the possible situations.

Previous work

In the development and improvement of UAV control design Uhrmann's concept of task-based guidance (Uhrmann & Schulte 2012) allowed a dramatic increase in pilot efficiency. An intelligent agent with the abilities to analyze commands, as well as the current situation, and to plan accordingly executes the tasks assigned to it by the human pilot. The pilot therefore can concentrate on what to achieve instead of how to achieve it, thus dramatically reducing the workload. The agent does this by breaking down tasks into subtasks. (Clauss & Schulte 2014) augmented this approach by allowing the pilot to take over the execution of certain subtasks, e.g. the sensor guidance, thus creating a first implementation of variable automation.

Another approach to increase pilot effectiveness is the playbook concept by (Miller et al. 2004). The difference to a task-based guidance system is mainly the focus on pre-scripted actions instead of dynamic goal oriented planning, which is able to adjust to the current situation. In combination with a plan execution system the playbook approach is also able to provide different levels of automation by specifying additional constrains (Miller 2014), starting from the planning level down to manipulating single waypoints. There is no restriction for the pilot to prevent accessing each level. The problem with this unlimited access to all automation levels is the inability of the available cognitive resources of the software to support the pilot, while he is using them. No information about his intentions or goals is communicated to the system and therefore the pool of available cognitive capacity is not used effectively. A better concept than the naive approach of direct access to automation functions is necessary. The content of this article is to present the scalable autonomy with guided access approach and the comparison to existing concepts.

Concept

The concept of guided access was presented in (Rudnick & Schulte 2016). Instead of allowing the usage of all automation functions on every level, it restricts access to automation functions. Prior to being able to use lower levels the pilot is forced to communicate his intent by giving a task to the system. The system thus knows the objectives and can create a plan to reach them. After the plan is calculated, the pilot is able to manipulate automation functions inside the scope of the plan. The software agent in turn can analyze these changes, compare them to the objectives and goals, and warn about conflicts. It can also suggest alternatives, incorporating the constraints given by the pilot, as the objectives are known. This combination allows full usage of the cognitive resources of the software agent, while still offering full control to the human pilot on lower automation levels. For example in a reconnaissance mission of a certain location, the pilot gives a recon task to the UAV, which in turn calculates a flight path taking minimum distance to the target into account in order to not be detected. If the pilot wants to change that route, since the target is only visible from one side, he can give constraints to the automation or just change the flight path to fit his needs. The agent can then warn, if the newly created path is to close to the target, since it knows about the intention of the pilot.

The concept works in principle with every kind of plan structure, e.g. just an ordered list of tasks, but especially well with tree structured plans. Human access to lower levels can be attributed to the parent tasks of the manipulated subtasks and therefore the software agent is able to do local changes on the plan and construct alternative suggestions with minimal modifications to the overall plan. For example, the agent created the plan displayed in Figure 1.

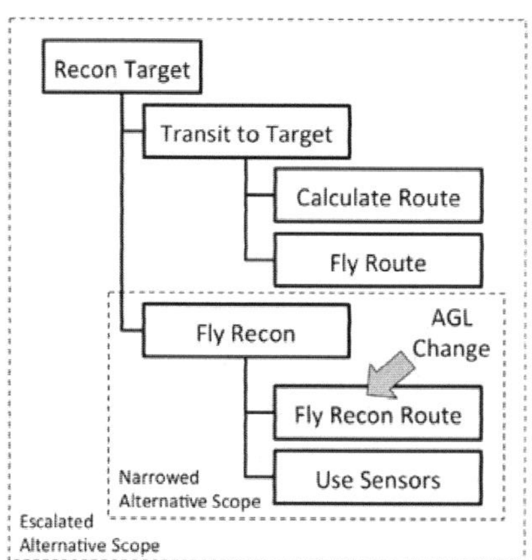

Figure 1. Agent plan with pilot modification and escalating alternative scopes

The pilot wants to reduce the altitude above ground level (AGL) for better reconnaissance and changes it for the Fly-Recon task. This endangers the UAV, since at a certain point the AGL is too low over a power supply line. The agent can now suggest a different route with the set AGL and avoid the danger, while maintaining the constraints, without modifying the rest of the plan. It is also possible to suggest escalating options, when the pilot declines a suggestion, and increase the scope of affected tasks, by taking more parent tasks into the calculation. In the previous example, the agent could change the transit towards the target instead, to open up new approach directions.

Implementation

To implement a fair comparison between the guided access concept and the naive approach, the same intelligent agent software was used for both cases. Only the user interface was modified for the different configurations. As described before, the naive approach configuration allowed unrestricted access to automation functions, while the guided access configuration only allowed access via a previously given task. The configurations are described in Table 1.

Table 1.

Available automation access for compared configurations.

	Naïve approach configuration	Guided access configuration
Flight guidance	Task-based guidance	Task-based guidance
	Manual waypoint guidance	Waypoint guidance by editing planned routes
		Waypoint guidance by editing planned recon routes
	FMS altitude	FMS altitude as constraints for tasks or subtasks
Sensor operation	Gimbal lock on position	Gimbal lock on position, during recon tasks
	Gimbal scan automation	

The task-based guidance configuration only allowed task-based guidance for flight guidance and sensor operation.

Evaluation

For the evaluation, the three configurations were compared during flight experiments with a single UAV, controlled from a ground control station, with the task to do several reconnaissance mission vignettes. Because the guided access concept is only feasible, were unexpected situations are probable, all mission vignettes were chosen to be out of the implementation scope of the task-based guidance system. To increase the amount of gathered data, slightly different variations of the same vignette were tasked in a row. This reduced complexity and increased flight safety, since the safety pilot was prepared for the current type of mission vignette. The main measurements taken were mission performance, objective workload via a secondary task, and subjective workload via a NASA-TLX (NASA 2010) questionnaire. Additional questionnaires captured usability and acceptance of the system. Mission performance was measured in three facets, mission completion, execution time and RoE violations. Mission completion captured the degree to which each mission vignette was successfully executed, for example, if a reconnaissance target was successfully and completely captured on camera. Planning time was measured between receiving the order for a task and commanding the UAV to execute the plan. Execution time was measured between the execute command and the completion of the plan. RoE violations consisted of flying through no-fly-zones, having a flight path to enter the target range of a SAM or actually entering it with the UAV and approaching a reconnaissance target closer than the briefed detection level allowed.

The secondary task was chosen to be in the realm of UAV operations, as well as affecting the same visual and thought resources as the main task, thereby adhering to the principles found by (Ogden et al. 1979). Therefore a position report, using the angle and distance between predefined reporting points and the current UAV position was used. Figure 2 shows the reporting points on the map (left) and the reporting tool UI (right) with the selected point (alpha), the direction (NE) and the distance (200m).

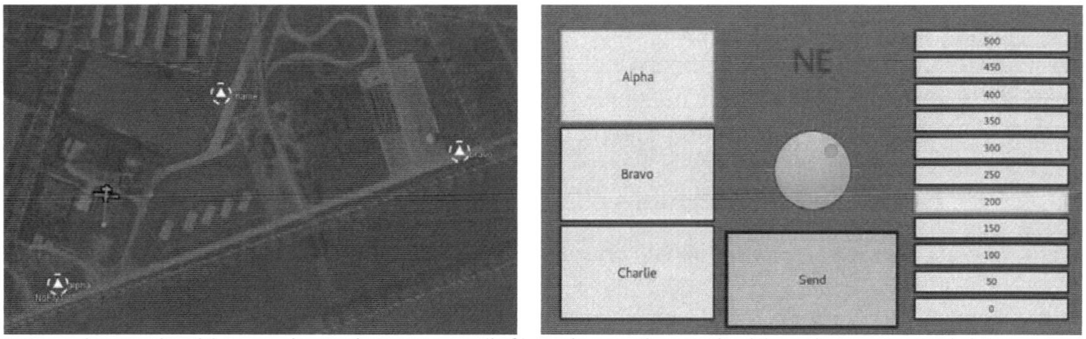

Figure 2. Secondary task with reporting points on map (left) and reporting tool with active request (right)

After a periodic random time between 6 and 10 seconds, the background of the reporting tool turned to light red, alerting the pilot of a position request, as depicted in Figure 1 (right). The red color changed to a dark red in the next 6 seconds, thus indicating the currentness of the request. This way the attention of the pilot can be drawn to a new request, even if he did not process the previous one. After filling in the data, the "Send" button transmitted the report and the background of the UI changed to grey, indicating a successful transmission.

The experiments were conducted with 5 test subjects from the German Armed Forces. After an introduction, the subjects started with a basic tutorial for the operation of the system. This included an introduction to the task-based guidance and its application. The subjects then executed a mission block for each configuration. The order of the configuration was selected in advance by chance. Each mission block consisted of a tutorial for the specific configuration, explaining the abilities of, as well as functions not available with, this configuration. After that, the subjects received a short briefing for the following mission vignettes. Then the aircraft was launched and the mission started. Due to weather and safety constraints, around 60% of the missions had to be flown with a simulation environment. This did not observably influence workload or other performance measures. The following vignettes were tasked in this order, with the number in braces indicating the variation count.

Cave (3): A reconnaissance target had to be sensed from a briefed direction for a certain time. The automatic reconnaissance route calculation was unable to provide this kind of view, which forced the pilot to manually create or edit the flight path.

Low altitude flight (2): A show of force action in order to frighten away enemy forces was tasked at a certain location. The automatic flight path planner was not allowed to reduce altitude under a certain safety level, but the pilot was able to either manipulate the created route or to influence the flight management system in order to achieve a sufficiently low altitude.

False SAM (3): Due to a wrong report, a false SAM was entered into the tactical situation. The pilot was aware of this mistake, while the software agent recognized it as threat and acted accordingly. To achieve a better mission performance, thus not flying around the false SAM, the pilot had to edit the planned flight routes.

Target count (2): The pilot was tasked to count targets on a given route. A SAM threatened parts of that route. Due to the implementation of the reconnaissance route planner, the agent was not able to avoid the threat. The pilot had to manually prevent the UAV from entering the threat range.

During the mission, the subjects had to work on the secondary task. A mission concluded with landing the aircraft, while the subjects filled out a NASA-TLX questionnaire, as well as a rating questionnaire for the current configuration. After a short break, the next configuration was loaded and the new mission block started again. Having completed all three mission-blocks the subjects were asked to fill out a general usability and acceptance questionnaire.

Results

The figures in this section display the minimum and maximum values as diamonds and the average as a box. Figure 3 (top left) shows the workload comparison between the three configurations, measured with the NASA Task Load Index and a normalized secondary task (ST) score. The workload decrease between the naïve approach and the guided access configuration is not significant (Wilcoxon signed rank test), but a trend can be observed. The test subjects also vocalized especially the subjective workload decrease, indicated by the NASA-TLX value. The large spread in the TBG-TLX value derives from larger frustration values of some subjects, as the missions were deliberately chosen outside the scope of the TBG design.

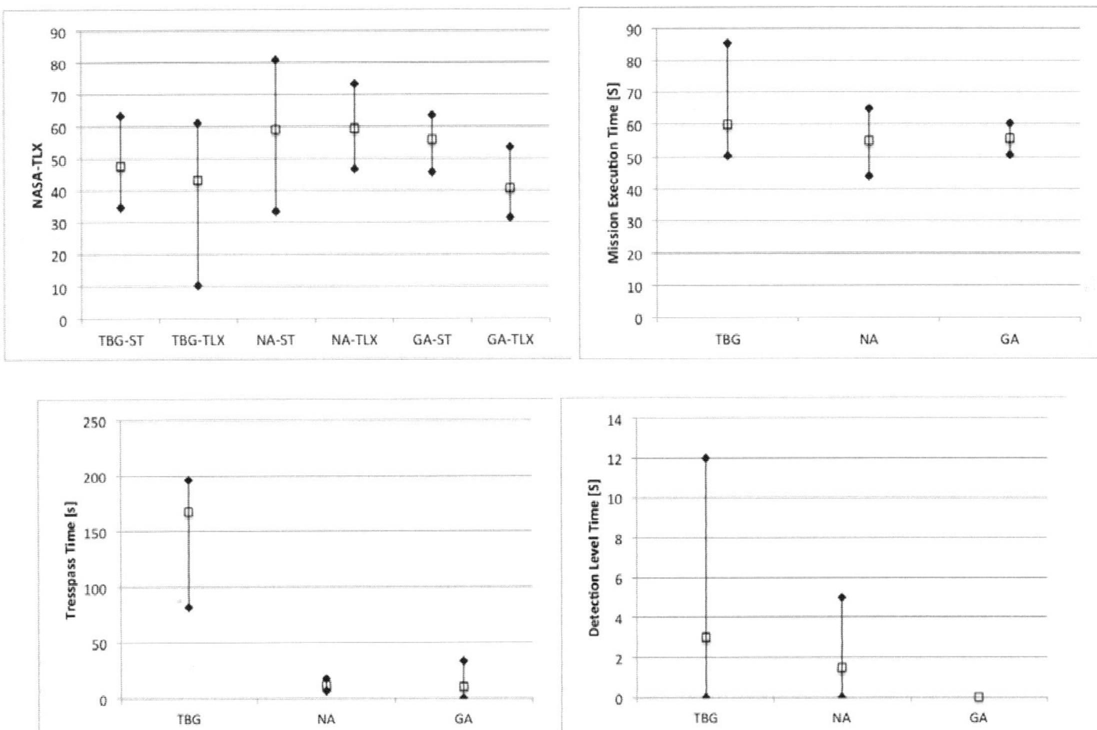

Figure 3 Comparison between configurations: Workload, determined by side task and questionaire (top left) and mission performance with execution time (top right), trespass time (bottom left) and detection level violation time (bottom right)

The subjects were able to complete the missions successfully with all three configurations and retrieve the needed reconnaissance data, although the TBG configuration only allowed diminished results, as the required angles for the camera in the Cave missions, as well as the low flight altitudes could not be achieved. Figure 3 (top right) shows the reduction in mission execution time, as unnecessary detours in the False SAM vignettes can be avoided with low-level access. Figure 3 (bottom left) displays the time in which the aircraft was on a flight path into an enemy threat, without the pilot reacting (Trespass). The difference between the TBG configuration and the other two is significant (Wilcoxon signed rank test, 5%, W=15). This results from the ability to edit the flight path around threats in the lower levels, while the recon route generator determined the flight path of the high level system. Finally Figure 3 (bottom right) illustrates the reduction in approaching to close to a reconnaissance target (Detection Level). The GA configuration can issue warnings before the critical distance is reached, as it is aware of the pilot's intent, thus reducing this kind of error to zero.

Conclusion

This article presented the experimental evaluation of a scalable autonomy concept with guided access to lower level functions. The experiments consisted of several mission vignettes, which were deliberately chosen to be out of the design scope of a task-based guidance system without low-level access (TBG). In comparison to the naïve approach configuration providing lower level inputs (VA), the guided access configuration (GA) offered a reduced workload. This was especially observed for the subjective workload. The mission error rate was reduced significantly, while the overall mission performance improved slightly. The guided access configuration offers a solution for low-level access to automation functions without increasing the workload severely and adds the benefit of better support to fulfill the pilot's intentions.

References

Clauss, S. & Schulte, A., 2014. Implications for operator interactions in an agent supervisory control relationship. *2014 International Conference on Unmanned Aircraft Systems, ICUAS 2014 - Conference Proceedings*, pp.703–714.

Miller, C.A., 2014. Delegation and transparency: Coordinating interactions so information exchange is no surprise. *Lecture Notes in Computer Science (including subseries Lecture Notes in Artificial Intelligence and Lecture Notes in Bioinformatics)*, 8525 LNCS(PART 1), pp.191–202.

Miller, C., Goldman, R. & Funk, H., 2004. A playbook approach to variable autonomy control: Application for control of multiple, heterogeneous unmanned air vehicles. *Proceedings of FORUM 60, the Annual Meeting of the American Helicopter Society.*

NASA, 2010. NASA TLX: Task Load Index. *Planta Medica*, 35(4), pp.308–315.

Ogden, M.G.D., Levine, M.J.M. & Eisner, M.E.J., 1979. Measurement of Workload by Secondary Tasks. *Human Factors*, 21(5), pp.529–548.

Rudnick, G. & Schulte, A., 2016. Scalable autonomy concept for reconnaissance UAVs on the basis of an HTN agent architecture. In *2016 International Conference on Unmanned Aircraft Systems (ICUAS)*. pp. 40–46.

Schulte, A. & Meitinger, C., 2010. Introducing Cognitive and Co-operative Automation into UAV Guidance Work Systems. In M. Barnes & F. Jentsch, eds. *Human-Robot Interaction in Future Military Operations*. Ashgate, pp. 145–170.

Uhrmann, J. & Schulte, A., 2012. Concept, Design and Evaluation of Cognitive Task-based UAV Guidance. *International Journal On Advances in Intelligent Systems*.

WARHORSE: A NEW APPROACH TO MANNED/UNMANNED TEAMING

Paul Schutte
US Army - Aviation Development Directorate
Fort Eustis, Virginia, USA

Ralph Williams
Analytical Mechanics Associates, Inc.
Hampton, VA, USA

Ronald Daiker
NASA Langley Research Center
Hampton, VA, USA

This paper presents a multi-human/multi-vehicle control and integration concept called Warhorse. Warhorse is a control paradigm that is applied to all vehicles (air or ground) that are used in conducting operations. The Warhorse concept aids humans in developing operational plans, enacting a plan using single or multiple vehicles (manned and/or unmanned), and individually controlling a single vehicle (i.e., one-on-one). It synergistically integrates the human with the machine so that the warfighter can bring all their skills and experience to bear on accomplishing the mission.

Introduction

The warfighter of the future will be called upon to command several manned or unmanned vehicles of differing types at the same time to complete a mission. It may be the case that the warfighter himself occupies one of these vehicles. It is a goal of the Army's Synergistic Unmanned Manned Intelligent Teaming (SUMIT) program to develop onboard and remote interfaces that will enable collaboration and/or control of multiple air and/or ground vehicles. These vehicles will, by definition, have different and unique capabilities in the form of payload, range, speed, stealth, and versatility. In addition, these vehicles are likely to be in a dynamic, hostile environment where the enemy, weather, and terrain may inflict restrictions and/or harm on them. It is important to provide robustness and adaptability, and allow for performing the mission with degraded or inoperable equipment (including the automation itself) on multiple vehicles. Vehicles and assets may not always be present or may be disabled during the mission, causing the warfighter to swiftly replan and reconfigure for the mission.

This paper presents a multi-human/multi-vehicle control and integration concept called Warhorse. Warhorse is a control paradigm that is applied to all vehicles (air or ground) that are used in conducting operations. The Warhorse concept aids humans in developing operational plans, enacting a plan using single or multiple vehicles (manned and/or unmanned), and individually controlling a single vehicle (i.e., one-on-one). It synergistically integrates the human with the machine so that the warfighter can bring all their skills and experience to bear on accomplishing the mission. Using the Warhorse concept, the warfighter can quickly command vehicles as part of an operation, even if the warfighter does not have a great deal of experience with some vehicles. This is accomplished through the use of common command paradigm used for all vehicles so that they all seem familiar. The command paradigm uses a hierarchical set of behaviors – plays, missions, and maneuvers. It is important to note that this is not a common interface – inceptors and actual control devices may vary greatly in order to best accommodate and enable the vehicle being commanded. The warfighter can command behaviors on any vehicle in a similar manner, just as most computer programs offer the "File, Edit, View, Format, Help, etc." interface paradigm; the programs themselves may be very different, but the user already has a sense of where certain commands are located and how to interact with the program. This allows the warfighter to bring their human capabilities, experiences, and training to bear on the mission rather than being relegated to managing and monitoring the automation.

The human being has a number of unique and valuable capabilities that are difficult to encode in a machine, and yet often provide a significant advantage over the enemy. These are flexibility, adaptability, comprehension of overall goals and the 'big picture', and an 'all-purpose' nature that allows one individual to perform a wide variety of disparate tasks (serially). The human can manually control a vehicle or conduct communications or replan a mission or troubleshoot and repair a damaged system. No single automation agent can be this versatile. Indeed, even multiple automation agents rarely work together to form a notional single agent that can perform all of these tasks as well as a human. The human being is creative, innovative, and brings a poorly understood and yet extremely valuable trait called intuition to the mission. Rather than attempting to imbue these attributes (many of which are extremely difficult and risky) into machines, the Warhorse concept allows both human and machine to do what they do best using a process called complementary automation or Complemation (Schutte, 1999). This design approach runs counter to the philosophy of automating as much as possible and then relegating the remaining tasks and functions to the human.

Warhorse combines two metaphor-based command and control interaction concepts. The first is the Playbook concept (Miller, 2004). The Playbook concept uses the metaphor of a sports team's playbook, similar to those used extensively by American football teams. They represent strategies or behavioral patterns that lay out what each member of the team will do on the play. Each play is generally named and practiced so that in the game, the quarterback or team captain can just call out the name of the play and all of the team members will know what to do. A quarterback can quickly change the play based on the situation. The Playbook approach has been used to create a control system in which a warfighter with little or no knowledge about the operation of unmanned aerial vehicles (UAVs) can call for a play and count on the UAVs involved to perform that play. Examples of plays are tracking a target, area reconnaissance, and sustained surveillance of a target.

The Playbook approach focuses on a minimally trained operator, but there are times where unique human capabilities are necessary, such as flexibility and opportunistic reactions to changes on the battlefield. There are times when the warfighter needs to make dynamic changes in the play and even times when he or she wants to take a more direct control of a vehicle or its payload. In the Playbook metaphor, the warfighter might want to "pass" to a player who is unexpectedly "open" or may want to "run the ball" his or her self. In these cases, another metaphor is needed – the Horse metaphor (Flemisch, 2003). In the Horse metaphor, the vehicle is assumed to have a certain level of intelligence that is strictly limited to transportation. It responds to the rider's commands, but if the rider offers no command, the horse will stay on its current path. The horse can respond to changes in the environment and it can respond automatically to certain threats. But the horse has no higher level sense of the mission – that is left to the rider. The horse has a number of behaviors or maneuvers that it can perform on its own, such as trot, jump, and gallop. For horses used in work or sport, there are more complex behaviors such as following a calf so the rider can rope it in ranching or maintaining safety and safe distances in polo while the player concentrates on the ball. The Horse metaphor has been applied to aviation (Schutte, et al, 2007; Schutte, Goodrich, & Williams, 2017) and automobiles (Altendorf, E., et al, 2015). Examples of Horse-metaphor maneuvers for a helicopter are takeoff, climb (direction, rate), level off (altitude), cruise (destination, speed), hover (destination), pop-up (altitude), laze (target), and mask (altitude). The warfighter can take more detailed control of a particular asset in a play and utilize the unique capabilities of that asset. As mentioned earlier, the warfighter might be in an aircraft during the mission and the warfighter might need to take 'manual' control of his or her vehicle (due to loss of vehicle automation or to perform more detailed maneuvers commensurate for the current situation). The Horse-metaphor allows for varying levels of automation assisted 'manual' control (see Schutte, et al, 2007 and Schutte, Goodrich, & Williams, 2017 for more detail).

In the Warhorse concept, the warfighter can play three roles: Planner, Commander, and Player. Each of these will be described below. The Warhorse concept is defined in relation to the warfighter as opposed to describing it solely as the system.

Planning Using Warhorse

German military strategist Helmuth Von Moltke once said, "No battle plan survives first contact with the enemy." The battlefield is highly dynamic and unpredictable: for example, the enemy can change tactics or have assets in locations not detected by intelligence; the weather and ground conditions can change; and systems and machinery can fail. That said, it is still important to have a robust plan before going into battle. The warfighter needs to be able to quickly plan using the best intelligence available; but perhaps more importantly, the warfighter needs to be able to replan as the situation changes. In the Warhorse concept, each and every vehicle (ground/air, manned/unmanned) is treated as a semi-autonomous agent that can be commanded in a common manner. Each vehicle has a list of maneuvers that it can perform. These maneuvers are meaningful, high-level descriptions of scripted maneuvers such as takeoff, max climb, hover, orbit, or fly Nap-Of-The-Earth (NOE). Maneuvers will vary from vehicle to vehicle based on their capabilities; however, the manner in which the warfighter assigns maneuvers to each vehicle is the same. Maneuvers can be temporally and spatially linked together to create simple Plays. A surveillance Unmanned Air Vehicle (UAV) can be given a string of maneuvers to execute autonomously in a Play. As the warfighter creates these simple plays, the performance characteristics of the vehicle are automatically represented in the play. For example, assigning a climb maneuver to an attack helicopter will graphically represent the performance angles and speeds associated with that helicopter. Assigning a climb maneuver to a UAV will present a much different range of climb profiles than those of the attack helicopter. This allows the warfighter to quickly assess the capabilities of the vehicle.

On top of maneuvers are missions. Missions are goal-directed behaviors. Missions use maneuvers but also use sensor data and other additional information. A mission is generally performed by a single vehicle. For example, a mission could be to track an enemy asset. The mission is highly dependent upon the enemy's movements. Another mission would be to guard a perimeter. Again, each vehicle has its own unique capabilities and therefore its own available missions. An unarmed UAV is not capable of performing a "Guard the perimeter" mission. It may be that different vehicles have similar missions that are conducted in very different ways. For example, a stealth approach mission for a helicopter may be flying NOE, whereas a stealth approach for a ground vehicle may mean keeping out of the line-of-sight of a particular position.

Missions and maneuvers can be used in conjunction with additional information to run plays. Plays are the scripted behaviors of assets mentioned earlier, usually for several vehicles in a coordinated fashion. Plays can have timelines where the warfighter can set events such as time elapsed events, trigger events, or synchronization events. Thus, early arriving assets might hold their position until all assets have assembled and then proceed to the threat zone to perform the rest of the play. Plays can be highly developed, and are usually created by the warfighter well ahead of their performance. This does not preclude the use of previously created (templated) plays that a warfighter can suddenly call.

When creating a plan, the warfighter has a map display, a timeline, and available assets (vehicles). The timeline contains a 'play head' that can be scrubbed through time, showing the positions of all assets at different stages of the plan. These positions are based on the performance characteristics of each vehicle. The warfighter is given the objective, gathers the necessary assets, and locates them at their desired locations on the map. They can then assign behaviors to the assets. When a behavior is assigned, it is given a start time and then projects its progress on the timeline. For example, if an aircraft is assigned a climb maneuver, as the warfighter advances through the time line, the aircraft will move on the map based on its predicted performance characteristics. This assists the warfighter in creating and coordinating the plan. These predictive characteristics can be propagated backwards in time as well. For example, the warfighter may want a surveillance UAV to be in position at a particular time. They can place the UAV there at the time on the timeline and the system will project backwards when the UAV should be dispatched. Not all behaviors (e.g., goal directed missions) can be accurately placed on the timeline. However events can still be used to set start or stop times. Timeless events such as trigger and synchronization events can be created on the timeline without anchoring to a specific time.

Commanding/Replanning Using Warhorse

After the plan has been created using the timeline, the warfighter moves to a mission commander role. Here is where the flexibility of behaviors comes into play. As the warfighter watches the plan unfold in real-time, they can monitor it using whatever sensor information and intelligence available. If something in the plan needs to be changed or modified, the warfighter can simply select an asset or group of assets and apply maneuvers, missions, or plays to those assets. The warfighter can preview these changes to see the outcome before they are implemented. However, if something needs to be done immediately, for example, the need to clear assets out of the area, the warfighter can quickly assign and command maneuvers or missions without waiting to review. The warfighter could call on a vehicle to execute a maximum climb to escape an area that contains newly discovered hostiles. For another example, a new threat may pop up and the mission commander can basically pick one of the available assets, assign an attack mission to the asset and specify the target, and then return to monitoring the battle.

Recall that assets may be manned, unmanned (remotely controlled), or unmanned autonomous. The same behavior command structure (maneuvers, missions, and plays) still applies regardless of asset type. The exact protocol and procedure for these assets will vary based on the capability of the asset. For example, an unmanned armed fixed-wing UAV will likely have fewer options for attack patterns available due to its trajectory and missile launch capability (e.g., a fixed wing UAV must launch a missile in the direction of its flight and may not be able to launch a missile at a target in time).

Riding the Warhorse

It may be the case that the warfighter will actually be controlling their own Warhorse vehicle or may have to take over direct control of another Warhorse vehicle. The warfighter may be in a ground control station operating an unmanned asset remotely. In these cases, the warfighter is considered to be 'riding' the Warhorse vehicle. The interface for one-to-one vehicle control in this ownship is unique because it is based on the manual inceptors unique to that vehicle. In the case of an aircraft, the warfighter becomes the pilot. These inceptors are not necessarily linked directly to associated effectors, but will likely have an intermediate level of automation between the inceptor and effector. As such, the pilot is still commanding the automation, rather than flying 'manually. For example, if the pilot has to do a climb in a fixed wing aircraft, they pull back on the stick. This sends a signal to the automation of the pilot's intent to climb. The automation initiates a climb based on the pilot's inputs but also presents options for climb maneuvers on a screen or through some tactile interface. The pilot selects one of these options, and then commands the aircraft to climb. At this point, the automation will control the aircraft and climb according to the parameters previously selected by the pilot. Virtually all Warhorse maneuvers are engaged through the use of manual inceptors. Warhorse missions and plays are engaged using selections on a multifunction map display. Again, the pilot selects the mission or play, designates the parameters (e.g., the target, the role in the playbook) and then engages the automation. This control paradigm is consistent across all Warhorse vehicles.

There are several reasons that manual inceptors are used for maneuvers (as opposed to touch screens or voice). The primary reason is the robustness of operation in the case of automation failure or some other system failure. The automation cannot be guaranteed foolproof nor invulnerable to programming errors, system anomalies, spoofed data, enemy fire, etc. There will be cases (especially when a human is on-board) where the operator will need to become a pilot and 'manually' fly the vehicle with degraded or no automation. Additionally, the pilot may want to take control because they see a tactical opportunity that does not allow for setting up a behavior. In the Warhorse concept, the pilot already has their hands on the controls. There is no need to revert to a completely different interface (e.g., from touch screens back to physical inceptors). The added benefit is that the warfighter's mental model requires less modification. A second reason for having the manual inceptors used for commanding the automation is to reduce skill loss due to constant use of automation. A third reason is to engage the warfighter in the battle in a meaningful way without creating extremes in workload. This helps the warfighter fight complacency and loss of situational awareness. A fourth reason is to promote a synergistic relationship between the warfighter and the Warhorse in order to achieve Complemation – optimal performance in human/automation teaming.

Summary

The Warhorse concept is designed to offer the trained warfighter a decisive advantage by capitalizing on the unique skills of the human being and using the automation to do what it does best. It allows for strategic planning and provides the ability to tailor and modify the mission as it is underway (i.e., tactical execution). It is important to note that the Warhorse concept is not designed for an individual with minimal training although the automation may be capable of adapting to less experienced users. It is designed to make the best use of human expertise and experience while minimizing the training required to learn system specific information. This allows the warfighter to spend valuable training time learning how to successfully complete the mission instead of learning how to operate the system.

References

Altendorf, E.; Baltzer, M.; Heesen, M.; Kienle, M.; Weißgerber, T.; Flemisch, F.: H-Mode, a Haptic-Multimodal Interaction Concept for Cooperative Guidance and Control of Partially and Highly Automated Vehicles; Winner et. al. Eds.: Handbook of Driver Assistance Systems; Springer, 2015

Flemisch, F. O., Adams, C. A., Conway, S. R., Goodrich, K. H., Palmer, M. T., & Schutte, M. C. (2003). The H-Metaphor as a Guideline for Vehicle Automation and Interaction. Report No. NASA/TM-2003-212672. NASA Langley, Langley.

Miller, C., Goldman, R., Funk, H., Wu, P., & Pate, B. (2004, June). A playbook approach to variable autonomy control: Application for control of multiple, heterogeneous unmanned air vehicles. In Proceedings of FORUM 60, the Annual Meeting of the American Helicopter Society (pp. 7-10). Baltimore: American Helicopter Society International, Inc.

Schutte, P. (1999). Complemation: An alternative to automation. Journal of Information Technology Impact, 1(3), 113-118.

Schutte, P. C., Goodrich, K. H., Cox, D. E., Jackson, B., Palmer, M. T., Pope, A. T., ... & Kinney, J. B. (2007). The naturalistic flight deck system: An integrated system concept for improved single-pilot operations. Report No. NASA/TM-2007-215090. NASA Langley, Langley.

Schutte, P., Goodrich, K., & Williams, R. (2017). Synergistic Allocation of Flight Expertise on the Flight Deck (SAFEdeck): A design concept to combat mode confusion, complacency, and skill loss in the flight deck. In Advances in Human Aspects of Transportation (pp. 899-911). Springer International Publishing.

EFFECTS OF RADAR SIDE CONFLICT PROBE FUNCTIONALITY ON ATC CONFLICT DETECTION PERFORMANCE AND EFFICIENCY

Anthony J. Masalonis
Spectrum Software Technology
Atlantic City International Airport, NJ
Ben Willems
Federal Aviation Administration
Atlantic City International Airport, NJ
Thomas Fincannon
Applied Research Associates
Atlantic City International Airport, NJ

We investigated integrating Conflict Probe (CP) on air traffic controllers' Radar Side (R-Side) displays. Eight controllers worked realistic, high-traffic simulation scenarios alone, using both R-Side and Radar Associate Side (RA-Side) displays. We manipulated CP presence on the R-Side—like today, it always appeared on the RA-Side—and the presence of yellow alerts for near-conflicts. We used established controller performance and workload metrics, plus novel operational analyses not used in past studies. R-Side CP had few workload effects, but increased voice communications when we included yellow alerts. It improved the efficiency of correcting conflict-inducing clearances, and seemed to facilitate proactive control to avert more urgent alerts. Though our simulated CP was less reliable than the current operational version, it showed evidence of benefiting performance and acceptance. Participants commented that R-Side yellow alerts were desirable in moderation. Future research should assess the appropriate alerting criterion.

In Air Traffic Control (ATC), a conflict occurs when two aircraft are closer than the minimum separation standard or an aircraft violates an unauthorized airspace volume. For many years, ATC workstations have included a short-term, tactical Conflict Alert (CA), which alerts controllers to conflicts predicted within about the next two minutes. Each conflicting aircraft's datablock—textual information near the airplane's location symbol—flashes, and the callsigns of the pair appear in a list. The Conflict Probe (CP) helps controllers detect and resolve aircraft conflicts earlier, to enhance safety and efficiency and help controllers manage their own workload. The CP grew out of research and development activities by the Federal Aviation Administration (FAA) and its contactors, who developed the Automated En Route ATC (AERA) concept (Goldmuntz et al., 1981). MITRE's User Request Evaluation Tool (URET, Brudnicki & McFarland, 1997), and the National Aeronautics and Space Administration (NASA)'s Conflict Prediction and Trial Planning (CPTP) tool (Paielli & Erzberger, 1997), extended AERA's capabilities. CP is now an integral part of the operational en route ATC system. It alerts sooner than CA—up to 20 minutes for aircraft conflicts and 40 minutes for airspace—and uses flight plan information, rather than current speed and heading alone. CP and CA are each useful in different settings and are complementary.

A typical en route ATC workstation includes two displays: the Radar Side (R-Side), and the Radar Associate Side (RA-Side), also known as the Data Side (D-side). The R-Side shows information essential to keeping aircraft separated, including each controlled aircraft's position as derived from radar, textual information about each flight, and the aforementioned CA alerts. The RA-Side includes a list of current and future controlled aircraft, with further information about each flight, and its own trajectory-based position display of aircraft.

Currently CP appears only on the RA-Side display. The RA-Side is sometimes operated by a separate controller in a two-person sector team, necessitating communication between the controllers when dealing with conflicts. At other times, a controller works a sector alone and uses both the R- and RA-Side displays, but must attend primarily to the R-Side to support the primary task, separating aircraft. In these settings, controllers often cannot make productive use of the RA-Side, including the CP. Therefore, integrating the CP onto the R-Side should enhance controllers' conflict resolution ability.

The present study was part of a succession of Human-in the-Loop (HITL) experiments for the FAA's Separation Management/Modern Procedures (SepMan) program. Under this program, Zingale, Willems, Schulz, and Higgins (2012) tested the implementation of CP on the R-Side, comparing various display methods and alerting criteria in a HITL simulation using current and retired controllers. They found evidence that controllers working alone and using both displays chose to view the details of the alert for more of the CP alerts when CP was on the R-

Side than when it was only on the RA-Side. Their data also suggested that a rule presenting only conflicts predicted within six minutes on the R-Side resulted in more frequent viewing of notifications by the R-Side controller in a two-person team. Their questionnaire data showed a preference for displaying CP information on the R-Side.

Zingale et al. (2012) analyzed objective safety and efficiency measures, such as losses of separation where controllers did not prevent a pair of aircraft from violating the prescribed separation minima, and efficiency of working aircraft through the sector in terms of time and distance traveled. None of these measures showed significant benefits of R-Side CP, so they recommended further CP research using more varied traffic scenarios. Therefore, we further investigated the CP location question in a HITL study analyzing a wider variety of objective performance data, including controller behaviors not assessed in previous studies of this research program. We also introduced more variety between the scenarios run in each condition to lessen predictability.

Another question regarding CP that had not been previously explored was the degree of reliability required for this automation to support performance and to be accepted by controllers when integrated on the R-Side (Masalonis, Rein, Messina, & Willems, 2013; Rein, Masalonis, Messina, & Willems, 2013). We defined reliability as a combination of Hit and False Alert rates, after Wickens and Dixon (2007). We set out to study whether improvements over the reliability of the currently-fielded algorithms would make R-Side CP functionality operationally acceptable. Here, we will focus mostly on how integration of CP information on the R-side affected operational performance, perceptions, and behaviors. We reported other analyses and results for the same simulation in Willems, Masalonis, Fincannon, Puzen, & Bastholm (2016). We predicted that CP information on the R-Side would improve controllers' conflict resolution performance and reduce workload. We also hypothesized that controllers would perceive locating CP on the R-Side as beneficial.

Method

Participants

Eight current, en route, Certified Professional Controllers—seven male, one female; mean age 50.13 years (SD = 3.94), ranging in age from 42 to 54 years—from the En Route Automation Modernization (ERAM) National User Team participated in a HITL experiment at the Research Development and Human Factors Laboratory at the FAA's William J. Hughes Technical Center (WJHTC). All reported having worked traffic at their facility in the preceding 12 months. Their mean ATC experience was 26.04 years (SD = 2.88), ranging from 21.83 to 30.50 years.

Apparatus

We conducted a high-fidelity ATC simulation using tools developed at WJHTC and/or regularly used there for research and testing: the Distributed Environment for Simulation, Rapid Engineering, and Experimentation (DESIREE), the Simulation Driver and Radar Recorder (SDRR), the Target Generation Facility (TGF), and the ERAM Evaluation System (EES). This was the first HITL simulation to interface EES to DESIREE.

Each controller used an R- and RA-Side workstation. The R-side workstations included a radar display (BARCO 29″ LCD, resolution 2048 x 2048); and a Cortron keyboard, Keypad Selection Device (KSD), and trackball. The RA-side workstations contained an EIZO 30″ LCD monitor, resolution 2560 x 1600, showing the Aircraft List [ACL] View) and Graphic Plan Display (GPD) depicting trajectory-based data, and a Cortron keyboard and trackball. The controllers and the simulation pilots used push-to-talk (PTT) communications through a voice switching and control system that simulated the current operational system. The simulation pilots made requests and responded to clearances as real-life pilots would. They worked in rooms separated from the experiment rooms. DESIREE collected real-time subjective workload assessments every two minutes via the custom-made Workload Assessment Keypad (WAK), based on the Air Traffic Workload Input Technique (ATWIT; Stein, 1985). The WAK has 10 buttons labeled 1 through 10. Participants may press the 1 to indicate very low workload, the 10 for very high workload, or any button in between. We provided a handout detailing the anchors for the scale.

We used a variety of questionnaires to collect participants' demographic information, and their subjective opinions about the CP functionality and concept and about other aspects of the simulation. Over-the-shoulder observers, who were also experienced controllers, used forms to rate participants' performance during each scenario.

In all conditions, CP information appeared on the RA-Side. The system identified aircraft involved in one or more conflicts with a color-coded square(s) depicting the number and type(s) of alert (aircraft or airspace). The color code for airspace alerts was orange. Aircraft alerts were red, meaning that CP predicts the centerlines of the two trajectories to violate separation minima, or yellow, meaning that CP predicts the adherence bounds—buffer zones surrounding each centerline to allow for prediction uncertainty—to get closer than the separation minima. In the conditions where CP was also available on the R-Side, a color-coded square in each conflicting aircraft's data block showed the total number of alerts and the color of the most severe alert; controllers could click it to show the trajectories. CP also provided a list of all alerted aircraft, with the same color-coded square for each flight and an indication of the number of minutes until the conflict.

We simulated a high-altitude sector in Indianapolis Center (ZID). Traffic ranged from about 7 to 26 aircraft over the course of each scenario. ZID personnel consider the real-life sector's capacity to be 19 aircraft.

Design and Procedure

In this paper, we mainly discuss the manipulation of two variables: CP Location, with two levels, Present and Absent (on R-Side); and Algorithm, with two levels (Improved and Legacy). The Legacy condition used the fielded algorithms; the Improved condition introduced features that engineering studies had shown to improve reliability (Crowell, Fabian, Young, Musialek, & Paglione, 2011, 2012). However, most participants reported not noticing or being sensitive to these differences, and objective data showed little evidence that the enhancements affected controller preferences or performance (Willems et al., 2016). The main difference between Legacy and Improved Algorithms of interest to this paper is that in the CP Present, Legacy Algorithm condition, the R-Side displayed CP alerts not only for red and orange alerts, but also for yellow alerts.

Depending on the analysis, we either employed a 2 x 2 design or focused on one of the independent variables. The overall experiment contained other manipulations, in particular a CP Reliability variable which we were not able to manipulate correctly due to simulation errors. The result of this error was that all the conditions had a lower CP reliability than that available in the currently fielded system. We took advantage of this fact to see if R-Side CP would provide benefit at the lower reliability presented to our participants; if so, it should also do so with the current or an improved algorithm.

Four controllers participated at a time. In each run, two participants worked side by side in each of two experimental rooms. The simulations were independent; controllers did not interact with each other. We did not present identical scenarios simultaneously to multiple participants. Each participant worked alone, operating both the R- and RA-Sides. Each group of four spent four days at WJHTC, a full day of training followed by about two and one-half days of testing. The testing sessions comprised twelve 50-minute scenarios. We conducted a short debrief session at the end of most experimental days, and a longer one at experiment end on the fourth day.

Results

Workload

We conducted multiple regression on the individual WAK ratings, with aircraft count and time-on-task as covariates, and unique run nested within Participant as a random effect. The ratings did not vary according to Location, $F(1, 28.1) = 0.55$, $p < .47$; Algorithm, $F(1, 28.02) = 0.069$, $p < .80$; or their interaction, $F(1, 28.02) = 0.092$, $p < .77$. We conducted a similar multiple regression on the communication workload data, except that the only random effect was Participant. In preliminary analysis, this model was more reliable for this model than the model used for WAK. The Location x Algorithm interaction significantly affected the number of voice transmissions, $F(1, 536.1) = 4.83$, $p < .03$. A post-hoc Tukey test showed that the significance resulted from 19 more transmissions per hour in the R-Side Present, Legacy condition than the R-Side Absent, Legacy condition. Similar regression analyses on the number of various types of commands issued showed no meaningful effects of Location or Algorithm.

Subjective Assessments

The results of the subjective assessments are further covered in Willems et al. (2016). To summarize a few results relevant to the present paper, CP on the R-Side increased controller ratings of CP usefulness, and increased over-the-shoulder observer ratings of participants' ability to use the CP in appropriate situations and in a timely and effective manner. Controllers also said that they were more likely to believe and respond to CP alerts when the information was on the R-Side. Observer ratings did, however, suggest reduced Situation Awareness (SA) with R-Side CP due to controllers reacting to the CP information instead of detecting conflicts on their own.

Proactive Altitude Clearances

We identified all cases where the controller instructed a pilot to "expedite" a climb or descent. Controllers do not use "expedite" often, reserving it for urgent situations such as impending conflicts. For this analysis, we determined whether each expedited clearance occurred after a CA activation on the given aircraft. This sequence of events would indicate that the highly tactical CA automation detected a conflict event before the participant, and the expedite clearance was a purely reactive response to something the controller had not known about. We classified each event according to whether a CP alert had activated for that particular aircraft. Therefore, if an event happened during an R-Side CP Present condition but there was no CP alert for the aircraft in question, it was classed as "No Alert." In four of the 15 cases where the expedited aircraft had not received a CP Alert on the R-Side, there was a CA associated with the aircraft (see Figure 1). In the seven situations where the aircraft did receive a CP alert, the controller always proactively gave an altitude clearance before an imminent conflict triggering a CA could develop.

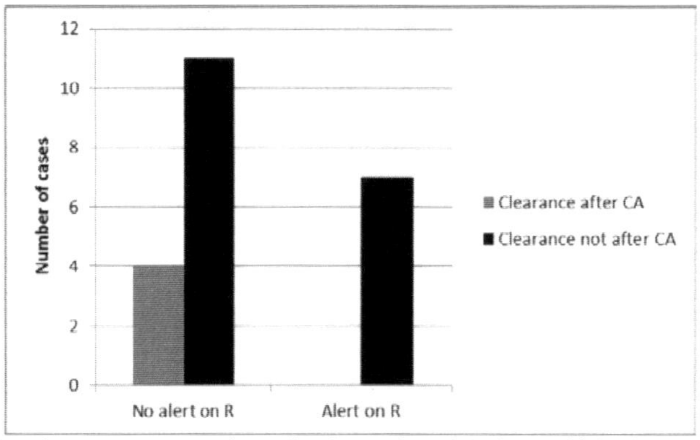

Figure 1. Number of Expedite clearances occurring after/not after CA, by R-Side CP Presence.

We did not have enough observations for statistical significance; with Yates' correction for low expected frequencies (Yates, 1934), $\chi^2(1, N = 22) = 0.85, p < .36$. However, it is operationally notable that in the entire experiment, it was never the case that a situation where the controller received a CP alert on the R-Side escalated to the point where it seemed necessary to expedite an altitude change due to a CA. This finding suggests that CP on the R-Side facilitated proactive conflict resolution behavior.

Modified Clearances

Clearances requiring an amendment sometimes indicate that the initial clearance was inappropriate for some reason, such as being the cause of a potential conflict. However, quickly amending these indicates good performance: the controller has corrected a less-than-ideal decision before a conflict or other problem resulted.

For this analysis we focused on cases where the participant gave and then later revised an instruction. We identified 22 relevant cases for analysis, all of which involved altitude clearances. Of these, 10 had an R-Side CP alert on the given aircraft, and 12 did not—either by virtue of being a CP Absent condition or because there was no alert for that particular aircraft.

We compared the time elapsed between the initial and corrected/rescinded clearances. Participants took 37 seconds longer to correct altitude clearances in R-Side Absent than in R-Side Present events (see Figure 2). A within-subjects t test showed the difference to be statistically significant, $t(19) = 2.50, p < .03$.

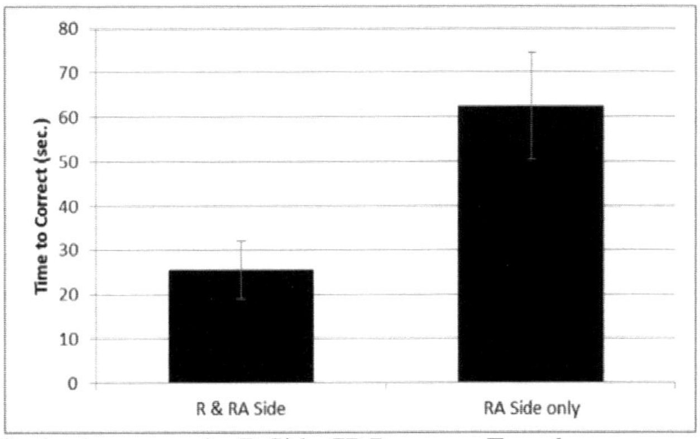

Figure 2. Time to correct altitude clearances, by R-Side CP Presence. Error bars represent standard error of the mean.

Behavioral Observations and Participant Comments

Observers and experimenters witnessed productive use of the R-Side display during the simulations, with controllers issuing clearances to aircraft with yellow CP alerts on the RA-Side in R-Side CP Absent runs. Conversely, situations occurred where the lack of a yellow alert on the R-Side resulted in a delay in acting, and aircraft nearing the separation minima—sometimes resulting in a CA—followed by the controller expressing discomfort with the lack of an R-Side alert. Cases such as these serve as anecdotal evidence that some yellow alerting may be beneficial on the R-Side. Therefore, during the debriefing sessions, experimenters raised the topic of presenting yellow CP alerts on the R-Side. Some participants held that R-Side yellow alerts could represent too much information: two of the eight controllers explicitly articulated a benefit to only showing red alerts, and a third participant called running with yellow alerts on the R-Side a "waste."

Comments in the debriefing in favor of R-Side yellow alerts focused on the notion that these alerts might be acceptable if the reliability were higher. Five of the eight controllers mentioned the potential benefit of presenting yellow alerts defined according to a more intuitive criterion. As mentioned earlier, CP's current definition of a yellow alert is when the conformance boxes surrounding the trajectory line, rather than the centerlines themselves, are predicted to violate separation standards. This rule sometimes misleads controllers as to why a given situation resulted in a yellow vs. red vs. no alert. Therefore, in the debriefing we discussed the concept of coloring an alert yellow based on the predicted distance between the centerlines: if this distance was greater than 5 nautical miles for aircraft at the same altitude, it would result in a yellow alert, regardless of the proximity between the conformance boxes. The question then arises as to what the largest separation should be that would still generate a yellow alert. Not all participants commented on this topic, but those who did exhibited a range on the exact distance they would prefer, with three explicitly stating preferred distances: 5.5, 6, and 8 nautical miles.

Discussion

Analyzing additional types of operational performance not addressed previously in this research program or, in some cases, any previous ATC HITLs, allowed the derivation of a fuller picture of CP's effects on controller performance. The CP display integration reduced the time to correct suboptimal altitude clearances by more than 50%, a result both statistically and operationally significant. Analysis of expedited clearances, while not powerful enough for statistical significance, showed that R-Side CP alerts always prevented controllers from having to issue an expedited clearance in response to a CA. We recommend using these performance measurements in future work.

Integrating CP on the R-Side might cause controllers to react too often to alerts, especially those that do not necessarily require action, resulting in higher workload and lower performance. The present results provide a mixed answer to this concern. The lack of increase in subjective workload or most objective workload measures with CP on the R-Side suggests that this is not an issue. However, R-Side CP increased communications workload, and hindered performance on tasks like timely verbal handoffs to the next frequency. Our operationally-experienced observers also indicated that R-Side CP might compromise SA. This latter finding corroborates previous research, such as Endsley and Kaber (1999), who stated that depending on automation can reduce SA.

The debrief discussions of yellow alerts show mixed opinions about whether these should appear on the R-Side, and how the answer to this question is affected by reliability. The consensus among participants appeared to be that with an acceptable reliability level, yellow alerts would be beneficial especially if defined in an intuitive manner, such as the number of miles of predicted separation between trajectory centerlines. The present study set out partly to establish the overall level of reliability needed for CP acceptance on the R-Side, but did not systematically investigate how red versus yellow alerts should be defined, whether yellows should appear on the R-Side, and how reliability affects these answers. These are all important future research areas.

The experimental design introduced a potential confound. The R-Side yellow alerts appeared only in the Legacy Algorithm condition, where CP had a higher false alert rate, so that the effects of yellow alerts cannot be totally removed from the lower reliability experienced in this condition. However, as mentioned earlier, several participants commented that they did not notice the reliability differences introduced by the algorithmic manipulations, and this factor should not seriously compromise the conclusions. Furthermore, the fact that the CP reliability was lower in the same conditions where R-Side yellow alerts were present served as a catalyst for the debriefing discussion and revealed useful insights.

CP's reliability in the runs analyzed for this paper was lower than in the fielded system, enabling us to conclude that the observed benefits of CP, and the fact that controllers were generally willing to accept the R-Side integration as evaluated in this experiment, would extend to the field, even without improving the fielded algorithm.

References

Brudnicki, D. J., & McFarland, A. L. (1997). *User Request Evaluation Tool (URET) conflict probe performance and benefits assessment*. McLean, VA: The MITRE Corporation.

Crowell, A., Fabian, A., Young, C., Musialek, B., & Paglione, M. (2011). *Evaluation of prototype enhancements to the En Route Automation Modernization's conflict probe* (DOT/FAA/TC-TN-12/3). Atlantic City International Airport, NJ: Federal Aviation Administration William J. Hughes Technical Center.

Crowell, A., Fabian, A., Young, C., Musialek, B., & Paglione, M. (2012). *Evaluation of prototype enhancements to the En Route Automation Modernization's conflict probe with an updated traffic sample from Chicago* (DOT/FAA/TC-TN-12/56). Atlantic City International Airport, NJ: Federal Aviation Administration William J. Hughes Technical Center.

Endsley, M., & Kaber, D. (1999). Level of automation effects on performance, situation awareness, and workload in a dynamic control task. *Ergonomics, 42*(3), 462–492.

Goldmuntz, L., Kefaliotis, J. T., Weathers, D., Kleiman, L. A., Rucker, R. A., & Schuchman, L. (1981). *The AERA Concept* (FAA-EM-81-3). Washington, DC: Federal Aviation Administration.

Masalonis, A. J., Rein, J. R., Messina, J., & Willems, B. (2013). Accuracy assessment of air traffic conflict probe prototype for operational evaluations. In *Proceedings of the Human Factors and Ergonomics Society Annual Meeting, 57*(1), 66–70.

Paielli, R. A., & Erzberger, H. (1997). Conflict probability estimation for Free Flight. *Journal of Guidance, Control, and Dynamics, 20,* 588–596.

Rein, J. R., Masalonis, A. J., Messina, J., & Willems, B. (2013). Meta-analysis of the effect of imperfect alert automation on system performance. In *Proceedings of the Human Factors and Ergonomics Society Annual Meeting, 57*(1), 280–284.

Stein, E. (1985). *Air traffic controller workload: An examination of workload probe* (DOT/FAA/CT-TN84/24). Atlantic City International Airport, NJ: Federal Aviation Administration William J. Hughes Technical Center.

Wickens, C. D., & Dixon, S. R. (2007). The benefits of imperfect diagnostic automation: A synthesis of the literature. *Theoretical Issues in Ergonomics Science, 8*(3), 201–212.

Willems, B., Masalonis, A. J., Fincannon, T., Puzen, G. M., & Bastholm, R. A. (2016). *Separation management human-in-the-loop simulation: Evaluations of conflict probe alert reliability requirements and conflict probe algorithmic improvements* (DOT/FAA/TC-16/38). Atlantic City International Airport, NJ: Federal Aviation Administration William J. Hughes Technical Center.

Yates, F. (1934). Contingency table involving small numbers and the χ^2 test. Supplement to the *Journal of the Royal Statistical Society, 1*(2), 217–235.

Zingale, C. M., Willems, B., Schulz, K., & Higgins, J. S. (2012). *Separation management (SepMan 2): Human factors simulation: Evaluation of conflict probe location and format, display monitors, and pointing devices* (DOT/FAA/TC-12/62). Atlantic City International Airport, NJ: Federal Aviation Administration William J. Hughes Technical Center.

A COMPARISON OF HELMET-MOUNTED DISPLAY SYMBOLOGIES DURING LIVE FLIGHT OPERATIONAL TASKS

Thomas Schnell, Christopher Reichlen, Christopher Reuter
University of Iowa Operator Performance Laboratory (OPL), Iowa City, IA 52242
Eric Geiselman and Jonathan Knox
United States Air Force Research Laboratory, 711th HPW/RHCV, WPAFB, OH
Henry Williams, Naval Medical Research Unit Dayton, OH
Bill Ercoline, KBR-Wyle, Brooks City Base, San Antonio, TX

Spatial disorientation continues to be one of the most costly problems in military aviation, as measured by both life and equipment loss. The unique Helmet-Mounted Display (HMD) centric interface within 5th generation aircraft has the potential to increase tactical capability when compared to previous similar-role aircraft. This study investigated the addition of off-axis ownship attitude information within the HMD field-of-view when the operator looks away from the virtual Head-Up Display (vHUD). In some 5th generation aircraft, traditional HUD symbology is presented via the HMD as there is no aircraft-fixed combiner. In some instances, the only attitude information included via the HMD is part of the vHUD symbology and is only available when the operator looks forward. For this study, a comparison was performed between a baseline representative symbology design and two other interfaces which included variations of off-boresight attitude information symbology. Air-to-ground tactical tasks of varying complexity were performed in live flight by evaluation pilots seated in the rear cockpit of an L-29 aircraft while donning a 5th generation representative HMD system. In addition to the HMD symbology, the visual scene presented was a virtual depiction of a mountainous terrain area. The real outside world was occluded by an opaque hood affixed to the canopy glass. Qualified pilots (n = 10) participated in the study and each flew three approximately one-hour flights. Data collection included quantitative performance, physiological response, and subjective feedback, and preliminary results are presented here.

Background

Historically, the objective of new cockpit technology development has been to enhance pilot performance (such as situation awareness) without causing problems such as Spatial Disorientation (SD). However, when improperly designed or poorly integrated, such technologies may actually reduce performance and increase the likelihood of unintended consequences. SD continues to be a serious problem in the military flight domain and it is critical that both the potential to cause problems as well as support effective defensive mitigation strategies be considered early during the development of new technologies. Past research has shown that new technologies change operator behaviors. For example, the availability of visual information provided via Helmet-Mounted Displays (HMDs) results in pilots looking farther off-axis for longer durations than when the information is not provided (Geiselman, 1999; Geiselman & Havig, 2011; Geiselman, Havig, & Brewer, 2000; Geiselman & Osgood, 1995; Post, Geiselman, & Goodyear, 1999). Presently, we are not able to accurately predict and characterize the potential traps of these behavioral changes, especially under operational conditions. There are two important usability questions which follow: 1) Applied to the 5th generation fighter representative environment, what are the potential effects of the technology on the causation of SD and are the resulting effects predictable and, 2) can effective mitigation strategies be designed into the system to minimize unintended consequences of the technology use?

Test Objectives

The overall aim of this project was to develop and test symbologies that support prevention of Spatial Disorientation (SD) during tactical off-boresight (OBS) use of an HMD in a fighter aircraft platform. Specific aims of this effort included the development of scenarios that are anticipated to cause SD in a 5th generation fighter platform using an HMD and evaluation/refinement of OBS HMD symbology configurations subjected to a high dynamics airborne evaluation.

Experimental Apparatus

A 5th generation fighter aircraft representative HMD was integrated in an OPL L-29 instrumented flight test aircraft and connected to a head-tracked graphics processor that served as a simulated Distributed Aperture System (DAS) for use in actual flight. While wearing the HMD the Evaluation Pilots (EPs) experienced a highly realistic nighttime Close Air Support (CAS) scenario while operating the L-29 aircraft from the back seat crew station as if they were in a single seat 5th generation HMD fighter environment (see Figure 1).

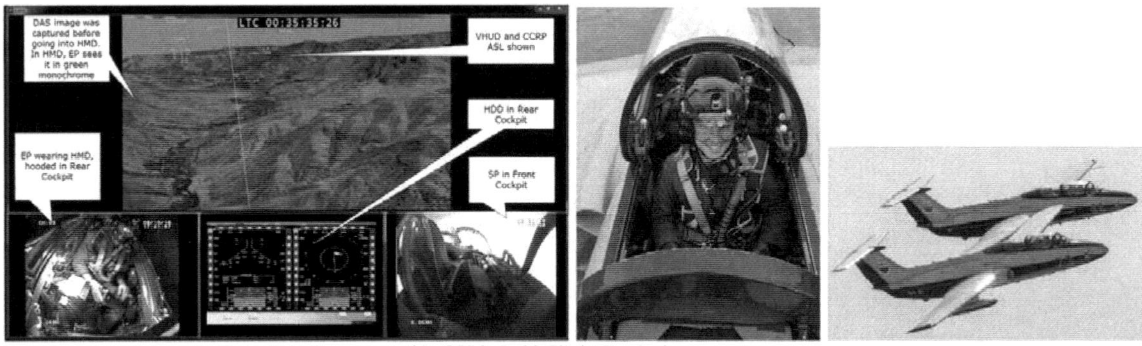

Figure 1. Fifth Generation Fighter HMD Integrated in OPL's L-29 Flight Testbed

The two OPL L-29 aircraft are single engine, tandem-seat fighter jet trainers with pressurized cockpits. These aircraft are fully acrobatic and capable of performing high dynamic maneuvers up to +8/-4gz at speeds up to Mach 0.7. These testbeds are highly instrumented aircraft that use state of the art avionics that incorporate onboard and netcentric airwarfare simulation capabilities, weapon models, Fire Control Radar simulation, and HMD capabilities. Additionally, the aircraft are equipped with human performance state assessment equipment which allows for monitoring of EP physiological based cognitive workload parameters, control inputs, and 6 channel audio/visual recording for human factors assessments. The aircraft are instrumented in such a way that they can also serve as aircraft-in-the-loop (AIL) simulators. The AIL capability was extensively used on the ground to train the EPs on the symbology and L-29 EP crew duties. For live flight operations, the Safety Pilot (SP) performed the taxi-operations, takeoff, and landings from the front seat. The EP crew station canopy was covered with a sliding cloth hood to eliminate the view of the "real-world" outside. A lateral and vertical position proxy mechanism allowed operation of the L-29 in Iowa airspace at mid-teen flight levels, while the EP experienced a nighttime, low-level DAS (monochrome shades of green) environment corresponding to an operations area in a mountainous region of Afghanistan. A ground based Joint Terminal Attack Controller (JTAC) used an immersive graphics environment of the same Afghan battlespace as seen from a ground soldier perspective to coordinate simulated airstrikes on a variety of target areas.

Experimental Procedures and Symbologies

Both the airborne EP fighter pilot and the ground based JTAC used a local area map-like product that we referred to as a placemat. It showed numbered buildings of tactical interest and road names for standardized situation awareness in the "keyhole" CAS procedures. Keyhole CAS uses a standardized template of the target area and initial points. EPs were given a coordinate designation of the Echo point (general target area) that could be visualized in the HMD as a superimposed target diamond and Azimuth Steering Line (ASL). A talk-on to the target selected by the JTAC followed the issuance of a standardized nine-line brief. The talk-on made frequent reference to features on the placemat product carried onboard by the EP and generally worked from large visible features such as the main river and highways to smaller features such as numbered buildings. During the talk-on, the EP was given altitude block assignments of increasing tightness to require attentional division between airmanship and weaponeering with a large percentage of the time spent looking OBS to visually identify target features in the DAS. The talk-on, requiring long and frequent OBS head movements, provided the majority of data of interest to our team. Specifically, we were interested in assessing EP airmanship, weaponeering, situation awareness, and cognitive workload, as a function of three different OBS symbology formats in the HMD. We assessed the comparative potential of those symbologies to prevent loss of spatial orientation. The airborne and ground based battlespaces were synchronized through a High Level Architecture Distributed Interactive Simulation (HLA/DIS) data protocol carried on a tactical utility data link from a ground station to the aircraft. A Rockwell Collins Live Virtual Constructive (LVC) avionics load was used to simulate the weapon flyout models that provided realistic ground attack weapons cueing on the Virtual Head Up Display (vHUD) in the HMD and tactical situation awareness as well as weapons Stores Management System on the Head Down Display. The three different OBS HMD test symbologies assessed were: 1). Current Display Format (CDF, see Figure 2), 2). Distributed Flight Path Reference (DFR, see Figure 3), and 3). Non-Distributed Flight Path Reference (NDFR, see Figure 4). It is important to note that the graphics generator output was in full color (see Figure 1, picture with mountains) and was fed into the fighter HMD that had green

Organic LED imagers. Figures 2-4 were manipulated to dim the background image to highlight the symbology and the background was tinted green to illustrate what the color image would have looked like through the green monochrome HMD. The EPs therefore saw the image on the HMD's combiner as a biocular, fully overlapped monochrome green picture of 1280 x 1024 pixels on a 30 x 40 degrees field of view. The OBS symbologies were shown when the EP turned his head more than 15 degrees laterally or tilted his head by more than 25 degrees vertically from the aircraft forward center line. In the CDF (Figure 2) symbology, there was no attitude (climb/dive/roll) information available. Only speed, head heading, and altitude were shown. Pilots thus had to look in the forward direction to the vHUD or interpret the rate of change in speed, heading, and altitude readouts to obtain aircraft attitude information, which is a difficult and error prone process. The DFR symbology (Figure 3) added aircraft attitude information in the upper right corner of the HMD field of view. This feature had a fixed aircraft symbol with a movable earth reference circle that rotated around the symbol center with regard to bank angle and which grew or shrank with regard to flightpath angle (climb/dive vs. pitch). The earth reference circle had two end-tick marks that referenced the nearest horizon on each side. Thus, for a flight path that pointed straight up or nearly so, the earth circle perimeter was small to non-existent, with only the end-tick marks left, thus indicating that the aircraft was in a climb attitude. For a flight path that pointed straight down or nearly so, the earth reference was a nearly full circle, thus indicating that there was no sky left around the forward direction and that the aircraft was in a dive attitude. The end-tick marks indicated the direction of the nearest horizon. In a level flight-path attitude, the earth reference was a semi-circle. For a full description of the symbology mechanization, see (Geiselman, 1999).

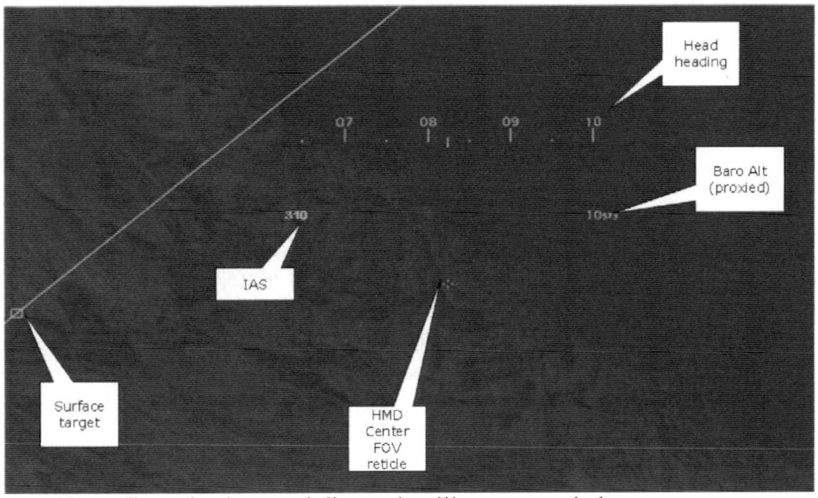

Note: Green background dimmed to illustrate symbology.

Figure 2. *Current Display Format (CDF) Seen During Off-Boresight Head movements*

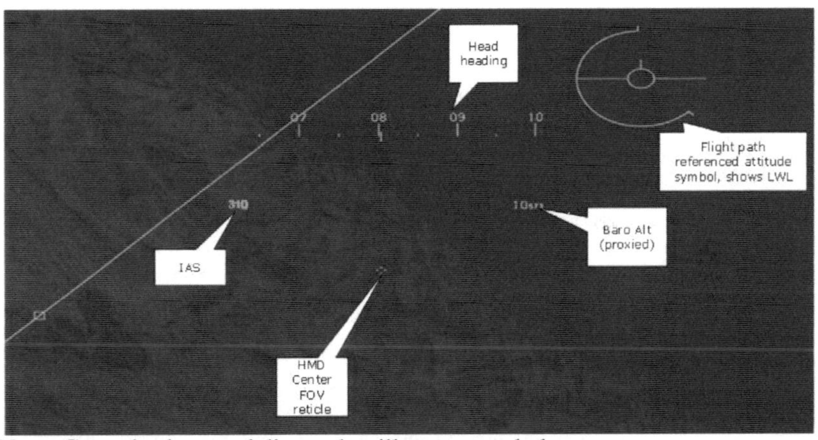

Note: Green background dimmed to illustrate symbology.

Figure 3. *Distributed Flight Path Reference (DFR) seen During Off-Boresight Head movements*

In the NDFR symbology case (see Figure 4), the flight path reference symbol in the upper right corner of the OBS HMD field of view (FOV) was furnished with flight information readouts. These included airspeed on the left wing of the aircraft symbol, aircraft heading as a two-digit number in the center circle, and altitude (MSL) on the right wing of the aircraft symbol. In this configuration, the corresponding flight information readouts were removed from the central section of the HMD FOV. The design of experiments (DOE) plan involved a total of ten current military pilots with jet aircraft experience and training or combat experience with air-to-ground (A/G) attack doctrine. We planned to enroll these ten EPs from two equally sized strata with five having prior HMD experience (e.g. Joint Helmet, F-35, or Scorpion) and five EPs having no such experience. In executing this project, we ended up with a sample of three EPs who had prior HMD experience and seven EPs who did not. The DOE also planned for three sorties per EP. In executing the flights, we ended up with nine EPs who flew all three sorties, one EP who flew only one sortie but then contracted a head cold and elected not to continue, and one EP who made up the remaining two sorties. Since these sorties were representative of nighttime CAS, we refer to them as sorties N1, N2, and N3, in increasing order of intended difficulty. Within each sortie, three attack scenarios were executed so that all three symbologies (CDF, DFR, NDFR) were used for a full A/G attack profile each. During the first scenario of each sortie, the EP checked in with the JTAC, call-sign SWIFT 06 as fragged (as stated by the briefed simulated air tasking order). The JTAC then provided a situation report (SITREP), issued an altitude clearance limit, and provided a nine-line for the first attack. Following issuance of the nine-line, the JTAC provided a visual talk-on to the intended first target using visual references that were available on the placemat product and which the EP had to identify visually using the HMD DAS. The CAS left-orbit stack altitude clearance limits were as follows N1: Remain above 9,000 ft, N2: Remain above 9,000 ft and clear of clouds, with an overcast cloud deck at 13,000, and N3: Remain in a block of 9,000 ft to 11,000 ft. During the talk-on, the EP had to maintain the altitude block in the CAS stack and maneuver the aircraft in such a way as to facilitate visual identification of the target. Once the target was identified, the JTAC requested immediate time-on-target (TOT) for either a show-of-force (SOF) or a bomb-on-target (BOT) delivery. This request also cleared the EP off the CAS stack. The EP then performed the necessary weaponeering and maneuvering to execute the SOF or employ a Mk-82 bomb on the target using a Continuous Computed Impact Point delivery method. Following completion of the necessary attacks, the JTAC asked the EP to provide subjective workload ratings on the Bedford rating scale (1=low, 10=very high) and a 3D Situation Awareness Rating Technique (SART) rating.

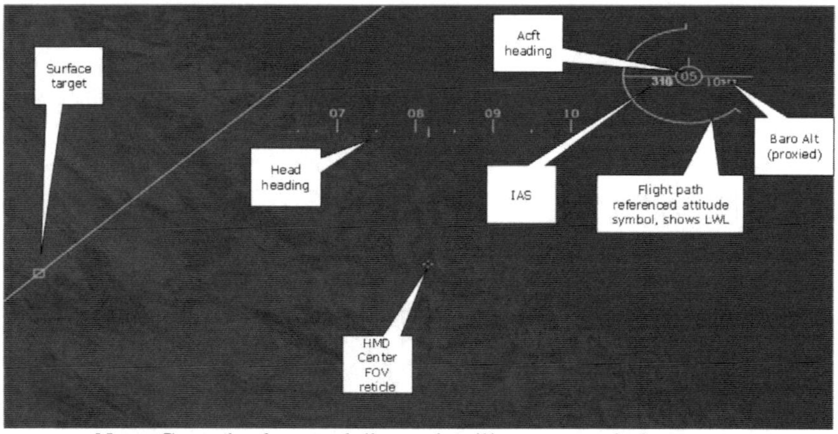

Note: Green background dimmed to illustrate symbology.

Figure 4. Non-Distributed Flight Path Reference (NDFR) seen During Off-Boresight Head movements

Results & Discussion

The following is a small sample of preliminary results as data analysis is still ongoing. This section is intended to provide an initial look at potentially important trends driving further analysis. We specifically chose to include head-tracking and workload data to illustrate the value of off-boresight attitude information. We examined data in three specific time periods of interest within individual attack scenarios based on the likelihood of OBS head movements: 1) SITREP/Nine-Line – the period when the JTAC issued the situation report or Nine-Line, 2) Talk-On start to Talk-On complete, and 3) Total Attack – from Talk-On start to weapons release or SOF complete. The Total Attack time period is thus basically a summary of the tactical (non-administrative) part of the CAS. We noticed that during the SITREP/Nine-Line, EPs were often focused on kneeboard and chart products to the apparent detriment of fine control of the aircraft. A look into the cockpit toward a kneeboard is an OBS look and the DFR and NDFR symbologies are believed to be beneficial in that regime. Additionally, during the Talk-On, EPs are looking frequently OBS to the left to identify landmarks and features in the DAS image. We believe that the DFR and NDFR attitude symbols are very useful for that phase of operation as well. Figure 5 is a collection of "heat maps" produced from head-tracker data. These show aggregated head positions for all subjects by aforementioned time periods of interest and symbology format. The higher density pixels indicate a higher number of head counts in the 1x1 degree region. Each map extends +/- 90 degrees in x (lateral) and y (vertical) axes. The red outlined box in the center of each map indicates the vHUD FOV (15 degrees lateral by 25 degrees vertical). We characterized EP cross-check behavior with a series of measures relating to the head movements OBS (see Table 1). Specifically, we analyzed cross-check rate, which in Table 1, we define as the number of times per minute the EP transitions from inside, to outside, to back inside of the vHUD FOV. Further, we include the total time spent OBS and the average duration of each single "look" during the time period of interest. From the heat maps, we notice a qualitative difference between CDF, DFR, and NDFR in the spread of the head-center orientation toward the kneeboard (4 o'clock position on the heat maps) during the SITREP/Nine-Line phase. There appears to be a tighter focus (better concentration ability) on the kneeboard, presumably better supported by the NDFR symbology when compared to the CDF.

Table 1. *Off-Boresight Look Metrics for Time Periods of Interest.*

SITREP/9-Line OBS Look Metrics - Aggregate for All Sorties						
Symbology	CDF		DFR		NDFR	
	Mean	SD	Mean	SD	Mean	SD
OBS Looks/Minute	12.20	5.08	10.19	5.18	11.28	6.53
%Time OBS	37.33	15.85	37.01	17.19	36.19	19.15
Avg. OBS Look Dur.	2.00	1.34	2.51	1.75	2.02	1.28
Talk-On OBS Look Metrics - Aggregate for All Sorties						
Symbology	CDF		DFR		NDFR	
	Mean	SD	Mean	SD	Mean	SD
OBS Looks/Minute	8.38	3.30	7.35	4.76	7.53	4.25
%Time OBS	47.12	25.03	41.30	23.63	47.66	25.20
Avg. OBS Look Dur.	3.46	2.27	4.14	3.78	4.12	3.05
Total Attack OBS Look Metrics - Aggregate for All Sorties						
Symbology	CDF		DFR		NDFR	
	Mean	SD	Mean	SD	Mean	SD
OBS Looks/Minute	6.84	2.69	5.55	2.50	5.95	2.87
%Time OBS	34.31	17.80	33.01	19.00	37.02	21.44
Avg. OBS Look Dur.	3.04	1.60	3.54	2.27	3.96	2.68

The heat maps for the Talk-On phase clearly show that EPs were able to venture OBS farther more often, and in a more organized (stratified) way with the NDFR when compared to the CDF. The DFR was in the middle of that trend, meaning that EPs went OBS a little more often and somewhat farther under the DFR condition when compared to the CDF condition. This is confirmed with the basic statistics in Table 1 for the Talk-On phase in that the DFR and NDFR symbology enabled the EPs to get the Talk-On job done with fewer OBS head movements of longer duration when compared to the CDF. This enabling capability of the DFR and NDFR is practically significant as the more frequent disruption of the search task in the CDF costs the EPs extra time to re-acquire the Talk On targets. The large spread of the OBS head movements in the CDF illustrate the struggle that EPs faced in re-acquisition of the Talk-On targets after a check-look to the vHUD (forward direction).

Table 2. *Mean Bedford Workload and SART Situation Awareness Scores*

	Bedford	SART
CDF	4.75	7.0
DFR	4.0	7.75
NDFR	4.0	8.0*

* denotes statistically significant difference to CDF

Bedford workload and situation awareness ratings were provided by the 10 EPs after each attack, and the mean scores are shown in Table 2. A Kolmogorov-Smirnov test indicated that the normality assumption could be made for the Bedford data (KS=0.03, p>0.15) and the SART data (KS=0.021, p>0.15). ANOVA did not indicate any statistically significant differences in the mean Bedford workload ratings. The ANOVA on the SART ratings indicated a statistically significant ($F_{2,78}$=5.99, p=0.004) effect for the symbology factor. A pairwise post-hoc t-test on the differences in effect means indicated that the NDFR ratings were significantly higher (t=3.43, p=0.0027) than their CDF counterparts. That same trend for the DFR was not quite significant (t=2.03, p=0.11), nor was the difference between DFR and NDFR.

While certainly preliminary, the results of this current live-flight study lend support to previous findings that HMD use results in pilots looking OBS for longer durations (Geiselman et al., 2000). Within the heat maps, there is evidence of an improved structure to the Talk-On scan behavior in the DFR and NDFR conditions. The SART ratings indicate that an advantage was provided by the NDFR format as well. Future analyses of this dataset will look for any instances of SD, examine the objective airmanship and weaponeering data, and apply metrics that better characterize the spread of the data in the heat maps. The results of those analyses will inform empirically based recommendations for the design of HMD symbology that improves pilot performance and mission effectiveness during tactically challenging operations with 5th generation fighter HMDs.

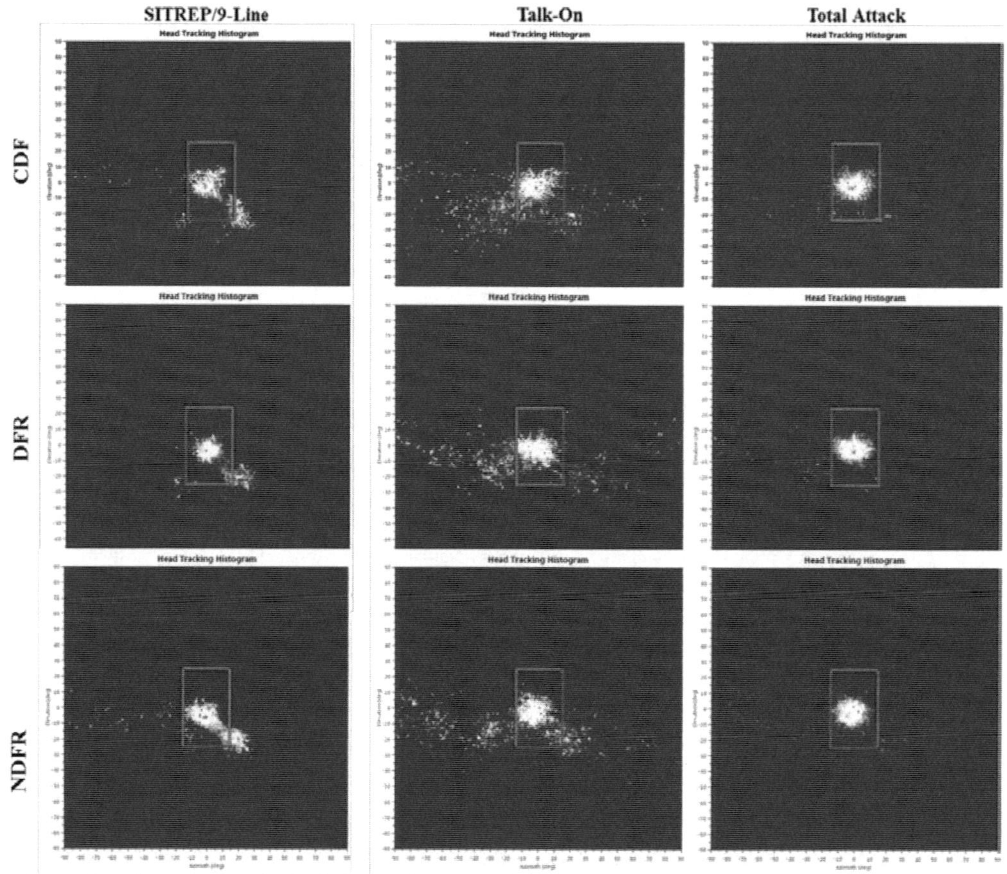

Figure 5. Heat Maps of Head Tracker Data by Symbology and Time period of Interest, N=10 Pilots.

Disclaimers & Acknowledgments

The views expressed in this article are those of the author and do not necessarily reflect the official policy or position of the Department of the Navy, Department of the Air Force, Department of Defense, nor the U.S. Government. This project was funded by the USAF 711th HPW using the Wyle Quick Reaction USAFSAM Assessments, Studies, Analysis, Evaluation, and Research (QUASAR) contract vehicle, FA 8650-12-D-6280 TO 33. I am an employee of the U.S. Government. This work was prepared as part of my official duties. Title 17 U.S.C. §105 provides that 'Copyright protection under this title is not available for any work of the United States Government.' Title 17 U.S.C. §101 defines a U.S. Government work as a work prepared by a military service member or employee of the U.S. Government as part of that person's official duties. The flights were performed under Public Aircraft Operations (PAO) Military Flight Release MFR-1216-012. The authors would like to thank USAF AFRL and Edwards Flight Test Center for their help and support in achieving the military flight release.

References

Geiselman, E. (1999). *Practical considerations for fixed wing helmet-mounted display symbology design.* Paper presented at the Proceedings of the human factors and ergonomics society 43rd annual meeting, Santa Monica, CA.

Geiselman, E., & Havig, P. R. (2011). *Rise of the HMD: the need to review our human factors guidelines.* Paper presented at the SPIE, Bellingham, WA.

Geiselman, E., Havig, P. R., & Brewer, M. T. (2000). *A non distributed flight reference symbology for helmet-mounted display use during off-boresight viewing: development and evaluation.* Paper presented at the The International Society for Optical Engineering, Bellingham, WA.

Geiselman, E., & Osgood, R. K. (1995). Helmet-mounted display attitude symbology: an evaluation of compression ratio. *International Journal of Industrial Ergonomics, 15*, 111-121.

Post, D. L., Geiselman, E., & Goodyear, C. D. (1999). Benefits of Color Coding Weapons Symbology for an Airborne Helmet-Mounted Display. *Human Factors, 41*(4), pp. 515-523.

USAF Spatial Disorientation Prevention: A Meta-Analytical Human Systems Integration Perspective

Patrick O'Neill
Booz Allen Hamilton
Wright-Patterson AFB, OH
Michael Miller
Air Force Institute of Technology
Wright-Patterson AFB, OH

Human Systems Integration involves the systematic consideration of tradeoffs in system structure or behavior, which affect seven human-centered domains to optimize total system performance and life cycle cost. All too often, HSI is overlooked or poorly practiced, despite the specific directive within DoDI 5000.02 for program managers to plan for and conduct HSI. In the worst of cases, poor consideration of human capabilities and limitations leads to errors, mishaps, and death or serious injury. One such human limitation in aviation is the inability of the pilot to maintain proper spatial orientation during flight, as mismatches between the stimuli present during flight create an erroneous perception of aircraft attitude with respect to the horizon. The consequences include preventable departures from controlled flight and unusual attitudes, unnecessary aircraft ejections, and controlled flight into terrain. Between 1993 and 2013, spatial disorientation contributed to 12% of all Class A Mishaps, resulting in the loss of 65 aircraft and 101 lives. With a fatality rate of 24.9%, disorientation leads all Class A mishap causal factors. While it may not be feasible to prevent all spatial disorientation mishaps, it may be possible to significantly reduce mishap rates through proper tradeoff analysis, resulting in better total system performance and reduced life cycle costs. To that end, this paper will discuss the HSI domains applicable to spatial disorientation and provide a meta-analytical perspective on practicing HSI and its implications for disorientation prevention.

Air Force Instruction (AFI) 63-1201, *Life Cycle Systems Engineering*, defines Human Systems Integration (HSI) as "a disciplined, unified, and interactive systems engineering approach to integrate human considerations into the system development, design, and life cycle management to improve total system performance and reduce total costs of ownership". The recent change to Department of Defense (DoD) Instruction 5000.02 reduced the recognized HSI domains to seven: Manpower, Personnel, Training (commonly MPT), Safety and Occupational Health, Human Factors Engineering (HFE), Survivability, and Habitability. Tradeoffs are made between system features balance total system performance (including the operator) with total life cycle costs (LCC). The instruction also requires the program manager (PM) to plan for and implement HSI beginning early and throughout a system's life cycle in order to optimize total system performance and total ownership costs, while ensuring that the system effectively provides the user with the ability to complete their mission.

Early and iterative consideration of HSI during system design is believed to offer significant payoffs to the total system, including but not limited to increased system availability, reliability, safety, and survivability. However, the intent of this guidance is not always realized, and HSI is often overlooked or poorly practiced. The results can lead to errors, mishaps, even death or serious injury. One such human limitation in aviation is spatial disorientation (SD), in which mismatches between the available environmental stimuli and the ability of the human physiological sensory pathways to accurately perceive these stimuli create an erroneous perception of aircraft attitude with respect to the horizon. The consequences of such orientation illusions can precipitate preventable in-flight loss of control (LOC-I), unusual attitudes (UA), unnecessary aircraft ejections, controlled flight into terrain (CFIT) or other undesirable consequences.

Spatial Disorientation

Spatial disorientation is often indicated as a causal factor in USAF Class A mishaps, particularly in fighter/attack aircraft, and is considered a leading cause of pilot fatalities (R. Gibb, Ercoline, & Scharff, 2011). Recent analysis suggests that, from 1993 to 2013, SD was involved in 12% of USAF Class A mishaps, of which 61.1% resulted in the loss of life (Poisson & Miller, 2014). Estimates in financial terms suggest that these mishaps have generated over $2-billion in property loss and medical costs. However, the true prevalence of SD may be even greater than 25% of all aviation mishaps, possibly as high as 33%, due to inaccuracies and underreporting (R. Gibb et al., 2011). Figure 1 provides a graphical summary of the statistics and trends over the last 35 years based on a meta-analysis of data from five separate publications (R. W. Gibb & Olson, 2008; Gillingham, 1992; Matthews, Previc, & Bunting, 2002; Poisson & Miller, 2014; Sundstrom, 2004). Of concern is the notion that the frequency and severity statistics have historically remained unchanged (R. Gibb et al., 2011); however, best-fit (greatest R^2) trend lines in Figure 1 suggest SD-related Class A mishap frequency peaked in the mid-1990s before returning to historical levels. Of note, a number of factors may be a play, evolving definitions of SD, relative emphasis on education and reporting requirements, or technology improvements.

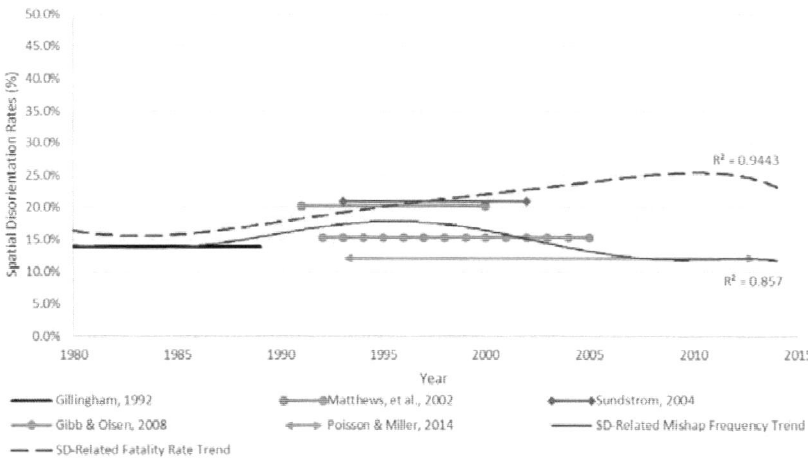

Figure 1. Frequency of SD-Related Class A Mishaps, 1980-2013.

While the primary causes are physiological (mismatches between the visual and vestibular senses), based largely on visual information, certain psychological factors contribute to the experience of SD (R. Gibb et al., 2011). These include cognitive tunneling, loss of situational awareness (SA), visual capture and clutter, and task saturation. For example, "pilots who have lost SA through being out-of-the-loop may be both slower to detect problems," (Endsley, 1997). Effectively, these psychological factors increase the risk of physiological modality conflicts: the wrong combinations of which ultimately set the stage for experiencing one of the three types of SD: Unrecognized, Recognized, and Incapacitating. Unrecognized (Type I), reflects a state where the pilot is unaware of the disorientation and believes their aircraft is responding well to inputs. Type II, Recognized, involves the pilot's awareness of conflicting orientation cues. Finally, Type III or Incapacitating, occurs when the pilot is aware of the disorientation, but cannot correctly adjust or recover.

HSI Considerations for Spatial Disorientation Mishap Reduction

Some researchers have argued that technology has made no effect in mitigating SD incidents; rather, technologies such as night vision goggles, heads-up displays, helmet/head-mounted displays (HMDs), or other technology has rather changed the types of errors committed by increasing perceptual and cognitive demands (R. Gibb et al., 2011). The result is an increased likelihood of SD due to visual clutter, reduction of peripheral vision cues, cognitive tunneling, task saturation, or other cognitive factors. However, the systems in question were designed for purposes other than SD mitigation, primarily to enhance pilot SA. Arguments against these technologies suggest that they fail to improve some aspects of SA and contribute to SD. However, this research suggests that technology solutions rely on the HSI plan and its human-centered considerations, primarily from the HFE, MPT, and Survivability domains to assist pilots in maintaining normal expected flight parameters.

Manpower, Personnel, and Training

The Manpower, Personnel, and Training HSI domains are often interlinked due to inherent domain interdependencies and the direct impact of tradeoffs among these domains. In HSI terms, the Personnel domain is concerned with the types of people and their required knowledge, skills, and abilities (KSAs). Requirements definition for this domain is driven by performing task analyses and descriptions within existent career fields. Conversely, Manpower focuses on the number of each type of personnel possessing the specified KSAs. Finally, Training covers development of the KSAs required to operate and maintain the system of interest (SOI). Thus, once a certain set of KSAs to perform their duties, an example tradeoff concerns the economics of hiring and training.

According to the USAF Officer Classification Directory, entry into the Pilot career field as a Pilot Trainee with Air Force Specialty Code (AFSC) 92T0 has two KSA requirements beyond those necessary to obtain a commission in the USAF: the abilities to pass a Flying Class I physical and obtain the required security clearance via initiation of a background investigation. The candidate selection process generally assesses medical fitness, anthropometry, and educational achievement (Carretta, 2000). The progression toward becoming a fully qualified pilot in a particular major weapon system (MWS) varies with each airframe, but follows the same basic path: Initial Flight Screening (IFS), Specialized Undergraduate Pilot Training, MWS-specific transition and operational training, and MWS-specific upgrade training.

Today, preventative SD training strategies focus on familiarization with the causes of SD, defining specific illusions, and recognizing these illusions through classroom instruction. Per AFI 11-403 *Aerospace Physiological Training Program*, aircrew receive their first exposure to SD during their initial Aerospace Physiology course, which covers the characteristics of SD types, basic physiology and contributing factors, and related illusions. AFI 11-403 also requires all fighter/attack students be trained in an SD-demonstrator. Refresher training is required every five years and includes review of these topics and further demonstration as necessary. Alternative methods include SD-specific simulators, in-service simulators, and in-flight demonstration (R. Gibb et al., 2011). SD-specific simulators are capable of producing motion and visual cues necessary for inducing SD under representative workloads. In-service simulators can be tailored to provide training for a limited set of SD-producing scenarios, but are incapable of replicating many realistic experiences. Finally, in-flight demonstration promotes the most realistic scenarios and can be applied to develop both recognition and recovery skills. However, enhanced aircrew training does not guarantee significant improvements to SD-related mishap trends; it is merely one contributing perspective.

In summary, trades among the MPT domains could include enhanced manpower and personnel analyses towards the goal of reducing task saturation and unrecognized SD. Alternately, improvements in training may increase the likelihood of recognizing SD onset before situations grow out of control or SD becomes debilitating.

Human Factors Engineering

The HFE domain seeks to account for human capabilities and limitations in the SOI's development and evolution in order to optimize human-machine performance in SOI operation and maintenance. These design considerations are governed by MIL-STD-1472G, which addresses human factors design specifications and principles. The goal of the presented criteria and principles is to achieve required user performance, manpower readiness, personnel-equipment reliability, and design standardization. SD results as a consequence of our limited capability to resolve visual, vestibular and proprioceptive cue conflicts, which can often be mediated or induced by other cognitive limitations as noted. Research designed to investigate SD prevention strategies cover a wide array of topics, including alternative attitude references, multimodal cueing, or synthetic vision systems. In each case, the primary goal is to provide pilots with intuitive information via alternative presentation and/or supplemental cues.

Traditional attitude information has been presented using the Sperry design, which is an inside-out attitude reference. Alternatives to this design include an outside-in attitude indicator (AI) and the Arc-Segmented Attitude Reference. Research has somewhat confirmed that the two alternative designs are more intuitive than inside-out AI leading to performance improvements (Poisson, Miller, Haas, & Williams, 2014; Self, Breun, Feldt, Perry, & Ercoline, 2002); yet these have not been reliably demonstrated in-flight. Additionally, researchers developed a spatial disorientation geometric command icon (a diamond) designed to be easily interpreted. Evaluations of this icon demonstrated increased performance on UA recovery tasks (Small, Fisher, & Keller, 2005).

Other technologies incorporate multimodal (auditory and tactile) cueing and synthetic vision systems (SVS) to help maintain orientation. Multiple Resource Theory suggests that auditory cueing (including three-dimensional) can improve SA when visual attention is under high workload, but the advantages may be limited. Tactile cues offer a variety of strategies for conveying various information to the operator such as signal localization, cue pulse rates, varying the fundamental vibration frequency, and linear and radial spatial flow (Lawson, Cholewiak, Brill, Rupert, & Thompson, 2015), which might prove useful to convey orientation information. Synthetic vision systems provide an artificial view of the environment by combining computer-generated imagery, guidance displays, and on-board sensors. Integration of these technologies is believed to improve flight safety, SA, and orientation (Prinzel & Kramer, 2006) through augmentation of basic flight information.

It should be noted that for SD to become recognized, multiple cues must be present and visual cues must be directly observed. Therefore, each of these visual displays suffer when breakdowns occur in the pilot's scan path under low visibility conditions, preventing the gathering of visual cues which might result in the cue conflict necessary for SD to be recognized. These cues may provide the pilot with early information that their perception of spatial orientation is degrading, encouraging the inclusion of the attitude indicator in their scan path.

Table 1.
Summary of Mishap and Fatality Rates per Million Flight Hours.

Study	Years (FY)	Aircraft	Flt Hrs	Mishap Rate	(#)	Fatality Rate	(#)
Holland, et al., 1995	1980-1989	A-10	2.09	3.83	(8)	6.22	(13)
		F-15	1.80	5.57	(10)	3.34	(6)
		F-16	2.06	7.75	(16)	6.30	(13)
		Overall	5.95	5.71	(34)	5.38	(32)
Sundstrom, 2004	1993-2002	A-10	1.19	5.89	(7)	4.21	(5)
		F-15	1.96	1.02	(2)	1.53	(3)
		F-16	3.70	4.05	(15)	2.70	(10)
		Overall	6.85	3.50	(24)	2.63	(18)
Poisson & Miller, 2014	1993-2013	A-10	2.35	3.83	(9)	2.56	(6)
		F-15	3.52	1.70	(6)	0.85	(3)
		F-16	6.78	3.69	(25)	2.65	(18)
		Overall	12.84	3.19	(41)	2.18	(28)

Table 2.
Cost and SD Type Summary by Aircraft per Million Flight Hours.

Aircraft	Flt Hrs	Type I	(#)	Type II	(#)	Type III	(#)	Total	Cost
A-10	1.19	4.21	(5)	0.84	(1)	0.00	(0)	6	$66,647,399
F-15	1.96	0.51	(1)	0.51	(1)	0.51	(1)	3	$77,903,113
F-16	3.70	3.51	(13)	0.27	(1)	0.00	(0)	14	$259,427,142
Overall	6.85	2.77	(19)	0.44	(3)	0.15	(1)	23	$403,977,654
Frequency		83%		13%		4%			

Survivability

Survivability concentrates on the SOI characteristics that reduce risk of acute and/or chronic illness, disability, injury, or death in the event of fratricide, detection and attack, hostile or extreme environments, system faults, or other hazardous occurrence. In the aviation environment SD is one of the many physiological incident risks present. In fact, some suggest that SD-related Class A mishaps exhibit a probability of fatality that is 2.85 times greater than other Class A mishaps between 1993 and 2013 (Poisson & Miller, 2014). Table 1 summarizes extrapolated mishap and fatality rates for the three most often studied fighter/attack aircraft in the current USAF inventory (the A-10, F-15, and F-16) (Holland & Freeman, 1995; Poisson & Miller, 2014; Sundstrom, 2004). Table 2 breaks the data down by SD type and includes the mishap costs. Based on this, Sundstrom (2004) recommended re-evaluation of installing Automatic Ground Collision Avoidance Systems (Auto-GCAS) in USAF fighter aircraft.

Such systems promise stark impacts on reducing type I and III SD mishap rates. According to the data in Table 2, the types accounted for a combined 87.0% of the relevant mishaps. Furthermore, Sundstrom suggests that, CFIT resulted in 68% of the reported disorientation mishaps. By taking over for an unresponsive, possibly disorientated or incapacitated pilot, it is conceivable that Auto-GCAS might avoid as much as 59% of these mishaps, based on the combined probability of CFIT and Type I or III SD, depending on the MWS and its mission profile.

Most technology intended to mitigate aircraft mishaps has focused primarily on recovery, meaning the system's benefits are realized after a pilot has become disoriented. While the capability of systems such as Auto-GCAS can partially reduce the occurrence of specific types of SD and related mishaps through automated recovery, the need to react and recover in the first place might be eliminated by proactively preventing the effects of disorientation. If successful, a proactive approach might alert a pilot to potential impending SD scenarios. Effectively, Type I and III incidents are recognized (Type II) early or not experienced at all. If that is the case, where Auto-GCAS may avoid 59% of relevant SD-mishaps, a prevention strategy may increase the figure to 87% or more.

HSI Cost-Benefit Trades of Preventing Disorientation

From 2009 to 2013, Class A Mishaps totaled approximately $2.6B in financial losses based on data obtained from public USAF Accident Investigations Boards reports. Human factors were identified in 71.7% of these accounting for roughly $1.95B of the reported financial losses. In a perfect world, perfect HSI and HFE would eradicate all human error and subsequent mishaps. However, SD has a physiological component that cannot be completely eliminated, only reduced. If realized, a decrease in SD-related mishaps translates to fewer fatalities and lost aircraft, increased total system performance, and potentially reduced total LCC. Any reduction in these areas constitutes a strong argument for increased investment in HSI. For SD prevention, long term solutions will involve some combination of training and technology, with specific implications for manpower and personnel.

First, it is essential to understand that the human capability for orientation is fundamentally limited by our physiology: it did not evolve to traverse the atmosphere. Disorientation is thus an inherent in-flight risk that cannot be mitigated through design, but rather must accepted and minimized. Realizing this, it is incumbent upon the PM and HSI personnel to consider the HSI implications of each requirement. For example, certain generation five fighter capabilities (HMDs, thrust vectoring, etc.) pose significant concerns for orientation, specifically the ability to resolve cues mismatches during off-axis viewing or movement. Accounting for such activities in the use case and/or concept of operations (CONOPS) will drive their consideration within the solution space.

Specifically, the CONOPS and use cases shape the ensuing task analysis, which for aviation is increasingly important as the integration of cockpit automation has led to the evolution of the pilot's supervisory control role. The tasks with which the pilot will be assigned from this analysis also imply particular orientation solution strategies such as cockpit automation. Finally, integration of new technologies requires some level of operator training, in this case for both SD and the technologies. For the USAF, this could mean reduced mission readiness during upgrades.

These questions are normally answered early in development, assuming HSI activities are initiated as recommended. Addressing them early in the acquisition process often leads to the return on investment (ROI). According to the DoD Operating and Support Cost Estimation Guide, development costs on average account for 7% of the program lifecycle cost, of which early HSI activities are a small piece. For many programs, this can translate to millions of dollars annually for all system development activities. At a fraction of this budget, HSI activities require a modest investment above the necessary cost implications of the requirements and design tradeoffs. Unfortunately, HSI does not necessarily translate to direct system procurement savings, as additional development and integration costs are incurred to procure the system. The majority of a system's LCC is incurred during operations and sustainment, and the conventionally accepted "golden rule" is 70% (Jones, White, Ryan, & Ritchel, 2014), but the associated decisions are made early in development. Likewise, the HSI ROI is often overlooked.

The implication is that significant drivers of operations and sustainment costs need to be accounted for from the early stages of development, including HSI. Mishaps with human-related contributions pose an opportunity for to quantify the ROI. Summarizing the economic costs of SD (damages, lost aircraft, medical bills, etc.) of related Class A mishaps between 1993 and 2013, including those directly tied to individual programs (A-10, F-15, and F-16) led to interesting LCC savings estimates and implications based on the effect of Auto-GCAS and potential impacts of improved SD and CFIT prevention (Table 3). Per this analysis, the entire USAF fleet could see potential annual savings totaling approximately $96.1M, and fighter/attack aircraft account for $30-35M of this estimate.

These estimates are based on a number of assumptions. The first is a 100% reduction in preventable SD mishaps, which is likely not feasible. Secondly, the analysis only accounts for reported incidents leading to a Class A mishaps. Many SD incidents go unreported due to misidentification, lack of economic or casualty consequences, or pilot attitude and culture. Third, much of the data required transformation to arrive at these estimates, so they are only as good as the accuracy of the sources. Furthermore, aircraft specific estimates are based on the number of lost aircraft during the periods of interest and the reported flyaway costs. Finally, estimates for the entire fleet do not necessarily account for the relative contributions by individual MWS or fiscal year, potentially inflating the savings.

Table 3.
Summary of HSI Life Cycle Cost Impacts with Respect to Spatial Disorientation

Study	Aircraft	#Lost A/C	Flyaway Cost ($M)	Economic Cost ($M)	Observed CFIT Impact ($M)	Est. LCC Savings ($M)	Auto-GCAS Impact ($M)		Prevention Impact ($M)	
							Cumulative	Annual	Cumulative	Annual Avg
Sundstrom, 2004	A-10	7	$18.8	$66.6	$66.6	$12.8	$39.4	$3.9	$58.0	$5.8
FY1993-2002	F-15E	2	$31.1	$77.9	$38.0	$21.1	$46.1	$4.6	$67.7	$6.8
	F-16A	2	$14.6	$29.7	$29.7	$9.9	$17.5	$1.8	$25.8	$2.6
	F-16B	2	$18.8	$23.0	$15.9	$12.8	$13.6	$1.4	$20.0	$2.0
	F-16C	9	$14.6	$182.6	$166.7	$9.9	$108.0	$10.8	$158.8	$15.9
	F-16D	1	$18.8	$24.2	--	$12.8	$14.3	$1.4	$21.1	$2.1
	Fighter/Attack	23	--	$404.0	$317.0	$79.4	$238.9	$23.9	$351.3	$35.1
Poisson & Miller, 2014	A-10	9	$18.8	($169.2)		$12.8	$100.0	$4.8	$147.1	$7.0
FY1993-2013	F-15A/C	2	$27.9	($55.8)		$19.0	$33.0	$1.6	$48.5	$2.3
	F-15B/D/E	4	$29.9	($119.6)		$20.3	$70.7	$3.4	$104.0	$5.0
	F-16A/C	20	$14.6	($292.0)		$9.9	$172.7	$8.2	$253.9	$12.1
	F-16B/D	5	$18.8	($94.0)		$12.8	$55.6	$2.6	$81.7	$3.9
	Fighter/Attack	40	--	($730.6)		$74.8	$432.0	$20.6	$635.3	$30.3
	All Aircraft	65	--	$2,320		$1,578	$1,372	$65.3	$2,017	$96.1

Perhaps the greatest supposition is the inherent hypothesis that investment in HSI with respect to both SD and the larger efforts will actually realize potential LCC reductions associated with future mishaps. Only known risks can be mitigated or reduced, but it is impossible to identify all circumstances and failure modes during risk analysis. It is entirely likely that new, emergent failure modes may occur that were not yielded during the risk analysis. The lack of guarantee on total LCC reduction associated with HSI investment might sway the decision away from the venture. In this case, the effectiveness to which the need is communicated will drive the decision: the key might be to appeal for the additional HSI perspective, which might identify new, unique risks and mitigation strategies. Finally, the given savings estimates apply to only SD prevention, which is but one topic impacted by HSI.

Conclusion

At 2.85 times more likely to be fatal than all other Class A mishaps, those involving disorientation result in substantial financial costs and significantly impact the flight safety. While the manpower and personnel HSI domains exert little influence on the experience of SD, they bear some of the consequences through injury and/or loss of life. However, considerations for training, HFE, and survivability exhibit the most promise for mishap prevention. Technologies like Auto-GCAS are predicated upon first experiencing disorientation, UA, LOC-I, etc. A more proactive approach might identify impending scenarios so that a pilot can correct the situation prior to experiencing disorientation. This requires early investment in HSI efforts to affect the HFE and training domain tradeoffs. The SD considerations only account for a small piece of the overall effort, but the ROI can be described as of millions of dollars saved, aircraft and lives saved, or mission readiness and execution.

References

Carretta, T. R. (2000). U.S. Air Force Pilot Selection and Training Methods. *Aviation, Space and Environmental Medicine*, *71*(9), 950–956.

Endsley, M. R. (1997). Supporting situation awareness in aviation systems. In *1997 IEEE International Conference on Systems, Man, and Cybernetics. Computational Cybernetics and Simulation* (Vol. 5, pp. 257–276).

Gibb, R., Ercoline, W. R., & Scharff, L. (2011). Spatial disorientation: Decades of pilot fatalities. *Aviation Space and Environmental Medicine*, *82*(7), 717–724.

Gibb, R. W., & Olson, W. (2008). Classification of Air Force Aviation Accidents: Mishap Trends and Prevention. *The International Journal of Aviation Psychology*, *18*, 305–325.

Gillingham, K. K. (1992). The spatial disorientation problem in the United States Air Force. *Journal of Vestibular Research : Equilibrium & Orientation*, *2*(4), 297–306.

Holland, D. A., & Freeman, J. E. (1995). A ten-year overview of USAF F-16 mishpa attributes from 1980-89. In *Proceedings of the Human Factors and Ergonomics Society 39th Annual Meeting* (pp. 30–34).

Jones, G., White, E., Ryan, E. T., & Ritchel, J. D. (2014). Investigating into the ratio of operating and support costs to life-cycle costs for DoD weapon systems. *Defense ARJ*, *21*(1), 442–464.

Lawson, B. D., Cholewiak, R. C., Brill, J. C., Rupert, A. H., & Thompson, L. I. (2015). Tactile cueing strategies to convey aircraft motion or warn of collision. In *Proceedings of the 18th International Symposium on Aviation Psychology* (pp. 19–24). Dayton, OH: Wright State Universiy.

Matthews, R. S. J., Previc, F. H., & Bunting, A. (2002). USAF Spatial Disorientation Survey. In *RTO-MP-086: HFM Symposium on "Spatial Disorientation in Military Vehicles: Causes, Consequences and Cures"* (pp. 7–1-7–13).

Poisson, R. J., & Miller, M. E. (2014a). Spatial disorientation mishap trends in the U.S. Air Force 1993 -2013. *Aviation Space and Environmental Medicine*, *85*(9), 919–924.

Poisson, R. J., Miller, M. E., Haas, M. W., & Williams, H. P. (2014). Evaluation of a non-traditional aircaft attitude indicator. In Y. Guan & H. Liao (Eds.), *Proceedings of the 2014 Industrial and Systems Engineering Research Conference.*

Prinzel, L. J., & Kramer, L. J. (2006). Synthetic Vision Systems. In W. Karwowski (Ed.), *International Encyclopedia of Ergonomics and Human Factors* (pp. 25–1 – 25–10). Philadelphia, PA: Taylor & Francis.

Self, B. P., Breun, M., Feldt, B., Perry, C., & Ercoline, W. R. (2002). Assessment of pilot performane using a moving horizon (inside-out), a moving aircraft (outside-in), and an arc-segmented attitude reference display. In *Proceedings of RTO HFM Symposium on "Spatial Disorientation in Military Vehicles: Causes, Consequences and Cures* (pp. 30–1 – 30–7).

Small, R. L., Fisher, A. M., & Keller, J. W. (2005). A Pilot Spatial Orientation Aiding System. In *Proceedings of the AIAA 5th Aviation, Technology, Integration, and Operations Conference* (pp. 1–15). Arlington, VA.

Sundstrom, J. N. (2004). *Flight conditions leading to Class A spatial disorientation mishaps in U.S. Air Force fighter operations: FY93-02.* Wright-Patterson AFB, OH.

ROLE OF EARLY FLYING PERFORMANCE IN PILOT TRAINING TRACKS

Rana Ihara, Yumi Shibata
Aeromedical Laboratory, Japan Air-Self Defense Force
Tokyo, Japan
Yuko Aiba
Acquisition, Technology & Logistics Agency, Ministry of Defense
Tokyo, Japan
Takenori Nomiyama, Asao Kobayashi
Aeromedical Laboratory, Japan Air-Self Defense Force
Tokyo, Japan

Early flying skills have been reported to strongly influence future flying skills. Few published studies have evaluated the comprehensive relationship among all flying training skills. This study evaluated three causal path models for flying skills that influenced flying performance rated by instructor pilots in five training phases. A covariance structure analysis showed that a sequential model with a connection between the first and last phases was optimum for representing flying training performance and skills. The flying skill acquired in the first training phase in a primary propeller aircraft influenced the last training skill in a fighter jet directly as well as indirectly. Furthermore, the influence of the first training phase on the fighter training skill was the strongest among all the training phases. These results suggest that the skill of flying a primary propeller aircraft is important in predicting fighter pilot skill and estimating the validity of pilot aptitude tests.

Long-term stability of individual diffferences in skill acquisition has been concerned during multi-year long training. There has been much research for both applied and basic situations (Ackerman, 1988; Hofmann et al., 1992, 1993; Alessandri et al., 2015). In the research, task configuration, for example, of task complexity or consistency, was key for determining individual differences over practice. When a complex task was repeated, individual differences remained constant or increased. However, when a task was consistently repeated, they decreased.

Tasks during pilot training could be defined as complex. In the Japan Air-Self Defense Forse (JASDF), pilot training consists of five progressive phases that are conducted using both propeller and jet aircraft: the first training phase is conducted using a primary propeller aircraft, T-7; the second to fourth training phases are conducted using a jet trainer, T-4; and the fifth training phase conducted using a fighter jet, F-15 or F-2. In the Japan Civil Aviation College, there are three flight training phases conducted utilizing both single-engine and dual-engine propeller aircraft. Pilot candidates are initially trained for basic flight skills in low-speed, simple aircraft. Later, advanced skills are acquired using high-speed, high-performance aircraft.

To acquire both skills, candidates are likely to practice multiple maneuvers for more than two or three years and join multiple training phases. It could be possible that tasks performed during pilot training are different among training phases rather than repetition of complex tasks.

Therefore, student pilots might need new abilities for subsequent training and individual difference in early training would not influence future training.

Several studies have shown the relationship of flying performance between various pilot training phases. Ree et al. (1995) demonstrated that the flying performance in a subsonic primary aircraft strongly influenced the subsequent flying performance in a supersonic advanced aircraft in the U.S. Air Force. The finding was true for both male and female pilots (Carretta and Ree, 1997). Okaue et al. (1973) suggested that the relationships between the four training phases in a previous version of JASDF pilot training were strong. Furthermore, the research indicated that the initial performance in a primary aircraft correlated with later performance in an advanced aircraft.

Prior research has suggested that early flying performance should predict subsequent flying performance. In other words, it was possible that individual differences remained over flying practice. Therefore, it was considered that tasks performed during pilot training would be complex and not change. However, there is no finding about comprehensive relationships among all flying training phases. Can we say that tasks performed during all flying trainings are complex and do not change?

Three Hypotheses of Flying Training Performance

Three hypotheses were proposed regarding the role of primary pilot performance in a sequential flying training environment in JASDF. To deal with the many flight training performance factors, such as landing, navigation, and formation, all hypotheses considered that the flying skill acquired in each training phase influenced flying performance.

First, we hypothesized that a flying skill acquired in a previous training phase would directly influence a flying skill in a subsequent phase only. Therefore, the first training skill would immediately affect only the second training skill. This straight model was based on the findings of Ree et al. (1995), which suggested that skills learned in flying a primary aircraft were required in flying an advanced aircraft. We expected that the influence of early skill was not so strong in later skills because it only indirectly influenced them. This hypothesis was called "Straight Model I (Figure 1a)."

The second model extended the Straight Model I, with the first training skill directly influencing the third training skill. The first and third training phases include mastering navigation. Navigation is considered to be a key maneuver because it is difficult to master it (Yang et al., 2013; Sullivan et al., 2011). Therefore, we hypothesized that the first training skill would strongly influence the third training skill. This model was called "Skip Model P (Figure 1b)."

The third model also extended the Straight Model I, with the first training skill directly influencing the fifth training skill, like a circle. This model, called the "Circle Model O," was proposed considering that student pilots are exposed to different aircraft types. They are introduced to and trained in new aircraft types in the first, second, and fifth training phases. In these three phases, students might be required to master an unfamiliar aircraft. Therefore, we hypothesized that the first training skill would directly influence not only the second training but also the fifth training skill. Our expectation was that a skill acquired early strongly influenced the last skill because the early skill influenced the last skill directly as well as indirectly (Figure 1c).

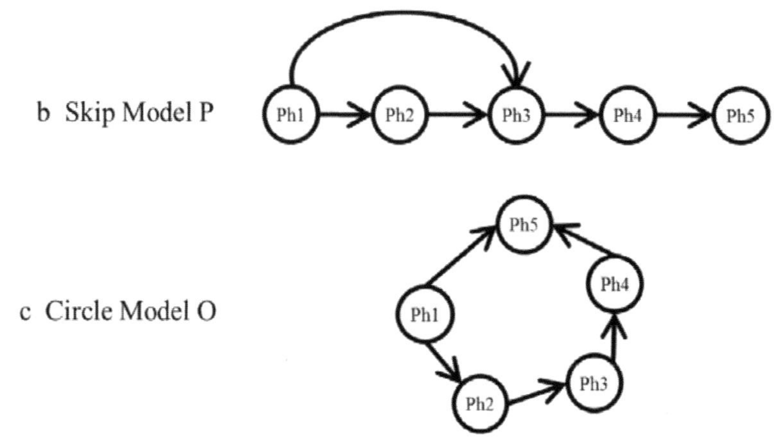

Figure 1. Hypothesized models.

Note. Ph1 through ph5 are the flying skill factors for the first through fifth pilot training phases.

Method

Participants

The participants were 270 JASDF male student pilots, who completed five pilot training phases. The age of the participants ranged between 20 and 29 years when the first phase began. These participants were either from the National Defense Academy, an accredited college or university, or the JASDF Aviation Cadet Corps. The participants who graduated from the National Defense Academy or an accredited college or university completed the JASDF Officer Candidate School before entering the pilot training course.

Flying Performance and Flying Skills

Student pilots were graded on maneuvers in each training phase. The maneuvers were different according to each phase; the first training phase, in a primary propeller trainer (T-7), included takeoff, landing, and basic navigation; the second training phase, in a jet trainer (T-4), included takeoff, landing, and formation; the third training phase, using the same aircraft as the second phase, included navigation for commercial pilot license; the fourth training phase again in the T-4, included elementary tactical maneuvers; the fifth training phase, in a fighter jet (F-15 or F-2), included takeoff, landing, and tactical maneuvers. We considered the flying skill of a student pilot in each phase as a factor affecting affecting their execution of these maneuvers, which was considered to be equivalent to flying performance in this study.

Data Analysis

A covariance structure analysis was performed on maneuver scores using PASW Statistics 18 and Amos 18 statistical analysis software. According to Toyoda (1998), an adequate fit is indicated if the value of root mean square error of approximation (RMSEA) is less than 0.05. With regard to the comparative fit index (CFI), values closer to 1.0 indicate better models. Akaike's information criterion was used for model comparisons, and the lowest scoring model was recommended.

Results

Goodness-of-fit indices of models are shown in Table. The RMSEA values of all the models were found to be acceptable. Circle Model O appeared to have the best fit among the competition models (CFI = 0.933; AIC = 658.364). The standardized estimates of the effects between flying skill factors among Circle Model O's variables are shown in Figure 2. The influence from the first training skill to the last training skill was directly 0.31 and indirectly 0.28 (0.81 × 0.79 × 0.90 × 0.49).

Table.
Fit indices of models of flying performance structure in pilot training phases.

Model	RMSEA	CFI	AIC	Description
Straight Model I	0.046	0.928	668.262	Model with indirect effects.
Skip Model P	0.045	0.930	666.513	Straight Model I with direct effect of a flying skill factor from ph1 to ph3.
Circle Model O	0.044	0.933	658.364	Straight Model I with direct effect of a flying skill factor from ph1 to ph5.

Note. RMSEA = root mean square error of approximation; CFI = comparative fit index; AIC = Akaike's information criterion. Ph1, ph3, and ph5 are the first, third, and fifth pilot training phases, respectively.

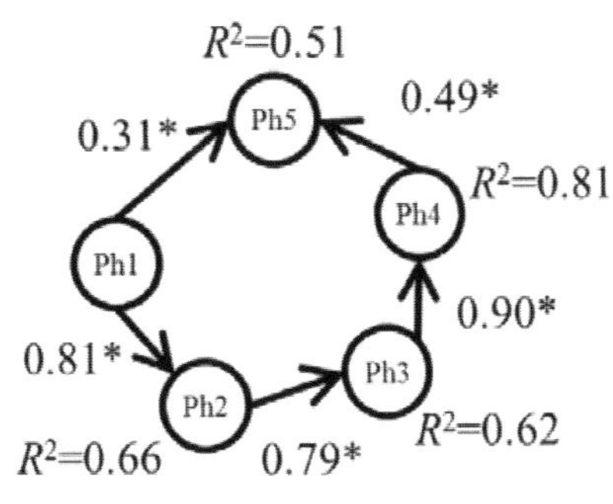

Figure 2. Path coefficients of Circle Model O.
$*P < 0.01$.

Note. Ph1 through ph5 are the flying skill factors for the first through fifth pilot training phases.

Discussion

The results showed that previous training skills influenced subsequent training skills. Moreover, the first training skill was confirmed to influence the last training skill both directly and indirectly. This was because student pilots may need similar skills in order to master an unfamiliar aircraft in both the first and fifth phases.

The effects of each phase on the fifth training skill were calculated: first phase 0.59 (0.31 + 0.28), second phase 0.35 (0.79 × 0.90 × 0.49), third phase 0.44 (0.90 × 0.49), and fourth phase .49. Among all the training phases, the first training phase had the most substantial effect on the fifth training phase. The first training performance was an important predictor of the last training performance.

The first training phase was strongly connected to the last phase. This means that the tasks during pilot training might involve complex repetitions and individual differences are retained throughout practice. From another point of view, however, the coefficient of determination (R^2) of the last training skill, as shown in Figure 2, was 0.51 and variances that we could not explain remained. It may be possible that the remaining variances were caused by new abilities in new tasks. We must continue to study the possible causes of these remaining variances.

In longitudinal studies on achievement tests, it was reported that relationships existed between the test scores near grades (Bloom,1964; Bracht & Hopkins,1972). In addition, the scores between far grades correlated (e.g., between third and eleventh grades, $r = 0.82$). First-year scores had direct and indirect effects on fourth-year scores in medical college (Harada & Nakamoto, 1997). These findings indicate that individual differences in early performance were maintained and the findings of the pilot training were consistent with those obtained by previous studies.

The findings of this study have a great impact on the development of pilot aptitude tests. The primary aim of these tests is to select persons who have the potential to become good pilots. When we evaluate pilot aptitude tests, in the truth, we want to use the flight duty performance scores of the pilot candidates who took the tests. Waiting a few years—which is usually the time required for a pilot candidate to become a duty pilot—to estimate the validity of pilot aptitude tests is undesirable. Therefore, it is helpful if the first training performance can predict the last training performance, and can be used as a criterion variable for evaluating future performance.

Interpreting the results from another perspective, enabling student pilots to further improve at each training phase might help them attain improved performance in all subsequent training phases. In particular, when the first training performance improves, the last training performance can also improve despite the use of different aircraft. The training values can be helpful in training student pilots now and in the future.

The findings of this study can be helpful in evaluating pilot aptitude tests and determining training values. However, these findings may lead to students having a fixed mindset, which might affect training performance. Blackwell et al. (2007) suggested that the students with a growth mindset were able to raise their scores, whereas the students with a fixed mindset werenot. Ratten et al. (2012) showed that instructors with a fixed mindset were likely to comfort low-performing students, and students respond to such comfort-oriented feedback with lower expectations of their own performance. In particular, in training or educational situations, exaggeration of the effect of early training might have a negative influence.

References

Ackerman, P. L. (1988). Determinants of individual differences during skill acquisition: Cognitive abilities and information processing. *Journal of Experimental Psychology: General*, 117, 288-318.

Alessandri, G., Borgogni, L., & Truxillo, D. M. (2015). Tracking job performance trajectories over time: A six-year longitudinal study. *European Journal of Work and Organizational Psychology*, 24, 560-577.

Blackwell, L. S., Trzesniewski, K. H., & Dweck, C. S. (2007). Implicit theories of intelligence predict achievement across an adolescent transition: A longitudinal study and an intervention. *Child Development*, 78, 246-263.

Bloom, B.S. (1964). Stability and change in human characteristics. NewYork: Wiley.

Bracht, G. H., & Hopkins, K. D. (1972). Stability of general academic achievement. In: Bracht, G. H., Hopkins, K. D., Stanley, J. C., editors. Perspectives in educational and psychological measurement. New Jersey: Prentice Hall.

Carretta, T. R., & Ree, M. J. (1997). A preliminary evaluation of causal models of male and female acquisition of pilot skills. *The International Journal of Aviation Psychology*, 7, 353-364.

Harada, N. & Nakamoto, M. (1997). Student selection methods for medical school and progress after admission-follow-up study at Yamaguchi University. 2. Multivariate analyses of factors influencing progress after admission. *Journal of the Japanese Society for Medical Education*, 28, 77-83. (In Japanese.)

Hofmann, D. A., Jacobs, R., & Gerras, S. J. (1992). Mapping individual performance over time. *Journal of Applied Psychology*, 77, 185-195.

Hofmann, D. A., Jacobs, R., & Baratta, J. E. (1993). Dynamic criteria and the measurement of change. *Journal of Applied Psychology*, 78, 194-204.

Okaue, M., Nakamura, M., & Hori, H. (1973). The analysis of aviation training evaluations (2): Similarities of aviation performances in flying training courses. The Reports of Aeromedical Laboratory, 14, 122-131. (In Japanese.)

Rattan, A., Good, C., & Dweck, C. S. (2012). "It's ok – Not everyone can be good at math": Instructors with an entity theory comfort (and demotivate) students. *Journal of Experimental Social Psychology*, 48, 731-737.

Ree, M. J., Carretta, T. R., & Teachout, M. S. (1995). Role of ability and prior job knowledge in complex training performance. *Journal of Applied Psychology*, 80, 721-730.

Sullivan, J., Yang J. H., Day, M., & Kennedy, Q. (2011). Training simulation for helicopter navigation by characterizing visual scan patterns. *Aviation, Space, and Environmental Medicine*, 82, 871-878.

Toyoda, H. (1998). Covariance Structure Analysis for Beginners. Asakura, Tokyo. (In Japanese.)

Yang, J. H., Kennedy, Q., Sullivan, J., & Fricker, R. D. (2013). Pilot performance: Assessing how scan patterns & navigational assessments vary by flight expertise. *Aviation, Space, and Environmental Medicine*, 84, 116-124.

HUMAN PERFORMANCE ASSESSMENT: EVALUATION AND EXPERIMENTAL USE OF WEARABLE SENSORS FOR BRAIN ACTIVITY MEASURES

Kurtulus Izzetoglu[1]
Drexel University

Dale Richards
Coventry University

Lei Ding
University of Oklahoma

Chen Ling
University of Akron

Ben Willems
Federal Aviation Administration

The emerging wearable human performance monitoring technologies can help evaluate the cognitive status and capacities of the crew in the cockpit as well as those operating ground control stations. Traditionally the use of behavioral measures and subjective metrics has been used to address cognitive factors associated with pilots or operators of safety critical systems. However, the advance in wearable physiology technologies could provide additional performance metrics directly driven from brain based measures, potentially validating subjective assessments and ultimately bringing us closer towards maintaining safe and effective performance. Furthermore, these techniques may also aid the design and evaluation of new technologies that are being presented as increasing operational capacity, efficiency and safety across the aerospace domain. The measurement of real time brain activity from the operator can help evaluate decision making, and reliably compare workload burden of next generation system versus legacy systems in the air transportation domain. This paper outlines key cognitive areas of interest when attempting to explore the correlation between physiological state changes and psychological constructs. A number of studies are described whereby wearable systems, namely electroencephalography (EEG), and functional near infrared spectroscopy (fNIRS), are used to evaluate human performance. The potential advantages and challenges are discussed in relation to implementing these sensors in real operational settings.

Civilian pilots, air traffic controllers, ground controllers are all increasingly required to utilize larger amounts of data and more complex systems. Hence, we are likely to observe an increase in the information-processing load and decision-making demands on aviation personnel. Many of these issues have been symbiotic with initiatives being developed under initiatives such as Next Generation Air Transportation System (NextGen) and the Single European Sky Air Traffic Management Research (SESAR) programmes. The human element within any future concept still represents a critical point that may either be seen as a point of failure or a means by which these new technologies are optimized. It is therefore important to consider how we not only assess such technologies, but the way in which the human interacts with them and ultimately arrives at making decisions.

The last decade has seen significant advances in physiological monitoring techniques, and in particular their integration into ubiquitous devices. One aspect of this has been the increase in wearable human performance monitoring technologies that can be used to evaluate the cognitive status and capacities of the crew on the flight deck, as well as on the ground (such as the ground control station or air traffic terminals). Non-invasive wearable technologies offer the potential to observe human cognitive performance directly driven from brain-based measures, which would be an important asset in evaluating (and maintaining) safe and effective operational performance. Further, such sensory input from the operator can help evaluate decision making, and reliably compare the cognitive workload burden of future versus legacy systems in the air transportation domain. Currently the most widely used brain activity measures are functional magnetic resonance imaging (fMRI), magnetoencephalography (MEG), electroencephalography (EEG), and functional near infrared spectroscopy (fNIRS).

This paper introduces some key theoretical aspects of cognition that are prevalent within aerospace, with particular attention to cognitive workload and human performance in safety critical environments; with a view to *bridging the gap* between cognition and measurement. Following this, a number of operational views are outlined through the description of field use cases: including ATC human-in-the-loop studies and the nature of human

[1] Corresponding author. E-mail: ki25@drexel.edu

performance in weather decision making. Principles of EEG and fNIRS are discussed in relation to application and calibration, before highlighting their potential contribution in providing reliable and objective assessment of pilot' and operator' cognitive performance.

Maintaining the Objective: Assessing Pilot and Operator Cognitive State

The Aerospace Industry is regarded as one of the safest transport domains, with a constantly improving safety record (Harris, 2014). However, when we consider the different roles and responsibilities that we ask of the humans that operate across the national airspace system (NAS) we can appreciate the diversity of tasks and systems that users of those systems have to utilize. When tasks become complex, laborious or dramatically increase, automation is commonly (and effectively) applied. Although we have traditionally seen a rise in the use of automation within aerospace applications, it is fair to say that the human will remain responsible for making critical decisions based on the information they are presented with.

Human Factors (HF) within aviation has provided us with a good understanding of the cognitive processes involved in aviation operations, predominantly focused on manned and unmanned aviation and the critical management task provided by Air Traffic Control Operations (ATCO). It is of little surprise, therefore, that we can identify a number of key cognitive components that play a role in human performance. In order to understand how an individual processes and acts on information it is critical that we define two important aspects that underpin Aviation HF; that of human information processing (HIP), mental workload (MW) and situation awareness (SA).

We must first consider the nature of a number of theoretical constructs that we need to understand when discussing these cognitive constructs. Without descending into an essay on the many different theories and approaches to understanding cognition, it is best to approach this by outlining the way in which humans process information. To start at the beginning we can describe, in general terms, the core aspects of HIP as related to how information travels from the environment to the human, and subsequently how he/she acts on that information. This in turn can be further deconstructed into three key factors: (1) **Encoding** data from the environment, (2) **Processing** the data into meaningful information we can use, and (3) **Executing** actions as a result of the first two steps. Although this sounds like a simple mechanistic approach we must remember that all of this activity must take place rapidly across different dynamic models of memory; namely sensory, short term (often referred to as working memory), and long term memory (Atkinson & Shiffrin, 1968, 1971). These distinct models of memory allow us to understand the processing of information in terms of how we attend to sensory stimuli, before we move on to register and encode aspects of the information, see Figure 1.

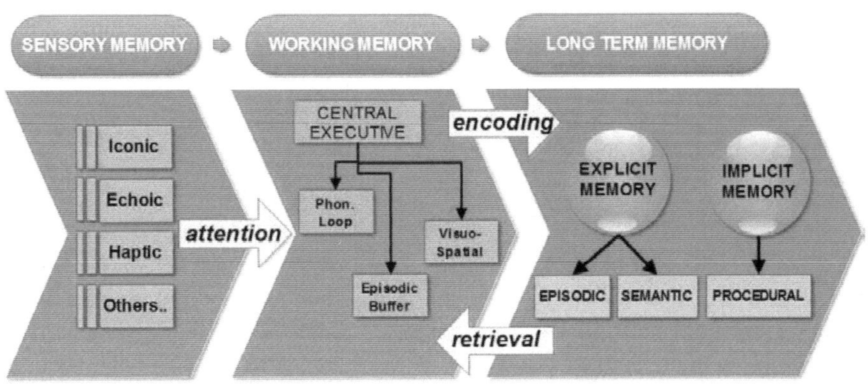

Figure 1 - Three component memory model of Information Processing (adapted from Atkinson & Shiffrin)

Of course the manner by which we process information is somewhat dependent on the characteristics of the information being attended to and the specific requirements of the task. This will further determine how attentional resource is utilized during the context of the task demand and which stimuli are attended to (Baddeley & Hitch, 1974; Baddeley, 2003). Inevitably this represents a constraint in terms of how humans process and store information, more so when confronted with dynamic and complex tasks to perform. Unsurprisingly there are many instances where this constraint of HIP can sometimes lead to *bottlenecks* whereby information will compete for the attention of the individual to process. The human brain adapts to this by selectively attending to certain information (very much dependent on the task context) whilst filtering out less salient information (Moran & Desimone, 1985).

If we focus on the processing of information within working memory, then it has been suggested that this represents our understanding of the external environment - or, to put it another way, our SA (Bell & Lyon, 2000). Endsley (1995) however views SA more as a cognitive product of information processing, and developed perhaps the most influential model of this construct. In essence Endsley (1995, 2000) suggests that SA is an active and ongoing process of achieving a state of knowledge of a given situation. But, in order to achieve this it is necessary to process information sequentially through the stages of (1) **Perceiving** the attributes and state of the elements within the environment, before beginning to (2) **Comprehend** what is being perceived, and then finally understanding this information by (3) **Projecting** ahead what is likely to happen in the future. While there are many different interpretations on the nature of SA as a theoretical construct, what it all boils down to is the nature of what it is we are attempting to measure; trying to make the intangible tangible. Stanton et al (2005) provide an overview of different methods employed to measure SA, which can be categorised into different techniques such as freeze probe recall, real-time probe techniques, post-trial subjective ratings, observer ratings and process indices (Salmon et al., 2006). All of these techniques have both good and bad points and may be used to claim a measure of SA (depending on which technique and definition you ascribe to). Indeed, Endsely (2015) concludes that the very nature of the construct makes it difficult in itself to measure it.

We can all think back to an instant where we have felt overwhelmed by a situation that has affected our ability to act efficiently and in a timely manner. Regardless of whether that experience was within an aviation context or not, it is likely that this increase in MW could also raise the likelihood of inducing human error and ultimately reducing your effectiveness (Moray, 1988). As with many cognitive constructs there is no single agreed definition of MW, but we can broadly agree that it is composed of a number of features that require: an input (or task load), a specified amount of effort required by the human to satisfy the task, and the actual performance of the human in *doing* the task (Jahns, 1973). Clearly the ability to assess an individual's MW during critical tasks can provide important details as to the manner of the task demand, which may then assist in the future design and integration of that system into an operational context.

The key element to consider here is that the assessment of MW requires a tangible value that can be assessed by employing a range of techniques. Primarily we can use observation and measurement to determine whether the task has been successfully completed, which may further be constructed of behavioral markers assigned to primary or secondary tasks. Thus, quantifiable measures (such as holding altitude or maintaining safe separation) may be used to determine whether the individual is operating under a higher or lower amount of MW. Either way we would witness an effect that could be translated as having an impact on the individual completing these tasks. Measuring behavioral response aligned to a particular task does not directly involve direct interaction with the participant, but an observation of the task with which they are engaged. However, it is almost impossible to enforce a completely *sterile* condition whereby we can state that any behavioral effect is solely attributed to MW. A more direct perception of what the participant may report in terms of their perceived effort can be gathered by a large number of available subjective MW measures.

Assessment and Measurement

Both MW and SA are cognitive processes that are theoretical constructs and somewhat illusive to direct measurement. However, all is not lost, as we may use a number of methods to assess human performance. We can see that there are several techniques that can be used, and these broadly fall into three categories: (1) Rating scales, (2) Performance Measures associated with primary and secondary tasks, and (3) Psychophysiological measures. In the past the most widely used of these methods would center on the first two methods of gathering data; due to their ease of use and lack of physiological techniques that can be readily applied and interpreted.

However, when selecting (or developing) an appropriate measurement technique there are a number of factors we must take into account. These factors are outlined in Figure 2, and show that the context within which we conduct human performance assessments plays a pivotal role in how we attempt to measure cognitive processes.

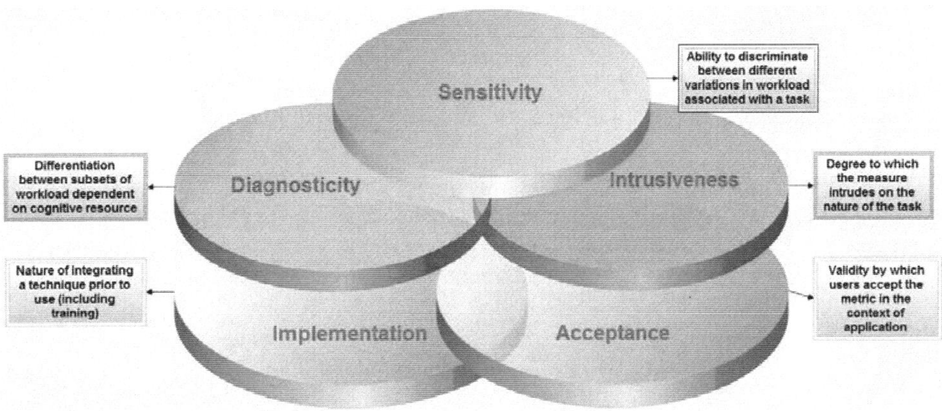

Figure 2 - Factors that should be considered within the selection criteria for metrics

While behavioral measures are largely non-intrusive and possess high participant acceptance, they are poor in terms of their sensitivity and diagnosticity for measuring mental workload. And consequently the validity of such approaches must often be examined.

Operational View of Human Performance Assessment

To assess the impact of changes being adopted under programmes such as NextGen and SESAR, we often run high fidelity Human-In-The-Loop (HITL) simulation experiments. Research efforts commonly examine the the impact of new technologies on human performance, with a particular focus on pilot/operator cognitive processing. Changes to mental state of the operator will directly affect the safety and efficiency of the NAS. One of the challenges with HITL experiments is to use an objective measure that is unobtrusive, real-time, and sensitive enough to detect changes due to human-automation interaction or procedural changes.

We have seen that the cognitive theories discussed in this paper are complex multi-dimensional constructs that are by their very nature difficult to quantify using any one single metric. By adopting a range of metrics, and choosing those that suit the nature of the task being examined, it brings us closer to a clearer picture of what an individual's cognitive response is within a given context. We have used several physiological measures in conjunction with system-derived as well as subjective measures. Here we present our experience across several studies as well as the pitfalls of using subjective measures to assess new technologies. The studies were conducted at the FAA's William J Hughes Technical Center and some reported by Willems (2002), Ayaz et al., (2011; 2012), and Harrison, et al. (2014). These provide a number of contexts which have shown promising results that appear to benefit from the application of neuropsychological measurement.

Context One: Decision Making and Significant Weather for Air Traffic Controllers and Pilots

About 70% of aviation delays are related to weather. To enhance NAS efficiency and safety, it is important that air traffic controllers and pilots work together to make sound decisions when encountering severe weather. Decision making and communication for air traffic controllers and pilots during severe weather situations could cause excessive MW for both controllers and pilots.

Severe weather creates challenges in decision making and communications for both controller and pilots in the complex sociotechnical system. Air traffic controllers need to make quick assessment about the weather scenarios and understand the current situation as well as future progress of the severe weather phenomenon. Additionally, they need to disseminate relevant weather information to pilots in the most effective way (Ahlstrom, 2005). For the pilots, with challenges created by severe weather, they have to make decisions about whether to divert from their original flight path with the help from air traffic controllers (Chamberlain & Latorella, 2001; Delaura & Evans, 2006). Effective and timely communication between controllers and pilots is critical to ensure safety and efficiency. To maintain a common weather picture and allow for shared SA, which facilitates collaborative decision making between controllers and pilots, communication protocol and channel (via Data Comm, ADS-B weather display enabled by NextGen technologies, voice) should be carefully designed.

As all weather forecasts are probabilistic in nature, controller and pilots also need to be trained to deal with inherent uncertainty in weather. National Severe Storm Laboratory (NSSL) from NOAA is developing Probabilistic Hazard Information (PHI) system, part of the vision of Forecasting a Continuum of Environmental Threats (FACETs; Rothfusz, Karstens, & Hilderbrand, 2014.). This new system provides dynamically updated probabilistic information of areas being impacted by severe weather threats, using graphical design methods to convey the likelihood of threat occurrence (Karstens, Stumpf, Ling et al., 2015). A graphical probabilistic weather display may become a useful tool to enhance decision making and communication for controllers and pilots.

Context Two: To Improve Safety and Evaluation of Training in ATM Using Neuroscience-Based technology

Air traffic management (ATM) is an essential part of air transportation and aviation, connecting cities and people citizens as well as boosting jobs and growth. However, worldwide ATM systems are based on aging technology and procedures and needs updating particularly in light of the expected traffic growth in the near future. The future ATM scenarios describe a system where high levels of automation should be deployed to support humans. However, automation brings a range of new challenges. A series of problems concerning the interaction between human and automation that have been reported are: deficiencies in human operator states, including vigilance decrements, complacency and out-of-the-loop problems, and training deficiencies.

We reviewed the state-of-the-art in assessing human performance and training under the advancement of aviation automation. Such technology capacities have been reflected in documented publications on MW assessment, alertness, training in air transportation management (ATM) with realistic environments and testers. We examined the state-of-the-art portable sensor technologies that are adaptable and inexpensive. This allowed us to identify a number of neurophysiologic conditions that can be associated with the levels of cognitive control (Astolfi et al., 2011; Shou et al., 2012; Borghini et al., 2014b; Kong et al., 2015). Further to this, we obtained information about the level of MW of ATM operators, through a combination of neurometrics and other physiologic measures (Arico et al., 2015; Borghini et al., 2015), in a realistic ATM context (Arico et al., 2014, 2016; Dasari et al., 2015). This allows us to recommend a number of safety measures. Finally, we gathered valuable data on the use of neurometrics that can assess the current learning level of trainees (Borghini et al., 2013, 2014a, 2016; Krishnan et al., 2014).

Context Three: An Investigation of Optical Brain Imaging Sensor in Performance Assessment

The safe and effective performance of aviation personnel depends on their ability to manage and maintain high levels of cognitive performance. A field-deployable optical brain imaging device can provide team member's cognitive state and relative level of expertise for a given level of performance by monitoring cortical areas that are known to be associated with MW, learning and the development of expertise.

Near-infrared spectroscopy (NIRS) has been widely used in brain studies as a noninvasive tool to study changes in the concentration of oxygenated hemoglobin (oxy-Hb) and deoxygenated hemoglobin (deoxy-Hb). Based on the NIRS technique, a functional brain activity assessment (fNIRS: functional Near InfraRed Spectroscopy) system has been deployed as a means to monitor cognitive functions, particularly during attention and working memory tasks as well as for complex tasks such as pilot training and air traffic control scenarios performed by healthy volunteers under operational conditions. The fNIRS is a field-deployable non-invasive optical brain monitoring technology that provides a direct measure of cerebral hemodynamics from the forehead in response to sensory, motor, or cognitive activation. This study also allowed us to progress brain based measures and biometrics across different human roles in aviation.

Our work utilizing fNIRS has allowed us to progress this technique towards deploying this device in the field; whereby operators can be assessed in their normal working condition and have included multiple studies with the Federal Aviation Administration (FAA) as well as with the Department of Defense (DoD). In the first study, we explored the impact of the different Conflict Resolution Advisory (CRA) conditions on air traffic control operator's behavior and MW. The fNIRS sensor was utilized to monitor the MW of the 12 operators using this new CRA system across 3-day human experimentation sessions (Harrison et al., 2014). Further to this, a HITL study was conducted using fNIRS to evaluate MW within a NextGen air traffic system that examined the difference between Data communication (DataCom) and Voice communication (VoiceCom) between pilot and air traffic controllers (Ayaz et al., 2012). Finally, we also adopted fNIRS to assess human performance unmanned aerial vehicle (UAV) operators (Izzetoglu et al., 2015). The results provided within these studies revealed that such fNIRS can be used to monitor true MW changes during aerospace operations. It also proved to be an objective measure of expertise development, i.e., the transition from novice to expert during operator training (Ayaz et al., 2012).

Discussion

Advances in neurophysiology and neuro-monitoring technologies have demonstrated that changes in physiology can explicitly be assessed and correlated with different tasks. These may relate to instances where the human is confronted with high cognitive loading, or events that can be identified as leading to a change in situation awareness. It may also be used to develop adaptive, personalized training regimes and provide indicative markers that are associated with expertise development. It is therefore essential that before we start to decide which metric to use, we must consider the context within which the measurement is to be applied, what we are exactly attempting to measure, and so on. Once we can establish these requirements we can begin to address the robustness of these neurophysiological biometrics in terms of reliability: does it produce the same results in similar situations? and validity: does it actually measure what it says it does?

It is worth noting that the sensitivity of these metrics may only provide one side of the story, in that they are perceived measures and sometimes do not reveal the full picture. Both subjective and objective metrics clearly have a role to play here, but we must exercise caution in not placing all our EEGs in one basket. Indeed, some studies have revealed contrasting results when we compare subjective versus physiological metrics in terms of MW (Richards et al, 2016). There has also been observations that suggest that subjective metrics, such as the NASA-TLX, can be limited by the nature of individual differences in introspection skills (Paulhus & Vazire, 2005). Chen et al (1995) even go so far to suggest that this limitation may even be observed at a cultural level, whereby instructing an individual to report perceived feelings of cognitive state are difficult to articulate.

We have shown that the advances in wearable sensors can be used to measure physiological state changes, and they represent an exciting opportunity to explore the psychology-physiology divide. Brain imaging measures allow us to add to our growing human performance toolkit, and when used with a battery of other metrics (including both behavioral and subjective), it provides us with a more robust understanding of cognitive performance.

References

Ahlstrom, U. (2005) Work domain analysis for air traffic controller weather displays. *Journal of Safety Research*, 36(2), 159-169.

Arico P, Borghini G, Di Flumeri G, Colosimo A, Pozzi S, Babiloni F. A passive Brain Computer Interface application for the mental workload assessment on professional air traffic controllers during realistic ATM tasks. *Progress in Brain Research*, 2016 (in press).

Arico P, Borghini G, Di Flumeri G, Colosimo A, Graziani I, Imbert JP, Granger G, Benhacene R, Terenzi M, Pozzi S, Babiloni F. Reliability over time of EEG-based mental workload evaluation during Air Traffic Management (ATM) tasks. *Conf Proc IEEE Eng Med Biol Soc. 2015* Aug;2015:7242-5.

Aricò P, Borghini G, Graziani I, Taya F, Sun Y, Bezerianos A, Thakor NV, Cincotti F, Babiloni F. Towards a multimodal bioelectrical framework for the online mental workload evaluation. Conf *Proc IEEE Eng Med Biol Soc. 2014*;2014:3001-4.

Astolfi L, De Vico Fallani F, Toppi J, Cincotti F, Salinari S, Vecchiato G, Wilke C, Yuan H, He B, Babiloni F: Imaging the social brain by simultaneous "hyperscanning" of different subjects during their mutual interactions. *IEEE Intelligent Systems, 26(5)*, 38-45, 2011.

Atkinson, R. C. & Shiffrin, R. M. (1968) Human memory: A proposed system and its control processes. In Spence, K. W., & Spence, J. T. *The psychology of learning and motivation* (Volume 2). New York: Academic Press. pp. 89–195.

Ayaz, H., Willems, B., Bunce, S., Shewokis, P. A., Izzetoglu, K., Hah, S., ... & Onaral, B. (2011). Estimation of cognitive workload during simulated air traffic control using optical brain imaging sensors. In *International Conference on Foundations of Augmented Cognition* (pp. 549-558). Springer Berlin Heidelberg.

Ayaz, H., Shewokis, P. A., Bunce, S., Izzetoglu, K., Willems, B., & Onaral, B. (2012). Optical brain monitoring for operator training and mental workload assessment. *Neuroimage, 59*(1), 36-47.

Baddeley, A. (2003) Working memory: looking back and looking forward. *Nature Reviews Neuroscience,* 4 (10), 829–39.

Baddeley, A.D. & Hitch, G.J. (1974) Working memory. In Bower, G.H. (ed.), *Recent Advances in Learning and Motivation (Volume 8)*, New York: Academic Press.

Bell, H. H. & Lyon, D. R. (2000) Using observer ratings to assess situation awareness. In M. R. Endsley & D. J. Garland (Eds.), *Situation Awareness Analysis and Measurement* (pp. 129-146). Mahwah, NJ: Lawrence Erlbaum Associates.

Borghini G, Arico P, Astolfi L, Toppi J, Cincotti F, Mattia D, Cherubino P, Vecchiato G, Maglione AG, Graziani I, Babiloni F. Frontal EEG theta changes assess the training improvements of novices in flight simulation tasks. *Conf Proc IEEE Eng Med Biol Soc. 2013*;2013:6619-22.

Borghini G, Aricò P, Di Flumeri G, Salinari S, Colosimo A, Bonelli S, Napoletano L, Ferreira A, Babiloni F. Avionic technology testing by using a cognitive neurometric index: A study with professional helicopter pilots. *Conf Proc IEEE Eng Med Biol Soc. 2015 Aug*;2015:6182-5.

Borghini G, Aricò P, Ferri F, Graziani I, Pozzi S, Napoletano L, Imbert JP, Granger G, Benhacene R, Babiloni F. A neurophysiological training evaluation metric for air traffic management. *Conf Proc IEEE Eng Med Biol Soc. 2014a*; 3005-8.

Borghini G, Aricò P, Graziani I, Salinari S, Sun Y, Taya F, Bezerianos A, Thakor NV, Babiloni F. Quantitative Assessment of the Training Improvement in a Motor-Cognitive Task by Using EEG, ECG and EOG Signals. *Brain Topogr. 2016 Jan*;29(1):149-61.

Borghini G, Astolfi L, Vecchiato G, Mattia D, Babiloni F. Measuring neurophysiological signals in aircraft pilots and car drivers for the assessment of mental workload, fatigue and drowsiness. *Neurosci Biobehav Rev. 2014b* Jul;44:58-75.

Chamberlain, J. P. & Latorella, K. A. (2001). Convective weather detection by general aviation pilots with conventional and data-linked graphical weather information sources. *20th Digital Avionics Systems Conference* October 2001.

Dasari D., Shou G. F., and Ding L. (2015). Monitoring Mental States of the Human Brain in Action: from Cognitive Test to Real-world Simulations, Foundations of Augmented Cognition, *Proceedings of 9th International Conference, AC 2015, held as part of 17th International Conference on Human-Computer Interaction*, pp. 178-186.

Dasari D., Shou G. F., and Ding L. (2013). Investigation of Independent Components based EEG Metrics for Mental Fatigue in Simulated ATC Task. *Proc. of the 6th International IEEE EMBS Neural Engineering Conference,* pp. 1331-1334.

Endsley, M. R. (1995) Toward a theory of situation awareness in dynamic systems. *Human Factors*, 37, 32–64.

Endsley M. R. (2000) Theoretical underpinnings of situation awareness: A critical review In Endsley M. R., Garland D. J. (Eds.), *Situation awareness analysis and measurement* (pp. 3–32). Mahwah, NJ: Erlbaum

Endsley, M.R. (2015) Final Reflections: Situation Awareness Models and Measures. Journal of Cognitive Engineering and Decision Making, 9(1), 4-32.

Harris, D. (2014) Improving Aircraft Safety. *The Psychologist*, February 2014, Vol.27, pp.90-95.

Harrison, J., Izzetoglu, K., Ayaz, H., Willems, B., Hah, S., Ahlstrom, U, ... & Onaral, B. (2014). Cognitive workload and learning assessment during the implementation of a next-generation air traffic control technology using functional near-infrared spectroscopy. *IEEE Transactions on Human-Machine Systems, 44*(4), 429-440.

Izzetoglu, K., Ayaz, H., Hing, J. T., Shewokis, P. A., Bunce, S. C., Oh, P., & Onaral, B. (2015). UAV operators workload assessment by optical brain imaging technology (fNIR). In *Handbook of Unmanned Aerial Vehicles* (pp. 2475-2500). Springer Netherlands.

Jahns, D.W. (1973) A Concept of Operator Workload in Manual Vehicle Operations. *Forschungsingstitut Anthropotechnik: Meckenheim.* Bericht Nr. 14.

Karstens, C., Stumpf, G, Ling, C., Hua, L., Kingfield, D., Smith, T., Correia, J., Calhoun, K., Ortega, K., Melick, C., Rothfusz, L. (2015) .Evaluation of a probabilistic forecasting methodology for severe convective weather in the 2014 Hazardous Weather Testbed. *Weather and Forecasting, 30(6),* 1551-1570.

Kong W, Lin W, Babiloni F, Hu S, Borghini G. Investigating Driver Fatigue versus Alertness Using the Granger Causality Network. *Sensors (Basel). 2015 Aug 5*;15(8):19181-98.

Kopardekar, P., Rhodes, J., Schwartz, A., Magyarits, S. and Willems, B., 2008, September. Relationship of maximum manageable air traffic control complexity and sector capacity. In *26th International Congress of the Aeronautical Sciences (ICAS 2008), and AIAA-ATIO-2008-8885, Anchorage, Alaska, Sept* (pp. 15-19).

Krishnan V. K., Dasari D., and Ding L. (2014). EEG Correlates of Fluctuations in Cognitive Performance in an Air Traffic Control Task. *Federal Aviation Administration.* DOT/FAA/AM-14/12.

Moran, J. & Desimone, R. (1985) Selective attention gates visual processing in the extrastriate cortex. *Science*, 229, 782-784.

Moray, N. (1988) Mental workload since 1979. In D. J. Osborne (Ed.), *International Review of Ergonomics* (pp. 123-150). Taylor & Francis.

Pribram, Karl H., Miller, George A., Galanter, E. *(1960) Plans and the structure of behavior.* New York: Holt, Rinehart and Winston. *p. 65. ISBN 0-03-010075-5. OCLC 190675.*

Richards, D. Scott, S., Furness, J., Lamb, P., Jordan, D., & Moore, D. (2016) Adaptive Symbology: Evaluation of Task-Specific Head-Up Display Information for use on a Commercial Flight Deck. Aviation 2016, Washington: June 2016.

Rothfusz, L., Karstens, C. D., & Hilderbrand, D. (2014). Forecasting a Continuum of Environmental Threats: Exploring Next-Generation Forecasting of High Impact Weather, *EOS Transactions, American Geophysical Union, 95*, 325-326.

Salmon, P., Stanton, N., Walker, G., Green, D. (2006). Situation Awareness measurement: a review of applicability for C4i environments. *Journal of Applied Ergonomics 37,* 225–238

Shou G. F., Ding L, and Dasari D. (2012). Probing neural activations from continuous EEG in a real-world task: Time-frequency independent component analysis. *Journal of Neuroscience Method, 209(1)*: 22-34.

Stanton, N.A., Salmon, P.M., Walker, G.H., Baber, C., Jenkins, D. (2005). *Human Factors Methods: A Practical Guide for Engineering and Design.* Ashgate, Aldershot.

Willems, B. (2002). *Decision support automation research in the en route air traffic control environment* (No. DOT/FAA/CT-TN02/10). FAA Technical Center, Atlantic City, NJ.

INTERACTIVE TEAM COGNITION: DO GAZE DATA ALSO TELL THE STORY?

Dirk Schulze Kissing
German Aerospace Center (DLR)
Hamburg, Germany
Carmen Bruder
German Aerospace Center (DLR)
Hamburg, Germany

Gaze-data could be feasible to assess interactions within small groups, and provide added value for the assessment of 'team cognition' (c.f., Cooke et al., 2013). In the synthetic Control Center Task Environment (ConCenT) 18 teams of three are collectively monitoring an array of displays and predicting malfunctions, indicated by a setpoint kickbacks. To locate potential malfunctions they have to collaboratively determine system dynamics patterns. Within-team expectations are measured by gaze parameters related to situational relevant areas of interest (AoI). Within-team standard errors (SE) in fixation frequencies are utilized for post-hoc classifications of teams according to gaze-behavior homogenity. The amount of attentional resources teams allocated to relevant AoIs increases during critical phases, with a change of emphasis between the three relevant task elements. Post hoc groups do not differ in their way of monitoring relevant elements. It is concluded that gaze-data provide promising measures of interaction-patterns for team cognition analysis.

The concept of "team cognition" (see Cooke, Gorman, Myers, & Duran, 2013) provides a system-theoretical approach to the analysis of cognitive requirements in collaborative working environments, like airline operational control center (OCC) or area control center (ACC) of air navigation service providers (ANSPs). While the methodological approach to the study of shared mental models (e.g., Cannon-Bowers, Salas, & Converse, 1993) lies in the reproduction of the individual knowledge representations of the team members, whose consistency is subsequently examined, the research on the team cognition starts with the recording of the interactions developing over task performance. The basic theoretical assumption is that teams can be viewed as systems in which the phenomenon of team cognition emerges. This phenomenon does not take place at the level of the representations of individual agents, but rather materializes in the interaction patterns of individuals. The individual behavior, in turn, is affected by these emergent behavioral patterns at the team level.

The phenomenon of team cognition, as Cooke et al (2013) continues, unfolds over time, with interaction patterns changing within shorter periods of time than individual behavior patterns do, which are supposed to be more strongly determined by individual knowledge structures. Cooke et al (2013) therefore assume that individual behavior patterns change more slowly than team behavior patterns.

Cooke et al. (2013) postulate that team cognition should be measured in the context of unfolding task processing, with a focus on the processes running at the team level rather than on relatively static representations. Eye-tracking technology can provide such within-task objective behavioral measures for determining the quality of team performance (see e.g., Hauland, 2008). Cooke and Gorman (2009) also discuss eye movement measurement as a method for recording event data that can be used to systematically describe interactions in teams.

The main goal the presented work is dedicated to is the recording of team cognition as an emergent dynamic activity (Cooke et al., 2013) using integrated visual data measurement. We are looking for a metric to visualize coordination patterns within groups and to separate them from individual behavior patterns. For this purpose a synthetic task environment is developed.

The Synthetic Control-Center Task Environment ConCenT

According to Hess et al. (2005), the main features of a synthetic task environment should be analyzed both in the field of task-work as well as from a cognitive perspective (see Cooke & Shope, 2004). Therefore the design of the *Synthetic <u>Con</u>trol <u>Cen</u>ter <u>T</u>ask Environment* (*ConCenT*) was well-based on field observations of task-work in operational control centers, and by the outcomes of subsequent counseling workshops with operational experts from various fields, but in particular from the aviation domain (cf. Schulze Kissing & Eissfeldt, 2015 for further information).

ConCenT (Schulze Kissing & Eissfeldt, 2015; Schulze Kissing & Bruder, 2016) was designed to assess coordinative behavior within small groups (of N= 3 members) collectively working on management by exception scenarios (cf., Dekker & Woods 1999). The main task is to monitor an array of displays and to detect a malfunction, indicated by a setpoint kickback, intime (i.e., within 4 seconds after kickback-occurrence). Before the critical phase of potential malfunction the location of its potential occurrence can be predicted by the team if it exchanges and interprets relevant information on certain system dynamics. During the upcoming critical phase the individuals' reliance on where to expect a malfunction-event is then measured by gaze parameters related to situative relevant areas of interest (AoI). *ConCenT* emulates an operations-control center (OCC) at a company headquarter where at three working positions the output of three production plants (denoted by numbers one to three) at different company sites (denoted as Alfa, Bravo and Charlie) are remotely supervised. At the fields, direct control of the plants, plant supervisory, production control and production scheduling is assumed to be under fully automatic control. The task of the human operators in the OCC is to manage by exception. Each company site features the same layout (compare Figure 1): one power-station is providing energy for the three plants. Each plant represents a production system consisting of seven units, three assembly lines and four production units providing the components they need. In the three assembly lines different combinations of three out of the four components are assembled to dissimilar end-products. In total the the team has to supervise the outputs of three assembly lines at each of the three plants at each of the three sites (i.e., 3x3x3=27 production-process outputs). To enhance coordination demands, and thus promote interaction within the OCC team, responsibilities are subdivided. A human operator of the OCC is in charge to monitor that output-quantities match the presets for one out of the three assembly lines at each of the plants at each site (so each human operator has to supervise 1x3x3=9 production-process outputs; compare Figure 1).

Working positions

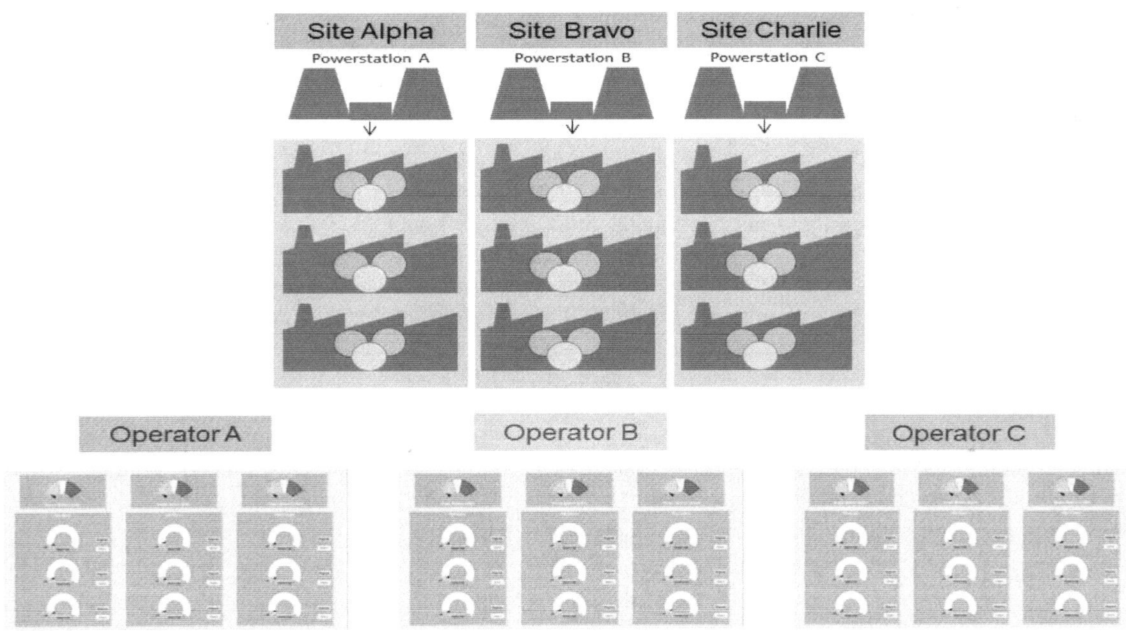

Figure 1: *Top:* Logical structure of ConCenT; three sites with three plants containing a production system with overlapping assembly lines controlled by different human operators; *Below:* User Inferfaces; each participant monitors 12 gauges, nine production-output and three power-supply displays, in a spatially corresponding array; at the output-gauges red arrows designate the actual setpoint, black arrows the actual production output, and green fields the range of tolerance for mismatches. At the power gauges black arrows display the current energy demand, and the coloured fields designate criticality.

Figure 1 presents the workstation for an operator with 9 gauges displaying the actual production output. The gauges are arranged in a 4x3 matrix correspondingly, with each column displaying the incoming information from only one site (Alfa, Bravo or Charlie), with the upper line showing the gauges displaying the actual energy consumption at the site, and each line below displaying the incoming information from the plants according to their numbers (1 to 3). Next to the right of each gauge a related response button labeled "Diagnose" is displayed.
The information stemming from the power-plants is identical at all working positions. The information stemming from the plants differ between working positions, because each gauge is displaying the output of another assembly line of the production system that is set up at that plant. So each operator has a different window to the production systems the OCC supervises. The operators therefore need to exchange their distinct information to get the whole picture.

Scenario

Sequence of phases. The whole system shows repetitive dynamics, shaped by assumed automatic control loops recurring every 30 seconds. During this cycle participants observe the events unfolding in the following sequence of phases: 1) *Notification* about a production schedule event: beginning with changing set-points at one or more gauges; this can be considered to be a notification to each team member where in his or her area of control a new set point is triggered and thus automation control is about to set in (duration: 6 seconds); 2) *Adjustment*: The remote automation adjusts production levels to the new presets; this can be observed only by the human operator in charge; energy is expended to perform this control process; thus the automated adjustment is accompanied by a temporary rise in energy demand. (duration from start of change: 9 seconds); 3) *Tension*: The temporary energy rise outlives the adjustment-process by several seconds. This is the critical phase of heightened tension within the technical system when malfunctions can occur (duration: 8 seconds); 4) *Relaxation*: The tension is removed from the production system as the energy level resets down to normal (duration: 5 seconds); 5) *Pause*: Phase before next interval starts (duration: 2 seconds).

Rules. The team task is to watch out for symptoms of malfunctioning. For the participants knowledge about constraints in the system dynamics mitigates the uncertainty about the location of malfunction occurence. The following rules apply: I.) a malfunction is (and only is) indicated by a setpoint kickback; II.) a setpoint kickback only occurs after an event of automatic adaption; III.) there are two necessary preconditions for a malfunction to occur: a) a malfunction can only be expected at sites where only one event of adaptive automation took place; b) The automated adjustment must be accompanied by an abnormal high energy demand of the automation control process; IV.) a setpoint kickback will be observable for all team members at exactly the gauge where the one team member observed the setpoint change; V.) a malfunction only occurs exactly during a phase of high tension within the production-system, when the output-quantities match the new presets but energy demand level is still high (critical interval; duration: 8 seconds); VI) when the energy level resets to normal a malfunction no longer is to be expected.

Task steps. So with the beginning of each interval the task-steps for the team are to: A) exchange information about set-point changes and collectively narrowing down at which site the first precondition for a malfunction is given; B) collectively observing, if the following adaptation process is accompanied by an abnormal high energy demand; C) distribute the information of the gauge position where the setpoint-change was observed so all team members can build up an expectation and closely monitor the relevant gauge during the critical interval; D) In case of a setpoint kickback perception: Respond with pressing the "diagnosis" button within 4 seconds. A scenario consists of N=144 intervals of 30 seconds durations, with 89 intervals showing relevant events, from that six intervals showing a malfunction event.

Research Question

The purpose of the reported exploratory experiment is to have a manipulation check whether a collective state of expectation can be induced by interactions related with a monitoring-task, and if this can be showed by team gaze indicators. Barzantny et al. (2017) provide indication that individual differences in frequencies of fixations on the relevant AoI during phases of expectancy discriminate between higher and lower performers in the malfunction detection task. It is an open question if these results can be replicated for patterns of gaze-behavior on the team level.

Material and Setup

A team of three participants took part in each experimental session. They were located within the same room, each one sitting in front of a 60Hz 1920 x 1080 Pixel 21''LCD display. During the session mobile dividing walls to visually separate the working positions. The eye movements were recorded with the Remote Eye Follower System from LC Technologies, Inc. The system worked at 120Hz and was linked to the simulation ConCenT using a common time stamp. The raw data were processed using the Nyan software. The fixation detection algorithm has been set to a certain point on the screen with a minimum threshold of six eye movements with a deviation threshold of 25 pixels. All successive fixations that fell into an area of interest (AOI) were categorized as "dwell time".

Participants

Among the 63 participants were 41 applicants for a air traffic controller trainingat DFS (Deutsche Flugsicherung GmbH), the others were students from local universities. The participants were between 18 and 34 years of age (M = 21.57, SD = 3.39). 47.6% were females, 52.4% were male. The participants received 25 to 35 € as compensation for the two to three-hour experiment.

Experimental Session

After filling out a demographic questionnaire the participants received a written instruction. After performing a guided tutorial they performed the experimental scenario in one team trial of 72 inutes duration. A session ended with a task-related questionnaire.

Measures

During the critical phase of uncertainty there are three types of AoIs that are relevant for successful task performance: a) process gauges where the rules apply so that setpoint-kickbacks can be expected as an AoI for gaze measures indicating the close monitoring of the critical task element; b) the energy gauge as an AoI for gaze measures indicating the participants' expectations for the offset of the critical phase; c) the response button to report setpoint-kickback as an AoI indicating action preparation. This report focuses on gaze measures related to these situational relevant AoI. The team was chosen as the entity of analysis. Only gaze data emitted during the critical phases were analyzed. For the purpose of this exploratoy analysis gaze-data registered for 30 intervals were prepared. The criterion for their selection was: all intervals showing a malfunction (N=6) plus the preceding four intervals in each case. All intervals were then excluded from the analysis that featured no critical event (N= 10), or more than one critical event (N= 9), so 11 of the prepared intervals remained for data analysis. The 11 intervals that were analyzed are representants of six exemplary temporal sections, covering points in scenario time from 0h 0min. 12sec. to 1h 2min.30sec.. Fixation counts and dwell times were analysed for the class of relevant AoI. This was constituted by all interactive elements on the screen, i.e. displays of relevant information (energy-demand and production-output) and the input element (response task) relevant for task performance at a given critical phase. After data clearing 18 complete teams with N=54 participants datasets were considered for the analysis. Individual data were aggregated over teams. The percentage of mean dwell times of a team on the relevant elements (3 AoIs) during critical intervals were calculated. Also the mean of the individual fixation frequencies measured for the relevant elements (3 AoIs) during each critical intervals were calculated for each team.

Results

Malfunction identification rate was generally high and increased with the third event in sequence (Events 1: 48.1%; 2: 48.1%; 3: 84.6%; 4: 80.8%; 5: 80.8; 6: 90.4%).

Posthoc Grouping of Teams

Teams were split into posthoc groups according to their collective fixation frequency measured for the relevant AoI during the critical phases (above or below mean), with higher values reflecting adequate resource allocation, and the standard deviation of this measure within a team (above or below mean) with low values reflecting the homogenity of resource allocation within a team. The rationale behind is that a recurrence in gazes between team members should result in relative low standad errors of fixation frequencies, may frequency be on a higher or lower level. These standard errors could serve as a metric for patterns of gaze-behavior, but without reflecing sequence information. This resulted in four categories with teams that were a) homogeneous in fixation frequencies for relevant objects on a high level (N=3), heterogenious in fixation frequencies for relevant objects on a high level (N=5), Homogeneous in fixation frequencies for relevant objects on a low level (N=7) and Heterogeneous in fixation frequencies for relevant objects on a low level (N=3). Class of teams with homogeneous high fixation counts represents only teams with at least two participants that show a malfunction identification rate above mean.

Team Monitoring Performance

A 19 (Interval in sequence) x 3 (Relevant AoI: Relevant Power-Gauge or Relevant Output-Gauge or Relevant Response Button) ANOVA with repeated measures with gaze-based posthoc team classification as a between subjects variable was performed for percentage of dwell times registered for a team during the interval of uncertainty.

During critical phases team ressource allocated to Relevant AoI accroding to the *Interval Position in Sequence* changes over the scenario (Sum of Means for Intervals 1-11 = 37,73%, 40,14%, 41,61% [*Section 1*] 66,35%, 54,27% [*Section 2*], 51,42%, 52,11% [*Section 3*], 67,29% [*Section 4*], 75,92%, 49,82% [*Section 5*], and 64,03% [*Section 6*] respectively; $F(5, 10) = 8.195$, $p < .05$; $\eta^2 = 0.94$). In the course of the scenario the amount of attentional resources teams allocated to relevant task elements during critical phases increased (polynomial contrast for a linear trend: $F=40.35$, $df= 1$ $p < .0001$, $\eta^2=0.74$). However, there were no *Interval Position in Sequence* by *Team Classification* [$F(21, 30) = 1.68$, $p < .11$, $\eta^2=0.71$] interactions (cf. Figure 2), indicating that team-classes do not differ in their trend to allocate more attentional resources to relevant AoIs over time.

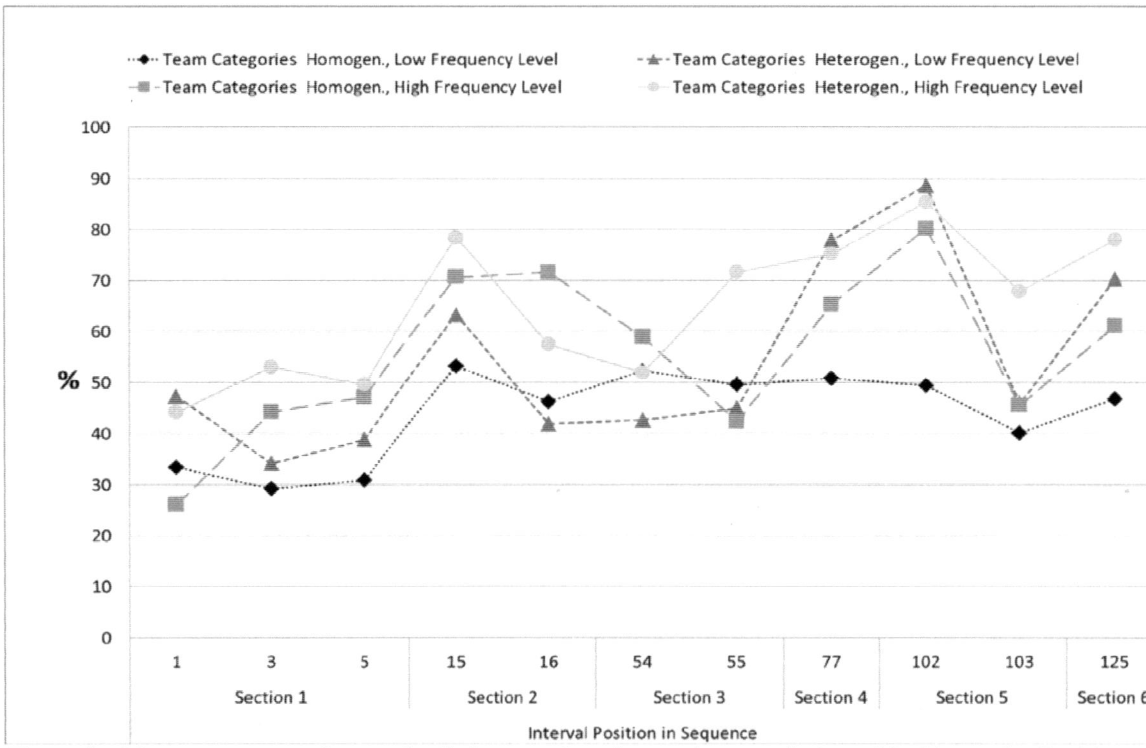

Figure 2: : Mean percentage of teams-dwell times on the relevant elements during critical intervals: *Position of Critical Interval in Seqence* by *Team Categories*

The allocation of resource of attentional resources to focussed monitoring also differed between the three relevant AoI (Means for AoI 1-3: 12.79% *[Power-Gauge], 33.85% [Output-Gauge] ,7.97% [Response Button]*); $F(2, 13) = 60.51$, $p < .0001$; $\eta^2= 0,90$) [in sum, more attention was allocated to relevant AoIs than to non-relevant AoI, which part of the 44.61% residual data can be assigned to non-relevant AoI, supposedly reflecting the allocation of attentional resources for distributed monitoring]. Tests of innersubject-effects indicate a *Relevant AoI* by *Interval Position in Sequence* interaction, F=3.04, df= 16.08 p < .0001, η^2=0.18, showing that teams were also changing their ressource allocation between the relevant AoI over time (cf. *Figure 2*). However there were no *Relevant AoI* by *Team Classification* [F(6, 28) = 1.29, p = .30, η^2=0.22] (cf. Figure 2), interactions, indicating that team-classes do not differ in their the way of monitoring relevant elements.

Discussion & Conclusions

The findings of the current study provide evidence that the quality of interactions, operationalized by measuring the prediction of correct locations using gaze data analysis, and the patterns in collective states of expectations can be induced by the chosen experimental manipulation. There was a trend that teams allocated more resources for providing focal attention and less resources for providing distributed attentions in the course of the scenario. This may be attributable to an increase in understanding of the task dynamics, paired with an increase in trust over time that events are predictable (decreases in the focal attention observable to the end of a scenario might be attributable to fatigue). Teams also were changing their ressource allocation between the relevant AoI over time. However, the chosen method for post-hoc grouping of teams based on fixation- frequency, and its within-group standar errors might not have led to a significant discrimmination between teams with different cognitive states. Although the high effect size observed for the no *Interval Position in Sequence* by *Team Classification* interaction might fuel the assumption that insignificant results were attributable to the small sample size. Nevertheless, the use of standard errors of fixation frequencies might be a too simplistic indicator for patterns of gaze behavior on team level, since this parameter also does not account for sequence information. Therfore a next step to search for team characterising patterns would be to apply the more complex method of gaze-recurrence analysis to data within teams of three. Furthermore, applying an design with experimental grous is the way to proceed. Especially as integrated gaze data analysis promises to provide an advanced method to test for Cooke et al`s (2013) assumption that team cognition patterns are more agile than individual performance patterns. Maybe gaze data will continue to tell this story, for the best of future ATM design.

References

Cannon-Bowers, J. A., Salas, E., & Converse, S. (1993). Shared mental models in expert team decision making. In Individual and Group Decision Making: Current Issues (pp. 221-245). Hillsdale, NJ: Lawrence Erlbaum Associates.

Cooke, N. J., Gorman, J. C., Myers, C. W., & Duran, J. L. (2013). Interactive team cognition. Cognitive Science, 37, 255-285.

Cooke, N. J. & Gorman, J. C. (2009). Interaction-based measures of cognitive systems. Journal of Cognitive Engineering and Decision Making, 3, 27-46.

Dekker, S. W. A. & Woods, D. D. (1999). To intervene or not to intervene: the dilemma of management by exception. Cognition, Technology & Work, 1, 86-96.

Hauland, G. (2008). Measuring individual and team situation awareness during planning tasks in training of en route air traffic control. The International Journal of Aviation Psychology, 18, 290-304.

Hess, S. M., MacMillan, J., Serfaty, D., & Elliott, L. (2005). From cognitive task analysis to simulation: Developing a synthetic team task for AWACS weapons directors. DTIC Document.

Schulze Kissing, D. & Eissfeldt, H. (2015). ConCenT: Eine Simulationsplattform zur Untersuchung kollaborativer Entscheidungsprozesse in Leitzentralen. In M. Grandt & S. Schmerwitz (Eds.), 57. Fachausschusssitzung Anthropotechnik: Kooperation und kooperative Systeme in der Fahrzeug- und Prozessführung (pp. 157-170). Bonn: Deutsche Gesellschaft für Luft- und Raumfahrt - Lilienthal-Oberth e.V.

Schulze-Kissing, D. & Bruder, C. (2016). Der Einsatz synthetischer Aufgabenumgebungen zur Untersuchung kollaborativer Prozesse in Leitzentralen am Beispiel der „Generic Control Center Task Environnment" (ConCenT). Proceedings of the Workshop Kognitive Systeme 2016, Bochum. Retrieved from http://elib.dlr.de/109031/1/63_215_1_SM.pdf

TRACKING WORKLOAD AND ENGAGEMENT IN AIR TRAFFIC CONTROL STUDENTS USING ELECTROENCEPHALOGRAPHY COGNITIVE STATE METRICS

Kyle A. Bernhardt
University of North Dakota Department of Psychology
Grand Forks, North Dakota
Terra Jorgenson, Craig Carlson, Paul Drechsel, Colt Iseminger
University of North Dakota Department of Aviation
Grand Forks, North Dakota
Dmitri Poltavski, F. Richard Ferraro, Thomas Petros
University of North Dakota Department of Psychology
Grand Forks, North Dakota

The current study evaluated the utility of electroencephalography (EEG) cognitive state to track workload and engagement changes in air traffic control students of differing experience during a Terminal Radar Approach Control (TRACON) scenario. EEG recordings were collected from *47* air traffic control students (27 with high and 20 with low experience) during a five phase TRACON scenario. The scenario fluctuated in the number of aircraft released per phase and the presence or absence of uncontrolled departures/arrivals. EEG workload probabilities were higher during the phase with uncontrolled departures/arrivals and maximum number of aircraft compared to phases with no uncontrolled arrivals/departures and fewer aircraft. Metrics of engagement did not vary throughout the scenario. Trends toward experience level differences in EEG metrics were observed, with less experienced students displaying slightly higher workload and engagement probabilities compared to their more experienced counterparts. Both experience groups made the most errors after the highest workload period.

The use of psychophysiological measurements to monitor operator cognitive states in the workplace is central to the field of neuroergonomics. Over the past two decades, a significant amount of research in the applied sector has focused on psychophysiological measures (e.g., electroencephalography) to monitor operators and evaluate their cognitive state (Parasuraman, 2015). In the profession of air traffic control (ATC), workload and engagement are two cognitive states critically tied to performance (Desmond & Hoyes, 1996; Signal, Gander, Anderson, & Brash, 2009). The complexity of ATC operations fosters an environment where overload situations can rapidly occur, potentially impairing controller decision making and placing aircraft in dangerous situations (Wickens, Mavor, & McGee, 1997). Moreover, low engagement, underload situations may lead to performance decrements as well (Desmond & Hoyes, 1996; Hancock & Warm, 1989; Saxby, Matthews, Warm, Hitchcock, & Neubauer, 2013). Psychophysiological measures of these constructs have been proposed as an alternative to performance-based metrics for evaluating an operator's skill level. Harrison et al. (2014) showed cerebral hemodynamic changes in ATCs as they learned to utilize Next-Generation technology. Moreover, Johnson et al. (2014) demonstrated electroencephalographic (EEG) differences in expert and novice shooters during a deadly force shooting simulator, with experts showing greater frontal theta activation and overall alpha suppression compared to novices. Experts also showed lower levels of EEG metrics of engagement compared to novices. Experts and novices did not differ with respect to EEG metrics of workload; however, the group size for each skill level was small ($n = 6$).

EEG provides several advantages for operator state assessment such as low cost, ease of application (Parasuraman, 2015), and the availability of turn-key systems that offer validated metrics of workload and engagement (e.g., Berka et al., 2007). Thus, EEG provides a suitable means for assessing ATC skill acquisition in ATC training institutions in addition to performance metrics, which only provide a unidimensional assessment of operator skill. That is, although two individuals may show the same task performance, one individual may experience significantly more workload and elevated task engagement than the other. The former individual would then be less adept at handling emergency situations than the latter due to reduced resource allocation capacity (Wickens & Tsang, 2015). More training may be necessary to ensure successful adaptation to increased workplace demands. The current study investigated the utility of using EEG derived cognitive state metrics of workload and engagement to evaluate ATC students of differing experience levels during a dynamic Terminal Radar Approach Control (TRACON) scenario. It was predicted that less experienced students would exhibit higher workload and engagement as well as worse performance than more experienced students.

Method

Participants

A total of 49 undergraduate air traffic control training students (two women and 47 men) from the University of North Dakota Department of Air Traffic Control participated in this study (M_{age} = 20.97, SD_{age} = 1.01). At the time of the study, all participants were enrolled in air traffic control training courses. From their prior coursework and advanced standing, all participants had adequate and on-going experience with the general procedures used with the TRACON simulator; however, they remained naïve with respect to the scenario to be performed. Two participants were excluded because of simulator technical difficulty, resulting in a final sample size of 47. Participants were split into low (n = 20) and high (n = 27) experience levels based on the highest ATC course completed that allowed them to perform all aspects of the scenario adequately.

Simulator and Scenario

This study utilized a high-fidelity TRACON simulator running ATCoach Global (V.4.32.5) Air Traffic Software with a Ubuntu Linux 11.04 (32-bit) operating system. The simulated RADAR scope was presented on a computer monitor measuring 20.1 in diagonal viewable area (1600 x 1200 resolution) with a 4:3 aspect ratio. Controllers utilized a keyboard and trackball mouse during the scenario to accept/initiate aircraft handoffs and manipulate aircraft data tags. During the scenario, controllers gave verbal instructions (e.g., altitude changes) communicated via a headset connected to a local radio connection to two research assistants (graduates of the air traffic control program) designated as "pseudo-pilots" located out of view from the controller's station. The pseudo-pilots utilized stations similar to the controller's station and inputted the commands for the aircraft in accordance with the controller's instructions. Pseudo-pilots gave verbal readbacks of controller commands.

This study utilized Academy Airspace, a fictitious training airspace used by the Federal Aviation Administration (FAA) to train new controllers. The Academy airspace consists of centrally located Academy Airport (KAAC) and five surrounding airports (one towered military airport and four uncontrolled airports). All inbound and outbound aircraft entered/departed the airspace through established arrival/departure gates. The scenario consisted of three types of aircraft: jets (both light and heavy), turboprop, and reciprocating. Participants gained control of inbound aircraft 40 nautical miles out from KAAC and were responsible for controlling the aircraft to their designated airport and clearing the aircraft for published instrument approaches (ILS, NDB, or GPS approach). All airports were operating under single runway operations.

Participants performed a single 1-hr and 15-min continuous scenario composed of five, 15-min phases. The phases were developed during prior work to create a multidimensional, variable task demand TRACON scenario. All simulated aircraft were operating under instrument flight rules (IFR). Arrivals and departures occurred at both KAAC and the five satellite airports. Phases were distinguished via two workload-contributing factors: (1) number of aircraft arriving or departing the airspace within a given 15-min phase and (2) the presence or absence of arriving and departing uncontrolled aircraft. Phases were designated 1, 2, 3, 4, 5 and consisted of 8, 11, 16, 8, and 3 departing/arriving aircraft, respectively. Uncontrolled aircraft occurred during Phase 3. From these phase parameters, it was predicted that workload and engagement would follow a negative quadratic function shape, with peak values occurring at Phase 3. Manipulations of air traffic volume have shown to reliably elicit workload changes in ATCs (Vogt, Hagemann, & Kastner, 2006). Moreover, handling uncontrolled aircraft strains working memory capacity. Working memory strain has also been shown to be a reliable manipulation of operator workload (Wickens & Tsang, 2015). To add realism to the scenario, controllers were also required to complete flight strip marking procedures for each aircraft. Paper strips housed in plastic holders were located to the right of the controller in two bays for active and non-active aircraft.

Scenario scoring. Scenario runs were recorded by the TRACON simulator program and later scored by a team of University of North Dakota Department of Air Traffic Control faculty. Combining for over 40 years of experience, these faculty members have backgrounds in either controlling professionally or were graduates of the University of North Dakota Air Traffic Control program. Participant performance was evaluated by tracking the number of errors made in phraseology, violation of aircraft separation minimums, and airspace procedural violations for each 15-min phase. Errors were identified according to violations of standards set by FAA Order JO 7110.65W and letters of agreement (LOAs) established by Academy Airspace for training purposes.

EEG Recording

EEG was recorded using the Advanced Brain Monitoring (ABM) B-Alert X24 wireless Bluetooth system sampling at 256 Hz. The system incorporates 20 electrodes placed according to the international 10/20 system: Fp1, Fp2, F7, F3, Fz, F4, F8, T3, C3, Cz, C4, T4, T5, P3, Pz, P4, T6, O1, POz, and O2. Reference electrodes were placed at the left and right mastoids. Data were filtered with 50, 60, 100, and 120 Hz notch filters with Low Pass FIR filters online during data collection. Proprietary ABM artifact decontamination algorithms removed artifacts resulting from electromyography, eye blinks, excursions, saturations, and spikes. Power spectral density (PSD) was calculated on a second by second epoch frequency by applying a 50% overlapping Kaiser window for data smoothing to three data point windows consisting of 256 decontaminated data points each. These data were then subject to Fast Fourier Transformation resulting in four standard bandwidths (delta, alpha, theta, and beta). ABM proprietary algorithms (Berka et al., 2007) then computed cognitive state metric probabilities of high engagement and average workload ranging from 0.00 to 1.00 based on PSD values. Higher values indicate a higher probability of being in an engaged or overload state, respectively. EEG metrics were averaged within each 15-min phase using 5% trimmed means to eliminate extreme values.

Procedure

Participants first arrived at the TRACON simulation room on the University of North Dakota campus and provided written consent. Then, participants completed individual difference measures (results not reported here) and were fitted with the B-Alert EEG headset. To obtain ABM cognitive state metrics, participants performed three computerized cognitive benchmark tasks: 3-choice vigilance, visual psychomotor vigilance test, and an eyes-closed auditory psychomotor vigilance test. One of the trained pseudo-pilots then briefed the participant on the scenario and gave instructions regarding strip marking procedures. The simulation was then commenced and ran continuously for 1.25-hr.

Data Analytics

Series of 5 (scenario phase: 1, 2, 3, 4, 5) x 2 (experience: low, high) mixed model ANOVAs were conducted to evaluate scenario performance and EEG cognitive state metrics. Scenario phase served as the within-subjects factor and experience served as the between-subjects factor. Analyses were evaluated at an $\alpha = .05$. If the assumption of sphericity was violated, degrees of freedom corrections were employed. When ε values were less than .75, the Greenhouse-Geisser correction was used and when ε values were greater than .75, the Huynh-Feldt correction was used (Field, 2009). Sidak corrections were utilized for multiple comparisons if pairwise comparisons were warranted.

Results

Simulation Performance

Phraseology. Graphical representations of performance data are displayed in Figure 1. The results of the mixed model ANOVA for phraseology errors revealed a significant main effect for phase, $F(4, 180) = 19.72$, $p < .001$, $\eta_p^2 = .31$. Overall, the number of phraseology errors significantly increased from Phase 1 to Phase 2 and from Phase 3 to Phase 4. Phraseology errors then decreased during Phase 5, but not significantly from Phase 4 (see Figure 1). Neither the main effect for experience nor the experience by phase interaction were significant, $Fs < 1$, $ps > .05$.

Separation. Analysis of separation errors revealed a significant main effect for phase, $F(2.31, 103.75) = 43.70$, $p < .001$, $\eta_p^2 = .49$. Separation errors significantly increased after Phase 2 and reached a maximum at Phase 4, and then significantly decreased during Phase 5 (see Figure 1). The main effect for experience and the experience by phase interaction were not significant, $Fs < 1$, $ps > .05$.

Procedures. The results of a mixed ANOVA analyzing procedural errors revealed a significant main effect for phase, $F(2.71, 121.91) = 108.69$, $p < .001$, $\eta_p^2 = .71$. Similar to phraseology and separation errors, procedural errors significantly increased from Phase 1 onward until reaching a maximum at Phase 4. Procedural errors then decreased from Phase 4 to Phase 5 (see Figure 1). The main effect for experience, $F < 1$, $p > .05$, and the interaction between phase and experience, $F(2.71, 121.91) = 1.89$, $p = .14$, $\eta_p^2 = .04$, were not significant.

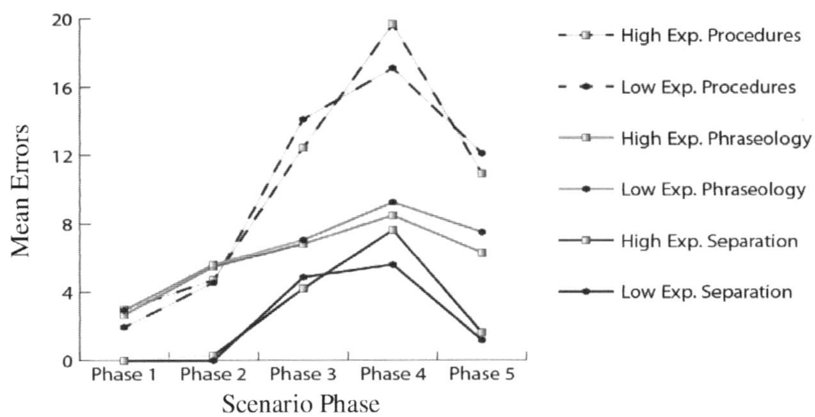

Figure 1. Scenario phase performance by controller experience (Exp.).

EEG Metrics

Workload. Results of EEG metrics are displayed in Figure 2. The results of a mixed ANOVA analyzing the B-Alert average workload metric revealed a main effect for phase, $F(2.29, 102.90) = 12.79$, $p < .001$, $\eta_p^2 = .22$. Examination of the group means demonstrated that workload increased across the phases, peaking at Phase 3 (see Figure 2 left panel). Phase 3 was significantly higher in workload than Phase 1, Phase 2, and Phase 5. Moreover, Phase 4 was significantly higher in workload than Phase 1 and Phase 5. The main effect for experience failed to reach significance, $F < 1$, $p > .05$. Additionally, the interaction between phase and experience was not significant, $F(2.29, 102.90) = 1.72$, $p = .18$, $\eta_p^2 = .04$.

Engagement. The analysis of the B-Alert high engagement metric (see Figure 2 right panel) revealed a nonsignificant main effect for phase, $F(2.62, 118.00) = 2.40$, $p = .80$, $\eta_p^2 = .05$, and a nonsignificant main effect for experience, $F(1, 45) = 2.59$, $p = .11$, $\eta_p^2 = .05$. The phase by experience interaction was also not significant, $F(2.26, 118.00) = 1.19$, $p = .31$, $\eta_p^2 = .03$.

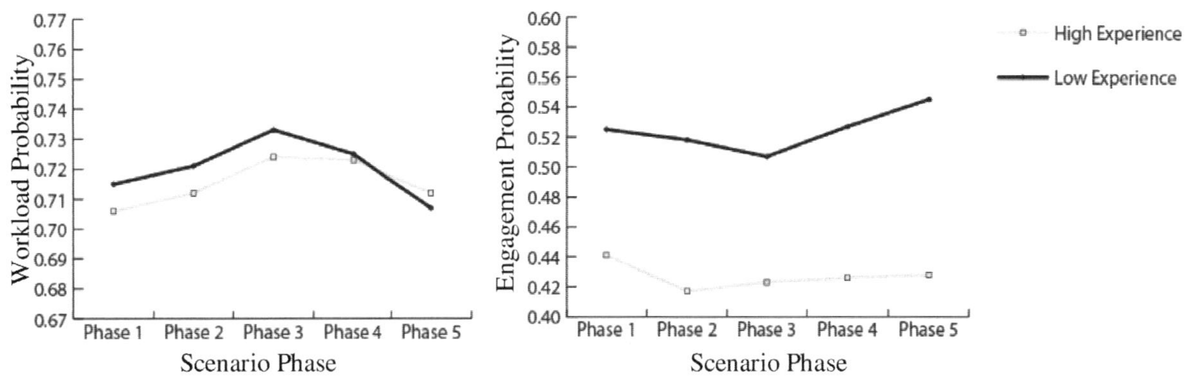

Figure 2. Scenario phase EEG workload (left panel) and engagement (right panel) probabilities by experience.

Discussion

The current study examined the utility of EEG based cognitive metrics for tracking workload and engagement in differentially experienced ATC students over the course of a dynamic, five phase high-fidelity TRACON simulation. Performance results indicated that both low and high experienced students performed similarly on the scenario, with the most errors occurring during Phase 4. Phase 3 had the most aircraft released along with uncontrolled aircraft, making Phase 3 more likely to induce errors. However, the most errors occurred during Phase 4. This was likely the result of a progressive buildup of traffic from the high workload of Phase 3, causing more errors further along the line in the scenario. The B-Alert workload metric demonstrated a trend that mirrored the predicted workload of the scenario. That is, the metric increased from Phases 1 through 3 and then decreased after Phase 3. Thus, this metric was able to track changes in cognitive workload throughout the simulation. This study provides further evidence for potential dissociations that can occur with performance and other measures of workload. The most errors occurred during Phase 4, however, peak EEG workload was observed in Phase 3. This finding underscores the importance of assessing operator workload with different measures to acquire a holistic picture of the operator's state. The B-Alert engagement metric did not significantly fluctuate throughout the scenario for either experience group, indicating that controllers remained consistently engaged throughout the scenario.

Although statistical analysis did not demonstrate experience main effects or phase by experience interactions with respect to the EEG workload and engagement metrics, visual analysis of the data shows a trend toward lower experience students exhibiting higher workload during Phases 1 through 3, as well as higher engagement throughout the scenario. The slightly increased workload observed in lower experienced students likely reflects less of an ability to mobilize cognitive resources efficiently (Berka et al., 2007). Moreover, the elevated engagement in lower experienced students is likely the antecedent to increased workload. Low experience controllers must maintain a higher degree of focused attention to maintain performance and allocate attentional resources, a cost not revealed by simulator performance metrics. The trend in engagement is similar to previous research demonstrating higher EEG indices of engagement in novices compared to experts during task performance (Johnson et al., 2014). Future studies should utilize stronger manipulations of controller experience (e.g., professional controllers vs. students). In using upper and lower leveled students, the skill level difference between the two groups may not have been strong enough to produce statistically significant differences between groups.

References

Berka, C., Levendowski, D. J., Lumicao, M. N., Yau, A., Davis, G., Olmstead, R. E., … Craven, P. L. (2007). EEG correlates of task engagement and mental workload in vigilance, learning, and memory. *Aviation, Space, and Environmental Medicine*, *78*(5), B231- B243

Desmond, P. A., & Hoyes, T. W. (1996). Workload variation, intrinsic risk and utility in a simulated air traffic control task: Evidence for compensatory effects. *Safety Science*, *22*(1-3), 87-101.

Field, A. (2009). *Discovering statistics using SPSS.* Thousand Oaks, CA: SAGE Publications Inc.

Hancock, P. A., & Warm, J. S. (1989). A dynamic model of stress and sustained attention. *Human Factors, 31*(5), 519-537.

Johnson, R. R., Stone, B. T., Miranda, C. M., Vila, B., James, L., James, S. M., … Berka, C. (2014). Identifying psychophysiological indices of expert vs. novice performance in deadly force judgment and decision making. *Frontiers in Human Neuroscience*, *8*, 1-13. doi: 10.3389/fnhum.2014.00512

Loft, S., Sanderson, P., Neal, A., & Mooij, M. (2007). Modeling and predicting mental workload in en route air traffic control: Critical review and broader implications. *Human Factors*, *49*(3), 376-399. doi: 10.1518/001872007X197017

Parasuraman, R. (2015). Neuroergonomic perspectives on human systems integration: Mental workload, vigilance, adaptive automation, and training. In D. A. Boehm-Davis, F. T. Durso, J. D. Lee. (Eds.), *APA handbook of human systems integration* (pp. 163-176). Washington, DC: American Psychological Association.

Saxby, D. J., Matthews, G., Warm, J. S., Hitchcock, E. M., & Neubauer, C. (2013). Active and passive fatigue in simulated driving: discriminating styles of workload regulation and their safety impacts. *Journal of Experimental Psychology: Applied*, *19*(4), 287-300.

Signal, T. L., Gander, P. H., Anderson, H., Brash, S. (2009). Scheduled napping as a countermeasure to sleepiness in air traffic controllers. *Journal of Sleep Research*, *18*, 11-19. doi: 10.1111/j.1365-2869.2008.00702.x

Vogt, J., Hagemann, T., & Kastner, M. (2006). The impact of workload on heart rate and blood pressure in en-route and tower air traffic control. *Journal of Psychophysiology*, *20*(4), 297-314.

Wickens, C. D., Mavor, A. S., & McGee, J. P. (Eds.). (1997). *Flight to the future: Human factors in air traffic control*. Washington, DC, US: National Academy Press.

Wickens, C., & Tsang, P. (2015). Workload. In D. A. Boehm-Davis, F. T. Durso, & J. D. Lee (Eds.), *APA handbook of human systems integration* (pp. 277-292). Washington, DC: American Psychological Association.

PERIODIC BLINK MEASURES USING DYNAMIC WINDOWING

Matt Middendorf
Middendorf Scientific Services
Chelsey Credlebaugh
Ball Aerospace and Technologies Corp.
Christina Gruenwald
Oak Ridge Institute for Science and Education
Jonathan Mead
Oak Ridge Institute for Science and Education
Scott Galster
Air Force Research Laboratory
Wright-Patterson AFB, Ohio

Artificial neural network (ANN) models are a common tool for cognitive state assessment. It is best if the inputs to the model are periodic. Typically, these inputs are extracted from physiological signals such as the electroencephalogram (EEG), electrocardiogram (ECG), electrooculogram (EOG), and others. Spectral measures derived from EEG data are periodic due to the signal processing. Features based on heart activity and respiration are quasi-periodic by nature. Features extracted from EOG, such as blink rate, can be especially non-periodic and can contain outliers. One approach to deal with this problem is to use static windows to compute average blink rate. This approach has some shortcomings. A new approach that uses dynamic windowing, filtering, and sampling is presented here. This new approach produces periodic data that are dynamic, adaptive to the individual, and well suited for ANN model use.

Being able to assess a person's cognitive state as it relates to stress and cognitive workload has great appeal in many domains. A high level of workload can have detrimental effects in real-world operations. For example, air traffic controllers should avoid exceeding a moderate level of workload because they have the responsibility of keeping airplane passengers safe (Brookings, Wilson, & Swain, 1996). Medical physicians performing lifesaving operations cannot afford to slip into a state of cognitive overload and risk harm to the patient. There are many other examples of when accurate cognitive state assessment is paramount.

One technique for performing a workload assessment is the use of artificial neural network (ANN) models. ANN models are biologically inspired computer programs designed to simulate the way in which the human processes information (Agatonovic-Kustrin & Beresford, 2000). Wilson (2003), for example, used ANNs and physiological measures to assess workload in a well-controlled laboratory task. Borghini, Astolfi, Vecchiato, Mattia, & Babiloni (2014) provides a review of research that use neurophysiological signals for workload assessment of aircraft pilots and car drivers. Physiological measures that are periodic are best suited for ANN models used for assessment. Specifically, the physiological inputs are regularly (e.g., one Hz) provided to the model.

The electroencephalogram (EEG) is an electrical signal associated with brain activity and is often used in ANNs for cognitive assessment. Typically, the EEG signals are processed using windows (i.e., four seconds with 75% overlap) to generate spectral measures. Because of the signal processing, the EEG measures are periodic.

The electrocardiogram (ECG) is an electrical signal associated with heart activity. The measures (heart rate and heat rate variability) determined from the ECG are quasi-periodic. Specifically, the heart beats are somewhat consistent, but are not perfectly regular.

The vertical electrooculogram (VEOG) can be used to detect eye blinks. Blink rate and blink duration have been shown to be sensitive to changes in workload (Hoepf et al., 2016) and are useful inputs to ANN models. However, blink rate is very non-periodic (Figure 1) and varies substantially between individuals. Instantaneous blink rate is computed by taking the inverse (reciprocal) of the time (period) between two consecutive blinks. The resulting value is multiplied by 60 to convert it to blinks per minute. As shown in Figure 1, instantaneous blink rate will not represent a long term average when double blinks occur. This 30 second window contains four blinks, corresponding to a blink rate of eight blinks per minute (BPM). The blinks occur at 7.6, 15.0, 20.7, and 21.5 seconds. The associated instantaneous blink rates for blinks 2, 3, and 4 are 8.12, 10.5, and 74.63 respectively. The blink rate for the fourth blink is an outlier. Because the instantaneous values are not periodic and may contain outliers, they are not well-suited for use in ANN models.

One approach to overcome these problems is to compute blink rate using static (i.e., fixed length) windows. This approach produces periodic data, but has potential pitfalls associated with lag and accuracy. A new approach using dynamic (i.e., varying length) windowing is presented that overcomes these pitfalls. This new approach is coupled with a filtering and sampling technique to produce periodic blink measures. Both the static and dynamic approaches rely on algorithms that can detect blinks based on eye activity.

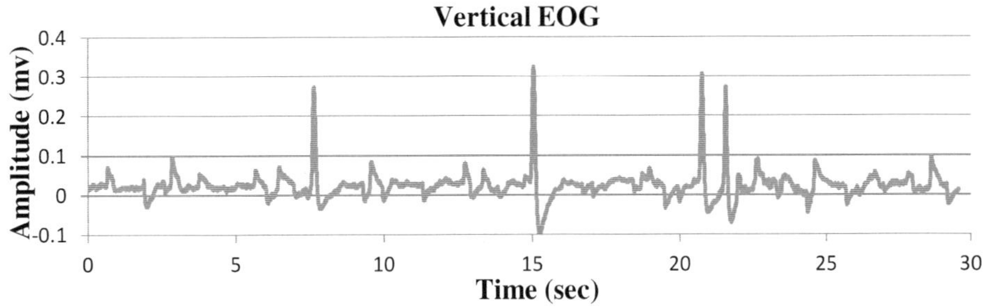

Figure 1. Vertical EOG data collected with blinks occurring at 7.6, 15.0, 20.7, and 21.5 seconds.

Background

Blink frequency and duration, can be acquired using various techniques. A common choice employs camera-based systems to measure eyelid position. The type of camera can vary, using technology such as laptop-based webcams (Krolak & Strumillo, 2011) or head-mounted eye trackers (Jiang, Tien, Huang, Zheng, & Atkins, 2012). These systems obtain video sequences, which can then be used in various algorithms to detect blinks.

Another common technique uses VEOG data to detect eye blinks (Pedrotti, Lei, Dzaack, & Rötting, 2011; Hu & Zheng, 2009). The VEOG data is processed by algorithms to detect blinks, commonly using thresholds and algorithm criteria. These thresholds can remain constant across trials (Ebrahim, Stolzmann, & Yang, 2013) or for the duration of the eye closure (Hu & Zheng, 2009). For a higher degree of specificity, the thresholds can be set specifically for each subject. An alternative route involves training the algorithm to adjust the detection parameters individually for each subject, allowing for more precision in eye blink detection (Kong & Wilson, 1998).

Because most camera-based detection methods depend on the stability of the subject's head for tracking position, they require the subject to remain still in an often unnatural position. In contrast, EOG recordings allow the participant to move around, since the electrodes are attached to the participant. This provides more reliable data, often with a higher sampling rate than camera-based methods.

Method

The intent of this paper is to present a new technique for generating periodic blink data using dynamic windowing. Therefore, the method sub-sections are presented concisely, with just enough detail to support the results section.

Participants

Ten individuals were recruited from the Midwest region to participate in this study. Eight participants were male and two were female, with an age range from 18-33 and a mean of 21.9. They read and signed the informed consent document before participating and were compensated for their time. All study procedures were reviewed and approved by the Air Force Research Laboratory Institutional Review Board.

Task Description

Two separate primary tasks (surveillance or tracking) were presented using a remotely piloted aircraft (RPA) simulation. A secondary communications task was also present.

Primary - Surveillance Task. The surveillance task required the participants to search a market place to find four high value targets (HVTs).

Primary - Tracking Task. The tracking task required participants to track HVTs travelling by motorcycle.

Secondary - Communications Task. A secondary task was presented concurrently with the primary tasks. Participants verbally answered a variety of mental math questions asked over a headset.

Apparatus

The VEOG data were acquired using two electrodes placed above and below the right eye. The VEOG data were sampled at 480 Hz using the Cleveland Medical Devices BioRadio 150. This device has hardware high pass filters with break frequencies of 0.5 Hz.

Procedure

Participants were brought into the laboratory for two days of task training and six days of data collection. During the course of the six data collection days, participants completed 24 surveillance trials and 24 tracking trials. VEOG acquisition hardware was present only on the six data collection sessions.

Static Window Approach

As reported earlier, researchers have used static windows to overcome the outlier issue associated with double blinks (Estepp & Christensen, 2015). In the current effort, static windows were explored. Blink rate was initially computed using a 30 second window with 29 second overlap. This approach deals well with the double blink outliers and produces periodic data. One concern with this approach is that the resulting blink rate measure will have a lag. This may not be ideal as an ANN input when a timely assessment is needed.

In an attempt to mitigate the lag concern, another approach was developed that uses a five second static window with four seconds of overlap. The output from this approach was often zero, quite granular, and was somewhat susceptible to the double blink effect. To address these issues, the output was low pass filtered. The filter was updated at a rate of 31.25 Hz so the output would transition smoothly from one value to the next. The output of the filter was sampled to generate periodic data.

The New Approach

The new approach is dependent on a blink detection algorithm that has a high degree of accuracy (Epling et al., 2015). The times of the detected blinks are buffered and supplied to the dynamic windowing algorithm to generate the periodic blink data (Figure 2). The dynamic windowing algorithm has three steps, including dynamic window calculation, filtering, and sampling.

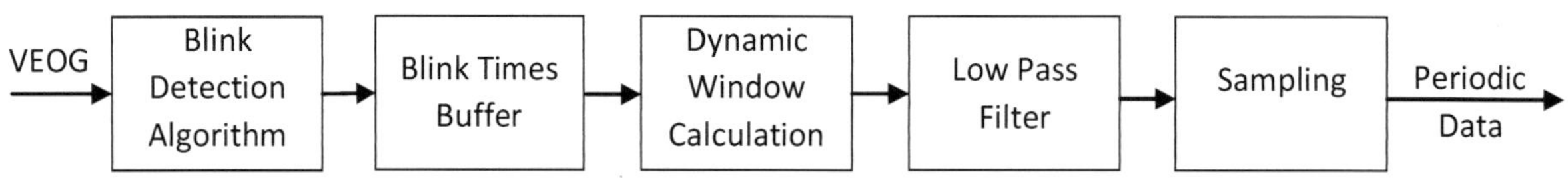

Figure 2. Steps in the periodic data calculation.

The dynamic window size (in seconds) is calculated based on the running average blink rate, which is computed by dividing the total number of blinks detected since the software was started is divided by the length of time the software has been running. The window size must be large enough to contain two blinks. This requirement (as opposed to one blink) was necessary to prevent several windows from having zero blinks in them. An individual with a blink rate of 12 BPM (an average of 5 seconds per blink) will have a window size of 10 seconds. The computed window size is limited to a maximum value of 30 seconds. The dynamic windowing approach results in window sizes that are adaptive between individuals, and dynamic within the individual.

The software that performs all of the calculations updates at 31.25 Hz. During each update, the dynamic window size is computed, the number of blinks in the window are counted, and the resulting blink rate is computed by dividing the count by the window size. The result is multiplied by 60 to convert the blink rate from blinks per second to blinks per minute (BPM). The blink rate is then processed by a first order Butterworth low pass filter with a cutoff frequency of 0.1 Hz. The output of the filter is sampled at regular intervals (i.e., one second) to produce the periodic blink data. Figure 3 shows instantaneous blink rate data that has been processed by the dynamic blink algorithm. The outlier (74.6 BPM) due to a double blink has been substantially reduced.

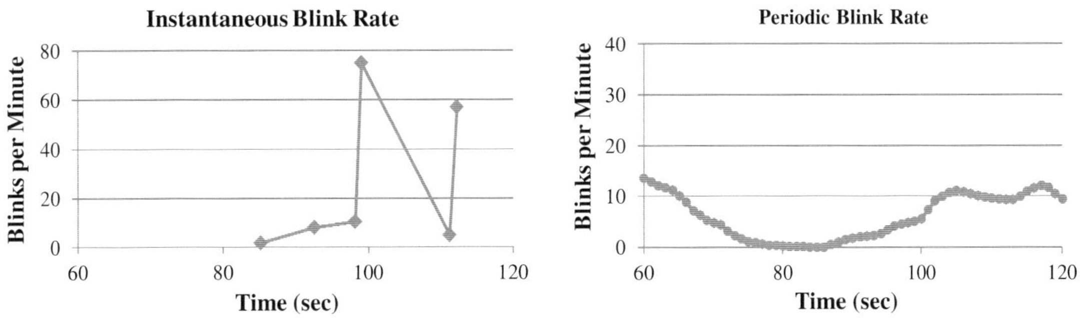

Figure 3. The left panel shows instantaneous blink rate produced by the blink detection algorithm. The right panel shows the same data after it has been processed by the dynamic windowing algorithm. The data in the right panel is sampled at 60 Hz.

Results

Actual blink rates were computed *post hoc* by dividing the number of blinks in an experimental trial by the length of the trial. This results in 24 values for each participant and task (surveillance & tracking). This serves as truth data for evaluating the dynamic blink rate data. The dynamic blink rate data were averaged across trials for each participant and task. Correlations were performed between the actual blink rates and the average dynamic blink rates. For the surveillance task, the correlations ranged from 0.91 to 1.0 with a mean of 0.97. For tracking, the range was from 0.97 to 1.0 with a mean of 0.99. These correlations indicate that the periodic blink rate based on dynamic windowing accurately reflects the actual blink rate. Figure 4 shows the reported correlations.

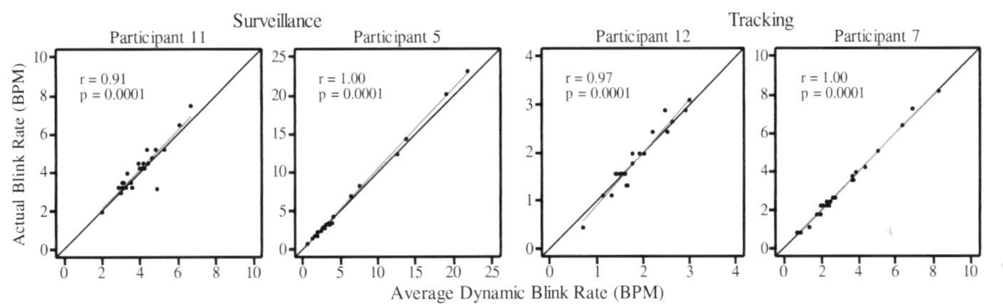

Figure 4. This figure shows the lowest and highest correlations between actual blink rate and average dynamic blink rate for the surveillance and tracking tasks. For surveillance, the lowest correlation is from participant 11 and the highest is from participant 5. For tracking, the lowest is from participant 12 and highest is from participant 7.

The instantaneous blink rate data were also averaged and correlated with the actual blink rate data. The correlations were lower than the correlations between average dynamic blink rate and actual blink rate. For the surveillance task, the correlations ranged from -0.2 to 0.89 with a mean of 0.5. For tracking, the range was from 0.02 to 0.93 with a mean of 0.63. These results indicate that instantaneous blink rate is inferior to dynamic blink rate because instantaneous blink rate contains huge outliers due to double blinks. Figure 5 shows examples of these correlations.

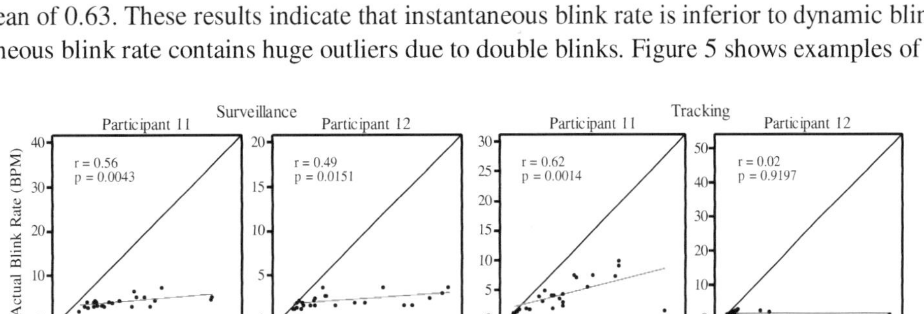

Figure 5. These examples show correlations between actual blink rates and average instantaneous blink rates.

The blink rate data from the two static windowing approaches were also averaged across trials and correlated with actual blink rate. The correlations for the 30 second window were generally high with the exception of participants with low blink rates. The correlations ranged from 0.75 to 0.99 with a mean of 0.86 for the surveillance task and 0.48 to 0.99 with a mean of 0.86 for the tracking task.

The correlations for the five second window were generally high, but were slightly lower than the 30 second window. The correlations ranged from 0.67 to 0.99 with a mean of 0.92 for the surveillance task, and 0.42 to 0.99 with a mean of 0.87 for the tracking task.

The above correlations are based on averages across trials. Additional insights are realized when the data are examined as a time series. Figure 6 shows time series data from a participant with a low blink rate (~3 BPM). The data from a static 30 second windowing approach closely matches the data from the dynamic windowing approach (left panel). Data from the five second windowing approach does not follow the data from dynamic windowing approach very closely (right panel), and appears to be "noisy."

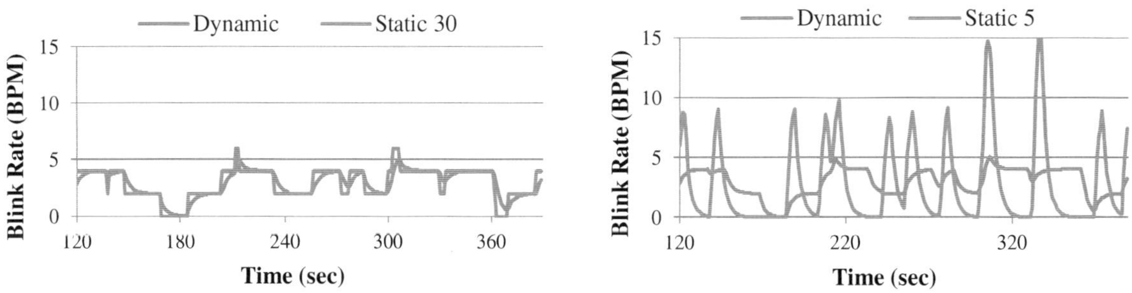

Figure 6. Time series data from a participant with a low blink rate (~3 BPM). Both panels have the periodic blink rate data computed using dynamic windows. The left panel includes blink rate computed using a static 30 second window, while the right panel includes data using a static five second window.

Figure 7 shows data from a participant with a high blink rate (~21 BPM). In this case, the five second window matches the dynamically windowed data better than the 30 second window. The two time series on the right panel (Figure 7) are nearly identical, making it nearly impossible to observe a difference.

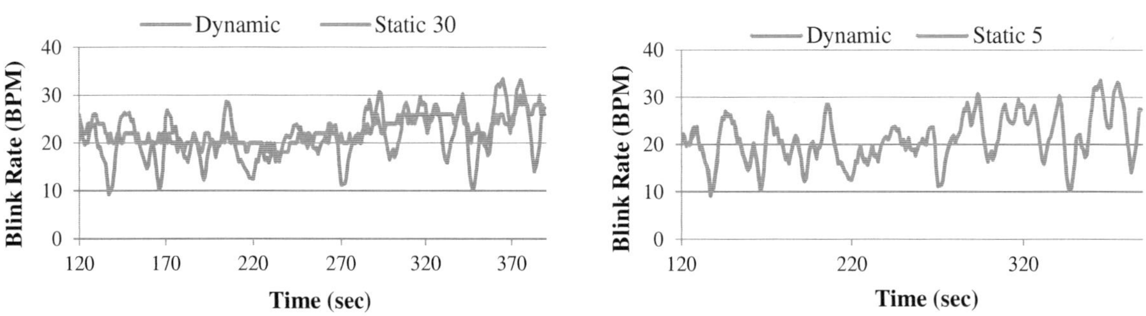

Figure 7. Time series data from a participant with a high blink rate (~21 BPM).

Discussion

The new method of generating periodic data using dynamic windowing is superior to static windowing because the window size is adaptive to the individual. This is in contrast to a one-size-fits-all approach. Figures 6 and 7 show that different static window sizes are needed dependent on the participants' blink rate. The new approach is also dynamic within an individual because the window size can change over time. The dynamic windowing algorithm is effective (Figure 3, right panel) in dealing with instantaneous outlier values due to double blinks. The filtering and sampling works well for creating periodic data.

The high correlations between actual blink rate and average dynamic blink rate indicate that these two quantities are measuring the same thing (i.e., blink rate). One advantage of the new dynamic blink rate approach is that the data is generated in real-time, whereas the actual blink rate is *post hoc*. Having periodic data available in real-time, with minimal lag and free from outliers, is optimal for use with ANN models.

One limitation of the current approach was in the calculation of the dynamic window size. Currently, it is based on rolling blink rate, which is computed from the time the software is started. Over a long period of time, the rolling blink rate becomes very stable and the dynamic aspect of the new approach is diminished. In future work, a long window (on the order of minutes) will be used to compute rolling blink rate to address this concern.

Acknowledgements

The authors would like to thank Chuck Goodyear for assistance with data analysis. The views expressed in this report are solely those of the authors and do not necessarily reflect the views of their organizations.

This research was supported in part by an appointment to the Student Research Participation Program at the U.S. Air Force Research Laboratory (USAFRL) administered by the Oak Ridge Institute for Science and Education through an interagency agreement between the U.S. Department of Energy and USAFRL. The views expressed in this report are solely those of the authors and do not necessarily reflect the views of the employers or granting organizations.

References

Agatonovic-Kustrin, S., & Beresford, R. (2000). Basic concepts of artificial neural network (ANN) modeling and its application in pharmaceutical research. *Journal of Pharmaceutical and Biomedical Analysis*, *22*(5), 717-727. doi: 10.1016/s0731-7085(99)00272-1

Borghini, G., Astolfi, L., Vecchiato, G., Mattia, D., & Babiloni, F. (2014). Measuring neurophysiological signals in aircraft pilots and car drivers for the assessment of mental workload, fatigue and drowsiness. *Neuroscience & Biobehavioral Reviews, 44*, 58-75. doi: 10.1016/j.neubiorev.2012.10.003

Brookings, J. B., Wilson, G. F., & Swain, C. R. (1996). Psychophysiological responses to changes in workload during simulated air traffic control. *Biological Psychology, 42,* 361-377.

Ebrahim, P., Stolzmann, W., & Yang, B. (2013). Eye movement detection for assessing driver drowsiness by electrooculography. In *Proceedings of the 2013 IEEE International Conference on Systems, Man, and Cybernetics.* doi: 10.1109/smc.2013.706

Epling, S., Middendorf, M., Hoepf, M., Gruenwald, C., Stork, L., & Galster, S. (2015). The electrooculogram and a new blink detection algorithm. In *Proceedings of the International Symposium on Aviation Psychology, 18*, 512-517.

Estepp, J., & Christensen, J. (2015). Electrode replacement does not affect classification accuracy in dual-session use of a passive brain-computer interface for assessing cognitive workload. *Frontiers in Neuroscience,* 9(54), 285-314. doi: 10.3389/fnins.2015.00054

Hoepf, M., Middendorf, M., Mead, J., Gruenwald, C., Credlebaugh, C., Middendorf, P., & Galster, S. (2016). Evaluation of Physiologically – Based Artificial Neural Network Models to Detect Operator Workload in Remotely Piloted Aircraft Operations (Report No. AFRL-RH-WP-TR-2016-0075). Wright-Patterson Air Force Base: Air Force Research Laboratory, Airman Systems Directorate.

Hu, S., & Zheng, G. (2009). Driver drowsiness detection with eyelid related parameters by Support Vector Machine. *Expert Systems with Applications, 36*(4), 7651-7658. doi: 10.1016/j.eswa.2008.09.030

Jiang, X., Tien, G., Huang, D., Zheng, B., & Atkins, M. S. (2013). Capturing and evaluating blinks from video-based eyetrackers. *Behavior Research* Methods, *45*(3), 656-663. doi: 10.3758/s13428-012-0294-x

Kong, X., & Wilson, G. (1998). A new EOG-based eyeblink detection algorithm. *Behavioral Research Methods, Instruments, & Computers, 30*(4), 713-719.

Krolak, A., & Strumillo, P. (2011). Eye-blink detection system for human-computer interaction. *Universal Access in the Information Society, 11*(4), 409-419. doi:10.1007/s10209-011-0256-6

Pedrotti, M., Lei, S., Dzaack, J., & Rotting, M. (2011). A data-driven algorithm for offline pupil signal preprocessing and eyeblink detection in low-speed eye-tracking protocols. *Behavioral Research Methods, 43*(2), 372-383. doi: 10.3758/s13428-010-0055-7

Wilson, G., & Russell, C. (2003). Real-time assessment of mental workload using psychophysiological measures and artificial neural networks. *Human Factors: The Journal of the Human Factors and Ergonomics Society, 45*(4), 635-643.

DATA-DRIVEN STAFFING RECOMMENDATIONS FOR AIR TRAFFIC CONTROL TOWERS

Linda G. Pierce
FAA Aviation Safety Civil Aerospace Medical Institute
Oklahoma City, OK
Terry L. Craft
FAA Air Traffic Organization Management Services
Washington, D.C.

The FAA is interested in an optimal strategy for placing air traffic controllers into high-level Terminal facilities. Our research question is whether new hire trainees (developmentals) should begin field training at lower-level facilities and transfer to higher-level facilities later, if successful at the lower-level facility, or begin training at a higher-level facility and transfer to a lower-level facility if unsuccessful? We compared the success rates of developmentals placed into medium- and high-level Terminal facilities after completing Academy training to the success rates of certified professional controllers (CPCs) allowed to transfer after completing field training at a lower-level facility. We found that the CPCs who began training at a lower-level facility succeeded in training at medium-level Combined Tower/TRACON facilities at a significantly higher rate than did developmentals at the same facility type and level. We recommended that the FAA staff higher-level facilities with CPC transfers rather than with new hires.

Should the Federal Aviation Administration (FAA) place newly hired air traffic controllers, with no prior experience in air traffic control (ATC), into a high, medium, or low-level ATC facility as their first facility? The FAA assigns a numerical level to a facility based on the volume, complexity, and sustainability of air traffic at that facility (FAA, 2016). Facility levels range between 4 and 12. In general, controlling air traffic is more challenging at higher than at lower-level facilities. It also takes most trainees longer to complete training and more trainees fail training at higher-level than at lower-level facilities (FAA, 2014a; FAA, 2016). However, because of the challenge and the pay (i.e., controllers are paid more at high-level than at low-level facilities), many controllers express a desire to be assigned to a high-level facility as early as possible. Our research objective was to determine if new hire trainees (called "developmentals") should be allowed to proceed directly to a medium- or high-level facility for field qualification training after they complete training at the FAA ATC Academy or if they should have to demonstrate their proficiency by becoming a successful certified professional controller (CPC; i.e., successfully completing all ATC training) at a less complex facility prior to moving to a higher-level, more complex facility and completing the additional training there as a CPC-In-Training (CPC-IT). Which placement strategy will result in a greater rate of successful training completion and qualification as CPC at higher-level facilities?

Terminal Operations

This question is especially important within the Terminal option of ATC, because unlike En Route centers, in which most are classified as high-level facilities (Levels 10–12), Terminal facility levels vary from low (Levels 4–6), to medium (Levels 7–9), to high (Levels 10–12) levels. Controllers at Terminal facilities work in airport towers, terminal radar approach control (TRACON) facilities or a combined tower and TRACON facility to "watch over" the aircraft traveling through the airspace of the airport or airports assigned (FAA, 2015). Terminal facilities vary in the extent to which air traffic is controlled using visual observation or radar and the number of airports for which the facility is responsible. There are 314 Terminal facilities within the National Airspace System (NAS) and the FAA manages the hiring and placement of new air traffic controllers at these facilities. A way to assess placement strategies of air traffic controllers is to compare the success rates at medium- and high-level facilities of developmentals trained as new hires with the success rates of those who transfer to a higher-level ATC facility after attaining CPC status at a lower-level facility.

National Training Database

Training outcomes for developmentals and CPC-ITs at FAA facilities are contained in the FAA's National Training Database (NTD; FAA, 2011). Researchers at the FAA's Civil Aerospace Medical Institute (CAMI) Aerospace Human Factors Research Division extract data from the NTD to develop a longitudinal ATC training database. Researchers regularly update and use the longitudinal ATC training database to respond to questions about air traffic controllers and to conduct human factors research to develop recommendations for improving controller placement and training practices.

Possible training outcomes that are stored in the NTD are as follows: *Completed, In Progress, Facility Fail, Transfer Lower, Transfer, and Separated – Other Reasons.* The outcomes of *Facility Fail* and *Transfer Lower* reflect unsuccessful completion of field qualification training. Developmentals coded as *Transfer Lower* failed field qualification training at their first facility but demonstrated the potential for being successful in training at a lower-level ATC facility, and thus were allowed to transfer to a less complex (lower-level) facility following FAA policies (FAA, 2013; Pierce, Byrne, & Manning, 2016). Records in the NTD allow analyses to be conducted based on training outcomes by option (Terminal or En Route) and by Terminal facility type and level. To determine which placement method produced the highest success rates in training at higher-level facilities, we compared the success rates in field training of new hires and CPC-ITs at medium- and high-level Terminal facilities. The type of Terminal facilities analyzed were Tower with Radar (Medium-level facilities = Levels 7–9 and High-level facilities = Levels 10–12) and Combined Tower with Radar and Terminal Radar Approach Control (Combined Tower/TRACON; Medium-level facilities = Levels 7–9 and High-level facilities = Levels 10–12). In general, Tower with Radar facilities rely more heavily on visual observation to control air traffic at one airport and Combined Tower/TRACON facilities rely on both visual observation and radar procedures to control air traffic at more than one airport (FAA, 2016). TRACON-only facilities were not included in the analysis because, in current practice, new hires no longer begin training at a TRACON-only facility as most of the facilities are high-level due to combinations of tower and TRACON facilities that occurred since 1995 and the relatively low number of low-volume TRACON facilities that remain. The placement of new hires at the remaining high-level TRACON facilities results in an exceptionally low success rate.

ATC Placement

Trainees with no prior experience in ATC attend the FAA ATC Academy to receive initial training that is germane for all facilities in either the En Route or Terminal option before receiving a facility assignment at which they receive site-specific training. The process for assigning ATC Academy graduates to a field facility has varied over time, but currently, graduating classes are offered a list of facilities from which to choose based on the current needs of the Air Traffic Organization (ATO). The number of facilities to be included on the list (facilities from which graduates in a class may choose) is based on the number of students in the class who successfully completed the Initial Qualification training course. The majority of classes begin with 18 trainees and graduation rates typically vary between 50% and 75%. Academy graduates are allowed to make their selections based on class rank. The trainee earning the highest overall point total in the initial qualification training course chooses first from among the facilities offered. Facility selection proceeds through the class in overall point rank order such that those whose scores rank them lower in the class have fewer options from which to choose. The list is generated by the FAA's ATO Management Services, Technical Requirements and Forecasting Group, Air Traffic Services Team (AJG-P21) and is based on the needs of the FAA to fill controller vacancies at specific facilities.

Management Services may use the results of the current research as input in developing policy for making field assignments for Academy graduates. This data-driven approach is in line with FAA efforts to improve safety and identify hazards and risks based on continuous analysis of data (FAA, Destination 2025) and the FAA's current strategic initiatives, Risk-Based Decision Making and Workforce of the Future.

Method

Database

From the longitudinal ATC training database, we extracted records for controllers who had trained at medium- and high-level Tower with Radar and Combined Tower/TRACON facilities as new hires from 2004 to 2015. Our sample included developmentals and CPC-ITs who had *Completed* training and were either *Successful* or *Unsuccessful* in training. Controllers with training outcomes of *Completed* were considered *Successful*. Controllers with training outcomes of *Facility Fail* or *Transfer Lower* were considered *Unsuccessful.*

We created two datasets. The first dataset included the training outcomes of developmentals at medium- (Levels 7–9) and high-level (Levels 10–12) Towers with Radar or Combined Tower/TRACONs as their first facility for training from 2004 to 2015. There were 1,997 records in the first dataset. Of those, 379 records were excluded because the developmental was still *In Progress* (n=176, 8.8%) and had not completed training or had *Transferred* (n=105, 5.3%) or left training for other reasons (*Other* (n=98, 4.9%)). To ensure independence of our groups, we excluded an additional 201 (12.4%) records because the new hires were also included in our second group of CPC-ITs. As shown in Table 1, there were 1,417 records remaining in the dataset. The number and percentage of developmentals categorized as either *Successful* or *Unsuccessful* in training are also shown.

Table 1.
Sample Characteristics for New Hires.

		Tower with Radar (7–9)	Tower with Radar (10–12)	Combined Tower/ TRACON (7–9)	Combined Tower/ TRACON (10–12)	Totals
Successful	Number	306	147	622	71	1,146
	Percent	(88.2)	(79.5)	(77.9)	(81.6)	(80.9)
Unsuccessful	Number	41	38	176	16	271
	Percent	(11.8)	(20.5)	(22.1)	(18.4)	(19.1)
Totals		347	185	798	87	1,417

The second dataset extracted from the longitudinal ATC training database was for the CPC-ITs, the comparison group. The CPC-IT group (n = 797) included controllers who were new hires at their first facility, had made CPC at that facility, and then transferred and began training at a second facility from 2004 to 2015. We only included records for those controllers who had transferred to a medium- or high-level Tower with Radar or Tower/TRACON facility after reaching CPC-IT at a lower level facility (of any type) and had completed training (Successfully or Unsuccessfully) at the second facility. We excluded 171 records with training outcomes listed as *In Progress* (n=139, 17.4%), *Transferred* (n=23, 2.9%), *or Other* (n=9, 1.1%). There were 626 records remaining in the CPC-IT dataset (see Table 2).

Table 2.
Sample Characteristics for CPC-ITs at a 2nd Facility.

		Tower with Radar (7–9)	Tower with Radar (10–12)	Combined Tower/ TRACON (7–9)	Combined Tower/ TRACON (10–12)	Totals
Successful	Number	149	163	203	44	559
	Percent	(93.7)	(86.2)	(89.4)	(86.3)	(89.3)
Unsuccessful	Number	10	26	24	7	67
	Percent	(6.3)	(13.8)	(10.6)	(13.7)	(10.7)
Totals		159	189	227	51	626

Procedure

To determine which group was more successful in training at medium- and high-level Tower with Radar and Combined Tower/TRACON facilities, the percentage of new hires who successfully completed training at medium- and high-level facilities was compared to the percentage of successful CPC-ITs at medium- and high level Terminal facilities of the same level.

Results

The percentage of successful new hires at a first facility and CPC-ITs at a second facility are shown by facility type and level grouping (Medium-Level 7–9 and High-Level 10–12) in Figure 1. Across all medium-level (7–9) and high-level (10–12) Tower with Radar and Combined Tower/TRACON facilities, 80.9% of the new hires were successful. The success rate for CPC-ITs at medium- and high-level Tower with Radar and Combined Tower/TRACON facilities was 89.3%.

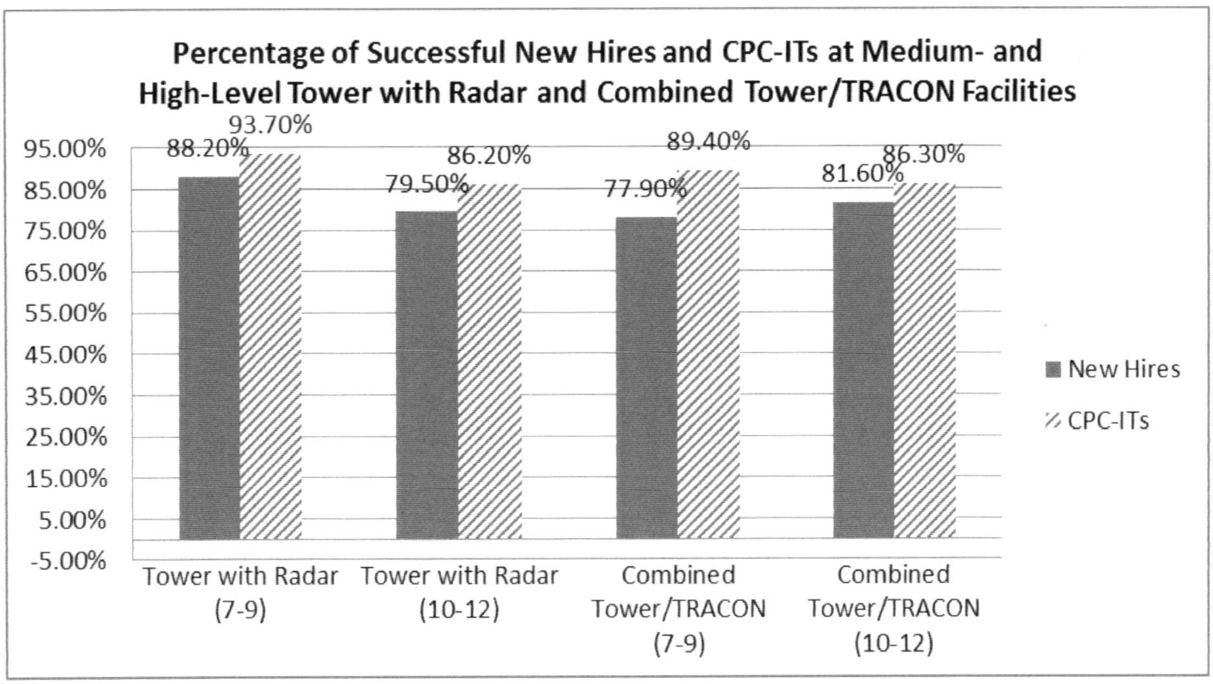

Figure 1.

We tested the significance of the difference using a Z-test to compare the proportion of successful developmentals and CPC-ITs at each facility type and level. We used the $p < .05$ value to determine if the difference between the two groups was statistically significant. The Z-test statistic and the p-value for each comparison are shown in Table 3. We found that the CPC-ITs assigned to medium-level Combined Tower/TRACON facilities completed training successfully at a significantly higher proportion than did new hires assigned to the same type and level facility. The difference between CPC-ITs and new hires at medium-level Tower with Radar facilities was marginally significant, but comparisons at high-level facilities, Tower with Radar and Combined Tower/TRACON, were non-significant.

Table 3.
Z-Test of the Difference Between Group Proportions.

	Tower with Radar (7–9)	Tower with Radar (10–12)	Combined Tower/ TRACON (7–9)	Combined Tower/ TRACON (10–12)
Z-score	-1.917	-1.742	-3.852	-0.710
p-value	.055	.081	.001	.478

Discussion

The FAA's Management Services, Technical Requirements and Forecasting Group, Air Traffic Services Team (AJG-P21) is responsible for technical workforce planning, prioritization, and hiring plan development for the ATO, as well as onboarding and placement of newly hired controllers. The average cost to train one developmental is approximately $139,207 per year, based on training costs reported from 2009 through 2013 (FAA, 2014b). On average, developmentals spend from 18 to 36 months in training, depending on facility type and level (FAA, 2014a; FAA, 2016). Thus, the FAA strives to place developmental controllers for field qualification training in ATC facilities in which they have the highest probability of success.

The results of the current effort clearly indicate that success rates at medium-level Combined Tower/TRACON facilities can be increased by staffing those facilities with CPC-ITs. The time for CPC-ITs to certify at a medium-level Combined Tower/TRACON facility is approximately 1 year (FAA, 2014a). Thus, the overall benefit to certification rates at medium-level Combined Tower/TRACON facilities of having new hires certify at a lower-level facility prior to transferring may be slightly diminished by higher training costs.

While the other comparisons were not statistically significant, the practical importance of the results is worth considering. More CPC-ITs were successful than new hires at the same type and level facility. While our results do not allow us to predict that there will continue to be a difference in success rates at these facilities, we have no reason to believe that the differences will not continue. It is likely that the small number of controllers, especially the CPC-ITs trained at the higher-level facilities, as well as the dichotomous outcome measure used in these analyses influenced our inability to find a significant difference. Future research will need to increase the sample size and consider other outcome measures to verify these findings. However, we believe that this information is useful in developing future practices and policies in ATC placement and training.

Limitations

Although we constrained our groups to developmentals and CPC-ITs at two facility types (Tower with Radar and Combined Tower/TRACON) there are approximately 130 independent facilities in each group. It is possible that variability in training methods at the facilities and differences in training methods used over time could differentially affect eventual training outcomes. A second limitation is the number of developmentals and CPC-ITs *In Progress,* who were excluded from the assessment. Seventeen percent of the CPC-ITs in the group assessed were still *In Progress* and due to the time needed to complete training, the majority of the developmentals and CPC-ITs *In Progress* and excluded from the assessment were from the 2014-2015 timeframe, which may also have differentially affected the results. We recommend updating the assessment as the developmentals and CPC-ITs currently *In Progress* complete training. We further recommend evaluating the total cost to achieve training success in field facilities including a comparison of the cost of training developmentals at higher-level facilities as compared to the recommended path of requiring developmentals to be trained to initial CPC at lower-level facilities followed by training them as CPC-ITs to achieve CPC at higher-level facilities.

Acknowledgments

Research reported in this paper was conducted under the Air Traffic Program Directive/Level of Effort Agreement between the Human Factors Division (ANG-C1), FAA Headquarters, and the Aerospace Human Factors Research Division (AAM-500) of the Civil Aerospace Medical Institute.

References

Federal Aviation Administration (2011, May). *National Training Database Guidelines.* Washington, DC: FAA.

Federal Aviation Administration (2013). Human Resources Policy Manual, EMP-1.14a (ATCS) Employment Policy for Air Traffic Control Specialists in Training. Washington, DC: FAA.

Federal Aviation Administration (2014a). *Air Traffic Control Specialist New Hire Training Performance Semi-Annual Report.* Washington, DC; FAA.

Federal Aviation Administration. (2014b). *Post-hearing Questions for the Record submitted to Ms. Patricia McNall from Senator Claire McCaskill.* Hearings before the Subcommittee on Financial and Contracting Over-sight of the Senate Committee on Homeland Security and Government Affairs, 113th Congress, 2nd Session, January 14, 2014. Retrieved from http://www.hsgac.senate.gov/download/?id=2B7 09C2C-DB78-4B00-9B7F-325B940B8EF7

Federal Aviation Administration (2015). *JO 3120.4P Air Traffic Technical Training*. Washington, DC: FAA.

Federal Aviation Administration (2016). *A plan for the future: 10-year strategy for the air traffic control workforce 2016-2025.* Washington, DC: FAA.

Federal Aviation Administration. (n.d.). *Destination 2025.* Retrieved from https://www.faa.gov/about/plans_reports/media/Destination2025.pdf.

Pierce, L.G., Byrne, C.L., & Manning, C.A. (2016). *An examination of the success rates of developmental controllers transferring to lower level facilities* (Report No. DOT/FAA/AM-16/2). Washington, DC: FAA Office of Aerospace Medicine.

PREDICTING AIR TRAFFIC CONTROL STUDENT SUCCESS AND IMPROVING PLACEMENT THROUGH SKILL APTITUDE TESTING

Kerry Hutson
The MITRE Corporation
McLean, VA
Keith Baldwin
The MITRE Corporation
McLean, VA
Susan Schultheis
The MITRE Corporation
McLean, VA

In order to improve Air Traffic Control (ATC) training efficiency and reduce the risk of student failure, the FAA needs tools that can predict student success and inform student placement in training. By considering student aptitude, the FAA could reduce training costs by reducing student failure rates and transfers between facilities. In order to explore the benefits of early aptitude assessment, The MITRE Corporation (MITRE) created a prototype Radar Vectoring Aptitude Test designed to be administered to students before they begin their training to assess their aptitude for the skill of radar vectoring. The FAA office of Safety and Technical Training (AJI-2) in Air Traffic Operations (ATO), the FAA Civil Aerospace Medical Institute (CAMI), and MITRE are working in collaboration to evaluate the prototype with trainees at the FAA Academy.

Introduction

The Federal Aviation Administration (FAA) plans to hire more than 6,300 air traffic controllers over the next 5 years (FAA, 2015) and the need to improve training to increase efficiency and effectiveness have been well documented (Barr, Brady, Koleszar, New, and Pounds, 2011; Hutson, et al., 2014). A number of factors contribute to inefficiency in Air Traffic Control (ATC) training including imperfect assessment of student aptitude early in the training process. After a candidate is hired by the FAA, they attend the FAA Academy at the Mike Monroney Aeronautical Center in Oklahoma City, OK. At that time, they are assigned a training option (en route or terminal). The FAA bases assignments primarily on facility staffing needs but students may indicate an interest in a particular facility or geographical preference. At the completion of their training, top performers at the FAA Academy may be offered a choice among openings at various facilities. Towers, TRACONS and en route facilities differ in operations, complexity, and required skills (Pierce, et al., 2016). Because student assignment to facilities is not based on an assessment of student strengths and aptitude for the unique skills required by different facility types, there can be a mismatch between student aptitude and facility placement. This mismatch can contribute to students failing to complete training at their first assigned facility. The FAA has a need for enhanced tools that can predict student success in training and help place students in either an en route or terminal environment based on an assessment of their aptitude for specific fundamental ATC skills. By using aptitude assessment to inform student placement in training, the FAA could possibly decrease training cost and risk;

more appropriate student placement in training could reduce the overall amount of time students spend in training and the student failure rate.

In order to explore the validity and benefits of early aptitude assessment, The MITRE Corporation (MITRE) created a prototype Radar Vectoring Aptitude Test capability. Radar vectoring is one of the critical tasks performed by terminal and en route controllers to ensure safe separation, to space aircraft, to sequence traffic, and to facilitate the efficient flow of traffic. The aptitude test is designed to be administered to students before they begin their training at the FAA Academy to assess their aptitude for the skills required for efficient and effective radar vectoring. The prototype is currently being evaluated with developmental trainees at the FAA Academy (referred to as students throughout this report) at the beginning of their training. Those students will be followed over the course of their training so that the relationship between aptitude test performance, training performance, and, ultimately, their success in achieving certification as an Air Traffic Controller can be assessed. If the evaluation indicates a relationship between aptitude test performance and student success in training, then additional skill aptitude tests, beyond vectoring, could aid the FAA in predicting student success and further support student placement in training.

Radar Vector Aptitude Test Prototype Description

The following is an overview description of the Radar Vector Aptitude Test Prototype. Radar vectoring is one of the critical tasks performed by terminal and en route controllers. The knowledge, skills, abilities, and other characteristics (KSAOs) necessary to perform an air traffic controller's job have been determined and documented by the American Institute of Research (AIR) (Krokos, et al., 2011; Krokos, et al., 2011; Krokos, et al., 2011). Throughout this document, the term aptitude is used to encompass the set of KSAOs that are needed to successfully perform the task of radar vectoring. The Radar Vector Aptitude Test prototype is designed to objectively assess those KSAOs. Specifically, the Radar Vector Aptitude Test will assess student aptitude for the following:

- Basic and advanced compass use
- Phraseology for issuing a vector clearance and oral communication
- Interpreting a data block
- Vectoring Skill
- Scanning, Prioritization, and Planning
- Situation Awareness
- Tolerance for increased/high workload

The aptitude test is composed of 4 sections. Each section has multiple subtests, allowing for varying levels of difficulty and an opportunity to adequately test for basic knowledge and skill. Each subsequent section is more difficult than the last and later sections are designed to assess more operationally comprehensive aptitudes and skills. The test takes approximately 3 hours to complete. Additionally, there is a participant survey at the end of the test. The survey captures feedback on the completeness of instructions and practice allowed, as well as data about previous student knowledge.

Radar Vector Aptitude Test Sections

The goal of Section 1 is to familiarize the student with the prototype and to test the student's understanding or knowledge of the compass, data block, and phraseology, as well as their aptitude to use this knowledge to effectively vector aircraft. Section 1 consists of three tests: ***Practice Test, Basic Compass Test,*** and ***Advanced Compass Test***. Figure 1 is a screen shot from the ***Practice Test:***

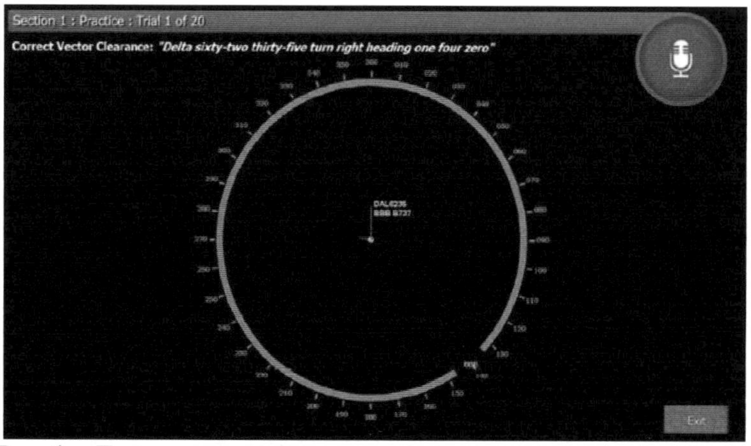

Figure 1. Section 1 Practice Test

The goal of Section 2 is to test the student's aptitude for compass use and proper phraseology in order to vector more than one aircraft at a time. Additionally, since multiple aircraft are moving, situation awareness, prioritization, planning, scanning, and oral communication are also assessed in Section 2. There are two tests in Section 2, the ***Basic Shapes Test*** and the ***Advanced Shape Test.*** Figure 2 is screen shot of the ***Advanced Shape Test***:

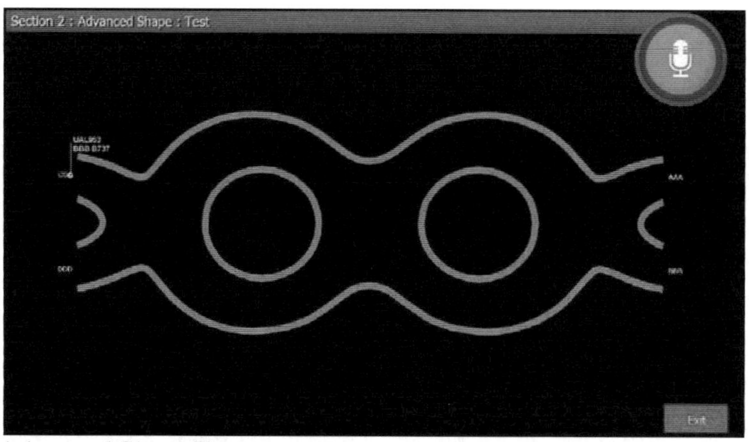

Figure 2. Section 2 Advanced Shape Test

The goal of Section 3 is to continue assessing a student's vectoring skills, as well as situation awareness, prioritization, and scanning. The following tests are administered in Section 3: ***Simple Shape Test, Scenario 1 Test,*** and ***Scenario 2 Test.*** Figure 3 is screen shot of the ***Scenario 1 Test***:

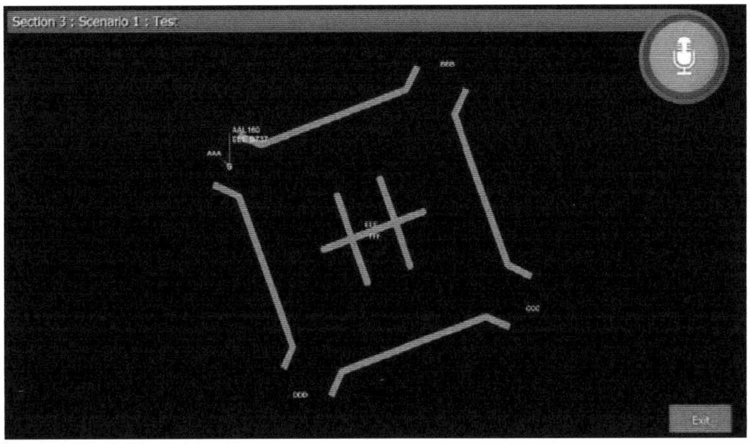

Figure 3. Section 3 Scenario 1 Test

The goal of Section 4, which is the most complex section, is to continue assessing a student's vectoring skills, as well as situation awareness, prioritization, and scanning in more complicated situations, all with an increased workload. Section 4 consists of 2 tests, ***Scenario 3 Test*** and ***Scenario 4 Test***, which aid in determining a student's ability to vector in a terminal and en route environment. Figure 4 is a screen shot from the ***Scenario 3 Test.***

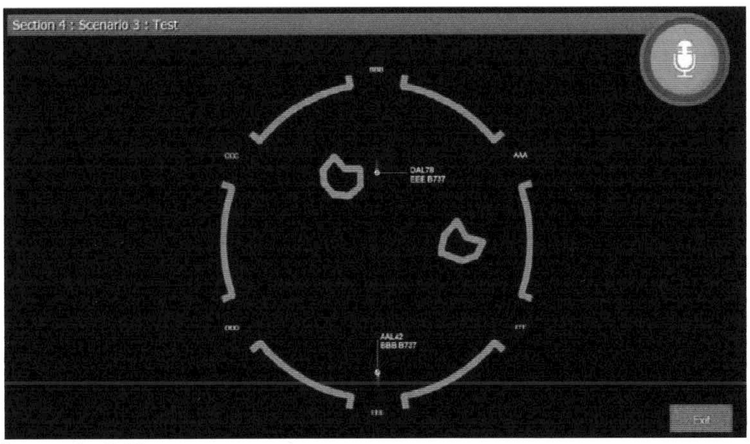

Figure 1. Section 4 Scenario 3 Test

Evalution Conduct

These next two sections present an overview of the evaluation conduct and data analysis. The evaluation is being conducted in partnership with CAMI in their lab facilities in Oklahoma City, OK. Before each test, A CAMI Principal Investigator (PI) gives students a 30-minute overview of the purpose and specifics of the evaluation. A CAMI appointed proctor is present to monitor system performance and student usage. Additionally, the students complete a demographics questionnaire. The students are asked to sign a voluntary consent form that describes the purpose, goals, risks, benefits, voluntary nature, and data collection/storage procedures of the study.

Data Collection and Assessment

MITRE and CAMI have analyzed the data to ensure that there is enough variation between students' scores, verifying that the test is neither too hard nor too easy and validating that the test is capturing differences in student aptitude. Using the first six months of data collection (N = 594), analyses are being conducted to determine which metrics generated by the aptitude test software will be used to create an overall scoring algorithm. These will be completed using correlation and regression methods with FAA Academy performance as the predicted outcome variable. In other words, each metric available will be examined separately and in combination with the other metrics to create a usable and predictive score.

Data from the ***Basic*** and ***Advanced Compass Tests*** will be studied to determine student success based on the number of correct vector clearances the student issued. Data from the ***Shapes*** and ***Scenario Tests*** will be used to calculate how many aircraft students successfully guided to the destination gate. For those aircraft that exited successfully, the number of vectors issued will be tallied. Additionally, data such as how many aircraft exited incorrect gates, how many times did the countdown clock reach zero, and how many times did the system respond to a student issued clearance with "say again" will be reviewed. Further analysis may include a determination of how far off incorrect vectors were and calculations of how close students came to the boundary. For shapes that included multiple aircraft, data may be studied to see in cases where the first aircraft failed to successfully reach the exit gate (i.e. the aircraft hit the boundary), were adjustments made to successfully vector subsequent aircraft? Data collected from later scenario tests, such as students' ability to control multiple aircraft and reaction to gate changes and moving objects may be indicators of a tolerance for higher workload and situation awareness. Data will also be examined to determine if students' performance improved over the course of the test, possibly indicative of aptitude.

The data collected from January 2017 – June 2017 (approximate sample size of 732) will be used to cross-validate the scoring algorithm determined using the data from first six months of the evaluation and make modifications, if needed. Then combining all data collected (N = 1,326), the overall predictive validity, utility, and fairness of the aptitude test for placement purposes will be evaluated. Utility will be assessed by comparing the cross-tabulations of actual placements versus indicated test placements. If those who would have been placed, based on the placement indicated by the Radar Vectoring Aptitude Test prototype score, into the option to which they were actually placed succeed at a higher rate than those who were placed in a different option than the one indicated by the aptitude test placement, then it is possible that the aptitude test will help increase the pass rate at the FAA Academy and be operationally useful. Fairness, as defined by the U.S. Equal Employment Opportunity Commission's Uniform Guidelines on Employee Selection Procedures, will be assessed to determine likelihood of adverse impact against protected groups ("Adoption of Questions and Answers to Clarify and Provide a Common Interpretation of the Uniform Guidelines on Employee Selection Procedures", 1979.)

Next Steps

MITRE, CAMI, and AJI-2 will continue to collaborate on the prototype evaluation and validation. MITRE and CAMI will continue the data analysis with data collected from students through FY2017 to first determine an overall Radar Vectoring Aptitude score that will be used to predict performance. A report on the results of the first year of the evaluation will be delivered to the FAA at the end of FY17. Also in FY17 detailed plans for the longitudinal study of students' performance will be developed in order to assess the test's ability to predict field training performance. The students will be followed over the course of their training to assess the relationships between score, training performance at the FAA Academy and at their first facility and, ultimately, their achievement of CPC status.

References

"Adoption of Questions and Answers to Clarify and Provide a Common Interpretation of the Uniform Guidelines on Employee Selection Procedures". 1979. [Online]. Available: https://www.eeoc.gov/policy/docs/qanda_clarify_procedures.html. [Accessed] September 23, 2016

Barr, M., Brady, T., Koleszar, G., New, M., & Pounds, J. 2011, FAA Independent Review Panel on the Selection, Assignment and Training of Air Traffic Control Specialists, *Independent Panel Review Report*, Washington, DC.

FAA, 2015, A Plan for the Future, 10-Year Strategy for the Air Traffic Control Workforce 2015-2024, Washington, DC.

Hutson, K., et al., 2014, Technical Training Reports 2008 – 2014: Summarization and Recommendations, McLean, VA: The MITRE Corporation

Krokos, K. J., et al., 2011, *Job Description for the NextGen Mid-Term ARTCC Controller,* Appendix A, American Institutes for Research, Washington, DC.

Krokos, K. J., et al., 2011, *Job Description for the NextGen Mid-Term TRACON Controller,* Appendix A, American Institutes for Research, Washington, DC.

Krokos, K. J., et al., 2011, *Job Description for the NextGen Mid-Term ATCT Controller,* Appendix A, American Institutes for Research, Washington, DC.

Pierce, L.G. et al., 2016, An Examination of the Success of Developmentals Air Traffic Controllers Transferring to Lower-Level Facilities, Washington, DC.

ATTRITION IN U.S. AIR TRAFFIC CONTROL SPECIALIST (ATCS) TRAINING: A REVIEW OF 50 YEARS OF DATA

Dana Broach
FAA Civil Aerospace Medical Institute
Oklahoma City, Oklahoma, USA

Aptitude testing and "screening" at the FAA Academy have been viewed as keys to reducing field ATCS training attrition. To what extent have ATCS field training attrition rates changed over time with testing and screening? Historical data on training outcomes were extracted from FAA reports and other documented sources for controllers hired in five non-overlapping cohorts spanning 50 years. *Academy Attrition Rate* averaged 26% (SD=18%) over the 50 years and across options, compared to 25% (SD=4%) in field training. Lower *Field Training Attrition Rates* coincided with no screening (22%, 1968–1970) and intensive screening (19%, 1986–1992). Elimination of screening did not result in an increase in the En Route *Field Training Attrition Rate* in 2005–2010, but the Terminal *Field Training Attrition Rate* tripled (29%) from the 1986–1992 low of 9%. The lack of a consistent pattern suggests that field training itself warrants investigation to better understand the dynamics of attrition.

Air Traffic Control Specialist (ATCS) training in the U.S. averages two to three years to achieve Certified Professional Controller (CPC) status. Attrition in that expensive and extensive training has long been a concern for the Federal Aviation Administration (FAA) and its stakeholders. For example, researchers in 1960 noted that "[P]roviding training for those employees who eventually will either drop out or be washed out of the training program has become expensive in time and money. The current attrition rate indicated that has become a serious problem" (Davis, Kerle, Silvestro, & Wallace, 1960). Similar concerns were noted in the 1970 report of the Air Traffic Controller Career Committee (Corson, Bernhard, Catterson, Fleming, Lewis, Mitchell, & Ruttenberg, 1970). High field attrition rates caught the attention of the U.S. Congress in 1975, resulting in a hearing and Congressional recommendations on how to reduce such losses (*Selection and Training of FAA Air Traffic Controllers*, 1975). Training attrition rates were a significant concern all through the 1980s as the FAA rebuilt the controller workforce following the 1981 strike by the Professional Air Traffic Controller Organization (PATCO) (U.S. General Accounting Office, 1986, 1987).

A central idea over the past 50 years is that training attrition rates can be "...solved or reduced by developing a realistic selection program for controllers" (Davis, et al.). Similarly, the U.S. Congress found in 1975 that the "[T]he selection process for admission to the ATC program is inadequate to predict with reasonable accuracy the selectees' potential for successfully completing the training program" (*Selection and Training of FAA Air Traffic Controllers*). From 1961 through 1975, the FAA used a one-stage selection process based on prior experience and education and placed new controllers into Academy training. Aptitude testing was incorporated into the selection process in 1963. Initial training was conducted at the FAA Academy on a pass/fail basis but without any explicit intention to eliminate or "screen out" new controllers. From 1976 through 1992, the FAA used a two-stage selection process with the expectation of a lower attrition rate in field training. The first stage was aptitude testing of applicants. The second stage of selection was "screening" at the FAA Academy where the explicit intent was to "screen out" those new controllers unlikely to succeed in field training. The "screening" component was incorporated into FAA Academy training in 1976 at the specific direction of the U.S. Congress (see the recommendations in *Selection and Training of FAA air traffic controllers*, 1975) and was especially prominent during the post-strike recovery period (see Broach, 1998). In the period 2005 to 2010, FAA reverted to a one-stage selection process based on a computerized aptitude test battery and training at the FAA Academy was conducted on a "pass/pass" basis. The question addressed in this review is to what degree attrition rates in field training varied over this 50-year interval (1960–2010) as the controller selection process changed. It is important to note that the training for each cohort reflected the technology, procedures, and traffic of that time period.

Method

Historical data on selection and training were extracted from FAA reports and databases maintained for research purposes at the FAA's Civil Aerospace Medical Institute (CAMI) for five non-overlapping cohorts (Table 1). Attrition rates in Academy and new hire field training were calculated from these primary sources. Descriptions of the selection process used for each cohort and FAA Academy training programs (Table 2) were also extracted from these and other sources such as training documentation.

Table 1
Primary ATCS data sources by cohort

Cohort	N	Source
1960 – 1963	1,741	Cobb, et al. 1972
1968 –1970	4,094	Cobb, et al, 1972
1981 –1985	13,533	CAMI Post-Strike ATCS Tracking Database
1986 –1992	14,392	CAMI Post-Strike ATCS Tracking Database
2005 – 2010	6,158	CAMI Next Generation ATCS Tracking Database

Attrition and retention rates were computed as follows for each cohort. Attrition from FAA Academy training (*Academy Attrition Rate*) was computed as the ratio of Academy losses (failures and withdrawals) to total entrants into the Academy for a given cohort. Attrition in field training (*Field Training Attrition Rate*) was computed as the ratio of losses from new hire field training (excluding deaths) to the number of persons (developmentals) entering new hire field training after completion of the FAA Academy. Total attrition (*Net Attrition Rate*) was computed as the ratio of the sum of Academy and field training losses to the total number of entrants into the Academy at the start of the training process. Persons with prior ATC experience hired at higher grade levels and placed directly into field ATC facilities (bypassing the FAA Academy) were excluded from this analysis of attrition rates.

Results

Historical Academy, field training, and total attrition data and rates by option and combined are presented in Table 3 for persons hired into the FAA Academy by year (or time period) and cohort. The combined (both options) Academy and field training attrition rates are illustrated in Figure 1.

Table 2
Summary description of selection and Academy training 1960–2010

Cohort	Selection	Academy
1960–1963	Prior education & experience No aptitude testing No maximum age at entry	By option, pass/fail 8-weeks (for both options) No explicit screening
1968–1970	Prior education & experience Aptitude testing for GS-5/7 No maximum age at entry	By option, pass/fail 8.5 weeks No explicit screening
1981–1985	Aptitude testing (OPM test) Top-down hiring based on score Maximum age at entry of 31	By option, pass/fail 11 weeks En Route (Fundamentals & Non-radar) 15 weeks Terminal (Fundamentals, Tower, Non-radar) Explicit screening
1986–1992	Aptitude testing (OPM test) Top-down hiring based on score Maximum age at entry of 31	Combined, pass/fail 9 weeks Explicit screening
2005–2010	Aptitude testing (AT-SAT) Hiring based on score bands (Qualified, Well Qualified, determined by AT-SAT score) Maximum age at entry of 31	By option, pass/pass 17 weeks En Route (Basics, En Route) 13 weeks Terminal (Basics, Tower) No explicit screening

Inspection of the data in Table 3 and as illustrated in Figure 1 suggests that the *Field Training Attrition Rate* varied less across time than did the *Academy Attrition Rate*. The *Academy Attrition Rate* spiked at over 40% for the 1981–1985 and 1986–1992 cohorts hired after the 1981 PATCO strike. In contrast, the *Field Training Attrition Rate* is flatter across years, varying 19 to 32% across both options and cohorts, even during the post-strike recovery period. The attrition rates by option (Table 3) follow the same pattern with large variations in *Academy Attrition Rate* as second-stage "screening" was introduced for the post-strike cohorts and then eliminated for the 2005–2010 cohort.

One might expect that removal of the "screening" component of the Academy training program might result in a higher *Field Training Attrition Rate* in subsequent years. But as shown in Figure 1 (and in Table 3 by option), the combined *Field Training Attrition Rate* did not dramatically increase for the 2005–2010 cohort following removal of the "screening" element in FAA Academy training. The En Route *Field Training Attrition Rate* for 2005–2010 (28%) is very comparable to the 1986–1992 En Route *Field Training Attrition Rate* of 27%. However, the Terminal *Field Training Attrition Rate* for the 2005–2010 cohort of 29% is approximately triple the 1986–1992 Terminal attrition rate of 9%. The increase in Terminal *Field Training Attrition Rate* might be attributable to the elimination of "screening" at the FAA Academy. However, other explanations such as changes in new hire aptitude, prior ATC experience and education and changes in field training rigor might be possible and should be evaluated.

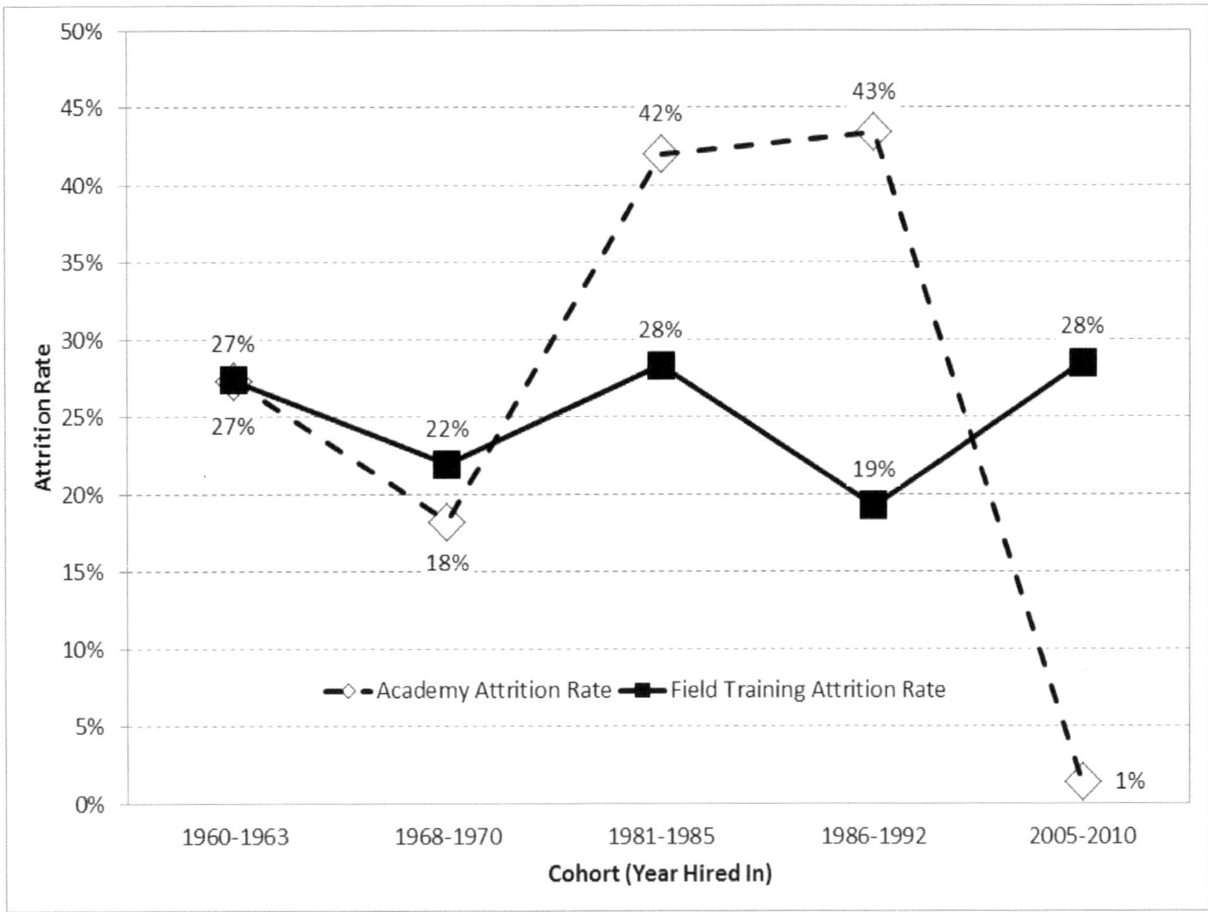

Figure 1. Academy and field training attrition rates for combined options by cohort

Discussion

Attrition in field training is a significant and persistent concern for the FAA and its stakeholders. For many years, "better" first-stage selection and explicit second-stage "screening" at the FAA Academy were held out as key methods for reducing field attrition. One might expect field training attrition rates to decrease with the introduction of second-stage "screening" at the FAA Academy over time. No such consistent decline is apparent. On the other hand, elimination of second-stage "screening" might be expected to result in higher field training attrition rates. This seems to be the case in the Terminal but not in En Route option for the 2005–2010 cohort.

While a selection process is needed for practical (and legal) reasons, it does not appear that first-stage selection and second-stage screening will *necessarily* reduce the new hire *Field Training Attrition Rate*. Rather, the relationship appears to be complex, and organizational circumstances, candidate characteristics, technology, and traffic might play significant roles. Furthermore, while selection and screening processes are reasonably well documented, the field training process itself is less well documented. Field training is conducted within a framework established by the ATCS technical training order (FAA, 2015) but is necessarily decentralized and facility-specific. Future research should explore in greater depth historical circumstances for each cohort and the interplay between selection, screening, and especially the field training process itself to better understand the dynamics of controller attrition.

Table 3

Historical ATCS hire, attrition, and retention data by cohort and option

Cohort[1]	N Enter Academy[2]	N Academy Loss	Academy Attrition Rate	N Academy Pass	N to Field[3]	N Field Training Loss	Field Training Attrition Rate	N Retentions[4]	N Losses	Net Retention Rate	Net Attrition Rate
					En Route Option						
60–62	1,008	323	23%	685	685	229	33%	456	552	45%	55%
68–70	3,159	565	18%	2,594	2,594	640	25%	1,954	1,205	62%	38%
81–85	8,536	4,073	48%	4,463	4,461	1,629	37%	2,832	5,702	34%	66%
86–92					4,732	1,237	26%	3,495			
05–10	2,753	49	2%	2,704	2,704	763	28%	1,941	812	71%	29%
					Terminal Option						
60–63	733	153	21%	580	580	117	20%	463	270	63%	37%
68–70	935	180	19%	755	755	94	12%	661	274	71%	29%
81–85	4,997	1,607	32%	3,390	3,384	590	17%	2,794	2,198	56%	44%
86–92					3,298	308	9%	2,990			
05–10	3,405	35	1%	3,370	3,370	967	29%	2,403	1,002	71%	29%
					Combined Options						
60–63	1,741	476	27%	1,265	1,265	346	27%	919	822	53%	47%
68–70	4,094	745	18%	3,349	3,349	734	22%	2,615	1,479	64%	36%
81–85	13,533	5,680	42%	7,853	7,844	2,478	32%	5,373	8,160	40%	60%
86–92	14,392	6,243	43%	8,149	8,030	1,545	19%	6,485	7,788	45%	54%
05–10	6,158	84	1%	6,074	6,074	1,730	28%	4,344	1,814	71%	29%

Notes: [1]60–62=1960–1962; 68–70=1968–1970; 81–85=1981–1985; 86–92=1986–1992; 05–10=2005–2010

[2]Hires into FAA Academy only, excludes hires direct to facilities; Losses are withdrawals and failures

[3]Numbers passing Academy and number reporting to field facilities are sometimes less due to no shows at the facility. "No shows" are not included in the calculation of field training attrition and net retention and loss rates

[4]Number of retentions (achieved Full Performance Level or Certified Professional Controller or still in training to be consistent with Cobb, et al., 1972) at 1st facility only; Losses are those that failed or transferred before completing field training at the 1st facility (excluding only deaths)

References

Broach, D. (Ed.). (1998). *Recovery of the FAA air traffic control specialist workforce, 1981-1992* (pp. 23–32). (Report No. DOT/FAA/AM-98/25). Washington, DC: Federal Aviation Administration Office of Aviation Medicine.

Cobb, B.B., Mathews, J.J., & Nelson, P.L. (1972, November). *Attrition-retention rates of air traffic control trainees recruited during 1960–1963 and 1968–1970.* (Report No. DOT/FAA/AM-72/33). Washington, DC: Federal Aviation Administration Office of Aviation Medicine.

Corson, J.J., Bernhard, P.W., Catterson, A.D., Fleming, R.W., Lewis, A.D., Mitchell, J.M., & Ruttenberg, S.H. (1970, January). *The career of the air traffic controller – A course of action.* (Report of the Air Traffic Controller Career Committee.) Washington, DC: U.S. Department of Transportation, Office of the Secretary of Transportation.

Davis, C. G., Kerle, R. H., Silvestro, A. W., & Wallace, W. H. (1960, March). *The air traffic control training program as viewed by training supervisors.* (Technical report No. 33, Project O under FAA contract FAA/BRD-40). Philadelphia, PA: Courtney and Company.

Federal Aviation Administration. (2015). *Air Traffic Technical Training* (FAA Joint Order No. 3120.24P). Washington, DC: Author. https://employees.faa.gov/tools_resources/orders_notices/index.cfm/go/document.information/documentID/1028512

Selection and training of FAA air traffic controllers. Hearing before the House Government Activities and Transportation Subcommittee of the Committee on Government Operations, First Session, 94th Congress (June 21, 1975).

U.S. General Accounting Office (GAO). (1986). *Aviation safety: Serious problems concerning the air traffic control workforce.* (RCED-86-121). Washington, DC: Author.

U.S. General Accounting Office (GAO). (1987). *FAA air traffic controller staffing and related issues.* (Report No. GAO/T-RCED-87-42). Washington, DC: Author.

Acknowledgments

Research reported in this paper was conducted under the Air Traffic/Technical Operations Program Directive/Level of Effort Agreement between the NextGen Human Factors Research Division (ANG-C1), FAA Headquarters, and the Aerospace Human Factors Research Division (AA-500) of the FAA Civil Aerospace Medical Institute. The opinions expressed are those of the author alone and do not necessarily reflect those of the Federal Aviation Administration, the U.S. Department of Transportation, or the government of the United States of America.

ENGINEERING FOR HUMANS: A NEW EXTENSION TO SYSTEMS THEORETIC PROCESS ANALYSIS

Megan France & John Thomas
Massachusetts Institute of Technology
Cambridge, MA

Systems Theoretic Process Analysis (STPA) is a new hazard analysis method developed at MIT to address a broad range of accident causal factors including dysfunctional interactions among components, design flaws, and requirements problems. This paper presents a new extension for analyzing human interactions with automation and understanding why unsafe behaviors may appear appropriate in the operational context. The extension is demonstrated by applying it to pilot control of aircraft pitch control during stall recovery using scenarios from the Air France 447 accident.

On May 31, 2009, Air France flight 447 was scheduled to fly to Paris from Rio de Janeiro. Tragically, the A330's pitot tubes became clogged with ice during the transatlantic flight, causing inconsistent airspeed indications and the disconnection of the autopilot system. These events in the cockpit, combined with environmental conditions, led to intense pilot confusion. The ensuing interactions between the pilots and the aircraft sent the plane into an aerodynamic stall, which went undetected until the plane plunged into the ocean. Two hundred sixteen passengers and twelve crewmembers were killed.

Following the accident, the French Accident Investigation Bureau (BEA, 2012) released a thorough investigation into the causes of the accident, including a human factors perspective into the pilots' behavior. This analysis provided the aviation industry with valuable information; however, it cannot undo the tragedy that occurred. In order to effectively prevent future accidents, it is necessary to perform both hazard analyses and human factors investigations during design and early development. In this paper, we demonstrate a method for incorporating human factors into the hazard analysis process by expanding upon an existing technique.

Systems-Theoretic Process Analysis (STPA) is a hazard analaysis technique designed to capture not only accidents which result from component failures, but also accidents which result from design flaws and unsafe interactions (Leveson, 2012). STPA is well-suited for analyzing complex systems, but it does not provide guidance specific to humans. A new extension to the STPA method, "Engineering for Humans", was recently developed to provide guidance early in the design process and address human interactions in the system (Thomas and France, 2016). This paper demonstrates how the new extension can be applied in an aviation context to understand pilot behavior. To demonstrate the relevance of this method to the aviation domain, we apply STPA to the process of aircraft pitch control during an aerodynamic stall and show how the Engineering for Humans extension could be used to identify factors involved in the fatal Air France 447 accident.

Systems Theoretic Process Analysis

STPA begins with the identification of relevant high-level *system accidents:* any undesired or unplanned event that results in a loss. Next, high-level *system hazards* are identified: the system states or set of conditions that, together with a particular worst-case environmental conditions, will lead to an accident (loss). For example "aircraft collides with terrain" is the system accident we will be concerned with in this paper, and the system hazard which could lead to that accident is "loss of lift during flight," which may occur due to inadequate speed or excessive angle of attack.

Once the system accidents and hazards are identified, the analyst must draw the *safety control structure.* Figure 1 below shows the safety control structure for this system, which includes the control actions and feedback between entities in the system.

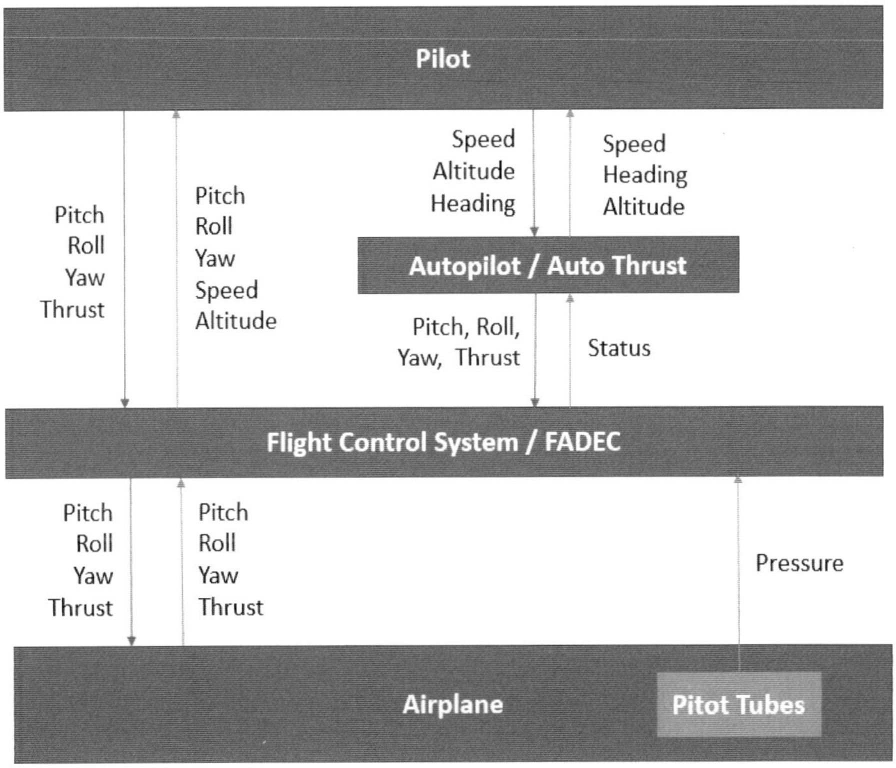

Figure 1.
Safety control structure for pilot interactions with aircraft. Note that for the purposes of this paper, a simplified and abstracted depiction of basic pilot controls can be used.

STPA has two main analysis steps. The first step examines how each control action in the system could cause a hazard. These *unsafe control actions* or UCAs, shown in Table 1, must fall under one of the four categories included as column headings.

Table 1.
Examples of Unsafe Control Actions (UCAs) for the Control Action "Increase Pitch."

Control Action	Not Providing Control Action Causes Hazard	Providing Control Action Causes Hazard	Wrong Timing or Order	Stopped Too Soon or Applied Too Long
Increase Pitch	UCA-1: PF does not increase pitch when aircraft is at risk of collision with terrain.	UCA-2: PF increases pitch while the aircraft is in a stall or approaching a stall.	-	UCA-3: PF increases pitch, but stops too soon before reaching the target pitch. UCA-4: PF continues to increase pitch too long when doing so exceeds the safe flight envelope.

Notes. "PF" refers to the pilot flying. UCA-2 is used for additional examples later in this paper.

The second step of STPA is where causal scenarios related to the UCAs are identified. The new STPA extension, Engineering for Humans, provides a process to anticipate and explain why humans might provide these unsafe control actions. This process is summarized in the next section.

Method

The new extension uses the human controller model depicted in Figure 2 below.

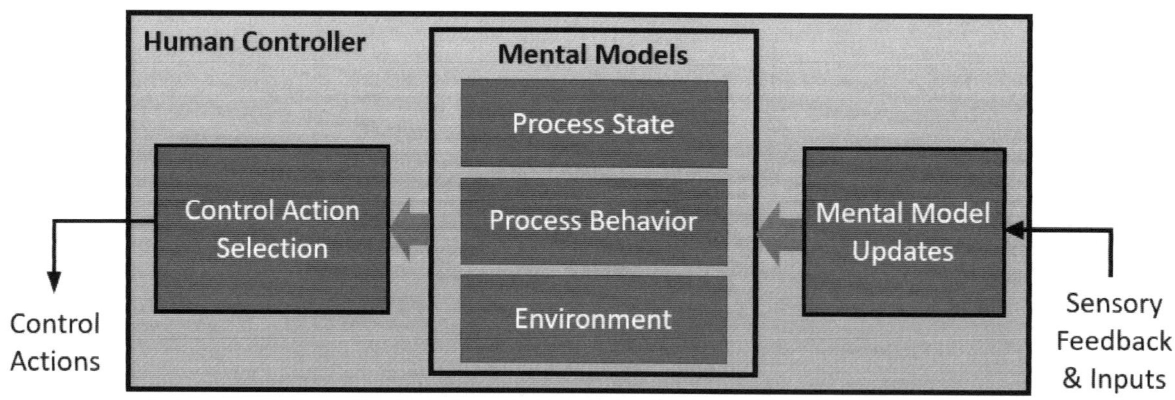

Figure 2.
Extended model of the human controller.

In this model, the *Control Action Selection* stage explains why a particular control action may be chosen by considering factors such as the operator's goals and other tasks that may compete for priority. Whether an action is skill, rule, or knowledge-based (Rasmussen, 1982) is also an important aspect of this stage of the model.

The *Mental Models* stage captures various human beliefs about the outside world. First, the *Mental Model of Process State* reflects the operator's awareness of software modes and the current state of operation. Incorrect mental models of process state may result from automatic mode changes, or progression of a controlled process to the next stage without feedback to the operator. The *Mental Model of Process Behavior* describes the operator's expectations for how the system will behave in a particular mode or stage of operations, and includes cause and effect relationships between the operator's actions and the system's behavior. Lastly, the *Mental Model of the Environment* includes factors outside the operator's control, including the behavior of other controllers and the novelty or variability of the environment.

Finally, this model requires the analyst to consider the source of mental models, including both how they are formed and how they are updated in response to change. Factors such as the salience of the change and the operator's expectations influence how likely that change is to be sensed (Wickens, Helleberg, Goh, Xu, & Horrey, 2001). In this stage we may also consider how factors such as time pressures and limitations of human attention may lead to the formation of incomplete or incorrect mental models.

Results

The output of the new extension is a set of scenarios for each UCA that explain why that action may have appeared logical to the human operator in context. These scenarios can be written in a paragraph or outline format and summarize the systemic factors that could contribute to the operator's behavior. The new extension also provides a way to illustrate the scenarios in a graphical format, at shown in Figure 4.

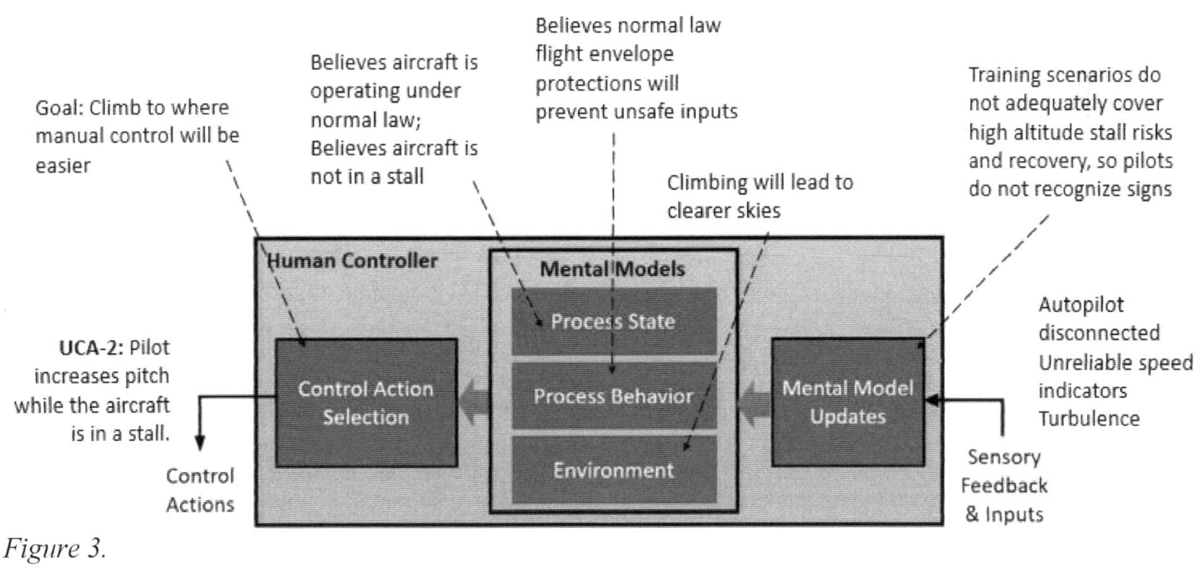

Figure 3.
Graphical depiction of a scenario based on events of the Air France 447 crash.

In Figure 3 above, a scenario is depicted graphically to show its relationship to each aspect of the new human controller model. The Unsafe Control Action in this figure is UCA-2: PF increases pitch while the aircraft is in a stall. Why in the world would a pilot do that? The first part of the model, the Control Action Selection, explores factors like pilot goals—as explained in the previous section. For example, the pilot may believe manual control will be easier at a higher altitude, where he knows the skies are clearer. The next stage, the Mental Models, explores pilot beliefs. For exmaple, the pilot may believe it is safe because he thinks the aircraft is not in a stall, is operating under normal law, and flight envelope protections will prevent him from stalling the plane. The last stage, the Mental Model Updates, explores how the pilot interprets information and updates (or doesn't update) their mental model. For example, due to inadequate training in high altitude stalls, he does not expect one to occur, and thus does not recognize the turbulence he is experiencing as part of a stall.

In the case of Air France flight 447, the pilots were faced with the sudden disconnection of the autopilot system, as shown in the scenario above. Lacking accurate speed information, they did not realize that they were at risk for a stall, proceeding to climb and even decrease speed. They realized too late that normal law, which provides flight envelope protections, was no longer in effect and the aircraft was operating under alternate law, which permitted the unsafe pitch inputs. The pilots may have been able to recover from an incident involving any of these factors alone, but it was the combination of the cockpit and stimuli, their beliefs about the system, and the circumstances influencing their decisions that ultimately led to the accident.

The scenario shown in Figure 3 is just one of many potential ways that an accident could occur. Other scenarios related to pitch input could lead to different accidents, and the method demonstrated here provides a systematic way of identifying such scenarios so that proactive efforts can be made to eliminate the factors that contribute to accidents. For example, the scenario in Figure 3 may suggest a need to more conspicuously indicate shifts from normal law to alternate law, or to improve training in high altitude stall procedures. Using STPA as a design tool, rather than a means of understanding causes of an accident after the fact, allows engineers and company management to make proactive decisions to improve safety.

The advantage of this extended STPA method is that it prompts the analyst to consider not only the different components of the operator's mental model, but also how that model is formed and what impact it has on decisions. While other models may provide a more nuanced view into human cognition (eg. Rasmussen, 1982; Endsley, 1995; Wickens, 1992), this model is deliberately simplified so that it can be used by industry practitioners without an extensive background in psychology or human factors. Those who do have such a background can also use this method to elicit their knowledge and experience in detailed topics such as sensation, perception, learning, and decision making while ensuring their explanations are structured in a way that is accessible to engineers and practitioners of all backgrounds. This model can thus be used as a common framework to talk about human automation interactions during system design and early development efforts.

Conclusions

STPA is a valuable hazard analysis technique with applicability across domains, and the new Engineering for Humans extension proposed in this paper provides additional guidance for analyzing human-automation interactions within the context of the larger system. When used to examine aircraft pitch control during an aerodynamic stall, the new extension can be used to model scenarios from the tragic Air France 447 crash. This new model provides a framework for understanding and anticipating human behavior during the design process, which is necessary to prevent such tragic accidents in the future.

Acknowledgements

This work is based on original research sponsored by General Motors (Thomas and France, 2016). The authors thank General Motors for their ongoing support in this area.

References

BEA (2012). *Final report on the accident on 1st June 2009 to the Airbus A330-203 registered F-GZCP operated by Air France, flight AF 447 Rio de Janeiro – Paris*. Paris, France: Bureau d'Enquêtes et d'Analyses pour la sécurité de l'aviation civile.

Endsley (1995). Toward a theory of situation awareness in dynamic systems. *Human Factors, 37(1),* 32-64.

Leveson, N. G. (2012). *Engineering a safer world: Systems thinking applied to safety.* Cambridge, MA: The MIT Press.

Rasmussen, J. (1982). Human errors. A taxonomy for describing human malfunction in industrial installations. *Journal of Occupational Accidents*, *4,* 311-333.

Thomas, J., France, M. (2016). *Engineering for Humans: STPA Analysis of an Automated Parking System*. Fifth MIT STAMP Conference, Cambridge, MA.

Wickens, C. D., Helleberg, J., Goh, J., Xu, X., & Horrey, W. J. (2001). Pilot Task Management: Testing an Attentional Expected Value Model of Visual Scanning (Technical Report No. ARL-01-14/NASA-01-7). NASA Ames Research Center. Retrieved from http://www.aviation.illinois.edu/avimain/papers/research/pub_pdfs/techreports/01-14.pdf.

Wickens, C.D. (1992). Engineering Psychology and Human Performance, New York: Harper Collins.

WORK DYNAMICS OF TASKWORK AND TEAMWORK IN FUNCTION ALLOCATION FOR MANNED SPACEFLIGHT OPERATIONS

Martijn IJtsma, Lanssie M. Ma, Amy R. Pritchett and Karen M. Feigh
Georgia Institute of Technology
Atlanta GA USA

This paper proposes a methodology for human-robot function allocation for future manned space exploration missions that uses fast-time computational simulation. Dynamics of taskwork and teamwork often result in emergent work patterns that are difficult to predict from static analysis of function allocations. We model the dynamics of taskwork and teamwork and demonstrate our approach through a case study that explores the function allocation design space for an on-orbit maintenance mission involving humans and various robots. The case study highlights the method's ability to predict possible concerns associated with limited availability of physical resources, action interdependencies, and communication requirements with possible time delays, and shows the influence of work dynamics on mission performance.

Communication delays associated with future manned exploration missions are on the order of tens of minutes. These delays no longer allow for ground-centered concepts of operation in which the locus of control lies with the mission control center; instead, there is a requirement to shift autonomy from the ground to astronaut crewmembers. In this context, crew autonomy refers to the crew's ability to make decisions and execute tasks independently from ground control. NASA envisions robots to play a large role in this higher crew autonomy.

Function allocation is the process of distributing tasks over multiple agents and comprises two dimensions: allocation of authority and responsibility. Authority is the notion of which agent is executing a task. Responsibility denotes which agent is accountable for the outcome of a task. Function allocation for manned space flight operations poses several unique challenges: (1) due to communication delays the mission crew can no longer rely on ground support; (2) there is limited availability of resources (i.e., consumables and tools); and (3) the environment is unfamiliar and extremely hazardous. Possible concerns associated with these challenges are, amongst others, long idle times for some agents as they wait for others to complete tasks, excessive communication required between agents, and high taskloads. Thus, there is a need for design methods that can objectively evaluate human-robot function allocations.

In the space domain, earlier work on evaluation of function allocations ranges from descriptions of a method for optimizing cost or reliability in human-robot teaming (Shah, Saleh, Hoffman, 2007) to 1-g full-scale testing of different function allocations in a space assembly task (Rehnmark, Currie, Ambrose, Culbert, 2004). One of the difficulties in evaluating function allocation is that the interplay of availability of resources, interdependencies of tasks, and communication requirements with possible time delays results in emergent work patterns that are difficult to predict from static analysis of function allocations. Therefore, function allocation evaluation methods should account for the dynamics of both the taskwork conducted by a team and the teamwork within the team.

This paper applies a fast-time computational simulation framework called Work Models that Compute (WMC) to evaluate human-robot function allocations for manned space flight operations. The focus of this paper is on the modeling and simulation of the dynamics of the team's taskwork and teamwork. A case study demonstrates simulation of the work dynamics for various function allocations for an on-orbit maintenance mission, highlighting the ability of this method to predict potential concerns.

Computational Simulation Framework

WMC is a computational simulation framework that can evaluate function allocation options. The framework has previously been used for analyzing and synthesizing function allocation in the air traffic management system (Pritchett, Bhattacharyya, & IJtsma, 2016) and between the pilot and the autoflight system in the flight deck (Pritchett, Feigh, Kim, 2014a; Pritchett, Feigh, Kim, 2014b). Work dynamics are modeled in WMC through three inter-acting types of models: agent models, resources, and actions.

Information and physical entities in the work environment are modeled as resources. Information resources, are represented computationally as variables. Physical resources, such as tools or spacecraft components, are modeled by computational structures defining attributes important to their use, including containing information on the resource's location, current "ownership" by an agent, and availability.

Actions are standalone descriptors of the work that is being performed. Each action has an attribute that defines the interaction with the environment through three types of relationships with resources: (1) an action *gets* an information resource, thereby retrieving information from the environment; (2) an action *sets* an information resource, corresponding to changing information in the environment; and (3) an action *uses* a physical resource, similar to using a tool in real-life.

Agent models do not contain descriptors of specific work activities. Thus, by modeling the actions outside of the agent models, new function allocations can be easily tested by differing assignments of actions to agents during run-time. The agent model then takes in the action it is assigned and executes it. WMC can use any type of agent model that is deemed appropriate for the analysis as long as it meets the computational interface standard of accepting calls from the simulation framework to execute the actions it is passed during run-time. Examples of agent models currently used in WMC are a perfect agent that can execute all tasks instantly and perfectly, and a more-extensive model that also adds elements of task management, delaying and interrupting actions when its assigned taskload reaches limits.

The interplay of the three types of models determines the work dynamics. Agents influence the work dynamics through how they manage them, such as methods for task management. The duration of an action can be contingent on the agent executing it (e.g., robots typically take longer to complete an action than a human). Physical resources can be used by only one action at a time: When an action requires a physical resource that is not available at the scheduled action time, WMC will delay the action until all the required resources are available. Finally, actions influence the work dynamics through their interdependencies, which are modeled through actions scheduling follow-up actions. Actions can be scheduled to occur immediately after one action has been completed, or can be scheduled at a later time. Additionally, the modeler can specify the action's own update cycle, through which the action can dynamically determine its own required next execution time.

This format allows simulation of both taskwork (i.e., the work inherently required to fulfill mission objectives, regardless of function allocation), and the teamwork that is implicitly or explicitly required to coordinate the taskwork for any given function allocation. When allocating the taskwork to agents in terms of authority and responsibility, WMC will automatically engender teamwork actions that are required by an authority-responsibility mismatches. This mismatch occurs when one agent is authorized to perform an action but a different agent is responsible for the outcome of that action (Woods, 1985). Mismatches are common when allocating authority to robotic agents, as they typically cannot be held accountable for action outcomes and a human is therefore required to verify whether the action has been completed to standard. This required teamwork is modeled in the form of monitoring and confirmation actions. Monitoring is a parallel action in which the responsible agent observes the authorized agent during the execution of the main action. Confirmation is a subsequent action in which the responsible agent confirms the successful execution of the main action after it has been executed. The authorized agent needs to wait for confirmation before it can continue with its next action. These monitoring and confirmation actions emerge as the work progresses, and themselves can impose significant taskload on agents.

WMC assesses several metrics of the function allocation. Logging of the action execution times and duration for each agent captures the total time required to perform the mission, as well as the taskload imposed on each agent. Likewise, logs of physical interaction capture when an agent performs an action that requires a physical resource that has last been used by another agent implying the need for an exchange of physical resources. Similarly, cognitive interaction is logged when one agent gets an information resource that has earlier been set by another agent, reflecting the transfer of information between the two agents.

Case Study

To demonstrate the methodology, we analyze several function allocation options for an on-orbit maintenance mission involving several robots and human astronauts. The goal of the on-orbit maintenance mission is to inspect and, whenever necessary, repair three exterior panels on the spacecraft. The agent included in the analysis are an extra-vehicular astronaut (EV), an intra-vehicular astronaut (IV), a Remote Manipulator System (RMS), two Robonaut robots, and a Mission Control Center (MCC) agent. The agents are modeled with a simple performance model that can only execute one action at a time, except for monitoring and confirmation actions for which there is no taskload limit. Physical resources in the scenario include, amongst others, toolsets to conduct the inspection and perform necessary repairs (usually two), five panels (three that need to be inspected and two backup panels that can be used for repair) and a Portable Life Support System (PLSS). Information resources include information on the panels that need to be repaired and information on confirmation by a responsible agent.

The first two columns of Table 1 show the decomposition of the required taskwork into functional blocks and actions. Columns 3-6 show four different function allocation (FA) options in terms of authority/responsibility; FA1 and FA2 are reasonably attainable with current robotic capabilities, and FA3 and FA4 may be possible with future robotic capabilities. To demonstrate the effect of resources on work dynamics we additionally simulate an altered version of option FA4 in which there is just one instead of two toolsets. Responsibility for actions that are executed by robots is usually allocated to the IV astronauts, engendering monitoring actions, although FA4-B examines the impact of giving responsibility to the MCC engendering confirmation actions, an allocation impacted by an assumed 10 second communication delay. The EV astronaut is responsible for his/her own actions.

Table 1. Candidate function allocations for the on-orbit maintenance scenario.

Functional blocks	Actions	Current capabilities		Future day capabilities		
		FA1	FA2	FA3	FA4-A	FA4-B
1. Exit dock	1.1 Prepare	EV/EV	EV/EV	EV/EV	Robo2/ IV	Robo2/MCC
	1.2 Leave dock	EV/EV	EV/EV	EV/EV	Robo2/ IV	Robo2/MCC
2. Traverse	2.1 Traverse	EV/EV	EV/EV	EV/EV	Robo2/ IV	Robo2/MCC
3. Inspect panel	3.1 Get inspection tools	EV/EV	EV/EV	Robo1/ IV	Robo2/ IV	Robo2/MCC
	3.2 Apply inspection tools	EV/EV	EV/EV	EV/EV	Robo2/ IV	Robo2/MCC
	3.3 Store inspection tools	EV/EV	EV/EV	Robo1/ IV	Robo2/ IV	Robo2/MCC
4. Repair panel	4.1 Get repair tools	EV/EV	EV/EV	Robo1/ IV	Robo1/ IV	Robo1/MCC
	4.2 Get new panel	EV/EV	RMS/IV	RMS/ IV	RMS/ IV	RMS/MCC
	4.3 Remove broken panel	EV/EV	EV/EV	EV/EV	Robo1/ IV	Robo1/MCC
	4.4 Emplace new panel	EV/EV	EV/EV	EV/EV	Robo1/ IV	Robo1/MCC
	4.5 Dispose of broken panel	EV/EV	RMS/ IV	RMS/ IV	RMS/ IV	RMS/MCC
	4.6 Store repair tools	EV/EV	EV/EV	Robo1/ IV	Robo1/ IV	Robo1/MCC
5. Enter dock	5.1 Enter dock	EV/EV	EV/EV	EV/EV	Robo2/ IV	Robo2/MCC

Results

Figure 1 shows the time trace of the actions for function allocation option FA1. With just one agent, the EV, performing all the work, this manifests as a linear sequence of the actions. From left to right, the agent first prepares and leaves the dock, then traverses to the first panel, inspects it and continues to the next panel. Inspection of the second panel shows it needs repair and the agent thus gets a new panel and swaps it with the damaged panel. After also repairing the third panel, the agent returns to the dock. This function allocation has zero idle time, no interaction between agents and no monitoring requirements.

Figure 2 shows the time trace for option FA2. Some actions can occur simultaneously; the RMS can get a backup panel while the EV prepares the tools and start removing the broken panel. Then, the RMS can dispose of the broken panel while the EV installs the backup panel. The parallel occurrence of the actions reduces the total mission time. However, because the two processes are not perfectly synchronized, the RMS frequently needs to wait for the EV, reflected in the increased idle time as compared to FA1. Also, divvying up the tasks in this way requires physical interaction between in the agents, in this case the backup panel is first being "used" by the RMS and then transferred to the EV. Similarly, the broken panel is transferred from the EV to the RMS. Finally, the IV is responsible for the RMS's operation and thus needs to monitor the task progress, with idle times in between.

The simulation results for FA3 are shown in Figure 3. Off-loading the tasks from EV to Robonaut does not result in a notable decrease in mission time, mostly because the interdependencies between their actions do not allow for parallel execution. For example, inspection cannot start before the inspection tools have been prepared. Additionally, this FA requires notable physical interaction between the Robonaut and EV to interchange tools.

Figure 4 shows the result for option FA4-A in which a Robonaut performs panel inspection and a second Robonaut performs repair actions, in coordination with RMS. The actions for inspection and repair are fairly independent, and, thus, having them executed by different agents can notably decrease mission duration. Additionally, these actions being standalone and comparable in duration results in a reduction in the total idle time and the required number of physical resource exchanges. However, with multiple robotic operations occurring in parallel, the IV experiences a high monitoring load.

The constraints imposed by physical resources are clearly shown in the differences between FA4-A with two toolsets (Figure 4) versus FA4-A with one toolset (Figure 5). With two toolsets, repair and inspection can be performed in parallel, whereas sharing one toolset between multiple agents results in agents waiting for each other. The sharing of the toolset also increases the required number of physical resource exchanges.

Finally, Figure 6 shows the time trace for FA4-B, which has MCC responsible for all robotic operations. Here, the authority-responsibility mismatch was assumed to be addressed through confirmation actions, since the 10 second delay in the MCC's receipt of any portrayal of the execution of the action would obstruct real-time monitoring. From the time trace it is clear that the required confirmation actions together with the communication delay result in an inefficient function allocation with long idle times. It does, however, alleviate the monitoring load of the IV, who is now available for other tasks. High confirmation taskload for MCC may not be a problem as MCC can easily increase capacity through (comparably) unlimited human resources.

Conclusions & Future Work

This paper described the modeling and method for evaluating function allocation options for future manned space flight. We argue that the benefit of using computational simulation is its ability to identify emergent work patterns that might go unnoticed when applying static analysis methods. Insight in the emergent interplay between agents, the availability of resources and interdependencies of actions can help the designer make more informed trade-offs in the function allocation process. The case study highlights our approach to simulating the dynamics of taskwork and teamwork and the major influence it can have on the system's performance.

Beyond the initial evaluation provided here, the simulation framework can be applied throughout the process of designing function allocations. Identifying the emergent patterns and steering design decisions based on a good understanding of the implications of function allocation options will ease the design process in later stages. Although we intend to extend our case study with more accurate and elaborate task decompositions, we also believe there is a benefit in modeling coarser tasks and simulating them in WMC to gain higher-level insights before breaking down the tasks into smaller subtasks. WMC can be used in an iterative process in which identification of the required taskwork and teamwork is updated based on results from the computational simulation and vice versa.

In the future we plan to extend the WMC architecture to simulate the dynamics of taskwork and teamwork in greater detail and at higher levels of fidelity, supporting the later, more detailed evaluations of function allocations and the procedures and mechanisms supporting them. We will additionally consider new teamwork actions between robots and humans that are of interest to function allocation designers, particularly focusing on the differences in capabilities of robots and humans and the consequences thereof on the required teamwork.

Acknowledgements

This work is sponsored by the NASA Human Research Program with Jessica Marquez serving as Technical Monitor under grant number NNJ15ZSA001N. The authors would also like to thank the other WMC developers.

References

Pritchett, A.R., Bhattacharyya, R.P. & IJtsma, M. (2016). Computational Assessment of Authority and Responsibility in Human-Automation Function Allocation. *Journal of Air Transportation,* 93–101.

Pritchett, A.R., Kim, S.Y., Feigh, K.M. (2014a). Modeling Human-Automation Function Allocation. *Journal of Cognitive Engineering and Decision Making, 8*(1), 33–51.

Pritchett, A.R., Kim, S.Y., Feigh, K.M. (2014b). Measuring Human-Automation Function Allocation. *Journal of Cognitive Engineering and Decision Making, 8*(1), 52–77.

Rehnmark, F., Currie, N., Ambrose, R.O., & Culbert, C. (2004). *Human-Centric Teaming in a Multi-Agent EVA Assembly Task* (No. 2004-01-2485). SAE Technical Paper.

Shah, J.A., Saleh, J.H., & Hoffman, J.A. (2007). Review and synthesis of considerations in architecting heterogeneous teams of humans and robots for optimal space exploration. *IEEE Transactions on Systems, Man and Cybernetics Part C: Applications and Reviews, 37*(5), 779–793.

Woods, D.D. (1985). Cognitive Technologies: The Design of Joint Human-Machine Cognitive Systems. *AI Magazine, 6*(4), 86–81.

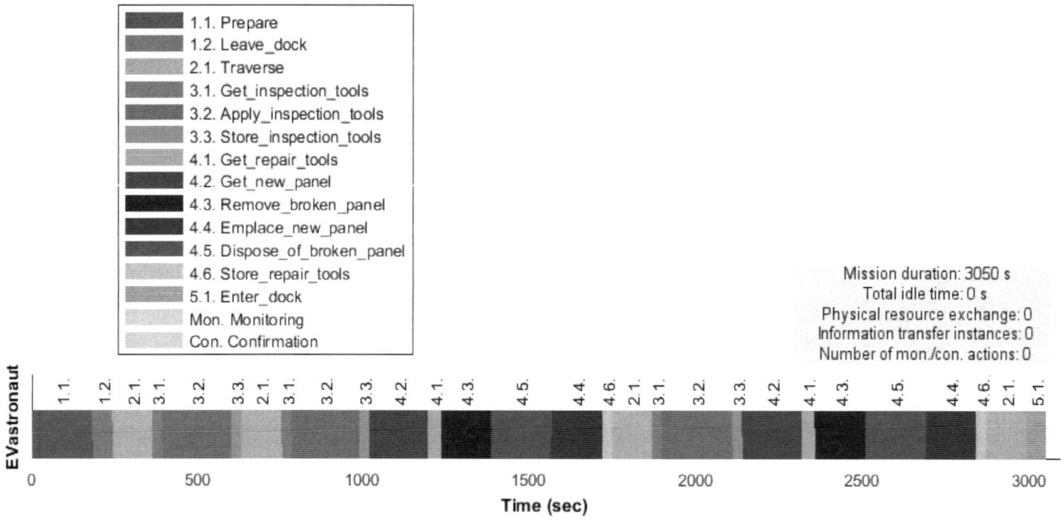

Figure 1. Function allocation option 1 (FA1) with EV astronaut performed all of the work.

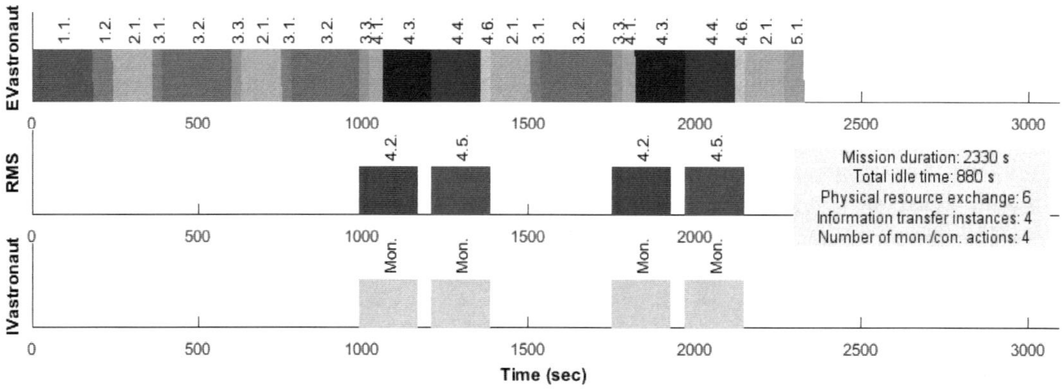

Figure 2. Function allocation option 2 (FA2) with RMS handling panels.

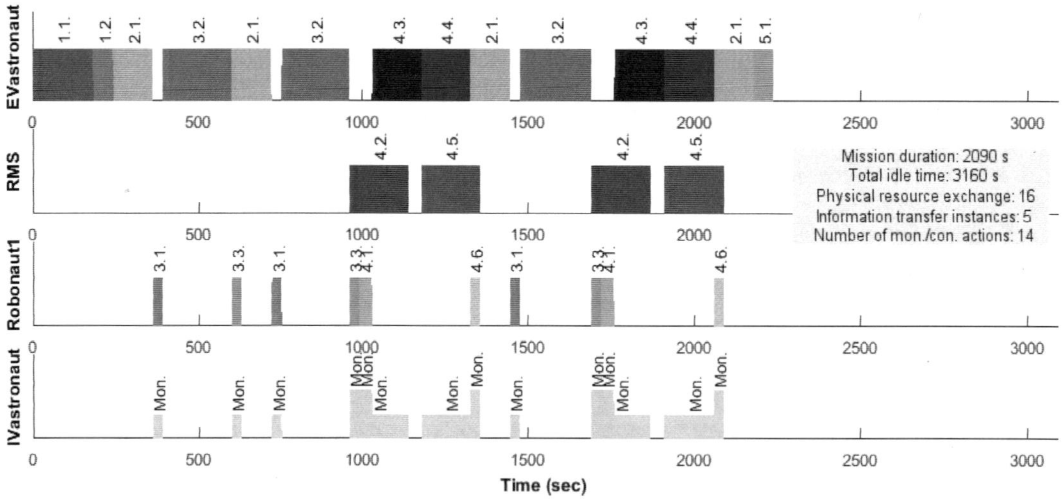

Figure 3. Function allocation option 3 (FA3) with RMS handling panels and Robonaut managing tools.

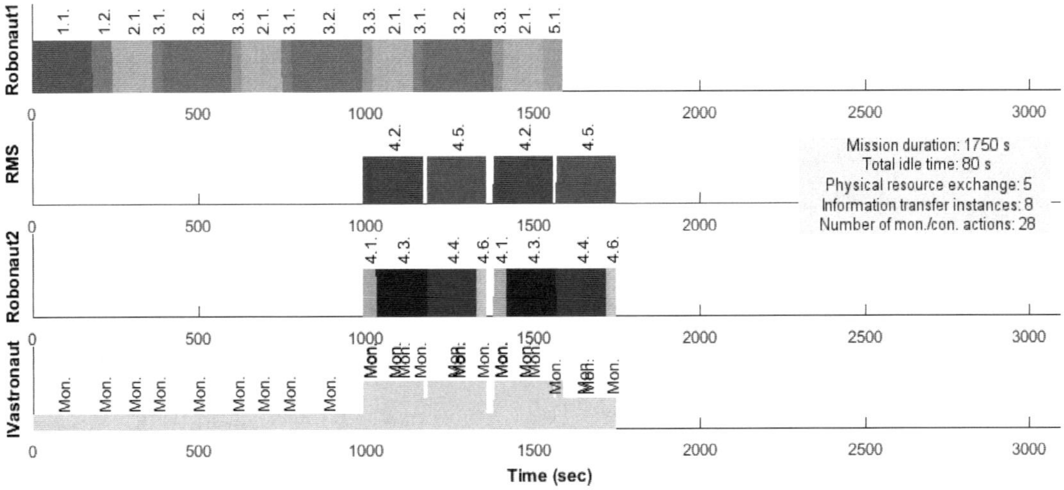

Figure 4. Function allocation option 4 (FA4-A) with Robonaut1 doing inspection, Robonaut2 & RMS doing repair.

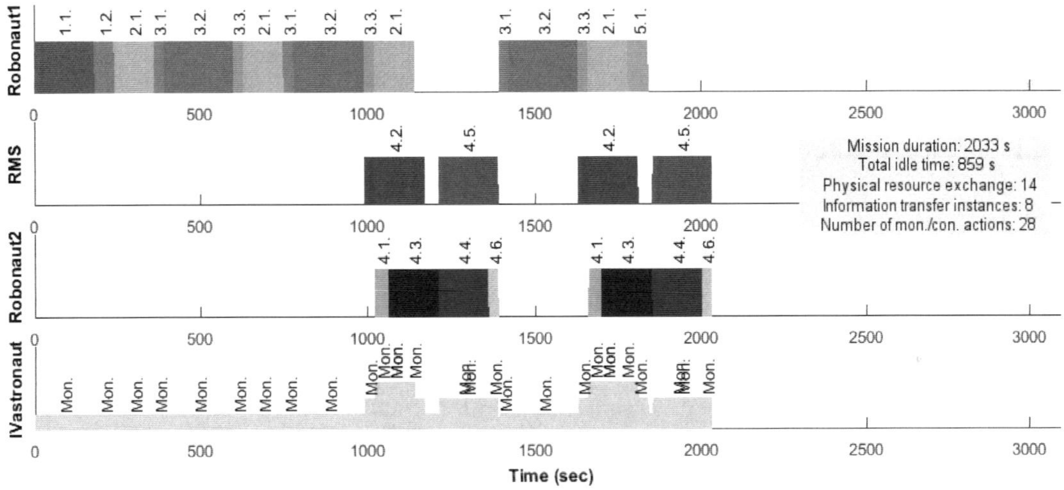

Figure 5. Function allocation option 4 (FA4-A) with only one toolset, shared by the agents, for inspection and repair.

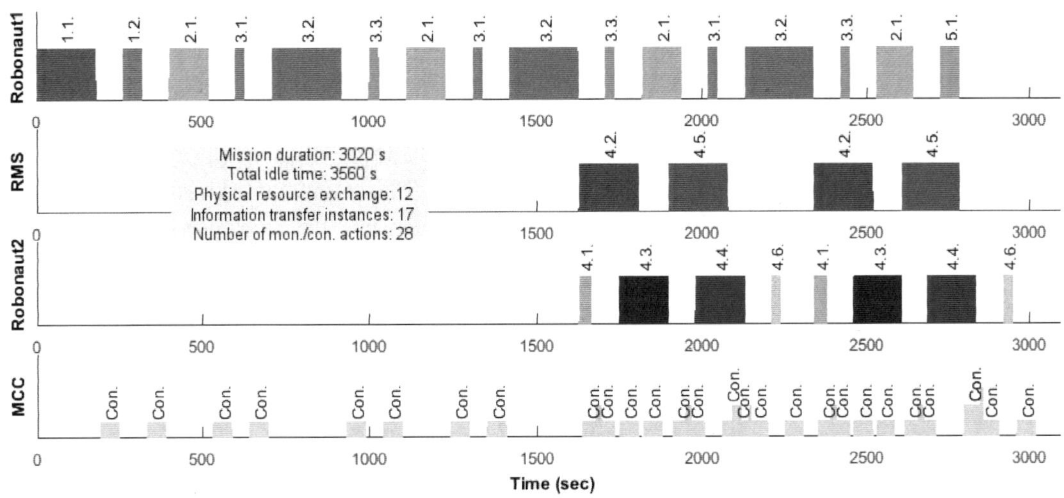

Figure 6. Function allocation option 4 (FA4-B) with MCC being responsible for robotic operations.

TAKING A CLOSER LOOK AT FLIGHT CREW HANDLING OF COMPLEX FAILURES – TEN CASE STUDIES

J.P.Reitsma*, L. Fucke†, C.Borst*, M.M.van Paassen*

* *Delft University of Technology, 2629 HS Delft, the Netherlands*

† *Boeing Research and Technology Europe, Madrid, Spain*

Non-normal events, in particular system failures with serious operational impact are rare in flight operations. These events are not always easy to handle by flight crews. The aim of the performed study is to determine where in this process potential issues may lie. Ten incident reports are studied using a newly developed operational issue analysis framework. The framework is used to determine whether and how the current interfaces communicate the initial *functional* impact and functional impact delayed in time. Additionally, results from pilot interviews are presented which identified three phases of non-normal event handling: fault detection, fault management and strategic planning. Analysis of the ten cases shows that current alert systems are mainly supporting the first two phases while the strategic planning phase, requiring higher level functional information integrated into the operational context as well as failure impact later in time, is relying almost entirely on pilot knowledge and reasoning.

Flight deck alerting systems have changed considerably in the past decades. The dials and warning lights used in the first generation airplanes were replaced by a centralized alphanumerical alert readout device, which presents descriptive text messages that are categorized by system and criticality (Veitengruber, 1978). More information and automation is provided than ever before. Although computerization changed the way alerts are presented, the fundamental concept of alerting has not changed and alert messages still largely refer to states of physical components or functions that were previously performed by a physical component. Several recent studies indicate that non-normal events are not always handled as desired, and procedures do not always provide sufficient guidance (Burian, et al., 2005). Difficulties can arise especially during failures involving interconnecting and automated systems (Singer & Dekker, 2000). The unchanged alerting approach, the shift of the pilot's role to an exception handler and manager of automated resources (Sarter, 1997) and the increased complexity of airplanes (Hasson & Crotty, 1997), may introduce various human performance issues. Current alerting systems present malfunctions to the flight crew as a list of messages that present physical information describing the status of individual systems, often physical components such as pumps, computers or valves. Lintern, Waite and Taller (1999) argue that the human-performance issues are not caused by the amount of presented information nor the complexity of the systems, but are mainly caused by the type of information that is presented. Multiple researchers confirm that flight decks lack functional information (Dinadis & Vicente, 1999) and that this lack of functional information can make handling non-normal events more taxing.

This paper will report on the results of an exploratory study with the goal to identify potential human-performance issues related to the current alerting systems and investigate where improvements could be possible. For this study 10 non-normal events were analyzed on how the current alerting systems present functional impact of a failure and what human performance issues might hinder understanding of the impact of the failure on the airplane.

Method

The Operational Issue Analysis (OIA) framework was developed to analyze incidents and accidents regarding potential human performance issues. The basic principle of the framework will be explained at the hand of Table 1.

Table 1
Operational issues analysis (OIA) framework.

			Human Performance Hinders					
			Detection			Understanding		...
Function	**Impact**	**Representation**	#1	#2	#3	#4	#5	
Generate Electricity	Lost	'Standby Bus Off'		x		x		
Distribute Fuel	Degraded	Fuel weight values	x	x	x			
...	...	...						

First, the initial functional impact of a system failure is determined from the incident reports and captured in the first two columns as presented in Table 1. The functions are obtained from a high-level functional decomposition, which enables the comparison across different alerting systems and airplane system architectures. A functional impact can be classified as either a loss, a degradation, a redundancy reduction or no impact. The function-specific flight deck effects that may indicate an impact are gathered and they are presented next to the functional impact. It can happen that a functional impact is not represented by any flight deck effect. The number of impacts without an indication is counted. The presented indications are then analyzed regarding the presence of any potential human performance hinders. The human performance hinders were obtained from the published Boeing in-house Cockpit-operations Reliability Evaluation Worksheet (CREW) (Fucke, et al., 2011). This worksheet is built around the Rasmussen's decision step ladder model. The CREW worksheet lists helps and hinders for all of the decision making steps. The OIA framework uses only hinders related to the first three steps, i.e. detection, understanding and prioritization, a total of 25 hinders. The latter decision making steps require more detailed procedural information, which is outside the focus of this study.

Next, the framework is used to determine how a functional impact delayed in time is communicated. This is done by evaluating each cascading failure step in the same fashion as before. This provides an indication whether the crew is able to detect repercussions at an early stage based on the presented flight deck effects.

The ten cases studied were selected from the Aviation Safety Reporting System (ASRS) and incident reports based on the following criteria: a system failure occurred in-flight on moderate to highly integrated airplanes and the malfunctions caused a severe operational impact delayed in time. The failures originate from a variety of systems.

Additionally, five experienced flight training instructors were interviewed to understand how non-normal events should be handled, what to consider during the event and what challenges are most typically encountered in operations and flight crew training. The results from the pilot interviews were used to assess validity of the findings of the operational issue analysis.

Results

The cases used for the OIA are categorized based on the initially affected function when the failure occurred: multiple failure scenarios, i.e. multiple functions are affected at the same moment, cases in which the "distribute fuel" function was initially affected, cases in which the "electric power generation" function was affected and cases in which the "hydraulic power generation" function was affected. The selected cases occurred on a variety of airplane models with varying system architectures. First is determined how many of the functional impacts are represented by any flight deck effect (FDE), which will be presented as a percentage of the total affected functions in Table 2. If there is an indication, the indication is evaluated on the number of hinders present. Finally, the cascading steps are analyzed and the percentage of impacts delayed in time that are communicated by a flight deck effect is calculated. The results are presented per category in Table 2, from which the following main observations can be extracted.

Table 2.
Hinders identified in showing the initial functional impact and impact delayed in time.

Case	Cases	Initial impacts with FDE per initially affected function [%]	Average # of hinders per initially affected function	Impacts delayed in time with FDE [%]
Multiple functions	2	51%	5	51%
Electric power generation	2	100%	1	-
Hydraulic power generation	2	100%	0	82%
Fuel distribution				
- (with alert)	1	100%	1	2%
- (without alert)	3	100%	8	2%

Not all initial functional impacts are presented by the interfaces. This was observed in particular for the multiple failure scenarios in which some impacts were not presented even though they were severe. The fact that the flight crew was surprised when detecting an uncontrollable engine (ATSB, 2013) (NTSB, 2010) and when there was no response after commanding reverse thrust, indicates that the impact of the failure was presented incompletely (NTSB, 2010). Furthermore, in one case the crew was not able to detect a loss of the fire extinguishing system (ATSB, 2013). The only way these effects could have been detected is by using detailed system knowledge.

Fuel leaks are hard to detect on planes that do not have an appropriate fuel alert. The crew can detect the initial impact on the fuel function, if no message is presented, only by comparing fuel on board figures with the flight plan, which requires mental effort, takes up additional time, is performed in large time intervals and is perceived as a lower priority in case other alerts are present, e.g., when symptoms appear in a different system. For example oil-related messages drew the attention of the crew away from routine tasks (GPIAA, 2004). The high number of average identified hinders in these cases confirm this. The planes that have a message in place such as 'FUEL DISAGREE' (ASRS #1184574), handled a fuel leak without difficulty. Hence, these cases are presented separately in Table 2.

The level of degradation can be difficult to determine. In the several cases, failure messages that include for example, 'monitoring fault', 'L/G CTL 1 FAULT' do not provide clarity if the system is still functioning. This might be because the messages present only physical states, e.g. 'HYD B+Y SYS LO PR','BRAKE TEMP'. The transformation to a higher level function needs to be done by the crew in order to determine if the system is still functioning.

Mode indications rely heavily on the pilots' system knowledge. The crew has to understand what functions are still supported in a specific mode. As an example the messages 'ALT LAW (PROT LOST)' and 'EMER ELEC CONFIG' shall be provided. In these cases, the crew may be required to remember what is covered by these modes and what is not, which increases mental effort. Difficulties exist in determining what is affected after an electric bus failure, since a lot of systems are dependent on these buses and often no clear overview is available to the crew.

The amount of presented messages during multiple failure scenarios can be overwhelming. In one case it took 50 minutes to obtain a clear overview on what systems were inoperative (ATSB, 2013). Hence it can be concluded that it can be difficult to obtain a clear overview of all the affected functionalities during failure scenarios that affect multiple sub-systems.

As Table 2 shows, determining the degradations delayed in time are almost entirely based on flight crew reasoning and procedural information. No indications are present that project future failure effects. These indications may be valuable for resource systems such as equipment cooling and depleting batteries, since they have a severe impact on connected systems.

Identifying the consequences of a system failure turns out to be challenging and often relies on flight crew reasoning. Examples of these are; difficulty in determining how the landing distance of the plane was affected while considering degraded braking capability and the higher approach speed (ATSB, 2013) (NTSB, 2010). Reduced range due to an extended landing gear (SUB, 2001). Additional fuel burn due to APU operation and a therefore limited range (ASRS #925795, ASRS #854044). And finally, the crosswind limitation due to hydraulic failure (ATSB 2001). This indicates that deriving functional, context specific information about the airplanes capabilities from alert messages can be difficult. The process of determining which parts of the mission can be performed without change and which not is one of the most challenging tasks the crews face. This was confirmed in the pilot interviews we performed.

Discussion

Some identified human performance issues could potentially be addressed by introducing new messages as was seen by the fuel cases. On the other hand, an increasing number of messages may also hinder detection and understanding.

The interviews and investigation reports show that handling of a non-normal event can be split up in three phases; 'manage the moment', 'fault management' and 'strategic planning' phase. The last phase is relatively unsupported by the current interface, this can be concluded from the observation that the impact delayed in time is not presented. This finding was confirmed in the interviews. Basic impacts on the mission can be difficult to extract from the alert messages, e.g. range or landing distance. This process is mainly based on the system knowledge and experience of the flight crew. While the majority of the tasks on the flight deck fall into the rule-based realm this process remains highly knowledge-based. The integration of the system effects into the operating context is complex. This can lead to interpretation errors, which can in turn lead to undesired consequences.

Making a diversion decision for example depends largely on aeronautical decision-making and can be very complex due to the many factors involved. Even clear procedural guidance stating that a diversion is needed, e.g. 'Land at nearest suitable airport' can be challenging to follow, as a lot of factors have to be considered to determine whether an airport in fact is suitable and what configuration is needed or available for landing. What the effects are of a changed configuration has to be determined by the crew, requiring additional interaction with on-board systems, performance tables and additional reasoning. In addition go-around performance may have to be considered, which leads again to an increase in workload.

As we determined in our pilot interviews, operational issues can often be detected using a step-by-step story-telling approach, in which each flight phase is briefed based on what the effects are and how to handle the plane differently from normal operating procedures. Obtaining information about failure effects, weather, performance data, level of available automation, airport navigational aids or other services, is often tedious and can take a lot of time.

While it is understandable that providing improved support of the strategic planning phase can be challenging due to the ever changing environment in which an aircraft operates, it may be worthwhile looking into better integration of a failure effect with the environment by making full use of the current computing capabilities and ways of information exchange. This may simplify information integration and decision-making, lead to a reduction in workload and the ability to evaluate more options. Further, this could reduce the potential of undesired consequences by moving some tasks from the knowledge-based to the rule-based realm. Also, by providing a better overview of failure consequences, unnecessary diversions might be reduced. These have a significant economic impact in flight operations.

Further, with advancing automation it is likely that the fault management phase may become less important altogether as automation will take over more and more of the associated reconfiguration tasks. A more integrated support of the strategic management tasks therefore appears to merit a priority.

Concluding Words

To date, transforming physical state information into functional availability as well as integration with the operational environment requires a high level of reasoning and system knowledge from the flight crews and hence considerable training. The current alerting and checklist systems may not always represent the operational effects of a system failure in a way that lends itself to ad-hoc understanding. This can lead to undesired consequences. Improvements can potentially be made by providing increased interface support for the information gathering and integration process. Automated processing of state information and relating it to the operational context can likely reduce the complexity of handling non-normal events.

References

ATSB (2001). Airbus A340-300, B-2380 Sydney, NSW 1 November 2000. *Investigation report 200005030.* Canberra, Australia: Australian Transport Safety Bureau (ATSB).

ATSB (2013). In-flight uncontained engine failure Airbus A380-842, VH-OQA. *Investigation report AO-2010-089.* Canberra, Australia: Australian Transport Safety Bureau (ATSB).

Burian, B.K., Barshi, I., & Dismukes, K. (2005). The challenge of aviation emergency and abnormal situations. *NASA Technical Memorandum, 213462*. Moffett Field, CA: NASA Ames Research Center.

Dinadis, N., & Vicente, K. J. (1999). Designing functional visualizations for aircraft systems status displays. *The International Journal of Aviation Psychology,* 9(3): 241–269.

Fucke, L., Mumaw, R. J., Kennedy, R. J., & Nicholson, R. K. (2011). Assessment of risk from human performance on the flight deck - Driving sound design decisions. *6th IET International Conference on System Safety,* (pp. 1–5). doi: 10.1049/cp.2011.0247

GPIAA (2001). All Engines-out landing due to fuel exhaustion, Air Transat, Airbus A330-243 marks C-GITS, Lajes, Azores, Portugal, 24 August 2001. *Investigation report 22/ACCID/2001.* Lisbon, Portugal: Gabinete de Prevenção e Investigação de Acidentes com Aeronaves (GPIAA).

Hasson, J., & Crotty, D. (1997). Boeing's safety assessment processes for commercial airplane designs. *16th DASC. AIAA/IEEE Digital Avionics Systems Conference*, vol.1(4.4): 1–7. doi: 10.1109/DASC.1997.635076

Lintern, G., Waite, T., & Talleur, D. A. (1999). Functional interface design for the modern aircraft cockpit. *The International Journal of Aviation Psychology*, 9(3): 225–240.

NTSB (2010). B752 Electrical Failure, Chicago IL, 22 Sept 2008. *NTSB Report CHI08IA292.* Washington, DC: National Transportation Safety Board (NTSB).

OTK (2009). Serious incident at Helsinki-Vantaa airport due to tread separation and hydraulic failure, 22 june 2009. *Investigation report C2/2009L.* Helsinki, Finland: Onnettomuustutkintakeskus (OTK).

Sarter, N. B., Woods, D. D., & Billings, C. E. (1997). Automation surprises. *Handbook of human factors and ergonomics*, 2:1926–1943.

Singer, G., & Dekker, S. W. A. (2000). Pilot performance during multiple failures: An empirical study of different warning systems. *Transportation human factors,* 2(1): 63–76.

SUB (2000). Flugunfall mit dem Motorflugzeug Type Airbus A310 am 12. Juli 2000 am Flughafen Wien-Schwechat, Niederösterreich. *Investigation report GZ. 85.007/0001-FUS/2006.* Vienna, Austria: Sicherheitsuntersuchungsstelle des Bundes (SUB).

Veitengruber, J. E. (1978). Design criteria for aircraft warning, caution, and advisory alerting systems. *Journal of Aircraft,* 15(9): 574–581.

TASK ANALYSIS OF TWO CREW OPERATIONS IN THE FLIGHT DECK: INVESTIGATING THE FEASIBILITY OF USING SINGLE PILOT

Paul Schutte
US Army - Aviation Development Directorate
Fort Eustis, Virginia, USA

This paper describes a task analysis of two crew operations in the flight deck in Part 121 Operations from gate to gate on a cross country flight. In addition, several non-normal scenarios were analyzed. Literature and operations materials were combined to develop the task lists. A primary source for this study is "An Exploration of Function Analysis and Function Allocation in the Commercial Flight Domain" by McGuire, et. al. (1991). Once the tasks were identified, they were set to an actually flight time schedule. Each task was assigned to either the Pilot Flying (PF) or the Pilot Monitoring (PM) in today's operations. Additionally, tasks were categorized as: discrete or continuous; heads up or heads down; simultaneous with other tasks or performed serially; and whether any tasks are redundant.

Introduction

Whether space, ground, or sea, whether military or civilian operations, automation is enabling the reduction of humans required to perform a task. While self-driving cars are currently most prominent in the headlines, efforts in commercial aviation and in military aviation to reduce personnel in the air and on the ground are marching forwards. In 2013, the National Aeronautics and Space Adminstration investigated reducing the flight crew on aircraft in Part 121 operations from 2 to 1. In the military, concepts of operation such as the Optionally Piloted Vehicle in the Future Vehicle Lift Army aircraft are looking to allow a two-, one-, or zero-crew complement to perform different pilotage and mission duties in a single aircraft platform. The Army's Synergistic Unmanned/Manned Intelligent Teaming effort looks to allow a single human mission commander to manage many unmanned and manned aviation assets during a mission.

What is the role of the human in the midst of all this automation? What is the best way to integrate increasing automation into current operational environments? The first step in answering these questions is understanding the role that the human and the automation currently hold. The study described in this paper begins to better understand these roles. Part 121 operations (see FAA 14 CFR Part 121), is a well documented operational environment that is covered extensively by government regulations, aircraft and avionics manufacturers' procedures, and operators (e.g., airlines, cargo handlers) procedures. For this reason, and because the commercial market is looking for ways to reduce crew complement, the Part 121, two-crew operations was the subject of this study.

The goal of this study was to enumerate all of the flight tasks currently conducted by the Pilot Flying (PF) and by the Pilot Monitoring (PM) in today's operations. They included whether the tasks are discrete or continuous, heads up or heads down, simultaneous with other tasks or performed serially, and whether any tasks are redundant.

To achieve this goal, literature and operations materials were combined to develop the task analysis. One source is worth mentioning here in the introduction since it provided the bulk of the material and served as the starting point for task list development. This is the report entitled, "An Exploration of Function Analysis and Function Allocation in the Commercial Flight Domain" by McGuire, et. al. (1991) This report provides a lengthy and detailed task description at a fairly low level of granularity, without being specific to any particular aircraft. While this report was created in 1991, it still reflects most of what goes on in the modern flight deck. In addition, the "USAirways Boeing B757//767 Pilot Handbook" (2006), Transport Canada's "Multicrew Aircraft Standard Operating Procedures" (Transport Canada, 2014) the Aeronautical Information Manual (FAA, 2008), and a task analysis of approach and landing (Leiden, 2002) were also used in creating the task analysis.

Four attributes were used to categorize each task in the task analysis. The first attribute was task duration: discrete or continuous. The second attribute was whether the crewmember performing the task had their 'heads-up' (looking at primary flight displays and/or out the window) or 'heads-down' (e.g., programming the FMS, tuning radios, dealing with systems). The third attribute was the type of task being performed: doing (e.g., select, retract, modify), communicating (e.g., talking or listening to ATC), observing (e.g., scan, monitor), or cognitive activity (e.g., evaluate, consider, compare). The fourth attribute assigned a mission management category (aviate, navigate, communicate, or manage systems) to each task.

Despite the thoroughness of these documents, there are still aspects of flight crew tasks that were not covered. These are cognitive aspects of a crew's duties, managing automation (automation is generally assumed to be fully functional), task management, and monitoring the other crew member.

Flight Crew Roles in Part 121 Operations

A two-person crew consists of a Captain and a First Officer. The Captain is the final authority in the flight deck and responsible for the flight and the First Officer is second in command. Two crewmembers on the flight deck provide redundancy, workload distribution, and increased monitoring (including monitoring each other). Prior to moving the aircraft from the gate and after the aircraft is parked at the gate, duties and tasks are assigned specifically to the Captain or the First Officer. Once the aircraft starts to move on the ramp, one crewmember will take on the role of Pilot Flying (PF) and the other that of Pilot Monitoring (PM). These roles are not assigned specifically to the Captain and First Officer (e.g., the Captain can be the PF or the PM) but at least one and only one of the crewmembers must be designated as PF. The PF is responsible for operating the controls for taxiing on the ground, and operating the flight controls (either manually or through automation) in the air. The PM is responsible for handling communications, monitoring the PF, assisting the PF where needed, and monitoring the overall situation of the aircraft. In addition, the PM generally handles systems management and contingency management.

Results

The first step was to develop a timeline of crew tasks. A flight plan from LAX to JFK was created for a midsize passenger jet. The flight phases were broken down into three segments: Departure, Cruise, and Arrival. Obviously, a significant portion of that is taken up in cruise (nearly 5 hours). During cruise, the duties were fairly constant and the number of tasks was relatively low. Based on the roles and responsibilities defined in the handbooks and operating procedures, tasks were assigned to either the pilot flying (PF) or the pilot monitoring (PM). In addition, several system failure contingencies were evaluated. Not surprisingly, the task load is very high during the departure and arrival phases of flight and relatively low during cruise. The PF has a significantly greater number of tasks than the PM. This is due to the fact that the PF is responsible for the closed continuous monitoring, evaluating, and managing of all flight parameters. Over a normal flight, the maximum number of tasks at a given time for the PF is 22, for the PM is 18. The maximum combined (both PF and PM) tasks occurring at the same time is 37. All of these maxima occur during departure. The maxima during arrival are 16 for the PF, 12 for the PM, and 27 for the combined tasks. The average combined tasks are 19, 16, and 15 across departure, cruise, and arrival, respective.

Task Categorization

For an additional perspective on what the crew members are doing, the tasks were categorized based on a number of factors: Verb type, Contiuous tasks, Mission Mangement type, and Heads Down tasks.

Verb Type. This perspective on task type has to do with the type of verb that the crewmember is performing – DOING, COMMUNICATING, COGNITIVE activity, or OBSERVING. All the verbs were placed into one of these four categories. There were 33 verbs in the DOING category and they consisted of verbs such as activate, select, configure, and open. There were 10 verbs in the COMMUNICATING category and they consisted of verbs such as report, request, announce, and acknowledge. There were 8 verbs in the COGNITIVE category and they consisted of verbs such as compare, consider, evaluate, and compute. There were 6 verbs in the OBSERVING category and they consisted of verbs such as monitor, detect, scan, and identify. The PF had nearly 300 DOING tasks, compared to a little over 100 for the PM. The PM had all of the approximatly 150 COMUNICATING tasks. The PF had approximatly 150 COGNITIVE tasks, as opposed to the PM's count of about 35. Finally, the PF had almost 200 OBSERVE tasks, while the PM had about 175.

Continuous Tasks. It is interesting to note that there were only 8 OBSERVING verbs in the previous categorization. However, the total number of OBSERVING tasks in a flight are second only to the DOING tasks. The reason for this is that OBSERVING tasks are contiouous for both PF and PF throughout the flight. Whereas some tasks are discrete and may only be performed a few times throughout the flight. To further explore this phenomena, the tasks were categorized into Contiuous and Discrete. Continuous tasks are those that the crewmember is supposed to constantly be monitoring/controlling. Some may actually be truly continuous such as manual control of the wheel and column. Most are intermittent such as a scanning pattern over a number of instruments. During the Departure phase of flight, nearly 60% of the PF's task are contiuous vs. 8% of the PM's tasks. While the total number of tasks decreases during the Cruise phase, over 80% of the PF's tasks are contiuous while the just over 40% of the PF's tasks are continuous. The Arrival phase reflects the Departure phase with the PF having over 40% continuous tasks and the PM having 8% again.

Mission Management Tasks. Tasks were also categorized according to the traditional mission management categories of Aviate, Navigate, Communicate, and Manage Systems. Not surprisingly, the PF has the lion's share of the Aviate tasks (~500) and Navigate tasks (~140) verses the PM's ~50 and ~20, respective. The reverse is true for Communicate and Manage Systems tasks where the PM has ~100 of each and the PF only has less than 10.

Heads Down Tasks. Both PF and PM are responsible to maintaining situational awareness regarding the flight. The PM is valued for being a second set of eyes outside of the cockpit to avoid obstacles and to identify hazards. Many tasks are called 'heads down' because they take the crewmember's focus away from the outside world and focus them on instrumentation and documentation in the cockpit. Heads down tasks do NOT include viewing the Primary Flight Display or the Navigation Display. Again, not surprisingly, very few PF tasks are heads down during any phase of flight. However, for the PM, nearly 50% of his or her tasks are heads down during departure, 80% during cruise, and 70% during arrival.

Contingencies

Contingencies. When a contingency or non-normal occurs, the crew task load not only increases, but some of the assignments shift from one crewmember to the other. Examples of these situations are system degradations or system failures, extreme weather, onboard emergencies, and fires. In these cases, the PM has a significant increase in the number of tasks. Non-normal checklists are almost exclusively performed by the PM. In these contingencies involving system failures, the PM will have an average of 30 additional tasks. In general, if multiple contingencies occur, these tasks are cumulative. For example, if an engine fire requires an emergency landing, fuel may need to be dumped if the aircraft is too heavy to land. Similarly, if all engines are inoperative, then it is likely that the loss of all generators checklist will be required (or the APU start). The PM's task workload can increase significantly. In these cases, the PF often takes over some of the PM's tasks such as communicating with ATC or dispatch. Not surprisingly most of these tasks that are assigned to the PM are heads-down. Because the PF assumes some of the PM's duties, there is an increase in the PF's heads-down time as well.

Discussion

Automation Issues and Opportunities

The question of what tasks and functions to allocate to the automation and what to allocate to the human – especially when looking to decrease the crew complement – is does not have an easy answer. Automation is quite amenable to repetitive, precise, deterministic, and long duration tasks. Similarly, any task that involves computations, memory (declarative, retrospective, prospective), or vigilance is probably best performed by automation instead of the human pilot. It is tempting to consider simply removing a crewmember and all the associated tasks that go with that role. This is essentially what occurred when commercial flightcrews replaced virtually the entire role of the flight engineer in the 1980's. However, the results of this study suggest caution in this approach.

Replacing the PM. The PM has a relatively low task count and based on this, it may seem easier to replace the PM. However, when contingencies occur, the PM is crucial. Humans have a unique capability for dealing with contingencies, especially when multiple contingencies occur. It is difficult to program automation for all contingencies. In addition, communications tasks are often best handled by humans given the vagaries of communication that will still exist if there are any humans in the loop (e.g., air traffic control, tactical operations center).

Replacing the PF. The PF has a number of factors that make that role attractive to automate. A majority of the PF's task are continuous tasks that involve, amoung many things, monitoring and observing. But humans are not good monitors when it comes to highly reliable systems (Parasuraman, 2010) and automation is excellent when it comes to continuous tasks that involve vigilance. Also, automation is quite capable of handling the DOING verbs and the PF performs an overwhelming percentage of those DOING tasks. The problem with replacing the PF is that it is a common understanding that some human is always responsible for 'flying the aircraft' (Wiener, et al, 2010). The human has qualities of personal responsibility and a strong will to live that automation does not posess.

Recommendation – New Roles for Humans and Automation

Rather than attempt to replace either the PF or the PM, it is likely that a new role be created for a single pilot and that the automation complement that role (Schutte, 1999). This is because human skills are necessary for some aspects of both the PF and PM's jobs. Creativity, situation awareness, responsibility and accountability are required in both positions. A new role should be defined for the pilot in the flight deck - perhaps a moniker of Pilot In Command (PIC). The pilot would be in command of all the automation resources. The PIC could safely say to the automation, "You two have got the airplane, while I look into why this checklist hasn't solved the problem," and rest assured that the automation will let the PIC know if there is any problem. However, the PIC could also intervene and 'manually fly' the aircraft. The PIC can take control of a difficult to handle aircraft (e.g., United 232 (NTSB, 1990)) and concentrate on flying and let the automation handle basic checklists. The PIC will remain the "last line of defense" for aircraft problems. And the automation must support that role.

The PIC concept requires that the PIC can perform a variety of tasks, can dynamically allocate tasks to the automation, and can easily do so. This represents a new design challenge that is more formidable than designing for two crew operations. Currently, the PF and the PM hand off tasks to each other with relative ease, knowing that the other crewmember is just as capable of performing those tasks and is just as situationally aware. There are likely to be significant differences in capability and situation awareness between the pilot and the automation. Protocols and procedures will need to be created for allocating tasks while ensuring that 'someone' is always flying the airplane. One way of viewing this design challenge is to consider it as CRM between two agents – pilot and automation. Interfaces and procedures will need to be developed to enable inter-agent communication, leadership and command, and decision-making. Automation is not especially communicative in the CRM sense, and ground support must communicate through the bandwidth 'keyhole' that can constrict communications. The SAFEdeck design approach (Schutte, et al, 2017) represents an attempt to design such a flight deck. SAFEdeck uses a single inceptor for both flying the aircraft and interacting with the automation so that the PIC can easily transition between high autonomy and low autonomy.

References

Federal Aviation Administration (2008). Aeronautical Information Manual.

Leiden, K., Keller, J., & French, J. (2002). Information to support the human performance modeling of a B757 flight crew during approach and landing. NASA Ames Research Center, Moffett Field, CA.

McGuire, J. C., Zich, J. A., Goins, R. T., Erickson, J. B., Dwyer, J. P., Cody, W. J., & Rouse, W. B. (1991). An Exploration of Function Analysis and Function Allocation in the Commercial Flight Domain (NASA CR-4374). *Long Beach, CA: McDonnell Douglas Aircraft Company.*

National Transportation Safely Board. (1990). Aircraft accident report: United Airlines Flight 232 McDonnell Douglas DC-10-10 Sioux Gateway Airport, Sioux City, Iowa, July 19, 1989 (NTSB- AAR-90-06). Washington, DC: National Transportation Safety Board.

Parasuraman, R., & Manzey, D. H. (2010). Complacency and bias in human use of automation: An attentional integration. Human Factors: The Journal of the Human Factors and Ergonomics Society, 52(3), 381-410.

Schutte, P. (1999). Complemation: An alternative to automation. Journal of Information Technology Impact, 1(3), 113-118.

Schutte, P., Goodrich, K., & Williams, R. (2017). Synergistic Allocation of Flight Expertise on the Flight Deck (SAFEdeck): A design concept to combat mode confusion, complacency, and skill loss in the flight deck. In Advances in Human Aspects of Transportation (pp. 899-911). Springer International Publishing.

Transport Canada (2014). Multicrew Aircraft Standard Operating Procedures. Retrieved from http://www.tc.gc.ca/eng/civilaviation/standards/commerce-manuals-multicrewsop-menu-1796.htm .

Wiener, E. L., Kanki, B. G., & Helmreich, R. L. (Eds.). (2010). Crew resource management. Academic Press.

BROWSE BY AUTHOR

AUTHOR NAME

Broach, Dana

Bronkhorst, Adelbert

Brown, Lori

Brown, Matthew

Bruder, Carmen

Cabral, Lisia Maria Espinola da Silva Pacheco

Calhoun, Gloria

Carlson, Craig

Castilho, Diogo

Catena, Andres

Chappell, Sheryl

Chen, Chun-Hsien

Cleanthous, Charalambos

Comstock, Ray

Craft, Terry

Craig, Paul

Crapanzano, Robert

Credlebaugh, Chelsey

Crews, Don

Crook, Brenda

Crossman, Jacob

Crutchfield, Jerry

Cutler, Victoria

Daiker, Ronald

Dattel, Andrew

Dattell, Andrew

Davis, Barry

Davis, Matthew

Deerenberg, Ronald

Dehais, Frédéric

Demas, Matthew

DeVivo, Sharon

Di Stasi, Leandro

Diaz-Piedra, Carolina

Ding, Lei

Dong, Ziyi

Donnot, Julien

Douglass, Scott

Draper, Mark

Drechsel, Paul

Dreher, Jean-Claude

Ellis, Kyle

Entzinger, Jorg

Ercoline, Bill

Estellita Lins, Marcos

Etherington, Timothy

Evans, Dakota

Fanjoy, Richard

Feigh, Karen

Fennell, Karl

Ferraro, Richard

Ferris, Thomas

Fincannon, Thomas

Flach, John

France, Megan

Frederiksen, Rich

Frost, Elizabeth

Fucke, L.

Fuentes, Luis

Fussell, Stephanie

Futrell, Mel

Galster, Scott

Gayraud, David

Geiselman, Eric

Georgiou, Andrea

Giombelli, Ermanno

Goldiez, Brian

Goode, William

Goodwin, Martin

Grigsby, Scott

Groen, Eric

Gruenwald, Christina

Hammack, Taleri

Hansen, Michael

Harrivel, Angela

Hauret, Daniel

Hein, Michael

Henneberry, Deb

Herdman, Chris

Holtaway, Lorraine

Holt, Timothy

Hourlier, Sylvain

Houston, Vincent

Ho, Victor

Howard, Michael

Howell, James

Howland, Jennifer

Huang, Chenyu

Hubbard, Sarah

Hubbs, Clay

Hutson, Kerry

Ihara, Rana

Iijima, Tomoko

IJtsma, Martijn

Iseminger, Colt

Izzetoglu, Kurtulus

Jackson, Kyle

Jin, Linfeng

Johnson, Ian

Jorgenson, Terra

Kay, Alison

Keller, Julius

Keller, Mikayla

Kennedy, Kellie

Kiggins, Daniel

Knecht, William

Knox, Jonathan

Kobayashi, Asao

Kozak, Brian

Kramer, Lynda

Krois, Paul

Landman, Annemarie

Last, Mary

Le Vie, Lisa

Lester, Evan

Liggett, Kristen

Ling, Chen

Li, Tianhua

Littlepage, Glenn

Liu, Dahai

Lo, Edmund

Lofaro, Ronald

Loukopoulou, Loukia

Lubner, Maxine

Luedtke, Jacqueline

Major, Wesley

Majumdar, Arnab

Ma, Lanssie

Martin, Matt

Masalonis, Anthony

Matayoshi, Naoki

Matsumoto, Jun

Matton, Nadine

Mauro, Robert

McCauley, Darendia

McDonald, Nick

McIntire, Stacey

Mead, Jonathan

Mehta, Rian

Mendonca, Flavio

Mercado-Velasco, G.A.

Merkt, Juan

Middendorf, Matt

Middendorf, Paul

Millan, Angel

Miller, Michael

Milner, Mattie

Moffett, Richard

Morrison, Rabea

Mulder, Max

Napoli, Nicholas

Neely, Jared

Nesthus, Thomas

Nichelli, Paolo

Nicholas, Stephanie

Nomiyama, Takenori

Ochieng, Washington

Oh, Chang-Geun

O'Neill, Patrick

Park, Seongmin

Patzek, Michael

Pavesi, Giacomo

Pederson, Eric

Petrenko, Oleg

Petrenko, Oleksandr

Petros, Thomas

Pierce, Linda

Pierce, Matt

Poisson, Jonathan

Poltavski, Dmitri

Pope, Alan

Post, David

Pounds, Julia

Prinzel, Lawrence

Pritchett, Amy

Pruchnicki, Shawn

Psylou, Elena

Purman, Ben

Qu, Xingda

Rantz, William

Rao, Arjun

Reichlen, Christopher

Reinerman-Jones, Lauren

Reitsma, J.P.

Reuter, Christopher

Revell, Saskia

Reynal, Maxine

Rice, Stephen

Richards, Dale

Rister, Frank

Risukhin, Vladimir

Roady, Trey

Rodriguez-Paras, Carolina

Roggow, Brian

Roll, Jason

Romigh, Griffin

Ropp, Tim

Rothwell, Clayton

Rowe, Allen

Rudnick, Georg

Ruff, Heath

Ryan, Margaret

Sain, Ryan

Scannella, Sébastien

Schmorrow, Dylan

Schnell, Thomas

Schulte, Axel

Schultheis, Susan

Schulze Kissing, Dirk

Schutte, Paul

Scoditti, Umberto

Sebok, Angelina

Servantie, Xavier

Sestito, Mariateresa

Shalin, Valerie

Shibata, Yumi

Sierra, Edmundo

Snodgrass, Megan

Spriggs, Sarah

Stanard, Terry

Stephens, Chad

Sutton, Jennifer

Suzuki, Shinji

Thomas, Gina

Thomas, John

Tinio, Raymart

Tippey, Kathryn

Trapsilawati, Fitri

Tricot, André

Trippe, Julia

Trippe, Ljulia

Uemura, Tsuneharu

Van Benthem, Kathleen

van Paassen, René

Venero, Peter

Vera, Steven

Verstynen, Harry

Villa, Jose

Wajon, Lloyd

Walters, Brett

Weedon, Sarah

Weldon, William

Wertheimer, Megan

Whitehurst, Geoff

Wickens, Christopher

Wilkerson, James

Willems, Ben

Williams, Henry

Williams, Kevin

Williams, Ralph

Winter, Scott

Wishy, Melissa

Wydoodt, Pierre

Yang, Qianru